JEE (Advanced)

Vectors and 3D Geometry

FOURTH EDITION

G. Tewani

Cengage

Australia • Brazil • Canada • India • Mexico • Singapore • United Kingdom • United States

Cengage

JEE (Advanced)
Vectors and 3D Geometry

Fourth Edition

G. Tewani

ISBN-13: 978-93-6660-304-9
ISBN-10: 93-6660-304-1

Cengage Learning India Pvt. Ltd.
T-303, 3rd Floor
DAV Complex, Plot 3
Mayur Vihar Phase 1
Delhi 110091

Printed by:
Rajiv Book Binding House
D-1518, DSIIDC Industrial Complex,
Narela, New Delhi 110040

Printed in India
Print Number: 02 Print Year: 2026

Preface

Engineering entrance examinations in India continue to evolve in structure, assessment methodology, and conceptual depth. As the primary gateway to premier institutions such as IITs, NITs, IIITs, ISM, IISERs, and other leading engineering colleges, the Joint Entrance Examination (JEE) emphasizes conceptual clarity, analytical reasoning, and application-based problem-solving. The **fourth edition** of this market-leading series has been developed with these objectives at its core, offering aspirants a structured, exam-oriented, and comprehensive learning framework.

The learning journey in this book begins with a carefully designed **Preliminary Section**, which serves as a preparation guide for students. A detailed **Table of Contents** further enables learners to navigate the subject systematically and plan their study schedule effectively. A **Road Map** has been introduced that further outlines the recommended learning pathway from an examination perspective, enabling students to approach their preparation in a strategic and organized manner.

The core of each chapter begins with the **Theory Section**, which focuses on building strong conceptual foundations. This section integrates **Concept Application Exercises** that are aligned with the concepts being taught, their corresponding difficulty levels, and application depth. These exercises are supported by **author-assessed time guidance**, helping students develop speed, accuracy, and exam temperament. The theory is further strengthened with **ample solved illustrations** that demonstrate step-by-step application of concepts and logical problem-solving approaches.

Following the theory, a dedicated set of advanced-level **Solved Examples** in subjective format has been included. These problems are designed to expose students to multi-step reasoning, deeper analytical thinking, and the type of structured solutions required for tackling high-difficulty examination problems, particularly at the JEE (Advanced) level.

The next stage of preparation is the **Exercise Section**, which provides extensive practice across multiple question formats. This includes **Single Correct Answer Type, Numerical Value Type, Integer Answer Type, Multiple Correct Answers Type, Linked Comprehension Type,** and **Matrix Match Type questions**. In this edition, the Single Correct Answer Type questions have been further refined and organized topic-wise, and are graded according to difficulty levels, beginning with easy problems, followed by medium and advanced-level questions. This structured progression enables students to gradually enhance their problem-solving ability and conceptual confidence. Collectively, the exercise sets are systematically designed to strengthen accuracy, improve speed, and help aspirants adapt to the diverse question patterns encountered in competitive examinations.

All books in this series comprise **latest solved JEE (Main) Previous Years' Questions (PYQs)** provided in a supplementary booklet. In addition, each chapter in the book concludes with a curated set of **solved JEE (Advanced) PYQs** updated up to the latest year, allowing aspirants to experience authentic examination-level problems and benchmark their preparation against actual exam standards. Together, these components form a comprehensive preparation structure designed to support progressive learning and exam readiness.

Comprehensive solutions to all concept application exercises, chapter-end exercises, and PYQs have been carefully revised and compiled at the end of the book. Special attention has been given to clarity, logical flow, and accuracy, ensuring that students can effectively analyze their performance and strengthen conceptual understanding. In addition, the entire series has undergone refinement, including correction of typographical inconsistencies and improvement in presentation quality, to enhance overall readability and learning effectiveness.

Beyond the main textbook, supplementary practice material has been provided to further support systematic preparation. A **Daily Practice Problems (DPP) booklet** offers topic-wise sets of fundamental questions, with each DPP focused on a specific concept area, enabling students to reinforce learning through short, targeted practice sessions. Another companion booklet includes the **solved JEE (Main) PYQs** asked in the latest JEE (Main) session, along with **one full-length mock test each aligned with JEE (Main)** and **JEE (Advanced) examination patterns**, helping students assess their preparedness under realistic test conditions.

In addition to the print content, learners are provided access to **free digital learning supplements** through **Cengage Digital** app, designed to extend and enrich the preparation experience. These include a **book-synchronized batch of self-learning video lectures** that closely follow the theory of the textbook, enabling concept reinforcement and flexible, self-paced study. The digital platform also offers **JEE (Advanced) PYQs (1979 onwards)** and JEE (Main and Advanced) level **chapter-wise assessments**, allowing students to assess their performance, identify strengths and gaps, and track progress in an exam-like environment.

Together, this integrated learning ecosystem—comprising structured theory, solved examples, graded practice, JEE PYQs, digital support resources, and supplementary booklets—aims to provide aspirants with a complete preparation solution. The book is designed to build strong fundamentals, enhance problem-solving proficiency, and instill the confidence required to excel in JEE (Main) and JEE (Advanced).

With best wishes for your success!

G. Tewani

Contents

Road Map

Build 3D Coordinate Geometry Foundation

Understand 3D coordinate system, position of points, distance formula, section formula, and geometrical interpretation of equations in three dimensions.

Study Spheres in Space

Understand centre–radius form and geometrical interpretation of sphere equations.

Learn Basic Vector Concepts

Understand vector types, direction ratios, angles between vectors, and geometric vector addition.

Use Component Form & Linear Combinations

Express vectors in component form and study linear dependence and independence.

Understand Vector Algebra

Learn dot product, cross product, and their properties and applications.

Apply Triple Products

Use scalar triple product for volume and vector triple product identities.

Study Lines in Space

Understand direction ratios, equations of lines, angles, and shortest distance problems.

Master Planes and Line–Plane Relations

Study plane equations, distance from plane, parallel and perpendicular planes, skew lines, and angle between planes.

Chapter

1 Introduction to Three-dimensional Geometry

Chapter Prerequisites

Before starting this chapter, ensure that you understand the following concepts thoroughly:

1. **Coordinate Geometry in Two Dimensions:** Familiarity with axes, the coordinate plane, plotting points, distance formula, and section formula in 2D, which directly extend into three dimensions.
2. **Basic Algebraic Manipulation:** Ability to simplify algebraic expressions and handle equations involving multiple variables.
3. **Understanding of Basic Three-dimensional Shapes and Spatial Visualization:** Familiarity with common three-dimensional objects such as cuboids, cubes, tetrahedrons, cylinders, cones, and spheres, and the ability to visualise their positions and dimensions in space.
4. **Logical and Step-wise Reasoning:** Ability to follow multi-step calculations and interpret geometric conditions clearly, important for deriving equations and locating points in 3D.
5. **Mid-point Formula in Two Dimensions:** Familiarity with finding the midpoint of a line segment joining two points in a plane.

Topic List

Learning Objectives

Introduction to 3D Coordinate System

Understand how three coordinate axes define space, interpret the coordinates of a point, and visualise its position relative to the coordinate planes.

Distance Formula in 3D

Learn to calculate the distance between two points in three-dimensional space and interpret the result geometrically.

Section Formula in 3D

Understand how to determine the coordinates of a point dividing a line segment in a given ratio using internal and external section formulas.

Sphere

Understand the concept of a sphere, form its equation in centre–radius form, and interpret the general equation geometrically.

1.1 THREE-DIMENSIONAL COORDINATE SYSTEM

Let us recall the two-dimensional coordinate system, which is determined by two perpendicular axes, the x-axis (horizontal axis) and the y-axis (vertical axis). These axes divide the plane into four regions called **quadrants.**

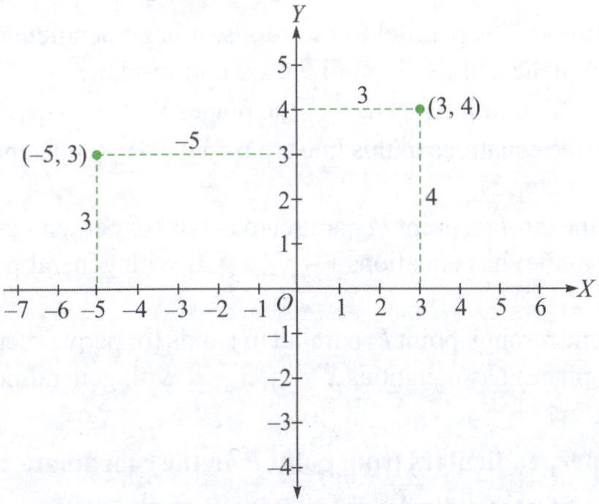

We know that for any point $P(x, y)$:

- The x-value (abscissa) represents the signed distance of the point from the y-axis, measured parallel to the x-axis.
- The y-value (ordinate) represents the signed distance of the point from the x-axis, measured parallel to the y-axis.

Thus, each point in the plane corresponds to an ordered pair (x, y) belonging to the Cartesian product $R \times R$, where R is the set of real numbers.

In the three-dimensional (3D) coordinate system, we insert one more axis through the origin, called the z-axis, which is perpendicular to both the x-axis and the y-axis.

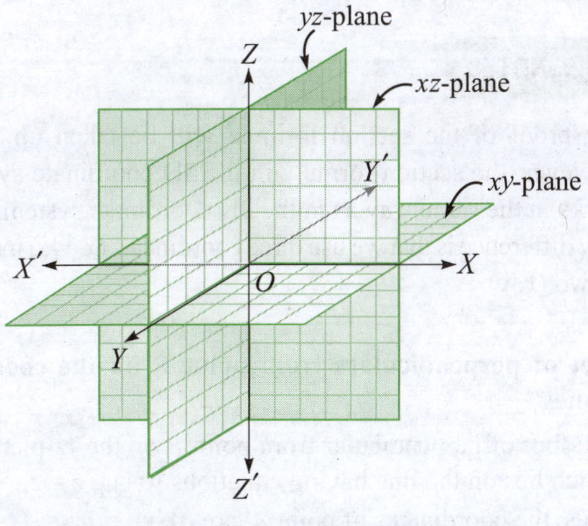

The three mutually perpendicular coordinate axes (x-, y-, z-axis) define 3D space.

- In the figure, OX, OY, OZ are the positive directions of the axes, and OX', OY', OZ' are the corresponding negative directions.
- These axes determine three mutually perpendicular coordinate planes: xy-plane, yz-plane, and xz-plane.
- These planes divide space into eight regions called **octants.**

The first four octants are numbered I to IV in the same way as the quadrants of the two-dimensional coordinate system, by following the signs of x and y and assigning a positive sign to z. The octants V to VIII follow the same order for x and y signs, but with z negative.

Octant / Coordinates	I	II	III	IV	V	VI	VII	VIII
x	+	–	–	+	+	–	–	+
y	+	+	–	–	+	+	–	–
z	+	+	+	+	–	–	–	–

Any point in this coordinate system has coordinates (x, y, z) which is an ordered triplet belonging to the Cartesian product $R \times R \times R$, where R is the set of real numbers.

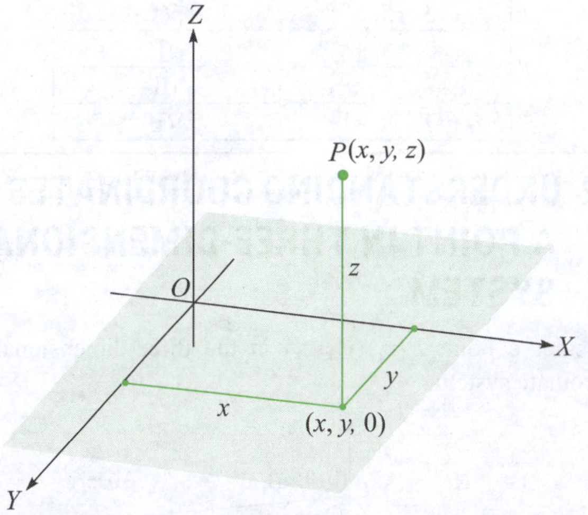

- The x-value and y-value have the same meaning in the xy-plane as in 2D.
- The z-value represents the *height* (or depth) of the point with respect to the xy-plane, i.e., how far above or below the xy-plane the point lies.

Since any point on the xy-plane has zero depth in the xy-plane, its z-coordinate is zero.

Hence, the equation of the xy-plane is $z = 0$, and a general point on it is $(x, y, 0)$.

Similarly, equations of coordinate axes and coordinate planes are as follows.

Plane/Axis	Equation(s)	General point
xy-plane	$z = 0$	$(x, y, 0)$
yz-plane	$x = 0$	$(0, y, z)$
xz-plane	$y = 0$	$(x, 0, z)$
x-axis (line of intersection of xy-plane and xz-plane)	$y = 0, z = 0$	$(x, 0, 0)$
y-axis (line of intersection of yz-plane and xy-plane)	$x = 0, z = 0$	$(0, y, 0)$
z-axis (line of intersection of yz-plane and xz-plane)	$x = 0, y = 0$	$(0, 0, z)$

Illustration 1.1

Name the octants in which the following points lie:
(2, 3, 4), (5, –3, 4), (2, –3, –7), (8, 6, –3), (–3, 5, –8), (–9, 3, 4), (–2, –5, 8), (–7, –8, –1)

Sol.

Point	Octant
(+2, +3, +4)	I
(+5, –3, +4)	IV
(+2, – 3, –7)	VIII
(+8, +6, –3)	V
(–3, +5, –8)	VI
(–9, +3, +4)	II
(–2, –5, +8)	III
(–7, –8, –1)	VII

1.2 UNDERSTANDING COORDINATES OF A POINT IN THREE-DIMENSIONAL SYSTEM

Consider a point $P(x_1, y_1, z_1)$ in the three-dimensional (3D) coordinate system.

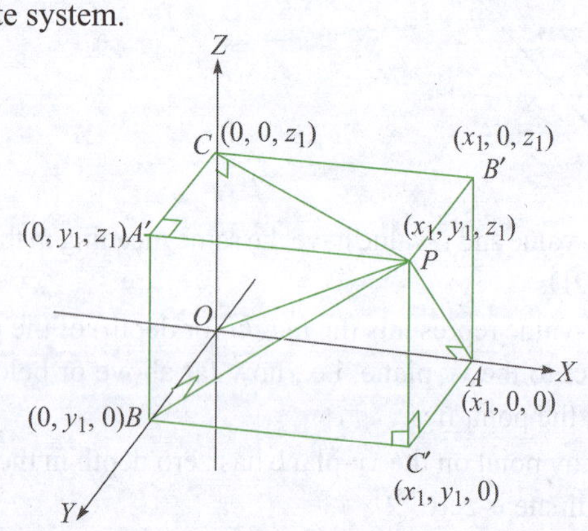

Here,

- x_1 is the signed distance of point P from the yz-plane, measured parallel to the x-axis,
- y_1 is the signed distance of point P from the xz-plane, measured parallel to the y-axis,
- z_1 is the signed distance of point P from the xy-plane, measured parallel to the z-axis.

Now, construct the a cuboid such that three faces $OAC'B$, $OBA'C$, $OAB'C$ through the origin lie on the coordinate planes, while the other three faces $PA'CB'$, $PB'AC'$ and $PC'BA'$ through point P are parallel to the xy, yz and xz-planes respectively.

1. The equations of planes through point P, parallel to the coordinate planes (or perpendicular to coordinate axes), are as follows.

(a) The depth of any point on the plane $PB'AC'$ from the yz-plane is x_1. Therefore, the equation of the plane through P, parallel to the yz-plane (perpendicular to the x-axis) is $x = x_1$.
This plane is perpendicular to both the xy-plane and the xz-plane.

(b) The equation of plane through P, parallel to the xz-plane (or perpendicular to y-axis) is $y = y_1$.
This plane is perpendicular to both xy-plane and yz-plane.

(c) The equation of plane through P, parallel to the xy-plane (or perpendicular to z-axis) is $z = z_1$.
This plane is perpendicular to both xz-plane and yz-plane.

2. The equations of lines through point P, parallel to the coordinate axes (or perpendicular to coordinate planes), are as follows.

(a) The line PC' is parallel to the z-axis; it is perpendicular to the xy-plane and lies in both the xz- and yz-plane.
It is the line of intersection of the planes $x = x_1$, $y = y_1$.
Thus, the equations of this line are $x = x_1$, $y = y_1$ with general point (x_1, y_1, z).

(b) The line through point P, parallel to x-axis (or perpendicular to yz-plane) has equations $y = y_1$, $z = z_1$ with general point (x, y_1, z_1).

(c) The line through point P, parallel to y-axis (or perpendicular to xz-plane) has equations $x = x_1$, $z = z_1$ with general point (x_1, y, z_1).

3. Feet of perpendiculars from point P on the coordinate axes

The foot of perpendicular from point P on the x-axis is the point A, which lies in the plane $x = x_1$.
Thus, the coordinates of point A are $(x_1, 0, 0)$.
Similarly, the feet of perpendiculars from point P on the y-axis and the z-axis are $B(0, y_1, 0)$ and $C(0, 0, z_1)$ respectively.

4. Reflections (images) of point P in the coordinate axes

Let P' be the reflection of P in the x-axis.
The foot of perpendicular from P onto the x-axis is $A(x_1, 0, 0)$.
Since A is the midpoint of PP', the reflection of P in the x-axis is $P' \equiv (x_1, -y_1, -z_1)$.
Similarly, the reflections of P in the y-axis is $(-x_1, y_1, -z_1)$ and in the z-axis is $(-x_1, -y_1, z_1)$.

> **NOTES**
>
> The proof of the section formula will be taken up later. However, the section formula in the 3D coordinate system works in the same way as in the 2D coordinate system. The only difference is that we use three coordinates (x, y, z) instead of two (x, y).

5. Feet of perpendiculars from point P on the coordinate planes

The foot of perpendicular from point P on the yz-plane is A', which lies on the line having equations $y = y_1$, $z = z_1$.
Thus, the coordinates of point A' are $(0, y_1, z_1)$.
Similarly, the feet of perpendiculars from P on the xz-plane and the xy-plane are $B'(x_1, 0, z_1)$ and $C'(x_1, y_1, 0)$ respectively.

6. Reflections (images) of point P in the coordinate planes

Let P'' be the reflection of P in the yz-plane.
The foot of perpendicular from P onto the yz-plane is $A'(0, y_1, z_1)$.
Since A' is the midpoint of PP'', the reflection of P in the yz-plane is $P'' \equiv (-x_1, y_1, z_1)$.
Similarly, the reflections of P in the xy-plane is $(x_1, y_1, -z_1)$ and in the xz-plane $(x_1, -y_1, z_1)$.

7. Distances of point *P* from the coordinate planes

In the figure, the depth of point *P* from the *yz*-plane is x_1.

Therefore, the distance of point *P* from the *yz*-plane is $|x_1|$ (= *PA*′).

Similarly, the distance of point *P* from the *xz*-plane is $|y_1|$ (= *PB*′) and the distance of point *P* from the *xy*-plane is $|z_1|$ (= *PC*′).

8. Distances of point *P* from the coordinate axes

In the figure, in a right-angled triangle *PB*′*A*,

$$PA^2 = PB'^2 + AB'^2 = y_1^2 + z_1^2$$

$\therefore \quad PA = \sqrt{y_1^2 + z_1^2}$, which is the distance of point *P* from the *x*-axis.

Similarly, the distance of point *P* from the *y*-axis is $\sqrt{x_1^2 + z_1^2}$ (= *PB*) and from the *z*-axis is $\sqrt{x_1^2 + y_1^2}$ (= *PC*).

9. Distance of point *P* from the origin

In the figure, in a right-angled triangle *OAP*,

$$OP^2 = OA^2 + AP^2 = x_1^2 + y_1^2 + z_1^2$$

$\therefore \quad OP = \sqrt{x_1^2 + y_1^2 + z_1^2}$, which is distance of point *P* from the origin.

Illustration 1.2

Consider the point *P*(–4, 5, 9). Find

a. the feet of perpendiculars from *P* on the coordinate axes.
b. the feet of perpendiculars from *P* on the coordinate planes.
c. the equations of planes through *P* which are parallel to the coordinates planes.
d. the equations of lines through *P* which are parallel to the coordinates axes.
e. the distances of *P* from the coordinate planes.
f. the distances of *P* from the coordinate axes.
g. the distance of *P* from the origin.
h. the reflections of *P* in the coordinate planes.
i. the reflections of *P* in the coordinate axes

Sol.

a. Feet of perpendicular from *P* on the *x*-, *y*- and *z*-axis are (–4, 0, 0), (0, 5, 0) and (0, 0, 9) respectively.

b. Feet of perpendicular from *P* on the *xy*-, *yz*- and *xz*-plane are (–4, 5, 0), (0, 5, 9) and (–4, 0, 9) respectively.

c. Equation of plane through *P*, parallel to, *yz* plane, is $x = -4$

Equation of plane through *P*, parallel to the *xz* plane, is $y = 5$

Equation of plane through *P*, parallel to the *xy* plane, is $z = 9$

d. Equations of line through *P*, parallel to the *z*-axis are $x = -4, y = 5$

Equations of line through *P*, parallel to the *y*-axis are $x = -4, z = 9$

Equations of line through *P*, parallel to the *x*-axis are $y = 5, z = 9$

e. Distances of point *P* from the *xy*-, *xz*-and *yz*-plane are respectively |9|, |5| and |–4| or 9, 5 and 4 units.

f. Distance of point *P* from the *x*-axis

$$= \sqrt{5^2 + 9^2} = \sqrt{106}$$

Distance of point *P* from the *y*-axis $= \sqrt{(-4)^2 + 9^2} = \sqrt{97}$

Distance of point *P* from the *z*-axis $= \sqrt{(-4)^2 + 5^2} = \sqrt{41}$

g. Distance of point *P* from the origin is

$$\sqrt{(-4)^2 + 5^2 + 9^2} = \sqrt{122}$$

h. The reflection of point *P* in the *xy*-plane lies on the line through *P*, parallel to the *z*-axis, having equations $x = -4$, $y = 5$. So, the coordinates of the reflection are (–4, 5, –9). Here the *z*-coordinate changes sign, since the given point and its reflection lie on opposite sides of the *xy*-plane. Similarly, the reflection of *P* in the *xz*-plane is (–4, –5, 9), and in the *yz*-plane is (4, 5, 9).

i. The reflection of *P* in the *x*-, *y*- and *z*-axis are (–4, –5, –9), (4, 5, –9) and (4, –5, 9) respectively.

Illustration 1.3

If the distances of point *P* from *x*, *y* and *z*-axis are respectively $\sqrt{41}$, $\sqrt{34}$ and 5, then find its distance from the origin and from the coordinate planes. Also find all possible coordinates of point *P*.

Sol. Let $P \equiv (x, y, z)$

The distances of point *P* from the *x*-, *y*-and *z*-axis are $\sqrt{y^2 + z^2}$, $\sqrt{x^2 + z^2}$ and $\sqrt{x^2 + y^2}$ respectively.

$\therefore \quad \sqrt{y^2 + z^2} = \sqrt{41}$, $\sqrt{x^2 + z^2} = \sqrt{34}$ and $\sqrt{x^2 + y^2} = 5$

Squaring and adding, we get

$$2(x^2 + y^2 + z^2) = (41 + 34 + 25) = 100$$

$$\Rightarrow \quad x^2 + y^2 + z^2 = 50$$

or, $\sqrt{x^2 + y^2 + z^2} = 5\sqrt{2}$, which is the distance of point *P* from the origin.

Now, $x^2 = (x^2 + y^2 + z^2) - (y^2 + z^2) = 50 - 41 = 9$

$y^2 = (x^2 + y^2 + z^2) - (x^2 + z^2) = 50 - 34 = 16$

and $z^2 = (x^2 + y^2 + z^2) - (x^2 + y^2) = 50 - 25 = 25$

$\therefore \quad |x| = 3, |y| = 4$ and $|z| = 5$.

So, the distances of point *P* from the *yz*-, *xz*- and *xy*-plane are 3, 4 and 5 respectively.

Also, there are eight possible points with coordinates (±3, ±4, ±5).

Illustration 1.4

Find the distance of the line having equations $x = 3, y = -4$ from the coordinate axes and the coordinate planes. Also, determine through which octants does this line pass.

Sol. Clearly equations $x = 3, y = -4$ represent a line parallel to the *z*-axis or perpendicular to the *xy*-plane.

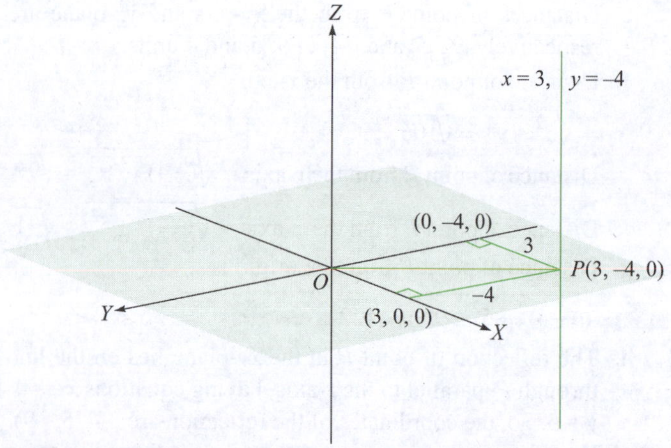

This line passes through the point $P(3, -4, 0)$ on the xy-plane.

The distance of the line from the axes is the same as the distance of point P from the axes.

Thus, the distances of the line from the x-axis and the y-axis are 4 units and 3 units respectively.

The distance of the line from the z-axis is $\sqrt{3^2+(-4)^2} = 5$ units.

Now, any point on the line can be taken as $(3, -4, z)$.

This point lies in the fourth octant if z is positive and in the eighth octant if z is negative.

Hence, the line passes through the fourth and eighth octants.

Illustration 1.5

Find the equations of line that passes through the point $(2, -4, -5)$ and is parallel to both the xy-plane and the xz-plane.

Sol. The intersection of the xy-plane and xz-plane is the x-axis. So, the required line is parallel to the x-axis.

The line passes through the point $(2, -4, -5)$.

Therefore, the equations of line are $y = -4, z = -5$.

Illustration 1.6

Find the equation of plane that passes through the point $(-3, 6, 7)$ and is perpendicular to both the xy-plane and the yz-plane.

Sol. The plane perpendicular to both the xy-plane and the yz-plane is parallel to the xz-plane.

Hence, the equation of plane through point $(-3, 6, 7)$ that is parallel to the xz-plane is $y = 6$.

Illustration 1.7

Explain the locus having equation is $x^2 + y^2 = 25$. Also, explain the locus having equations $x^2 + y^2 = 25, z = 0$.

Sol. Let the variable point $P(x, y, z)$ satisfy the equation $x^2 + y^2 = 25$.

$\therefore \sqrt{x^2+y^2} = 5$, i.e. the distance of point P from the z-axis is always '5'.

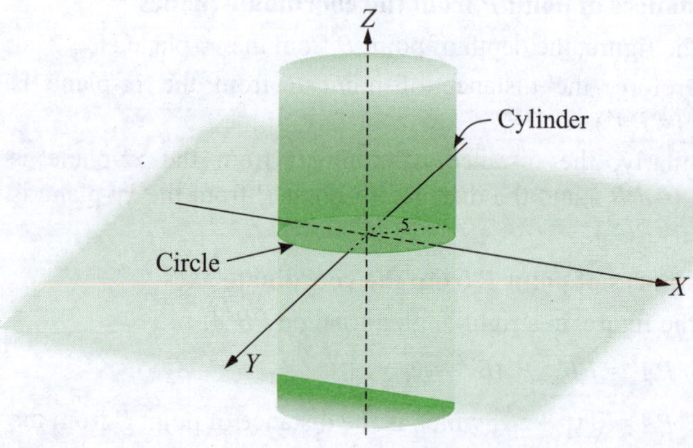

Clearly all such points lie on the curved surface of cylinder whose axis is the z-axis and the radius 5 units.

Now, planes parallel to the xy-plane cuts the cylinder in a circle.

Now, the equation $z = 0$ represents the xy-plane.

Hence, the equations $x^2 + y^2 = 25, z = 0$ represent a circle of radius 5 units having centre at the origin on the xy-plane.

Illustration 1.8

What do equations $y = 3, z = 4$ represent?

Sol. The equation $y = 3$ represents a plane parallel to the xz-plane and passing through the point $(0, 3, 0)$.

The equation $z = 4$ represents a plane parallel to the xy-plane and passing through the point $(0, 0, 4)$.

Thus, the equations $y = 3$, $z = 4$ together represent the line of intersection of planes $y = 3$ and $z = 4$, which is parallel to the x-axis. This line passes through point $(0, 3, 4)$ on the yz-plane.

Illustration 1.9

Let $P(x, y, z)$ be a point in the first octant, whose projection in the xy-plane is the point Q. Let $OP = \gamma$; the angle between OQ and the positive x-axis be θ; and the angle between OP and the positive z-axis be f, where O is the origin. Then find the distance of P from the x-axis.

Sol.

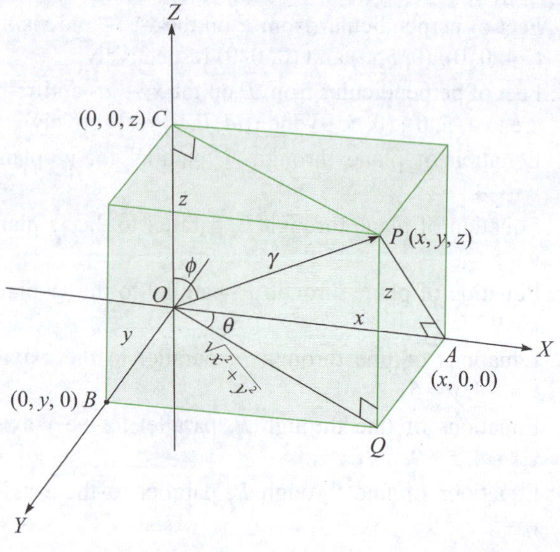

From the figure, the distance of P from the x-axis,

$$PA = \sqrt{y^2 + z^2}$$

Now, in triangle OCP, $OC = z = OP\cos\phi = \gamma\cos\phi$

Also, $\gamma^2 = x^2 + y^2 + z^2 = x^2 + y^2 + \gamma^2\cos^2\phi$

$\Rightarrow \quad x^2 + y^2 = \gamma^2(1 - \cos^2\phi) = \gamma^2\sin^2\phi \quad \ldots\text{(i)}$

Now, in triangle OAQ, $x = \sqrt{x^2 + y^2}\cos\theta$

$\Rightarrow \quad x^2 = (x^2 + y^2)\cos^2\theta$

$\Rightarrow \quad x^2\sin^2\theta = y^2\cos^2\theta$

$\Rightarrow \quad x^2 = y^2\cot^2\theta$

Putting in (i), we get

$$y^2\cot^2\theta + y^2 = \gamma^2\sin^2\theta$$

$\Rightarrow \quad y^2 = \gamma^2\sin^2\phi\sin^2\theta$

So, $PA = \sqrt{y^2 + z^2}$

$$= \sqrt{\gamma^2\sin^2\phi\sin^2\theta + \gamma^2\cos^2\phi}$$

$$= \sqrt{\gamma^2\sin^2\phi\sin^2\theta + \gamma^2(1 - \sin^2\phi)}$$

$$= \gamma\sqrt{1 - \sin^2\phi(1 - \sin^2\theta)}$$

$$= \gamma\sqrt{1 - \sin^2\phi\cos^2\theta}$$

Aim to complete within 10 minutes

Concept Application Exercise 1.1

1. Find the feet of perpendiculars from point $P(-4, 5, 9)$ on the coordinate axes and the coordinate planes.
2. Find the distances of a point $P(3, -5, 6)$ from the coordinate axes and the coordinate planes.
3. Find the equation of the plane passing through the point $(-2, 3, -5)$ and parallel to the
   a. xy-plane **b.** yz-plane **c.** zx-plane
4. Find the equations of a line passing through the point $P(2, 3, -4)$, parallel to the
   a. x-axis **b.** y-axis **c.** z-axis
5. What does equation $x^2 - 2x - 3 = 0$ represent in 3D coordinate system?
6. What is the distance of point $P(3, 4, -5)$ from
   a. the plane $z = 6$ **b.** the plane $x = 7$
7. Find the images of the point $P(-3, 2, 4)$ in the xy-, yz- and zx-planes. Also find the octant in which they lie.

Answers

1. Feet of perpendiculars on x, y and z-axes are respectively $(-4, 0, 0)$, $(0, 5, 0)$ and $C(0, 0, 9)$.
Feet of perpendiculars on xy, yz and zx-plane are respectively $(-4, 5, 0)$, $(0, 5, 9)$ and $(-4, 0, 9)$.

2. Distances of point P from x, y and z-axes are respectively $\sqrt{61}$, $\sqrt{45}$ and $\sqrt{34}$.
Distances of point P from xy, yz and zx-planes are respectively 6, 3 and 5.

3. **a.** $z = -5$ **b.** $x = -2$ **c.** $y = 3$

4.
a. equations of line parallel to x-axis is $y = 3$, $z = -4$
b. equations of line parallel to y-axis is $x = 2$, $z = -4$
c. equations of line parallel to z-axis is $x = 2$, $y = 3$

5. Combined equation of planes $x = -1$ and $x = 3$ which are parallel to yz-plane.

6. **a.** 11 units **b.** 4 units

7. Image of P in xy-plane is $(-3, 2, -4)$ which lies in the sixth octant.
Image of P in yz-plane is $(3, 2, 4)$ which lies in the first octant.
Image of P in xz-plane is $(-3, -2, 4)$ which lies in the third octant.

1.3 DISTANCE FORMULA

Distance between Two Points Lying on the Same Axis

The distance between two points $(x_1, 0, 0)$ and $(x_2, 0, 0)$ lying on the x-axis is $|x_1 - x_2|$.

The distance between two points $(0, y_1, 0)$ and $(0, y_2, 0)$ lying on the y-axis is $|y_1 - y_2|$.

The distance between two points $(0, 0, z_1)$ and $(0, 0, z_2)$ lying on the z-axis is $|z_1 - z_2|$.

Distance between Two Points Lying on the Line Parallel to One of the Coordinate Axes

Consider a line parallel to z-axis having equations $x = a$, $y = b$.

Now, consider two points $P(a, b, z_1)$ and $Q(a, b, z_2)$ on this line. In the figure, feet of perpendiculars from points P and Q on the z-axis are $P'(0, 0, z_1)$ and $Q'(0, 0, z_2)$ respectively.

Clearly, $PQ = P'Q' = |z_1 - z_2|$.

Similarly:

- On a line parallel to the x-axis distance $= |x_1 - x_2|$.
- On a line parallel to the y-axis distance $= |y_1 - y_2|$.

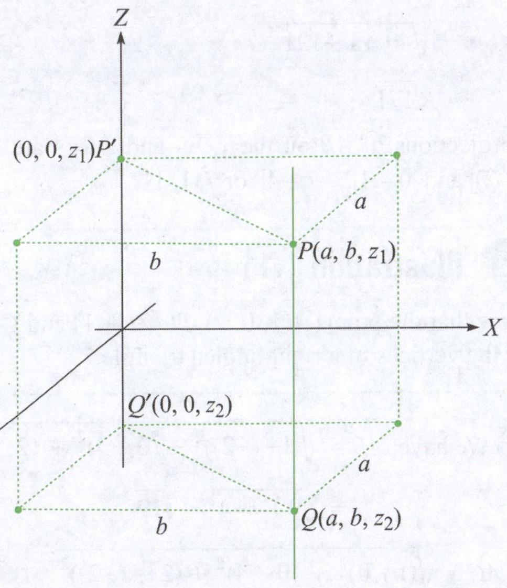

Distance between Two Points

Consider two points $P(x_1, y_1, z_1)$ and $Q(x_2, y_2, z_2)$. Construct a cuboid such that three faces through P and three faces through Q are parallel to the coordinate planes.

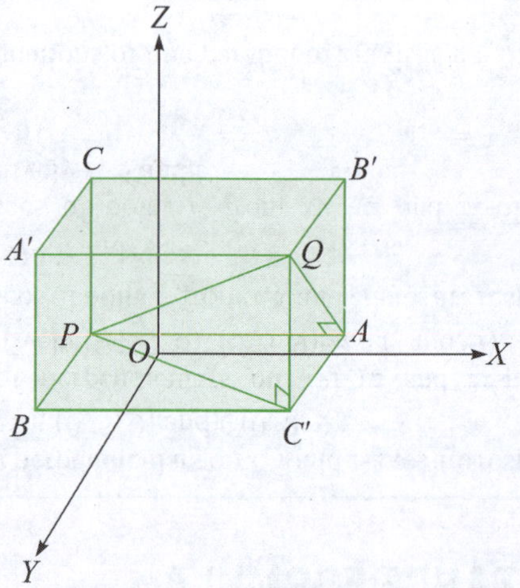

Equation of the plane $PBA'C$, which is parallel to the yz-plane is $x = x_1$.

Equation of the plane $QC'AB'$ which is parallel to the yz-plane is $x = x_2$.

So, the distance between these planes is $|x_1 - x_2|$

$\therefore \quad PA = |x_1 - x_2|$.

Similarly, $PB = |y_1 - y_2|$ and $PC = |z_1 - z_2|$.

$\therefore \quad PQ^2 = PA^2 + PB^2 + PC^2 = (x_1 - x_2)^2 + (z_1 - z_2)^2 + (y_1 - y_2)^2$

Thus, $PQ = \sqrt{(x_1 - x_2)^2 + (y_1 - y_2)^2 + (z_1 - z_2)^2}$

Illustration 1.10

Find the distance between points $A(-2, 1, 4)$ and $B(5, 0, -7)$. Also find the projections of AB on the coordinate axes.

Sol. The distance between points A and B is

$$AB = \sqrt{(5-(-2))^2 + (0-1)^2 + (-7-4)^2}$$
$$= \sqrt{49 + 1 + 121}$$
$$= \sqrt{171}$$

The projections of AB on the x-, y- and z-axis are respectively $|5 - (-2)|$ and $|0 - 1|$, $|-7 - 4|$ or 7, 1, 11.

Illustration 1.11

Show that the points $A(1, 0, 2)$, $B(-2, 3, 1)$ and $C(0, 4, -2)$ are the vertices of a right-angled triangle.

Sol. We have, $AB = \sqrt{(1-(-2))^2 + (0-3)^2 + (2-1)^2}$

$$= \sqrt{9+9+1} = \sqrt{19}$$

$$AC = \sqrt{(1-0)^2 + (0-4)^2 + (2-(-2))^2}$$
$$= \sqrt{1+16+16}$$
$$= \sqrt{33}$$

and $BC = \sqrt{(-2-0)^2 + (3-4)^2 + (1-(-2))^2}$

$$= \sqrt{4+1+9}$$
$$= \sqrt{14}$$

Clearly, $AC^2 = AB^2 + BC^2$

Hence, triangle ABC is a right-angled triangle with the right angle at vertex B.

Illustration 1.12

Show that the points $A(2, 3, 5)$, $B(6, 1, 6)$ and $C(14, -3, 8)$ are collinear.

Sol. We have,

$$AB = \sqrt{(6-2)^2 + (1-3)^2 + (6-5)^2}$$
$$= \sqrt{4^2 + (-2)^2 + 1^2} = \sqrt{21},$$
$$AC = \sqrt{(14-2)^2 + (-3-3)^2 + (8-5)^2}$$
$$= \sqrt{12^2 + (-6)^2 + 3^2} = \sqrt{189} = 3\sqrt{21}$$

and $BC = \sqrt{(14-6)^2 + (-3-1)^2 + (8-6)^2}$

$$= \sqrt{8^2 + (-4)^2 + 2^2}$$
$$= \sqrt{84} = 2\sqrt{21}$$

Hence, $AB + BC = AC$.

Therefore, points A, B and C are collinear such that point B divides AC internally in the ratio 1 : 2.

Illustration 1.13

Find the points on the z-axis which are at a distance 7 units from the point $P(3, -2, 5)$.

Sol. Let the point on the z-axis be $A(0, 0, z)$.

Given point is $P(3, -2, 5)$.

It is given that $AP = 7$

$\therefore \quad AP^2 = 49$

$\Rightarrow \quad (3-0)^2 + (-2-0)^2 + (5-z)^2 = 49$

$\Rightarrow \quad 9 + 4 + (5-z)^2 = 49$

$\Rightarrow \quad (5-z)^2 = 36$

$\Rightarrow \quad 5 - z = \pm 6$

$\Rightarrow \quad z = -1$ or 11

Thus, the coordinates of the required point A are $(0, 0, -1)$ and $(0, 0, 11)$.

Illustration 1.14

Find the locus of the point which is equidistant from the points $A(-2, 1, 4)$ and $B(5, 0, -7)$.

Sol. Let the point $P(x, y, z)$ be equidistant from the points A and B.

So, $PA = PB$ or $PA^2 = PB^2$

$\Rightarrow \quad (x+2)^2 + (y-1)^2 + (z-4)^2 = (x-5)^2 + y^2 + (z+7)^2.$

$\Rightarrow \quad x^2 + 4x + 4 + y^2 - 2y + 1 + z^2 - 8z + 16 = x^2 - 10x + 25 + y^2 + z^2 + 14z + 49.$

$\Rightarrow \quad 14x - 2y - 22z = 53$, which is the required locus.

Aim to complete within 15 minutes

Concept Application Exercise 1.2

1. A cuboid is formed by planes passing through the points $P(6, 8, 10)$ and $Q(3, 4, 8)$, parallel to the coordinate planes. Find the lengths of the edges of the cuboid. Also, find the length of its diagonal.
2. If $PQ = 10$ and the projections of PQ on the x-axis and the y-axis are 3 and 4 respectively, then find the projection of PQ on the z-axis.
3. Verify the following:
   a. (0, 7, –10), (1, 6, –6) and (4, 9, –6) are the vertices of an isosceles triangle.
   b. (–1, 2, 1), (1, –2, 5), (4, –7, 8) and (2, –3, 4) are the vertices of a parallelogram.
4. The distance of the point P from the origin is 7 units, from the x-axis is $\sqrt{13}$ units and from the y-axis is $\sqrt{40}$ units. Find the coordinates of point P.
5. Let $x^2 + 3y^2 = 3$ be the equation of an ellipse in the xy-plane. If $A(-\sqrt{3}, 0, 0)$ and $B(-\sqrt{3}, 0, 2)$ are two points, then find the point P on the ellipse such that $\angle APB = 45°$.

Answers

1. Length of edges are 3, 4, 2. Length of diagonal $= \sqrt{29}$
2. $5\sqrt{3}$
4. $(\pm 6, \pm 3, \pm 2)$
5. $(0, \pm 1, 0)$

1.4 SECTION FORMULA

1.4.1 Internal Section Formula

Consider two points $P(x_1, y_1, z_1)$ and $Q(x_2, y_2, z_2)$.

Let us find the coordinates of the point $R(x, y, z)$ which divides PQ internally in the ratio $m : n$.

Let all the points be in the first octant.

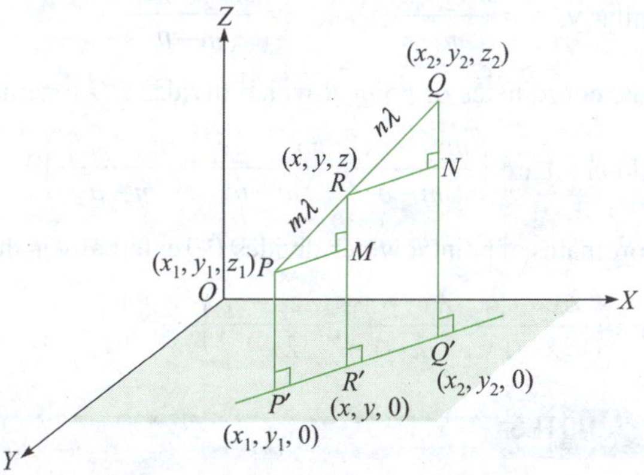

Here, P', Q' and R' are the feet of the perpendiculars from points P, Q and R on the xy-plane, respectively.

Points M and N are the feet of the perpendiculars from points P and R on lines RR' and QQ' respectively.

Clearly $PP' = z_1$, $QQ' = z_2$.

$\therefore \quad RM = RR' - MR' = RR' - PP' = z - z_1$

and $\quad QN = QQ' - NQ'$

$\quad = QQ' - RR' = z_2 - z$

Now, triangles PMR and RNQ are similar.

$$\therefore \quad \frac{PR}{QR} = \frac{RM}{QN}$$

$$\Rightarrow \quad \frac{m\lambda}{n\lambda} = \frac{z - z_1}{z_2 - z}$$

$$\Rightarrow \quad z = \frac{mz_2 + nz_1}{m+n}$$

Similarly, we can get $x = \dfrac{mx_2 + nx_1}{m+n}$ and $y = \dfrac{my_2 + ny_1}{m+n}$

Hence, the coordinates of the point R which divides PQ internally in the ratio $m : n$ are $\left(\dfrac{mx_2 + nx_1}{m+n}, \dfrac{my_2 + ny_1}{m+n}, \dfrac{mz_2 + nz_1}{m+n}\right)$.

Coordinates of point R which divides PQ internally in the ratio $\lambda : 1$ are $\left(\dfrac{\lambda x_2 + x_1}{\lambda+1}, \dfrac{\lambda y_2 + y_1}{\lambda+1}, \dfrac{\lambda z_2 + z_1}{\lambda+1}\right)$

1.4.2 External Section Formula

Now, let's find coordinates of point R which divides PQ externally in the ratio $m : n$.

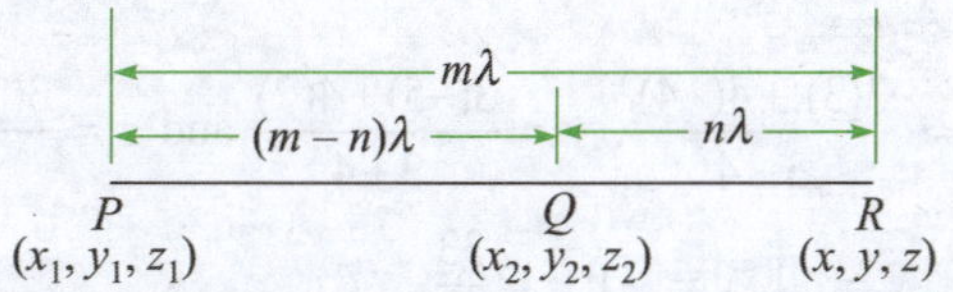

In the figure point R divides PQ externally in the ratio $m : n$.

We have $PR : QR = m : n$.

$\therefore \quad PQ : QR = (m-n) : n.$

Thus, point Q divides PR internally in the ratio $(m-n) : n$.

So, using internal section formula:

$$x_2 = \frac{(m-n)x + nx_1}{(m-n)+n}$$

$$\therefore \quad mx_2 = (m-n)x + nx_1 \text{ or } x = \frac{mx_2 - nx_1}{m-n}$$

Similarly, $y = \frac{my_2 - ny_1}{m-n}$ and $z = \frac{mz_2 - nz_1}{m-n}$

Thus, coordinates of point R which divides PQ externally in the ratio $m : n$ are $\left(\frac{mx_2 - nx_1}{m-n}, \frac{my_2 - ny_1}{m-n}, \frac{mz_2 - nz_1}{m-n}\right)$.

Coordinates of point R which divides PQ externally in the ratio $\lambda : 1$ are $\left(\frac{\lambda x_2 - x_1}{\lambda - 1}, \frac{\lambda y_2 - y_1}{\lambda - 1}, \frac{\lambda z_2 - z_1}{\lambda - 1}\right)$.

NOTES

- The coordinates of the mid-point of the line segment joining two points $P(x_1, y_1, z_1)$ and $Q(x_2, y_2, z_2)$ are $\left(\frac{x_1 + x_2}{2}, \frac{y_1 + y_2}{2}, \frac{z_1 + z_2}{2}\right)$.
- The coordinates of the centroid of the triangle, having vertices (x_1, y_1, z_1), (x_2, y_2, z_2) and (x_3, y_3, z_3) are $\left(\frac{x_1 + x_2 + x_3}{3}, \frac{y_1 + y_2 + y_3}{3}, \frac{z_1 + z_2 + z_3}{3}\right)$.
- Point R on the line joining points $P(x_1, y_1, z_1)$ and $Q(x_2, y_2, z_2)$ can be taken as $\left(\frac{tx_2 + x_1}{t+1}, \frac{ty_2 + y_1}{t+1}, \frac{tz_2 + z_1}{t+1}\right)$.

  If $t > 0$, then point R divides PQ internally in the ratio $t:1$.

  If $t < 0$, then point R divides PQ, externally in the ratio $|t|:1$.

Illustration 1.15

Find the coordinates of the point which divides the line segment joining the points (–4, 2, 3) and (3, –5, 7)

a. internally in the ratio 3 : 4

b. externally in the ratio 3 : 4.

Sol.

a. Let $R(x, y, z)$ be the point that divides the line segment joining points $P(-4, 2, 3)$ and $Q(3, -5, 7)$ internally in the ratio 3 : 4.

$$x = \frac{3(3) + 4(-4)}{3+4},\ y = \frac{3(-5) + 4(2)}{3+4} \text{ and } z = \frac{3(7) + 4(3)}{3+4}$$

i.e., $x = -1$, $y = -1$, $z = \frac{33}{7}$

Thus, $R \equiv \left(-1, -1, \frac{33}{7}\right)$.

b. Let $S(x, y, z)$ be the points that divides PQ externally in the ratio 3 : 4.

$$x = \frac{3(3) - 4(-4)}{3-4},\ y = \frac{3(-5) - 4(2)}{3-4} \text{ and } z = \frac{3(7) - 4(3)}{3-4}$$

i.e., $x = -25$, $y = 23$, $z = -9$

Thus, $S \equiv (-25, 23, -9)$.

Illustration 1.16

Prove that points $P(3, 2, -4)$, $Q(5, 4, -6)$ and $R(9, 8, -10)$ are collinear. Find the ratio in which Q divides PR.

Sol. Let the point $Q(5, 4, -6)$ divide the line segment joining $P(3, 2, -4)$ and $R(9, 8, -10)$ in the ratio $k : 1$.

Therefore, by section formula,

$$(5, 4, -6) \equiv \left(\frac{k(9) + 3}{k+1}, \frac{k(8) + 2}{k+1}, \frac{k(-10) - 4}{k+1}\right)$$

If points P, Q and R are collinear, we must get the same value of k from each comparison.

Comparing the x-coordinates, we have

$$\frac{9k + 3}{k+1} = 5$$

$$\Rightarrow\quad 9k + 3 = 5k + 5$$

$$\Rightarrow\quad 4k = 2 \Rightarrow k = \frac{2}{4} = \frac{1}{2}$$

Comparing the y-coordinates, we have

$$\frac{8k + 2}{k+1} = 4$$

$$\Rightarrow\quad 4k + 1 = 2k + 2$$

$$\Rightarrow\quad k = \frac{1}{2}$$

Similarly, comparing the z-coordinates we get $k = \frac{1}{2}$.

Thus, points P, Q and R are collinear.

Also, point Q divides PR internally in the ratio 1 : 2.

Illustration 1.17

Consider two points $P(-2, 3, 5)$ and $Q(1, -4, 3)$. Find the point where line PQ meets

a. yz-plane **b.** xz-plane and

c. xy-plane

Sol. Let the point R divide the line segment joining $P(-2, 3, 5)$ and $Q(1, -4, 3)$ in the ratio $\lambda_1 : 1$.

Hence, by section formula, the coordinates of point R are

$$\left(\frac{\lambda_1(1) - 2}{\lambda_1 + 1}, \frac{\lambda_1(-4) + 3}{\lambda_1 + 1}, \frac{\lambda_1(3) + 5}{\lambda_1 + 1}\right) \quad \ldots\text{(i)}$$

a. If point R lies on the yz-plane then the x-coordinate of R is zero.

$$\therefore\quad \frac{\lambda_1(1) - 2}{\lambda_1 + 1} = 0 \text{ or } \lambda_1 = 2$$

Putting the value of λ_1 in (i), we get

$$R \equiv \left(0, \frac{-5}{3}, \frac{11}{3}\right)$$

b. If point R lies on the xz-plane then the y-coordinate of R is zero.

$$\therefore\quad \frac{\lambda_1(-4) + 3}{\lambda_1 + 1} = 0 \text{ or } \lambda_1 = \frac{3}{4}$$

Putting the value of λ_1 in (i), we get

$$R \equiv \left(\frac{-5}{7}, 0, \frac{29}{7}\right)$$

c. If point R lies on the xy-plane then the z-coordinate of R is zero.

$$\therefore \quad \frac{\lambda_1(3)+5}{\lambda_1+1} = 0 \text{ or } \lambda_1 = -\frac{5}{3}$$

Putting the value of λ_1 in (i), we get

$$R \equiv \left(\frac{11}{2}, -\frac{29}{2}, 0\right)$$

Thus, the yz-plane and the xz-plane divide PQ internally while the xy-plane divides PQ externally.

Illustration 1.18

Prove that the line joining points $A(1, -2, -3)$ and $B(-3, 4, 6)$ intersect the x-axis. Also, find the point of intersection.

Sol. Let the line joining A and B intersects x-axis at $C(x, 0, 0)$.

Let $AC : CB = k : 1$

$$\therefore \quad C \equiv \left(\frac{-3k+1}{k+1}, \frac{4k-2}{k+1}, \frac{6k-3}{k+1}\right) \equiv (x, 0, 0)$$

Now, $4k - 2 = 0$ and $6k - 3 = 0$

From both the equations, we get the same value of k that is $k = \frac{1}{2}$.

Then: $x = \dfrac{-3\cdot\frac{1}{2}+1}{\frac{1}{2}+1} = \dfrac{-1}{3}$

Thus, the line AB meets the x-axis at C.

$$\therefore \quad C \equiv \left(\frac{-1}{3}, 0, 0\right)$$

Illustration 1.19

Prove that the three diagonals of a cube intersect. Also, find the acute angle between any two of them.

Sol. A cube of side length 'a', with three of its edges along the positive coordinate axes is as shown in the figure.

The four diagonals are OP, AD, BE and FC.

The midpoint of OP has coordinates $\left(\frac{a}{2}, \frac{a}{2}, \frac{a}{2}\right)$.

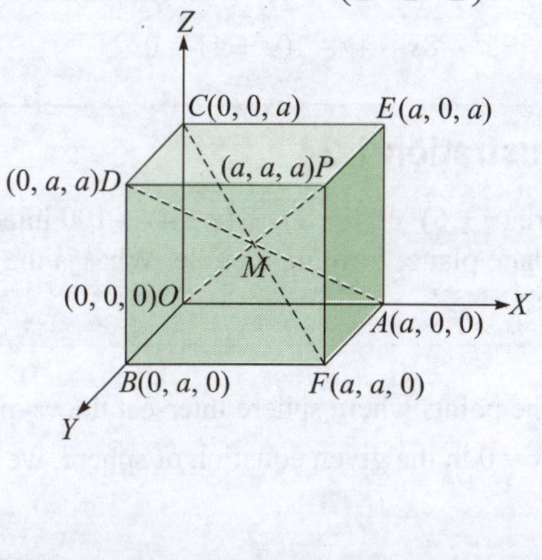

Also, midpoints of AD, BE and FC have coordinates $\left(\frac{a}{2}, \frac{a}{2}, \frac{a}{2}\right)$.

Thus, the diagonals are concurrent at $\left(\frac{a}{2}, \frac{a}{2}, \frac{a}{2}\right) \equiv M$.

Let us find the angle 'θ' between the diagonals OP and AD.

Consider triangle MPD.

$$OP = \sqrt{a^2+a^2+a^2} = \sqrt{3}a$$

$$\therefore \quad MP = MD = \frac{\sqrt{3}a}{2}$$

Also, $PD = a$

Now, in triangle MPD, using the cosine rule, we have

$$\cos\theta = \frac{(MP)^2+(MD)^2-(PD)^2}{2\times MP\times MD} = \frac{\frac{3}{4}a^2+\frac{3}{4}a^2-a^2}{2\times\frac{\sqrt{3}a}{2}\times\frac{\sqrt{3}a}{2}}$$

$$= \frac{\frac{1}{2}a^2}{\frac{3a^2}{2}} = \frac{1}{3}$$

$$\therefore \quad \theta = \cos^{-1}\frac{1}{3}$$

Thus, the acute angle between the diagonals is $\cos^{-1}\frac{1}{3}$.

Illustration 1.20

Prove that the line segments joining each vertex of a tetrahedron to the centroid of the opposite face are concurrent.

Sol. Consider tetrahedron $A_1A_2A_3A_4$ with $A_i \equiv (x_i, y_i, z_i)$; $i = 1, 2, 3, 4$.

The centroid of triangle $A_2A_3A_4$ is

$$G_1 \equiv \left(\frac{x_2+x_3+x_4}{3}, \frac{y_2+y_3+y_4}{3}, \frac{z_2+z_3+z_4}{3}\right)$$

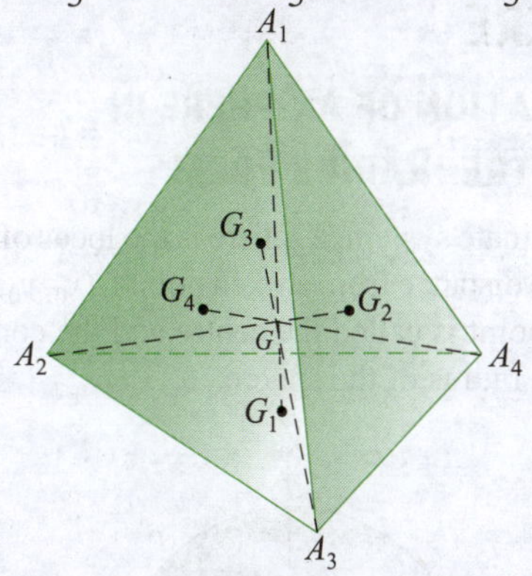

Now, we have a point G on A_1G_1 which divides it in the ratio 3 : 1.

$$\therefore G \equiv \left(\frac{3\left(\frac{x_2+x_3+x_4}{3}\right)+1(x_1)}{3+1}, \frac{3\left(\frac{y_2+y_3+y_4}{3}\right)+1(y_1)}{3+1}, \frac{3\left(\frac{z_2+z_3+z_4}{3}\right)+1(z_1)}{3+1}\right)$$

$$\equiv \left(\frac{x_1+x_2+x_3+x_4}{4}, \frac{y_1+y_2+y_3+y_4}{4}, \frac{z_1+z_2+z_3+z_4}{4}\right)$$

The centroid of triangle $A_1A_3A_4$ is

$$G_2 \equiv \left(\frac{x_1+x_3+x_4}{3}, \frac{y_1+y_3+y_4}{3}, \frac{z_1+z_3+z_4}{3}\right).$$

Now, we have a point on A_2G_2 which divides it in the ratio 3 : 1, having coordinates

$$\left(\frac{x_1+x_2+x_3+x_4}{4}, \frac{y_1+y_2+y_3+y_4}{4}, \frac{z_1+z_2+z_3+z_4}{4}\right) \equiv G$$

Thus, A_1G_1 and A_2G_2 intersect at G.

Similarly, A_3G_3 and A_4G_4 pass through the point G, where G_3 and G_4 are centroids of triangular faces opposite to vertices A_3 and A_4, respectively.

Aim to complete within 15 minutes

Concept Application Exercise 1.3

1. Using the section formula, prove that the points $A(3, 2, -4)$, $B(5, 4, -6)$ and $C(9, 8, -10)$ are collinear.
2. Find the coordinates of the points which trisect the line segment joining the points $P(4, 2, -6)$ and $Q(10, -16, 6)$.
3. Equation $ax + by + cz + d = 0$ represents the plane in the space. Prove that points $A(2, 3, -4)$ and $B(7, 2, 0)$ lie on the same side of the plane $x - y + 2z - 7 = 0$.
4. Prove that the line segments joining midpoints of opposite edges of a tetrahedron are concurrent.

Answers

2. $(6, -4, -2)$ and $(8, -10, 6)$

1.5 SPHERE

1.5.1 Equation of a Sphere in Centre–Radius Form

In the 3D coordinate system, a sphere is the locus of variable point $P(x, y, z)$ whose distance from a fixed point $C(x_0, y_0, z_0)$ is constant 'r'. The fixed point is called the centre and the constant distance 'r' is called the radius of the sphere.

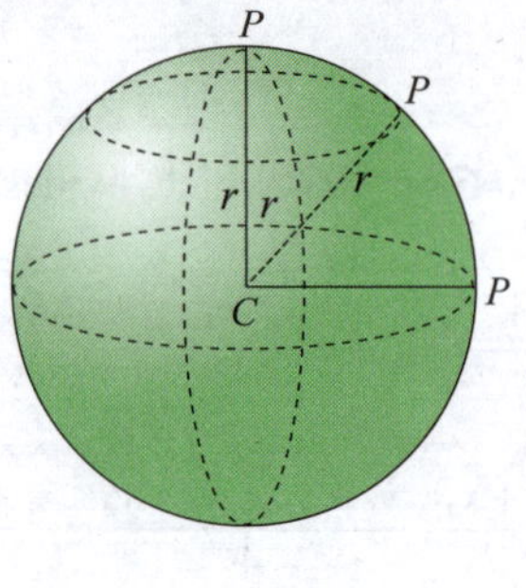

Thus, $CP = r$ or $CP^2 = r^2$

So, $(x - x_0)^2 + (y - y_0)^2 + (z - z_0)^2 = r^2$, which is the equation of the sphere in the centre-radius form.

Illustration 1.21

Find the equation of a sphere with centre $(2, -1, 3)$ and radius 4.

Sol. The equation of sphere is given by

$$(x-2)^2 + (y-(-1))^2 + (z-3)^2 = 4^2,$$

which simplifies to $x^2 + y^2 + z^2 - 4x + 2y - 6z = 2$

Illustration 1.22

Find the equation of the sphere for which two points $A(4, 1, -1)$ and $B(6, 7, 9)$ are diametrically opposite.

Sol. Clearly the centre is the midpoint of AB, which is

$$C \equiv \left(\frac{4+6}{2}, \frac{1+7}{2}, \frac{-1+9}{2}\right) \equiv (5, 4, 4).$$

Radius of the sphere is

$$CA = \sqrt{(5-4)^2 + (4-1)^2 + (4-(-1))^2}$$
$$= \sqrt{1^2 + 3^2 + 5^2} = \sqrt{35}$$

Therefore, the equation of the sphere is $(x-5)^2 + (y-4)^2 + (z-4)^2 = 35$, which simplifies to

$$x^2 + y^2 + z^2 - 10x - 8y - 8z + 22 = 0$$

Illustration 1.23

Find the equation of the largest possible sphere with centre (4, 2, 5) that lies entirely within the first octant. (tangency with a coordinate plane being permissible).

Sol. Clearly the largest possible sphere is the one that touches all the coordinate planes.

Since the y-coordinate of the centre is numerically the smallest among the three coordinates, it will be the radius.

Thus, the required equation of the sphere is

$(x-4)^2 + (y-2)^2 + (z-5)^2 = 4$, which simplifies to

$$x^2 + y^2 + z^2 - 8x - 4y - 10z + 41 = 0$$

Illustration 1.24

The sphere $(x+6)^2 + (y-1)^2 + (z-4)^2 = 100$ intersects the yz-coordinate plane, forming a circle. What is the radius of this circle?

Sol. For the points where sphere intersect the yz–plane, $x = 0$.

So, putting $x = 0$ in the given equation of sphere, we have

$$(0 + 6)^2 + (y - 1)^2 + (z - 4)^2 = 100$$

or $\quad (y - 1)^2 + (z - 4)^2 = 64$

Clearly this is the equation of circle in the yz–plane.

The centre of the circle is (0, 1, 4) and the radius is '8'.

1.5.2 General Equation of a Sphere

The equation of a sphere with centre (x_0, y_0, z_0) and radius 'r' is $(x - x_0)^2 + (y - y_0)^2 + (z - z_0)^2 = r^2$

This equation can be written in the form

$$x^2 + y^2 + z^2 + 2gx + 2fy + 2hz + c = 0.$$

This is the general equation of a sphere. It can be written in the centre-radius form as:

$$(x + g)^2 + (y + f)^2 + (z + h)^2 = g^2 + f^2 + h^2 - c$$

The centre of this sphere is $(-g, -f, -h)$ and the radius is $\sqrt{g^2 + f^2 + h^2 - c}$.

Illustration 1.25

Find the centre and radius of the sphere $x^2 + y^2 + z^2 - 2x + 6y - 4z = 22$.

Sol. The given equation can be rewritten as

$$(x - 1)^2 + (y + 3)^2 + (z - 2)^2 = 36$$

The sphere has centre (1, –3, 2) and radius $r = 6$.

Illustration 1.26

Find the intercept of sphere $(x + 1)^2 + (y - 4)^2 + (z - 6)^2 = 33$ on the z–axis.

Sol. For the points on the z-axis, $x = 0$ and $y = 0$.

So, from the equation of sphere, we have

$$(0 + 1)^2 + (0 - 4)^2 + (z - 6)^2 = 33$$

$\Rightarrow \quad 1 + 16 + (z - 6)^2 = 33$

$\Rightarrow \quad (z - 6)^2 = 16$

$\Rightarrow \quad z - 6 = \pm 4$

$\Rightarrow \quad z = 10$ or 2

Thus, the sphere intersects the z–axis at $A(0, 0, 2)$ and $B(0, 0, 10)$.

Therefore, $AB = 8$.

Illustration 1.27

Find the equation of the sphere passing through the points (1, 0, 0), (0, 1, 0), (0, 0, 1) and having the smallest possible radius.

Sol. Let the equation of the required sphere be

$$x^2 + y^2 + z^2 + 2ux + 2vy + 2wz + d = 0$$

Since the sphere passes through (1, 0, 0), (0, 1, 0) and (0, 0, 1), we get $1 + 2u + d = 0$, $1 + 2v + d = 0$, $1 + 2w + d = 0$

$\Rightarrow \quad u = v = w = -\dfrac{1}{2}(d + 1)$

If R is the radius of the sphere, then $R^2 = u^2 + v^2 + w^2 - d$

$$\Rightarrow \quad R^2 = \frac{3}{4}(d + 1)^2 - d$$

$$= \frac{3}{4}\left[d^2 + 2d + 1 - \frac{4}{3}d\right]$$

$$= \frac{3}{4}\left[d^2 + \frac{2}{3}d + 1\right]$$

$$= \frac{3}{4}[(d + 1/3)^2 + 1 - 1/9] = \frac{3}{4}[(d + 1/3)^2 + 8/9]$$

Clearly, the least value of R^2 occurs, when $d = -\dfrac{1}{3}$.

Therefore, $u = v = w = -\dfrac{1}{2}\left(1 - \dfrac{1}{3}\right) = -\dfrac{1}{3}$

Hence, the equation of the sphere is

$$x^2 + y^2 + z^2 - \frac{2}{3}(x + y + z) - \frac{1}{3} = 0$$

or $\quad 3(x^2 + y^2 + z^2) - 2(x + y + z) - 1 = 0$

Illustration 1.28

Prove that the equation of sphere having diametric end points $P(x_1, y_1, z_1)$ and $Q(x_2, y_2, z_2)$ is $(x - x_1)(x - x_2) + (y - y_1)(y - y_2) + (z - z_1)(z - z_2) = 0$

Sol. Let the variable point on the sphere be $R(x, y, z)$.

Since PQ is diameter, PQ subtends a right-angle at point R.

So, using the Pythagoras theorem, we have

$$PQ^2 = PR^2 + QR^2$$

$\therefore \quad (x_1 - x_2)^2 + (y_1 - y_2)^2 + (z_1 - z_2)^2 = (x - x_1)^2 + (y - y_1)^2 + (z - z_1)^2 + (x - x_2)^2 + (y - y_2)^2 + (z - z_2)^2$

$\Rightarrow \quad 2x^2 + 2y^2 + 2z^2 - 2(x_1 + x_2)x - 2(y_1 + y_2)y - 2(z_1 + z_2)z = -2x_1x_2 - 2y_1y_2 - 2z_1z_2$

$\Rightarrow \quad (x^2 - (x_1 + x_2)x + x_1x_2) + (y^2 - (y_1 + y_2)y + y_1y_2) + (z^2 - (z_1 + z_2)z + z_1z_2) = 0$

$\Rightarrow \quad (x - x_1)(x - x_2) + (y - y_1)(y - y_2) + (z - z_1)(z - z_2) = 0$

NOTES

In 2D coordinate system, the equation of circle in the xy-plane having $P(x_1, y_1)$ and $Q(x_2, y_2)$ as diametric end points is $(x - x_1)(x - x_2) + (y - y_1)(y - y_2) = 0$

Illustration 1.29

A sphere of constant radius k passes through the origin and meets the axes at A, B and C. Prove that the centroid of ΔABC lies on the sphere $9(x^2 + y^2 + z^2) = 4k^2$.

Sol. Let the equation of sphere passing through the origin and having radius k be

$$x^2 + y^2 + z^2 + 2ux + 2vy + 2wz = 0 \quad \text{...(i)}$$

Since, the radius of the sphere is k, we get

$$u^2 + v^2 + w^2 = k^2 \quad \text{...(ii)}$$

Now (i) meets x-axis at $O(0, 0, 0)$ and $A(-2u, 0, 0)$; y-axis at $O(0, 0, 0)$ and $B(0, -2v, 0)$ and z-axis at $O(0, 0, 0)$ and $C(0, 0, -2w)$.

Let the centroid of the triangle ABC be (α, β, γ).

Then $\alpha = -\dfrac{2u}{3}, \beta = -\dfrac{2v}{3}, \gamma = -\dfrac{2w}{3}$

or $\quad u = -\dfrac{3\alpha}{2}, v = -\dfrac{3\beta}{2}, w = -\dfrac{3\gamma}{2}$

Putting these values in (ii), we get

$$\left(\frac{-3}{2}\alpha\right)^2 + \left(\frac{-3}{2}\beta\right)^2 + \left(\frac{-3}{2}\gamma\right)^2 = k^2$$

or, $\quad \alpha^2 + \beta^2 + \gamma^2 = \dfrac{4}{9}k^2$.

Thus, the centroid of ΔABC lies on the sphere $x^2 + y^2 + z^2 = \dfrac{4}{9}k^2$.

Aim to complete within 15 minutes

Concept Application Exercise 1.4

1. Find the equation of the sphere with centre (3, 1, 2) and radius 5.
2. Find the equation of the sphere passing through the points (0, 0, 0), (1, 0, 0), (0, 1, 0) and (0, 0, 1).
3. Find the locus of a point which moves such that the sum of the squares of its distances from the points $A(1, 2, 3)$, $B(2, -3, 5)$ and $C(0, 7, 4)$ is 120.
4. A point $P(x, y, z)$ is such that $3PA = 2PB$, where A and B have the coordinates (1, 3, 4) and (1, –2, –1) respectively. Find the equation of the locus of P and verify that it represents a sphere. Also, find its centre and radius.
5. Find the equation of the sphere having diametric end-points (1, 2, 3) and (–2, 4, –1).

Answers

1. $x^2 + y^2 + z^2 - 6x - 2y - 4z - 11 = 0$
2. $x^2 + y^2 + z^2 - x - y - z = 0$
3. $x^2 + y^2 + z^2 - 2x - 4y - 8z - 1 = 0$
4. $x^2 + y^2 + z^2 - 2x - 14y - 16z + 42 = 0$; Centre $\equiv (1, 7, 8)$;
   Radius $= 6\sqrt{2}$
5. $x^2 + y^2 + z^2 + x - 6y - 2z + 3 = 0$

EXERCISES

Single Correct Answer Type

1. The distance of point $P(3, 4, 5)$ from the xy-plane is
(1) 3 units (2) 4 units
(3) 5 units (4) $5\sqrt{2}$ units

2. The length of perpendicular drawn from the point $P(-3, -4, 5)$ on the y-axis is
(1) $\sqrt{41}$ (2) $\sqrt{34}$
(3) 5 (4) 6

3. Let M be the foot of the perpendicular drawn from a point $P(-1, 7, -5)$ on the xy-plane. The coordinates of point M are
(1) (–1, 7, 0) (2) (1, –7, 0)
(3) (7, 5, 0) (4) (–1, 6, 0)

4. If the feet of perpendiculars from point $P(1, 2, 3)$ on the x, y and z-axis are A, B and C respectively, then the distance of the centroid of triangle ABC from the origin is
(1) $\dfrac{\sqrt{14}}{3}$ (2) $\dfrac{\sqrt{14}}{2}$
(3) $\sqrt{14}$ (4) $3\sqrt{14}$

5. If the feet of perpendiculars from point $P(1, 2, 3)$ on the xy, yz and zx-plane are A, B and C respectively, then the distance of the centroid of triangle ABC from the origin is
(1) $\dfrac{\sqrt{14}}{3}$ (2) $\dfrac{2\sqrt{14}}{3}$
(3) $\dfrac{3\sqrt{14}}{2}$ (4) $3\sqrt{14}$

6. A point R with x–coordinate 6 lies on the line segment joining the points $P(2, -3, 4)$ and $Q(8, 0, 10)$. Then distance of R from the origin is
(1) $\sqrt{47}$ (2) $\sqrt{41}$
(3) $\sqrt{101}$ (4) $3\sqrt{7}$

7. The volume of the cuboid having combined equations of opposite faces as $x^2 - 3x - 4 = 0$, $y^2 - 5y + 6 = 0$ and $z^2 + 6z + 8 = 0$ is
(1) 20 cu. units
(2) 10 cu. units
(3) 15 cu. units
(4) 24 cu. units

8. Number of spheres touching all the coordinate planes and having the fixed radius is
(1) 2 (2) 4
(3) 6 (4) 8

9. If the sum of the squares of the distances of a point from the three coordinate axes is 36, then its distance from the origin is
(1) 5 (2) $3\sqrt{2}$
(3) $\sqrt{21}$ (4) $3\sqrt{3}$

10. The xy-plane divides the line joining the points (–1, 3, 4) and (2, –5, 6)
(1) internally in the ratio 2 : 3
(2) externally in the ratio 2 : 3
(3) internally in the ratio 3 : 2
(4) externally in the ratio 4 : 3

11. If the points (–1, 3, 2), (–4, 2, –2) and (5, 5, λ) are collinear, then λ =
(1) 10 (2) –10
(3) –8 (4) 8

12. From which of the following is the distance of the point (1, 2, 3) equal to $\sqrt{10}$?
(1) origin (2) x-axis
(3) y-axis (4) z-axis

13. Distances of the point (1, 2, 3) from the x, y and z-axes are, respectively,
(1) 13, 10, 5 (2) $\dfrac{1}{\sqrt{13}}, \dfrac{1}{\sqrt{10}}, \dfrac{1}{\sqrt{5}}$
(3) $\sqrt{5}, \sqrt{13}, \sqrt{10}$ (4) $\sqrt{13}, \sqrt{10}, \sqrt{5}$

14. If the centroid of a triangle whose vertices are $(a, 1, 3)$, $(-2, b, -5)$ and $(4, 7, c)$ is the origin, then the values of a, b and c are
(1) –2, –8, –2 (2) 2, 8, –2
(3) –2, –8, 2 (4) 7, –1, 0

15. The point of intersection of the line joining the points (3, 4, 1) and (5, 1, 6) and the xy-plane is
(1) (13, 23, 0) (2) $\left(\dfrac{13}{5}, \dfrac{23}{5}, 0\right)$
(3) (–13, 23, 0) (4) $\left(-\dfrac{13}{5}, \dfrac{23}{5}, 0\right)$

16. The coordinates of a point which is equidistant from the points (0, 0, 0), $(a, 0, 0)$, $(0, b, 0)$ and $(0, 0, c)$ are given by
(1) $\left(\dfrac{a}{2}, \dfrac{b}{2}, \dfrac{c}{2}\right)$ (2) $\left(\dfrac{a}{3}, \dfrac{b}{3}, \dfrac{c}{3}\right)$
(3) $\left(\dfrac{a}{4}, \dfrac{b}{4}, \dfrac{c}{4}\right)$ (4) $\left(-\dfrac{a}{2}, -\dfrac{b}{2}, -\dfrac{c}{2}\right)$

17. The points $A(5, -1, 1)$, $B(7, -4, 7)$, $C(1, -6, 10)$ and $D(-1, -3, 4)$ are vertices of a
(1) square (2) rhombus
(3) rectangle (4) parallelogram

18. Which of the following points lies inside the sphere $x^2 + y^2 - 4x - 6y + 8z - 7 = 0$?
(1) (1, –2, 1) (2) (0, 0, 3)
(3) (0, 1, 2) (4) (1, 2, 0)

19. Number of points common to the spheres $x^2 + y^2 + z^2 = 36$ and $x^2 + y^2 + z^2 - 20x + 84 = 0$ is
(1) zero (2) one
(3) infinite (4) two

20. The xy-plane cuts the sphere $x^2 + y^2 + z^2 - 2x - 4y + 6z - 4 = 0$ in a circle, whose radius is

(1) 2 (2) 3
(3) 4 (4) 5

21. If the equation $6x + 3y - 2z = 18$ represents a plane in space, then the centroid of ΔABC where, A, B, C are the points of intersection of the plane with the coordinate axes is

(1) (1, 2, –3)
(2) (–1, 2, 3)
(3) (–1, –2, –3)
(4) (1, – 2, 3)

22. If the reflection of the point (2, 3, –4) in the x-axis is Q and in the xy-plane is R, then QR is

(1) $2\sqrt{29}$ (2) 6
(3) $2\sqrt{21}$ (4) 7

23. Consider the point $P(2, -3, -5)$, then the match the following lists.

	List I		List II
(P)	the reflection of P in the x-axis lies in the octant	(I)	2
(Q)	the reflection of P in the xy-plane lies in the octant	(II)	1
(R)	the reflection of P in the y-axis lies in the octant	(III)	3
(S)	the reflection of P in the yz-plane lies in the octant	(IV)	4
		(V)	7

(1) (P) → (II); (Q) → (III); (R) → (IV); (S) → (V)
(2) (P) → (III); (Q) → (II); (R) → (II); (S) → (IV)
(3) (P) → (II); (Q) → (IV); (R) → (III); (S) → (V)
(4) (P) → (II); (Q) → (IV); (R) → (I); (S) → (V)

ANSWER KEY

EXERCISES

Single Correct Answer Type

1. (3) **2.** (2) **3.** (1) **4.** (1) **5.** (2)
6. (3) **7.** (2) **8.** (4) **9.** (2) **10.** (2)
11. (1) **12.** (3) **13.** (4) **14.** (3) **15.** (2)
16. (1) **17.** (2) **18.** (4) **19.** (2) **20.** (2)
21. (1) **22.** (1) **23.** (3)

Chapter

2 Introduction to Vectors

Chapter Prerequisites

Before starting this chapter, ensure that you understand the following concepts thoroughly:

1. **Coordinate Geometry in Two and Three Dimensions:** Understanding of coordinate axes, plotting points, and the use of distance and section formulas, along with familiarity in representing points in space using ordered triples.
2. **Basic Spatial and Geometric Visualization:** Ability to visualise directions, line segments, and simple three-dimensional positions, and interpret geometric ideas such as angles, lengths, and parallelism.
3. **Basic Understanding of Scalars and Vectors:** Awareness of the difference between scalar and vector quantities in terms of magnitude and direction.
4. **Comfort with Algebraic and Step-wise Reasoning:** Ability to simplify algebraic expressions, work with equations, and follow multi-step procedures needed for computations involving vector magnitude, direction, and components.

Topic List

- **2.1** Vectors Definition
  - **2.1.1** Types of Vectors
  - **2.1.2** Angle Between Vectors
  - **2.1.3** Addition of Vectors
  - **2.1.4** Section Formula
- **2.2** Vectors in Components Form
  - **2.2.1** Direction Cosines of a Vector
  - **2.2.2** Angle Between Two Vectors
  - **2.2.3** Vectors Addition in Component Form
  - **2.2.4** Bisector of Angle Between Two Vectors
- **2.3** Linear Combination of Vectors
  - **2.3.1** Linearly Dependent Vectors
  - **2.3.2** Linearly Independent Vectors

Learning Objectives

Definition of Vectors & Basic Concepts

Understand vectors and their types, identify angles between vectors, distinguish collinear and non-collinear vectors, and apply geometric laws of vector addition.

Linear Combination of Vectors

Learn about linear combinations of vectors and distinguish between linearly dependent and linearly independent vectors.

Vectors in Component Form

Express vectors in component form, determine direction cosines and magnitude, and perform vector addition using components.

2.1 VECTORS DEFINITION

In day-to-day life and science, we come across two types of quantities. One in which only magnitude is required, such as mass, time, area, volume, distance, speed, etc. All we want to know about these quantities is how much, i.e., "What is the mass?", "What time is it?", What is the speed?" These quantities can be specified only by magnitude. Such quantities are called scalar quantities, which have magnitude only. On the other hand, we find quantities in which both magnitude and direction are required, such as displacement, velocity, force, acceleration etc. If some force is being applied to a body, we are interested in knowing how much force has been applied and in which direction. In the displacement of a particle, we are interested in knowing the distance between the initial and terminal position of the particle and the direction of the terminal position with respect to the original position. Such quantities are called vector quantities, which have both magnitude and direction.

In mathematics a vector is understood as a directed line segment. Let 'l' be any straight line in a plane or three-dimensional space. This line can be given two directions by means of arrowheads. A line with one of these directions prescribed is called a directed line (figure (a), (b)).

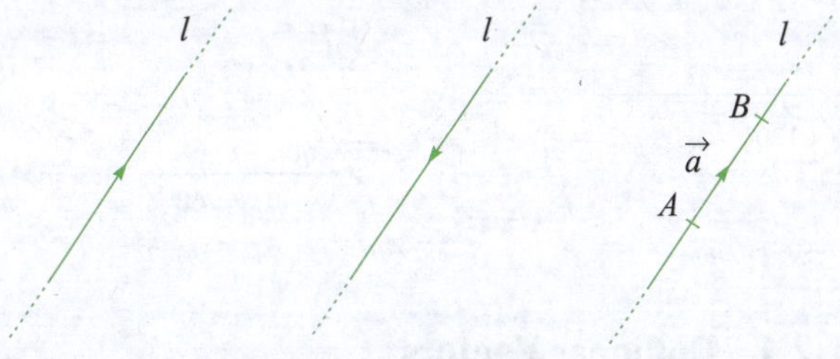

(a) (b) (c)

Now, if we restrict the line 'l' to the line segment AB, then a magnitude is prescribed on the line 'l' with one of the two directions, so that we obtain a directed line segment (figure (c)). Thus, a directed line segment has magnitude as well as direction, which is called a vector.

In figure (c), directed line segment is from point A to B, which is denoted by $\overrightarrow{AB}$, read as "vector AB". This vector can also be denoted by $\vec{a}$. The point A from where vector $\overrightarrow{AB}$ starts is called its initial point (tail) and the point B where it ends is called its terminal point (head). The distance between the initial and terminal points of a vector is called the magnitude (or length) of the vector, denoted as $|\overrightarrow{AB}|$ or $|\vec{a}|$ or a. The arrow indicates the direction of the vector.

2.1.1 Types of Vectors

Vectors can be classified into different types based on their magnitude, direction, and position. The main types are described below.

1. **Zero Vector (Null Vector):** A vector whose initial and terminal points coincide, is called a zero vector (or null vector), and it is denoted by $\vec{0}$ or simply 0. Clearly, the magnitude of the null vector is zero, and it cannot be assigned a definite direction since it has zero magnitude. The vectors $\overrightarrow{AA}$ and $\overrightarrow{BB}$ are examples of zero vectors.
2. **Unit Vector:** A vector of magnitude equal to unity is called a unit vector. A unit vector is denoted by a small letters with a cap on it, e.g., $\hat{a}$ is a unit vector and $|\hat{a}| = 1$. The unit vectors in the positive directions of the x-, y- and z-axes are denoted by $\hat{i}$, $\hat{j}$ and $\hat{k}$ respectively.
3. **Localized Vector:** A vector whose initial point is fixed is called a localised vector.
4. **Free Vector:** A vector whose initial point is not fixed is called a free vector or non-localised vector. This vector can move anywhere in space but has a fixed magnitude and direction. This is called parallel shifting of a vector, which means changing the position of a vector from one place to another without changing its direction and magnitude.
5. **Position Vector:** A position vector is a vector in a coordinate system that defines the position of a point in space with respect to reference point, usually the origin. Thus, the position vector of any point is a localized vector whose initial point is the origin. The position vector of point P with respect to origin O is written as $\overrightarrow{OP}$. If the coordinates of point P are (x, y, z), then $|\overrightarrow{OP}| = \sqrt{x^2 + y^2 + z^2}$. If $\overrightarrow{OP} = \vec{r}$, then the position vector of point P is $\vec{r}$ and point P is denoted as $P(\vec{r})$.
6. **Co-initial Vectors:** Two or more vectors having the same initial point are called co-initial vectors. As shown in the figure, all the vectors are originating or emanating from point P. These vectors are called co-initial vectors.

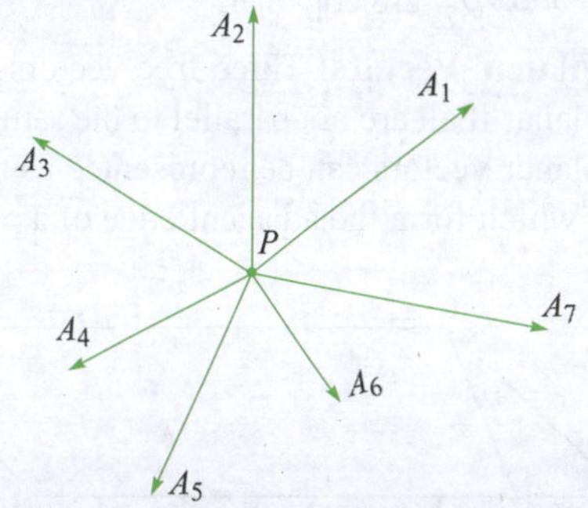

7. **Co-terminal (Coterminous) Vectors:** Two or more vectors having the same terminal point are called co-terminal vectors.

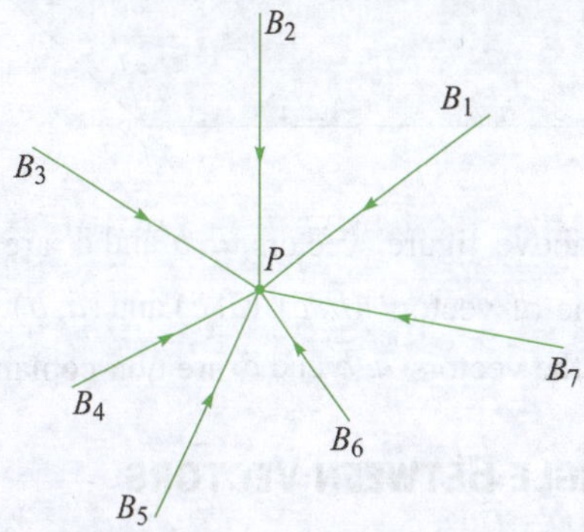

As shown in the figure, all vectors terminate at point P. These vectors are co-terminal vectors.

8. **Equal Vectors:** Since a vector has both magnitude and direction, two vectors $\vec{a}$ and $\vec{b}$ are said to be equal, if they have the same magnitude and direction, regardless of the positions of their initial points. If two vectors $\vec{a}$ and $\vec{b}$ are equal, we write $\vec{a} = \vec{b}$.

9. **Coplanar Vectors:** Three free vectors are said to be coplanar if they are parallel to the same plane. Similarly, three localized vectors are coplanar if they lie in the same plane. In the following figure, vectors $\vec{a}, \vec{b}$ and $\vec{c}$ are parallel to the plane P, hence they are coplanar.

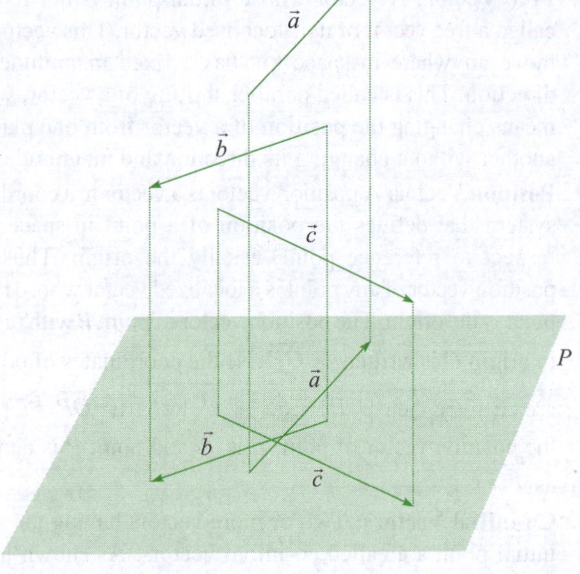

Four points A, B, C and D are coplanar iff vectors $\overrightarrow{AB}$, $\overrightarrow{AC}$ and $\overrightarrow{AD}$ are coplanar.

10. **Non-coplanar Vectors:** Three free vectors are said to be non-coplanar if all are not parallel to the same plane. Three non-coplanar vectors can be represented as three co-initial vectors, which form the adjacent edge of a parallelepiped.

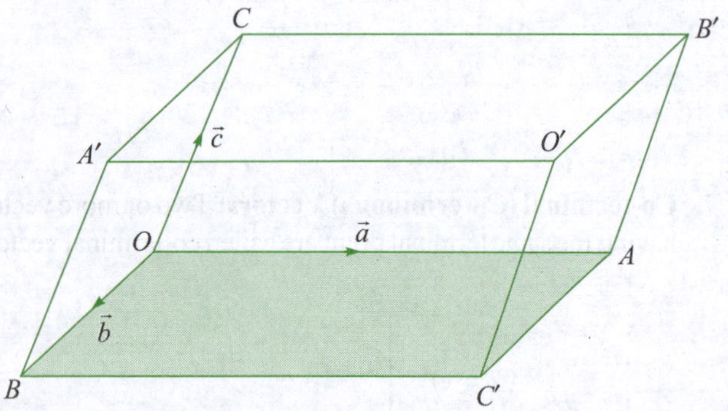

In the above figure, vectors $\vec{a}, \vec{b}$ and $\vec{c}$ are not lying in the plane of vectors $(\vec{b}, \vec{c}), (\vec{a}, \vec{c})$ and $(\vec{a}, \vec{b})$ respectively. Hence, the vectors $\vec{a}, \vec{b}$ and $\vec{c}$ are non-coplanar.

2.1.2 Angle Between Vectors

The **angle between two vectors** is the angle formed at the intersection of their tails or heads. If the vectors are NOT joined tail-to-tail, then one of them can be shifted (by parallel shifting) so that they are placed tail-to-tail or head-to-head.

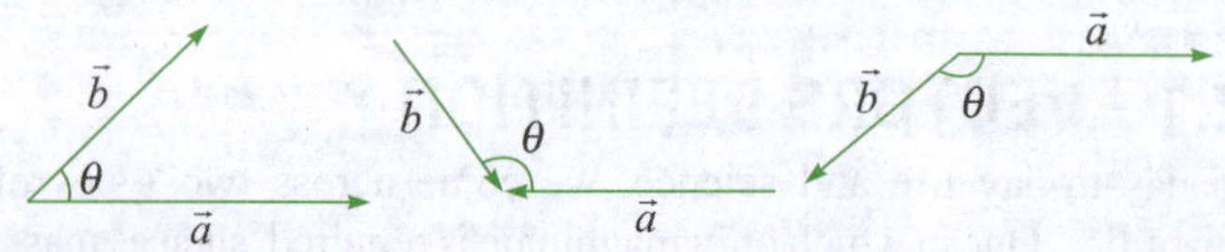

The angle between two vectors can be acute, right, or obtuse, depending their directions.

The Angle θ between two vectors lies in the interval $[0, \pi]$.

If two vectors are in the same direction, the angle between them is $0°$ and if two vectors are in the opposite directions, the angle between them is $180°$.

Here are some examples to illustrate how to find the angle between two vectors.

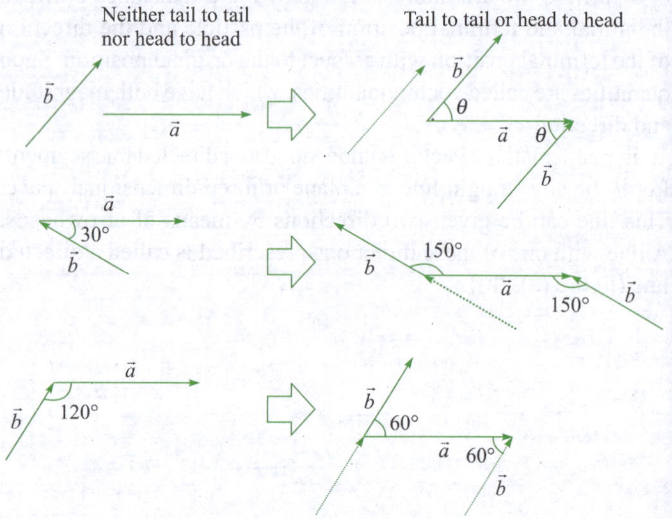

2.1.2.1 Collinear Vectors

Two or more vectors are said to be collinear if they are parallel to the same line, irrespective of their magnitudes and directions. Thus, two collinear vectors are either in the same direction (called like vectors) or in the opposite directions (called unlike vectors). Equal vectors are always collinear but collinear vectors need not be equal. The angle between collinear vectors is either $0°$ or $180°$.

In the following figure, vectors $\vec{a}, \vec{b}$ and $\vec{c}$ are collinear, as they are parallel to the same straight line.

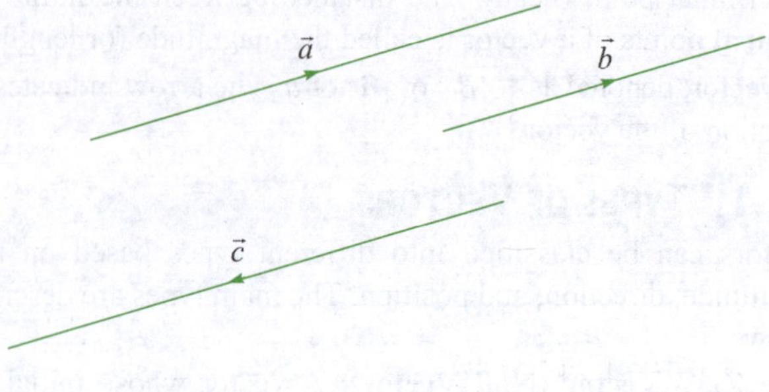

Points A, B and C are collinear it the vectors $\overrightarrow{AB}$ and $\overrightarrow{BC}$ or $\overrightarrow{AB}$ and $\overrightarrow{AC}$ or $\overrightarrow{BC}$ and $\overrightarrow{AC}$ are collinear.

2.1.2.2 Non-collinear Vectors

Two vectors which are not collinear are called non-collinear vectors. In other words, two vectors are said to be non-collinear if they are neither in the same direction nor in opposite directions. Two localized non-collinear vectors define a fixed plane, as shown in the figure.

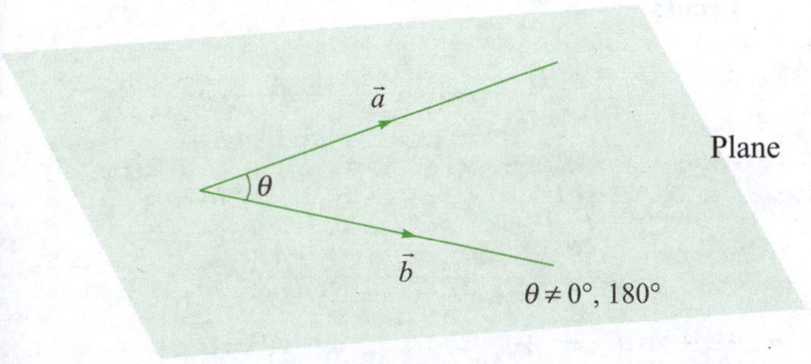

2.1.3 Addition of Vectors

We know that a vector $\overrightarrow{AB}$ simply means the displacement from a point A to the point B. Now consider a situation that a girl moves from point A to B and then from point B to C. The net displacement made by the girl from point A to the point C is given by the vector $\overrightarrow{AC}$ and expressed as $\overrightarrow{AC} = \overrightarrow{AB} + \overrightarrow{BC}$.

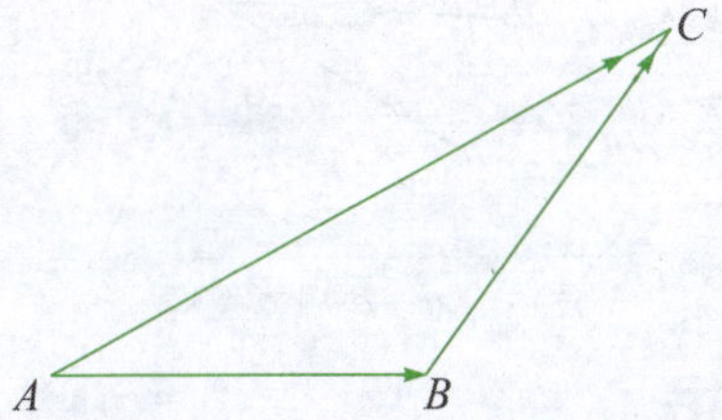

This is known as the triangle law of vector addition. Vector $\overrightarrow{AC}$ is called the resultant of vectors $\overrightarrow{AB}$ and $\overrightarrow{BC}$.

In general, if we have two free vectors $\vec{a}$ and $\vec{b}$ (figure, (a)), then to add them, they are positioned so that the initial point of one coincides with the terminal point of the other (figure, (b)).

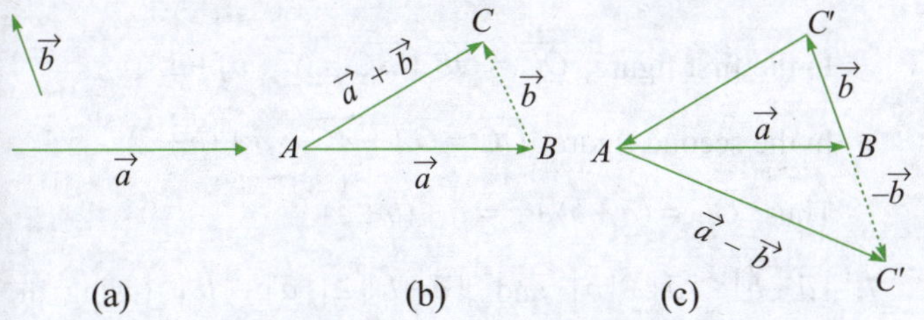

(a) (b) (c)

For example, in figure (b), we have shifted vector $\vec{b}$ without changing its magnitude and direction, so that its initial point coincides with the terminal point of $\vec{a}$. Then the vector $\vec{a}+\vec{b}$, represented by the third side AC of triangle ABC, gives the sum (or resultant) of the vectors $\vec{a}$ and $\vec{b}$, i.e. in triangle ABC [figure (b)], we have

$$\overrightarrow{AB} + \overrightarrow{BC} = \overrightarrow{AC}$$

Also, $\overrightarrow{AB} + \overrightarrow{BC} + \overrightarrow{CA} = \overrightarrow{AC} + \overrightarrow{CA} = \overrightarrow{AA} = \vec{0}$

This means that when vectors representing the sides of a triangle are taken in order, the resultant is zero, since the initial and terminal points coincide.

Now, construct a vector $\overrightarrow{BC'}$ such that its magnitude is the same as that of vector $\overrightarrow{BC}$, but its direction is opposite (figure (c)), i.e., $\overrightarrow{BC'} = -\overrightarrow{BC}$

Then, by applying the triangle law, (figure (c)), we have

$$\overrightarrow{AC'} = \overrightarrow{AB} + \overrightarrow{BC'} = \overrightarrow{AB} - \overrightarrow{BC} = \vec{a} - \vec{b}$$

Thus, the vector $\overrightarrow{AC'}$ represents the difference of $\vec{a}$ and $\vec{b}$.

Now, consider a boat in a river going from one bank to the other in a direction perpendicular to the flow of the river. It is acted upon by two velocity vectors – one is the velocity imparted to the boat by its engine and the other is the velocity of the flow of the river water. Under the simultaneous influence of these two velocities, the boat actually starts traveling with a different velocity. To have a precise idea about the effective speed and direction (i.e., the resultant velocity) of the boat, we use the following law of vector addition.

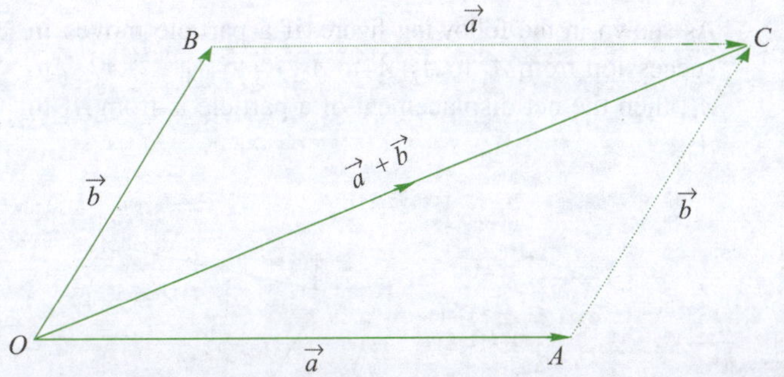

If we have two vectors $\vec{a}$ and $\vec{b}$, represented by the two adjacent sides of a parallelogram in magnitude and direction, then their sum $\vec{a}+\vec{b}$ is represented in both magnitude and direction by the diagonal of the parallelogram passing through their common point. This is known as the parallelogram law of vector addition.

Also, from the figure, using the triangle law, we have

$$\overrightarrow{OA} + \overrightarrow{AC} = \overrightarrow{OC}$$

or $\quad \overrightarrow{OA} + \overrightarrow{OB} = \overrightarrow{OC} \qquad$ (since $\overrightarrow{AC} = \overrightarrow{OB}$)

This is the parallelogram law of vector addition.

Thus, we may say that the triangle law and the parallelogram law of vector addition are equivalent.

NOTES

- If $\vec{a}$ and $\vec{b}$ are collinear vectors, then their resultant is still $\vec{a}+\vec{b}$, although in this case we do not obtain a triangle or a parallelogram in the usual sense.

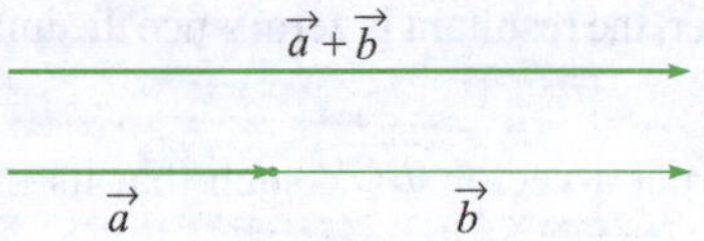

- **Polygon law of vector addition**
  The polygon law of vector addition states that if a number of vectors can be represented, in both magnitude and direction, by the sides of a polygon taken in order, then their resultant is represented in both magnitude and direction by the closing side of the polygon taken in the opposite order.

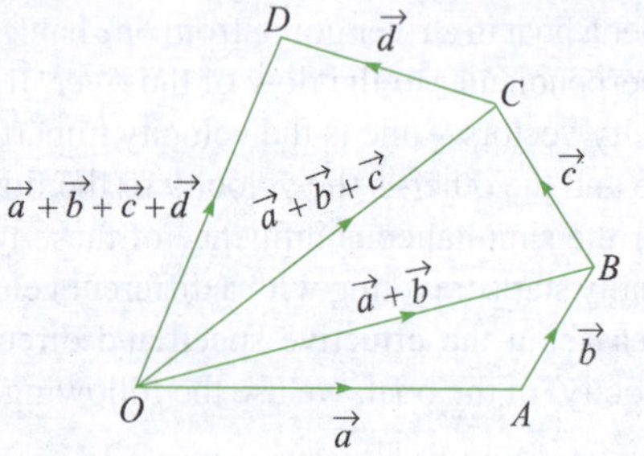

- **Resultant of 'n' number of vectors**
  As shown in the following figure, if a particle moves in succession from A_1 to A_2, A_2 to A_3, A_3 to A_4, …, A_{n-1} to A_n, then the net displacement of a particle is from A_1 to A_n.

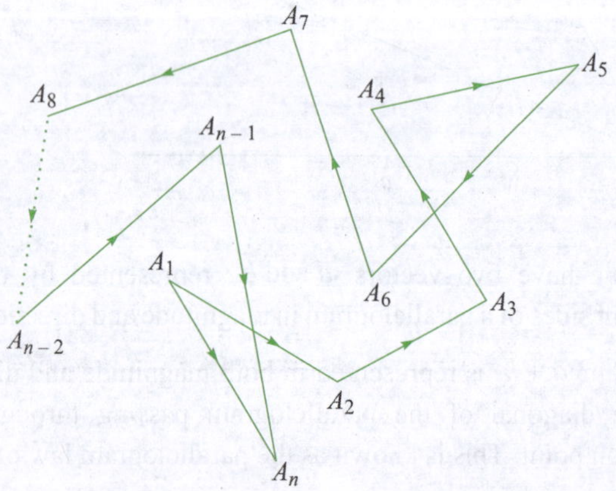

This can be written in vector-addition form as $\overline{A_1A_2}+\overline{A_2A_3}+\overline{A_3A_4}\ldots\overline{A_{n-2}A_{n-1}}+\overline{A_{n-1}A_n}=\overline{A_1A_n}$.

Thus, the resultant of vectors $\overline{A_1A_2}$, $\overline{A_2A_3}$, $\overline{A_3A_4}\ldots$, $\overline{A_{n-1}A_n}$ is $\overline{A_1A_n}$.

2.1.3.1 Properties of Vector Addition

The following are the fundamental properties of vector addition, which hold for all vectors in space.

1. $\vec{a}+\vec{0}=\vec{a}$ (additive identity)
2. $\vec{a}+(-\vec{a})=\vec{0}$ (additive inverse)
3. $\lambda\vec{a}\pm\mu\vec{a}=(\lambda\pm\mu)\vec{a}$, where λ and μ are scalars.
4. $\lambda(\vec{a}+\vec{b})=\lambda\vec{a}+\lambda\vec{b}$, where λ is a scalar.
5. Vector addition is commutative, i.e. $\vec{a}+\vec{b}=\vec{b}+\vec{a}$

Proof:

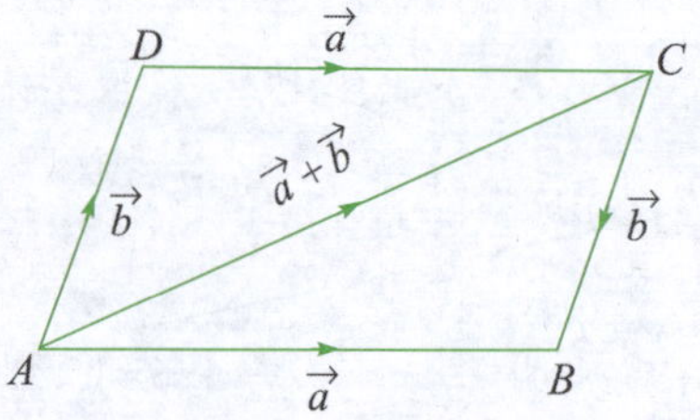

Since $\overline{AB}+\overline{BC}=\overline{AC} \Rightarrow \vec{a}+\vec{b}=\overline{AC}$

Also, $\overline{AD}+\overline{DC}=\overline{AC} \Rightarrow \vec{b}+\vec{a}=\overline{AC}$

Thus, $\vec{a}+\vec{b}=\vec{b}+\vec{a}$

6. Vector addition is associative, i.e. $(\vec{a}+\vec{b})+\vec{c}=\vec{a}+(\vec{b}+\vec{c})$

Proof:

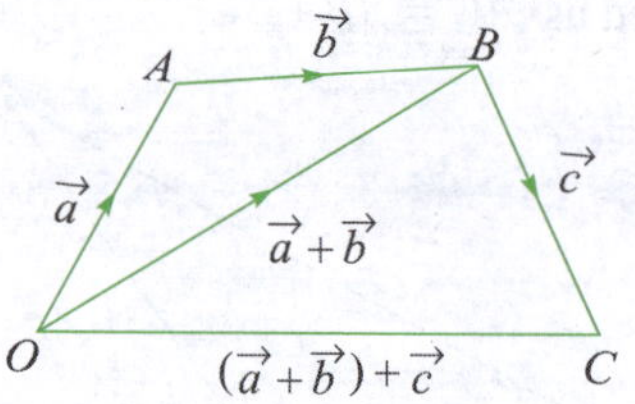

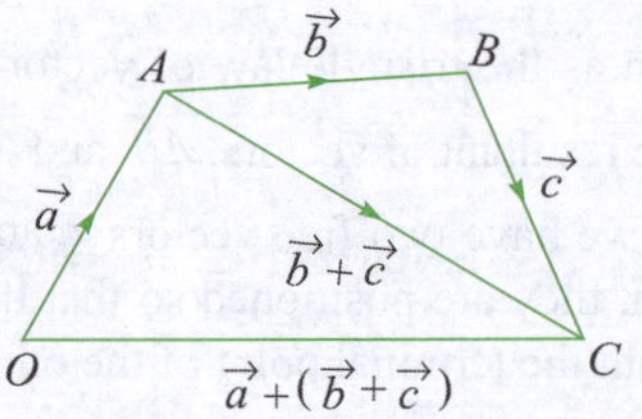

In the first figure, $\overline{OC}=\overline{OB}+\overline{BC}=(\vec{a}+\vec{b})+\vec{c}$

In the second figure, $\overline{OC}=\overline{OA}+\overline{AC}=\vec{a}+(\vec{b}+\vec{c})$

Thus, $\overline{OC}=(\vec{a}+\vec{b})+\vec{c}=\vec{a}+(\vec{b}+\vec{c})$

7. $|\vec{a}+\vec{b}|\le|\vec{a}|+|\vec{b}|$ and $||\vec{a}-\vec{b}||\ge||\vec{a}|-|\vec{b}||$ (using the triangular inequality).

   Equality holds when $\vec{a}$ and $\vec{b}$ are in the same direction.

   Also, $|\vec{a}-\vec{b}|\le|\vec{a}|+|\vec{b}|$ and $||\vec{a}+\vec{b}||\ge||\vec{a}|-|\vec{b}||$ (using the triangular inequality).

   Equality holds when vectors $\vec{a}$ and $\vec{b}$ are in the opposite directions.

8. $\overrightarrow{AB} = \overrightarrow{OB} - \overrightarrow{OA}$ = Position vector of B – Position vector of A, where O is the origin.

Proof:

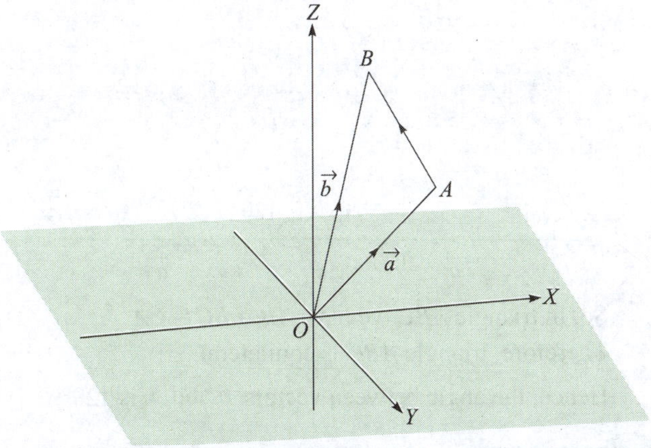

From the figure, using the triangle rule of vector addition, we have,

$$\overrightarrow{OB} = \overrightarrow{OA} + \overrightarrow{AB}$$

$\therefore \quad \overrightarrow{OB} - \overrightarrow{OA} = \overrightarrow{OA} - \overrightarrow{OA} + \overrightarrow{AB}$

$\Rightarrow \quad \overrightarrow{OB} - \overrightarrow{OA} = \vec{0} + \overrightarrow{AB}$

$\Rightarrow \quad \overrightarrow{AB} = \overrightarrow{OB} - \overrightarrow{OA}$ = Position vector of B – Position vector of A

$\Rightarrow \quad \overrightarrow{AB} = \vec{b} - \vec{a}$

9. If three vectors $\vec{a}, \vec{b}$ and $\vec{c}$ are coplanar then $\vec{c} = \lambda\vec{a} + \mu\vec{b}$, where λ and μ are scalars.

Proof:

Three coplanar vectors $\vec{a}, \vec{b}$ and $\vec{c}$ can be made coinitial at a point O.

We have $\overrightarrow{OA} = \vec{a}$, $\overrightarrow{OB} = \vec{b}$ and $\overrightarrow{OC} = \vec{c}$

Now, we draw parallelogram $OA'CB'$ by drawing vectors parallel to $\vec{a}$ and $\vec{b}$ through the terminal point C of the vector $\overrightarrow{OC}$.

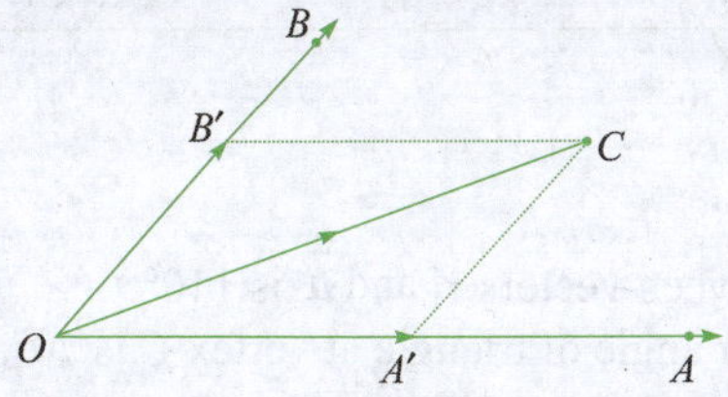

$\therefore \quad \overrightarrow{OA'} = \lambda\vec{a}$ and $\overrightarrow{OB'} = \mu\vec{b}$, where λ and μ are scalars.

Using the parallelogram law of addition, we have

$$\overrightarrow{OC} = \overrightarrow{OA'} + \overrightarrow{OB'} \text{ or } \vec{c} = \lambda\vec{a} + \mu\vec{b}$$

Similarly, for some scalars $\lambda_1, \mu_1, \lambda_2, \mu_2$, we can write

$$\vec{a} = \lambda_1\vec{b} + \mu_1\vec{c} \text{ and } \vec{b} = \lambda_2\vec{a} + \mu_2\vec{c}.$$

10. Three vectors $\vec{a}$, $\vec{b}$ and $\vec{c}$ represent the sides of a triangle (taken head-to-tail in order) if and only if $\vec{a} + \vec{b} + \vec{c} = \vec{0}$.

Proof:

Place the vectors head-to-tail to form a closed triangle: start at a point P, draw $\vec{a}$ from P to Q, then $\vec{b}$ from Q to R, then $\vec{c}$ from R back to P.

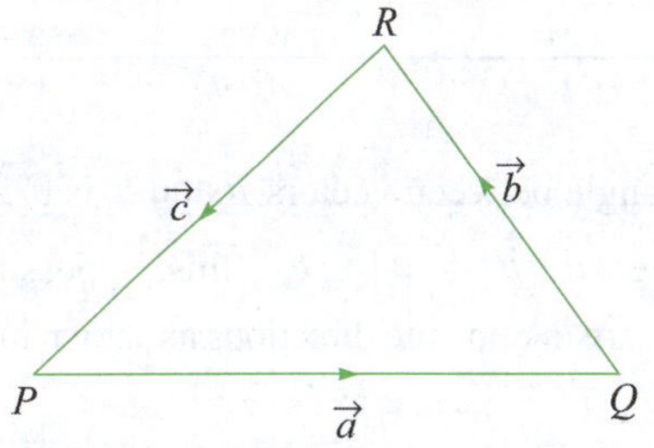

By the triangle (polygon) law of vector addition,

$$\vec{a} + \vec{b} = \overrightarrow{PR}$$

Since, $\vec{c} = \overrightarrow{RP} = -\overrightarrow{PR}$, we get

$$\vec{a} + \vec{b} + \vec{c} = \overrightarrow{PR} + \overrightarrow{RP} = \vec{0}$$

Now, let $\vec{a} + \vec{b} + \vec{c} = \vec{0}$

Then $\vec{c} = -(\vec{a} + \vec{b})$

Place $\vec{a}$ from P to Q, then $\vec{b}$ from Q to R. The resultant $\vec{a} + \vec{b}$ is $\overrightarrow{PR}$, so, $c = -\overrightarrow{PR} = \overrightarrow{RP}$.

Thus, drawing $\vec{c}$ from R to P closes the figure PQR, giving a triangle with successive sides $\vec{a}, \vec{b}, \vec{c}$.

Illustration 2.1

If $2\overrightarrow{AC} = 3\overrightarrow{CB}$, then prove that $2\overrightarrow{OA} + 3\overrightarrow{OB} = 5\overrightarrow{OC}$, where O is origin.

Sol. We have $2\overrightarrow{AC} = 3\overrightarrow{CB}$.

$\therefore \quad 2(\overrightarrow{OC} - \overrightarrow{OA}) = 3(\overrightarrow{OB} - \overrightarrow{OC})$

$\Rightarrow \quad 2\overrightarrow{OC} + 3\overrightarrow{OC} = 2\overrightarrow{OA} + 3\overrightarrow{OB}$

$\Rightarrow \quad 5\overrightarrow{OC} = 2\overrightarrow{OA} + 3\overrightarrow{OB}$

Illustration 2.2

In each of the following cases, find the angle between non-zero vectors $\vec{a}$ and $\vec{b}$.

a. $|\vec{a} + \vec{b}| = |\vec{a}| + |\vec{b}|$ **b.** $|\vec{a} - \vec{b}| = |\vec{a}| + |\vec{b}|$

c. $|\vec{a} + \vec{b}|^2 = |\vec{a}|^2 + |\vec{b}|^2$ **d.** $|\vec{a} + \vec{b}| = |\vec{a} - \vec{b}|$

e. $|\vec{a}| = |\vec{b}| = |\vec{a} + \vec{b}|$

Sol.

a. We have $|\vec{a} + \vec{b}| = |\vec{a}| + |\vec{b}|$, which is possible if vectors $\vec{a}$ and $\vec{b}$ are in the same direction as shown in the following figure.

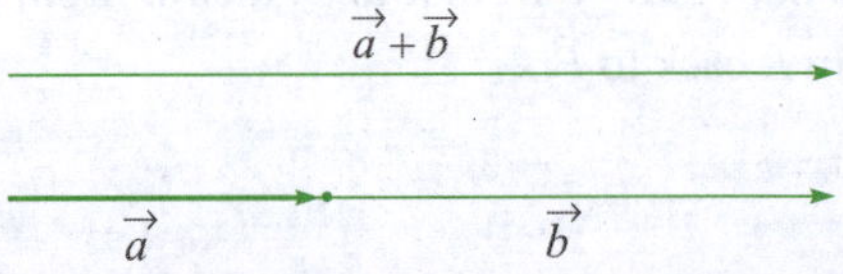

So, the angle between vectors $\vec{a}$ and $\vec{b}$ is 0°.

b. We have $|\vec{a}-\vec{b}|=|\vec{a}|+|\vec{b}|$, this is possible if vectors $\vec{a}$ and $\vec{b}$ are in opposite directions as shown in the following figure.

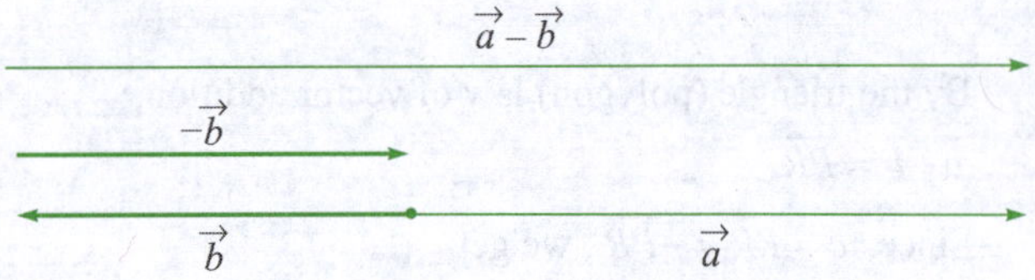

So, the angle between vectors $\vec{a}$ and $\vec{b}$ is 180°.

c. We have $|\vec{a}+\vec{b}|^2=|\vec{a}|^2+|\vec{b}|^2$.

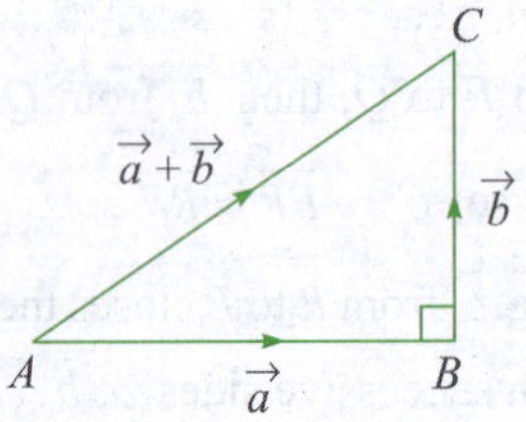

As shown in the figure this is possible if vectors $\vec{a}$ and $\vec{b}$ are the sides of a right-angled triangle.

So, the hypotenuse is $\vec{a}+\vec{b}$.

Using Pythagoras theorem, we get $|\vec{a}+\vec{b}|^2=|\vec{a}|^2+|\vec{b}|^2$

So, the angle between vectors $\vec{a}$ and $\vec{b}$ is 90°

d. We have $|\vec{a}+\vec{b}|=|\vec{a}-\vec{b}|$.

Vectors $\vec{a}+\vec{b}$ and $\vec{a}-\vec{b}$ are diagonal vectors of a parallelogram whose sides are vectors $\vec{a}$ and $\vec{b}$.

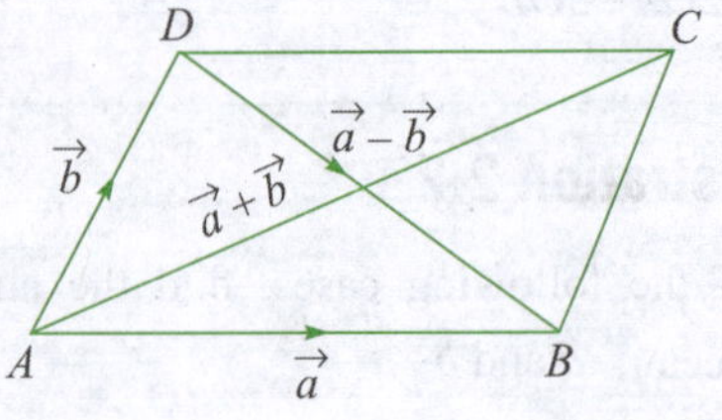

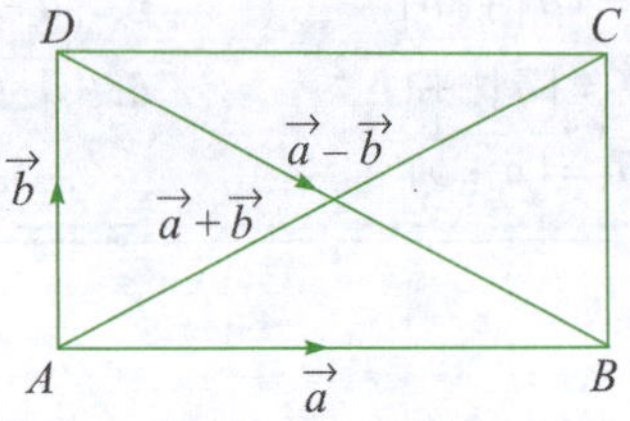

So, $|\vec{a}+\vec{b}|=|\vec{a}-\vec{b}| \Rightarrow$ diagonals have the same length.

Thus, parallelogram is a rectangle.

Hence, vectors $\vec{a}$ and $\vec{b}$ are perpendicular.

e. We have $|\vec{a}|=|\vec{b}|=|\vec{a}+\vec{b}|$.

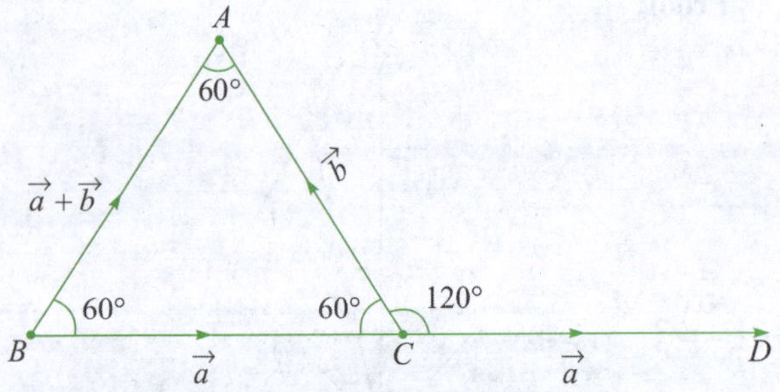

So, in triangle ABC we have $AB=BC=CA$.

Therefore, triangle ABC is equilateral.

Hence, the angle between vectors $\vec{a}$ and $\vec{b}$ is 120°.

Illustration 2.3

Let $\vec{a},\vec{b},\vec{c}$ be three vectors such that $\vec{a}+\vec{b}+\vec{c}=0$. If the angle between vectors $\vec{a}$ and $\vec{b}$ is 110° and the angle between vectors $\vec{b}$ and $\vec{c}$ is 100°, then find the angle between vectors $\vec{a}$ and $\vec{c}$.

Sol. Since $\vec{a}+\vec{b}+\vec{c}=0$; the vectors $\vec{a},\vec{b},\vec{c}$ can be taken as the sides of triangle ABC as shown in the following figure.

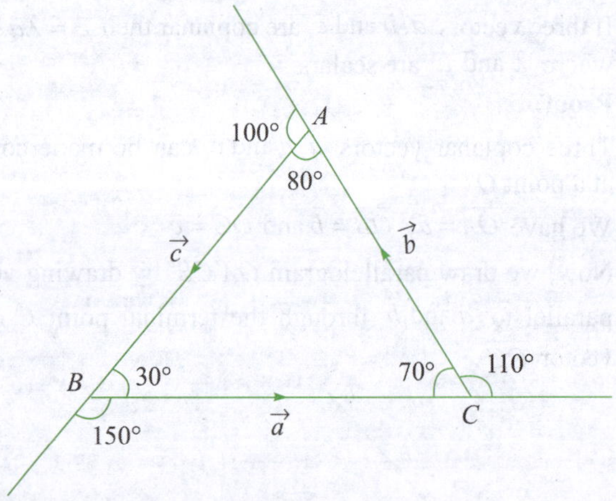

The angle between vectors $\vec{a}$ and $\vec{b}$ is 110°.

So, the interior angle of triangle at vertex C is 70°.

The angle between vectors $\vec{b}$ and $\vec{c}$ is 100°.

So, the interior angle of triangle at vertex A is 80°.

Therefore, the interior angle of triangle at vertex B is 30°.

Hence, the angle between vectors $\vec{a}$ and $\vec{c}$ is 150°.

Illustration 2.4

If vector $\vec{a}+\vec{b}$ bisects the angle between the vectors $\vec{a}$ and $\vec{b}$, then prove that $|\vec{a}|=|\vec{b}|$.

Sol. We know that vector $\vec{a}+\vec{b}$ is along the diagonal of the parallelogram whose adjacent sides are vectors $\vec{a}$ and $\vec{b}$. Now if $\vec{a}+\vec{b}$ bisects the angle between vectors $\vec{a}$ and $\vec{b}$, then the parallelogram must be a rhombus; hence, $|\vec{a}|=|\vec{b}|$.

Illustration 2.5

If $\overrightarrow{AO}+\overrightarrow{OB}=\overrightarrow{BO}+\overrightarrow{OC}$, then prove that B is the midpoint of AC.

Sol. We have, $\overrightarrow{AO}+\overrightarrow{OB}=\overrightarrow{BO}+\overrightarrow{OC}$

$\Rightarrow \quad \overrightarrow{AB}=\overrightarrow{BC}$

Thus, vectors $\overrightarrow{AB}$ and $\overrightarrow{BC}$ are collinear

Therefore, Points A, B, C are collinear

Also $|\overrightarrow{AB}|=|\overrightarrow{BC}|$

Hence, B is the midpoint of AC

Illustration 2.6

If $ABCDE$ is a pentagon then prove that the resultant of forces $\overrightarrow{AB}$, $\overrightarrow{AE}$, $\overrightarrow{BC}$, $\overrightarrow{DC}$, $\overrightarrow{ED}$ and $\overrightarrow{AC}$ is $3\overrightarrow{AC}$.

Sol. Let the resultant force be $\vec{R}$.

Then, $\vec{R}=\overrightarrow{AB}+\overrightarrow{AE}+\overrightarrow{BC}+\overrightarrow{DC}+\overrightarrow{ED}+\overrightarrow{AC}$

$=(\overrightarrow{AB}+\overrightarrow{BC})+(\overrightarrow{AE}+\overrightarrow{ED}+\overrightarrow{DC})+\overrightarrow{AC}$

$=\overrightarrow{AC}+\overrightarrow{AC}+\overrightarrow{AC}=3\overrightarrow{AC}$

Illustration 2.7

Prove that the resultant of two forces acting at point A represented by $\overrightarrow{AB}$ and $\overrightarrow{AC}$ is given by $2\overrightarrow{AD}$, where D is the mid-point of BC.

Sol. Let the resultant force be $\vec{R}$.

Then, $\vec{R}=\overrightarrow{AB}+\overrightarrow{AC}$

$=(\overrightarrow{AD}+\overrightarrow{DB})+(\overrightarrow{AD}+\overrightarrow{DC})$

$=2\overrightarrow{AD}+(\overrightarrow{DB}+\overrightarrow{DC})$

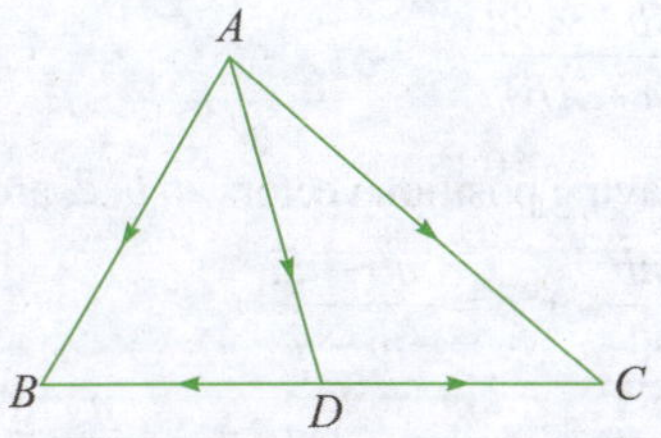

Since D is the mid-point of BC, we have

$\therefore \quad \overrightarrow{DB}=-\overrightarrow{DC}$

Thus, $\vec{R}=2\overrightarrow{AD}+\vec{0}=2\overrightarrow{AD}$

Illustration 2.8

Prove that the sum of three vectors determined by the medians of a triangle directed from the vertices is zero.

Sol. In the figure D, E and F are mid-points of sides BC, AC and AB respectively.

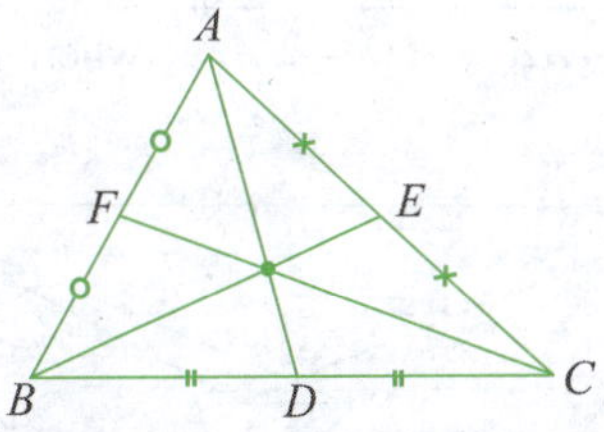

Using the result of the previous question, we have

$\overrightarrow{AB}+\overrightarrow{AC}=2\overrightarrow{AD}$,

$\overrightarrow{BC}+\overrightarrow{BA}=2\overrightarrow{BE}$ and

$\overrightarrow{CA}+\overrightarrow{CB}=2\overrightarrow{CF}$

Adding, we get

$\overrightarrow{AB}+\overrightarrow{AC}+\overrightarrow{BC}+\overrightarrow{BA}+\overrightarrow{CA}+\overrightarrow{CB}=2(\overrightarrow{AD}+\overrightarrow{BE}+\overrightarrow{CF})$

$\therefore \quad (\overrightarrow{AB}+\overrightarrow{BA})+(\overrightarrow{BC}+\overrightarrow{CB})+(\overrightarrow{CA}+\overrightarrow{AC})=2(\overrightarrow{AD}+\overrightarrow{BE}+\overrightarrow{CF})$

$\Rightarrow \quad \vec{0}+\vec{0}+\vec{0}=2(\overrightarrow{AD}+\overrightarrow{BE}+\overrightarrow{CF})$

$\Rightarrow \quad \overrightarrow{AD}+\overrightarrow{BE}+\overrightarrow{CF}=0$

Illustration 2.9

Let ABC be a triangle and P be any point on BC. If $\overrightarrow{PQ}$ is the sum of $\overrightarrow{AP}+\overrightarrow{PB}+\overrightarrow{PC}$, then show that $ABQC$ is a parallelogram and that Q is therefore, a fixed point.

Sol. We have $\overrightarrow{PQ}=\overrightarrow{AP}+\overrightarrow{PB}+\overrightarrow{PC}$

or $\quad \overrightarrow{PQ}-\overrightarrow{PC}=\overrightarrow{AP}+\overrightarrow{PB}$

$\Rightarrow \quad \overrightarrow{PQ}+\overrightarrow{CP}=\overrightarrow{AP}+\overrightarrow{PB}$

$\Rightarrow \quad \overrightarrow{CQ}=\overrightarrow{AB}\Rightarrow CQ=AB$ and $CQ \parallel AB$

Therefore, $ABQC$ is a parallelogram.

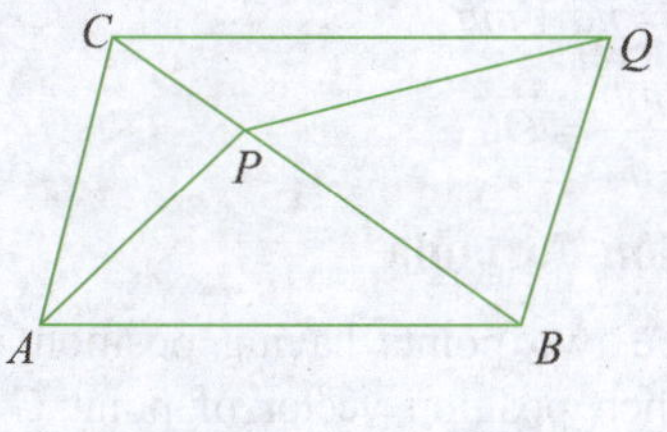

Now, A, B and C are fixed points and $ABQC$ is a parallelogram. Therefore, Q is a fixed point.

Illustration 2.10

Two forces $\overrightarrow{AB}$ and $\overrightarrow{AD}$ are acting at the vertex A of a quadrilateral $ABCD$ and two forces $\overrightarrow{CB}$ and $\overrightarrow{CD}$ at C. Prove that their resultant is given by $4\overrightarrow{EF}$, where E and F are the midpoints of AC and BD, respectively.

Sol. We have, $\overrightarrow{AB} + \overrightarrow{AD} = 2\overrightarrow{AF}$, where F is the midpoint of BD.

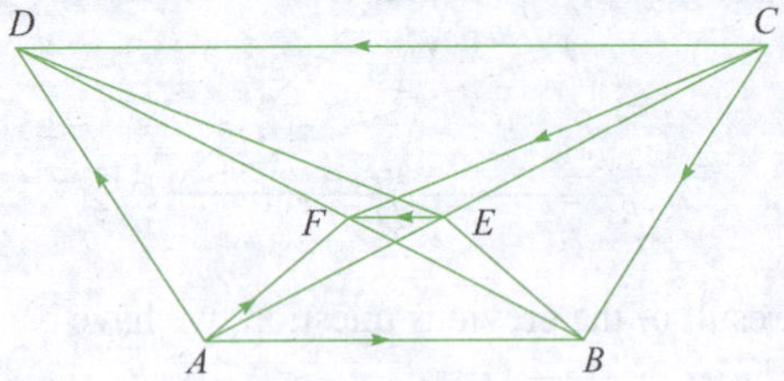

Similarly, $\overrightarrow{CB} + \overrightarrow{CD} = 2\overrightarrow{CF}$

$\therefore \quad \overrightarrow{AB} + \overrightarrow{AD} + \overrightarrow{CB} + \overrightarrow{CD} = 2(\overrightarrow{AF} + \overrightarrow{CF})$

$= 2(2\overrightarrow{EF})$, where E is the midpoint of AC

$= 4\overrightarrow{EF}$

2.1.4 Section Formula

Internal Section Formula

Let A and B be two points having position vectors $\vec{a}$ and $\vec{b}$ respectively. Then the position vector of point C that divides AB internally in the ratio $m : n$ is $\dfrac{m\vec{b}+n\vec{a}}{m+n}$.

Proof: We have $\dfrac{|\overrightarrow{AC}|}{|\overrightarrow{CB}|} = \dfrac{m}{n}$

Now, vectors $\overrightarrow{AC}$ and $\overrightarrow{CB}$ are collinear.

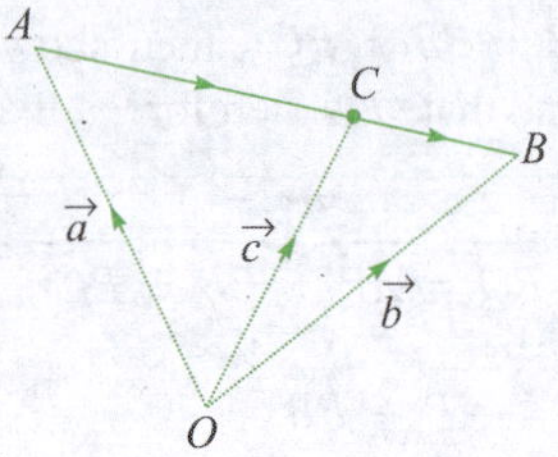

$\therefore \quad n\overrightarrow{AC} = m\overrightarrow{CB}$

$\Rightarrow \quad n(\overrightarrow{OC} - \overrightarrow{OA}) = m(\overrightarrow{OB} - \overrightarrow{OC})$

$\Rightarrow \quad n(\vec{c} - \vec{a}) = m(\vec{b} - \vec{c})$

$\Rightarrow \quad n\vec{c} - n\vec{a} = m\vec{b} - m\vec{c}$

$\Rightarrow \quad m\vec{c} + n\vec{c} = n\vec{a} + m\vec{b}$

$\Rightarrow \quad (m+n)\vec{c} = n\vec{a} + m\vec{b}$

$\Rightarrow \quad \vec{c} = \dfrac{n\vec{a}+m\vec{b}}{m+n} = \overrightarrow{OC}$

External Section Formula

Let A and B be two points having position vectors $\vec{a}$ and $\vec{b}$ respectively. Then position vector of point C that divides AB externally in the ratio $m : n$ is $\dfrac{m\vec{b}-n\vec{a}}{m-n}$.

Proof: We have $\dfrac{|\overrightarrow{AC}|}{|\overrightarrow{BC}|} = \dfrac{m}{n}$

Now, vectors $\overrightarrow{AC}$ and $\overrightarrow{BC}$ are collinear.

$\therefore \quad n\overrightarrow{AC} = m\overrightarrow{BC}$

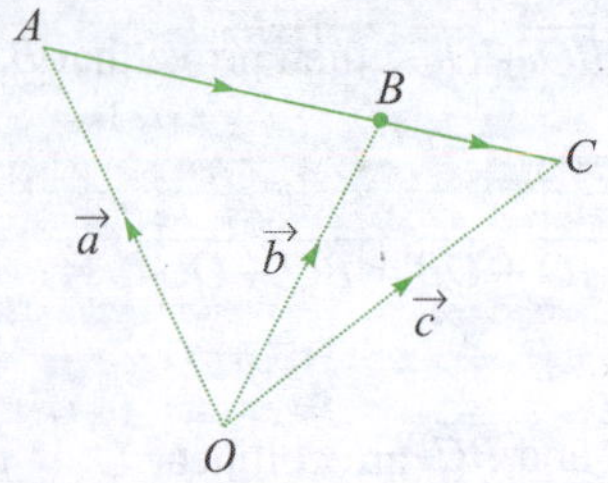

$\Rightarrow \quad n(\overrightarrow{OC} - \overrightarrow{OA}) = m(\overrightarrow{OC} - \overrightarrow{OB})$

$\Rightarrow \quad n(\vec{c} - \vec{a}) = m(\vec{c} - \vec{b})$

$\Rightarrow \quad n\vec{c} - n\vec{a} = m\vec{c} - m\vec{b}$

$\Rightarrow \quad m\vec{b} - n\vec{a} = m\vec{c} - n\vec{c}$

$\Rightarrow \quad (m-n)\vec{c} = m\vec{b} - n\vec{a}$

$\Rightarrow \quad \vec{c} = \dfrac{m\vec{b}-n\vec{a}}{m-n} = \overrightarrow{OC}$

NOTES

- If the position vectors of points A and B are $\vec{a}$ and $\vec{b}$ respectively, then the position vector of point dividing AB internally and externally in the ratio $\lambda : 1$ $(= AC : BC)$ are $\dfrac{\lambda\vec{b}+\vec{a}}{\lambda+1}$ and $\dfrac{\lambda\vec{b}-\vec{a}}{\lambda-1}$ respectively.
- If the position vectors of point A and B are $\vec{a}$ and $\vec{b}$ respectively, then the position vector of the mid-point of AB is $\dfrac{\vec{a}+\vec{b}}{2}$.
- In ΔABC, if the vertices A, B and C have position vectors $\vec{a}, \vec{b}$ and $\vec{c}$ respectively, then the positive vector of the centroid of triangle is $\dfrac{\vec{a}+\vec{b}+\vec{c}}{3}$.

  We know that in triangle (not equilateral), orthocentre(H), centroid (G) and circumcentre (O) are collinear and G divides HO internally in the ratio 2 : 1.

  So, if the circumcentre is at origin, then the position vector of the orthocentre is $(\vec{a}+\vec{b}+\vec{c})$.

  Also, the position vector of the incentre of the triangle is $\dfrac{BC\vec{a}+CA\vec{b}+AB\vec{c}}{BC+CA+AB}$.
- If points having position vectors $\vec{a}, \vec{b}, \vec{c}$ are collinear then

  $\vec{c} = \dfrac{n\vec{a}+m\vec{b}}{m+n}$ or $\vec{c} = \dfrac{m\vec{b}-n\vec{a}}{m-n}$.

  $\therefore \vec{c} = \dfrac{n}{m+n}\vec{a} + \dfrac{m}{m+n}\vec{b}$ or $\vec{c} = -\dfrac{n}{m-n}\vec{a} + \dfrac{m}{m-n}\vec{b}$

  $\Rightarrow \vec{c} = \lambda\vec{a} + \mu\vec{b}$, where $\lambda + \mu = 1$

  $\left(\because \dfrac{n}{m+n} + \dfrac{m}{m+n} = 1 \text{ and } -\dfrac{n}{m-n} + \dfrac{m}{m-n} = 1\right)$

Illustration 2.11

If a parallelepiped is formed by three co-initial and non-coplanar vectors $\overrightarrow{OA}, \overrightarrow{OB}$, and $\overrightarrow{OC}$, then prove that the diagonal vector of the parallelepiped through point O is $\overrightarrow{OA}+\overrightarrow{OB}+\overrightarrow{OC}$.

Sol. We have parallelepiped as shown in the following figure.

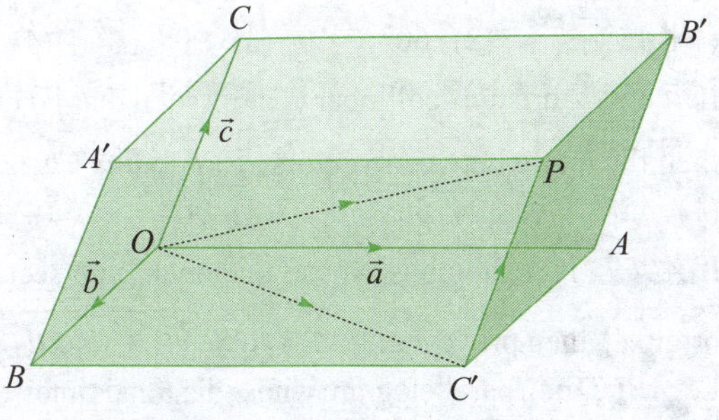

Diagonal through point O is OP.

Now, $\overrightarrow{OP}=\overrightarrow{OC'}+\overrightarrow{C'P}$

$$=(\overrightarrow{OA}+\overrightarrow{OB})+\overrightarrow{OC} \quad (\because \overrightarrow{C'P}=\overrightarrow{OC})$$

Thus, $\overrightarrow{OP}=\overrightarrow{OA}+\overrightarrow{OB}+\overrightarrow{OC}$

Hence, the diagonal vector is the resultant of vectors $\overrightarrow{OA}, \overrightarrow{OB}$, and $\overrightarrow{OC}$.

Illustration 2.12

If $\vec{a}, \vec{b}, \vec{c}, \vec{d}$ are the position vectors of the points A, B, C and D respectively such that no three points are collinear and $\vec{a}+\vec{c}=\vec{b}+\vec{d}$, then prove that the quadrilateral $ABCD$ is a parallelogram.

Sol. We have $\vec{a}+\vec{c}=\vec{b}+\vec{d}$

$$\therefore \quad \frac{\vec{a}+\vec{c}}{2}=\frac{\vec{b}+\vec{d}}{2}$$

Thus, the position vector of the mid-point of diagonal AC is equal to the position vector of the mid-point of diagonal BD.

Hence, the mid-points of the diagonals coincide.

Therefore, quadrilateral $ABCD$ is a parallelogram.

Illustration 2.13

The midpoints of two opposite sides of a quadrilateral, together with the midpoints of its diagonals, are the vertices of a parallelogram. Prove this using vectors.

Sol. Let $\vec{a}, \vec{b}, \vec{c}$ and $\vec{d}$ be the position vectors of vertices A, B, C and D, respectively.

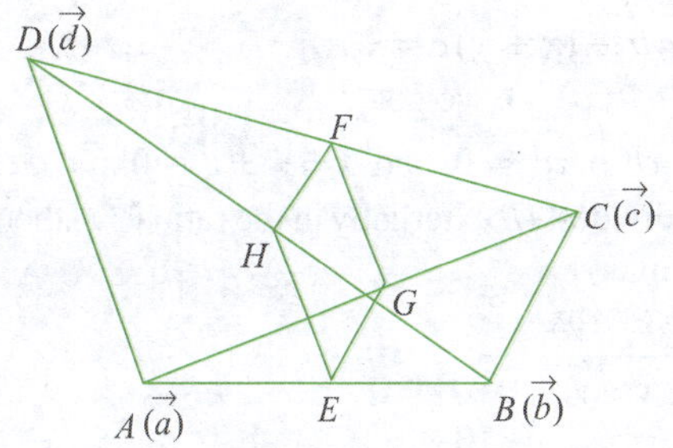

Let E, F, G and H be the midpoints of AB, CD, AC and BD, respectively.

$$\therefore \quad \text{P.V. of } E=\frac{\vec{a}+\vec{b}}{2},$$

$$\text{P.V. of } F=\frac{\vec{c}+\vec{d}}{2},$$

$$\text{P.V. of } G=\frac{\vec{a}+\vec{c}}{2}, \text{ and}$$

$$\text{P.V. of } H=\frac{\vec{b}+\vec{d}}{2}$$

Now, $\overrightarrow{EG}=\text{P.V. of G} - \text{P.V. of } E=\frac{\vec{a}+\vec{c}}{2}-\frac{\vec{a}+\vec{b}}{2}=\frac{\vec{c}-\vec{b}}{2}$

and $\overrightarrow{HF}=\text{P.V. of F} - \text{P.V. of } H=\frac{\vec{c}+\vec{d}}{2}-\frac{\vec{b}+\vec{d}}{2}=\frac{\vec{c}-\vec{b}}{2}$

$\therefore \quad \overrightarrow{EG}=\overrightarrow{HF} \Rightarrow EG \parallel HF$ and $EG=HF$

Hence, $EGHF$ is a parallelogram.

Illustration 2.14

Prove that the points having position vectors $\vec{a}, \vec{b}$ and $\vec{b}+t(\vec{a}-\vec{b})$ are collinear for all real values of t.

Sol. We have $\vec{b}+t(\vec{a}-\vec{b}) = t\vec{a}+(1-t)\vec{b} = \dfrac{t\vec{a}+(1-t)\vec{b}}{t+(1-t)}$

Therefore, by section formula, the points having position vectors $\vec{a}, \vec{b}$ and $\vec{b}+t(\vec{a}-\vec{b})$ are collinear.

Also, we observe that $\vec{b}+t(\vec{a}-\vec{b}) = t\vec{a}+(1-t)\vec{b} = \lambda\vec{a}+\mu\vec{b}$, where $\lambda+\mu=1$.

Illustration 2.15

Prove that three points with position vectors $\vec{a}, \vec{b}, \vec{c}$ are collinear if and only if there exist scalars x, y, z not all zero such that $x\vec{a}+y\vec{b}+z\vec{c}=0$ and $x+y+z=0$.

Sol. **Necessity**

Let A, B, C be collinear points with position vectors $\vec{a}, \vec{b}, \vec{c}$.

Therefore, point C must divide the line segment AB either internally or externally in some ratio.

Let C divides AB internally in the ratio $y : x$ where $x, y > 0$.

By the internal section formula, we have

$$\vec{c}=\frac{x\vec{a}+y\vec{b}}{x+y}.$$

or $\quad x\vec{a} + y\vec{b} - (x+y)\vec{c} = \vec{0}$.

Choosing $z = -(x+y)$, we get

$$x\vec{a} + y\vec{b} + z\vec{c} = \vec{0} \text{ and } x + y + z = 0.$$

If instead C divides AB externally in the ratio $y : x$, then by external section formula,

$$\vec{c} = \frac{x\vec{a} - y\vec{b}}{x - y}.$$

or $\quad x\vec{a} - y\vec{b} - (x-y)\vec{c} = \vec{0}$.

In this case also the sum of coefficients of $\vec{a}, \vec{b}, \vec{c}$ is zero.

Thus, in both internal and external division, the required relation holds.

Sufficiency

Conversely, suppose there exist scalars x, y, z, not all zero, such that

$$x\vec{a} + y\vec{b} + z\vec{c} = \vec{0} \text{ and } x + y + z = 0.$$

From $x + y + z = 0$, we obtain $z = -(x+y)$.

Substituting this into the first equation yields,

$$x\vec{a} + y\vec{b} - (x+y)\vec{c} = \vec{0}, \text{ or } \vec{c} = \frac{x\vec{a} + y\vec{b}}{x + y}.$$

This is precisely the section formula.

If x and y have the same sign, the division is internal; if they have opposite signs, the division is external.

In either case, C lies on the line through A and B.

Thus the three points are collinear.

Illustration 2.16

If $O(\vec{0})$ is the circumcentre and O' is the orthocentre of a triangle ABC then prove that

a. $\overrightarrow{OA} + \overrightarrow{OB} + \overrightarrow{OC} = \overrightarrow{OO'}$

b. $\overrightarrow{O'A} + \overrightarrow{O'B} + \overrightarrow{O'C} = 2\overrightarrow{O'O}$

c. $\overrightarrow{AO'} + \overrightarrow{O'B} + \overrightarrow{O'C} = \overrightarrow{AP}$, where AP is the diameter of the circumcircle through point A.

Sol.

a. Let the position vectors of vertices A, B and C be $\vec{a}, \vec{b}$ and $\vec{c}$ respectively.

Then, the position vector of the centroid G is $\dfrac{\vec{a}+\vec{b}+\vec{c}}{3}$.

Now, the orthocentre of triangle is O'.

We know that the centroid divides the join of the orthocentre and the circumcentre in the ratio 2 : 1.

Thus, using $O'G : GO = 2 : 1$, we get the position vector of the orthocentre as $\vec{a}+\vec{b}+\vec{c}$

So, $\overrightarrow{OO'} = \overrightarrow{OA} + \overrightarrow{OB} + \overrightarrow{OC}$.

b. $\overrightarrow{O'A} + \overrightarrow{O'B} + \overrightarrow{O'C}$

$$= (\overrightarrow{OA} - \overrightarrow{OO'}) + (\overrightarrow{OB} - \overrightarrow{OO'}) + (\overrightarrow{OC} - \overrightarrow{OO'})$$

$$= \overrightarrow{OA} + \overrightarrow{OB} + \overrightarrow{OC} - 3\overrightarrow{OO'}$$

$$= \vec{a} + \vec{b} + \vec{c} - 3(\vec{a} + \vec{b} + \vec{c})$$

$$= -2(\vec{a} + \vec{b} + \vec{c}) = -2\overrightarrow{OO'} = 2\overrightarrow{O'O}$$

c. $\overrightarrow{AO'} + \overrightarrow{O'B} + \overrightarrow{O'C}$

$$= 2\overrightarrow{AO'} + (\overrightarrow{O'A} + \overrightarrow{O'B} + \overrightarrow{O'C})$$

$$= 2\overrightarrow{AO'} + 2\overrightarrow{O'O}$$

$= 2\overrightarrow{AO}$ (2 × the radius vector from A to O)

$= \overrightarrow{AP}$ (the diameter vector from A to P)

Aim to complete within 30 minutes

Concept Application Exercise 2.1

1. If $|\vec{a}| = |\vec{b}| = |\vec{a} - \vec{b}|$, then the angle between vectors $\vec{a}$ and $\vec{b}$ is
   (1) 45° (2) 60° (3) 90° (4) 120°
2. If $\vec{a}$ and $\vec{b}$ are non-collinear vectors such that $|\vec{a}| = |\vec{b}|$, then the angle between vectors $\vec{a}+\vec{b}$ and $\vec{a}-\vec{b}$ is
   (1) 45° (2) 60° (3) 90° (4) 120°
3. If $ABCD$ is a rhombus whose diagonals intersect at the origin O, then prove that $\overrightarrow{OA} + \overrightarrow{OB} + \overrightarrow{OC} + \overrightarrow{OD} = \vec{0}$.
4. Let $ABCD$ be a parallelogram whose diagonals intersect at P and let O be the origin. If $\overrightarrow{OA} + \overrightarrow{OB} + \overrightarrow{OC} + \overrightarrow{OD} = \lambda\overrightarrow{OP}$, then $\lambda =$
   (1) 1 (2) 2 (3) 4 (4) 8
5. Let $ABCD$ be a parallelogram. If L and M are the mid-points of BC and CD, respectively, express $\overrightarrow{AL}$ and $\overrightarrow{AM}$ in the form of $\overrightarrow{AB}$ and $\overrightarrow{AD}$. Also show that $\overrightarrow{AL} + \overrightarrow{AM} = (3/2)\,\overrightarrow{AC}$.
6. Let $ABCD$ be a quadrilateral and E be the point of intersection of the lines joining the mid-points of opposite sides. Show that the resultant of vectors $\overrightarrow{OA}, \overrightarrow{OB}, \overrightarrow{OC}$ and $\overrightarrow{OD}$ is equal to $4\,\overrightarrow{OE}$, where O is any point.
7. If $\vec{a}, \vec{b}$ and $\vec{c}$ are mutually perpendicular vectors or equal magnitudes, then find the angle between the vectors $\vec{a}$ and $\vec{a}+\vec{b}+\vec{c}$.
8. If $\vec{a}$ and $\vec{b}$ are two vectors of magnitude 1 inclined at 120°, then the angle between vectors $\vec{b}$ and $\vec{b}-\vec{a}$ is
   (1) 30° (2) 45° (3) 60° (4) 120°
9. If $\vec{a}, \vec{b}, \vec{c}$ and $\vec{d}$ are four vectors in three-dimensional space with the same initial point such that $3\vec{a} - 2\vec{b} + \vec{c} - 2\vec{d} = \vec{0}$, show that terminals A, B, C and D of these vectors are coplanar. Find the point at which AC and BD meet. Find the ratio in which P divides AC and BD.

Answers

1. (2) **2.** (3) **4.** (3)

7. $\cos^{-1}\dfrac{1}{\sqrt{3}}$ **8.** (1)

9. AC and BD intersect at P whose P.V. is $\dfrac{3\vec{a}+\vec{c}}{4}$; P divides AC in the ratio 3 : 1 and BD in the ratio 1 : 1.

2.2 VECTORS IN COMPONENTS FORM

Consider points A (1, 0, 0), B (0, 1, 0) and C (0, 0, 1) on the x-axis, y-axis and z-axis, respectively.

Then, clearly $|\overrightarrow{OA}|=1, |\overrightarrow{OB}|=1$ and $|\overrightarrow{OC}|=1$.

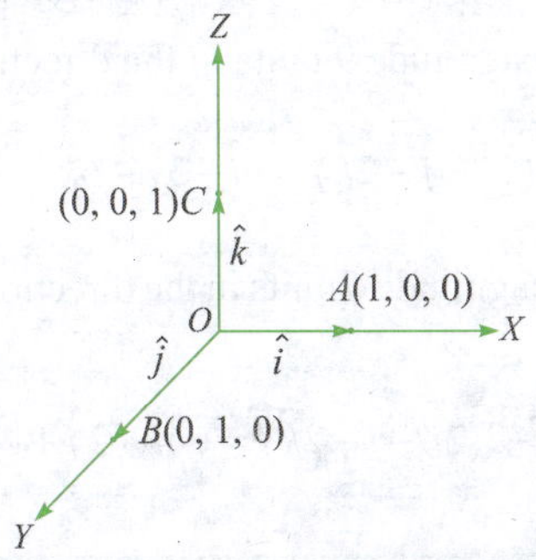

The vectors $\overrightarrow{OA}, \overrightarrow{OB}$ and $\overrightarrow{OC}$, each having magnitude 1, are called the unit vectors along the positive axes OX, OY and OZ, and are denoted by $\hat{i}$, $\hat{j}$ and $\hat{k}$, respectively. These vectors are said to form an orthonormal triad of vectors.

Now, consider a point $P(x, y, z)$. The position vector of point P is $\overrightarrow{OP}$.

From point P, draw three planes parallel to the coordinate planes (xy, yz and xz planes). In this way, we obtain a cuboid having three faces on the coordinate planes and three faces through point P, parallel to the coordinate planes.

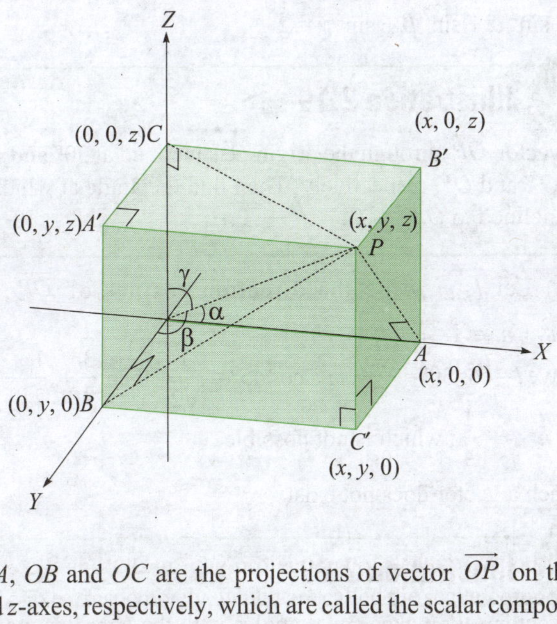

OA, OB and OC are the projections of vector $\overrightarrow{OP}$ on the x-, y, and z-axes, respectively, which are called the scalar components of vector $\overrightarrow{OP}$ on the axes.

Hence, the scalar components of vector $\overrightarrow{OP}$ on the x, y and z-axis are x, y, and z respectively.

Clearly, $\overrightarrow{OA}=x\hat{i}, \overrightarrow{OB}=y\hat{j}$ and $\overrightarrow{OC}=z\hat{k}$.

Now, $\overrightarrow{OC'}=\overrightarrow{OA}+\overrightarrow{OB}=x\hat{i}+y\hat{j}$

Then, $\overrightarrow{OP}=\overrightarrow{OC'}+\overrightarrow{C'P}=\overrightarrow{OC'}+\overrightarrow{OC}$ $\quad (\because \overrightarrow{C'P}=\overrightarrow{OC})$

or $\quad \overrightarrow{OP}=x\hat{i}+y\hat{j}+z\hat{k}$

This is the component form of vector $\overrightarrow{OP}$, where $x\hat{i}$, $y\hat{j}$ and $z\hat{k}$ are the vector components or the resolved parts of vector $\overrightarrow{OP}$ on the x-, y- and z-axes respectively.

NOTES

- $\lambda(x\hat{i}+y\hat{j}+z\hat{k})=(\lambda x)\hat{i}+(\lambda y)\hat{j}+(\lambda z)\hat{k}$, where λ is a scalar.
- We know that if $(x_1, y_1, z_1)=(x_2, y_2, z_2)$ then $x_1=x_2, y_1=y_2$ and $z_1=z_2$.
  So, $x_1\hat{i}+y_1\hat{j}+z_1\hat{k}=x_2\hat{i}+y_2\hat{j}+z_2\hat{k}$ iff $x_1=x_2, y_1=y_2$ and $z_1=z_2$

2.2.1 DIRECTION COSINES OF A VECTOR

Let the angles made by the vector $\overrightarrow{OP}$ with the positive directions of x-, y- and z-axes are α, β and γ respectively. These angles are called the direction angles of vector $\overrightarrow{OP}$. The cosine values of these angles, i.e., $\cos\alpha$, $\cos\beta$ and $\cos\gamma$ are called the direction cosines of the vector $\overrightarrow{OP}$, usually denoted by l, m and n respectively.

In the right angled triangle OAP, we have $\cos\alpha=\frac{OA}{OP}=\frac{x}{|\overrightarrow{OP}|}$.

Similarly, from the right angled triangles OBP and OCP, we get

$$\cos\beta=\frac{OB}{OP}=\frac{y}{|\overrightarrow{OP}|} \text{ and } \cos\gamma=\frac{OC}{OP}=\frac{z}{|\overrightarrow{OP}|}.$$

Thus, the coordinates of the point P can be expressed as $(l|\overrightarrow{OP}|, m|\overrightarrow{OP}|, n|\overrightarrow{OP}|)$.

Also, $l^2+m^2+n^2=\cos^2\alpha+\cos^2\beta+\cos^2\gamma$

$$=\frac{x^2}{|\overrightarrow{OP}|^2}+\frac{y^2}{|\overrightarrow{OP}|^2}+\frac{z^2}{|\overrightarrow{OP}|^2}$$

$$=\frac{x^2+y^2+z^2}{|\overrightarrow{OP}|^2}=\frac{|\overrightarrow{OP}|^2}{|\overrightarrow{OP}|^2}=1$$

Thus, $l^2+m^2+n^2=1$, i.e. the sum of the squares of the direction cosines is unity.

Also, the unit vector in the direction of $\overrightarrow{OP}=x\hat{i}+y\hat{j}+z\hat{k}$ is $l\hat{i}+m\hat{j}+n\hat{k}$ or $\frac{1}{\sqrt{x^2+y^2+z^2}}(x\hat{i}+y\hat{j}+z\hat{k})$, which verifies the fact that, by dividing a vector by its magnitude, we obtain the unit vector in the direction of the given vector.

If the initial point of a given vector $\vec{a}$ is not the origin, then we draw a vector $\vec{a}'$, equal to vector $\vec{a}$, through the origin. Thus, the direction angles, and hence the direction cosines, of vectors $\vec{a}$ and $\vec{a}'$ are the same.

Illustration 2.17

If O is origin and point P is (2, –1, –2), then find the following.

a. Octant in which point P lies

b. Vector $\overrightarrow{OP}$

c. scalar components of $\overrightarrow{OP}$ on the axes.

d. vector components of $\overrightarrow{OP}$ on the axes.

e. Direction cosines of $\overrightarrow{OP}$

f. Direction angles of $\overrightarrow{OP}$

g. Terminal point of the vector which is equal to $\overrightarrow{OP}$ and having the start point (1, –3, 4)

h. Initial point of the vector which is equal to $\overrightarrow{OP}$ having the terminal point (–1 , 5, –6)

i. Unit vector in the direction of $\overrightarrow{OP}$

j. Vector of magnitude 9 units in the direction of $\overrightarrow{OP}$

k. Vector of magnitude 6 units in the direction opposite to $\overrightarrow{OP}$.

Sol.

a. Since y- and z- coordinates are negative and x- coordinate is positive, point P lies is eighth octant.

b. $\overrightarrow{OP} = 2\hat{i} - \hat{j} - 2\hat{k}$

c. Scalar components of $\overrightarrow{OP}$ on x-axis, y-axis and z-axis are 2, –1, –2 respectively.

d. Vector components of $\overrightarrow{OP}$ on x-axis, y-axis and z-axis are $2\hat{i}, -\hat{j}$ and $-2\hat{k}$ respectively.

e. $|\overrightarrow{OP}| = \sqrt{2^2 + (-1)^2 + (-2)^2} = \sqrt{9} = 3$

So, the direction cosines are $l = \frac{2}{3}, m = -\frac{1}{3}, n = -\frac{2}{3}$

f. $l = \cos\alpha = \frac{2}{3}, m = \cos\beta = -\frac{1}{3}, n = \cos\gamma = -\frac{2}{3}$

So, the direction angles are

$$\alpha = \cos^{-1}\frac{2}{3},\ \beta = \cos^{-1}\left(-\frac{1}{3}\right) = \pi - \cos^{-1}\frac{1}{3} \text{ and}$$

$$\gamma = \cos^{-1}\left(-\frac{2}{3}\right) = \pi - \cos^{-1}\frac{2}{3}$$

g. Let $\overrightarrow{QR} = \overrightarrow{OP} = 2\hat{i} - \hat{j} - 2\hat{k}$, where $Q \equiv (1, -3, 4)$

Now, $\overrightarrow{QR} = \overrightarrow{OR} - \overrightarrow{OQ} = 2\hat{i} - \hat{j} - 2\hat{k}$

$\therefore\quad \overrightarrow{OR} - (\hat{i} - 3\hat{j} + 4\hat{k}) = 2\hat{i} - \hat{j} - 2\hat{k}$

$\Rightarrow\quad \overrightarrow{OR} = (\hat{i} - 3\hat{j} + 4\hat{k}) + 2\hat{i} - \hat{j} - 2\hat{k}$

$\Rightarrow\quad \overrightarrow{OR} = 3\hat{i} - 4\hat{j} + 2\hat{k}$

$\Rightarrow\quad R \equiv (3, -4, 2)$

h. Let $\overrightarrow{QR} = \overrightarrow{OP} = 2\hat{i} - \hat{j} - 2\hat{k}$, where $R \equiv (-1, 5, -6)$

Now, $\overrightarrow{QR} = \overrightarrow{OR} - \overrightarrow{OQ} = 2\hat{i} - \hat{j} - 2\hat{k}$

$\therefore\quad (-\hat{i} + 5\hat{j} - 6\hat{k}) - \overrightarrow{OQ} = 2\hat{i} - \hat{j} - 2\hat{k}$

$\Rightarrow\quad \overrightarrow{OQ} = (-\hat{i} + 5\hat{j} - 6\hat{k}) - (2\hat{i} - \hat{j} - 2\hat{k})$

$\Rightarrow\quad \overrightarrow{OQ} = -3\hat{i} + 6\hat{j} - 4\hat{k}$

$\Rightarrow\quad Q \equiv (-3, 6, -4)$

i. Unit vector in the direction of $\overrightarrow{OP}$ is $\frac{\overrightarrow{OP}}{|\overrightarrow{OP}|} = \frac{1}{3}(2\hat{i} - \hat{j} - 2\hat{k})$

j. Vector of magnitude 9 units in the direction of $\overrightarrow{OP}$ is

$$9\frac{\overrightarrow{OP}}{|\overrightarrow{OP}|} = 3(2\hat{i} - \hat{j} - 2\hat{k}) = 6\hat{i} - 3\hat{j} - 6\hat{k}$$

k. Vector of magnitude 6 units in the direction opposite to $\overrightarrow{OP}$ is

$$-6\frac{\overrightarrow{OP}}{|\overrightarrow{OP}|} = -2(2\hat{i} - \hat{j} - 2\hat{k}) = -4\hat{i} + 2\hat{j} + 4\hat{k}$$

Illustration 2.18

If α, β and γ are the angles which a directed line makes with the positive directions of the coordinate axes, then find the value of $\sin^2\alpha + \sin^2\beta + \sin^2\gamma$.

Sol. The DCs of the line are $l = \cos\alpha, m = \cos\beta, n = \cos\gamma$

We know that

$l^2 + m^2 + n^2 = 1$

$\therefore\quad \cos^2\alpha + \cos^2\beta + \cos^2\gamma = 1$

$\Rightarrow\quad (1 - \sin^2\alpha) + (1 - \sin^2\beta) + (1 - \sin^2\gamma) = 1$

$\Rightarrow\quad \sin^2\alpha + \sin^2\beta + \sin^2\gamma = 2$

Illustration 2.19

A vector $\overrightarrow{OP}$ through the origin O is inclined at 30° and 45° to OX and OY, respectively. Then find the angle at which it is inclined to OZ.

Sol. Let l, m, n be the direction cosines of $\overrightarrow{OP}$, then $l^2 + m^2 + n^2 = 1$.

We have $l = \cos 30° = \frac{\sqrt{3}}{2}, m = \cos 45° = \frac{1}{\sqrt{2}}$, then $\frac{3}{4} + \frac{1}{2} + n^2 = 1$

$\Rightarrow\quad n^2 = -\frac{1}{4}$, which is not possible.

So, such a vector does not exist.

Illustration 2.20

A vector $\overrightarrow{OP}$ makes equal angles with the coordinate axes then find its direction cosines.

Sol. Let l, m, n be the direction cosines of $\overrightarrow{OP}$, which makes an equal angle α with each of the coordinate axes.

$\therefore\quad l = m = n = \cos\alpha$

Now $l^2 + m^2 + n^2 = 1$

$\therefore\quad 3l^2 = 1$

$\Rightarrow\quad l = \pm\frac{1}{\sqrt{3}} = m = n$

Therefore, $l = m = n = \frac{1}{\sqrt{3}}$ or $l = m = n = -\frac{1}{\sqrt{3}}$

For $l = m = n = \frac{1}{\sqrt{3}}$, point P lies in the first octant, where $\overrightarrow{OP}$ makes an angle $\cos^{-1}\left(\frac{1}{\sqrt{3}}\right)$ with each coordinate axis.

For $l = m = n = -\frac{1}{\sqrt{3}}$, point P lies in the seventh octant, where $\overrightarrow{OP}$ makes an angle $(180° - \cos^{-1}\frac{1}{\sqrt{3}})$ with each coordinate axis.

Illustration 2.21

Vector $\overrightarrow{OP}$ has direction cosines, $\cos\alpha$, $\cos\beta$ and $\cos\gamma$. Assuming yz-plane as a mirror then what are the direction cosines of the reflected image of OP in the yz-plane?

Sol. Let point P lies in first octant.

$\therefore$ Directional angles of OP are α, β and γ.

If OP is reflected in yz-plane and OP' is reflected line segment then OP' will make same angle β and γ with y-axis and z-axis, respectively, but its inclination with x-axis will be $(180° - \alpha)$

So, direction cosines of OP' are $\cos(180° - \alpha)$, $\cos\beta$, $\cos\gamma$ or $-\cos\alpha$, $\cos\beta$ and $\cos\gamma$.

Illustration 2.22

A unit vector of modulus 2 is equally inclined to the x- and the y-axes at an angle $\pi/3$. Find the length of the projection of the vector on the z-axis.

Sol. Given that the vector is inclined at an angle $\pi/3$ with both the x- and the y-axes. Then

$$\cos\alpha = \cos\beta = \frac{1}{2}$$

Also, $\cos^2\alpha + \cos^2\beta + \cos^2\gamma = 1$. Therefore,

$$\cos^2\gamma = \frac{1}{2}$$

or $$\cos\gamma = \pm\frac{1}{\sqrt{2}}$$

Thus, the given vector is

$$2(\cos\alpha\hat{i} + \cos\beta\hat{j} + \cos\gamma\hat{k}) = 2\left(\frac{\hat{i}}{2} + \frac{\hat{j}}{2} \pm \frac{\hat{k}}{\sqrt{2}}\right) = \hat{i} + \hat{j} \pm \sqrt{2}\hat{k}$$

Hence, the length of projection of the vector on the z-axis is $\sqrt{2}$ units.

Illustration 2.23

If the projections of vector $\vec{a}$ on the x-, y- and z-axes are 2, 1 and 2 units, respectively, find the angle at which vector $\vec{a}$ is inclined to the z-axis.

Sol. Since the projections of vector $\vec{a}$ on the x-, y-, and z-axes are 2, 1 and 2 units, respectively, we have

$$\vec{a} = 2\hat{i} + \hat{j} + 2\hat{k}$$

$$\therefore \quad |\vec{a}| = \sqrt{2^2 + 1^2 + 2^2} = 3$$

Then, $\cos\gamma = \frac{2}{3}$ (where γ is the angle of vector $\vec{a}$ with the z-axis), i.e.,

$$\gamma = \cos^{-1}\frac{2}{3}$$

Illustration 2.24

Find the terminal point of the vector of magnitude 6 units having its initial point $(-1, 3, 5)$ and direction angles $\cos^{-1}\frac{1}{3}$, $\cos^{-1}\frac{2}{3}$, $\cos^{-1}\left(-\frac{2}{3}\right)$.

Sol. Let the required vector be $\overrightarrow{AB}$.

We have $A \equiv (-1, 3, 5)$, $|\overrightarrow{AB}| = 6$.

Also, the direction angles,

$$\alpha = \cos^{-1}\frac{1}{3},\ \beta = \cos^{-1}\frac{2}{3} \text{ and } \gamma = \cos^{-1}\left(-\frac{2}{3}\right)$$

Therefore, the direction cosines are

$$l = \cos\alpha = \frac{1}{3},\ m = \cos\beta = \frac{2}{3} \text{ and } n = \cos\gamma = -\frac{2}{3}$$

So,

$$\overrightarrow{AB} = |\overrightarrow{AB}|(l\hat{i} + m\hat{j} + n\hat{k}) = 6\left(\frac{1}{3}\hat{i} + \frac{2}{3}\hat{j} - \frac{2}{3}\hat{k}\right) = 2\hat{i} + 4\hat{j} - 4\hat{k}$$

Now, $\overrightarrow{AB} = \overrightarrow{OB} - \overrightarrow{OA}$

$$\therefore \quad 2\hat{i} + 4\hat{j} - 4\hat{k} = \overrightarrow{OB} - (-\hat{i} + 3\hat{j} + 5\hat{k})$$

$$\Rightarrow \quad \overrightarrow{OB} = \hat{i} + 7\hat{j} + \hat{k}$$

2.2.2 Angle between Two Vectors

Consider two vectors $\overrightarrow{AB} = x_1\hat{i} + y_1\hat{j} + z_1\hat{k}$ and $\overrightarrow{AC} = x_2\hat{i} + y_2\hat{j} + z_2\hat{k}$.

Let $A \equiv (0, 0, 0)$.

Then, $B \equiv (x_1, y_1, z_1)$ and $C \equiv (x_2, y_2, z_2)$.

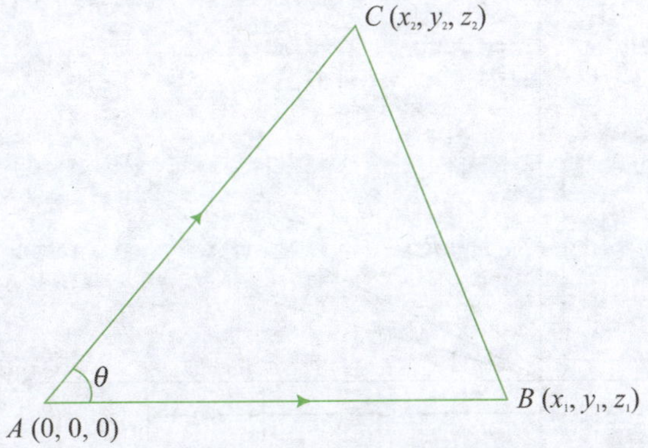

Let $\angle BAC = \theta$.

Now, using the cosine formula in the triangle ABC, we get

$$\cos\theta = \frac{AB^2 + AC^2 - BC^2}{2AB \cdot AC}$$

$$= \frac{(x_1^2+y_1^2+z_1^2)+(x_2^2+y_2^2+z_2^2)-[(x_1-x_2)^2+(y_1-y_2)^2+(z_1-z_2)^2]}{2\sqrt{x_1^2+y_1^2+z_1^2}\sqrt{x_2^2+y_2^2+z_2^2}}$$

$$= \frac{x_1x_2+y_1y_2+z_1z_2}{\sqrt{x_1^2+y_1^2+z_1^2}\sqrt{x_2^2+y_2^2+z_2^2}}$$

Thus, $\theta = \cos^{-1}\dfrac{x_1x_2+y_1y_2+z_1z_2}{\sqrt{x_1^2+y_1^2+z_1^2}\sqrt{x_2^2+y_2^2+z_2^2}}$

1. If θ is acute, then $\cos\theta > 0$, Therefore, $x_1x_2+y_1y_2+z_1z_2>0$.
2. If θ is obtuse, then $\cos\theta<0$. Therefore, $x_1x_2+y_1y_2+z_1z_2<0$.
3. If θ is a right angle, then $\cos\theta = 0$. Therefore, $x_1x_2+y_1y_2+z_1z_2=0$.

Illustration 2.25

Find the angle between the vectors $2\hat{i}+\hat{j}-3\hat{k}$ and $3\hat{i}-\hat{j}+2\hat{k}$.

Sol. Let the angle between the vectors be θ.

Then, $\cos\theta = \dfrac{(2)(3)+(1)(-1)+(-3)(2)}{\sqrt{4+1+9}\sqrt{9+1+4}} = -\dfrac{1}{14}$

Thus, the angle between the vectors is $\cos^{-1}\left(-\dfrac{1}{14}\right)$, which is obtuse.

Illustration 2.26

If vectors $3\hat{i}+4\hat{j}-5\hat{k}$ and $\hat{i}-4\hat{j}+\lambda\hat{k}$ are perpendicular then find the value of λ.

Sol. Since vectors are perpendicular, we have

$(3)(1)+(4)(-4)+(-5)(\lambda)=0$

$\Rightarrow \quad \lambda = -\dfrac{13}{5}$

2.2.3 Vector Addition in Component Form

Consider parallelogram $OPQR$ formed by vectors $\overrightarrow{OP} = x_1\hat{i}+y_1\hat{j}$ and $\overrightarrow{OR} = x_2\hat{i}+y_2\hat{j}$ in the xy-plane.

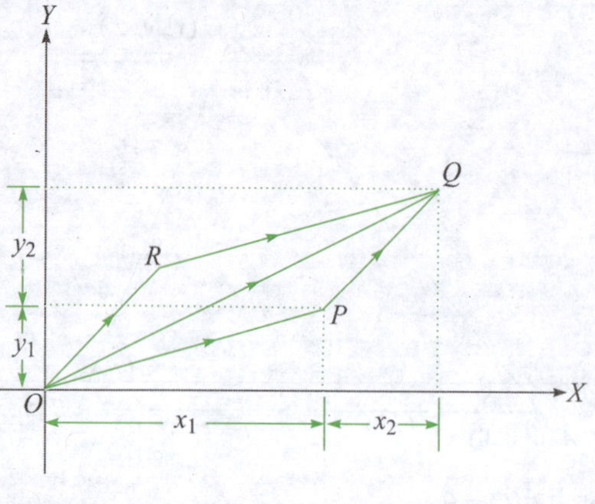

Now, $\overrightarrow{PQ} = \overrightarrow{OR} = x_2\hat{i}+y_2\hat{j}$

So, from the figure, we have

$$\overrightarrow{OQ} = \overrightarrow{OP}+\overrightarrow{OR} = \overrightarrow{OP}+\overrightarrow{PQ} = (x_1+x_2)\hat{i}+(y_1+y_2)\hat{j}$$

Thus, $(x_1\hat{i}+y_1\hat{j})+(x_2\hat{i}+y_2\hat{j}) = (x_1+x_2)\hat{i}+(y_1+y_2)\hat{j}$

Also, we have

$$(x_1\hat{i}+y_1\hat{j})-(x_2\hat{i}+y_2\hat{j}) = (x_1-x_2)\hat{i}+(y_1-y_2)\hat{j}$$

With similar reasoning, it can be proved that

$$(x_1\hat{i}+y_1\hat{j}+z_1\hat{k}) \pm (x_2\hat{i}+y_2\hat{j}+z_2\hat{k}) = (x_1\pm x_2)\hat{i}+(y_1\pm y_2)\hat{j}+(z_1\pm z_2)\hat{k}$$

2.2.3.1 Vector Joining Two Points

Consider two point $A(x_1, y_1, z_1)$ and $B(x_2, y_2, z_2)$.

So, $\overrightarrow{OA} = x_1\hat{i}+y_1\hat{j}+z_1\hat{k}$ and $\overrightarrow{OB} = x_2\hat{i}+y_2\hat{j}+z_2\hat{k}$.

Then,

$\overrightarrow{AB}$ = The position vector of point B – The position vector of point A

$= \overrightarrow{OB}-\overrightarrow{OA}$

$= (x_2\hat{i}+y_2\hat{j}+z_2\hat{k})-(x_1\hat{i}+y_1\hat{j}+z_1\hat{k})$

$= (x_2-x_1)\hat{i}+(y_2-y_1)\hat{j}+(z_2-z_1)\hat{k}$

Also, $\overrightarrow{BA} = (x_1-x_2)\hat{i}+(y_1-y_2)\hat{j}+(z_1-z_2)\hat{k}$

Illustration 2.27

Find the unit vector in the direction of vector $\overrightarrow{PQ}$, where P and Q are the points (1, 2, 3) and (4, 5, 6), respectively.

Sol. The given points are P (1, 2, 3) and Q (4, 5, 6). Therefore,

$\overrightarrow{PQ} = (4-1)\hat{i}+(5-2)\hat{j}+(6-3)\hat{k} = 3\hat{i}+3\hat{j}+3\hat{k}$

$\therefore \quad |\overrightarrow{PQ}| = \sqrt{3^2+3^2+3^2}$

$= \sqrt{9+9+9}$

$= \sqrt{27} = 3\sqrt{3}$

Hence, the unit vector in the direction of $\overrightarrow{PQ}$ is

$$\frac{\overrightarrow{PQ}}{|\overrightarrow{PQ}|} = \frac{3\hat{i}+3\hat{j}+3\hat{k}}{3\sqrt{3}} = \frac{1}{\sqrt{3}}\hat{i}+\frac{1}{\sqrt{3}}\hat{j}+\frac{1}{\sqrt{3}}\hat{k}$$

Illustration 2.28

If in a parallelogram $ABCD$, the diagonal vectors are $\overrightarrow{AC} = 2\hat{i}+3\hat{j}+4\hat{k}$ and $\overrightarrow{BD} = -6\hat{i}+7\hat{j}-2\hat{k}$.

Then find the vectors $\overrightarrow{AB}$ and $\overrightarrow{AD}$.

Sol. Let $\overrightarrow{AB} = \vec{a}$ and $\overrightarrow{AD} = \vec{b}$

Then $\overrightarrow{AC} = \vec{a}+\vec{b}$ and $\overrightarrow{BD} = \vec{b}-\vec{a}$

$\Rightarrow \quad \vec{a} = \dfrac{\overrightarrow{AC}-\overrightarrow{BD}}{2}$ and $\vec{b} = \dfrac{\overrightarrow{AC}+\overrightarrow{BD}}{2}$

$\Rightarrow \quad \vec{a} = \dfrac{2\hat{i}+3\hat{j}+4\hat{k}-(-6\hat{i}+7\hat{j}-2\hat{k})}{2}$ and

$\vec{b} = \dfrac{2\hat{i}+3\hat{j}+4\hat{k}+(-6\hat{i}+7\hat{j}-2\hat{k})}{2}$

$\Rightarrow \quad \vec{a} = \overrightarrow{AB} = 4\hat{i}-2\hat{j}+3\hat{k}$ and $\vec{b} = \overrightarrow{AD} = -2\hat{i}+5\hat{j}+\hat{k}$

Illustration 2.29

Find the resultant of the vectors $\vec{a} = \hat{i} - \hat{j} + 2\hat{k}$ and $\vec{b} = \hat{i} + 2\hat{j} - 4\hat{k}$. Find the unit vector in the direction of the resultant vector.

Sol. The resultant vector of $\vec{a}$ and $\vec{b}$ is

$\vec{a} + \vec{b} = 2\hat{i} + \hat{j} - 2\hat{k} = \vec{c}$ (say)

Now, unit vector in the direction of

$$\vec{c} \text{ is } \hat{c} = \frac{2\hat{i} + \hat{j} - 2\hat{k}}{\sqrt{(2)^2 + (1)^2 + (-2)^2}} = \frac{1}{3}(2\hat{i} + \hat{j} - 2\hat{k})$$

Illustration 2.30

Show that the points A, B and C with position vectors $\vec{a} = 3\hat{i} - 4\hat{j} - 4\hat{k}$, $\vec{b} = 2\hat{i} - \hat{j} + \hat{k}$ and $\vec{c} = \hat{i} - 3\hat{j} - 5\hat{k}$, respectively are the vertices of a right-angled triangle.

Sol. Position vectors of points A, B and C are respectively given as

$\vec{a} = 3\hat{i} - 4\hat{j} - 4\hat{k}$, $\vec{b} = 2\hat{i} - \hat{j} + \hat{k}$ and $\vec{c} = \hat{i} - 3\hat{j} - 5\hat{k}$

$\therefore \overrightarrow{AB} = \vec{b} - \vec{a} = (2-3)\hat{i} + (-1+4)\hat{j} + (1+4)\hat{k} = -\hat{i} + 3\hat{j} + 5\hat{k}$

$\overrightarrow{BC} = \vec{c} - \vec{b} = (1-2)\hat{i} + (-3+1)\hat{j} + (-5-1)\hat{k} = -\hat{i} - 2\hat{j} - 6\hat{k}$

$\overrightarrow{CA} = \vec{a} - \vec{c} = (3-1)\hat{i} + (-4+3)\hat{j} + (-4+5)\hat{k} = 2\hat{i} - \hat{j} + \hat{k}$

$\therefore \quad |\overrightarrow{AB}|^2 = (-1)^2 + 3^2 + 5^2 = 35$

$|\overrightarrow{BC}|^2 = (-1)^2 + (-2)^2 + (-6)^2 = 41$

$|\overrightarrow{CA}|^2 = 2^2 + (-1)^2 + 1^2 = 6$

$\therefore \quad |\overrightarrow{AB}|^2 + |\overrightarrow{CA}|^2 = 36 + 6 = 41 = |\overrightarrow{BC}|^2$

Hence, ΔABC is a right-angled triangle.

Illustration 2.31

Prove that the points $\hat{i} + 2\hat{j} - 3\hat{k}$, $2\hat{i} - \hat{j} + \hat{k}$ and $2\hat{i} + 5\hat{j} - \hat{k}$ form a triangle in space.

Sol. Given points are

$A(\hat{i} + 2\hat{j} - 3\hat{k})$, $B(2\hat{i} - \hat{j} + \hat{k})$, $C(2\hat{i} + 5\hat{j} - \hat{k})$

$\therefore \quad \overrightarrow{AB} = \hat{i} - 3\hat{j} + 4\hat{k}$ and $\overrightarrow{AC} = \hat{i} + 3\hat{j} + 2\hat{k}$

Clearly, the vectors $\overrightarrow{AB}$ and $\overrightarrow{AC}$ are non-collinear as there does not exist any real λ for which $\overrightarrow{AB} = \lambda\overrightarrow{AC}$.

Hence, the vectors $\overrightarrow{AB}$ and $\overrightarrow{AC}$ are non-collinear, therefore the given three points form a triangle.

Illustration 2.32

Find the position vector of a point R which divides the line segment joining two points P and Q whose position vectors are $\hat{i} + 2\hat{j} - \hat{k}$ and $-\hat{i} + \hat{j} + \hat{k}$, respectively, in the ratio 2 : 1.

a. Internally **b.** Externally

Sol. The position vector of point R dividing the line segment joining two points P and Q in the ratio $m : n$ is given by

1. Internally: $\dfrac{m\vec{b} + n\vec{a}}{m+n}$ **2.** Externally: $\dfrac{m\vec{b} - n\vec{a}}{m-n}$

Position vectors of P and Q are given as

$\overrightarrow{OP} = \hat{i} + 2\hat{j} - \hat{k}$ and $\overrightarrow{OQ} = -\hat{i} + \hat{j} + \hat{k}$

a. The position vector of point R which divides the line segment joining two points P and Q internally in the ratio 2 : 1 is given by

$$\overrightarrow{OR} = \frac{2(-\hat{i} + \hat{j} + \hat{k}) + 1(\hat{i} + 2\hat{j} - \hat{k})}{2+1}$$

$$= \frac{(-2\hat{i} + 2\hat{j} + 2\hat{k}) + (\hat{i} + 2\hat{j} - \hat{k})}{3}$$

$$= \frac{-\hat{i} + 4\hat{j} + \hat{k}}{3} = -\frac{1}{3}\hat{i} + \frac{4}{3}\hat{j} + \frac{1}{3}\hat{k}$$

b. The position vector of point R which divides the line segment joining two points P and Q externally in the ratio 2 : 1 is given by

$$\vec{OR} = \frac{2(-\hat{i} + \hat{j} + \hat{k}) - 1(\hat{i} + 2\hat{j} - \hat{k})}{2-1}$$

$$= (-2\hat{i} + 2\hat{j} + 2\hat{k}) - (\hat{i} + 2\hat{j} - \hat{k})$$

$$= -3\hat{i} + 3\hat{k}$$

Illustration 2.33

Check whether the three vectors $2\hat{i} + 2\hat{j} + 3\hat{k}$, $-3\hat{i} + 3\hat{j} + 2\hat{k}$ and $3\hat{i} + 4\hat{k}$ form a triangle or not.

Sol. If vectors $\vec{a} = 2\hat{i} + 2\hat{j} + 3\hat{k}$, $\vec{b} = -3\hat{i} + 3\hat{j} + 2\hat{k}$ and $\vec{c} = 3\hat{i} + 4\hat{k}$ form a triangle, then we must have $\vec{a} + \vec{b} + \vec{c} = 0$.

But for the given vectors, $\vec{a} + \vec{b} + \vec{c} \neq 0$. Hence, these vectors do not form a triangle.

Illustration 2.34

If two sides of a triangle ABC are $\overrightarrow{AB} = \hat{i} + 2\hat{j}$ and $\overrightarrow{BC} = \hat{i} + \hat{k}$, then find the length of the third side.

Sol. Given sides of the triangle are $\overrightarrow{AB} = \hat{i} + 2\hat{j}$ and $\overrightarrow{BC} = \hat{i} + \hat{k}$

Now $\overrightarrow{AB} + \overrightarrow{BC} + \overrightarrow{CA} = \vec{0}$

$\therefore \quad \overrightarrow{CA} = -\overrightarrow{AB} - \overrightarrow{BC} = -(\hat{i} + 2\hat{j}) - (\hat{i} + \hat{k})$

$= -2\hat{i} - 2\hat{j} - \hat{k}$

Therefore, the length of the third side,

$$|\overrightarrow{CA}| = \sqrt{(-2)^2 + (-2)^2 + (-1)^2} = 3.$$

Illustration 2.35

Find the lengths of the diagonals of the parallelepiped whose sides are vectors $\hat{i}+4\hat{j}-\hat{k},\ -3\hat{i}+\hat{j}+2\hat{k}, \hat{i}+2\hat{j}+\hat{k}$.

Sol. In the figure

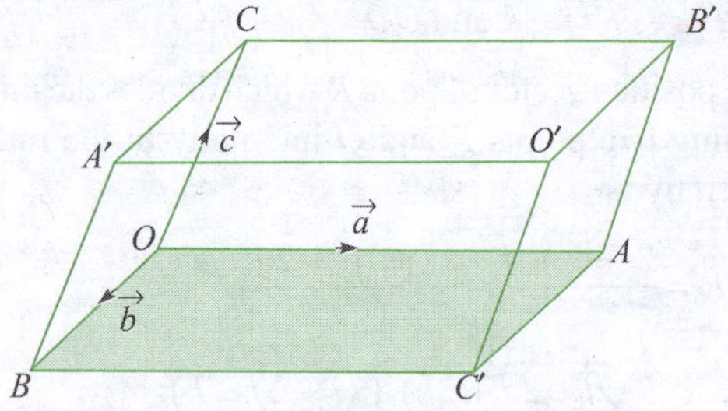

$\vec{a}=\overrightarrow{OA}=\hat{i}+4\hat{j}-\hat{k},\ \vec{b}=\overrightarrow{OB}=-3\hat{i}+\hat{j}+2\hat{k},$

$\vec{c}=\overrightarrow{OC}=\hat{i}+2\hat{j}+\hat{k}$

$\therefore$ Diagonal vectors are

$\overrightarrow{OO'}=\vec{a}+\vec{b}+\vec{c}=-\hat{i}+7\hat{j}+2\hat{k},$

$\overrightarrow{AA'}=-\vec{a}+\vec{b}+\vec{c}=-3\hat{i}-\hat{j}+4\hat{k},$

$\overrightarrow{BB'}=\vec{a}-\vec{b}+\vec{c}=5\hat{i}+5\hat{j}-2\hat{k},$

and $\overrightarrow{CC'}=\vec{a}+\vec{b}-\vec{c}=-3\hat{i}+3\hat{j}$

$\therefore$ $\overrightarrow{OO'}=\sqrt{1+49+4}=\sqrt{54}=3\sqrt{6}$

$|\overrightarrow{AA'}|=\sqrt{9+1+16}=\sqrt{26}$

$|\overrightarrow{BB'}|=\sqrt{25+25+4}=\sqrt{54}=3\sqrt{6}$

$|\overrightarrow{CC'}|=\sqrt{9+9}=3\sqrt{2}$

Illustration 2.36

Prove that the sum of the three vectors determined by the diagonals of three adjacent faces of a cube passing through the same corner, the vectors being directed from the corner, is twice the vector determined by the diagonal of the cube.

Sol. Consider a cube as shown in the following figure whose co-initial edges are vectors $\overrightarrow{OA}=a\hat{i}, \overrightarrow{OB}=a\hat{j}$ and $\overrightarrow{OC}=a\hat{k}$ through the origin.

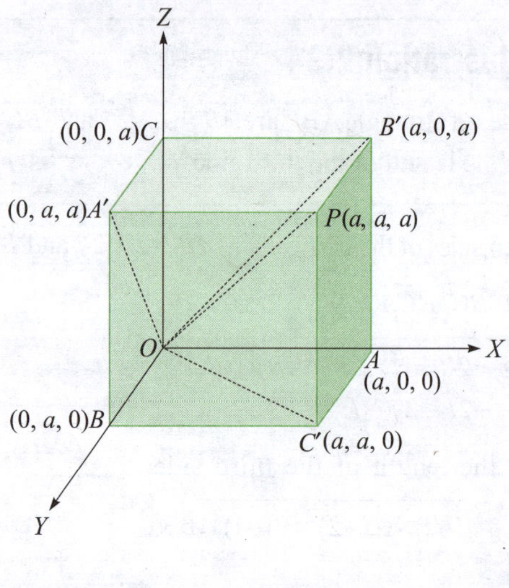

Vectors determined by the diagonals of the faces are

$\overrightarrow{OA'}=\overrightarrow{OB}+\overrightarrow{OC}=a\hat{j}+a\hat{k}$

$\overrightarrow{OB'}=\overrightarrow{OA}+\overrightarrow{OC}=a\hat{i}+a\hat{k}$

and $\overrightarrow{OC'}=\overrightarrow{OA}+\overrightarrow{OB}=a\hat{i}+a\hat{j}$

$\therefore$ $\overrightarrow{OA'}+\overrightarrow{OB'}+\overrightarrow{OC'}=2(a\hat{i}+a\hat{j}+a\hat{k})$

Now, the diagonal vector $\overrightarrow{OP}=a\hat{i}+a\hat{j}+a\hat{k}$

Therefore, $\overrightarrow{OA'}+\overrightarrow{OB'}+\overrightarrow{OC'}=2\overrightarrow{OP}$

Illustration 2.37

Three coinitial vectors of magnitudes a, $2a$ and $3a$ meet at a point and their directions are along the diagonals of the three adjacent faces of a cube. Determine the magnitude of their resultant R.

Sol. Let the three co-initial vectors $\overrightarrow{OA}=\hat{i}, \overrightarrow{OB}=\hat{j}, \overrightarrow{OC}=\hat{k}$ acting at the origin are the edges of a cube.

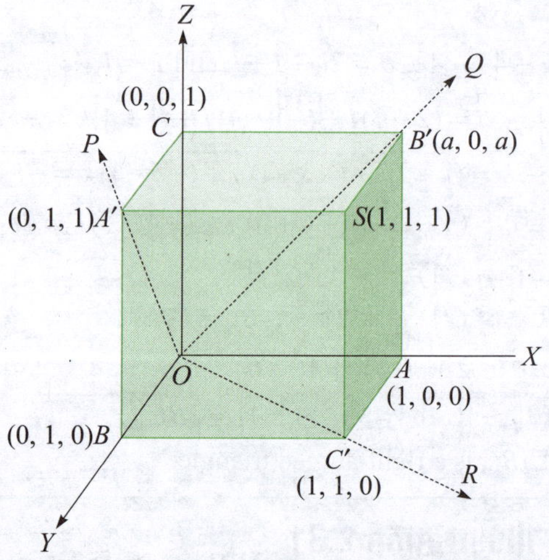

Let the vectors $\overrightarrow{OP}, \overrightarrow{OQ}, \overrightarrow{OR}$ in order are acting along the diagonals OA', OB' and OC' of the faces such that

$|\overrightarrow{OP}|=a, |\overrightarrow{OQ}|=2a$ and $|\overrightarrow{OR}|=3a$.

Now, $\overrightarrow{OA'}=\hat{j}+\hat{k}$, $\overrightarrow{OB'}=\hat{i}+\hat{k}$ and $\overrightarrow{OC'}=\hat{i}+\hat{j}$.

$\therefore$ $\overrightarrow{OP}=|\overrightarrow{OP}|\dfrac{\overrightarrow{OA'}}{|\overrightarrow{OA'}|}=\dfrac{a}{\sqrt{2}}(\hat{j}+\hat{k})$

$\overrightarrow{OQ}=|\overrightarrow{OQ}|\dfrac{\overrightarrow{OB'}}{|\overrightarrow{OB'}|}=\dfrac{2a}{\sqrt{2}}(\hat{i}+\hat{k})$ and

$\overrightarrow{OR}=|\overrightarrow{OR}|\dfrac{\overrightarrow{OC'}}{|\overrightarrow{OC'}|}=\dfrac{3a}{\sqrt{2}}(\hat{i}+\hat{j})$

So, the resultant, $\vec{R}=\overrightarrow{OP}+\overrightarrow{OQ}+\overrightarrow{OR}$

$$=\frac{a}{\sqrt{2}}(\hat{j}+\hat{k})+\frac{2a}{\sqrt{2}}(\hat{i}+\hat{k})+\frac{3a}{\sqrt{2}}(\hat{i}+\hat{j})=\frac{a}{\sqrt{2}}(5\hat{i}+4\hat{j}+3\hat{k})$$

Illustration 2.38

If the resultant of two forces is equal in magnitude to one of the components and is perpendicular to it in direction, find the other component using vector method.

Sol. Let one of the component forces is vector $\overrightarrow{OA} = P\hat{i}$ acting at the origin.

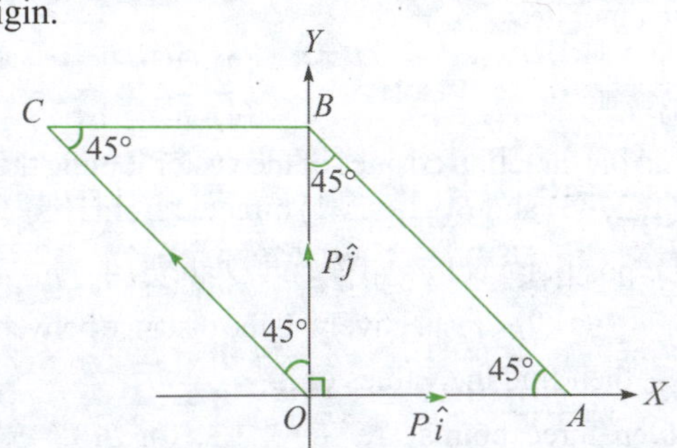

Given that the resultant of two forces is equal in magnitude to one of the components and is perpendicular to it. Let the resultant force $\overrightarrow{OB}$ is perpendicular to $P\hat{i}$ acting along the positive y-axis.

So, the resultant force is $\overrightarrow{OB} = P\hat{j}$.

Clearly, the other force $\overrightarrow{OC}$ will make an obtuse angle with the vector $\overrightarrow{OA}$.

Using the parallelogram law of vector addition, we have

$$\overrightarrow{OA} + \overrightarrow{OC} = \overrightarrow{OB}$$

$$\therefore \quad P\hat{i} + \overrightarrow{OC} = P\hat{j}$$

$$\Rightarrow \quad \overrightarrow{OC} = -P\hat{i} + P\hat{j}$$

From the geometry of the figure, vector $\overrightarrow{OC}$ makes an angle 135° with the vector $\overrightarrow{OA}$.

Also, $|\overrightarrow{OC}| = \sqrt{(-P)^2 + P^2} = \sqrt{2P^2} = \sqrt{2}P$

Thus, the other component force has magnitude $\sqrt{2}$ times the magnitude of the first force and makes an angle 135° with it.

Illustration 2.39

A man travelling east at 8 km per hour finds that the wind seems to blow directly from the north. On doubling the speed, he finds that it appears to come from North east. Find the velocity of the wind.

Sol.

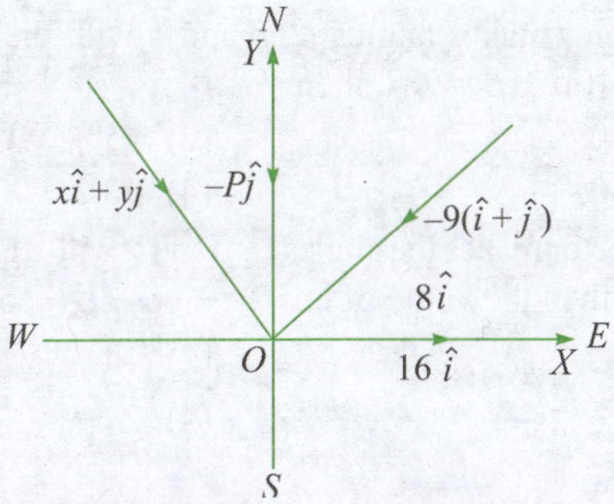

Let the actual velocity of the wind is $x\hat{i} + y\hat{j}$.

Man is travelling east.

If his speed is 8 k/h, he finds that wind is blowing from the north.

If the speed of the wind is p km/hr, then velocity of the wind is $-p\hat{j}$.

This is the velocity of the wind relative to the man.

So, we have $-p\hat{j} = x\hat{i} + y\hat{j} - (8\hat{i})$

Comparing coefficients of $\hat{i}$ and $\hat{j}$, we get

$$x - 8 = 0 \text{ and } y = -p \qquad \text{...(i)}$$

Now, when man doubles his speed, he finds that wind is blowing from the north-east.

If the speed of the wind is q km/hr, then velocity of the wind is $-q(\hat{i} + \hat{j})$.

Again, this is the velocity of the wind relative to the man.

So, we have $-q(\hat{i} + \hat{j}) = x\hat{i} + y\hat{j} - (16\hat{i})$

Comparing coefficients of $\hat{i}$ and $\hat{j}$, we get

$$-q = x - 16 \text{ and } -q = y \qquad \text{...(ii)}$$

From (i) and (ii), we get

$$x = 8,\ y = -8 \text{ and } p = q = 8$$

Therefore, actual velocity of the wind is $8\hat{i} - 8\hat{j}$, i.e. wind is blowing from North-west.

Also, speed is $\sqrt{8^2 + 8^2} = 8\sqrt{2}$ km/hr.

2.2.4 BISECTOR OF ANGLE BETWEEN TWO VECTORS

Consider a parallelogram $ABCD$ formed by vectors $\overrightarrow{AB} = \vec{a}$ and $\overrightarrow{AD} = \vec{b}$.

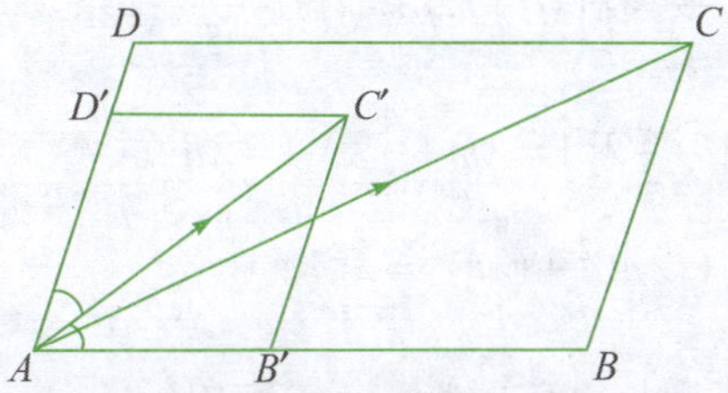

Clearly, the diagonal vector $\overrightarrow{AC} = \overrightarrow{AB} + \overrightarrow{AD} = \vec{a} + \vec{b}$.

We know that the rhombus is a parallelogram in which diagonals bisect the opposite angles.

So, we construct a rhombus $AB'C'D'$ by taking unit vectors $\overrightarrow{AB'} = \dfrac{\vec{a}}{|\vec{a}|}$ and $\overrightarrow{AD'} = \dfrac{\vec{b}}{|\vec{b}|}$.

Now, the diagonal vector of this rhombus is $\overrightarrow{AC'} = \dfrac{\vec{a}}{|\vec{a}|} + \dfrac{\vec{b}}{|\vec{b}|}$, which bisects the opposite angles.

Hence, the vector bisecting the angle between vectors $\vec{a}$ and $\vec{b}$ is $\lambda\left(\dfrac{\vec{a}}{|\vec{a}|} + \dfrac{\vec{b}}{|\vec{b}|}\right)$ or $\lambda(\hat{a} + \hat{b})$, where $\lambda \in R^+$.

Illustration 2.40

If $\vec{a} = 7\hat{i} - 4\hat{j} - 4\hat{k}$ and $\vec{b} = -2\hat{i} - \hat{j} + 2\hat{k}$, determine the vector $\vec{c}$ along the internal bisector of the angle between the vectors $\vec{a}$ and $\vec{b}$ such that $|\vec{c}| = 5\sqrt{6}$.

Sol. We have $|\vec{a}| = \sqrt{49 + 16 + 16} = \sqrt{81} = 9$ and

$|\vec{b}| = \sqrt{4+1+4} = \sqrt{9} = 3$

So, the vector bisecting the angle between $\vec{a}$ and $\vec{b}$ is

$$\vec{p} = \frac{\vec{a}}{|\vec{a}|} + \frac{\vec{b}}{|\vec{b}|}$$

$$= \frac{1}{9}(7\hat{i} - 4\hat{j} - 4\hat{k}) + \frac{1}{3}(-2\hat{i} - \hat{j} + 2\hat{k}) = \frac{1}{9}(\hat{i} - 7\hat{j} + 2\hat{k})$$

Now, the unit vector in the direction of $\vec{p}$ is

$$\frac{\vec{p}}{|\vec{p}|} = \frac{1}{3\sqrt{6}}(\hat{i} - 7\hat{j} + 2\hat{k})$$

So, the vector in the direction of $\vec{p}$ having magnitude $5\sqrt{6}$ is

$$5\sqrt{6}\frac{\vec{p}}{|\vec{p}|} = \frac{5}{3}(\hat{i} - 7\hat{j} + 2\hat{k}) = \vec{c}$$

Illustration 2.41

Find a unit vector $\vec{c}$ if $-\hat{i} + \hat{j} - \hat{k}$ bisects the angle between the vectors $\vec{c}$ and $3\hat{i} + 4\hat{j}$.

Sol. Let $\vec{c} = l\hat{i} + m\hat{j} + n\hat{k}$ where $l^2 + m^2 + n^2 = 1$

Given that the vector $-\hat{i} + \hat{j} - \hat{k}$ bisects the angle between vectors $\vec{c}$ and $3\hat{i} + 4\hat{j}$.

$$\therefore \quad -\hat{i} + \hat{j} - \hat{k} = \lambda\left((l\hat{i} + m\hat{j} + n\hat{k}) + \frac{3\hat{i} + 4\hat{j}}{5}\right), \lambda > 0 \qquad \text{...(i)}$$

$$\Rightarrow \quad -1 = \lambda l + \frac{3}{5}\lambda, 1 = \lambda m + \frac{4}{5}\lambda, -1 = \lambda n$$

Now, $(\lambda l)^2 + (\lambda m)^2 + (\lambda n)^2 = \lambda^2$

$$\therefore \quad \left(-1 - \frac{3}{5}\lambda\right)^2 + \left(1 - \frac{4}{5}\lambda\right)^2 + (-1)^2 = \lambda^2$$

$$\Rightarrow \quad \lambda = \frac{15}{2}$$

Putting this value in (i), we get $\vec{c} = \frac{1}{15}(-11\hat{i} - 10\hat{j} - 2\hat{k})$

Aim to complete within 40 minutes

Concept Application Exercise 2.2

1. Find the unit vector in the direction of the vector $\vec{a} = \hat{i} + \hat{j} + 2\hat{k}$.
2. Find the direction cosines of the vector $\hat{i} + 2\hat{j} + 3\hat{k}$.
3. The terminal point of a vector $-4\hat{i} - 2\hat{j} + 3\hat{k}$ having its initial point (2, –1, 4) lies in the octant
   (1) 3rd (2) 5th
   (3) 7th (4) 8th
4. The number of integral values of a for which the angle between the vectors $a\hat{i} - 2\hat{j} - 4\hat{k}$ and $a\hat{i} + 2\hat{j} + 3\hat{k}$ is obtuse is
   (1) 4 (2) 6
   (3) 7 (4) 9
5. If the vector $\overrightarrow{OP}$ (where O is origin) makes an obtuse angle the positive x-axis, then in which octant the point P cannot lie
   (1) 2nd (2) 3rd
   (3) 6th (4) 8th
6. Find the direction cosines of the vector joining the points $A(1, 2, -3)$ and $B(-1, -2, 1)$ directed from A to B.
7. The position vectors of P and Q are $5\hat{i} + 4\hat{j} + a\hat{k}$ and $-\hat{i} + 2\hat{j} - 2\hat{k}$, respectively. If the distance between them is 7, then find the values of a.
8. Given three points are $A(-3, -2, 0)$, $B(3, -3, 1)$ and $C(5, 0, 2)$. Then find a vector in the direction of $\overrightarrow{AB}$ with magnitude $|\overrightarrow{AC}|$.
9. Find a vector of magnitude 5 units, which is parallel to the resultant of the vectors $\vec{a} = 2\hat{i} + 3\hat{j} - \hat{k}$ and $\vec{b} = \hat{i} - 2\hat{j} + \hat{k}$.
10. What is the unit vector in the direction of the vector $\vec{a} = 3\hat{i} + 4\hat{j} - 2\hat{k}$? What vector should be added to $\vec{a}$ so that the resultant is the unit vector $\hat{i}$?
11. The position vectors of points A and B w.r.t. the origin are $\vec{a} = \hat{i} + 3\hat{j} - 2\hat{k}$ and $\vec{b} = 3\hat{i} + \hat{j} - 2\hat{k}$, respectively. Then the vector $\overrightarrow{OP}$ which bisects angle AOB, where P is a point on AB is
    (1) $2(\hat{i} - \hat{j} - \hat{k})$ (2) $2(\hat{i} + \hat{j} + \hat{k})$
    (3) $2(\hat{i} - \hat{j} + \hat{k})$ (4) $2(\hat{i} + \hat{j} - \hat{k})$
12. Find the vector of magnitude 3, bisecting the angle between the vectors $\vec{a} = 2\hat{i} + \hat{j} - \hat{k}$ and $\vec{b} = \hat{i} - 2\hat{j} + \hat{k}$.
13. If the vectors $\overrightarrow{AB} = 2\hat{i} + 3\hat{j} - \hat{k}$ and $\overrightarrow{AC} = -4\hat{i} - \hat{j} + 3\hat{k}$ are the sides of triangle ABC, then the length of the median through A is
    (1) $\sqrt{2}$ (2) $\sqrt{3}$
    (3) $2\sqrt{3}$ (4) $3\sqrt{3}$
14. A vector makes angles α, β and γ with the coordinate axes. If $\alpha + \beta = 90°$, then γ =
    (1) 30° (2) 45°
    (3) 60° (4) 90°
15. If a vector makes the angles α, β and γ with 3D coordinate axes, then the value of $\cos 2\alpha + \cos 2\beta + \cos 2\gamma$ is
    (1) –2 (2) –1
    (3) 1 (4) 2

16. Let the point P be $P(2, 3, 6)$ and O be the origin. Then with which axis the angle of $\overrightarrow{OP}$ is the greatest?

17. The position vectors of the vertices A, B and C of a triangle are $\hat{i}+\hat{j}$, $\hat{j}+\hat{k}$ and $\hat{i}+\hat{k}$, respectively. Find a unit vector $\hat{r}$ lying in the plane of ABC which is perpendicular to IA, where I is the incentre of the triangle.

Answers

1. $\frac{1}{\sqrt{6}}\hat{i}+\frac{1}{\sqrt{6}}\hat{j}+\frac{2}{\sqrt{6}}\hat{k}$ **2.** $\left(\frac{1}{\sqrt{14}}, \frac{2}{\sqrt{14}}, \frac{3}{\sqrt{14}}\right)$

3. (1) **4.** (3)

5. (4) **6.** $\left(-\frac{1}{3}, -\frac{2}{3}, \frac{2}{3}\right)$

7. $a=-5, 1$ **8.** $\frac{6}{\sqrt{19}}(\hat{i}-\hat{j}+\hat{k})$

9. $\pm\frac{3\sqrt{10}\hat{i}}{2}\pm\frac{\sqrt{10}}{2}\hat{j}$

10. $\frac{1}{\sqrt{29}}(3\hat{i}+4\hat{j}-2\hat{k}); -2\hat{i}-4\hat{j}+2\hat{k}$

11. (4) **12.** $\frac{3(3\hat{i}-\hat{j})}{\sqrt{10}}$

13. (2) **14.** (4)

15. (2) **16.** x-axis

17. $\hat{r}=\pm\frac{1}{\sqrt{2}}(\hat{i}-\hat{j})$

2.3 LINEAR COMBINATION OF VECTORS

Linear combination of a set of vectors $\vec{a_1}, \vec{a_2}, \vec{a_3}, \ldots, \vec{a_n}$ is the expression $m_1\vec{a_1}+m_2\vec{a_2}+m_3\vec{a_3}+\ldots+m_n\vec{a_n}$, where $m_1, m_2, m_3, \ldots, m_n$ are scalars.

2.3.1 Linearly Dependent Vectors

A set of vectors $\vec{a_1}, \vec{a_2}, \vec{a_3}, \ldots, \vec{a_n}$ is said to be linearly dependent if at least one of them can be expressed as a linear combination of the remaining vectors.

Alternatively, if the set of vectors $\vec{a_1}, \vec{a_2}, \vec{a_3}, \ldots, \vec{a_n}$ is linearly dependent then $m_1\vec{a_1}+m_2\vec{a_2}+m_3\vec{a_3}+\ldots+m_n\vec{a_n}=0$, where not all $m_1, m_2, m_3, \ldots, m_n$ are zero.

Theorem

A pair of collinear vectors is linearly dependent.

Verification

We know that if vectors $\vec{a_1}$ and $\vec{a_2}$ are collinear then for some scalar λ, $\vec{a_2}=\lambda\vec{a_1}$. This can be put as $m_1\vec{a_1}+m_2\vec{a_2}=0$, where m_1 and m_2 are non-zero scalars. This is the condition of linearly dependent vectors.

Theorem

Three coplanar vectors are linearly dependent.

Verification

We know that if vectors $\vec{a_1}$, $\vec{a_2}$ and $\vec{a_3}$ are coplanar then for some scalars λ and μ, $\vec{a_3}=\lambda\vec{a_1}+\mu\vec{a_2}$. This can be put as $m_1\vec{a_1}+m_2\vec{a_2}+m_3\vec{a_3}=0$, where m_1, m_2 and m_3 are non-zero scalars. This is the condition of linearly dependent vectors.

Theorem

Four vectors in space are linearly dependent.

Proof: Let the four vectors $\vec{a_1}, \vec{a_2}, \vec{a_3}$ and $\vec{a_4}$ are co-initial at point O.

We have $\overrightarrow{OA}=\vec{a_1}, \overrightarrow{OB}=\vec{a_2}, \overrightarrow{OC}=\vec{a_3}$ and $\overrightarrow{OD}=\vec{a_4}$

We construct a parallelepiped by drawing vectors through point D parallel to the vectors $\vec{a_1}, \vec{a_2}$ and $\vec{a_3}$ such that $\overrightarrow{OD}$ is the diagonal of the parallelepiped.

Now, for some scalars λ, μ and δ, we have $\overrightarrow{OA'}=\lambda\vec{a_1}$, $\overrightarrow{OB'}=\mu\vec{a_2}$ and $\overrightarrow{OC'}=\delta\vec{a_3}$

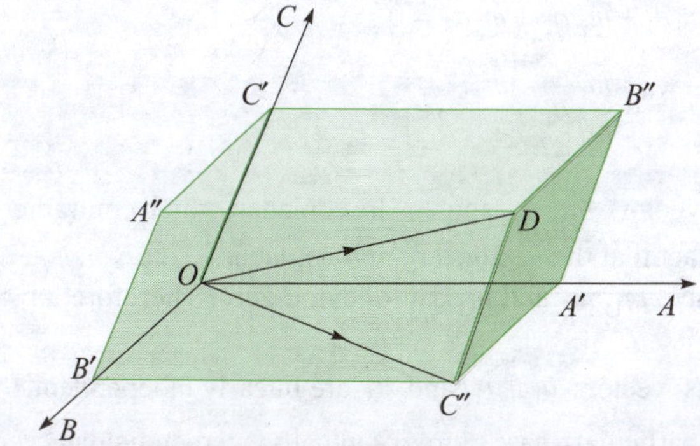

$$\therefore \quad \overrightarrow{OD}=\overrightarrow{OC''}+\overrightarrow{C''D}$$
$$=(\overrightarrow{OA'}+\overrightarrow{OB'})+\overrightarrow{OC'}$$
$$=(\lambda\vec{a_1}+\mu\vec{a_2})+\delta\vec{a_3}$$

So, $\vec{a_4}=\lambda\vec{a_1}+\mu\vec{a_2}+\delta\vec{a_3}$, i.e. vector $\vec{a_4}$ is expressible as a linear combination of the other three vectors.

Therefore, vectors $\vec{a_1}, \vec{a_2}, \vec{a_3}$ and $\vec{a_4}$ are linearly dependent.

2.3.2 Linearly Independent Vectors

A set of vectors $\vec{a_1}, \vec{a_2}, \vec{a_3}, \ldots, \vec{a_n}$ is said to be linearly independent if no vector of the set can be expressed as a linear combination of the others.

Alternatively, if the set of vectors $\vec{a_1}, \vec{a_2}, \vec{a_3}, \ldots, \vec{a_n}$ is linearly independent then, the equation

$m_1\vec{a_1}+m_2\vec{a_2}+m_3\vec{a_3}+\ldots+m_n\vec{a_n}=0$ has only the trivial solution,

$$m_1=m_2=m_3=\ldots=m_n=0$$

Theorem

A pair of non-collinear vectors is linearly independent.

Proof: Consider non-collinear vectors $\overrightarrow{a_1}$ and $\overrightarrow{a_2}$.

Linear combination of these vectors is $m_1\overrightarrow{a_1}+m_2\overrightarrow{a_2}$; where m_1 and m_2 are scalars.

Now, if m_1 and m_2 are non-zero, then we have

$$m_1\overrightarrow{a_1}+m_2\overrightarrow{a_2}=0 \Rightarrow \overrightarrow{a_1}=-\frac{m_2}{m_1}\overrightarrow{a_2}$$

So, vectors $\overrightarrow{a_1}$ and $\overrightarrow{a_2}$ are collinear, which contradicts the given fact that the vectors are non-collinear.

Hence, m_1 and m_2 can't be non-zero. Therefore, $m_1 = m_2 = 0$

Thus vectors $\overrightarrow{a_1}$ and $\overrightarrow{a_2}$ are linearly independent.

Theorem

Three non-coplanar vectors are linearly independent.

Proof: Consider three non-coplanar vectors $\overrightarrow{a_1}$, $\overrightarrow{a_2}$ and $\overrightarrow{a_3}$.

Linear combination of these vectors is $m_1\overrightarrow{a_1}+m_2\overrightarrow{a_2}+m_3\overrightarrow{a_3}$; where m_1, m_2 and m_3 are scalars.

Now, if m_1, m_2 and m_3 are non-zero, then we have

$$m_1\overrightarrow{a_1}+m_2\overrightarrow{a_2}+m_3\overrightarrow{a_3}=0$$

$$\Rightarrow \quad \overrightarrow{a_3}=-\frac{m_1}{m_3}\overrightarrow{a_1}-\frac{m_2}{m_3}\overrightarrow{a_2}$$

So, vectors $\overrightarrow{a_1}$, $\overrightarrow{a_2}$ and $\overrightarrow{a_3}$ are coplanar, which contradicts the given fact that the vectors are non-coplanar.

Hence, m_1, m_2 and m_3 can't be non-zero. Therefore, $m_1 = m_2 = m_3 = 0$.

Thus, vectors $\overrightarrow{a_1}$, $\overrightarrow{a_2}$ and $\overrightarrow{a_3}$ are linearly independent.

In particular case three mutually perpendicular vectors $\hat{i}$, $\hat{j}$ and $\hat{k}$ are linearly independent.

Theorem

If vectors $x_1\vec{a}+y_1\vec{b}+z_1\vec{c}$, $x_2\vec{a}+y_2\vec{b}+z_2\vec{c}$ are collinear, where $\vec{a}, \vec{b}, \vec{c}$ are non-coplanar then $\dfrac{x_1}{x_2}=\dfrac{y_1}{y_2}=\dfrac{z_1}{z_2}$.

Proof: Vectors $x_1\vec{a}+y_1\vec{b}+z_1\vec{c}$, $x_2\vec{a}+y_2\vec{b}+z_2\vec{c}$ are collinear.

$$\therefore \quad x_1\vec{a}+y_1\vec{b}+z_1\vec{c}=\lambda(x_2\vec{a}+y_2\vec{b}+z_2\vec{c}), \text{ where } \lambda \text{ is a scalar.}$$

$$\Rightarrow \quad (x_1-\lambda x_2)\vec{a}+(y_1-\lambda y_2)\vec{b}+(z_1-\lambda z_2)\vec{c}=0$$

But given that vectors $\vec{a}, \vec{b}, \vec{c}$ are non-coplanar, i.e. linearly independent.

$$\therefore \quad x_1-\lambda x_2=0,\ y_1-\lambda y_2=0 \text{ and } z_1-\lambda z_2=0$$

$$\Rightarrow \quad \frac{x_1}{x_2}=\frac{y_1}{y_2}=\frac{z_1}{z_2}=\lambda$$

Theorem

If vectors $x_1\vec{a}+y_1\vec{b}+z_1\vec{c}$, $x_2\vec{a}+y_2\vec{b}+z_2\vec{c}$ and $x_3\vec{a}+y_3\vec{b}+z_3\vec{c}$ are coplanar; where $\vec{a}, \vec{b}, \vec{c}$ are non-coplanar then $\begin{vmatrix} x_1 & y_1 & z_1 \\ x_2 & y_2 & z_2 \\ x_3 & y_3 & z_3 \end{vmatrix}=0$

Proof: Vectors $x_1\vec{a}+y_1\vec{b}+z_1\vec{c}$, $x_2\vec{a}+y_2\vec{b}+z_2\vec{c}$ and $x_3\vec{a}+y_3\vec{b}+z_3\vec{c}$ are coplanar.

So, there exist scalars λ and μ such that

$$x_1\vec{a}+y_1\vec{b}+z_1\vec{c}=\lambda(x_2\vec{a}+y_2\vec{b}+z_2\vec{c})+\mu(x_3\vec{a}+y_3\vec{b}+z_3\vec{c})$$

$$\therefore (x_1-\lambda x_2-\mu x_3)\vec{a}+(y_1-\lambda y_2-\mu y_3)\vec{b}+(z_1-\lambda z_2-\mu z_3)\vec{c}=0$$

But given that vectors $\vec{a}, \vec{b}, \vec{c}$ are non-coplanar, i.e. linearly independent.

So, $\quad x_1-\lambda x_2-\mu x_3=0 \quad$ ...(i)

$\quad y_1-\lambda y_2-\mu y_3=0 \quad$ ...(ii)

and $\quad z_1-\lambda z_2-\mu z_3=0 \quad$ ...(iii)

Here the system of equations in two unknowns λ and μ must be consistent, i.e. values of λ and μ obtained from any two equations must satisfy the third equation.

$$\therefore \quad \begin{vmatrix} x_1 & -x_2 & -x_3 \\ y_1 & -y_2 & -y_3 \\ z_1 & -z_2 & -z_3 \end{vmatrix}=0$$

$$\Rightarrow \quad \begin{vmatrix} x_1 & y_1 & z_1 \\ x_2 & y_2 & z_2 \\ x_3 & y_3 & z_3 \end{vmatrix}=0 \quad \text{(using properties of determinant)}$$

Thus, vectors $x_1\hat{i}+y_1\hat{j}+z_1\hat{k}$, $x_2\hat{i}+y_2\hat{j}+z_2\hat{k}$ and $x_3\hat{i}+y_3\hat{j}+z_3\hat{k}$ are coplanar, then $\begin{vmatrix} x_1 & y_1 & z_1 \\ x_2 & y_2 & z_2 \\ x_3 & y_3 & z_3 \end{vmatrix}=0$.

Illustration 2.42

Prove that points $\vec{a}-2\vec{b}+3\vec{c}$, $2\vec{a}+3\vec{b}-4\vec{c}$, $-7\vec{b}+10\vec{c}$ are collinear, where $\vec{a}, \vec{b}, \vec{c}$ are non-collinear vectors.

Sol. Let point A, B and C are

$\vec{a}-2\vec{b}+3\vec{c}$, $2\vec{a}+3\vec{b}-4\vec{c}$ and $-7\vec{b}+10\vec{c}$ respectively.

$$\therefore \quad \overrightarrow{AB} = \text{Position vector of } B - \text{Position vector of } A$$

$$= 2\vec{a}+3\vec{b}-4\vec{c}-(\vec{a}-2\vec{b}+3\vec{c})$$

$$= \vec{a}+5\vec{b}-7\vec{c}$$

$$\overrightarrow{AC} = \overrightarrow{OC}-\overrightarrow{OA}$$

$$= (-7\vec{b}+10\vec{c})-(\vec{a}-2\vec{b}+3\vec{c})$$

$$= -\vec{a}-5\vec{b}+7\vec{c}$$

Thus, $\overrightarrow{AB} = -\overrightarrow{AC}$

So, vectors $\overrightarrow{AB}$ and $\overrightarrow{AC}$ are collinear, and hence points *A*, *B* and *C* are collinear.

Illustration 2.43

Using vector method prove that points (1, –1, 2), (3, 3, 4), (4, 5, 11) are collinear.

Sol. Let $P \equiv (1, -1, 2)$, $Q \equiv (3, 3, 8)$ and $R \equiv (4, 5, 11)$.

$\therefore \quad \overrightarrow{PQ} = (3-1)\hat{i} + (3-(-1))\hat{j} + (8-2)\hat{k} = 2\hat{i} + 4\hat{j} + 6\hat{k}$

and $\quad \overrightarrow{QR} = (4-3)\hat{i} + (5-3)\hat{j} + (11-8)\hat{k} = \hat{i} + 2\hat{j} + 3\hat{k}$

Clearly, $\overrightarrow{PQ} = 2\overrightarrow{QR}$.

So, vectors $\overrightarrow{PQ}$ and $\overrightarrow{QR}$ are collinear.

Hence, points *P*, *Q* and *R* are collinear.

Illustration 2.44

Using the vector method, find the values of λ and μ if points $(2, \lambda, 3)$, $(-1, 2, 1)$ and $(4, 7, \mu)$ are collinear.

Sol. Let $P \equiv (2, \lambda, 3)$, $Q \equiv (-1, 2, 1)$ and $R \equiv (4, 7, \mu)$.

$\therefore \quad \overrightarrow{PQ} = (-1-2)\hat{i} + (2-\lambda)\hat{j} + (1-3)\hat{k} = -3\hat{i} + (2-\lambda)\hat{j} - 2\hat{k}$

and $\quad \overrightarrow{QR} = (4-(-1))\hat{i} + (7-2)\hat{j} + (\mu-1)\hat{k} = 5\hat{i} + 5\hat{j} + (\mu-1)\hat{k}$

Points *P*, *Q* and *R* are collinear.

So, vectors $\overrightarrow{PQ}$ and $\overrightarrow{QR}$ are collinear.

$$\therefore \quad \frac{-3}{5} = \frac{2-\lambda}{5} = \frac{-2}{\mu-1}$$

$$\Rightarrow \quad \lambda = 5, \ \mu = \frac{13}{3}$$

Illustration 2.45

Let $\vec{a}$, $\vec{b}$, $\vec{c}$ be three non-zero vectors, such that no two of them are collinear. If $\vec{a} + 2\vec{b}$ is collinear with $\vec{c}$ and $\vec{b} + 3\vec{c}$ is collinear with $\vec{a}$, then find the value of $|\vec{a} + 2\vec{b} + 6\vec{c}|$

Sol. Given that vectors $\vec{a} + 2\vec{b}$ and $\vec{c}$ are collinear.

$\therefore \quad \vec{a} + 2\vec{b} = \lambda\vec{c}$, where λ is a scalar ...(i)

Also, given that vectors $\vec{b} + 3\vec{c}$ and $\vec{a}$ are collinear.

$\therefore \quad \vec{b} + 3\vec{c} = \mu\vec{a}$, where μ is a scalar ...(ii)

Eliminating vector $\vec{b}$ from the above equations, we get

$$\vec{a} - 6\vec{c} = \lambda\vec{c} - 2\mu\vec{a}$$

or $\quad \vec{a}(1+2\mu) - (\lambda+6)\vec{c} = 0$

Now, vectors $\vec{a}$ and $\vec{c}$ are non-collinear, i.e. linearly independent.

$\therefore \quad 1 + 2\mu =$ and $\lambda + 6 = 0$

$\Rightarrow \quad \mu = -\dfrac{1}{2}$ and $\lambda = -6$.

Putting the value of λ in (i) or the value of μ in (ii), we get

$$\vec{a} + 2\vec{b} + 6\vec{c} = 0$$

Therefore, $|\vec{a} + 2\vec{b} + 6\vec{c}| = 0$

Illustration 2.46

If vectors $\vec{a}$ and $\vec{b}$ are non-collinear, then find the value of *x* for which vectors $\vec{c} = (x-2)\vec{a} + \vec{b}$ and $\vec{d} = (2x+1)\vec{a} - \vec{b}$ are collinear.

Sol. If vectors $\vec{c}$ and $\vec{d}$ are collinear, then

$\vec{c} = \lambda\vec{d}$, where λ is a scalar.

$\therefore \quad (x-2)\vec{a} + \vec{b} = \lambda[(2x+1)\vec{a} - \vec{b}]$

Now, given that vectors $\vec{a}$ and $\vec{b}$ are non-collinear.

So, comparing the coefficients of $\vec{a}$ and $\vec{b}$, we get

$\therefore \quad (x-2) = \lambda(2x+1)$ and $\lambda = -1$.

$\Rightarrow \quad (x-2) = -1(2x+1)$

$$\Rightarrow \quad x = \frac{1}{3}$$

Alternative method

Since the vectors $\vec{a}$ and $\vec{b}$ are non-collinear, vectors $\vec{c} = (x-2)\vec{a} + \vec{b}$ and $\vec{d} = (2x+1)\vec{a} - \vec{b}$ are collinear if

$$\frac{x-2}{2x+1} = \frac{1}{-1} \Rightarrow x = \frac{1}{3}$$

Illustration 2.47

If $\vec{a}$ and $\vec{b}$ are two non-collinear vectors, show that points $l_1\vec{a} + m_1\vec{b}$, $l_2\vec{a} + m_2\vec{b}$, $l_3\vec{a} + m_3\vec{b}$ are collinear if

$$\begin{vmatrix} l_1 & l_2 & l_3 \\ m_1 & m_2 & m_3 \\ 1 & 1 & 1 \end{vmatrix} = 0$$

Sol. Points $A(l_1\vec{a} + m_1\vec{b})$, $B(l_2\vec{a} + m_2\vec{b})$, $C(l_3\vec{a} + m_3\vec{b})$ are collinear.

Therefore, vectors $\overrightarrow{AB}$ and $\overrightarrow{AC}$ are collinear.

$\Rightarrow \quad (l_2\vec{a} + m_2\vec{b}) - (l_2\vec{a} + m_1\vec{b}) = \lambda[(l_3\vec{a} + m_3\vec{b}) - (l_1\vec{a} + m_1\vec{b})]$

$\Rightarrow \quad (l_2 - l_1)\vec{a} + (m_2 - m_1)\vec{b} = \lambda[(l_3 - l_1)\vec{a} + (m_3 - m_1)\vec{b}]$

$$\Rightarrow \quad \frac{l_1-l_2}{l_2-l_3}=\frac{m_1-m_2}{m_2-m_3}=\lambda$$

$$\Rightarrow \quad \begin{vmatrix} l_1 & l_2 & l_3 \\ m_1 & m_2 & m_3 \\ 1 & 1 & 1 \end{vmatrix}=0$$

Illustration 2.48

The vectors $2\hat{i}+3\hat{j}$, $5\hat{i}+6\hat{j}$ and $8\hat{i}+\lambda\hat{j}$ have their initial points at (1, 1). Find the value of λ so that the vectors terminate on a straight line.

Sol. Since the vectors $2\hat{i}+3\hat{j}$ and $5\hat{i}+6\hat{j}$ have (1, 1) as initial point, their terminal points are (3, 4) and (6, 7) respectively. The equation of the line joining these two points in the xy-plane is $x-y+1=0$.

Now, the terminal point of $8\hat{i}+\lambda\hat{j}$ is $(9, \lambda+1)$.

Since the vectors terminate on the same straight line, point $(9, \lambda+1)$ lies on the line $x-y+1=0$.

Therefore, $9-(\lambda+1)+1=0$

or $\quad \lambda=9$

Illustration 2.49

The median AD of the triangle ABC is bisected at E and BE meets AC at F. Find $AF:FC$

Sol. Let the vertex A is at origin and $\overrightarrow{AB}=\vec{b}$, $\overrightarrow{AC}=\vec{c}$.

Then, the position vectors of points B and C are $\vec{b}$ and $\vec{c}$ respectively.

Since AD is median, D is the midpoint of BC.

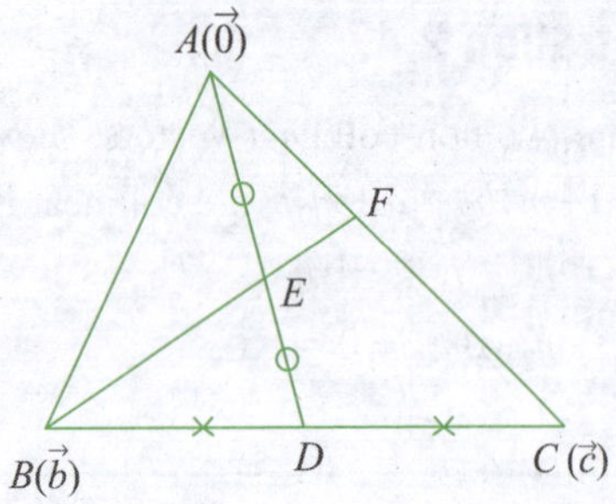

So, the position vector of D is $\dfrac{\vec{b}+\vec{c}}{2}$.

Now, AD is bisected at E.

So, the position vector of E is $\dfrac{\vec{0}+\dfrac{\vec{b}+\vec{c}}{2}}{2}=\dfrac{\vec{b}+\vec{c}}{4}$

Let $AF:FC=p:1$

Thus, the positive vector of F is $\dfrac{p\vec{c}+1(\vec{0})}{p+1}=\dfrac{p\vec{c}}{p+1}$ $\quad$ ...(i)

Let $BF:EF=q:1$

Therefore, the positive vector of F is

$$\frac{q\dfrac{\vec{b}+\vec{c}}{4}-1(\vec{b})}{q-1}=\frac{(q-4)\vec{b}+q\vec{c}}{4(q-1)} \quad \text{...(ii)}$$

From (i) and (ii), $\dfrac{p\vec{c}}{p+1}=\dfrac{(q-4)\vec{b}+q\vec{c}}{4(q-1)}$

Now, $\vec{b}$ and $\vec{c}$ are non-collinear (linearly independent) vectors.

$$\therefore \quad \frac{q-4}{4(q-1)}=0 \text{ and } \frac{p}{p+1}=\frac{q}{4(q-1)}$$

$$\Rightarrow \quad q=4 \text{ and } \frac{p}{p+1}=\frac{4}{4(4-1)}$$

$$\Rightarrow \quad q=4 \text{ and } 12p=4p+4$$

$$\Rightarrow \quad q=4 \text{ and } p=\frac{1}{2}$$

Thus, $AF:FC=1:2$

Illustration 2.50

Check the coplanarity of the vectors

$-2\hat{i}-2\hat{j}+4\hat{k}$, $-2\hat{i}+4\hat{j}-2\hat{k}$, $4\hat{i}-2\hat{j}-2\hat{k}$.

Sol. Given vectors are

$-2\hat{i}-2\hat{j}+4\hat{k}$, $-2\hat{i}+4\hat{j}-2\hat{k}$, $4\hat{i}-2\hat{j}-2\hat{k}$.

The determinant of the coefficients of the independent vectors $\hat{i}, \hat{j}$ and $\hat{k}$ is

$$\begin{vmatrix} -2 & -2 & 4 \\ -2 & 4 & -2 \\ 4 & -2 & -2 \end{vmatrix}=-2(-8-4)+2(4+8)+4(4-16)$$

$=24+24-48=0$

Therefore, vectors are coplanar.

Illustration 2.51

Prove that the four points

$6\hat{i}-7\hat{j}$, $16\hat{i}-19\hat{j}-4\hat{k}$, $3\hat{j}-6\hat{k}$ and $2\hat{i}+5\hat{j}+10\hat{k}$

form a tetrahedron in space.

Sol. Given points are

$A(6\hat{i}-7\hat{j})$, $B(16\hat{i}-19\hat{j}-4\hat{k})$, $C(3\hat{j}-6\hat{k})$, $D(2\hat{i}+5\hat{j}+10\hat{k})$

Hence, vectors

$$\overrightarrow{AB}=10\hat{i}-12\hat{j}-4\hat{k},\ \overrightarrow{AC}=-6\hat{i}+10\hat{j}-6\hat{k}$$

and $\overrightarrow{AD}=-4\hat{i}+12\hat{j}+10\hat{k}$

Now, the determinant of the coefficients of vectors $\overrightarrow{AB}, \overrightarrow{AC}, \overrightarrow{AD}$ is

$$\begin{vmatrix} 10 & -12 & -4 \\ -6 & 10 & -6 \\ -4 & 12 & 10 \end{vmatrix} = 10(100+72)+12(-60-24)-4(-72+40) \neq 0$$

Hence, the given points are non-coplanar and therefore form a tetrahedron in space.

Illustration 2.52

Find the value of m for which vectors $2\hat{i}-4\hat{j}+5\hat{k}$, $\hat{i}-m\hat{j}+3\hat{k}$ and $-\hat{i}+2\hat{j}-\hat{k}$ are coplanar (linearly independent).

or

Find the value of m for which there exists scalars λ and μ for which $2\hat{i}-4\hat{j}+5\hat{k} = \lambda(\hat{i}-m\hat{j}+3\hat{k})+(-\hat{i}+2\hat{j}-\hat{k})$

Sol. We must have $\begin{vmatrix} 2 & -4 & 5 \\ 1 & -m & 3 \\ -1 & 2 & -1 \end{vmatrix} = 0$

$\Rightarrow \quad 2(m-6)+4(-1+3)+5(2-m)=0$

$\Rightarrow \quad 3m = 6$

$\Rightarrow \quad m = 2$

Illustration 2.53

If $\vec{a}, \vec{b}, \vec{c}$ are non-coplanar vectors, prove that the four points $2\vec{a}+3\vec{b}-\vec{c}$, $\vec{a}-2\vec{b}+3\vec{c}$, $3\vec{a}+4\vec{b}-2\vec{c}$, $\vec{a}-6\vec{b}+6\vec{c}$ are coplanar.

Sol. Let $2\vec{a}+3\vec{b}-\vec{c}$, $\vec{a}-2\vec{b}+3\vec{c}$, $3\vec{a}+4\vec{b}-2\vec{c}$, $\vec{a}-6\vec{b}+6\vec{c}$ be the position vectors of points A, B, C and D respectively.

$\therefore \quad \overrightarrow{AB} = (\vec{a}-2\vec{b}+3\vec{c})-(2\vec{a}+3\vec{b}-\vec{c}) = -\vec{a}-5\vec{b}+4\vec{c}$

$\overrightarrow{AC} = (3\vec{a}+4\vec{b}-2\vec{c})-(2\vec{a}+3\vec{b}-\vec{c}) = \vec{a}+\vec{b}-\vec{c}$

and $\quad \overrightarrow{AD} = (\vec{a}-6\vec{b}+6\vec{c})-(2\vec{a}+3\vec{b}-\vec{c}) = -\vec{a}-9\vec{b}+7\vec{c}$

Now the determinant of the coefficients of vectors $\overrightarrow{AB}$, $\overrightarrow{AC}$ and $\overrightarrow{AD}$ is

$$\begin{vmatrix} -1 & -5 & 4 \\ 1 & 1 & -1 \\ -1 & -9 & 7 \end{vmatrix} = -1(7-9)+5(7-1)+4(-9+1)$$

$$= 2+30-32 = 0$$

So, the vectors $\overrightarrow{AB}$, $\overrightarrow{AC}$ and $\overrightarrow{AD}$ are coplanar.

Hence points A, B, C and D are coplanar.

Aim to complete within 30 minutes

Concept Application Exercise 2.3

1. Show that the vectors $2\vec{a}-\vec{b}+3\vec{c}$, $\vec{a}+\vec{b}-2\vec{c}$ and $\vec{a}+\vec{b}-3\vec{c}$ are non-coplanar vectors (where $\vec{a}, \vec{b}, \vec{c}$ are non-coplanar vectors).
2. Examine the following vectors for linear independence:
   a. $\hat{i}+\hat{j}+\hat{k}, 2\hat{i}+3\hat{j}-\hat{k}, -\hat{i}-2\hat{j}+2\hat{k}$
   b. $3\hat{i}+\hat{j}-\hat{k}, 2\hat{i}-\hat{j}+7\hat{k}, 7\hat{i}-\hat{j}+13\hat{k}$
3. If $\vec{a}$ and $\vec{b}$ are non-collinear vectors and $\vec{A} = (p+4q)\vec{a}+(2p+q+1)\vec{b}$ and $\vec{B} = (-2p+q+2)\vec{a}+(2p-3q-1)\vec{b}$. If $3\vec{A} = 2\vec{B}$, then the value of $p+q$ is
   (1) -1 (2) 0 (3) 1 (4) 2
4. If $\vec{a}, \vec{b}$ and $\vec{c}$ are three non-coplanar vectors, then prove that points $l_1\vec{a}+m_1\vec{b}+n_1\vec{c}$, $l_2\vec{a}+m_2\vec{b}+n_2\vec{c}$, $l_3\vec{a}+m_3\vec{b}+n_3\vec{c}$, $l_4\vec{a}+m_4\vec{b}+n_4\vec{c}$ are coplanar if $\begin{vmatrix} l_1 & l_2 & l_3 & l_4 \\ m_1 & m_2 & m_3 & m_4 \\ n_1 & n_2 & n_3 & n_4 \\ 1 & 1 & 1 & 1 \end{vmatrix} = 0$.
5. If $\vec{a}, \vec{b}$ and $\vec{c}$ are three non-zero, non-coplanar vectors, then find the linear relation between the following four vectors: $\vec{a}-2\vec{b}+3\vec{c}, 2\vec{a}-3\vec{b}+4\vec{c}; 3\vec{a}-4\vec{b}+5\vec{c}, 7\vec{a}-11\vec{b}+15\vec{c}$.
6. Let a, b, c be distinct non-negative numbers and the vectors $a\hat{i}+a\hat{j}+c\hat{k}, \hat{i}+\hat{k}, c\hat{i}+c\hat{j}+b\hat{k}$ lie in a plane, then the quadratic equation $ax^2+2cx+b=0$ has
   (1) equal roots (2) distinct real roots
   (3) non-real roots (4) can't say anything
7. In a triangle PQR, S and T are points on QR and PR, respectively, such that $QS = 3SR$ and $PT = 4TR$. Let M be the point of intersection of PS and QT. Then the ratio $QM : MT$ is
   (1) 3 : 2 (2) 4 : 3 (3) 15 : 4 (4) 13 : 5
8. Prove that vectors
   $\vec{u} = (al+a_1l_1)i+(am+a_1m_1)j+(an+a_1n_1)k$
   $\vec{v} = (bl+b_1l_1)i+(bm+b_1m_1)j+(bn+b_1n_1)k$
   $\vec{w} = (cl+c_1l_1)i+(cm+c_1m_1)j+(cn+c_1n_1)k$
   are coplanar.

Answers

2. a. Vectors are non-coplanar and linearly independent.
b. Vectors are coplanar and linearly dependent

3. (3)

5. $7\vec{a}-11\vec{b}+15\vec{c} = \vec{a}-2\vec{b}+3\vec{c}+3(2\vec{a}-3\vec{b}+4\vec{c}) + 0(3\vec{a}-4\vec{b}+5\vec{c})$

6. (1) **7.** (3)

SOLVED EXAMPLES

Example 2.1

Find all values of λ such that $(x, y, z) \neq (0, 0, 0)$ and $(\hat{i} + \hat{j} + 3\hat{k})x + (3\hat{i} - 3\hat{j} + \hat{k})y + (-4\hat{i} + 5\hat{j})z$ $= \lambda(x\hat{i} + y\hat{j} + z\hat{k})$, where, $\hat{i}, \hat{j}$ and $\hat{k}$ are unit vectors along the coordinate axes.

Sol. We have, $(\hat{i} + \hat{j} + 3\hat{k})x + (3\hat{i} - 3\hat{j} + \hat{k})y + (-4\hat{i} + 5\hat{j})z = \lambda(x\hat{i} + y\hat{j} + z\hat{k})$

Comparing coefficients of $\hat{i}$, $\hat{j}$ and $\hat{k}$, we get

$\Rightarrow \quad (1 - \lambda)x + 3y - 4z = 0$ ...(i)

$\Rightarrow \quad x - (3 + \lambda)y + 5z = 0$ ...(ii)

$3x + y - \lambda z = 0$ ...(iii)

Now, $(x, y, z) \neq (0, 0, 0)$

Therefore, the system of equations has non-trivial solutions

$$\therefore \begin{vmatrix} 1-\lambda & 3 & -4 \\ 1 & -(3+\lambda) & 5 \\ 3 & 1 & -\lambda \end{vmatrix} = 0$$

$\Rightarrow \quad (1 - \lambda)[3\lambda + \lambda^2 - 5] - 3[-\lambda - 15] - 4[1 + 9 + 3\lambda] = 0$

$\Rightarrow \quad \lambda^3 + 2\lambda^2 + \lambda = 0$

$\Rightarrow \quad \lambda(\lambda + 1)^2 = 0$

$\Rightarrow \quad \lambda = 0, -1$

Example 2.2

A vector has components A_1, A_2 and A_3 in a right-handed rectangular Cartesian coordinate system $OXYZ$. The coordinate system is rotated about the x-axis through an angle $\pi/2$. Find the components of A in the new coordinate system in terms of A_1, A_2 and A_3.

Sol. Since vector $\vec{A}$ has components A_1, A_2 and A_3, in the coordinate system $OXYZ$,

$$\vec{A} = \hat{i}A_1 + \hat{j}A_2 + \hat{k}A_3$$

When given system is rotated through $\pi/2$, the new x-axis is along the old y-axis and the new y-axis is along the old negative x-axis; z remains same as before.

Hence, the components of A in the new system are $A_2, -A_1$ and A_3.

Therefore, $\vec{A}$ becomes $A_2\hat{i} - A_1\hat{j} + A_3\hat{k}$.

Example 2.3

The position vectors of the vertices A, B and C of a triangle are $\hat{i} + \hat{j}$, $\hat{j} + \hat{k}$ and $\hat{i} + \hat{k}$, respectively. Find a unit vector $\hat{r}$ lying in the plane of ABC and perpendicular to IA, where I is the incentre of the triangle.

Sol. Since $|\overrightarrow{AB}| = |\overrightarrow{BC}| = |\overrightarrow{CA}|$, the triangle is equilateral and the incentre is same as the circumcentre, and hence IA is perpendicular to BC. Therefore, $\vec{r}$ is parallel to BC.

Thus, $\vec{r} = \lambda(\hat{i} - \hat{j})$

Hence, unit vector $\hat{r} = \pm\frac{1}{\sqrt{2}}(\hat{i} - \hat{j})$

Example 2.4

A ship is sailing towards the north at a speed of 1.25 m/s. The current is taking it towards the east at the rate of 1 m/s and a sailor is climbing a vertical pole on the ship at the rate of 0.5 m/s. Find the velocity of the sailor in space.

Sol. Let the unit vectors $\hat{i}, \hat{j}$ and $\hat{k}$ be parallel to the east, north and vertically upward in the direction of pole, respectively. Then the velocity vectors of the current, ship and the sailor are, respectively, $\hat{i}, 1.25\hat{j}$ and $0.5\hat{k}$. Therefore, the velocity $\vec{v}$ of the sailor in space is the resultant of these vectors. Hence,

$$\vec{v} = \hat{i} + 1.25\hat{j} + 0.5\hat{k}$$

Then, $|\vec{v}| = \sqrt{1 + (1.25)^2 + (0 \cdot 5)^2}$

$= \sqrt{1 + 1.5625 + .25}$

$= \sqrt{2.8125} = 1.677$ m/s

Example 2.5

Given four points P_1, P_2, P_3 and P_4 on the coordinate plane with origin O which satisfy the condition $\overrightarrow{OP}_{n-1} + \overrightarrow{OP}_{n+1} = \frac{3}{2}\overrightarrow{OP}_n$.

a. If P_1 and P_2 lie on the curve $xy = 1$, then prove that P_3 does not lie on the curve.

b. If P_1, P_2 and P_3 lie on the circle $x^2 + y^2 = 1$, then prove that P_4 also lies on this circle.

Sol. **a.** Putting $n = 2$ in $\overrightarrow{OP}_{n-1} + \overrightarrow{OP}_{n+1} = \frac{3}{2}\overrightarrow{OP}_n$, we get

$$\overrightarrow{OP}_3 = \frac{3}{2}\overrightarrow{OP}_2 - \overrightarrow{OP}_1 \quad \text{...(i)}$$

It points P_1 and P_2 lie on the curve $xy = 1$, then

$$\left.\begin{aligned} \overrightarrow{OP}_1 &= a\hat{i} + \frac{1}{a}\hat{j} \\ \overrightarrow{OP}_2 &= b\hat{i} + \frac{1}{b}\hat{j} \end{aligned}\right\} ab \neq 0$$

$$\therefore \quad \overrightarrow{OP_3} = \frac{3}{2}\left(b\hat{i} + \frac{1}{b}\hat{j}\right) - \left(a\hat{i} + \frac{1}{a}\hat{j}\right)$$

$$= \left(\frac{3b}{2} - a\right)\hat{i} + \left(\frac{3}{2b} - \frac{1}{a}\right)\hat{j}$$

Now, if P_3 lies on $xy = 1$, we must have

$$\left(\frac{3b}{2} - a\right)\left(\frac{3}{2b} - \frac{1}{a}\right) = 1$$

$$\Rightarrow \quad (3b - 2a)(3a - 2b) = 4ab$$

$$\Rightarrow \quad 2a^2 - 3ab + 2b^2 = 0$$

This is not possible as Discriminant < 0 ($a = 0$ and $b = 0$ not possible)

b. Let $\overrightarrow{OP_1} = \cos\alpha\,\hat{i} + \sin\alpha\,\hat{j}$

$$\overrightarrow{OP_2} = \cos\beta\,\hat{i} + \sin\beta\,\hat{j}$$

$$\therefore \quad \overrightarrow{OP_3} = \frac{3}{2}(\cos\beta\,\hat{i} + \sin\beta\,\hat{j}) - (\cos\alpha\,\hat{i} + \sin\alpha\,\hat{j})$$

$$= \left(\frac{3}{2}\cos\beta - \cos\alpha\right)\hat{i} + \left(\frac{3}{2}\sin\beta - \sin\alpha\right)\hat{j}$$

Since P_3 lies on $x^2 + y^2 = 1$, we have

$$\left(\frac{3}{2}\cos\beta - \cos\alpha\right)^2 + \left(\frac{3}{2}\sin\beta - \sin\alpha\right)^2 = 1$$

$$\text{or} \quad \frac{9}{4} + 1 - 3(\cos\beta\cos\alpha + \sin\beta\sin\alpha) = 1$$

$$\text{or} \quad \frac{9}{4} - 3\cos(\beta - \alpha) = 0 \Rightarrow \cos(\beta - \alpha) = \frac{3}{4} \quad \ldots\text{(ii)}$$

Putting $n = 3$ in the given relation, we have

$$\overrightarrow{OP_2} + \overrightarrow{OP_4} = \frac{3}{2}\overrightarrow{OP_3}$$

$$\text{or} \quad \overrightarrow{OP_4} = \frac{3}{2}\overrightarrow{OP_3} - \overrightarrow{OP_2}$$

$$= \frac{3}{2}\left(\frac{3}{2}\overrightarrow{OP_2} - \overrightarrow{OP_1}\right) - \overrightarrow{OP_2}$$

$$= \frac{5}{4}\overrightarrow{OP_2} - \frac{3}{2}\overrightarrow{OP_1}$$

$$= \frac{5}{4}(\cos\beta\,\hat{i} + \sin\beta\,\hat{j}) - \frac{3}{2}(\cos\alpha\,\hat{i} + \sin\alpha\,\hat{j}),$$

$$= \left(\frac{5}{4}\cos\beta - \frac{3}{2}\cos\alpha\right)\hat{i} + \left(\frac{5}{4}\sin\beta - \frac{3}{2}\sin\alpha\right)\hat{j}$$

$$\text{Now,} \quad \left(\frac{5}{4}\cos\beta - \frac{3}{2}\cos\alpha\right)^2 + \left(\frac{5}{4}\sin\beta - \frac{3}{2}\sin\alpha\right)^2$$

$$= \frac{25}{16} + \frac{9}{4} - \frac{15}{4}\cos(\beta - \alpha) = \frac{61}{16} - \frac{45}{16} = 1$$

Hence, P_2 lies on the curve.

Example 2.6

ABCD is a tetrahedron and *O* is any point. If the lines joining *O* to the vertices meet the opposite faces at *P*, *Q*, *R* and *S*, prove that $\frac{OP}{AP} + \frac{OQ}{BQ} + \frac{OR}{CR} + \frac{OS}{DS} = 1.$

Sol. Here *ABCD* is a tetrahedron. Let *O* be the origin and the P.V. of *A*, *B*, *C* and *D* be $\vec{a}, \vec{b}, \vec{c}$ and $\vec{d}$, respectively.

Since the four vectors $\vec{a}, \vec{b}, \vec{c}, \vec{d}$ are linearly dependent, there exist scalars x, y, z, t (not all zero) such that

$$x\vec{a} + y\vec{b} + z\vec{c} + t\vec{d} = 0 \quad \text{(where } x, y, z \text{ and } t \text{ are scalars)}$$

$$\text{or} \quad y\vec{b} + z\vec{c} + t\vec{d} = -x\vec{a}$$

$$\text{or} \quad \frac{y\vec{b} + z\vec{c} + t\vec{d}}{y + z + t} = -\frac{x\vec{a}}{y + z + t}$$ (assuming $y + z + t \neq 0$ as the case $y + z + t = 0$, is analogous with another vertex)

Here, the L.H.S. is the P.V. of a point in the plane *BCD* and R.H.S. is the P.V. of a point on the line *OA*.

Therefore, there must be a common point *P* common to both the plane and the straight line with is

$$\overrightarrow{OP} = \frac{-x\vec{a}}{y + z + t}$$

$$\text{But,} \quad \overrightarrow{AP} = \overrightarrow{OP} - \overrightarrow{OA} = -\frac{x\vec{a}}{y + z + t} - \vec{a}$$

$$= -\left(\frac{x + y + z + t}{y + z + t}\right)\vec{a}$$

$$\therefore \quad \overrightarrow{OP} = \frac{x}{y + z + t}\left(\frac{y + z + t}{x + y + z + t}\right)\overrightarrow{AP}$$

$$\text{or} \quad \overrightarrow{OP} = \left(\frac{x}{x + y + z + t}\right)\overrightarrow{AP}$$

$$\Rightarrow \quad \frac{OP}{AP} = \frac{x}{x + y + z + t}$$

$$\text{Similarly,} \quad \frac{OQ}{BQ} = \frac{y}{x + y + z + t},$$

$$\frac{OR}{CR} = \frac{z}{x + y + z + t} \quad \text{and} \quad \frac{OS}{DS} = \frac{t}{x + y + z + t}$$

$$\Rightarrow \quad \frac{OP}{AP} + \frac{OQ}{BQ} + \frac{OR}{CR} + \frac{OS}{DS} = 1$$

Example 2.7

A pyramid with vertex at point P has a regular hexagonal base $ABCDEF$. The position vectors of points A and B are $\hat{i}$ and $\hat{i}+2\hat{j}$, respectively. The centre of the base has the position vector $\hat{i}+\hat{j}+\sqrt{3}\hat{k}$. Altitude drawn from P on the base meets the diagonal AD at point G. Find all possible position vectors of G. It is given that the volume of the pyramid is $6\sqrt{3}$ cubic units and AP is 5 units.

Sol. Let the centre of the base be O. Therefore,

Now, $|\overrightarrow{AB}| = 2$

$$\therefore \quad \Delta OAB = \frac{\sqrt{3}}{4}(2)^2 = \sqrt{3} \text{ sq. units}$$

Also, base area $= 6\sqrt{3}$ sq. units

Let the height of the pyramid be h. Therefore,

$$\frac{1}{3}\times 6\sqrt{3}h = 6\sqrt{3} \text{ or } h = 3 \text{ units}$$

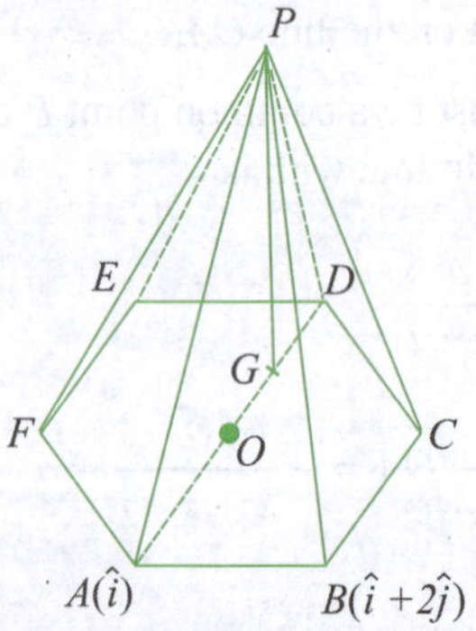

It is given that $|\overrightarrow{AP}| = 5$. Therefore,

$$AG = \sqrt{25-9} = 4 \text{ units}$$

$$\Rightarrow \quad |\overrightarrow{AG}| = 4 \text{ units}$$

Now $|\overrightarrow{AG}|$ and $|\overrightarrow{AO}|$ are collinear. Therefore,

$$\overrightarrow{AG} = \lambda \overrightarrow{AO} \Rightarrow |\overrightarrow{AG}| = |\lambda||\overrightarrow{AO}|$$

or $\quad 2|\lambda| = 4$ or $|\lambda| = 2$

$$\Rightarrow \quad \overrightarrow{AG} = \pm 2(\hat{i}+\hat{j}+\sqrt{3}\hat{k})$$

$$\Rightarrow \quad \overrightarrow{OG} = \pm 2(\hat{i}+\hat{j}+\sqrt{3}\hat{k}) + \hat{i}$$

$$= -(\hat{i}+2\hat{j}+2\sqrt{3}\hat{k}),\ 3\hat{i}+2\hat{j}+2\sqrt{3}\hat{k}$$

Example 2.8

A straight line L cuts the lines AB, AC and AD of a parallelogram $ABCD$ at points B_1, C_1 and D_1, respectively. If $\overrightarrow{AB_1} = \lambda_1 \overrightarrow{AB}$, $\overrightarrow{AD_1} = \lambda_2 \overrightarrow{AD}$ and $\overrightarrow{AC_1} = \lambda_3 \overrightarrow{AC}$, then prove that $\frac{1}{\lambda_3} = \frac{1}{\lambda_1} + \frac{1}{\lambda_2}$.

Sol. Let $\overrightarrow{AB} = \vec{a}$ and $\overrightarrow{AD} = \vec{b}$; then $\overrightarrow{AC} = \vec{a}+\vec{b}$.

Given $\overrightarrow{AB_1} = \lambda_1 \vec{a}$, $\overrightarrow{AD_1} = \lambda_2 \vec{b}$, $\overrightarrow{AC_1} = \lambda_3(\vec{a}+\vec{b})$

Since vectors $\overrightarrow{D_1C_1}$ and $\overrightarrow{B_1D_1}$ are collinear, we have

$$\overrightarrow{D_1C_1} = k\overrightarrow{B_1D_1} \text{ for some } k \in R$$

$$\Rightarrow \quad \overrightarrow{AC_1} - \overrightarrow{AD_1} = k(\overrightarrow{AD_1} - \overrightarrow{AB_1})$$

$$\Rightarrow \quad \lambda_3(\vec{a}+\vec{b}) - \lambda_2\vec{b} = k(\lambda_2\vec{b} - \lambda_1\vec{a})$$

or $\quad \lambda_3\vec{a} + (\lambda_3 - \lambda_2)\vec{b} = k\lambda_2\vec{b} - k\lambda_1\vec{a}$

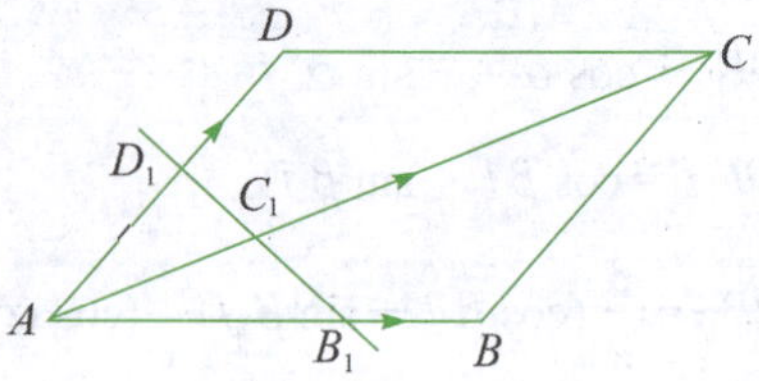

($\because$ $\vec{a}$ and $\vec{b}$ are non-collinear)

$$\Rightarrow \quad -\frac{\lambda_3}{\lambda_1} = \frac{\lambda_3 - \lambda_2}{\lambda_2}$$

$$\Rightarrow \quad \lambda_1\lambda_2 = \lambda_1\lambda_3 + \lambda_2\lambda_3$$

$$\Rightarrow \quad \frac{1}{\lambda_3} = \frac{1}{\lambda_1} + \frac{1}{\lambda_2}$$

Example 2.9

In a triangle ABC, D and E are points on BC and AC, respectively, such that $BD = 2DC$ and $AE = 3EC$. Let P be the point of intersection of AD and BE. Find BP/PE using the vector method.

Sol. Let the vertices of the triangle be $A(\vec{0})$, $B(\vec{b})$ and $C(\vec{c})$.

Given that D divides BC in the ratio 2 : 1.

Therefore, the position vector of D is $\frac{\vec{b}+2\vec{c}}{3}$.

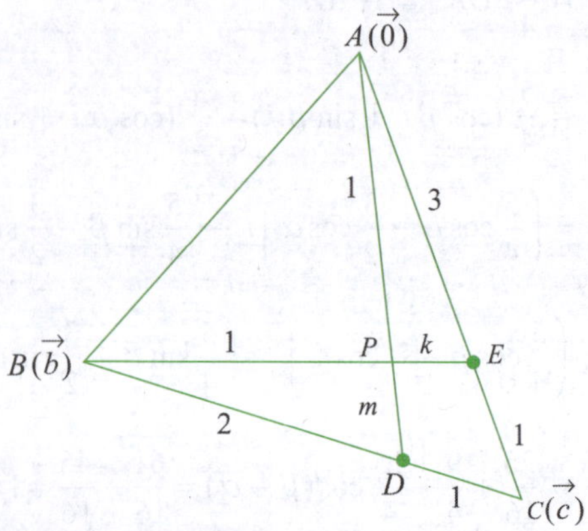

Also, E divides AC in the ratio 3 : 1.

Therefore, the position vector of E is $\frac{\vec{0}+3\vec{c}}{4} = \frac{3\vec{c}}{4}$.

Let the point of intersection P of AD and BE divide BE in the ratio $1 : k$ and AD in the ratio $1 : m$. Then the position vectors of P in these two cases are $\dfrac{k\vec{b} + 1(3\vec{c}/4)}{k+1}$ and $\dfrac{m\,0 + m((\vec{b} + 2\vec{c})/3)}{m+1}$, respectively.

Equating the position vectors of P, we get

$$\frac{k\vec{b}}{k+1} + \frac{3\vec{c}}{4(k+1)} = \frac{m\vec{b}}{3(m+1)} + \frac{2m\vec{c}}{3(m+1)}$$

$$\Rightarrow \frac{k}{k+1} = \frac{m}{3(m+1)} \text{ and } \frac{3}{4(k+1)} = \frac{2m}{3(m+1)}$$

($\because$ vectors $\vec{b}$ and $\vec{c}$ are non-collinear)

Dividing, we have $\dfrac{4k}{3} = \dfrac{1}{2}$ or $k = \dfrac{3}{8}$

Therefore, the required ratio is 8 : 3.

Example 2.10

A boat moves in still water with a velocity which is k times less than the river flow velocity. Find the angle to the stream direction at which the boat should be rowed to minimize drifting.

Sol. Let the flow velocity of river be u and the velocity of boat in still water be v. Thus,

$$v = \frac{u}{K}$$

Also, let the boat move at an angle θ with the direction of stream. Now the velocity of boat in the river is the vector resultant of the velocity of boat and flow velocity of river, which can be written as

$$\vec{v}_B = (u - v\cos\alpha)\hat{i} + (u\sin\alpha)\hat{j}$$

$$= (u + v\cos\theta)\hat{i} + (u\sin\theta)\hat{j}$$

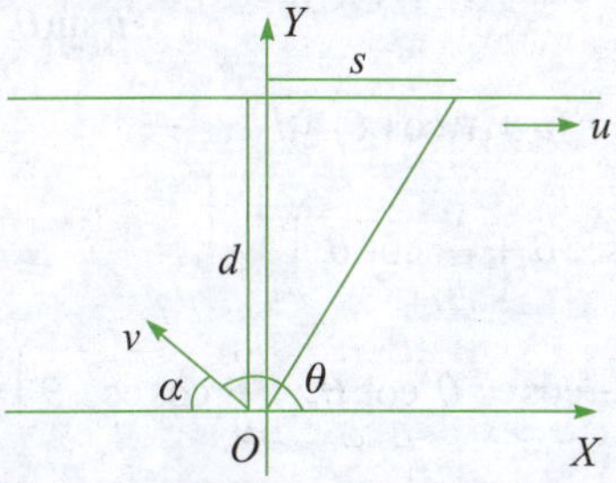

Hence, the time taken to cross the river $= \dfrac{d}{u\sin\theta}$ (d = width of the river)

Thus, the drift $s = (u + v\cos\theta)\cdot d$

or $\quad s = d\,(\operatorname{cosec}\theta + \dfrac{v}{u}\cot\theta)$

or $\quad \dfrac{ds}{d\theta} = d\,(\operatorname{cosec}\theta\cot\theta - \dfrac{v}{\mu}\operatorname{cosec}^2\theta) = 0$

or $\quad \dfrac{v}{u}\operatorname{cosec}^2\theta = \operatorname{cosec}\theta\cot\theta$

or $\quad \cos\theta = \dfrac{1}{k}$ or $\theta = \cos^{-1}\left(\dfrac{1}{k}\right)$

Example 2.11

In a quadrilateral $PQRS$, $\overrightarrow{PQ} = \vec{a}$, $\overrightarrow{QR} = \vec{b}$, $\overrightarrow{SP} = \vec{a} - \vec{b}$, M is the midpoint of $\overrightarrow{QR}$ and X is a point on SM such that $SX = \dfrac{4}{5}SM$. Prove that P, X and R are collinear.

Sol. We have, $\overrightarrow{QM} = \dfrac{\vec{b}}{2} \Rightarrow \overrightarrow{PM} = \vec{a} + \dfrac{\vec{b}}{2}$

Also, $\overrightarrow{SM} = \overrightarrow{PM} - \overrightarrow{PS} = 2\vec{a} - \dfrac{1}{2}\vec{b}$,

$$\overrightarrow{SX} = \frac{4}{5}\overrightarrow{SM} = \frac{8}{5}\vec{a} - \frac{2}{5}\vec{b},$$

and $\quad \overrightarrow{PX} = \overrightarrow{PS} + \overrightarrow{SX}$

$$= -\vec{a} + \vec{b} + \frac{8}{5}\vec{a} - \frac{2}{5}\vec{b} = \frac{3}{5}(\vec{a} + \vec{b})$$

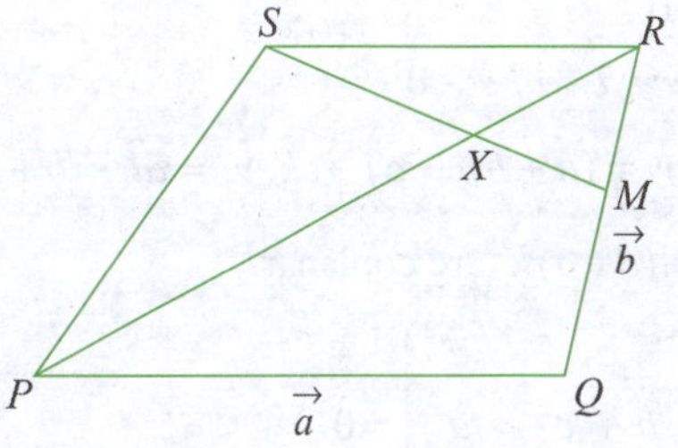

Also $\overrightarrow{PR} = \overrightarrow{PQ} + \overrightarrow{QR} = \vec{a} + \vec{b} = \dfrac{5}{3}\overrightarrow{PX}$

Hence P, X and R are collinear.

Example 2.12

Let $\vec{A}(t) = f_1(t)\hat{i} + f_2(t)\hat{j}$ and $\vec{B}(t) = g_1(t)\hat{i} + g_2(t)\hat{j}$, $t \in [0, 1]$, where f_1, f_2, g_1, g_2 are continuous functions. If $\vec{A}(t)$ and $\vec{B}(t)$ are non-zero vectors for all t and $\vec{A}(0) = 2\hat{i} + 3\hat{j}$, $\vec{A}(1) = 6\hat{i} + 2\hat{j}$, $\vec{B}(0) = 3\hat{i} + 2\hat{j}$ and $\vec{B}(1) = 2\hat{i} + 6\hat{j}$, then show that $\vec{A}(t)$ and $\vec{B}(t)$ are parallel for some t.

Sol. $\vec{A}(t)$ is parallel to $\vec{B}(t)$ for some $t \in [0, 1]$ if and only if

$\dfrac{f_1(t)}{g_1(t)} = \dfrac{f_2(t)}{g_2(t)}$ for some $t \in [0, 1]$

or $\quad f_1(t)\cdot g_2(t) = f_2(t)g_1(t)$ for some $t \in [0, 1]$

Let $\quad h(t) = f_1(t)\cdot g_2(t) - f_2(t)\cdot g_1(t)$

Then, $h(0) = f_1(0)\cdot g_2(0) - f_2(0)\cdot g_1(0)$

$= 2 \times 2 - 3 \times 3 = -5 < 0$

and $h(1) = f_1(1) \cdot g_2(1) - f_2(1) \cdot g_1(1)$

$= 6 \times 6 - 2 \times 2 = 32 > 0$

Since h is a continuous function, and $h(0) \cdot h(1) < 0$, there exists at least one $t \in [0, 1]$ for which $h(t) = 0$, i.e., $\vec{A}(t)$ and $\vec{B}(t)$ are parallel vectors for this t.

Example 2.13

Let the vectors $\vec{u}_1 = \hat{i} + \hat{j} + a\hat{k}$, $\vec{u}_2 = \hat{i} + b\hat{j} + \hat{k}$ and $\vec{u} = c\hat{i} + \hat{j} + \hat{k}$ be coplanar. If the vectors $\vec{v}_1 = (a+b)\hat{i} + c\hat{j} + c\hat{k}$, $\vec{v}_2 = a\hat{i} + (b+c)\hat{j} + a\hat{k}$ and $\vec{v}_3 = b\hat{i} + b\hat{j} + (c+a)\hat{k}$ are also coplanar, then find the value of $(a+b+c)$.

Sol. Vectors $\vec{u}_1 = \hat{i} + \hat{j} + a\hat{k}$, $\vec{u}_2 = \hat{i} + b\hat{j} + \hat{k}$ and $\vec{u} = c\hat{i} + \hat{j} + \hat{k}$ are coplanar

$$\therefore \quad \begin{vmatrix} 1 & 1 & a \\ 1 & b & 1 \\ c & 1 & 1 \end{vmatrix} = 0$$

$$\Rightarrow \quad a + b + c - 2 - abc = 0 \qquad \ldots\text{(i)}$$

Also, vectors $\vec{v}_1 = (a+b)\hat{i} + c\hat{j} + c\hat{k}$, $\vec{v}_2 = a\hat{i} + (b+c)\hat{j} + a\hat{k}$, and $\vec{v}_3 = b\hat{i} + b\hat{j} + (c+a)\hat{k}$ are coplanar.

$$\therefore \quad \begin{vmatrix} a+b & c & c \\ a & b+c & a \\ b & b & a+c \end{vmatrix} = 0$$

Applying $R_3 \to R_3 - (R_1 + R_2)$, we get

$$\begin{vmatrix} a+b & c & c \\ a & b+c & a \\ -2a & -2c & 0 \end{vmatrix} = 0$$

$$\Rightarrow \quad \begin{vmatrix} a+b & c & c \\ a & b+c & a \\ a & c & 0 \end{vmatrix} = 0$$

Applying $R_1 \to R_1 - R_3$ and $R_2 \to R_2 - R_3$ we get

$$\begin{vmatrix} b & 0 & c \\ 0 & b & a \\ a & c & 0 \end{vmatrix} = 0$$

$$\Rightarrow \quad abc = 0$$

So, from (i), $a + b + c = 2$.

Example 2.14

A boat moves in still water with a velocity which is k times less than the river flow velocity. Find the angle with the stream (downstream) at which the boat should be rowed to minimize drifting.

Sol. When a boat rows across a river, the water current pushes it downstream. This sideways push is called drift.

- If the boat points straight across, the river pushes it the most.
- If the boat points a little upstream, part of its velocity cancels the river's push, reducing drift.
- There is one special angle at which drift is the least possible.

We want to find that angle (with respect to the river flow direction).

Let the flow velocity of river be u and the velocity of boat in still water be v.

Thus, $v = \dfrac{u}{K}$

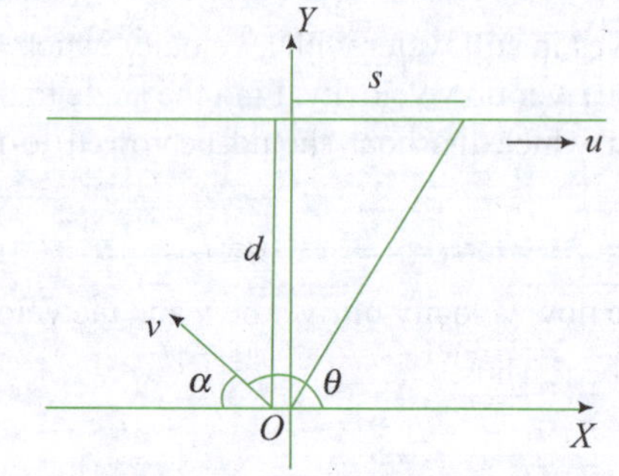

Also, let the boat move at an angle θ with the direction of stream. Now the velocity of boat in the river is the vector resultant of the velocity of boat an flow velocity of river, which can be written as

$$\vec{v}_B = (u - v\cos\alpha)\hat{i} + (u\sin\alpha)\hat{j}$$

$$= (u + v\cos\theta)\hat{i} + (u\sin\theta)\hat{j}$$

Hence, the time taken to cross the river $\dfrac{d}{u\sin\theta}$ (d = width of the river)

Thus, the drift $s = (u + v\cos\theta) \cdot d$

$$\Rightarrow \quad s = d\left(\operatorname{cosec}\theta + \frac{v}{u}\cot\theta\right)$$

$$\Rightarrow \quad \frac{ds}{d\theta} = d\left(-\operatorname{cosec}\theta\cot\theta - \frac{v}{\mu}\operatorname{cosec}^2\theta\right) = 0$$

$$\Rightarrow \quad \frac{v}{u}\operatorname{cosec}^2\theta = \operatorname{cosec}\theta\cot\theta$$

$$\Rightarrow \quad \cos\theta = -\frac{1}{k} \text{ or } \theta = \cos^{-1}\left(-\frac{1}{k}\right)$$

EXERCISES

[Note: Bulleted questions in Single Correct Answer Type, Numerical Value Type and Integer Answer Type sections are of high difficulty level.]

Single Correct Answer Type

Vectors Definition

1. Let $\vec{a}$, $\vec{b}$ and $\vec{c}$ be pairwise mutually perpendicular vectors, such that $|\vec{a}|=1$, $|\vec{b}|=2$, $|\vec{c}|=2$. Then the length of the vector $\vec{a}+\vec{b}+\vec{c}$ is

(1) 3 (2) $2\sqrt{3}$
(3) $3\sqrt{3}$ (4) 6

2. Let $\vec{a}, \vec{b}$ and $\vec{c}$ be three unit vectors such that $3\vec{a}+4\vec{b}+5\vec{c}=0$. Then which of the following statements is true?

(1) $\vec{a}$ is parallel to $\vec{b}$
(2) $\vec{a}$ is perpendicular to $\vec{b}$
(3) $\vec{b}$ bisects the angle between $\vec{a}$ and $\vec{c}$
(4) $\vec{a}, \vec{b}, \vec{c}$ are coplanar.

3. If G is the centroid of a triangle ABC, then $\overrightarrow{GA}+\overrightarrow{GB}+\overrightarrow{GC}$ is equal to

(1) $\vec{0}$ (2) $3\overrightarrow{GA}$
(3) $3\overrightarrow{GB}$ (4) $3\overrightarrow{GC}$

4. If $\vec{a}$ is a non-zero vector of magnitude a and m is a non-zero scalar, then $m\vec{a}$ is a unit vector if

(1) $m=\pm 1$ (2) $a=|m|$
(3) $a=1/|m|$ (4) $a=1/m$

5. It A, B, C and D have position vectors $\vec{a}, \vec{b}, \vec{c}$ and $\vec{d}$, respectively, such that $\vec{a}-\vec{b}=2(\vec{d}-\vec{c})$. Then

(1) AB and CD bisect each other
(2) BD and AC bisect each other
(3) AB and CD trisect each other
(4) BD and AC trisect each other

6. In a trapezium, $ABCD$, $\overrightarrow{BC}=\alpha\,\overrightarrow{AD}$. Also, vector $\vec{p}=\overrightarrow{AC}+\overrightarrow{BD}$ is collinear with the vector $\overrightarrow{AD}$. If $\vec{p}=\mu\,\overrightarrow{AD}$, then which of the following is true?

(1) $\mu=\alpha+2$ (2) $\mu+\alpha=1$
(3) $\alpha=\mu+1$ (4) $\mu=\alpha+1$

7. Let $\vec{A}$ be a vector with direction cosines cos α, cos β and cos γ. Assuming the yz-plane as a mirror, the direction cosines of the image of $\vec{A}$ in the yz-plane are

(1) $\cos\alpha, \cos\beta, \cos\gamma$
(2) $\cos\alpha, -\cos\beta, \cos\gamma$
(3) $-\cos\alpha, \cos\beta, \cos\gamma$
(4) $-\cos\alpha, -\cos\beta, -\cos\gamma$

• **8.** If $|\vec{a}+\vec{b}|<|\vec{a}-\vec{b}|$, then the angle between the vectors $\vec{a}$ and $\vec{b}$ lies in the interval

(1) $(0, \pi/2)$ (2) $(0, \pi)$
(3) $(\pi/2, \pi)$ (4) $(0, 2\pi)$

• **9.** If a point O is the centre of a circle circumscribed about a triangle ABC. Then $\overrightarrow{OA}$ sin $2A$ + $\overrightarrow{OB}$ sin $2B$ + $\overrightarrow{OC}$ sin $2C$ is equal to

(1) $(\overrightarrow{OA}+\overrightarrow{OB}+\overrightarrow{OC})\sin 2A$
(2) $3\overrightarrow{OG}$, where G is the centroid of triangle ABC
(3) $\vec{0}$
(4) $4\overrightarrow{OG}$

• **10.** If $\vec{a}$ and $\vec{b}$ are two unit vectors and θ is the angle between them, then the unit vector along the angle bisector of $\vec{a}$ and $\vec{b}$ is

(1) $\dfrac{\vec{a}-\vec{b}}{2\cos(\theta/2)}$ (2) $\dfrac{\vec{a}+\vec{b}}{2\cos(\theta/2)}$
(3) $\dfrac{\vec{a}-\vec{b}}{\cos(\theta/2)}$ (4) $\dfrac{\vec{a}+\vec{b}}{3\cos(\theta/2)}$

• **11.** Locus of the point P, for which $\overrightarrow{OP}$ represents a vector with the direction cosine $\cos\alpha=\dfrac{1}{2}$ (where O is the origin) is

(1) a circle parallel to the y–z plane with centre on the x-axis
(2) a cone concentric with the positive x-axis having vertex at the origin and the slant height equal to the magnitude of the vector
(3) a ray emanating from the origin and making an angle of 60° with the x-axis
(4) a disc parallel to the y–z plane with centre on the x-axis and radius equal to $|\overrightarrow{OP}|$ sin 60°

• **12.** Let be three non-collinear vectors in a plane such that $|\vec{a}|=1$, $|\vec{b}|=3$ and $|\vec{c}|=\sqrt{10}$. If the angle between $\vec{a}$ and $\vec{c}$ is α and that between $\vec{b}$ and $\vec{c}$ is β, where $\alpha, \beta \in \left[\dfrac{\pi}{2}, \pi\right]$ then $\alpha+\beta$ equals

(1) $\frac{7\pi}{6}$ (2) $\frac{4\pi}{3}$

(3) $\frac{3\pi}{2}$ (4) $\frac{7\pi}{4}$

Vectors in Components Form

13. If vectors $\overrightarrow{AB} = -3\hat{i} + 4\hat{k}$ and $\overrightarrow{AC} = 5\hat{i} - 2\hat{j} + 4\hat{k}$ are the sides of a ΔABC, then the length of the median through A is

(1) $\sqrt{14}$ (2) $\sqrt{18}$ (3) $\sqrt{29}$ (4) 5

14. Let us define the length of a vector $a\hat{i} + b\hat{j} + c\hat{k}$ as $|a| + |b| + |c|$. This definition coincides with the usual definition of length of a vector $a\hat{i} + b\hat{j} + c\hat{k}$ if and only if

(1) $a = b = c = 0$
(2) any two of a, b and c are zero
(3) any one of a, b and c is zero
(4) $a + b + c = 0$

15. Given three vectors $\vec{a} = 6\hat{i} - 3\hat{j}$, $\vec{b} = 2\hat{i} - 6\hat{j}$ and $\vec{c} = -2\hat{i} + 21\hat{j}$ are such that $\vec{\alpha} = \vec{a} + \vec{b} + \vec{c}$. Then the vector $\vec{\alpha}$ is given by

(1) $3\vec{a} - 2\vec{b}$ (2) $3\vec{b} - 2\vec{a}$

(3) $2\vec{a} - 3\vec{b}$ (4) $\vec{a} - 2\vec{b}$

16. Vectors $\vec{a} = \hat{i} + 2\hat{j} + 3\hat{k}$, $\vec{b} = 2\hat{i} - \hat{j} + \hat{k}$ and $\vec{c} = 3\hat{i} + \hat{j} + 4\hat{k}$ are

(1) not coplanar
(2) coplanar but cannot form a triangle
(3) coplanar and form a triangle
(4) coplanar and can form a right-angled triangle

17. Vectors $\vec{a} = -4\hat{i} + 3\hat{k}$; $\vec{b} = 14\hat{i} + 2\hat{j} - 5\hat{k}$ are laid off from one point. Then vector $\vec{d}$, which is being laid off from the same point dividing the angle between vectors $\vec{a}$ and $\vec{b}$ in equal halves and having the magnitude $\sqrt{6}$, is

(1) $\hat{i} + \hat{j} + 2\hat{k}$ (2) $\hat{i} - \hat{j} + 2\hat{k}$

(3) $\hat{i} + \hat{j} - 2\hat{k}$ (4) $2\hat{i} + \hat{j} - 2\hat{k}$

18. If $4\hat{i} + 7\hat{j} + 8\hat{k}$, $2\hat{i} + 3\hat{j} + 4\hat{k}$ and $2\hat{i} + 5\hat{j} + 7\hat{k}$ are the position vectors of the vertices A, B and C, respectively, of a triangle ABC, then the position vector of the point where the internal bisector of angle A meets BC is

(1) $\frac{2}{3}(-6\hat{i} - 8\hat{j} - 6\hat{k})$ (2) $\frac{2}{3}(6\hat{i} + 8\hat{j} + 6\hat{k})$

(3) $\frac{1}{3}(6\hat{i} + 13\hat{j} + 18\hat{k})$ (4) $\frac{1}{3}(5\hat{j} + 12\hat{k})$

19. If the vectors $\hat{i} - \hat{j}$, $\hat{j} + \hat{k}$ and $\vec{a}$ in order form a triangle, then $\vec{a} =$

(1) $-\hat{i} - \hat{k}$ (2) $\hat{i} - 2\hat{j} - \hat{k}$

(3) $2\hat{i} + \hat{j} + \hat{k}$ (4) $\hat{i} + \hat{k}$

20. The angle that the bisector of vectors $2\hat{i} + 2\hat{j} - \hat{k}$ and $-2\hat{i} - 3\hat{j} + 6\hat{k}$ makes with the x-axis is

(1) $\cos^{-1}\frac{8}{\sqrt{210}}$ (2) $\cos^{-1}\frac{5}{\sqrt{210}}$

(3) $\cos^{-1}\frac{11}{\sqrt{210}}$ (4) $\cos^{-1}\frac{4}{\sqrt{210}}$

21. If the angle between vectors $2\hat{i} - 3\hat{j} + a\hat{k}$ and $3\hat{i} + 2\hat{j} - 4\hat{k}$ is 60°, then the value of a is

(1) $\sqrt{\frac{377}{35}}$ (2) $-\sqrt{\frac{377}{35}}$

(3) $\sqrt{\frac{358}{33}}$ (4) $-\sqrt{\frac{358}{33}}$

22. If $\hat{i} - 3\hat{j} + 5\hat{k}$ bisects the angle between $\hat{a}$ and $-\hat{i} + 2\hat{j} + 2\hat{k}$, where $\hat{a}$ is a unit vector, then

(1) $\hat{a} = \frac{1}{105}(41\hat{i} + 88\hat{j} - 40\hat{k})$

(2) $\hat{a} = \frac{1}{105}(41\hat{i} + 88\hat{j} + 40\hat{k})$

(3) $\hat{a} = \frac{1}{105}(-41\hat{i} + 88\hat{j} - 40\hat{k})$

(4) $\hat{a} = \frac{1}{105}(41\hat{i} - 88\hat{j} - 40\hat{k})$

23. If $\vec{b}$ is a vector whose initial point divides the join of $5\hat{i}$ and $5\hat{j}$ in the ratio $k : 1$ and whose terminal point is the origin. If $|\vec{b}| \leq \sqrt{37}$, then k lies in the interval

(1) $[-6, -1/6]$
(2) $(-\infty, -6] \cup [-1/6, \infty)$
(3) $[0, 6]$
(4) none of these

24. Let $x^2 + 3y^2 = 3$ be the equation of an ellipse in the xy-plane. If A and B are two points whose position vectors are $-\sqrt{3}\hat{i}$ and $-\sqrt{3}\hat{i} + 2\hat{k}$ respectively. Then the position vectors of a point P on the ellipse such that $\angle APB = \pi/4$ are

(1) $\pm\hat{j}$ (2) $\pm(\hat{i} + \hat{j})$

(3) $\pm\hat{i}$ (4) $\vec{k}$

• **25.** A uni-modular tangent vector on the curve $x = t^2 + 2$, $y = 4t - 5$, $z = 2t^2 - 6t$ at $t = 2$ is

(1) $\frac{1}{3}(2\hat{i} + 2\hat{j} + \hat{k})$ (2) $\frac{1}{3}(\hat{i} - \hat{j} - \hat{k})$

(3) $\frac{1}{6}(2\hat{i} + \hat{j} + \hat{k})$ (4) $\frac{2}{3}(\hat{i} + \hat{j} + \hat{k})$

Linear Combination of Vectors

26. Let ABC be a triangle, the position vectors of whose vertices are respectively $\hat{i} + 2\hat{j} + 4\hat{k}$, $-2\hat{i} + 2\hat{j} + \hat{k}$ and $2\hat{i} + 4\hat{j} - 3\hat{k}$. Then ΔABC is

(1) isosceles (2) equilateral

(3) right angled (4) none of these

27. If $\vec{\alpha} + \vec{\beta} + \vec{\gamma} = a\vec{\delta}$ and $\vec{\beta} + \vec{\gamma} + \vec{\delta} = b\vec{\alpha}$, where $\vec{\alpha}$ and $\vec{\delta}$ are non-collinear, then $\vec{\alpha} + \vec{\beta} + \vec{\gamma} + \vec{\delta}$ equals

(1) $a\vec{\alpha}$ (2) $b\vec{\delta}$

(3) 0 (4) $(a + b)\vec{\gamma}$

28. Given three non-zero, non-coplanar vectors $\vec{a}$, $\vec{b}$ and $\vec{c}$. Let $\vec{r}_1 = p\vec{a} + q\vec{b} + \vec{c}$ and $\vec{r}_2 = \vec{a} + p\vec{b} + q\vec{c}$. If the vectors $\vec{r}_1 + 2\vec{r}_2$ and $2\vec{r}_1 + \vec{r}_2$ are collinear, then (p, q) is

(1) (0, 0) (2) (1, –1)

(3) (–1, 1) (4) (1, 1)

29. If the vectors $\vec{a}$ and $\vec{b}$ are linearly independent satisfying $(\sqrt{3}\tan\theta + 1)\vec{a} + (\sqrt{3}\sec\theta - 2)\vec{b} = 0$, then the values of θ are given by

(1) $n\pi - \frac{\pi}{6}$, $n \in Z$ (2) $2n\pi \pm \frac{11\pi}{6}$, $n \in Z$

(3) $n\pi \pm \frac{\pi}{6}$, $n \in Z$ (4) $2n\pi + \frac{11\pi}{6}$, $n \in Z$

30. If points $\hat{i} - \hat{j} + \hat{k}$, $2\hat{i} + \hat{j} - \hat{k}$, $3\hat{i} + 4\hat{k} + \hat{k}$ and $\alpha\hat{i} + 2\hat{j} + 3\hat{k}$ are coplanar then the value of α is

(1) 0 (2) 1 (3) 2 (4) 3

31. Vectors $\alpha\hat{i} + \hat{j} - \beta\hat{k}$, $\hat{i} - 2\hat{j} + 3\hat{k}$ and $2\hat{i} + \hat{j} - 4\hat{k}$, are linearly dependent for the point (α, β) lying on the line

(1) $x - y - 1 = 0$ (2) $x + y - 3 = 0$

(3) $x - y + 2 = 0$ (4) $2x - y + 4 = 0$

• **32.** Let $\vec{r}_1, \vec{r}_2, \vec{r}_3, \ldots, \vec{r}_n$ be the position vectors of points $P_1, P_2, P_3, \ldots, P_n$ with respect to the origin O. If the vector equation $a_1\vec{r}_1 + a_2\vec{r}_2 + \cdots + a_n\vec{r}_n = 0$ holds, then a similar equation will also hold w.r.t. to any other origin provided

(1) $a_1 + a_2 + \cdots + a_n = n$

(2) $a_1 + a_2 + \cdots + a_n = 1$

(3) $a_1 + a_2 + \cdots + a_n = 0$

(4) $a_1 = a_2 = a_3 = \cdots = a_n = 0$

• **33.** The value of λ so that the points P, Q, R and S on the sides OA, OB, OC and AB, respectively, of a regular tetrahedron $OABC$ are coplanar such that $\frac{OP}{OA} = \frac{1}{3}$, $\frac{OQ}{OB} = \frac{1}{2}$, $\frac{OR}{OC} = \frac{1}{3}$ and $\frac{OS}{AB} = \lambda$ is

(1) $\lambda = \frac{1}{2}$ (2) $\lambda = -1$

(3) $\lambda = 0$ (4) for no value of λ

• **34.** If points $\lambda\hat{i} - \hat{j} + 2\hat{k}$, $2\hat{i} + 4\hat{j} - 5\hat{k}$ and $\hat{i} - 2\hat{i} + \mu\hat{k}$ are collinear then length of the vector $\vec{p} = \frac{1}{\lambda}\hat{i} + \frac{1}{\mu}\hat{j} + \frac{2\sqrt{57}}{17}\hat{k}$ is

(1) 1 (2) 2 (3) 3 (4) 4

• **35.** The cartesian form of the vector equation $x\hat{i} + y\hat{j} + z\hat{k} = \hat{i} + \hat{j} + \hat{k} + \lambda(\hat{i} - \hat{j} + \hat{k}) + \mu(2\hat{i} + \hat{j} + \hat{k})$; $\lambda, \mu \in R$ is

(1) $2x - y - 3z + 2 = 0$ (2) $x - 2y + 3z - 2 = 0$

(3) $x - y + 2z - 3 = 0$ (4) $4x - y - 3z + 4 = 0$

Numerical Value Type

1. A vector $\vec{a}$ has components $3p$ and 1 with respect to a rectangular cartesian system. This system is rotated through a certain angle about the origin in the counter clockwise sense. If, with respect to new system, $\vec{a}$ has components $p + 1$ and $\sqrt{10}$, then the positive value of p is ________.

2. The value of m for which vectors $\hat{i} + 2\hat{j} + 3\hat{k}$, $3\hat{i} - \hat{j} + m\hat{k}$ and $\hat{i} - 2\hat{j} + 2\hat{k}$ are coplanar is ________

3. If the points (1, 5, 35), (7, 5, 5), (1, λ, 7) and (2λ, 1, 2) are coplanar, then the sum of all possible values of λ is ________.

4. Let for a triangle ABC, $\overrightarrow{AB} = -2\hat{i} + \hat{j} + 3\hat{k}$, $\overrightarrow{CB} = \alpha\hat{i} + \beta\hat{j} + 8\hat{k}$ and $\overrightarrow{CA} = 4\hat{i} + 3\hat{j} + \gamma\hat{k}$. Then the value of $\frac{\alpha\beta}{\gamma}$ is ________.

5. Let $\vec{a} = \hat{i} + \lambda\hat{j} + \hat{k}$ and $\vec{b} = 2\mu\hat{i} + \hat{j} + \hat{k}$. If $|\vec{a} + \vec{b}| = |\vec{a}| + |\vec{b}|$ then the value of $\lambda + \mu$ is ________.

• **6.** In a triangle ABC, if $|\overrightarrow{BC}| = 3$, $|\overrightarrow{CA}| = 5$ and $|\overrightarrow{BA}| = 7$, then the projection of the vector $\overrightarrow{BA}$ on $\overrightarrow{BC}$ is ________.

• **7.** Let $\vec{a}$, $\vec{b}$ and $\vec{c}$ be unit vectors such that $\vec{a} + \vec{b} - \vec{c} = 0$. If the area of triangle formed by vectors $\vec{a}$ and $\vec{c}$ is A, then the value of $4A^2$ is ________.

Integer Answer Type

1. Let ABC be a triangle whose centroid is G, orhtocentre is H and circumcentre is the origin 'O'. If D is any point in the plane of the triangle such that no three of O, A, C and D are collinear satisfying the relation $\overline{AD} + \overline{BD} + \overline{CH} + 3\overline{HG} = \lambda\, \overline{HD}$, then the value of the scalar 'λ' is_______.
2. If the resultant of three forces $\vec{F_1} = p\hat{i} + 3\hat{j} - \hat{k}, \vec{F_2} = -5\hat{i} + \hat{j} + 2\hat{k}$ and $\vec{F_3} = 6\hat{i} - \hat{k}$ acting on a particle has a magnitude equal to 5 units, then the difference in the values of p is________.
3. The least positive integral value of x for which the angle between vectors $\vec{a} = x\hat{i} - 3\hat{j} - \hat{k}$ and $\vec{b} = 2x\hat{i} + x\hat{j} - \hat{k}$ is acute is________.
4. Vectors along the adjacent sides of parallelogram are $\vec{a} = \hat{i} + 2\hat{j} + \hat{k}$ and $\vec{b} = 2\hat{i} + 4\hat{j} + \hat{k}$. Then the length of the longer diagonal of the parallelogram is________.
5. If vectors $\vec{a} = \hat{i} + 2\hat{j} - \hat{k}, \vec{b} = 2\hat{i} - \hat{j} + \hat{k}$ and $\vec{c} = \lambda\hat{i} + \hat{j} + 2\hat{k}$ are coplanar, then the value of λ is ________.
6. If the vectors $\vec{a} = \hat{i} + 2\hat{j} + 3\hat{k}, \vec{b} = 2\hat{i} + 3\hat{j} + \hat{k}, \vec{c} = 3\hat{i} + \hat{j} + 2\hat{k}$ are satisfying $\alpha\vec{a} + \beta\vec{b} + \gamma\vec{c} = -3(\hat{i} - \hat{k})$, then the value of $\alpha + \beta + \gamma$ is ________.

Multiple Correct Answers Type

1. The vectors $x\hat{i} + (x+1)\hat{j} + (x+2)\hat{k}, (x+3)\hat{i} + (x+4)\hat{j} + (x+5)\hat{k}$ and $(x+6)\hat{i} + (x+7)\hat{j} + (x+8)\hat{k}$ are coplanar if x is equal to
   (1) 1 (2) −3 (3) 4 (4) 0
2. The sides of a parallelogram are $2\hat{i} + 4\hat{j} - 5\hat{k}$ and $\hat{i} + 2\hat{j} + 3\hat{k}$. The unit vectors parallel to diagonals is/are
   (1) $\frac{1}{7}(3\hat{i} + 6\hat{j} - 2\hat{k})$ (2) $\frac{1}{7}(3\hat{i} - 6\hat{j} - 2\hat{k})$
   (3) $\frac{1}{\sqrt{69}}(\hat{i} + 2\hat{j} + 8\hat{k})$ (4) $\frac{1}{\sqrt{69}}(-\hat{i} - 2\hat{j} + 8\hat{k})$
3. A vector $\vec{a}$ has the components $2p$ and 1 w.r.t. a rectangular Cartesian system. This system is rotated through a certain angle about the origin in the counterclockwise sense. If, with respect to a new system, a has components $(p+1)$ and 1, then p is equal to
   (1) −1 (2) −1/3 (3) 1 (4) 2
4. If the points $\hat{i} + \hat{j}, \hat{i} - \hat{j}$ and $p\hat{i} + q\hat{j} + r\hat{k}$ are collinear, then
   (1) $p = 1$ (2) $r = 0$
   (3) $q \in R$ (4) $q \neq \pm 1$
5. If $\vec{a}, \vec{b}$ and $\vec{c}$ are non-coplanar vectors and λ is a real number, then the vectors $\vec{a} + 2\vec{b} + 3\vec{c}, \lambda\vec{b} + \mu\vec{c}$ and $(2\lambda - 1)\vec{c}$ are coplanar if
   (1) $\mu \in R$ (2) $\lambda = \frac{1}{2}$
   (3) $\lambda = 0$ (4) no value of λ
6. If the resultant of three forces $\vec{F_1} = p\hat{i} + 3\hat{j} - \hat{k}, \vec{F_2} = 6\hat{i}$ and $\vec{F_3} = -5\hat{i} + \hat{j} + 2\hat{k}$ acting on a particle has a magnitude equal to 5 units, then the value (s) of p is/are
   (1) −6 (2) −4 (3) 2 (4) 4
7. The vector $\hat{i} + x\hat{j} + 3\hat{k}$ is rotated through an angle θ and is doubled in magnitude. It now becomes $4\hat{i} + (4x - 2)\hat{j} + 2\hat{k}$. The value(s) of x is/are
   (1) 1 (2) −2/3 (3) 2 (4) 4/3
8. Let $\vec{a}, \vec{b}$ and $\vec{c}$ be three coplanar unit vectors such that $\vec{a} + \vec{b} + \vec{c} = 0$. If three vectors $\vec{p}, \vec{q}$ and $\vec{r}$ are parallel to $\vec{a}, \vec{b}$ and $\vec{c}$, respectively, and have integral but different magnitudes, then among the following options, $|\vec{p} + \vec{q} + \vec{r}|$ can take a value equal to
   (1) 1 (2) 0 (3) $\sqrt{3}$ (4) 2
9. If non-zero vectors $\vec{a}$ and $\vec{b}$ are equally inclined to the coplanar vector $\vec{c}$, then $\vec{c}$ can be
   (1) $\frac{|\vec{a}|}{|\vec{a}| + 2|\vec{b}|}\vec{a} + \frac{|\vec{b}|}{|\vec{a}| + |\vec{b}|}\vec{b}$
   (2) $\frac{|\vec{b}|}{|\vec{a}| + |\vec{b}|}\vec{a} + \frac{|\vec{a}|}{|\vec{a}| + |\vec{b}|}\vec{b}$
   (3) $\frac{|\vec{a}|}{|\vec{a}| + 2|\vec{b}|}\vec{a} + \frac{|\vec{b}|}{|\vec{a}| + 2|\vec{b}|}\vec{b}$
   (4) $\frac{|\vec{b}|}{2|\vec{a}| + |\vec{b}|}\vec{a} + \frac{|\vec{a}|}{2|\vec{a}| + |\vec{b}|}\vec{b}$
10. If $A(-4, 0, 3)$ and $B(14, 2, -5)$, then which of the following points can lie on the bisector of the angle between $\overrightarrow{OA}$ and $\overrightarrow{OB}$ (O is the origin of reference)?
   (1) (2, 2, 4) (2) (2, 11, 5)
   (3) (−3, −3, −6) (4) (1, 1, 2)

11. In a four-dimensional space where unit vectors along the axes are $\hat{i}, \hat{j}, \hat{k}$ and $\hat{l}$, and $\vec{a}_1, \vec{a}_2, \vec{a}_3, \vec{a}_4$ are four non-zero vectors such that no vector can be expressed as a linear combination of others and $(\lambda - 1)(\vec{a}_1 - \vec{a}_2)$ $+ \mu(\vec{a}_2 + \vec{a}_3) + \gamma(\vec{a}_3 + \vec{a}_4 - 2\vec{a}_2) + \vec{a}_3 + \delta\vec{a}_4 = \vec{0}$, then

(1) $\lambda = 1$ (2) $\mu = -2/3$

(3) $\gamma = 2/3$ (4) $\delta = 1/3$

12. Let ABC be a triangle, the position vectors of whose vertices are $7\hat{j} + 10\hat{k}$, $-\hat{i} + 6\hat{j} + 6\hat{k}$ and $-4\hat{i} + 9\hat{j} + 6\hat{k}$. Then ΔABC is

(1) isosceles (2) equilateral

(3) right angled (4) none of these

Linked Comprehension Type

For Questions 1–3

$ABCD$ is a parallelogram. L is a point on BC that divides BC in the ratio 1 : 2. AL intersects BD at P. M is a point on DC that divides DC in the ratio 1 : 2 and AM intersects BD in Q. Then

1. Point P divides AL in the ratio

(1) 1 : 2 (2) 1 : 3 (3) 3 : 1 (4) 2 : 1

2. Point Q divides DB in the ratio

(1) 1 : 2 (2) 1 : 3 (3) 3 : 1 (4) 2 : 1

3. $PQ : DB$ is equal to

(1) 2/3 (2) 1/3 (3) 1/2 (4) 3/4

For Questions 4–5

Let $OABCD$ be a pentagon in which the sides OA and CB are parallel and the sides OD and AB are parallel. It $OA : CB = 2 : 1$ and $OD : AB = 1 : 3$. Then,

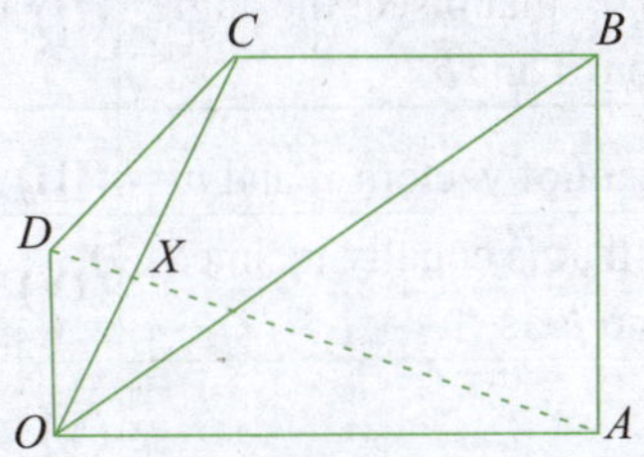

4. The ratio $\dfrac{OX}{XC}$ is

5. The ratio $\dfrac{AX}{XD}$ is

For Questions 6–7

Consider the regular hexagon $ABCDEF$ with centre at O (origin).

6. If $\overrightarrow{AD} + \overrightarrow{EB} + \overrightarrow{FC} = \lambda\,\overrightarrow{AB}$ then $\lambda =$

7. If the resultant of five forces $\overrightarrow{AB}$, $\overrightarrow{AC}$, $\overrightarrow{AD}$, $\overrightarrow{AE}$, $\overrightarrow{AF}$ acting at the vertex A is $\lambda\,\overrightarrow{AO}$, then $\lambda =$

For Questions 8–9

If $|\vec{a}| = 3$, $|\vec{b}| = 4$, and angle between $\vec{a}$ and $\vec{b}$ is 60°.

8. The value of $|\vec{a} + \vec{b}|^2$ is

9. The value of $|\vec{a} - \vec{b}|^2$ is

For Questions 10–11

Let A, B, C, D, E represent the vertices of a regular pentagon $ABCDE$. If the position vector of these vertices are $\vec{a}, \vec{a} + \vec{b}, \vec{b}, \lambda\vec{a}$ and $\lambda\vec{b}$, respectively. Then

10. The ratio $\dfrac{AD}{BC}$ is equal to

(1) $1 - \cos\dfrac{3\pi}{5} : \cos\dfrac{3\pi}{5}$

(2) $1 + 2\cos\dfrac{2\pi}{5} : \cos\dfrac{\pi}{5}$

(3) $1 + 2\cos\dfrac{\pi}{5} : 2\cos\dfrac{\pi}{5}$

(4) None of these

11. OE divides OC in the ratio

(1) $\cos\dfrac{2\pi}{5} : 1$ (2) $\cos\dfrac{3\pi}{5} : 1$

(3) $1 : 2\cos\dfrac{2\pi}{5}$ (4) 1 : 2

For Questions 12–13

In a parallelogram $OABC$, $\vec{a}, \vec{b}, \vec{c}$ are, respectively, the position vectors of the vertices A, B, C with reference to O as the origin. A point E is taken on the side BC that divides it in the ratio of 2 : 1. Also, the line segment AE intersects the line bisecting the angle $\angle AOC$ internally at point P. If CP when extended meets AB in point F, then

12. The position vector of point P is

(1) $\dfrac{|\vec{a}||\vec{c}|}{3|\vec{c}| + 2|\vec{a}|}\left(\dfrac{\vec{a}}{|\vec{a}|} + \dfrac{\vec{c}}{|\vec{c}|}\right)$ (2) $\dfrac{3|\vec{a}||\vec{c}|}{3|\vec{c}| + 2|\vec{a}|}\left(\dfrac{\vec{a}}{|\vec{a}|} + \dfrac{\vec{c}}{|\vec{c}|}\right)$

(3) $\dfrac{2|\vec{a}||\vec{c}|}{3|\vec{c}| + 2|\vec{a}|}\left(\dfrac{\vec{a}}{|\vec{a}|} + \dfrac{\vec{c}}{|\vec{c}|}\right)$ (4) None of these

13. The ratio in which F divides AB is

(1) $\dfrac{2|\vec{a}|}{\left||\vec{a}| - 3|\vec{c}|\right|}$ (2) $\dfrac{|\vec{a}|}{\left||\vec{a}| - 3|\vec{c}|\right|}$

(3) $\dfrac{3|\vec{a}|}{\left||\vec{a}| - 3|\vec{c}|\right|}$ (4) $\dfrac{3|\vec{c}|}{3|\vec{c}| - |\vec{a}|}$

Matrix Match Type

1. If $ABCDEF$ is a regular hexagon formed by vectors, $\overrightarrow{AB} = \vec{a}$ and $\overrightarrow{BC} = \vec{b}$, then match the following lists.

List-I	List-II
(P) If $\overrightarrow{CD} = x\vec{a} + y\vec{b}$, then the value of $x + y$ is	**(I)** -2
(Q) If $\overrightarrow{CE} = x\vec{a} + y\vec{b}$, then the value of $x + y$ is	**(II)** -1
(R) $\overrightarrow{AE} = x\vec{a} + y\vec{b}$, then the value of $(-x - y)$ is	**(III)** $y = 0$
(S) $\overrightarrow{AD} = x\vec{a} + y\vec{b}$, then the value of $x + y$ is	**(IV)** 1
	(V) 2

(1) (P) → (III); (Q) → (I); (R) → (II); (S) → (IV)
(2) (P) → (I); (Q) → (III); (R) → (II); (S) → (V)
(3) (P) → (II); (Q) → (V); (R) → (IV); (S) → (III)
(4) (P) → (III); (Q) → (II); (R) → (I); (S) → (V)

2. In triangle ABC, $\overrightarrow{AB} = 2\hat{i} + 2\hat{j} + \hat{k}$ and $\overrightarrow{AC} = 2\hat{i} + 4\hat{j} + 4\hat{k}$. Then match the following columns.

List-I	List-II
(P) Length of median through A	**(I)** 3
(Q) length of angle bisector through A	**(II)** $\frac{\sqrt{77}}{2}$
(R) Length of side BC	**(III)** $\frac{\sqrt{17}}{3}$
(S) length of altitude through B	**(IV)** $\frac{2\sqrt{34}}{3}$
	(V) $\sqrt{13}$

(1) (P) → (II); (Q) → (V); (R) → (IV); (S) → (I)
(2) (P) → (II); (Q) → (IV); (R) → (V); (S) → (III)
(3) (P) → (III); (Q) → (V); (R) → (IV); (S) → (II)
(4) (P) → (V); (Q) → (I); (R) → (II); (S) → (IV)

3. Match the following lists.

List-I	List-II
(P) If $\vec{a}$ and $\vec{b}$ are non–collinear vectors then the least value of λ for which vectors $\vec{\alpha} = (\lambda - 2)\,\vec{a} + \vec{b}$ and $\vec{\beta} = (4\lambda - 9)\,\vec{a} + 3\vec{b}$ are collinear is	**(I)** 1
(Q) The greatest value of μ, for which the vectors, $\mu\hat{i} + \hat{j} + \hat{k}$, $\hat{i} + \mu\hat{j} + \hat{k}$, $\hat{i} + \hat{j} + \mu\hat{k}$ are coplanar, is	**(II)** 2
(R) The smallest positive value of α for which the points A, B, C whose position vectors respectively are $\vec{a} = \hat{i} + 4\hat{j} + 3\hat{k}$, $\vec{b} = 2\hat{i} + \alpha\hat{j} + 4\hat{k}$, $\vec{c} = 3\hat{i} - 2\hat{j} + 5\hat{k}$ are non-collinear, is	**(III)** 3
(S) The greatest value of α, for which the points whose position vectors $\hat{i} - 2\hat{j} + 2\hat{k}$, $2\hat{i} - 3\hat{j} + 4\hat{k}$, $(\alpha + 1)i + 2k$ and $9\hat{i} + (\alpha - 8)\hat{j} + 6\hat{k}$ are coplanar, is	**(IV)** 4
	(V) 5

(1) (P) → (III); (Q) → (IV); (R) → (II); (S) → (I)
(2) (P) → (III); (Q) → (II); (R) → (I); (S) → (V)
(3) (P) → (III); (Q) → (I); (R) → (II); (S) → (IV)
(4) (P) → (II); (Q) → (I); (R) → (V); (S) → (IV)

4. Let $\vec{a} = 2\hat{i} + 2\hat{j} - \hat{k}$, $\vec{b} = 2\hat{i} - 4\hat{j} + 4\hat{k}$. Then match the following lists

List-I	List-II
(P) Vector $\vec{c}$ such that $\vec{a}, \vec{b}$ and $\vec{c}$ from a triangle	**(I)** $4\hat{i} - 2\hat{j} + 3\hat{k}$
(Q) Vector $\vec{c}$ that bisect the angle between $\vec{a}$ and $\vec{b}$	**(II)** $-4\hat{i} + 2\hat{j} - 3\hat{k}$
(R) Resultant of vectors $\vec{a}$ and $\vec{b}$	**(III)** $3\hat{i} - 2\hat{k}$
(S) Vector that is equally inclined to $\vec{a}$ and $\vec{b}$ is	**(IV)** $3\hat{i} + \hat{k}$
	(V) $3\hat{j} + 4\hat{k}$

(1) (P) → (II); (Q) → (V); (R) → (I); (S) → (IV)
(2) (P) → (I); (Q) → (IV); (R) → (II); (S) → (V)
(3) (P) → (II); (Q) → (IV); (R) → (I); (S) → (V)
(4) (P) → (I); (Q) → (V); (R) → (II); (S) → (IV)

ARCHIVES

Matrix Match Type

1. Match the following lists:

List I	List II
(P) In a triangle ΔXYZ, let a, b and c be the lengths of the sides opposite to the angles X, Y and Z respectively. If $2(a^2 - b^2) = c^2$ and $\lambda = \dfrac{\sin(X - Y)}{\sin Z}$ then possible values of n for which $\cos(n\pi\lambda) = 0$ is (are)	**(1)** 1
(Q) In a triangle ΔXYZ, let a, b and c be the lengths of the sides opposite to the angles X, Y and Z, respectively. If $1 + \cos 2X - 2\cos 2Y = 2\sin X \sin Y$, then possible value(s) of $\dfrac{a}{b}$ is (are)	**(II)** 2
(R) In R^2, let $\sqrt{3}\hat{i} + \hat{j}, \hat{i} + \sqrt{3}\hat{j}$ and $\beta\hat{i} + (1 - \beta)\hat{j}$ be the position vectors of X, Y and Z with respect of the origin O, respectively. If the distance of Z from the bisector of the acute angle of $\overrightarrow{OX}$ and $\overrightarrow{OY}$ is $\dfrac{3}{\sqrt{2}}$, then possible value(s) of $\lvert\beta\rvert$ is (are)	**(III)** 3
(S) Suppose that $F(\alpha)$ denotes the area of the region bounded by $x = 0$, $x = 2$, $y^2 = 4x$ and $y = \lvert\alpha x - 1\rvert + \lvert\alpha x - 2\rvert + \alpha x$, where $\alpha \in \{0, 1\}$. Then the value(s) of $F(\alpha) + \dfrac{8}{3}\sqrt{2}$, when $\alpha = 0$ and $\alpha = 1$, is (are)	**(IV)** 5
	(V) 6

(JEE Advanced 2015)

Numerical Value Type

1. For any two points M and N in the XY-plane, let $\overrightarrow{MN}$ denote the vector from M to N, and $\vec{O}$ denote the zero vector. Let P, Q and R be three distinct points in the XY-plane. Let S be a point inside the triangle ΔPQR such that

$$\overrightarrow{SP} + 5\overrightarrow{SQ} + 6\overrightarrow{SR} = \vec{0}.$$

Let E and F be the mid-points of the sides PR and QR, respectively. Then the value of $\dfrac{\text{length of the line segment } EF}{\text{length of the line segment } ES}$ is ________. **(JEE Advanced 2025)**

Integer Answer Type

1. Consider the set of eight vectors $V = \{a\hat{i} + b\hat{j} + c\hat{k}\,;\, a, b, c \in \{-1, 1\}\}$. Three non-coplanar vectors can be chosen from V in 2^p ways. Then p is ________. **(JEE Advanced 2013)**

2. Suppose that $\vec{p}, \vec{q}$ and $\vec{r}$ are three non-coplanar vectors in R^3. Let the components of a vector $\vec{s}$ along $\vec{p}, \vec{q}$ and $\vec{r}$ be 4, 3 and 5, respectively. If the components of this vector $\vec{s}$ along $(-\vec{p} + \vec{q} + \vec{r})$, $(\vec{p} - \vec{q} + \vec{r})$ and $(-\vec{p} - \vec{q} + \vec{r})$ are x, y and z, respectively, then the value of $2x + y + z$ is **(JEE Advanced 2015)**

3. Consider the vectors $\vec{x} = \hat{i} + 2\hat{j} + 3\hat{k}$, $\vec{y} = 2\hat{i} + 3\hat{j} + \hat{k}$, and $\vec{z} = 3\hat{i} + \hat{j} + 2\hat{k}$. For two distinct positive real numbers α and β, define $\vec{X} = \alpha\vec{x} + \beta\vec{y} - \vec{z}$, $\vec{Y} = \alpha\vec{y} + \beta\vec{z} - \vec{x}$, and $\vec{Z} = \alpha\vec{z} + \beta\vec{x} - \vec{y}$. If the vectors $\vec{X}, \vec{Y}$ and $\vec{Z}$ lie in plane, then the value of $\alpha + \beta - 3$ is ________. **(JEE Advanced 2025)**

ANSWER KEY

EXERCISES

Single Correct Answer Type

1. (1)	**2.** (4)	**3.** (1)	**4.** (3)	**5.** (4)
6. (4)	**7.** (3)	**8.** (3)	**9.** (3)	**10.** (2)
11. (2)	**12.** (3)	**13.** (2)	**14.** (2)	**15.** (3)
16. (2)	**17.** (1)	**18.** (3)	**19.** (1)	**20.** (1)
21. (2)	**22.** (4)	**23.** (2)	**24.** (1)	**25.** (1)
26. (3)	**27.** (3)	**28.** (4)	**29.** (4)	**30.** (3)
31. (3)	**32.** (3)	**33.** (2)	**34.** (1)	**35.** (1)

Numerical Value Type

1. (1.25)	**2.** (7.25)	**3.** (8.8)	**4.** (1.6)	**5.** (1.5)
6. (5.5)	**7.** (0.75)			

Integer Answer Type

1. (2)	**2.** (6)	**3.** (2)	**4.** (7)	**5.** (13)
6. (0)				

Multiple Correct Answers Type

1. (1), (2), (3), (4)	**2.** (1), (4)
3. (2), (3)	**4.** (1), (2), (4)
5. (1), (2), (3)	**6.** (2), (3)
7. (2), (3)	**8.** (3), (4)
9. (2), (4)	**10.** (1), (3), (4)
11. (1), (2), (4)	**12.** (1), (3)

Linked Comprehension Type

1. (3)	**2.** (2)	**3.** (3)	**4.** (0.4)	**5.** (6)
6. (4)	**7.** (6)	**8.** (13)	**9.** (37)	**10.** (3)
11. (3)	**12.** (2)	**13.** (4)		

Matrix Match Type

1. (4)
2. (2)
3. (3)
4. (3)

ARCHIVES

Matrix Match Type

1. R → I, II

Numerical Value Type

1. (1.2)

Integer Answer Type

1. (5) **2.** (9) **3.** (–2)

Chapter

3 Products of Vectors

Chapter Prerequisites

Before starting this chapter, ensure that you understand the following concepts thoroughly:

1. **Basic Vector Algebra:** Understanding of vector representation in three dimensions, types of vectors, vector magnitude, direction, and vector addition and subtraction in both geometric and component forms.
2. **Coordinate Geometry in Three Dimensions:** Familiarity with three-dimensional axes, ordered triples, the distance between points, and the basic spatial interpretation of directions and planes.
3. **Trigonometric Basics:** Understanding of sine, cosine, and angle relationships needed for interpreting the dot product and angles between vectors.
4. **Comfort with Multi-step Vector Calculations:** Ability to manipulate algebraic expressions and follow multi-step procedures involved in computing magnitudes, verifying perpendicularity, evaluating vector products, and interpreting vector equations.

Topic List

Learning Objectives

Foundations of Vector Operations

Understand the need for algebraic and geometric operations on vectors and how they extend the study of direction and magnitude in geometry.

Dot (Scalar) Product of Vectors

Learn the dot product and its properties, find vector components and squared magnitudes, and apply it to angles, projections, and perpendicularity.

Cross (Vector) Product of Vectors

Learn the cross product and its properties, solve vector equations, and apply it to find perpendicular vectors and areas of geometric figures.

Scalar Triple Product

Learn the scalar triple product, interpret its geometric meaning, and apply it to determine volumes and orientation in three dimensions.

Vector Triple Product

Learn the vector triple product, use standard identities, and interpret related concepts such as four-vector products and reciprocal systems.

3.1 INTRODUCTION

After studying the fundamental operations on vectors, we now proceed to the study of another type of operation known as vector products. These are introduced primarily to address and model various real-life problems encountered in physics and engineering. In this chapter, we shall discuss the different types of vector products and their properties, which will enable us to analyze and understand three-dimensional geometry and space more effectively.

3.2 DOT (SCALAR) PRODUCT OF VECTORS

3.2.1 Definition of Dot Product

To introduce the idea of the dot product, let us first consider the concept of **work done** in physics.

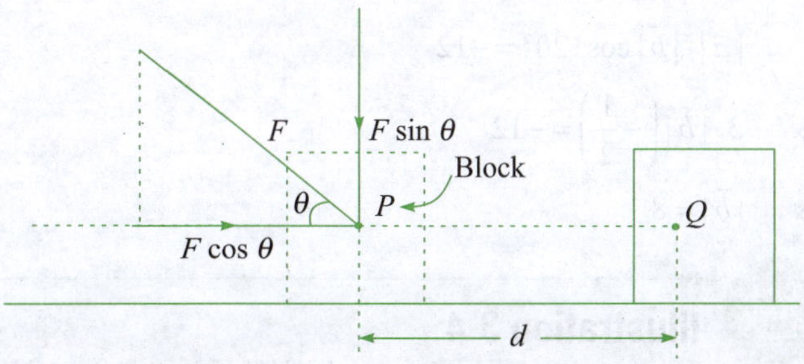

Suppose a block is placed on a horizontal surface at point P. A force F is applied at an angle θ with the horizontal, and the block is displaced to point Q. Thus,

$PQ = d$ units

Work done is defined as the product of the component of the force in the direction of displacement and the magnitude of displacement. From the figure, the component of the force along the displacement is $F \cos \theta$. Therefore,

$$W = (F \cos \theta) d = F d \cos \theta$$

Here, displacement is caused by the horizontal component of the force, which is in the direction of the displacement. The vertical component ($F \sin \theta$) does not cause any displacement.

This shows that the calculation of work involves multiplying two vectors: force and displacement, by taking their magnitudes and the cosine of the angle between them. This motivates us to define a new kind of product of two vectors.

Since force and displacement are both vector quantities, this expression can be written as:

$$\vec{F} \cdot \vec{d} = Fd \cos \theta$$

which is called the **dot product** of the force vectors $\vec{F}$ and the displacement vector $\vec{d}$.

Therefore, in general, the dot product (or scalar product) of two vectors $\vec{a}$ and $\vec{b}$ is defined as

$$\vec{a} \cdot \vec{b} = |\vec{a}||\vec{b}| \cos \theta$$

where θ is the angle between $\vec{a}$ and $\vec{b}$.

It is called the scalar product because the result is a scalar (a single number), not a vector.

3.2.2 Properties of Dot Product

The dot product has the following fundamental properties:

1. Since $\vec{a} \cdot \vec{b} = |\vec{a}||\vec{b}| \cos\theta$, we have $\vec{a} \cdot \vec{b} > 0$ if 'θ' is an acute angle and $\vec{a} \cdot \vec{b} < 0$ if 'θ' is an obtuse angle.

Also, if $\vec{a} \perp \vec{b}$, then $\vec{a} \cdot \vec{b} = 0$

Thus, $\hat{i} \cdot \hat{j} = \hat{j} \cdot \hat{k} = \hat{i} \cdot \hat{k} = 0$

2. $\vec{a} \cdot \vec{a} = |\vec{a}| \cdot |\vec{a}| \cos 0° = |\vec{a}|^2$

Thus, $\hat{i} \cdot \hat{i} = \hat{j} \cdot \hat{j} = \hat{k} \cdot \hat{k} = 1$.

Also, $\vec{a} \cdot \vec{a}$ is denoted as $\vec{a}^2$.

3. $\vec{a} \cdot \vec{b} = \vec{b} \cdot \vec{a}$ (Commutative property)

Since $\vec{a} \cdot \vec{b} = |\vec{a}||\vec{b}| \cos\theta$ and $\vec{b} \cdot \vec{a} = |\vec{b}||\vec{a}| \cos\theta$, we have $\vec{a} \cdot \vec{b} = \vec{b} \cdot \vec{a}$.

4. $(m\vec{a}) \cdot (n\vec{b}) = mn(\vec{a} \cdot \vec{b})$, where m and n are scalars.

5. $\vec{a} \cdot (\vec{b} + \vec{c}) = \vec{a} \cdot \vec{b} + \vec{a} \cdot \vec{c}$ (dot product is distributive over addition of vectors).

Proof:

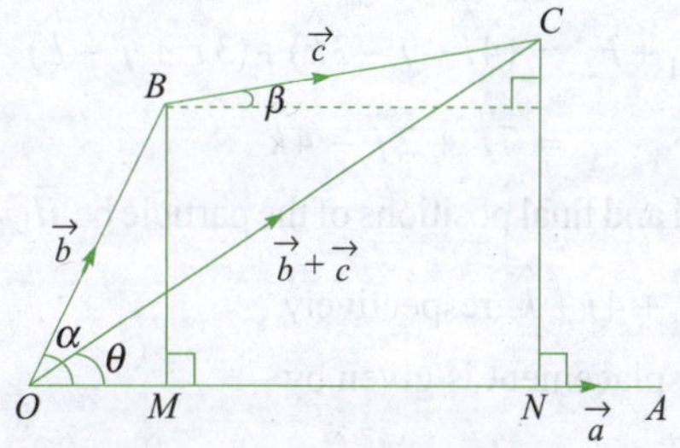

In the figure, $\overrightarrow{OA} = \vec{a}$, $\overrightarrow{OB} = \vec{b}$, $\overrightarrow{BC} = \vec{c}$

$\therefore \quad \overrightarrow{OC} = \overrightarrow{OB} + \overrightarrow{BC} = \vec{b} + \vec{c}$

Now, $BM \perp OA$ and $CN \perp OA$.

$$\begin{aligned}
\text{L.H.S.} &= \vec{a} \cdot (\vec{b} + \vec{c}) \\
&= \overrightarrow{OA} \cdot \overrightarrow{OC} \\
&= (OA)(OC) \cos \theta \text{ (where } \theta = \angle CON) \\
&= (OA)(ON) = (OA)(OM + MN) \\
&= (OA)(OM) + (OA)(MN) \\
&= (OA)(OB \cos \alpha) + (OA)(BC \cos \beta) \\
&= \overrightarrow{OA} \cdot \overrightarrow{OB} + \overrightarrow{OA} \cdot \overrightarrow{BC} \\
&= \vec{a} \cdot \vec{b} + \vec{a} \cdot \vec{c} = \text{R.H.S.}
\end{aligned}$$

6. $\vec{a} \cdot (\vec{b} - \vec{c}) = \vec{a} \cdot \vec{b} - \vec{a} \cdot \vec{c}$

7. If vectors $\vec{a} = a_1\hat{i} + a_2\hat{j} + a_3\hat{k}$ and $\vec{b} = b_1\hat{i} + b_2\hat{j} + b_3\hat{k}$ then $\vec{a} \cdot \vec{b} = a_1b_1 + a_2b_2 + a_3b_3$

Proof:

$$\begin{aligned}
\vec{a} \cdot \vec{b} &= (a_1\hat{i} + a_2\hat{j} + a_3\hat{k}) \cdot (b_1\hat{i} + b_2\hat{j} + b_3\hat{k}) \\
&= (a_1\hat{i}) \cdot (b_1\hat{i} + b_2\hat{j} + b_3\hat{k}) + (a_2\hat{j}) \cdot (b_1\hat{i} + b_2\hat{j} + b_3\hat{k}) \\
&\quad + a_3\hat{k} \cdot (b_1\hat{i} + b_2\hat{j} + b_3\hat{k}) \quad \text{(Distributive property)}
\end{aligned}$$

$$= a_1b_1\hat{i} \cdot \hat{i} + a_1b_2\hat{i} \cdot \hat{j} + a_1b_3\hat{i} \cdot \hat{k} + a_2b_1\hat{j} \cdot \hat{i} + a_2b_2\hat{j} \cdot \hat{j} + a_2b_3\hat{j} \cdot \hat{k} + a_3b_1\hat{k} \cdot \hat{i} + a_3b_2\hat{k} \cdot \hat{j} + a_3b_3\hat{k} \cdot \hat{k}$$

$$= a_1b_1 + a_2b_2 + a_3b_3 \quad (\because \hat{i} \cdot \hat{j} = \hat{j} \cdot \hat{k} = \hat{i} \cdot \hat{k} = 0, \ \hat{i} \cdot \hat{i} = \hat{j} \cdot \hat{j} = \hat{k} \cdot \hat{k} = 1)$$

8. $\vec{r} = (\vec{r}\cdot\hat{i})\hat{i} + (\vec{r}\cdot\hat{j})\hat{j} + (\vec{r}\cdot\hat{k})\hat{k}$

Proof:

Let $\vec{r} = x\hat{i} + y\hat{j} + z\hat{k}$.

Taking dot product with $\hat{i}$, $\hat{j}$ and $\hat{k}$ successively,

we have $x = \vec{r}\cdot\hat{i}$, $y = \vec{r}\cdot\hat{j}$ and $z = \vec{r}\cdot\hat{k}$.

$\therefore \quad \vec{r} = (\vec{r}\cdot\hat{i})\hat{i} + (\vec{r}\cdot\hat{j})\hat{j} + (\vec{r}\cdot\hat{k})\hat{k}$

Illustration 3.1

A particle acted by constant forces $4\hat{i} + \hat{j} - 3\hat{k}$ and $3\hat{i} + \hat{j} - \hat{k}$ is displaced from point $\hat{i} + 2\hat{j} + 3\hat{k}$ to the point $5\hat{i} + 4\hat{j} + \hat{k}$. Find the total work done by the forces.

Sol. Let the forces are $F_1 = 4\hat{i} + \hat{j} - 3\hat{k}$ and $F_2 = 3\hat{i} + \hat{j} - \hat{k}$.

Then the total (resultant) force is:

$$\vec{F} = \vec{F_1} + \vec{F_2} = (4\hat{i} + \hat{j} - 3\hat{k}) + (3\hat{i} + \hat{j} - \hat{k})$$

$$= 7\hat{i} + 2\hat{j} - 4\hat{k}$$

Let the initial and final positions of the particle be $\vec{d}_1 = \hat{i} + 2\hat{j} + 3\hat{k}$ and $\vec{d}_2 = 5\hat{i} + 4\hat{j} + \hat{k}$ respectively.

Then, the displacement is given by

$$\vec{d} = \vec{d}_2 - \vec{d}_1 = (5\hat{i} + 4\hat{j} + \hat{k}) - (\hat{i} + 2\hat{j} + 3\hat{k})$$

$$= 4\hat{i} + 2\hat{j} - 2\hat{k}$$

Therefore, the work done $= \vec{F}\cdot\vec{d}$

$$= (7\hat{i} + 2\hat{j} - 4\hat{k})\cdot(4\hat{i} + 2\hat{j} - 2\hat{k})$$

$$= (7)(4) + (2)(2) + (-4)(-2)$$

$$= 28 + 4 + 8 = 40 \text{ units}$$

Illustration 3.2

Find the angle between the vectors $\hat{i} - 2\hat{j} + 3\hat{k}$ and $3\hat{i} - 2\hat{j} + \hat{k}$.

Sol. The given vectors are $\vec{a} = \hat{i} - 2\hat{j} + 3\hat{k}$ and $\vec{b} = 3\hat{i} - 2\hat{j} + \hat{k}$.

$$|\vec{a}| = \sqrt{1^2 + (-2)^2 + 3^2} = \sqrt{1+4+9} = \sqrt{14}$$

and $\quad |\vec{b}| = \sqrt{3^2 + (-2)^2 + 1^2} = \sqrt{9+4+1} = \sqrt{14}$

Now, $\vec{a}\cdot\vec{b} = (\hat{i} - 2\hat{j} + 3\hat{k})\cdot(3\hat{i} - 2\hat{j} + \hat{k})$

$$= 1\cdot3 + (-2)(-2) + 3\cdot1$$

$$= 3 + 4 + 3$$

$$= 10$$

Thus, $\vec{a}\cdot\vec{b} = |\vec{a}||\vec{b}|\cos\theta$

or $\quad 10 = \sqrt{14}\sqrt{14}\cos\theta$

$\Rightarrow \quad \cos\theta = \dfrac{10}{14}$

$\Rightarrow \quad \theta = \cos^{-1}\left(\dfrac{5}{7}\right)$

Illustration 3.3

If $\vec{a} = \hat{i} - 2\hat{j} + 2\hat{k}$, $\vec{a}\cdot\vec{b} = -12$ and angle between vectors $\vec{a}$ and $\vec{b}$ is 120°, then find $|\vec{b}|$.

Sol. We have, $\vec{a} = \hat{i} - 2\hat{j} + 2\hat{k}$

$\therefore \quad |\vec{a}| = \sqrt{1+4+4} = \sqrt{9} = 3$

Now, $\vec{a}\cdot\vec{b} = -10$

$\therefore \quad |\vec{a}|\cdot|\vec{b}|\cos 120° = -12$

$\Rightarrow \quad 3\cdot|\vec{b}|\left(-\dfrac{1}{2}\right) = -12$

$\Rightarrow \quad |\vec{b}| = 8$

Illustration 3.4

If vectors $\vec{a} = x\hat{i} + (x-1)\hat{j} + \hat{k}$ and $\vec{b} = (x+1)\hat{i} + \hat{j} + a\hat{k}$ make an acute angle $\forall\, x \in R$, then find the values of a.

Sol. We have, $\vec{a}\cdot\vec{b} = (x\hat{i} + (x-1)\hat{j} + \hat{k})\cdot((x+1)\hat{i} + \hat{j} + a\hat{k})$

$$= x(x+1) + x - 1 + a$$

$$= x^2 + 2x + a - 1$$

Now, the angle between vectors is acute.

Then, we must have $\vec{a}\cdot\vec{b} > 0\ \forall\, x \in R$

$\Rightarrow \quad x^2 + 2x + a - 1 > 0\ \forall\, x \in R$

$\Rightarrow \quad 4 - 4(a-1) < 0 \qquad (\because \text{Discriminant} < 0)$

$\Rightarrow \quad a > 2$

Illustration 3.5

a. Using the dot product prove that for real numbers a_1, a_2, a_3 and b_1, b_2, b_3, $(a_1b_1 + a_2b_2 + a_3b_3)^2 \le (a_1^2 + a_2^2 + a_3^2)(b_1^2 + b_2^2 + b_3^2)$.

b. If $a + 2b + 3c = 4$ then, using the dot product find the least value of $a^2 + b^2 + c^2$.

Sol.

a. Consider vectors $\vec{a} = a_1\hat{i} + a_2\hat{j} + a_3\hat{k}$

and $\quad \vec{b} = b_1\hat{i} + b_2\hat{j} + b_3\hat{k}$

If the angle between vectors is θ, then

$$\cos\theta = \frac{a_1b_1 + a_2b_2 + a_3b_3}{\sqrt{a_1^2 + a_2^2 + a_3^2}\sqrt{b_1^2 + b_2^2 + b_3^2}}$$

Now, $|\cos\theta| \le 1$

$$\Rightarrow \frac{|a_1b_1 + a_2b_2 + a_3b_3|}{\sqrt{a_1^2 + a_2^2 + a_3^2}\sqrt{b_1^2 + b_2^2 + b_3^2}} \le 1$$

$$\Rightarrow (a_1b_1 + a_2b_2 + a_3b_3)^2 \le (a_1^2 + a_2^2 + a_3^2)(b_1^2 + b_2^2 + b_3^2)$$

b. Consider vectors $\vec{p} = a\hat{i} + b\hat{j} + c\hat{k}$

and $\vec{q} = \hat{i} + 2\hat{j} + 3\hat{k}$

Then, using the above result, we have

$$(a + 2b + 3c)^2 \le (1^2 + 2^2 + 3^2)(a^2 + b^2 + c^2)$$

Given that $a + 2b + 3c = 4$.

So, $a^2 + b^2 + c^2 \ge \frac{8}{7}$

Therefore, the least value of $a^2 + b^2 + c^2$ is $\frac{8}{7}$.

Illustration 3.6

If $\vec{a}, \vec{b}, \vec{c}$ are non-zero vectors such that $\vec{a}\cdot\vec{b} = \vec{a}\cdot\vec{c}$, then find the geometrical relation between the vectors.

Sol. We have, $\vec{a}\cdot\vec{b} = \vec{a}\cdot\vec{c}$

$\Rightarrow \quad \vec{a}\cdot\vec{b} - \vec{a}\cdot\vec{c} = \vec{0}$

$\Rightarrow \quad \vec{a}\cdot(\vec{b} - \vec{c}) = \vec{0}$

$\Rightarrow \quad \vec{b} - \vec{c} = \vec{0}$ or $\vec{a} \perp (\vec{b} - \vec{c})$

$\Rightarrow \quad \vec{b} = \vec{c}$ or $\vec{a} \perp (\vec{b} - \vec{c})$

Illustration 3.7

If $\vec{r}\cdot\hat{i} = \vec{r}\cdot\hat{j} = \vec{r}\cdot\hat{k}$ and $|\vec{r}| = 3$, then find vector $\vec{r}$

Sol. We know that $\vec{r} = (\vec{r}\cdot\hat{i})\hat{i} + (\vec{r}\cdot\hat{j})\hat{j} + (\vec{r}\cdot\hat{k})\hat{k}$

Given that $\vec{r}\cdot\hat{i} = \vec{r}\cdot\hat{j} = \vec{r}\cdot\hat{k} = \lambda$ (say)

$\Rightarrow \quad \vec{r} = \lambda\hat{i} + \lambda\hat{j} + \lambda\hat{k}$

Now given that $|\vec{r}| = 3$

$\Rightarrow \quad \sqrt{\lambda^2 + \lambda^2 + \lambda^2} = 3$

$\Rightarrow \quad \lambda = \pm\sqrt{3}$

Hence, the required vectors are $\vec{r} = \pm\sqrt{3}(\hat{i} + \hat{j} + \hat{k})$

Illustration 3.8

If $\vec{a}\cdot\hat{i} = \vec{a}\cdot(\hat{i} + \hat{j}) = \vec{a}\cdot(\hat{i} + \hat{j} + \hat{k})$, then find the unit vector $\vec{a}$.

Sol. Let $\vec{a} = x\hat{i} + y\hat{j} + z\hat{k}$

Then, $\vec{a}\cdot\hat{i} = (x\hat{i} + y\hat{j} + z\hat{k})\cdot\hat{i} = x$, $\vec{a}\cdot(\hat{i} + \hat{j}) = x + y$

and $\vec{a}\cdot(\hat{i} + \hat{j} + \hat{k}) = x + y + z$

Now, $x = x + y \Rightarrow y = 0$ and $x + y = x + y + z \Rightarrow z = 0$

Hence, $x = 1$ (since $\vec{a}$ is a unit vector)

Therefore, $\vec{a} = \hat{i}$

Illustration 3.9

If $\vec{a}, \vec{b}, \vec{c}$ are mutually perpendicular vectors of equal magnitudes, show that the vector $\vec{a} + \vec{b} + \vec{c}$ is equally inclined to $\vec{a}, \vec{b}$ and $\vec{c}$.

Sol. Since $\vec{a}, \vec{b}$, and $\vec{c}$ are mutually perpendicular vectors, we have $\vec{a}\cdot\vec{b} = \vec{b}\cdot\vec{c} = \vec{c}\cdot\vec{a} = 0$.

It is given that

$$|\vec{a}| = |\vec{b}| = |\vec{c}|.$$

Let the vector $\vec{a} + \vec{b} + \vec{c}$ be inclined to $\vec{a}, \vec{b}$ and $\vec{c}$ at angles θ_1, θ_2 and θ_3, respectively. Then, we have

$$\cos\theta_1 = \frac{(\vec{a} + \vec{b} + \vec{c})\cdot\vec{a}}{|\vec{a} + \vec{b} + \vec{c}||\vec{a}|} = \frac{\vec{a}\cdot\vec{a} + \vec{b}\cdot\vec{a} + \vec{c}\cdot\vec{a}}{|\vec{a} + \vec{b} + \vec{c}||\vec{a}|}$$

$$= \frac{|\vec{a}|^2}{|\vec{a} + \vec{b} + \vec{c}||\vec{a}|} [\vec{b}\cdot\vec{a} = \vec{c}\cdot\vec{a} = 0]$$

$$= \frac{|\vec{a}|}{|\vec{a} + \vec{b} + \vec{c}|}$$

Similarly, $\cos\theta_2 = \frac{|\vec{b}|}{|\vec{a} + \vec{b} + \vec{c}|}$ and $\cos\theta_3 = \frac{|\vec{c}|}{|\vec{a} + \vec{b} + \vec{c}|}$

Now, $|\vec{a}| = |\vec{b}| = |\vec{c}|$. Therefore, $\cos\theta_1 = \cos\theta_2 = \cos\theta_3$

or $\quad \theta_1 = \theta_2 = \theta_3$

Hence, the vector $(\vec{a} + \vec{b} + \vec{c})$ is equally inclined to $\vec{a}, \vec{b}$ and $\vec{c}$.

Illustration 3.10

Vectors $\vec{a}$, $\vec{b}$ and $\vec{c}$ are of the same length and when taken pair-wise they form equal angles. If $\vec{a} = \hat{i} + \hat{j}$ and $\vec{b} = \hat{j} + \hat{k}$, then find vector $\vec{c}$.

Sol. Let $\vec{c} = x\hat{i} + y\hat{j} + z\hat{k}$. Since $|\vec{a}| = |\vec{b}| = |\vec{c}|$, we have $x^2 + y^2 + z^2 = 2$.

It is given that the angles between the vectors taken in pairs are equal, say θ.

Then, $\cos\theta = \frac{\vec{a}\cdot\vec{b}}{|\vec{a}||\vec{b}|} = \frac{0+1+0}{\sqrt{2}\sqrt{2}} = \frac{1}{2}$

Therefore, $\frac{\vec{a}\cdot\vec{c}}{|\vec{a}||\vec{c}|}=\frac{1}{2}$ and $\frac{\vec{b}\cdot\vec{c}}{|\vec{b}||\vec{c}|}=\frac{1}{2}$

$\Rightarrow \quad \frac{x+y}{\sqrt{2}\sqrt{2}}=\frac{1}{2}$ and $\frac{y+z}{\sqrt{2}\sqrt{2}}=\frac{1}{2}$

$\Rightarrow \quad x+y=1$ and $y+z=1$

$\Rightarrow \quad y=1-x$ and $z=1-y=1-(1-x)=x$

Also $x^2+y^2+z^2=2 \Rightarrow x^2+(1-x)^2+x^2=2$

$\Rightarrow \quad (3x+1)(x-1)=0 \Rightarrow x=1, -1/3$

Now, $y=1-x \Rightarrow y=0$ for $x=1$ and $y=4/3$ for $x=-1/3$

Hence, $\vec{c}=\hat{i}+0\hat{j}+\hat{k}$ or $\vec{c}=-\frac{1}{3}\hat{i}+\frac{4}{3}\hat{j}-\frac{1}{3}\hat{k}$

Illustration 3.11

Let the plane $x+3y-2z+6=0$ meet the co–ordinate axes at the points A, B, C. If the orthocentre of the triangle ABC is $\left(\alpha, \beta, \frac{6}{7}\right)$, then find the value of α and β.

Sol. We have $A(-6, 0, 0)$, $B(0, -2, 0)$, $C=(0, 0, 3)$

$\therefore \quad \overrightarrow{AB}=6\hat{i}-2\hat{j},\ \overrightarrow{BC}=2\hat{j}+3\hat{k},\ \overrightarrow{AC}=6\hat{i}+3\hat{k}$

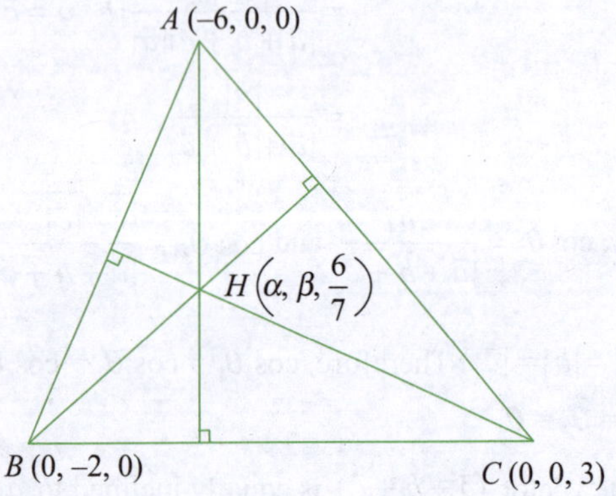

Now, $\overrightarrow{AH}\cdot\overrightarrow{BC}=0$

$\Rightarrow \quad 2\beta+\frac{18}{7}=0 \Rightarrow \beta=\frac{-9}{7}$

Also, $\overrightarrow{CH}\cdot\overrightarrow{AB}=0$

$\Rightarrow \quad 6\alpha-2\beta=0$

$\Rightarrow \quad \alpha=\frac{-3}{7}$

Illustration 3.12

Find a unit vector which makes an angle of 60° with $2\hat{i}+2\hat{j}-\hat{k}$ and an angle of 45° with $\hat{i}-\hat{k}$.

Sol. Let the required vector be $\vec{a}=x\hat{i}+y\hat{j}+z\hat{k}$

Since $|\vec{a}|=1$, we have $x^2+y^2+z^2=1$ …(i)

Now, $\vec{a}$ makes an angle 60° with $2\hat{i}+2\hat{j}-\hat{k}$.

$\therefore \quad \vec{a}\cdot(2\hat{i}+2\hat{j}-\hat{k})=|\vec{a}|(3)\cos 60°$

$\Rightarrow \quad 2x+2y-z=\frac{3}{2}$ …(ii)

Also, $\vec{a}$ makes an angle 45° with $\hat{i}-\hat{k}$.

$\therefore \quad \vec{a}\cdot(\hat{i}-\hat{k})=|\vec{a}|(\sqrt{2})\cos 45°$

$\Rightarrow \quad x-z=1$ …(iii)

From (ii) – (iii), we get $x+2y=\frac{1}{2}$ or $x=\frac{1}{2}-2y$.

So, from (iii), we get $z=-\frac{1}{2}-2y$

Putting the expressions for x and z in terms of y in (i), we get

$$\left(\frac{1}{2}-2y\right)^2+y^2+\left(-\frac{1}{2}-2y\right)^2=1$$

$\Rightarrow \quad 9y^2=\frac{1}{2}$ or $y=\pm\frac{1}{3\sqrt{2}}$

Hence, the corresponding values of x and z are $\frac{1}{2}\mp\frac{\sqrt{2}}{3}$ and $-\frac{1}{2}\mp\frac{\sqrt{2}}{3}$.

Therefore, the required vectors are

$$\left(\frac{1}{2}+\frac{\sqrt{2}}{3}\right)\hat{i}-\frac{1}{3\sqrt{2}}\hat{j}+\left(-\frac{1}{2}+\frac{\sqrt{2}}{3}\right)\hat{k}$$ and

$$\left(\frac{1}{2}-\frac{\sqrt{2}}{3}\right)\hat{i}+\frac{1}{3\sqrt{2}}\hat{j}+\left(-\frac{1}{2}-\frac{\sqrt{2}}{3}\right)\hat{k}.$$

Illustration 3.13

If $\vec{a}=-\hat{i}+\hat{j}+\hat{k}$ and $\vec{b}=2\hat{i}+0\hat{j}+\hat{k}$, then find vector $\vec{c}$ such that: (a) it is coplanar with $\vec{a}$ and $\vec{b}$, (b) it is ⊥ to $\vec{b}$ and (c) $\vec{a}\cdot\vec{c}=7$.

Sol. Let $\vec{c}=x\hat{i}+y\hat{j}+z\hat{k}$

Then from the condition (a),

$$\begin{vmatrix} x & y & z \\ -1 & 1 & 1 \\ 2 & 0 & 1 \end{vmatrix}=0 \text{ or } x+3y-2z=0 \quad \text{…(i)}$$

From the condition (b),

$2x+z=0$ …(ii)

From the condition (c),

$-x+y+z=7$ …(iii)

Solving (i), (ii) and (iii), we get the values of x, y and z,

Hence, $\vec{c}=\frac{1}{2}(-3\hat{i}+5\hat{j}+6\hat{k})$

Illustration 3.14

Vector $\overrightarrow{OA} = \hat{i}+2\hat{j}+2\hat{k}$ turns through a right angle passing through the positive x-axis on the way. Show that the vector in its new position is $\frac{4\hat{i}-\hat{j}-\hat{k}}{\sqrt{2}}$.

Sol. Let the vector in the new position be $\overrightarrow{OB} = x\hat{i}+y\hat{j}+z\hat{k}$. According to the given condition, we have

$$|\overrightarrow{OB}| = |\overrightarrow{OA}| \Rightarrow x^2+y^2+z^2=9 \quad \text{...(i)}$$

Also $\overrightarrow{OA} \perp \overrightarrow{OB} \Rightarrow \overrightarrow{OA}\cdot\overrightarrow{OB}=0 \Rightarrow x+2y+2z=0 \quad \text{...(ii)}$

Since while turning $\overrightarrow{OA}$, it passes through the positive x-axis on the way, vectors $\overrightarrow{OA}$, $\overrightarrow{OB}$ and $\lambda\hat{i}$ are coplanar. Thus,

$$\begin{vmatrix} x & y & z \\ 1 & 2 & 2 \\ \lambda & 0 & 0 \end{vmatrix} = 0$$

or $y - z = 0 \quad \text{...(iii)}$

Solving (i), (ii) and (iii) for x, y and z, we get $x = -4y$ and $y = z$.

$\Rightarrow \quad 16y^2+y^2+y^2=9 \quad$ (from (i))

$\Rightarrow \quad y = \pm\frac{1}{\sqrt{2}}, z = \pm\frac{1}{\sqrt{2}}$ and $x = \mp\frac{1}{\sqrt{2}}$

$\Rightarrow \quad \overrightarrow{OB} = \pm\left(\frac{4}{\sqrt{2}}\hat{i}-\frac{1}{\sqrt{2}}\hat{j}-\frac{1}{\sqrt{2}}\hat{k}\right)$

Since angle between $\overrightarrow{OB}$ and $\hat{i}$ is acute,

$$\overrightarrow{OB} = \frac{4}{\sqrt{2}}\hat{i}-\frac{1}{\sqrt{2}}\hat{j}-\frac{1}{\sqrt{2}}\hat{k}$$

3.2.3 Component of One Vector in the Direction of Another

In the following figure $\overline{OA} = \vec{a}$ and $\overline{OB} = \vec{b}$.

Angle between $\overline{OA}$ and $\overline{OB}$ is θ.

The foot of the perpendicular from B to OA is M.

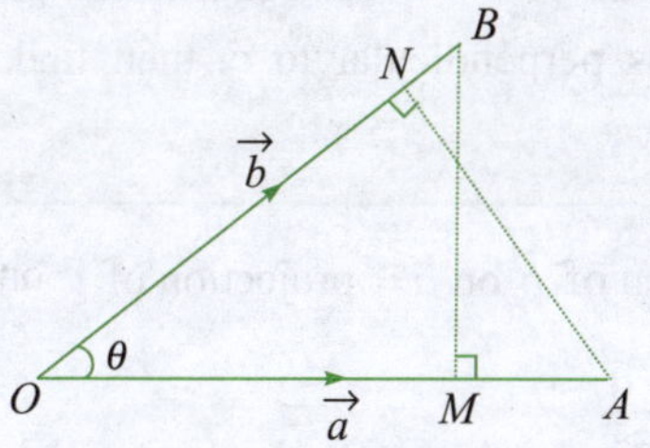

Here, OM is called the scalar projection of vector $\vec{b}$ in the direction of vector $\vec{a}$.

Now, $OM = OB\cos\theta = |\vec{b}|\cos\theta = |\vec{b}|\frac{\vec{a}\cdot\vec{b}}{|\vec{a}||\vec{b}|}$

$$= \frac{\vec{a}\cdot\vec{b}}{|\vec{a}|} = \left(\frac{\vec{a}}{|\vec{a}|}\right)\cdot\vec{b}$$

$\therefore \quad OM = \hat{a}\cdot\vec{b}$

Similarly, the scalar projection of vector $\vec{a}$ in the direction of vector $\vec{b}$ is $\hat{b}\cdot\vec{a}\,(=ON)$

Now, $\overline{OM} = (OM)\frac{\vec{a}}{|\vec{a}|} = \left(\frac{\vec{a}\cdot\vec{b}}{|\vec{a}|}\right)\frac{\vec{a}}{|\vec{a}|} = \left(\frac{\vec{a}\cdot\vec{b}}{|\vec{a}|^2}\right)\vec{a}$

Also, $\vec{b} = \overline{OM} + \overline{MB}$

$\Rightarrow \quad \overline{MB} = \vec{b} - \overline{OM} = \vec{b} - \left(\frac{\vec{a}\cdot\vec{b}}{|\vec{a}|^2}\right)\vec{a}$

Thus, the components of $\vec{b}$ along $\vec{a}$ and along the direction perpendicular to $\vec{a}$ are $\left(\frac{\vec{a}\cdot\vec{b}}{|\vec{a}|^2}\right)\vec{a}$ and $\vec{b} - \left(\frac{\vec{a}\cdot\vec{b}}{|\vec{a}|^2}\right)\vec{a}$ respectively.

Similarly, the components of $\vec{a}$ along $\vec{b}$ and along the direction perpendicular to $\vec{b}$ are $\left(\frac{\vec{a}\cdot\vec{b}}{|\vec{b}|^2}\right)\vec{b}$ and $\vec{a} - \left(\frac{\vec{a}\cdot\vec{b}}{|\vec{b}|^2}\right)\vec{b}$ respectively.

Illustration 3.15

Find the projection of the line segment joining the points (1, –1, 3) and (2, –4, 11) on the line joining the points (–1, 2, 3) and (3, –2, 10).

Sol. The given points are $P(1, -1, 3)$, $Q(2, -4, 11)$, $R(-1, 2, 3)$ and $S(3, -2, 10)$.

Then $\overline{PQ} = \hat{i}-3\hat{j}+8\hat{k}$ and $\overline{RS} = 4\hat{i}-4\hat{j}+7\hat{k}$

Therefore, the projection of $\overline{PQ}$ on $\overline{RS}$ is

$$= \frac{|\overline{PQ}\cdot\overline{RS}|}{|\overline{RS}|}$$

$$= \frac{|(\hat{i}-3\hat{j}+8\hat{k})\cdot(4\hat{i}-4\hat{j}+7\hat{k})|}{9}$$

$$= \frac{4+12+56}{9} = 8$$

Illustration 3.16

If the scalar projection of vector $x\hat{i}-\hat{j}+\hat{k}$ on the vector $2\hat{i}-\hat{j}+5\hat{k}$ is $\frac{1}{\sqrt{30}}$, then find the value of x.

Sol. Projection of $x\hat{i}-\hat{j}+\hat{k}$ on $2\hat{i}-\hat{j}+5\hat{k}$

$$=\frac{(x\hat{i}-\hat{j}+\hat{k})\cdot(2\hat{i}-\hat{j}+5\hat{k})}{\sqrt{4+1+25}}$$

$$=\frac{2x+1+5}{\sqrt{30}}$$

But, given $\frac{2x+6}{\sqrt{30}}=\frac{1}{\sqrt{30}}$ $\Rightarrow 2x+6=1$ or $x=\frac{-5}{2}$

Illustration 3.17

If $\vec{a}=4\hat{i}+6\hat{j}$ and $\vec{b}=3\hat{j}+4\hat{k}$, then find the component of $\vec{a}$ along $\vec{b}$ and the component of $\vec{a}$ perpendicular to $\vec{b}$.

Sol. The component of vector $\vec{a}$ along $\vec{b}$ is

$$\frac{(\vec{a}\cdot\vec{b})\vec{b}}{|\vec{b}|^2}=\frac{18}{25}(3\hat{j}+4\hat{k})$$

The component of vector $\vec{a}$ in the direction perpendicular to $\vec{b}$ is

$$\vec{a}-\frac{(\vec{a}\cdot\vec{b})\vec{b}}{|\vec{b}|^2}=4\hat{i}+6\hat{j}-\frac{18}{25}(3\hat{j}+4\hat{k})=4\hat{i}+\frac{96}{25}\hat{j}-\frac{72}{25}\hat{k}$$

Illustration 3.18

Let $\vec{a}=\hat{i}+\hat{j}+2\hat{k}$, $\vec{b}=2\hat{i}-3\hat{j}+\hat{k}$ and $\vec{c}=\hat{i}-\hat{j}+\hat{k}$ be three given vectors. Find vector $\vec{v}$ in the plane of $\vec{a}$ and $\vec{b}$ whose projection on $\vec{c}$ is $\frac{2}{\sqrt{3}}$ and $\vec{v}.\hat{j}=7$.

Sol. Since vector $\vec{v}$ lies in the plane of vectors $\vec{a}$ and $\vec{b}$, we have, $\vec{v}=\lambda\vec{a}+\mu\vec{b}$.

Now, the projection of $\vec{v}$ on $\vec{c}$ is $\frac{2}{\sqrt{3}}$

$$\therefore \quad \frac{\vec{v}\cdot\vec{c}}{|\vec{c}|}=\frac{2}{\sqrt{3}}$$

$$\Rightarrow \quad \frac{\lambda\vec{a}\cdot\vec{c}+\mu\vec{b}\cdot\vec{c}}{\sqrt{3}}=\frac{2}{\sqrt{3}}$$

$\Rightarrow \quad 2\lambda+6\mu=2$

$\Rightarrow \quad \lambda+3\mu=1$ ...(i)

Also, $\vec{v}\cdot\hat{j}=7$

$\therefore \quad \lambda\vec{a}\cdot\hat{j}+\mu\vec{b}\cdot\hat{j}=7$

$\Rightarrow \quad \lambda-3\mu=7$ ...(ii)

Solving (i) and (ii), we get $\lambda=4$, $\mu=-1$

$\therefore \quad \vec{v}=2\hat{i}+7\hat{j}+7\hat{k}$

3.2.4 Square of the Sum of Vectors

We have the following results.

1. $(\vec{a}+\vec{b})^2=a^2+b^2+2ab\cos\theta=|\vec{a}+\vec{b}|^2$, where $a=|\vec{a}|$ and $b=|\vec{b}|$.

Proof:

$(\vec{a}+\vec{b})^2=(\vec{a}+\vec{b})\cdot(\vec{a}+\vec{b})$

$=\vec{a}\cdot(\vec{a}+\vec{b})+\vec{b}\cdot(\vec{a}+\vec{b})$ (Distributive property)

$=\vec{a}\cdot\vec{a}+\vec{a}\cdot\vec{b}+\vec{b}\cdot\vec{a}+\vec{b}\cdot\vec{b}$

$=a^2+b^2+2\vec{a}\cdot\vec{b}$ $(\because \vec{a}\cdot\vec{b}=\vec{b}\cdot\vec{a})$

$=a^2+b^2+2ab\cos\theta$

Also, $(\vec{a}+\vec{b})^2=(\vec{a}+\vec{b})\cdot(\vec{a}+\vec{b})$

$=|\vec{a}+\vec{b}||\vec{a}+\vec{b}|\cos 0°=|\vec{a}+\vec{b}|^2$

Similarly, $(\vec{a}-\vec{b})^2=|\vec{a}-\vec{b}|^2=a^2+b^2-2ab\cos\theta$

Also, $(\vec{a}+\vec{b}+\vec{c})^2=|\vec{a}+\vec{b}+\vec{c}|^2$

$=a^2+b^2+c^2+2\vec{a}\cdot\vec{b}+2\vec{b}\cdot\vec{c}+2\vec{a}\cdot\vec{c}$

2. $(\vec{a}+\vec{b})\cdot(\vec{a}-\vec{b})=\vec{a}^2-\vec{b}^2$

Proof:

$(\vec{a}+\vec{b})\cdot(\vec{a}-\vec{b})=\vec{a}\cdot(\vec{a}+\vec{b})-\vec{b}\cdot(\vec{a}+\vec{b})$

$=\vec{a}\cdot\vec{a}+\vec{a}\cdot\vec{b}-\vec{b}\cdot\vec{a}-\vec{b}\cdot\vec{b}$

$=\vec{a}^2-\vec{b}^2$ $(\because \vec{a}\cdot\vec{b}=\vec{b}\cdot\vec{a})$

Illustration 3.19

If $|\vec{a}|=|\vec{b}|=|\vec{a}+\vec{b}|=1$, then find the value of $|\vec{a}-\vec{b}|$.

Sol. We have,

$|\vec{a}+\vec{b}|^2+|\vec{a}-\vec{b}|^2=(a^2+b^2+2\vec{a}\cdot\vec{b})+(a^2+b^2-2\vec{a}\cdot\vec{b})$

$=2(a^2+b^2)$

$\Rightarrow \quad 1+|\vec{a}-\vec{b}|^2=4$

$\Rightarrow \quad |\vec{a}-\vec{b}|=\sqrt{3}$

Illustration 3.20

Let the vectors $\vec{a},\vec{b},\vec{c}$ be such that $|\vec{a}|=2$, $|\vec{b}|=4$ and $|\vec{c}|=4$. If the projection of $\vec{b}$ on $\vec{a}$ is equal to the projection of $\vec{c}$ on $\vec{a}$ and $\vec{b}$ is perpendicular to $\vec{c}$, then find the value of $|\vec{a}+\vec{b}-\vec{c}|$.

Sol. Projection of $\vec{b}$ on $\vec{a}=$ projection of $\vec{c}$ on $\vec{a}$

$$\therefore \quad \frac{\vec{b}\cdot\vec{a}}{|\vec{a}|}=\frac{\vec{c}\cdot\vec{a}}{|\vec{a}|}$$

$\Rightarrow \quad \vec{b}\cdot\vec{a} = \vec{c}\cdot\vec{a}$

$\vec{b}$ is perpendicular to $\vec{c}$

$\Rightarrow \quad \vec{b}\cdot\vec{c} = 0$

Now, $|\vec{a}+\vec{b}-\vec{c}|^2$

$= \vec{a}^2 + \vec{b}^2 + \vec{c}^2 + 2\vec{a}\cdot\vec{b} - 2\vec{a}\cdot\vec{c} - 2\vec{b}\cdot\vec{c}$

$= \vec{a}^2 + \vec{b}^2 + \vec{c}^2 = 36$

$|\vec{a}+\vec{b}-\vec{c}| = 6$

Illustration 3.21

If $|\vec{a}|+|\vec{b}| = |\vec{c}|$ and $\vec{a}+\vec{b} = \vec{c}$, then find the angle between a and $\vec{b}$.

Sol. We have $\vec{a}+\vec{b} = \vec{c}$

$\therefore \quad |\vec{a}+\vec{b}|^2 = |\vec{c}|^2$

$\Rightarrow \quad |\vec{a}|^2 + |\vec{b}|^2 + 2\vec{a}\cdot\vec{b} = |\vec{c}|^2 \qquad \text{...(i)}$

Given that $|\vec{a}|+|\vec{b}| = |\vec{c}|$

$\therefore \quad |\vec{a}|^2 + |\vec{b}|^2 + 2|\vec{a}||\vec{b}| = |\vec{c}|^2 \qquad \text{...(ii)}$

$\Rightarrow \quad \vec{a}\cdot\vec{b} = |\vec{a}||\vec{b}| \qquad$ [from (i) and (ii)]

$\Rightarrow \quad \cos\theta = 1$ or $\theta = 0°$

Illustration 3.22

If three unit vectors $\vec{a}, \vec{b}$ and $\vec{c}$ satisfy $\vec{a}+\vec{b}+\vec{c} = \vec{0}$, then find the angle between $\vec{a}$ and $\vec{b}$.

Sol. $\vec{a}+\vec{b} = -\vec{c}$

$\Rightarrow \quad |\vec{a}+\vec{b}|^2 = |\vec{c}|^2 = 1 \qquad (\because |\vec{c}| = 1)$

$\Rightarrow \quad |\vec{a}|^2 + |\vec{b}|^2 + 2\vec{a}\cdot\vec{b} = 1$

$\Rightarrow \quad \vec{a}\cdot\vec{b} = -\dfrac{1}{2} \qquad (\because |\vec{a}| = |\vec{b}| = 1)$

$\Rightarrow \quad |\vec{a}||\vec{b}|\cos\theta = -\dfrac{1}{2}$

$\Rightarrow \quad \cos\theta = -\dfrac{1}{2}$

$\Rightarrow \quad \theta = \dfrac{2\pi}{3}$

Illustration 3.23

If θ is the angle between two unit vectors $\vec{a}$ and $\vec{b}$, then prove that

a. $\cos\dfrac{\theta}{2} = \dfrac{1}{2}|\vec{a}+\vec{b}|$

b. $\sin\dfrac{\theta}{2} = \dfrac{1}{2}|\vec{a}-\vec{b}|$

Sol.

a. $(\vec{a}+\vec{b})\cdot(\vec{a}+\vec{b}) = |\vec{a}|^2 + |\vec{b}|^2 + 2\vec{a}\cdot\vec{b}$

$= 1 + 1 + 2(1)(1)\cos\theta$

$= 2 + 2\cos\theta$

$= 2(1+\cos\theta)$

$\Rightarrow \quad |\vec{a}+\vec{b}|^2 = 2\cdot 2\cos^2\dfrac{\theta}{2}$

or $\quad \cos\dfrac{\theta}{2} = \dfrac{1}{2}|\vec{a}+\vec{b}|$

b. $(\vec{a}-\vec{b})\cdot(\vec{a}-\vec{b}) = |\vec{a}|^2 + |\vec{b}|^2 - 2\vec{a}\cdot\vec{b}$

$= 1 + 1 - 2(1)(1)\cos\theta$

$= 2 - 2\cos\theta$

$= 2(1-\cos\theta)$

$\Rightarrow \quad |\vec{a}-\vec{b}|^2 = 2\cdot 2\sin^2\dfrac{\theta}{2}$

or $\quad \sin\dfrac{\theta}{2} = \dfrac{1}{2}|\vec{a}-\vec{b}|$

Illustration 3.24

If $\vec{a}$ and $\vec{b}$ are unit vectors, then find the greatest value of $\sqrt{3}|\vec{a}+\vec{b}| + |\vec{a}-\vec{b}|$.

Sol. We have, $\sqrt{3}|\vec{a}+\vec{b}| + |\vec{a}-\vec{b}|$

$= 2\sqrt{3}\left|\cos\dfrac{\theta}{2}\right| + 2\left|\sin\dfrac{\theta}{2}\right|$

$\le \sqrt{(2\sqrt{3})^2 + (2)^2} = 4$

Therefore, the required maximum value is 4.

Illustration 3.25

If $\vec{a}, \vec{b}$ and $\vec{c}$ are three mutually perpendicular unit vectors and $\vec{d}$ is a unit vector which makes equal angles with $\vec{a}, \vec{b}$ and $\vec{c}$, then find the value of $|\vec{a}+\vec{b}+\vec{c}+\vec{d}|^2$.

Sol. We have, $|\vec{a}+\vec{b}+\vec{c}+\vec{d}|^2 = \Sigma|\vec{a}|^2 + 2\Sigma\vec{a}\cdot\vec{b}$

$= 4 + 2\vec{d}\cdot(\vec{a}+\vec{b}+\vec{c}) \quad (\because \vec{a}, \vec{b}, \vec{c}$ are mutually perpendicular).

Let $\vec{d} = \lambda\vec{a} + \mu\vec{b} + \nu\vec{c}$. Then $\vec{d}\cdot\vec{a} = \vec{d}\cdot\vec{b} = \vec{d}\cdot\vec{c} = \cos\theta$. Therefore,

$\lambda = \mu = \nu = \cos\theta$

Also $\lambda^2 + \mu^2 + \nu^2 = 1 \Rightarrow 3\cos^2\theta = 1$ or $\cos\theta = \pm\dfrac{1}{\sqrt{3}}$

$\therefore \quad |\vec{a}+\vec{b}+\vec{c}+\vec{d}|^2 = 4 \pm \dfrac{2\cdot 3}{\sqrt{3}} = 4 \pm 2\sqrt{3}$

Illustration 3.26

Let $\vec{a}$, $\vec{b}$ and $\vec{c}$ are vectors such that $|\vec{a}| = 3$, $|\vec{b}| = 4$ and $|\vec{c}| = 5$, and $(\vec{a}+\vec{b})$ is perpendicular to $\vec{c}$, $(\vec{b}+\vec{c})$ is perpendicular to $\vec{a}$ and $(\vec{c}+\vec{a})$ is perpendicular to $\vec{b}$. Then find the value of $|\vec{a}+\vec{b}+\vec{c}|$.

Sol. Given, $(\vec{a}+\vec{b})\cdot\vec{c} = 0 \Rightarrow \vec{a}\cdot\vec{c}+\vec{b}\cdot\vec{c} = 0$

$(\vec{b}+\vec{c})\cdot\vec{a} = 0 \Rightarrow \vec{a}\cdot\vec{b}+\vec{c}\cdot\vec{a} = 0$

$(\vec{c}+\vec{a})\cdot\vec{b} = 0 \Rightarrow \vec{b}\cdot\vec{c}+\vec{a}\cdot\vec{b} = 0$

$\therefore \quad 2(\vec{a}\cdot\vec{b}+\vec{b}\cdot\vec{c}+\vec{c}\cdot\vec{a}) = 0$

Now,

$$|\vec{a}+\vec{b}+\vec{c}|^2 = |\vec{a}|^2+|\vec{b}|^2+|\vec{c}|^2+2(\vec{a}\cdot\vec{b}+\vec{b}\cdot\vec{c}+\vec{c}\cdot\vec{a})$$
$$= 3^2+4^2+5^2 = 50$$

$\Rightarrow \quad |\vec{a}+\vec{b}+\vec{c}| = 5\sqrt{2}$

Illustration 3.27

In an isosceles triangle ABC, $|\overrightarrow{AB}| = |\overrightarrow{BC}| = 8$, a point E divides AB internally in the ratio 1 : 3. Then find the angle between $\overrightarrow{CE}$ and $\overrightarrow{CA}$, where $|\overrightarrow{CA}| = 12$.

Sol. Given, $|\vec{c}| = 12$ and $|\vec{b}| = |\vec{b}-\vec{c}| = 8$

$\Rightarrow \quad b^2 = b^2+c^2-2\,\vec{b}\cdot\vec{c}$

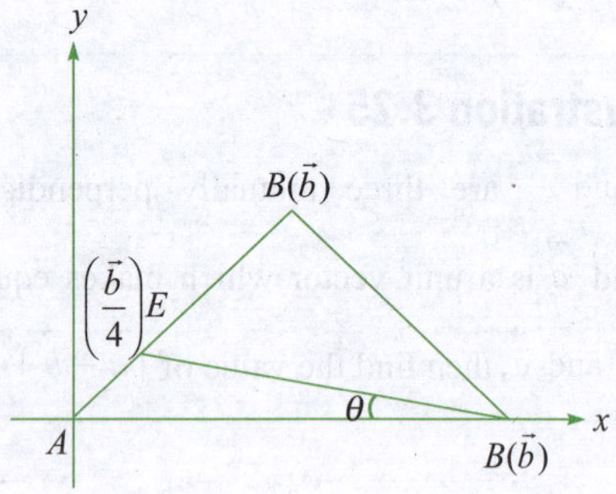

$\Rightarrow \quad \vec{b}\cdot\vec{c} = 72$

E divides AB in the ratio 1:3.

So, P.V. of E is $\dfrac{\vec{b}}{4}$.

$$\therefore \quad \cos\theta = \frac{\vec{c}\cdot\left(\vec{c}-\dfrac{\vec{b}}{4}\right)}{|\vec{c}|\left|\vec{c}-\dfrac{\vec{b}}{4}\right|} = \frac{\vec{c}\cdot\vec{c}-\dfrac{\vec{c}\cdot\vec{b}}{4}}{12\left|\vec{c}-\dfrac{\vec{b}}{4}\right|} = \frac{144-18}{12\left|\vec{c}-\dfrac{\vec{b}}{4}\right|}$$

Now $\left|\vec{c}-\dfrac{\vec{b}}{4}\right|^2 = |\vec{c}|^2+\dfrac{|\vec{b}|^2}{16}-\dfrac{\vec{b}\cdot\vec{c}}{2} = 144+4-36 = 112$

$$\Rightarrow \quad \cos\theta = \frac{21}{2\times\sqrt{112}} = \frac{21}{2\times4\sqrt{7}} = \frac{3\sqrt{7}}{8}$$

3.2.5 Applications of Dot Product in Geometry

The dot product of vectors is a powerful tool in geometry, as it helps both to prove new results and to verify known ones. Since it directly relates vectors to the angle between them, it offers a simple and systematic method for establishing important geometric properties. In this way, the dot product provides a natural bridge between algebra and geometry. In the following illustrations, we shall see how it can be applied to problems involving lines, triangles, and quadrilaterals."

Illustration 3.28

Prove by vector method that cos $(A + B)$ = cos A cos B – sin A sin B.

Sol. Let $\hat{i}$ and $\hat{j}$ be unit vectors along OX and OY, respectively.

Let $\overrightarrow{OP}, \overrightarrow{OQ}$ be two unit vectors drawn in the plane XOY such that

$\angle XOP = A,\ \angle XOQ = B$

$\therefore \quad \angle POQ = A + B$

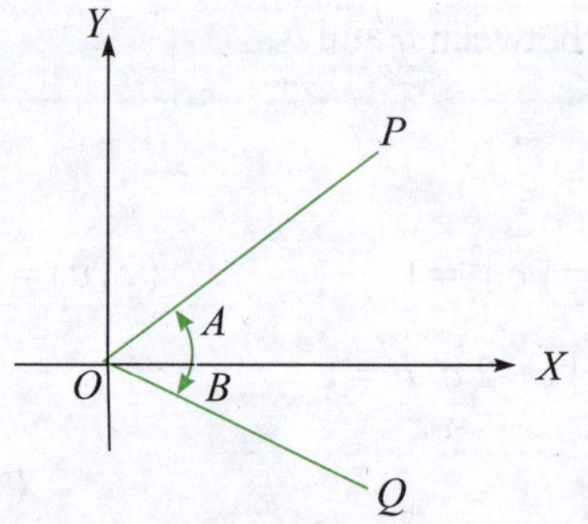

Now, $\overrightarrow{OP} = \hat{i}\cos A + \hat{j}\sin A$

$\overrightarrow{OQ} = \hat{i}\cos B - \hat{j}\sin B$

$\therefore \quad \overrightarrow{OP}\cdot\overrightarrow{OQ} = \cos A\cos B - \sin A\sin B$

$\Rightarrow \quad (1)(1)\cos(A+B) = \cos A\cos B - \sin A\sin B$

$\Rightarrow \quad \cos(A+B) = \cos A\cos B - \sin A\sin B$

Illustration 3.29

Using the vector method, prove the cosine rule in a triangle.

Sol. Let us prove that in triangle ABC, $a^2 = b^2+c^2-2bc\cos A$, where a, b, c are sides opposite to the angles A, B and C.

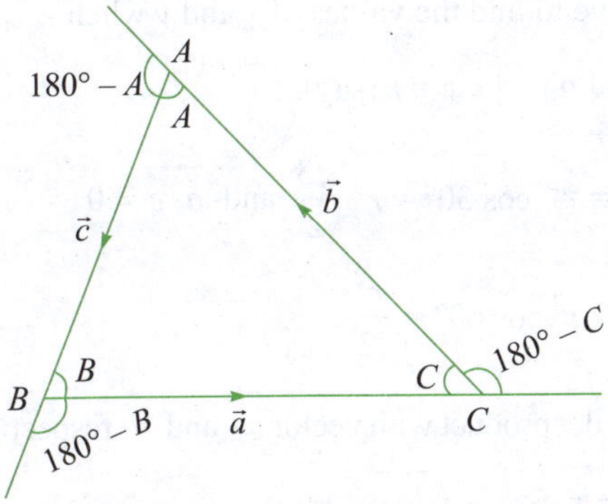

In a $\Delta\, ABC$,

Let, $\overrightarrow{AB} = \vec{c}, \overrightarrow{BC} = \vec{a}, \overrightarrow{CA} = \vec{b}$,

Since, $\vec{a}+\vec{b}+\vec{c} = 0$, we have $\vec{b}+\vec{c} = -\vec{a}$

$\Rightarrow \quad (\vec{b}+\vec{c})^2 = (-\vec{a})^2$

$\Rightarrow \quad b^2 + c^2 + 2\vec{b}.\vec{c} = a^2$

$\Rightarrow \quad b^2 + c^2 + 2bc\cos(180° - A) = a^2$

$\Rightarrow \quad a^2 = b^2 + c^2 - 2bc\cos A$

Illustration 3.30

Using the dot product of vectors prove that a parallelogram, whose diagonals are equal in length, is a rectangle.

Sol. Let the vectors along the adjacent sides of a parallelogram are $\vec{a}$ and $\vec{b}$.

Also, let the angle between $\vec{a}$ and $\vec{b}$ is θ.

The vectors along the diagonals are $(\vec{a}+\vec{b})$ and $(\vec{a}-\vec{b})$.

If the length of the diagonals are equal, we have

$|\vec{a}+\vec{b}| = |\vec{a}-\vec{b}|$

$\Rightarrow \quad |\vec{a}+\vec{b}|^2 = |\vec{a}-\vec{b}|^2$

$\Rightarrow \quad (\vec{a}+\vec{b})^2 = (\vec{a}-\vec{b})^2$

$\Rightarrow \quad a^2 + b^2 + 2ab\cos\theta = a^2 + b^2 - 2ab\cos\theta$

$\Rightarrow \quad 4ab\cos\theta = 0$

$\Rightarrow \quad \cos\theta = 0$

$\Rightarrow \quad \theta = 90°$

So, $\vec{a} \perp \vec{b}$.

Therefore, the parallelogram is a rectangle.

Illustration 3.31

In the triangle *ABC*, using the vector method prove the projection formula that is $a = b\cos C + c\cos B$.

Sol. Let $\vec{BC} = \vec{a}, \vec{CA} = \vec{b}, \vec{AB} = \vec{c}$, so that

$BC = a,\ CA = b,\ AB = c$

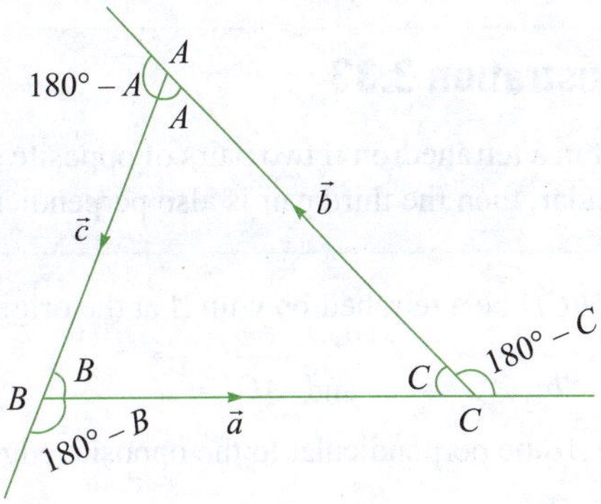

Now, $\vec{a}+\vec{b}+\vec{c} = \vec{0}$

$\therefore \quad \vec{a}\cdot(\vec{a}+\vec{b}+\vec{c}) = 0$

$\Rightarrow \quad \vec{a}\cdot\vec{a}+\vec{a}\cdot\vec{b}+\vec{a}\cdot\vec{c} = 0$

$\Rightarrow \quad a^2 + ab\cos(180° - C) + ac\cos(180° - B) = 0$

$\Rightarrow \quad a^2 - ab\cos C - ac\cos B = 0$

$\Rightarrow \quad a - b\cos C - c\cos B = 0$

$\Rightarrow \quad a = b\cos C + c\cos B$

Illustration 3.32

Prove that an angle inscribed in a semi-circle is a right angle using the vector method.

Sol. Let *O* be the centre of the semi-circle and *BA* be the diameter. Let *P* be any point on the circumference of the semi-circle.

Let $\overrightarrow{OA} = \vec{a}$, then $\overrightarrow{OB} = -\vec{a}$

Let $\overrightarrow{OP} = \vec{r}$

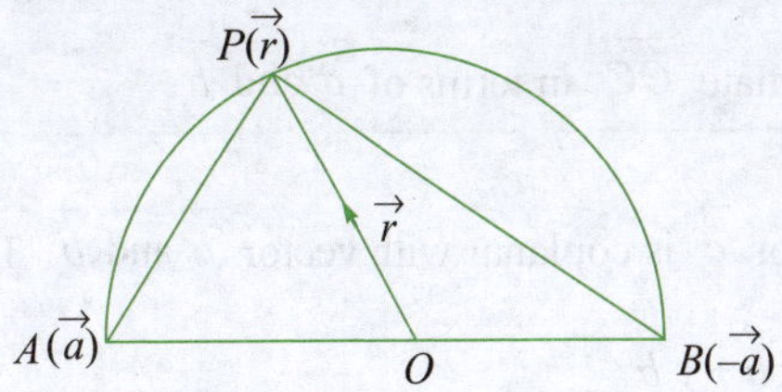

$\therefore \quad \overrightarrow{AP} = \overrightarrow{OP} - \overrightarrow{OA} = \vec{r} - \vec{a}$

and $\quad \overrightarrow{BP} = \overrightarrow{OP} - \overrightarrow{OB} = \vec{r} - (-\vec{a}) = \vec{r} + \vec{a}$

Now, $\overrightarrow{AP}\cdot\overrightarrow{BP} = (\vec{r}-\vec{a})\cdot(\vec{r}+\vec{a})$

$= \vec{r}^2 - \vec{a}^2$

$= a^2 - a^2 \qquad [\because r = a \text{ since } OP = OA]$

$= 0$

Therefore, $\overrightarrow{AP}$ is perpendicular to $\overrightarrow{BP}$.

$\therefore \quad \angle APB = 90°$

Illustration 3.33

Prove that in a tetrahedron if two pairs of opposite edges are perpendicular, then the third pair is also perpendicular.

Sol. Let $ABCD$ be a tetrahedron with A at the origin.

Let $\overrightarrow{AB} = \vec{b}$, $\overrightarrow{AC} = \vec{c}$ and $\overrightarrow{AD} = \vec{d}$

Let the edge AB be perpendicular to the opposite edge CD. Then

$$\overrightarrow{AB} \cdot \overrightarrow{CD} = 0$$

or $\vec{b} \cdot (\vec{d} - \vec{c}) = 0$

or $\vec{b} \cdot \vec{d} = \vec{b} \cdot \vec{c}$ ...(i)

Also let AC be perpendicular to the opposite edge BD. Therefore,

$$\overrightarrow{AC} \cdot \overrightarrow{BD} = 0$$

or $\vec{c} \cdot (\vec{d} - \vec{b}) = 0$

or $\vec{c} \cdot \vec{d} = \vec{b} \cdot \vec{c}$ ...(ii)

Now, from (i) and (ii), we have

$$\vec{b} \cdot \vec{d} = \vec{c} \cdot \vec{d}$$

or $(\vec{c} - \vec{b}) \cdot \vec{d} = 0$

or $\overrightarrow{BC} \cdot \overrightarrow{AD} = 0$

Hence, AD is perpendicular to the opposite edge BC.

Illustration 3.34

Arc AC of a circle subtends a right angle at the centre O. Point B divides the arc in the ratio 1 : 2. If $\overrightarrow{OA} = \vec{a}$ and $\overrightarrow{OB} = \vec{b}$, then calculate $\overrightarrow{OC}$ in terms of $\vec{a}$ and $\vec{b}$.

Sol. Vector $\vec{c}$ is coplanar with vector $\vec{a}$ and $\vec{b}$. Therefore,

$$\vec{c} = x\vec{a} + y\vec{b} \quad \text{...(i)}$$

Point B divides arc AC in the ratio 1 : 2 so that $\angle AOB = 30°$ and $\angle BOC = 60°$.

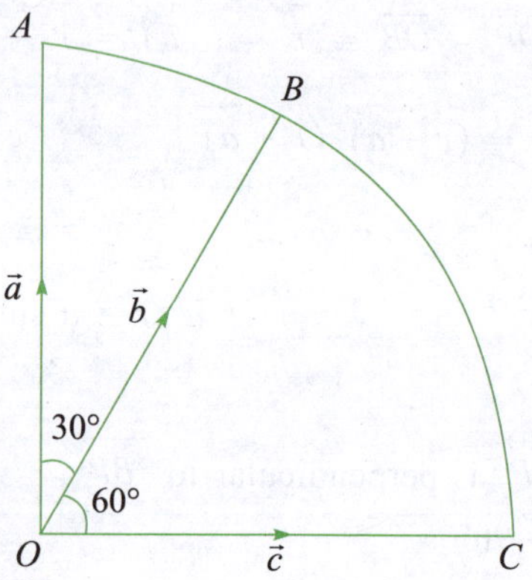

Now, we have to find the values of x and y when

$$|\vec{a}| = |\vec{b}| = |\vec{c}| = r \text{ (say)}.$$

$\therefore$ $\vec{a} \cdot \vec{b} = r^2 \cos 30° = r^2 \dfrac{\sqrt{3}}{2}$ and $\vec{a} \cdot \vec{c} = 0$

Also, $\vec{b} \cdot \vec{c} = r^2 \cos 60° = \dfrac{r^2}{2}$

In (i), taking dot product with vector $\vec{c}$ and $\vec{a}$ respectively, we get

$$\vec{c} \cdot \vec{c} = x\,\vec{a} \cdot \vec{c} + y\,\vec{b} \cdot \vec{c}$$

and $\vec{c} \cdot \vec{a} = x\,\vec{a} \cdot \vec{a} + y\,\vec{b} \cdot \vec{a}$

$\therefore$ $r^2 = 0 + \dfrac{r^2}{2} y$, or $y = 2$

and $0 = xr^2 + yr^2 \dfrac{\sqrt{3}}{2}$

Putting $y = 2$ we get $x = -\sqrt{3}$

$\therefore$ $\vec{c} = -\sqrt{3}\,\vec{a} + 2\vec{b}$

Aim to complete within 50 minutes

Concept Application Exercise 3.1

1. Constant forces $P_1 = \hat{i} - \hat{j} + \hat{k}$, $P_2 = -\hat{i} + 2\hat{j} - \hat{k}$ and $P_3 = \hat{j} - \hat{k}$ act on a particle at a point A. Then the work done when particle is displaced from position $A(4\hat{i} - 3\hat{j} - 2\hat{k})$ to $B(6\hat{i} + \hat{j} - 3\hat{k})$ is
   (1) 6 units (2) 8 units
   (3) 9 units (4) 12 units
2. Find $|\vec{a}|$ and $|\vec{b}|$, if $(\vec{a} + \vec{b}) \cdot (\vec{a} - \vec{b}) = 8$ and $|\vec{a}| = 8|\vec{b}|$.
3. Show that $|\vec{a}|\vec{b} + |\vec{b}|\vec{a}$ is a perpendicular to $|\vec{a}|\vec{b} - |\vec{b}|\vec{a}$, for any two non-zero vectors $\vec{a}$ and $\vec{b}$.
4. Let $\vec{u} = \hat{i} + \hat{j}$, $\vec{v} = \hat{i} - \hat{j}$ and $\vec{w} = \hat{i} + 2\hat{j} + 3\hat{k}$. If $\hat{n}$ is a unit vector perpendicular to both $\vec{u}$ and $\vec{v}$, then the value of $|\vec{w} \cdot \hat{n}|$ is
   (1) 3 (2) 4 (3) 5 (4) 6
5. A, B, C, D are any four points, prove that $\overrightarrow{AB} \cdot \overrightarrow{CD} + \overrightarrow{BC} \cdot \overrightarrow{AD} + \overrightarrow{CA} \cdot \overrightarrow{BD} = 0$.
6. If the vectors $3\vec{p} + \vec{q}$; $5\vec{p} - 3\vec{q}$ and $2\vec{p} + \vec{q}$; $4\vec{p} - 2\vec{q}$ are pairs of mutually perpendicular vectors, then the angle between vectors $\vec{p}$ and $\vec{q}$ is
   (1) $\cos^{-1}\dfrac{4}{5}$ (2) $\cos^{-1}\dfrac{3}{4}$
   (3) $\cos^{-1}\dfrac{3}{8}$ (4) $\cos^{-1}\dfrac{7}{8}$

7. Let $\vec{A}$ and $\vec{B}$ be two non-parallel unit vectors in a plane. If $(\alpha\vec{A} + \vec{B})$ bisects the internal angle between $\vec{A}$ and $\vec{B}$, then find the value of α, using the dot product.

8. The scalar projection of the vector $\hat{i} + 3\hat{j} + 7\hat{k}$ on the vector $7\hat{i} - \hat{j} + 8\hat{k}$ is

(1) $\frac{60}{\sqrt{114}}$ (2) $\frac{-60}{\sqrt{114}}$ (3) $\frac{60}{\sqrt{59}}$ (4) $\frac{-60}{\sqrt{59}}$

9. If $\vec{a}, \vec{b}, \vec{c}$ are unit vectors such that $\vec{a} + \vec{b} + \vec{c} = 0$, then find the value of $\vec{a}\cdot\vec{b} + \vec{b}\cdot\vec{c} + \vec{c}\cdot\vec{a}$.

10. If $|\vec{a}| = 3, |\vec{b}| = 4$ and the angle between $\vec{a}$ and $\vec{b}$ is $120°$, then find the value of $|4\vec{a} + 3\vec{b}|$.

11. If $\vec{a} + \vec{b} + \vec{c} = 0, |\vec{a}| = 3, |\vec{b}| = 5, |\vec{c}| = 7$, then find the angle between $\vec{a}$ and $\vec{b}$.

12. Let $\vec{a}, \vec{b}$ and $\vec{c}$ be unit vectors, such that $\vec{a} + \vec{b} + \vec{c} = \vec{x}$, $\vec{a}\cdot\vec{x} = 1$, $\vec{b}\cdot\vec{x} = \frac{3}{2}$, $|\vec{x}| = 2$. Then the angle between $\vec{c}$ and $\vec{x}$ is

(1) $\cos^{-1}\frac{4}{5}$ (2) $\cos^{-1}\frac{3}{4}$ (3) $\cos^{-1}\frac{3}{8}$ (4) $\cos^{-1}\frac{5}{8}$

13. If $\vec{a}$ and $\vec{b}$ are unit vectors, then the greatest value of $|\vec{a} + \vec{b}| + |\vec{a} - \vec{b}|$ is

(1) $\sqrt{2}$ (2) 2 (3) $2\sqrt{2}$ (4) 4

14. If $|\vec{a}| = 5, |\vec{a} - \vec{b}| = 8$ and $|\vec{a} + \vec{b}| = 10$, then find $|\vec{b}|$.

15. If A, B, C, D are four distinct points in space such that $\overline{AB}$ is not perpendicular to $\overline{CD}$ and satisfies $\overline{AB}\cdot\overline{CD} = k(|\overline{AD}|^2 + |\overline{BC}|^2 - |\overline{AC}|^2 - |\overline{BD}|^2)$, then the value of k is

(1) $\frac{1}{4}$ (2) $\frac{1}{2}$ (3) 1 (4) 2

16. Let $\vec{a}, \vec{b}$ and $\vec{c}$ be three unit vectors such that $|\vec{a} - \vec{b}|^2 + |\vec{a} - \vec{c}|^2 = 8$. Then $|\vec{a} + 2\vec{b}|^2 + |\vec{a} + 2\vec{c}|^2$ is equal to

(1) 1 (2) 2 (3) 4 (4) 8

17. Let $\vec{\alpha} = 4\hat{i} + 3\hat{j} + 5\hat{k}$ and $\vec{\beta} = \hat{i} + 2\hat{j} - 4\hat{k}$. Let $\vec{\beta}_1$ be parallel to $\vec{\alpha}$ and $\vec{\beta}_2$ be perpendicular to $\vec{\alpha}$. If $\vec{\beta} = \vec{\beta}_1 + \vec{\beta}_2$, then the value of $5\vec{\beta}_2\cdot(\hat{i} + \hat{j} + \hat{k})$ is

(1) 9 (2) 11 (3) 7 (4) 6

18. Let $\vec{x}$ be a vector in the plane containing vectors $\vec{a} = 2\hat{i} - \hat{j} + \hat{k}$ and $\vec{b} = \hat{i} + 2\hat{j} - \hat{k}$. If the vector $\vec{x}$ is perpendicular to $(3\hat{i} + 2\hat{j} - \hat{k})$ and its projection on $\vec{a}$ is $\frac{17\sqrt{6}}{2}$, then find the value of $|\vec{x}|^2$

Answers

1. (3)
2. $|\vec{a}| = \frac{16\sqrt{2}}{3\sqrt{7}}, |\vec{b}| = \sqrt{\frac{8}{63}}$ **4.** (1)
6. (3) **7.** 1
8. (1) **9.** $\frac{-3}{2}$
10. 12 **11.** $\pi/3$
12. (2) **13.** (3)
14. $\sqrt{57}$ **15.** $k = 1/2$
16. (2) **17.** (3)
18. 486

3.3 CROSS (VECTOR) PRODUCT OF VECTORS

3.3.1 Definition of Cross Product

Another product of vectors is the cross product or vector product. To understand the definition of this product, we consider its physical interpretation in Physics and Mechanics through the concept of torque.

Torque is the measure of the force that can cause an object to rotate about an axis. It is defined as the product of the distance of the point of application of the force from the axis of rotation and the component of the force perpendicular to the lever arm. In simple words, torque (also called the moment of force) is the tendency of a force to cause rotation.

Force causes an object to accelerate in linear motion; in the same way, torque causes angular acceleration. Thus, torque is the rotational analogue of linear force. The straight line about which the object rotates is called the axis of rotation.

A good and familiar example of applying a torque is using a wrench to turn a bolt. The wrench helps to amplify the twisting force applied on the bolt.

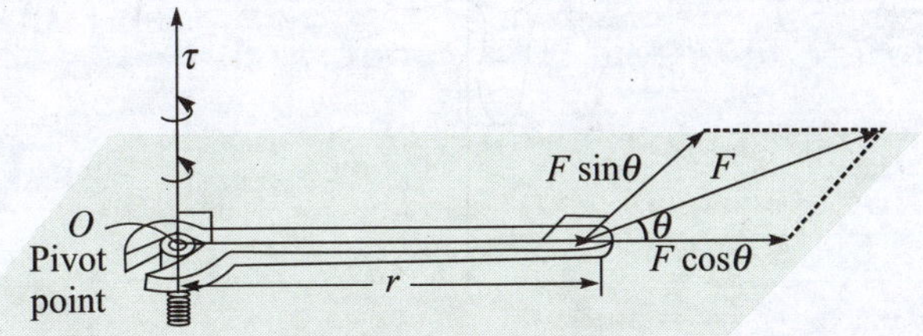

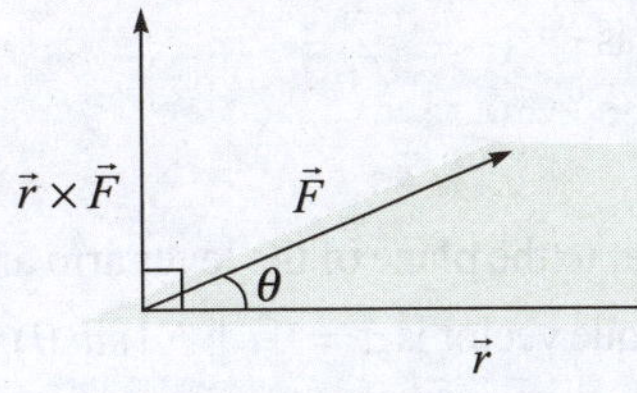

Free body diagram

(A Free Body Diagram (FBD) is a simplified sketch of an object, isolated from its surroundings, showing all the external forces acting on it as vectors.)

In the figure, a force **F** making an angle θ with the lever is applied at point P on the lever, at distance r from the pivot point O (about which the bolt rotates).

- The component of the force along the lever, $F\cos\theta$, does not produce any turning effect.
- The twisting effect is caused by the perpendicular component, $F\sin\theta$.

Hence, by definition, the magnitude of torque is

$$\tau = r\cdot(F\sin\theta).$$

If the force is applied so as to rotate the lever in the anticlockwise sense, the torque vector points upward (out of the plane of the paper). If the force rotates it in the clockwise sense, the torque vector points downward (into the plane).

Thus, the direction of torque is perpendicular to the plane determined by the lever arm vector $\vec{r}=\overrightarrow{OP}$ and the force vector $\vec{F}$. Its direction is decided by the right-hand rule: curl the fingers of the right hand from $\vec{r}$ to $\vec{F}$; the extended thumb points in the direction of torque.

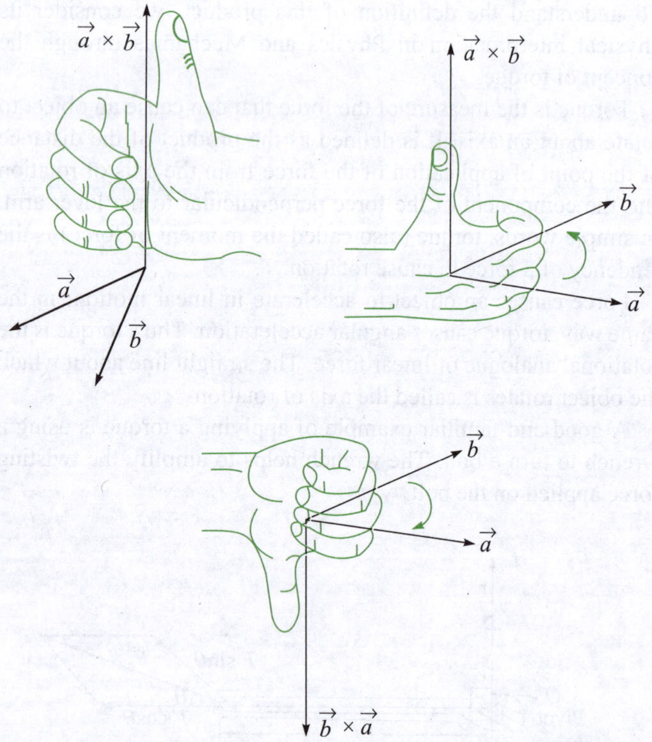

Therefore, torque has both magnitude and direction, and is a **vector quantity**. Its magnitude is

$$|\vec{\tau}| = |\vec{r}||\vec{F}|\sin\theta.$$

If the unit vector perpendicular to the plane of the lever arm and the force vector is $\hat{n}$, then the torque vector is $\vec{\tau}=(|\vec{r}||\vec{F}|\sin\theta)\hat{n}$, which is denoted as $\vec{\tau}=\vec{r}\times\vec{F}$. Here, $\vec{r}\times\vec{F}$ is called cross product (vector product) of the vectors $\vec{r}$ and $\vec{F}$.

In general, the cross product of two vectors $\vec{a}$ and $\vec{b}$ is defined as

$$\vec{a}\times\vec{b} = |\vec{a}||\vec{b}|\sin\theta\,\hat{n},$$

where θ is the angle between $\vec{a}$ and $\vec{b}$, and $\hat{n}$ is a unit vector perpendicular to the plane of $\vec{a}$ and $\vec{b}$, pointing in the direction given by the **right-hand rule**.

If instead we curl the fingers from $\vec{b}$ to $\vec{a}$, the direction of the cross product $\vec{b}\times\vec{a}$ will be opposite, i.e.,

$$\vec{b}\times\vec{a} = -|\vec{a}||\vec{b}|\sin\theta\,\hat{n} = -\vec{a}\times\vec{b}$$

3.3.2 Properties of Cross Product

1. $\vec{a}\times\vec{b} = -\vec{b}\times\vec{a}$
2. $\vec{a}\times\vec{a} = 0$, as angle between $\vec{a}$ and $\vec{a}$ is $0°$
3. $\hat{i}\times\hat{i} = \hat{j}\times\hat{j} = \hat{k}\times\hat{k} = 0$
4. Two non-zero vectors $\vec{a}$ and $\vec{b}$ are collinear iff $\vec{a}\times\vec{b}=0$
5. $\hat{i}\times\hat{j}=\hat{k}$, since $\hat{i}\perp\hat{j}$ and unit vector perpendicular to the plane of $\hat{i}$ and $\hat{j}$ is $\hat{k}$. Similarly, $\hat{j}\times\hat{k}=\hat{i}$ and $\hat{k}\times\hat{i}=\hat{j}$.
6. $(m\vec{a})\times(n\vec{b}) = mn(\vec{a}\times\vec{b})$; in particular

   $m(\vec{a}\times\vec{b}) = (m\vec{a})\times\vec{b} = \vec{a}\times(m\vec{b})$
7. $\vec{a}\times(\vec{b}+\vec{c}) = \vec{a}\times\vec{b}+\vec{a}\times\vec{c}$ (Cross-product is distributive over addition of vectors)

   Proof:

   Consider the following figure.

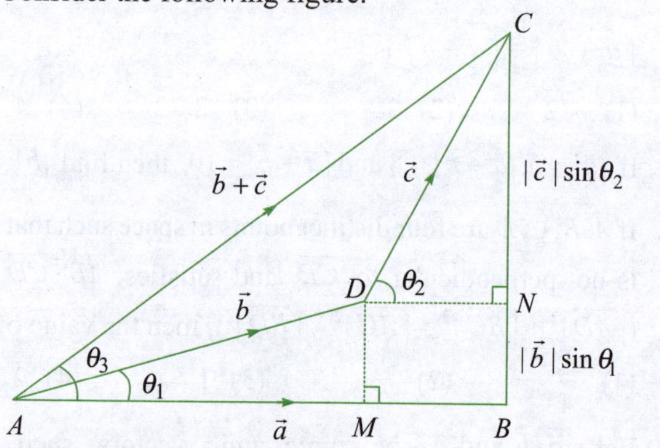

We have, $\overrightarrow{AB}=\vec{a}$, $\overrightarrow{AD}=\vec{b}$, $\overrightarrow{DC}=\vec{c}$

and $\overrightarrow{AC}=\overrightarrow{AD}+\overrightarrow{DC}=\vec{b}+\vec{c}$

In ΔABC, $BC = AC\sin\theta_3 = |\vec{b}+\vec{c}|\sin\theta_3$

In ΔAMD, $DM = AD\sin\theta_1 = |\vec{b}|\sin\theta_1$

$\therefore\quad BN = DM = |\vec{b}|\sin\theta_1$

In ΔCND, $CN = DC\sin\theta_2 = |\vec{c}|\sin\theta_2$

Now, $\vec{a}\times(\vec{b}+\vec{c}) = |\vec{a}||\vec{b}+\vec{c}|\sin\theta_3\,\hat{n}$

$= |\vec{a}|\left(BC\,\hat{n}\right)\ (\because BC = |\vec{b}+\vec{c}|\sin\theta_3)$

$= |\vec{a}|[(BN+NC)\hat{n}]$

$= |\vec{a}|[(|\vec{b}|\sin\theta_1 + |\vec{c}|\sin\theta_2)\hat{n}]$

$= |\vec{a}||\vec{b}|\sin\theta_1\hat{n} + |\vec{a}||\vec{c}|\sin\theta_2\hat{n}$

$= \vec{a}\times\vec{b}+\vec{a}\times\vec{c}$

8. If $\vec{a} = a_1\hat{i} + a_2\hat{j} + a_3\hat{k}$ and $\vec{b} = b_1\hat{i} + b_2\hat{j} + b_3\hat{k}$,

then $\vec{a}\times\vec{b} = \begin{vmatrix} \hat{i} & \hat{j} & \hat{k} \\ a_1 & a_2 & a_3 \\ b_1 & b_2 & b_3 \end{vmatrix}$

Proof:

$$\vec{a}\times\vec{b} = (a_1\hat{i} + a_2\hat{j} + a_3\hat{k})\times(b_1\hat{i} + b_2\hat{j} + b_3\hat{k})$$

$$= a_1\hat{i}\times b_1\hat{i} + a_1\hat{i}\times b_2\hat{j} + a_1\hat{i}\times b_3\hat{k}$$

$$+ a_2\hat{j}\times b_1\hat{i} + a_2\hat{j}\times b_2\hat{j} + a_2\hat{j}\times b_3\hat{k}$$

$$+ a_3\hat{k}\times b_1\hat{i} + a_3\hat{k}\times b_2\hat{j} + a_3\hat{k}\times b_3\hat{k}$$

$$= 0 + a_1b_2\hat{k} - a_1b_3\hat{j} - a_2b_1\hat{k} + 0 + a_2b_3\hat{i} + a_3b_1\hat{j} - a_3b_2\hat{i} + 0$$

$$= (a_2b_3 - a_3b_2)\hat{i} - (a_1b_3 - a_3b_1)\hat{j} + (a_1b_2 - a_2b_1)\hat{k}$$

$$\therefore \quad \vec{a}\times\vec{b} = \begin{vmatrix} \hat{i} & \hat{j} & \hat{k} \\ a_1 & a_2 & a_3 \\ b_1 & b_2 & b_3 \end{vmatrix}$$

Interchanging R_2 and R_3, we get

$$\vec{a}\times\vec{b} = -\begin{vmatrix} \hat{i} & \hat{j} & \hat{k} \\ b_1 & b_2 & b_3 \\ a_1 & a_2 & a_3 \end{vmatrix} = -\vec{b}\times\vec{a}$$

9. Unit vectors perpendicular to the plane of vectors $\vec{a}$ and $\vec{b}$ are $\pm\dfrac{\vec{a}\times\vec{b}}{|\vec{a}\times\vec{b}|}$.

10. $(\vec{a}\times\vec{b})^2 + (\vec{a}\cdot\vec{b})^2 = |\vec{a}|^2|\vec{b}|^2$

Proof:

$$(\vec{a}\times\vec{b})^2 + (\vec{a}\cdot\vec{b})^2 = |\vec{a}\times\vec{b}|^2 + (\vec{a}\cdot\vec{b})^2$$

$$= |\vec{a}|^2|\vec{b}|^2\sin^2\theta + |\vec{a}|^2|\vec{b}|^2\cos^2\theta$$

$$= |\vec{a}|^2|\vec{b}|^2$$

11. If $\vec{a}+\vec{b}+\vec{c} = 0$ then $\vec{a}\times\vec{b} = \vec{b}\times\vec{c} = \vec{c}\times\vec{a}$

Proof:

We have $\vec{a}+\vec{b}+\vec{c} = 0$ ...(i)

Taking cross product with vector $\vec{a}$, we get

$$\vec{a}\times\vec{a} + \vec{a}\times\vec{b} + \vec{a}\times\vec{c} = 0$$

$$\Rightarrow \quad 0 + \vec{a}\times\vec{b} - \vec{c}\times\vec{a} = 0$$

$$\Rightarrow \quad \vec{a}\times\vec{b} = \vec{c}\times\vec{a} \qquad \text{...(ii)}$$

Similarly, in (i) taking cross product with vector $\vec{b}$, we get

$$\vec{a}\times\vec{b} = \vec{b}\times\vec{c} \qquad \text{...(iii)}$$

From (ii) and (iii), we get

$$\vec{a}\times\vec{b} = \vec{b}\times\vec{c} = \vec{c}\times\vec{a}$$

Illustration 3.35

Find the moment (torque) about the point (1, –1, –1) of the force $\vec{F} = 3\hat{i} + 4\hat{j} - 5\hat{k}$, acting at the point (1, 0, –2).

Sol. The moment (torque) of a force $\vec{F}$ about the point A is given by

$$\vec{\tau}_A = \overrightarrow{AP}\times\vec{F}$$

where, A(1, –1, –1) is the point about which moment is required, P(1, 0, –2) is the point of application of force, and $\overrightarrow{AP}$ is the position vector of P relative to A.

Now, $\overrightarrow{AP} = (1-1)\hat{i} + (0-(-1))\hat{j} + (-2-(-1))\hat{k}$

or $\quad \overrightarrow{AP} = \hat{j} - \hat{k}$.

Therefore, $\vec{\tau}_A = (\hat{j} - \hat{k})\times(3\hat{i} + 4\hat{j} - 5\hat{k})$.

$$= \begin{vmatrix} \hat{i} & \hat{j} & \hat{k} \\ 0 & 1 & -1 \\ 3 & 4 & -5 \end{vmatrix}.$$

$$= -\hat{i} - 3\hat{j} - 3\hat{k}$$

Illustration 3.36

Find $|\vec{a}\times\vec{b}|$, if $\vec{a} = \hat{i} - 7\hat{j} + 7\hat{k}$ and $\vec{b} = 3\hat{i} - 2\hat{j} + 2\hat{k}$.

Sol. We have, $\vec{a} = \hat{i} - 7\hat{j} + 7\hat{k}$ and $\vec{b} = 3\hat{i} - 2\hat{j} + 2\hat{k}$

$$\therefore \quad \vec{a}\times\vec{b} = \begin{vmatrix} \hat{i} & \hat{j} & \hat{k} \\ 1 & -7 & 7 \\ 3 & -2 & 2 \end{vmatrix}$$

$$= \hat{i}(-14+14) - \hat{j}(2-21) + \hat{k}(-2+21) = 19\hat{j} + 19\hat{k}$$

$$\therefore \quad |\vec{a}\times\vec{b}| = \sqrt{(19)^2 + (19)^2} = \sqrt{2\times(19)^2} = 19\sqrt{2}$$

Illustration 3.37

Find a unit vector perpendicular to the plane determined by the points (1, –1, 2), (2, 0, –1) and (0, 2, 1).

Sol. Given points are A(1, –1, 2), B(2, 0, –1) and C(0, 2, 1)

$$\Rightarrow \quad \overrightarrow{AB} = \vec{a} = \hat{i} + \hat{j} - 3\hat{k},$$

and $\quad \overrightarrow{BC} = \vec{b} = -2\hat{i} + 2\hat{j} + 2\hat{k}$

$$\therefore \quad \vec{a}\times\vec{b} = \begin{vmatrix} \hat{i} & \hat{j} & \hat{k} \\ 1 & 1 & -3 \\ -2 & 2 & 2 \end{vmatrix} = 8\hat{i} + 4\hat{j} + 4\hat{k}$$

Hence, the required unit vector is $\dfrac{2\hat{i} + \hat{j} + \hat{k}}{\sqrt{6}}$ or $-\dfrac{2\hat{i} + \hat{j} + \hat{k}}{\sqrt{6}}$

Illustration 3.38

Let the vectors $\vec{a}$ and $\vec{b}$ be such that $|\vec{a}| = 3$ and $|\vec{b}| = \frac{\sqrt{2}}{3}$. Find the angle between $\vec{a}$ and $\vec{b}$ if $\vec{a} \times \vec{b}$ is a unit vector.

Sol. It is given that $|\vec{a}| = 3$ and $|\vec{b}| = \frac{\sqrt{2}}{3}$.

We know that $\vec{a} \times \vec{b} = |\vec{a}||\vec{b}| \sin\theta\, \hat{n}$, where $\hat{n}$ is a unit vector perpendicular to both $\vec{a}$ and $\vec{b}$ and θ is the angle between $\vec{a}$ and $\vec{b}$.

Since $\vec{a} \times \vec{b}$ is a unit vector, $|\vec{a} \times \vec{b}| = 1$.

$\Rightarrow \quad ||\vec{a}||\vec{b}| \sin\theta\, \hat{n}| = 1$

$\Rightarrow \quad ||\vec{a}||\vec{b}| \sin\theta| = 1$

$\Rightarrow \quad 3 \times \frac{\sqrt{2}}{3} \times \sin\theta = 1$

$\Rightarrow \quad \theta = \frac{\pi}{4}$

Hence, $\vec{a} \times \vec{b}$ is a unit vector if the angle between $\vec{a}$ and $\vec{b}$ is $\frac{\pi}{4}$.

Illustration 3.39

Show that $(\vec{a} - \vec{b}) \times (\vec{a} + \vec{b}) = 2(\vec{a} \times \vec{b})$.

Sol. We have, $(\vec{a} - \vec{b}) \times (\vec{a} + \vec{b}) = (\vec{a} - \vec{b}) \times \vec{a} + (\vec{a} - \vec{b}) \times \vec{b}$

$= \vec{a} \times \vec{a} - \vec{b} \times \vec{a} + \vec{a} \times \vec{b} - \vec{b} \times \vec{b}$

$= \vec{0} + \vec{a} \times \vec{b} + \vec{a} \times \vec{b} - \vec{0}$

$= 2\vec{a} \times \vec{b}$

Illustration 3.40

If $|\vec{a}| = 2$, then find the value of $|\vec{a} \times \hat{i}|^2 + |\vec{a} \times \hat{j}|^2 + |\vec{a} \times \hat{k}|^2$.

Sol. Let $\vec{a} = a_1\hat{i} + a_2\hat{j} + a_3\hat{k}$

$$\therefore \quad |\vec{a} \times \hat{i}|^2 = \begin{vmatrix} \hat{i} & \hat{j} & \hat{k} \\ a_1 & a_2 & a_3 \\ 1 & 0 & 0 \end{vmatrix}^2$$

$= |a_3\hat{j} - a_2\hat{k}|^2 = a_3^2 + a_2^2$

Similarly, $|\vec{a} \times \hat{j}|^2 = a_1^2 + a_3^2$ and $|\vec{a} \times \hat{k}|^2 = a_1^2 + a_2^2$

Hence, $|\vec{a} \times \hat{i}|^2 + |\vec{a} \times \hat{j}|^2 + |\vec{a} \times \hat{k}|^2$

$= 2(a_1^2 + a_2^2 + a_3^2) = 2|\vec{a}|^2 = 8$

Illustration 3.41

If $\vec{a} = \hat{i} + \hat{j} + \hat{k}$, $\vec{a} \cdot \vec{b} = 1$ and $\vec{a} \times \vec{b} = \hat{j} - \hat{k}$ then find vector $\vec{b}$.

Sol. Let $\vec{b} = \alpha\hat{i} + \beta\hat{j} + \gamma\hat{k}$

Now, $\vec{a} \times \vec{b} = \hat{j} - \hat{k}$

$$\Rightarrow \begin{vmatrix} \hat{i} & \hat{j} & \hat{k} \\ 1 & 1 & 1 \\ \alpha & \beta & \gamma \end{vmatrix} = \hat{j} - \hat{k}$$

$\Rightarrow \quad \beta - \gamma = 0, \alpha - \gamma = 1, \alpha - \beta = 1$

$\Rightarrow \quad \beta = \gamma, \alpha = 1 + \gamma$ and $\alpha = 1 + \beta$

Now, $\vec{a} \cdot \vec{b} = 1$

$\Rightarrow \quad \alpha + \beta + \gamma = 1$

$\Rightarrow \quad 1 + \beta + \beta + \beta = 1$

$\Rightarrow \quad \beta = 0$

$\therefore \alpha = 1$ and $\gamma = 0$

Thus, $\vec{b} = \hat{i}$

Illustration 3.42

If $\vec{a} \times \vec{b} = \vec{c} \times \vec{d}$ and $\vec{a} \times \vec{c} = \vec{b} \times \vec{d}$, then show that $\vec{a} - \vec{d}$, is parallel to $\vec{b} - \vec{c}$ provided $\vec{a} \neq \vec{d}$ and $\vec{b} \neq \vec{c}$.

Sol. We have $\vec{a} \times \vec{b} = \vec{c} \times \vec{d}$

and $\vec{a} \times \vec{c} = \vec{b} \times \vec{d}$ ...(i)

Now, $(\vec{a} - \vec{d}) \times (\vec{b} - \vec{c})$

$= \vec{a} \times \vec{b} - \vec{a} \times \vec{c} - \vec{d} \times \vec{b} + \vec{d} \times \vec{c}$

$= (\vec{a} \times \vec{b} + \vec{d} \times \vec{c}) - (\vec{a} \times \vec{c} + \vec{d} \times \vec{b})$

$= (\vec{a} \times \vec{b} - \vec{c} \times \vec{d}) - (\vec{a} \times \vec{c} - \vec{b} \times \vec{d})$

$= \vec{0} - \vec{0} = \vec{0}$ [from (i)]

Hence, $\vec{a} - \vec{d}$ and $\vec{b} - \vec{c}$ are parallel.

Illustration 3.43

Let $\hat{a}$ and $\hat{b}$ be two unit vectors such that the angle between them is $\frac{\pi}{4}$. If the angle between the vectors $(\hat{a} + \hat{b})$ and $(\hat{a} + 2\hat{b} + 2(\hat{a} \times \hat{b}))$ is θ, then find the value of $\cos\theta$.

Sol. We have, $|\hat{a} + \hat{b}|^2 = 1 + 1 + 2\cos\frac{\pi}{4} = 2 + \sqrt{2}$

and $|\hat{a}\times\hat{b}|^2=|\hat{a}|^2|\hat{b}|^2-|\hat{a}\cdot\hat{b}|^2=1-\frac{1}{2}=\frac{1}{2}$

$\therefore \quad |\hat{a}+2\hat{b}+2(\hat{a}\times\hat{b})|^2$

$=1+4+4|\hat{a}\times\hat{b}|^2+4\hat{a}\cdot\hat{b}+0+0 \quad (\because \hat{a}\cdot(\hat{a}\times\hat{b})=\hat{b}\cdot(\hat{a}\times\hat{b})=0)$

$=1+4+4\times\frac{1}{2}+4\frac{1}{\sqrt{2}}$

$=7+2\sqrt{2}$

Now, $\cos\theta=\dfrac{(\hat{a}+\hat{b})\cdot(\hat{a}+2\hat{b}+2(\hat{a}\times\hat{b}))}{|\hat{a}+\hat{b}||\hat{a}+2\hat{b}+2(\hat{a}\times\hat{b})|}$

$=\dfrac{1+2\hat{a}\cdot\hat{b}+0+\hat{b}\cdot\hat{a}+2+0}{\sqrt{2+\sqrt{2}}\sqrt{7+2\sqrt{2}}}$

$=\dfrac{3+\frac{3}{\sqrt{2}}}{\sqrt{2+\sqrt{2}}\sqrt{7+2\sqrt{2}}}$

$=\dfrac{3(1+\sqrt{2})}{\sqrt{4+2\sqrt{2}}\sqrt{7+2\sqrt{2}}}=\dfrac{3(1+\sqrt{2})}{\sqrt{36+22\sqrt{2}}}$

Illustration 3.44

Let $\vec{a}, \vec{b}, \vec{c}$ be three non-coplanar vectors such that $\vec{a}\times\vec{b}=4\vec{c}$, $\vec{b}\times\vec{c}=9\vec{a}$ and $\vec{c}\times\vec{a}=\alpha\vec{b}$, $\alpha>0$. If $|\vec{a}|+|\vec{b}|+|\vec{c}|=36$, then find the value of α.

Sol. Since $\vec{a}\times\vec{b}=4\vec{c}$, $\vec{c}$ is perpendicular to both $\vec{a}$ and $\vec{b}$. Also, $\vec{b}\times\vec{c}=9\vec{a}$, then $\vec{a}$ is perpendicular to both $\vec{b}$ and $\vec{c}$. Thus, vectors $\vec{a}, \vec{b}$ and $\vec{c}$ are mutually perpendicular.

Therefore, $|\vec{a}||\vec{b}|=4|\vec{c}|, |\vec{b}||\vec{c}|=9|\vec{a}|$ and $|\vec{c}||\vec{a}|=\alpha|\vec{b}|$

$\Rightarrow \quad (|\vec{a}||\vec{b}|)(|\vec{b}||\vec{c}|)=(4|\vec{c}|)(9|\vec{a}|)$

$\Rightarrow \quad |\vec{b}|=6$

$\Rightarrow \quad 6|\vec{a}|=4|\vec{c}|$ or $|\vec{c}|=\frac{3}{2}|\vec{a}|$

Now, $|\vec{a}|+|\vec{b}|+|\vec{c}|=36$

$\Rightarrow \quad |\vec{a}|+\frac{3}{2}|\vec{a}|+6=36$

$\Rightarrow \quad |\vec{a}|=12$ and $|\vec{c}|=18$

$\Rightarrow \quad \alpha=\dfrac{|\vec{c}||\vec{a}|}{|\vec{b}|}=\dfrac{18\times12}{6}=36$

Illustration 3.45

Let $\hat{u}$ and $\hat{v}$ be two non-collinear unit vectors such that $\left|\dfrac{\hat{u}+\hat{v}}{2}+\hat{u}\times\hat{v}\right|=1$. Prove that $|\hat{u}\times\hat{v}|=\left|\dfrac{\hat{u}-\hat{v}}{2}\right|$

Sol. Given that $\left|\dfrac{\hat{u}+\hat{v}}{2}+\hat{u}\times\hat{v}\right|=1$

or $\left|\dfrac{\hat{u}+\hat{v}}{2}+\hat{u}\times\hat{v}\right|^2=1$

$\Rightarrow \quad \left|\dfrac{\hat{u}+\hat{v}}{2}\right|^2+|\hat{u}\times\hat{v}|^2=1 \quad (\because \hat{u}\cdot(\hat{u}\times\hat{v})=\hat{v}\cdot(\hat{u}\times\hat{v})=0)$

$\Rightarrow \quad \dfrac{|\hat{u}|^2+|\hat{v}|^2+2\hat{u}\cdot\hat{v}}{4}+|\hat{u}\times\hat{v}|^2=1$

$\Rightarrow \quad \dfrac{2+2\cos\theta}{4}+\sin^2\theta=1$

$\Rightarrow \quad \cos^2\frac{\theta}{2}=\cos^2\theta$

$\Rightarrow \quad \theta=n\pi\pm\frac{\theta}{2}, n\in Z$

$\Rightarrow \quad \theta=\frac{2\pi}{3}$

$\Rightarrow \quad |\hat{u}\times\hat{v}|^2=\sin^2\frac{\pi}{3}=\frac{3}{4}$

and $\left|\dfrac{\hat{u}-\hat{v}}{2}\right|^2=\frac{1}{4}\left[1+1-2\cos\frac{2\pi}{3}\right]=\frac{3}{4}$

Illustration 3.46

Let $\vec{a}$ and $\vec{b}$ are two vectors. Such that $|\vec{a}|=1, |\vec{b}|=2$ and $\vec{a}\cdot\vec{b}=1$. If $\vec{c}=(\vec{a}\times\vec{b})-2\vec{b}$, then find the value of $\vec{b}\cdot\vec{c}$.

Sol. We have $\vec{c}=(\vec{a}\times\vec{b})-2b$

$\Rightarrow \quad |\vec{c}|^2=|\vec{a}\times\vec{b}|^2+4|\vec{b}|^2$

$(\because \vec{a}\times\vec{b}\perp\vec{b})$

$=|\vec{a}|^2|\vec{b}|^2-(\vec{a}\cdot\vec{b})^2+4|\vec{b}|^2$

$=4-1+4.4=19$

Now, $\vec{c}+2\vec{b}=\vec{a}\times\vec{b}$

$\therefore \quad |\vec{c}+2\vec{b}|^2=|\vec{a}\times\vec{b}|^2$

$\Rightarrow \quad |\vec{c}|^2+4|\vec{b}|^2+4\vec{b}\cdot\vec{c}=3$

$\Rightarrow \quad 19+16+4\vec{b}\cdot\vec{c}=3$

$\Rightarrow \quad \vec{b}\cdot\vec{c}=-8$

Illustration 3.47

Let $\vec{a}, \vec{b}$ and $\vec{c}$ be three vectors such that $\vec{a}\neq0$, $|\vec{a}|=|\vec{c}|=1$, $|\vec{b}|=4$ and $|\vec{b}\times\vec{c}|=\sqrt{15}$. If $\vec{b}-2\vec{c}=\lambda\vec{a}$, then find the value of λ.

Sol. Let the angle between $\vec{b}$ and $\vec{c}$ be α. Then

$|\vec{b}\times\vec{c}|=\sqrt{15}$

$\Rightarrow \quad |\vec{b}||\vec{c}| \sin\alpha = \sqrt{15}$

$\Rightarrow \quad \sin\alpha = \dfrac{\sqrt{15}}{4}$

$\Rightarrow \quad \cos\alpha = \dfrac{1}{4}$

Now, $\vec{b} - 2\vec{c} = \lambda \vec{a}$

$\Rightarrow \quad |\vec{b} - 2\vec{c}|^2 = \lambda^2 |\vec{a}|^2$

$\Rightarrow \quad |\vec{b}|^2 + 4|\vec{c}|^2 - 4\cdot\vec{b}\cdot\vec{c} = \lambda^2|\vec{a}|^2$

$\Rightarrow \quad 16 + 4 - 4\{|\vec{b}||\vec{c}|\cos\alpha\} = \lambda^2$

$\Rightarrow \quad 16 + 4 - 4 \times 4 \times 1 \times \dfrac{1}{4} = \lambda^2$

$\Rightarrow \quad \lambda^2 = 16$ or $\lambda = \pm 4$

Illustration 3.48

Given $|\vec{a}| = |\vec{b}| = 1$ and $|\vec{a} + \vec{b}| = \sqrt{3}$. If $\vec{c}$ is a vector such that $\vec{c} - \vec{a} - 2\vec{b} = 3(\vec{a} \times \vec{b})$, then find the value of $\vec{c}\,.\,\vec{b}$.

Sol. We have, $|\vec{a} + \vec{b}| = \sqrt{3}$

$\Rightarrow |\vec{a} + \vec{b}|^2 = 3$

$\Rightarrow \quad 1 + 1 + 2\vec{a}\cdot\vec{b} = 3$

$\Rightarrow \quad \vec{a}\cdot\vec{b} = \dfrac{1}{2}$

Now, $\vec{c} - \vec{a} - 2\vec{b} = 3(\vec{a} \times \vec{b})$

$\Rightarrow \quad (\vec{c} - \vec{a} - 2\vec{b})\cdot\vec{b} = 3(\vec{a} \times \vec{b})\cdot\vec{b}$

$\Rightarrow \quad \vec{c}\cdot\vec{b} - \vec{a}\cdot\vec{b} - 2\vec{b}\cdot\vec{b} = 0$

$\Rightarrow \quad \vec{c}\cdot\vec{b} - \dfrac{1}{2} - 2 = 0$

$\Rightarrow \quad \vec{c}\,.\,\vec{b} = \dfrac{5}{2}$

Illustration 3.49

Let $\vec{a}, \vec{b}, \vec{c}$ be three coplanar vectors such that $|\vec{a}| = 1, |\vec{b}| = 2, \vec{b}\cdot\vec{c} = 8$. If the angle between $\vec{b}$ and $\vec{c}$ is 45°, then find the value of $|\vec{a} \times (\vec{b} \times \vec{c})|$.

Sol. Vector $\vec{a}, \vec{b}, \vec{c}$ are coplanar.

$\therefore \quad \vec{a} \perp (\vec{b} \times \vec{c})$

$\Rightarrow \quad |\vec{a} \times (\vec{b} \times \vec{c})| = |\vec{a}||\vec{b} \times \vec{c}| \qquad \ldots\text{(i)}$

Now $\vec{b}\cdot\vec{c} = 8$

$\Rightarrow \quad |\vec{b}||\vec{c}| \cos 45° = 8$

$\Rightarrow \quad 2|\vec{c}| \dfrac{1}{\sqrt{2}} = 8$

$\Rightarrow \quad |\vec{c}| = 4\sqrt{2}$

$\therefore \quad |\vec{b} \times \vec{c}| = |\vec{b}||\vec{c}| \sin 45° = 2 \times 4\sqrt{2} \times \dfrac{1}{\sqrt{2}} = 8$

$\therefore \quad |\vec{a} \times (\vec{b} \times \vec{c})| = 1 \times 8 = 8 \qquad$ (from (i))

3.3.3 Solving Vector Equations Using Cross Product

Many vector equations involve the cross product of vectors. Using the definition and properties of the cross product, we can solve such equations. Let us examine the following illustrations.

Illustration 3.50

Let $\vec{a} = \hat{i} + 4\hat{j} + 2\hat{k}, \vec{b} = 3\hat{i} - 2\hat{j} + 7\hat{k}$ and $\vec{c} = 2\hat{i} - \hat{j} + 4\hat{k}$. Find a vector $\vec{d}$ which is perpendicular to both $\vec{a}$ and $\vec{b}$ and $\vec{c}\cdot\vec{d} = 15$.

Sol. Vector $\vec{d}$ is perpendicular to vectors

$\vec{a} = \hat{i} + 4\hat{j} + 2\hat{k}, \vec{b} = 3\hat{i} - 2\hat{j} + 7\hat{k}$

$$\therefore \quad \vec{d} = \lambda \begin{vmatrix} i & j & k \\ 1 & 4 & 2 \\ 3 & -2 & 7 \end{vmatrix} = \lambda(32i - j - 14\hat{k})$$

Also $\vec{c}\cdot\vec{d} = 15$

$\therefore \quad \lambda(2\hat{i} - \hat{j} + 4\hat{k})\cdot(32\hat{i} - \hat{j} - 14\hat{k}) = 15$

$\Rightarrow \quad 9\lambda = 15$

$\Rightarrow \quad \lambda = \dfrac{5}{3}$

Hence, the required vector $\vec{d} = \dfrac{1}{3}(160\hat{i} - 5\hat{j} - 70\hat{k})$

Illustration 3.51

Let $\vec{a} = \hat{i} - \hat{j} + 2\hat{k}$ and $\vec{b} = 2\hat{i} + \hat{j} - 3\hat{k}$. If $\vec{r} \times \vec{a} = \vec{r} \times \vec{b}, \vec{r}\cdot(\hat{i} + \hat{j} + \hat{k}) = -4$ then find vector $\vec{r}$.

Sol. We have $\vec{r} \times \vec{a} = \vec{r} \times \vec{b}$

$\Rightarrow \quad \vec{r} \times \vec{a} - \vec{r} \times \vec{b} = 0$

$\Rightarrow \quad \vec{r} \times (\vec{a} - \vec{b}) = 0$

$\Rightarrow \quad \vec{r} = \lambda(\vec{a} - \vec{b})$

$\Rightarrow \quad \vec{r} = \lambda(-\hat{i} - 2\hat{j} + 5\hat{k})$

Now, $\vec{r}\cdot(\hat{i} + \hat{j} + \hat{k}) = -4$

$\therefore \quad \lambda(-1 - 2 + 5) = -4$ or $\lambda = -2$

So, $\vec{r} = 2\hat{k} + 4\hat{j} - 10\hat{k}$

Illustration 3.52

If $\vec{b}$ is not perpendicular to $\vec{c}$, then find the vector $\vec{r}$ satisfying the equations $\vec{r} \times \vec{b} = \vec{a} \times \vec{b}$ and $\vec{r} \cdot \vec{c} = 0$.

Sol. Given $\vec{r} \times \vec{b} = \vec{a} \times \vec{b} \Rightarrow (\vec{r} - \vec{a}) \times \vec{b} = 0$

Hence, $(\vec{r} - \vec{a})$ and $\vec{b}$ are parallel.

$\Rightarrow \quad \vec{r} - \vec{a} = t\vec{b}$ ...(i)

Also $\vec{r} \cdot \vec{c} = 0$

Therefore, taking dot product of (i) by $\vec{c}$, we get

$$\vec{r} \cdot \vec{c} - \vec{a} \cdot \vec{c} = t(\vec{b} \cdot \vec{c})$$

or $\quad 0 - \vec{a} \cdot \vec{c} = t(\vec{b} \cdot \vec{c})$ or $t = -\left(\dfrac{\vec{a} \cdot \vec{c}}{\vec{b} \cdot \vec{c}}\right)$ ...(ii)

From (i) and (ii), $\vec{r} = \vec{a} - \left(\dfrac{\vec{a} \cdot \vec{c}}{\vec{b} \cdot \vec{c}}\right)\vec{b}$

3.3.4 Applications of Cross Product in Geometry

Just like the dot product, the cross product is also a powerful tool in geometry, as it helps both in proving new results and in verifying known ones. Let us examine the following illustrations.

Illustration 3.53

In a triangle ABC, prove that $\dfrac{a}{\sin A} = \dfrac{b}{\sin B} = \dfrac{c}{\sin C}$, where a, b, c are sides opposite to vertices A, B and C respectively.

Sol. Let $\overrightarrow{BC} = \vec{a}$, $\overrightarrow{CA} = \vec{b}$, $\overrightarrow{AB} = \vec{c}$, so that $\vec{a} + \vec{b} = -\vec{c}$. Therefore,

$$\vec{a} \times \vec{a} + \vec{a} \times \vec{b} = -\vec{a} \times \vec{c}$$

$\Rightarrow \quad 0 + \vec{a} \times \vec{b} = \vec{c} \times \vec{a}$

$\Rightarrow \quad |\vec{a} \times \vec{b}| = |\vec{c} \times \vec{a}|$

$\Rightarrow \quad ab \sin(180° - C) = ca \sin(180° - B)$

$\Rightarrow \quad ab \sin C = ca \sin B$

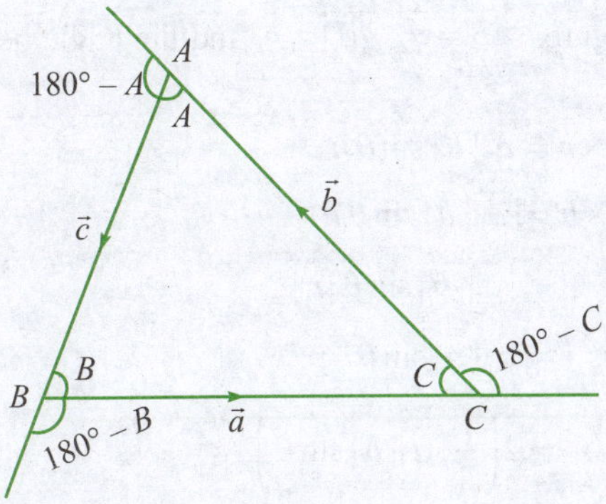

Dividing both sides by abc, we get

$$\frac{\sin C}{c} = \frac{\sin B}{b}$$

$\therefore \quad \dfrac{b}{\sin B} = \dfrac{c}{\sin C}$ ...(i)

Similarly, we get $\dfrac{c}{\sin C} = \dfrac{a}{\sin A}$ ...(ii)

From (i) and (ii), we have

$$\frac{a}{\sin A} = \frac{b}{\sin B} = \frac{c}{\sin C}$$

Illustration 3.54

Using cross product of vectors, prove that $\sin(A + B) = \sin A \cos B + \cos A \sin B$.

Sol. Let OP and OQ be unit vectors making angles A and B with the X-axis such that

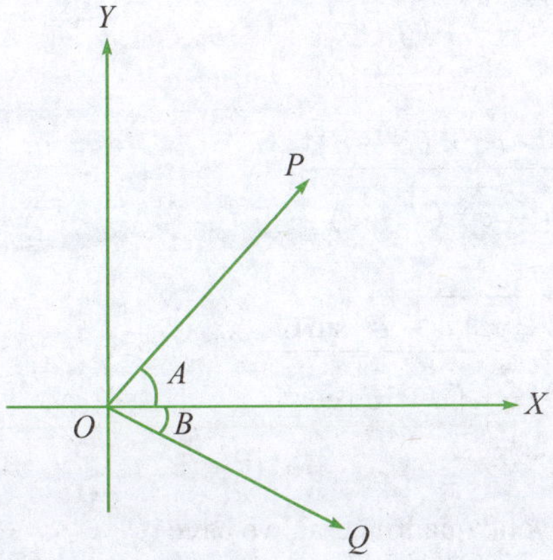

$\angle POQ = A + B$

$\therefore \quad \overline{OP} = \hat{i} \cos A + \hat{j} \sin A$

$\overline{OQ} = \hat{i} \cos B - \hat{j} \sin B$

Now, $\overline{OP} \times \overline{OQ} = (1)(1) \sin(A + B)(-\hat{k})$

$= -\sin(A + B)\hat{k}$ ...(i)

Also, $\overline{OP} \times \overline{OQ} = \begin{vmatrix} \hat{i} & \hat{j} & \hat{k} \\ \cos A & \sin A & 0 \\ \cos B & -\sin B & 0 \end{vmatrix}$

$= (-\cos A \sin B - \sin A \cos B)\hat{k}$

$= -(\sin A \cos B + \cos A \sin B)\hat{k}$ ...(ii)

From (i) and (ii), we get

$$\sin(A + B) = \sin A \cos B + \cos A \sin B$$

Illustration 3.55

If $\vec{a}, \vec{b}, \vec{c}$ and $\vec{d}$ are the position vectors of the vertices of a cyclic quadrilateral $ABCD$, prove that

$$\frac{|\vec{a}\times\vec{b}+\vec{b}\times\vec{d}+\vec{d}\times\vec{a}|}{(\vec{b}-\vec{a})\cdot(\vec{d}-\vec{a})}+\frac{|\vec{b}\times\vec{c}+\vec{c}\times\vec{d}+\vec{d}\times\vec{b}|}{(\vec{b}-\vec{c})\cdot(\vec{d}-\vec{c})}$$

$= 0.$

Sol. We have,

$$\frac{|\vec{a}\times\vec{b}+\vec{b}\times\vec{d}+\vec{d}\times\vec{a}|}{(\vec{b}-\vec{a})\cdot(\vec{d}-\vec{a})}=\frac{|(\vec{a}-\vec{d})\times(\vec{b}-\vec{a})|}{(\vec{b}-\vec{a})\cdot(\vec{d}-\vec{a})}$$

$$=\frac{|\vec{a}-\vec{d}||\vec{b}-\vec{a}|\sin A}{|\vec{b}-\vec{a}||\vec{d}-\vec{a}|\cos A}$$

$$= \tan A \qquad \ldots\text{(i)}$$

Also $$\frac{|\vec{b}\times\vec{c}+\vec{c}\times\vec{d}+\vec{d}\times\vec{b}|}{(\vec{b}-\vec{c})\cdot(\vec{d}-\vec{c})}$$

$$=\frac{|(\vec{b}-\vec{c})\times(\vec{c}-\vec{d})|}{(\vec{b}-\vec{c})\cdot(\vec{d}-\vec{c})}$$

$$=\frac{|\vec{b}-\vec{c}||\vec{c}-\vec{d}|\sin C}{|\vec{b}-\vec{c}|\cdot|\vec{d}-\vec{c}|\cos C}$$

$$= \tan C \qquad \ldots\text{(iii)}$$

$ABCD$ is a cyclic quadrilateral, we have

$A = 180° - C$

or $\tan A = \tan(180° - C)$

or $\tan A + \tan C = 0$

$\Rightarrow$

$$\frac{|\vec{a}\times\vec{b}+\vec{b}\times\vec{d}+\vec{d}\times\vec{a}|}{(\vec{b}-\vec{a})\cdot(\vec{d}-\vec{a})}+\frac{|\vec{b}\times\vec{c}+\vec{c}\times\vec{d}+\vec{d}\times\vec{b}|}{(\vec{b}-\vec{c})\cdot(\vec{d}-\vec{c})}=0$$

Illustration 3.56

Let A, B, C be points with position vectors $2\hat{i}-\hat{j}+\hat{k}, \hat{i}+2\hat{j}+\hat{k}$ and $3\hat{i}+\hat{j}+2\hat{k}$ respectively. Find the shortest distance between point B and plane OAC.

Sol. Shortest distance between B and plane OAC is the perpendicular distance of point B from the plane which is BM, where M is the foot of perpendicular from B on plane OAC.

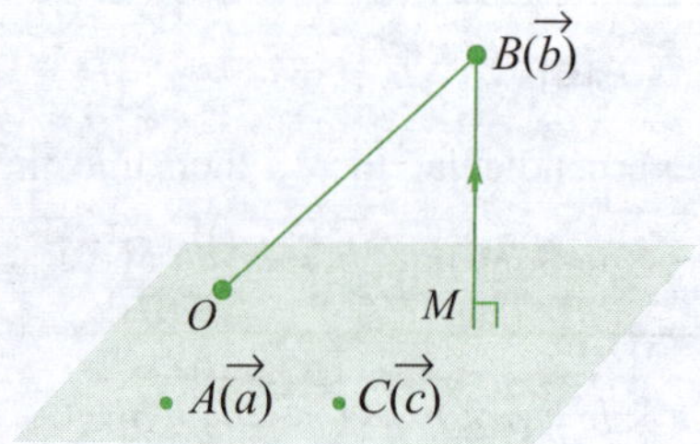

BM = projection vector $\overrightarrow{OB}$ on vector perpendicular to the plane.

Now vector perpendicular to the plane is $\overrightarrow{OA}\times\overrightarrow{OC}=\vec{a}\times\vec{c}$. Therefore,

$$BM=\frac{\overrightarrow{OB}\cdot(\vec{a}\times\vec{c})}{|\vec{a}\times\vec{c}|}$$

$$=\frac{\vec{b}\cdot(\vec{a}\times\vec{c})}{|\vec{a}\times\vec{c}|}$$

Now $\vec{a}\times\vec{c}=\begin{vmatrix}\hat{i}&\hat{j}&\hat{k}\\2&-1&1\\3&1&2\end{vmatrix}=-3\hat{i}-\hat{j}+5\hat{k}$

$\therefore\ \vec{b}\cdot(\vec{a}\times\vec{c})=(\hat{i}+2\hat{j}+3\hat{k})\cdot(-3\hat{i}-\hat{j}+5\hat{k})=-3-2+15=10$

and $|\vec{a}\times\vec{c}|=|-3\hat{i}-\hat{j}+5\hat{k}|=\sqrt{9+1+25}=\sqrt{35}$

$\therefore\quad BM=\dfrac{10}{\sqrt{35}}=2\sqrt{\dfrac{5}{7}}$

3.3.5 Area of a Triangle and a Quadrilateral Using Cross Product

We have the following results.

1. Area of a parallelogram whose adjacent edges are vectors $\vec{a}$ and $\vec{b}$ is $|\vec{a}\times\vec{b}|$

Proof:

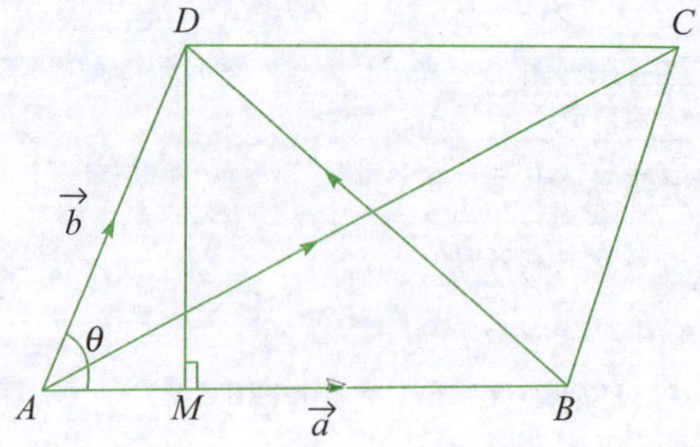

In the figure, $\overrightarrow{AB}=\vec{a}$, $\overrightarrow{AD}=\vec{b}$ and the angle between them is θ.

Now, $\vec{a}\times\vec{b}=|\vec{a}||\vec{b}|\sin\theta\,\hat{n}$

$\therefore\quad |\vec{a}\times\vec{b}|=|\,|\vec{a}||\vec{b}|\sin\theta\,\hat{n}\,|$

$= |\vec{a}||\vec{b}|\sin\theta\,|\hat{n}|$

$=|\vec{a}||\vec{b}|\sin\theta$

$=2\left(\dfrac{1}{2}|\vec{a}||\vec{b}|\sin\theta\right)$

$= 2\left(\frac{1}{2}AB \times DM\right)$

$= 2(\text{Area of } \Delta ABD)$

$=$ Area of parallelogram $ABCD$

2. Area of a triangle formed by the vectors $\vec{a}$ and $\vec{b}$ is $\frac{1}{2}|\vec{a} \times \vec{b}|$

Proof:

In the above figure, ΔABD is formed by the vectors $\vec{a}$ and $\vec{b}$.

So, the area of $\Delta ABD = \frac{1}{2} \times$ (area of parallelogram)

$= \frac{1}{2} \times |\vec{a} \times \vec{b}|$

3. Area of a parallelogram formed by the diagonal vectors $\vec{p}$ and $\vec{q}$ is $\frac{1}{2}|\vec{p} \times \vec{q}|$

Proof:

In the above figure, let $\overrightarrow{AC} = \vec{a} + \vec{b} = \vec{p}$ (*let*)

and $\overrightarrow{BD} = \vec{b} - \vec{a} = \vec{q}$ (*let*)

$\therefore \quad \frac{1}{2}|\vec{p} \times \vec{q}| = \frac{1}{2}|\overrightarrow{AC} \times \overrightarrow{BD}|$

$= \frac{1}{2}|(\vec{a} + \vec{b}) \times (\vec{b} - \vec{a})|$

$= \frac{1}{2}|\vec{a} \times \vec{b} - \vec{a} \times \vec{a} + \vec{b} \times \vec{b} - \vec{b} \times \vec{a}|$

$= \frac{1}{2}|\vec{a} \times \vec{b} + \vec{a} \times \vec{b}|$

$= |\vec{a} \times \vec{b}|$

$=$ Area of the parallelogram

4. Area of a quadrilateral formed by the diagonal vectors $\vec{p}$ and $\vec{q}$ is $\frac{1}{2}|\vec{p} \times \vec{q}|$

Proof:

Consider quadrilateral $ABCD$ whose diagonal vectors are $\overrightarrow{AC} = \vec{p}$ and $\overrightarrow{BD} = \vec{q}$.

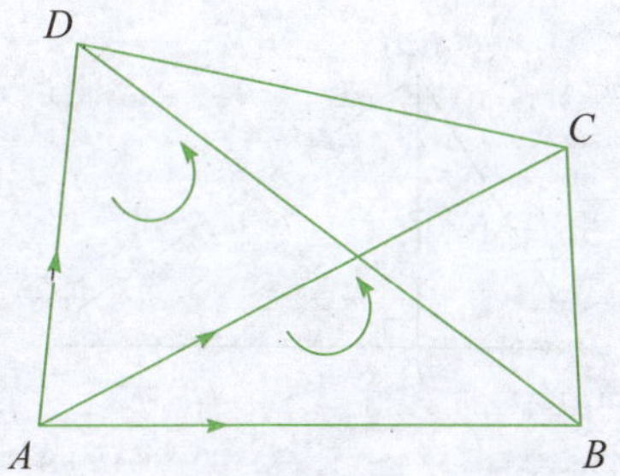

Now, vector area of the quadrilateral $ABCD$

$=$ vector area of ΔABC + vector area of ΔACD

$= \frac{1}{2}\overrightarrow{AB} \times \overrightarrow{AC} + \frac{1}{2}\overrightarrow{AC} \times \overrightarrow{AD}$

(as both these cross products are in the same direction)

$= -\frac{1}{2}\overrightarrow{AC} \times \overrightarrow{AB} + \frac{1}{2}\overrightarrow{AC} \times \overrightarrow{AD}$

$= \frac{1}{2}\overrightarrow{AC} \times (\overrightarrow{AD} - \overrightarrow{AB})$

$= \frac{1}{2}\overrightarrow{AC} \times \overrightarrow{BD} = \frac{1}{2}\vec{p} \times \vec{q}$

Thus, the area of quadrilateral is $\frac{1}{2}|\vec{p} \times \vec{q}|$

5. Area of triangle with position vectors of vertices as $\vec{a}$, $\vec{b}$ and $\vec{c}$ is $\frac{1}{2}|\vec{a} \times \vec{b} + \vec{b} \times \vec{c} + \vec{c} \times \vec{a}|$.

Proof:

Let in ΔABC, the position vectors of the vertices A, B, C are $\vec{a}$, $\vec{b}$ and $\vec{c}$ respectively.

$\therefore \quad$ Area of triangle $= \frac{1}{2}|\overrightarrow{AB} \times \overrightarrow{AC}|$

$= \frac{1}{2}|(\vec{b} - \vec{a}) \times (\vec{c} - \vec{a})|$

$= \frac{1}{2}|\vec{b} \times \vec{c} - \vec{b} \times \vec{a} - \vec{a} \times \vec{c} + \vec{a} \times \vec{a}|$

$= \frac{1}{2}|\vec{a} \times \vec{b} + \vec{b} \times \vec{c} + \vec{c} \times \vec{a}|$

Illustration 3.57

Find the area of the triangle having vertices A (1, 1, 2), B (2, 3, 5) and C (1, 5, 5).

Sol. The vertices of triangle ABC are given as A (1, 1, 2), B (2, 3, 5), and C (1, 5, 5).

The adjacent sides $\overrightarrow{AB}$ and $\overrightarrow{BC}$ of ΔABC are given as

$$\overrightarrow{AB} = (2-1)\hat{i} + (3-1)\hat{j} + (5-2)\hat{k} = \hat{i} + 2\hat{j} + 3\hat{k}$$

$$\overrightarrow{BC} = (1-2)\hat{i} + (5-3)\hat{j} + (5-5)\hat{k} = -\hat{i} + 2\hat{j}$$

Now, Area of $\Delta ABC = \frac{1}{2}|\overrightarrow{AB} \times \overrightarrow{BC}|$

$$\overrightarrow{AB} \times \overrightarrow{BC} = \begin{vmatrix} \hat{i} & \hat{j} & \hat{k} \\ 1 & 2 & 3 \\ -1 & 2 & 0 \end{vmatrix} = \hat{i}(-6) - \hat{j}(3) + \hat{k}(2+2) = -6\hat{i} - 3\hat{j} + 4\hat{k}$$

$$\therefore \quad |\overrightarrow{AB} \times \overrightarrow{BC}| = \sqrt{(-6)^2 + (-3)^2 + 4^2} = \sqrt{36 + 9 + 16} = \sqrt{61}$$

Hence, the area of ΔABC is $\frac{\sqrt{61}}{2}$ square units.

Illustration 3.58

Find the area of the parallelogram whose adjacent sides are determined by the vectors $\vec{a} = \hat{i} - \hat{j} + 3\hat{k}$ and $\vec{b} = 2\hat{i} - 7\hat{j} + \hat{k}$.

Sol. The area of the parallelogram whose adjacent sides are $\vec{a}$ and $\vec{b}$ is $|\vec{a} \times \vec{b}|$.

Adjacent sides are given as

$$\vec{a} = \hat{i} - \hat{j} + 3\hat{k} \text{ and } \vec{b} = 2\hat{i} - 7\hat{j} + \hat{k}$$

$$\text{Now, } \vec{a} \times \vec{b} = \begin{vmatrix} \hat{i} & \hat{j} & \hat{k} \\ 1 & -1 & 3 \\ 2 & -7 & 1 \end{vmatrix} = \hat{i}(-1+21) - \hat{j}(1-6) + \hat{k}(-7+2)$$

$$= 20\hat{i} + 5\hat{j} - 5\hat{k}$$

$$\therefore \quad |\vec{a} \times \vec{b}| = \sqrt{20^2 + 5^2 + (-5)^2} = \sqrt{400 + 25 + 25} = 15\sqrt{2}$$

Hence, the area of the parallelogram is $15\sqrt{2}$ square units.

Illustration 3.59

Find the area of a parallelogram whose diagonals are $\vec{a} = 3\hat{i} + \hat{j} - 2\hat{k}$ and $\vec{b} = \hat{i} - 3\hat{j} + 4\hat{k}$.

Sol. Area of the parallelogram, $\Delta = \frac{1}{2}|\vec{a} \times \vec{b}|$

$$\text{Now, } \vec{a} \times \vec{b} = \begin{vmatrix} \hat{i} & \hat{j} & \hat{k} \\ 3 & 1 & -2 \\ 1 & -3 & 4 \end{vmatrix} = -2\hat{i} - 14\hat{j} - 10\hat{k}$$

$$\text{Hence, } \Delta = \frac{1}{2}|\vec{a} \times \vec{b}| = \frac{1}{2}\sqrt{4+196+100} = 5\sqrt{3}$$

Illustration 3.60

The position vectors of the vertices of a quadrilateral with A at the origin are $B(\vec{b})$, $D(\vec{d})$ and $C(l\vec{b} + m\vec{d})$. Prove that the area of the quadrilateral is $\frac{1}{2}(l+m)|\vec{b} \times \vec{d}|$.

Sol. Area of the quadrilateral $= \frac{1}{2}|\overrightarrow{AC} \times \overrightarrow{BD}|$

$$= \frac{1}{2}|(l\vec{b} + m\vec{d}) \times (\vec{d} - \vec{b})|$$

$$= \frac{1}{2}|l\vec{b} \times \vec{d} - m\vec{d} \times \vec{b}|$$

$$= \frac{1}{2}(l+m)|\vec{b} \times \vec{d}|$$

Illustration 3.61

If A, B, C and D be any four non-coplanar points in the space, then prove that $|\overrightarrow{AB} \times \overrightarrow{CD} + \overrightarrow{BC} \times \overrightarrow{AD} + \overrightarrow{CA} \times \overrightarrow{BD}|$ = 4(area of ΔABC).

Sol. Let the position vectors of A, B, C and D be $\vec{a}$, $\vec{b}$, $\vec{c}$ and $\vec{0}$, respectively. Then

$$\overrightarrow{AB} \times \overrightarrow{CD} = (\vec{b} - \vec{a}) \times (-\vec{c}),$$

$$\overrightarrow{BC} \times \overrightarrow{AD} = (\vec{c} - \vec{b}) \times (-\vec{a})$$

and $\overrightarrow{CA} \times \overrightarrow{BD} = (\vec{a} - \vec{c}) \times (-\vec{b})$

$$\therefore \quad \overrightarrow{AB} \times \overrightarrow{CD} + \overrightarrow{BC} \times \overrightarrow{AD} + \overrightarrow{CA} \times \overrightarrow{BD}$$

$$= \vec{c} \times \vec{b} + \vec{a} \times \vec{c} + \vec{a} \times \vec{c} + \vec{b} \times \vec{a} + \vec{b} \times \vec{a} + \vec{c} \times \vec{b}$$

$$= 2(\vec{c} \times \vec{b} + \vec{b} \times \vec{a} + \vec{a} \times \vec{c})$$

$$= 2(\vec{c} \times (\vec{b} - \vec{a}) - \vec{a} \times (\vec{b} - \vec{a}))$$

$$= 2((\vec{c} - \vec{a}) \times (\vec{b} - \vec{a}))$$

$$= 2(\overrightarrow{AC} \times \overrightarrow{AB})$$

$$\therefore \quad |\overrightarrow{AB} \times \overrightarrow{CD} + \overrightarrow{BC} \times \overrightarrow{AD} + \overrightarrow{CA} \times \overrightarrow{BD}| = 4\left|\frac{1}{2}(\overrightarrow{AC} \times \overrightarrow{AB})\right|$$

= 4(area of ΔABC)

Illustration 3.62

In ΔABC, if $\vec{a}$, $\vec{b}$ and $\vec{c}$ are the position vectors of the vertices A, B and C, respectively then prove that the perpendicular distance of the vertex A from the base BC of the triangle is $\frac{|\vec{a} \times \vec{b} + \vec{b} \times \vec{c} + \vec{c} \times \vec{a}|}{|\vec{c} - \vec{b}|}$.

Sol. We have, $|\overrightarrow{BC} \times \overrightarrow{BA}| = |\vec{a} \times \vec{b} + \vec{b} \times \vec{c} + \vec{c} \times \vec{a}|$

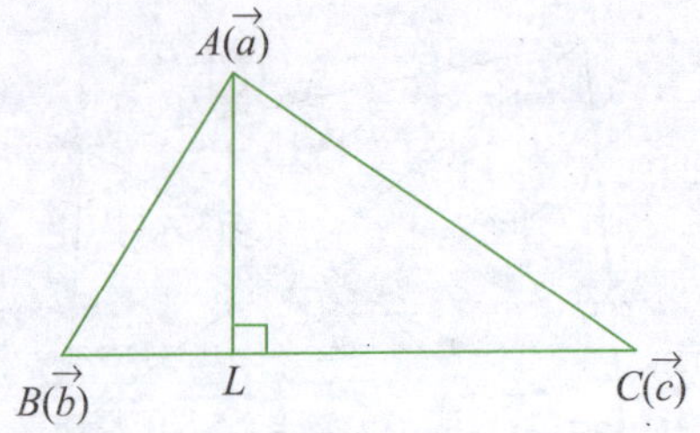

$$\therefore \quad |\overrightarrow{BC}||\overrightarrow{BA}|\sin B = |\vec{a} \times \vec{b} + \vec{b} \times \vec{c} + \vec{c} \times \vec{a}|$$

or $\quad |\vec{c} - \vec{b}|(AB \sin B) = |\vec{a} \times \vec{b} + \vec{b} \times \vec{c} + \vec{c} \times \vec{a}|$

Therefore, the length of perpendicular from A on BC is

$$AL = AB \sin B = \frac{|\vec{a} \times \vec{b} + \vec{b} \times \vec{c} + \vec{c} \times \vec{a}|}{|\vec{b} - \vec{c}|}.$$

Illustration 3.63

In triangle ABC, points D, E and F are taken on the sides BC, CA and AB, respectively, such that $\dfrac{BD}{DC}=\dfrac{CE}{EA}=\dfrac{AF}{FB}=n.$

Prove that $\Delta_{DEF}=\dfrac{n^2-n+1}{(n+1)^2}\Delta_{ABC}$.

Sol. Let A be the origin and let the position vectors of points B and C be $\vec{b}$ and $\vec{c}$, respectively.

Therefore, the position vectors of D, E and F are

$\dfrac{n\vec{c}+\vec{b}}{n+1}$, $\dfrac{\vec{c}}{n+1}$ and $\dfrac{n\vec{b}}{n+1}$ respectively. Hence,

$$\overrightarrow{ED}=\overrightarrow{AD}-\overrightarrow{AE}=\frac{(n-1)\vec{c}+\vec{b}}{n+1} \text{ and } \overrightarrow{EF}=\frac{n\vec{b}-\vec{c}}{n+1}$$

Now, the vector area of $\Delta ABC=\dfrac{1}{2}(\vec{b}\times\vec{c})$

and the vector area of $\Delta DEF=\dfrac{1}{2}(\overrightarrow{EF}\times\overrightarrow{ED})$

$$=\frac{1}{2(n+1)^2}[(n\vec{b}-\vec{c})\times\{(n-1)\vec{c}+\vec{b}\}]$$

$$=\frac{1}{2(n+1)^2}[(n^2-n)\vec{b}\times\vec{c}+\vec{b}\times\vec{c}]$$

$$=\frac{1}{2(n+1)^2}[(n^2-n+1)(\vec{b}\times\vec{c})]=\frac{n^2-n+1}{(n+1)^2}\Delta_{ABC}$$

Illustration 3.64

Show by a numerical example and geometrically also that $\vec{a}\times\vec{b}=\vec{a}\times\vec{c}$ does not imply $\vec{b}=\vec{c}$.

Sol. Let $\vec{a}=3\hat{i}+2\hat{j}+5\hat{k}$, $\vec{b}=6\hat{i}+5\hat{j}+8\hat{k}$, $\vec{c}=3\hat{i}+3\hat{j}+3\hat{k}$

$$\therefore\quad \vec{a}\times\vec{b}=\begin{vmatrix}\hat{i}&\hat{j}&\hat{k}\\3&2&5\\6&5&8\end{vmatrix}$$

$$=(16-25)\hat{i}-(24-30)\hat{j}+(15-12)\hat{k}$$
$$=-9\hat{i}+6\hat{j}+3\hat{k}$$

$$\text{Also, }\vec{a}\times\vec{c}=\begin{vmatrix}\hat{i}&\hat{j}&\hat{k}\\3&2&5\\3&3&3\end{vmatrix}$$

$$=(6-15)\hat{i}-(9-15)\hat{j}+(9-6)\hat{k}=-9\hat{i}+6\hat{j}+3\hat{k}$$

$\therefore\quad \vec{a}\times\vec{b}=\vec{a}\times\vec{c}$, but $\vec{b}\neq\vec{c}$.

Geometrically

Let $\overrightarrow{AB}=\vec{a}$, $\overrightarrow{AD}=\vec{b}$, $\overrightarrow{AD'}=\vec{c}$. Then

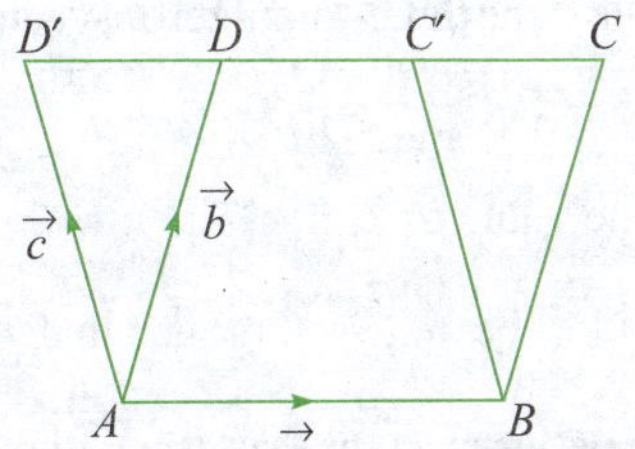

Vector area of parallelogram $ABCD=\vec{a}\times\vec{b}$

Vector area of parallelogram $ABC'D'=\vec{a}\times\vec{c}$

Now, Vector area of parallelogram $ABCD$

= Vector area of parallelogram $ABC'D'$

($\because$ both parallelograms have same base and same height)

$\therefore\quad \vec{a}\times\vec{b}=\vec{a}\times\vec{c}$ but $\vec{b}\neq\vec{c}$

Aim to complete within 65 minutes

Concept Application Exercise 3.2

1. Find the moment of $\vec{F}$ about point (2, –1, 3), when force $\vec{F}=3\hat{i}+2\hat{j}-4\hat{k}$ is acting on point (1, –1 , 2).
2. If $\vec{a}=2\hat{i}+3\hat{j}-5\hat{k}$, $\vec{b}=m\hat{i}+n\hat{j}+12\hat{k}$ and $\vec{a}\times\vec{b}=\vec{0}$, then the value of $(4m-n)$ is
   (1) 12 (2) –12 (3) 6 (4) –6
3. Let $\vec{a}=3\hat{i}+2\hat{j}+x\hat{k}$ and $\vec{b}=\hat{i}-\hat{j}+\hat{k}$, for some real x. Then find the values of r for which $|\vec{a}\times\vec{b}|=r$ is possible.
4. Let $\vec{u}=\hat{i}-\hat{j}-2\hat{k}$, $\vec{v}=2\hat{i}+\hat{j}-\hat{k}$, $\vec{v}\cdot\vec{w}=2$ and $\vec{v}\times\vec{w}=\vec{u}+\lambda\vec{v}$. Then the value of $\vec{u}.\vec{w}$ is
   (1) 1 (2) 2 (3) 3 (4) 1/2
5. If $\vec{a}\times\vec{b}=\vec{b}\times\vec{c}\neq 0$, where $\vec{a},\vec{b}$ and $\vec{c}$ are coplanar vectors, then for some scalar k prove that $\vec{a}+\vec{c}=k\vec{b}$.
6. If $\vec{a}=2\hat{i}+3\hat{j}-\hat{k}$, $\vec{b}=-\hat{i}+2\hat{j}-4\hat{k}$ and $\vec{c}=\hat{i}+\hat{j}+\hat{k}$, then the value of $(\vec{a}\times\vec{b})\cdot(\vec{a}\times\vec{c})$ is
   (1) –70 (2) –72 (3) –74 (4) –80
7. Given that $\vec{a}\cdot\vec{b}=\vec{a}\cdot\vec{c}$, $\vec{a}\times\vec{b}=\vec{a}\times\vec{c}$ and $\vec{a}$ is not a zero vector. Show that $\vec{b}=\vec{c}$.
8. Show that $(\vec{a}-\vec{b})\times(\vec{a}+\vec{b})=2\vec{a}\times\vec{b}$ and give a geometrical interpretation of it.

9. If $\vec{x}$ and $\vec{y}$ are unit vectors and $|\vec{z}| = \frac{2}{\sqrt{7}}$ such that $\vec{z} + \vec{z} \times \vec{x} = \vec{y}$, then the angle θ between $\vec{x}$ and $\vec{z}$ is

(1) 30° (2) 45° (3) 60° (4) 90°

10. Prove that $(\vec{a}\cdot\hat{i})(\vec{a}\times\hat{i})+(\vec{a}\cdot\hat{j})(\vec{a}\times\hat{j})+(\vec{a}\cdot\hat{k})(\vec{a}\times\hat{k}) = \vec{0}$.

11. Let $\vec{a}, \vec{b}$ and $\vec{c}$ be three non-zero vectors such that $\vec{a} + \vec{b} + \vec{c} = \vec{0}$ and $\lambda \vec{b} \times \vec{a} + \vec{b} \times \vec{c} + \vec{c} \times \vec{a} = \vec{0}$. Then find the value of λ.

12. Let $\vec{a}, \vec{b}$ and $\vec{c}$ be unit vectors such that $\vec{a}\cdot\vec{b} = 0 = \vec{a}\cdot\vec{c}$. If the angle between $\vec{b}$ and $\vec{c}$ is $\frac{\pi}{6}$, then find $\vec{a}$.

13. If $(\vec{a}\times\vec{b})^2 + (\vec{a}\cdot\vec{b})^2 = 144$ and $|\vec{a}| = 4$, then the value of $|\vec{b}|$ is

(1) 3 (2) 4 (3) 6 (4) 8

14. Given $|\vec{a}| = |\vec{b}| = 1$ and $|\vec{a} + \vec{b}| = \sqrt{3}$. If $\vec{c}$ is a vector such that $\vec{c} - \vec{a} - 2\vec{b} = 3(\vec{a} \times \vec{b})$, then the value of $\vec{c}\cdot\vec{b}$ is

(1) 1.5 (2) 2.5 (3) 4.5 (4) 5

15. The magnitude of the projection of the vector $2\hat{i} + 3\hat{j} + \hat{k}$ on the vector perpendicular to the plane containing the vectors $\hat{i} + \hat{j} + \hat{k}$ and $\hat{i} + 2\hat{j} + 3\hat{k}$, is

(1) $\frac{\sqrt{3}}{2}$ (2) $3\sqrt{6}$ (3) $\sqrt{\frac{3}{2}}$ (4) $\sqrt{6}$

16. The area of the quadrilateral $ABCD$ with vertices $A(2, 1, 1)$, $B(1, 2, 5)$, $C(-2, -3, 5)$ and $D(\lambda, -6, -7)$ is equal to

(1) $9\sqrt{38}$ (2) 54 (3) $8\sqrt{38}$ (4) 48

17. Let $\vec{a} = -5\hat{i} + \hat{j} - 3\hat{k}$, $\vec{b} = \hat{i} + 2\hat{j} - 4\hat{k}$ and $\vec{c} = (((\vec{a}\times\vec{b})\times\hat{i})\times\hat{i})\times\hat{i}$. Then $\vec{c}\cdot(-\hat{i}+\hat{j}+\hat{k})$ is equal to

(1) −12 (2) −10 (3) −13 (4) −15

18. If the area of parallelogram whose diagonals are $\hat{i} - \hat{j} + 2\hat{k}$ and $2\hat{i} + 3\hat{j} + \alpha\hat{k}$ is $\frac{\sqrt{93}}{2}$ square units then the greatest value of α is

(1) 2 (2) 4 (3) 6 (4) 8

19. In ΔABC, centroid O is the origin. If $\overrightarrow{OB} = \hat{i} - 2\hat{j} + 3\hat{k}$ and $\overrightarrow{OC} = 2\hat{i} + \hat{j} - 4\hat{k}$ then find the area of ΔABC.

20. If area of quadrilateral having diagonals as vectors $\vec{a}$ and $\vec{b}$ is 10 sq. units, then find the area of parallelogram formed by vectors $\vec{a}$ and $\vec{a} \times \vec{b}$, where $|\vec{a}| = 4$ is

(1) 40 sq. units (2) 60 sq. units
(3) 80 sq. units (4) 92 sq. units

Answers

1. $2\hat{i} - 7\hat{j} - 2\hat{k}$ **2.** (2)
3. $5\sqrt{\frac{3}{2}}$ **4.** (1)
6. (3) **9.** (3)
11. 2 **12.** $\pm 2(\vec{b} \times \vec{c})$
13. (1) **14.** (2)
15. (3) **16.** (3)
17. (1) **18.** (2)
19. $\frac{15}{2}\sqrt{6}$ sq. units **20.** (3)

3.4 SCALAR TRIPLE PRODUCT

3.4.1 Definition of the Scalar Triple Product

The scalar triple product or mixed product of three vectors is the dot product of one of the vectors with the cross product of the other two vectors.

The scalar triple product of three vectors $\vec{a}, \vec{b}, \vec{c}$ is defined as $\vec{a}\cdot(\vec{b}\times\vec{c})$ and denoted by $[\vec{a}\,\vec{b}\,\vec{c}]$. It is also called the box product of vectors.

The product $\vec{a}\cdot(\vec{b}\times\vec{c})$ can also be written without brackets as $\vec{a}\cdot\vec{b}\times\vec{c}$, because it is meaningful only when the cross-product is evaluated first.

If $\vec{a} = a_1\hat{i} + a_2\hat{j} + a_3\hat{k}$, $\vec{b} = b_1\hat{i} + b_2\hat{j} + b_3\hat{k}$ and $\vec{c} = c_1\hat{i} + c_2\hat{j} + c_3\hat{k}$, then

$$\vec{a}\cdot(\vec{b}\times\vec{c}) = (a_1\hat{i}+a_2\hat{j}+a_3\hat{k})\cdot\begin{vmatrix} \hat{i} & \hat{j} & \hat{k} \\ b_1 & b_2 & b_3 \\ c_1 & c_2 & c_3 \end{vmatrix}$$

$$= (a_1\hat{i}+a_2\hat{j}+a_3\hat{k})\cdot((b_2c_3 - b_3c_2)\hat{i} - (b_1c_3 - b_3c_1)\hat{j} + (b_1c_2 - b_2c_1)\hat{k})$$

$$= a_1(b_2c_3 - b_3c_2) - a_2(b_1c_3 - b_3c_1) + a_3(b_1c_2 - b_2c_1)$$

$$= \begin{vmatrix} a_1 & a_2 & a_3 \\ b_1 & b_2 & b_3 \\ c_1 & c_2 & c_3 \end{vmatrix}$$

Also, it can be proved that $(\vec{a}\times\vec{b})\cdot\vec{c} = \begin{vmatrix} a_1 & a_2 & a_3 \\ b_1 & b_2 & b_3 \\ c_1 & c_2 & c_3 \end{vmatrix}$ ...(i)

Thus, $\vec{a}\cdot(\vec{b}\times\vec{c}) = (\vec{a}\times\vec{b})\cdot\vec{c} = [\vec{a}\,\vec{b}\,\vec{c}]$.

Thus, the scalar triple product remains unchanged if the dot and cross product operators are interchanged.

Now, using the row-interchange property of determinant, we have $[\vec{a}\,\vec{b}\,\vec{c}]=[\vec{b}\,\vec{c}\,\vec{a}]=[\vec{c}\,\vec{a}\,\vec{b}]$ ...(ii)

Thus, we can begin writing the box product with any of the vectors, provided that the vectors $\vec{a}, \vec{b}$ and $\vec{c}$ are in cyclic order.

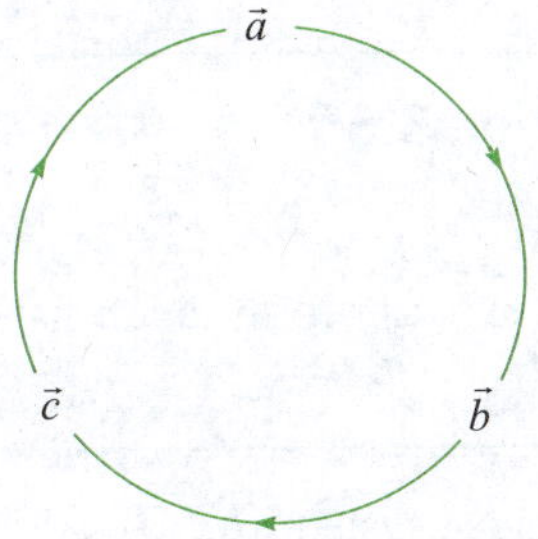

Further, if we interchange any two vectors in any of the box products in (ii), the sign of the product gets reversed, i.e. $[\vec{a}\,\vec{b}\,\vec{c}]=-[\vec{a}\,\vec{c}\,\vec{b}]=-[\vec{c}\,\vec{b}\,\vec{a}]=-[\vec{b}\,\vec{a}\,\vec{c}]$ ...(iii)

3.4.1.1 Handedness of Coordinate Systems and Scalar Triple Product

Typically, there are two types of Cartesian coordinate systems: left-handed and right-handed. In both systems, the positive x-axis is taken to the right and the positive y-axis upward. The orientation of the system determines whether the positive z-axis points outward (toward the observer) or inward (away from the observer). To see this, place the fingers of the left or right hand along the positive x-axis and curl them toward the positive y-axis. The direction of the thumb then gives the positive z-axis for that coordinate system. If the right hand is used, the system is right-handed; if the left hand is used, the system is left-handed.

Left-handed Cartesian Coordinates | Right-handed Cartesian Coordinates

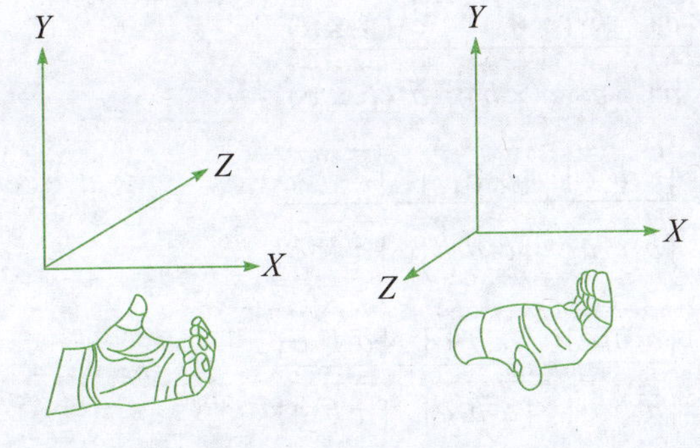

In summary, if the positive x-axis points to the right and the positive y-axis points upward, then the positive z-axis can either point toward the observer (right-handed system) or away from the observer (left-handed system).

Similarly, in a two-dimensional system, if the positive x-axis points to the right, then the positive y-axis may be taken either upward (right-handed system) or downward (left-handed system). The two-dimensional analogue of the cross product between two vectors is positive if the vectors defining the parallelogram are taken in the same order (that is, the second vector is obtained by a counterclockwise rotation from the first, just as the y-axis is from the x-axis).

In general, the ordered set of unit vectors $(\hat{i}, \hat{j}, \hat{k})$ is taken to form a right-handed system. This implies that $\hat{i}\cdot(\hat{j}\times\hat{k})>0$, and in fact, $\hat{i}\cdot(\hat{j}\times\hat{k})=1$

The scalar triple product (STP) in three dimensions, and the two-dimensional analogue of the cross product, do not by themselves fix the handedness of the coordinate system. They only indicate whether the orientation of the vectors involved agrees with the handedness of the chosen coordinate system.

For example, if the vectors are ordered as $\hat{i}, \hat{j}, \hat{k}$, then $\hat{i}\cdot(\hat{j}\times\hat{k})=1$

However, if we reorder them so that no rotation can put them in the same order as the coordinate axes of the system—for instance, $\hat{j}, \hat{i}, \hat{k}$ - then their scalar triple product becomes -1.

Conclusion

In general, the sign of the scalar triple product tells us about the handedness of the three vectors relative to the coordinate system:

- If $[\vec{a}\,\vec{b}\,\vec{c}]>0$, the three vectors are in right-handed order.
- If $[\vec{a}\,\vec{b}\,\vec{c}]<0$, the three vectors are in left-handed order.

Since the coordinate system is usually assumed to be right-handed, a negative scalar triple product indicates that the three vectors form a left-handed system. Conversely, if the coordinate system itself were left-handed, then a negative scalar triple product would indicate that the three vectors are arranged in right-handed order.

3.4.2 Properties of Scalar Triple Product

1. $[k\,\vec{a}\,\vec{b}\,\vec{c}]=k[\vec{a}\,\vec{b}\,\vec{c}]$, where k is a scalar,

Since $[k\vec{a}\,\vec{b}\,\vec{c}]=(k\vec{a})\cdot(\vec{b}\times\vec{c})=k(\vec{a}\cdot(\vec{b}\times\vec{c}))=k[\vec{a}\,\vec{b}\,\vec{c}]$

2. $[\vec{a}+\vec{d}\,\vec{b}\,\vec{c}]=[\vec{a}\,\vec{b}\,\vec{c}]+[\vec{d}\,\vec{b}\,\vec{c}]$,

Since $[\vec{a}+\vec{d}\,\vec{b}\,\vec{c}]=(\vec{a}+\vec{d})\cdot(\vec{b}\times\vec{c})$

$=\vec{a}\cdot(\vec{b}\times\vec{c})+\vec{d}\cdot(\vec{b}\times\vec{c})$

$=[\vec{a}\,\vec{b}\,\vec{c}]+[\vec{d}\,\vec{b}\,\vec{c}]$

3. If any two of the three vectors $\vec{a}, \vec{b}, \vec{c}$ are collinear, then $[\vec{a}\,\vec{b}\,\vec{c}]=0$

For example, if $\vec{b}=\lambda\vec{a}$, then $[\vec{a}\,\vec{b}\,\vec{c}]=[\vec{a}\,\lambda\vec{a}\,\vec{c}]=(\vec{a}\times\lambda\vec{a})\cdot\vec{c}=0$

Thus, $[\vec{a}\,\vec{a}\,\vec{c}]=0$, that is if any two of the three vectors are equal, then the box product is zero. This is because $\vec{a}\perp(\vec{a}\times\vec{c})$ or equivalenty, two rows in the determinant of $\vec{a}\cdot(\vec{a}\times\vec{c})$ become identical.

4. For three non-zero distinct vectors $\vec{a}, \vec{b}, \vec{c}$, no two of which are collinear, $[\vec{a}\,\vec{b}\,\vec{c}]=0$, if vectors are coplanar.

Proof:

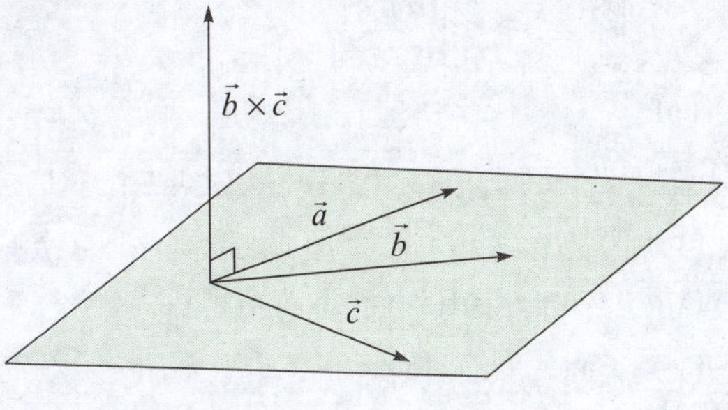

Vector $\vec{b}\times\vec{c}$ is perpendicular to the plane of vectors $\vec{b}$ and $\vec{c}$.
Since vector $\vec{a}$ lies in the plane of vectors $\vec{b}$ and $\vec{c}$, $\vec{a}\perp(\vec{b}\times\vec{c})$.

$\therefore \quad [\vec{a}\,\vec{b}\,\vec{c}]=\vec{a}\cdot(\vec{b}\times\vec{c})=0$

Also, if $[\vec{a}\,\vec{b}\,\vec{c}]=0$,
then $\vec{a}\cdot(\vec{b}\times\vec{c})=0$

$\Rightarrow \quad \vec{a}\perp(\vec{b}\times\vec{c})$

$\Rightarrow \quad \vec{a}$ is parallel to the plane of vectors $\vec{b}$ and $\vec{c}$.

$\Rightarrow \quad \vec{a}, \vec{b}, \vec{c}$ are coplanar.

5. $[\vec{a}\,\vec{b}\,\vec{c}][\vec{p}\,\vec{q}\,\vec{r}]=\begin{vmatrix}\vec{a}\cdot\vec{p} & \vec{a}\cdot\vec{q} & \vec{a}\cdot\vec{r}\\ \vec{b}\cdot\vec{p} & \vec{b}\cdot\vec{q} & \vec{a}\cdot\vec{r}\\ \vec{c}\cdot\vec{p} & \vec{c}\cdot\vec{q} & \vec{c}\cdot\vec{r}\end{vmatrix}$

Proof:

Let $\vec{a}=a_1\hat{i}+a_2\hat{j}+a_3\hat{k}$, $\vec{b}=b_1\hat{i}+b_2\hat{j}+b_3\hat{k}$,
$\vec{c}=c_1\hat{i}+c_2\hat{j}+c_3\hat{k}$

and $\vec{p}=p_1\hat{i}+p_2\hat{j}+p_3\hat{k}$, $\vec{q}=q_1\hat{i}+q_2\hat{j}+q_3\hat{k}$,
$\vec{r}=r_1\hat{i}+r_2\hat{j}+r_3\hat{k}$

$$\therefore \quad [\vec{a}\,\vec{b}\,\vec{c}][\vec{p}\,\vec{q}\,\vec{r}]=\begin{vmatrix}a_1 & a_2 & a_3\\ b_1 & b_2 & b_3\\ c_1 & c_2 & c_3\end{vmatrix}\begin{vmatrix}p_1 & p_2 & p_3\\ q_1 & q_2 & q_3\\ r_1 & r_2 & r_3\end{vmatrix}$$

$$=\begin{vmatrix}\sum a_1p_1 & \sum a_1q_1 & \sum a_1r_1\\ \sum b_1p_1 & \sum b_1q_1 & \sum b_1r_1\\ \sum c_1p_1 & \sum c_1q_1 & \sum c_1r_1\end{vmatrix},$$

where $\sum a_1p_1=a_1p_1+a_2p_2+a_3p_3=\vec{a}\cdot\vec{p}$ etc.

$$=\begin{vmatrix}\vec{a}\cdot\vec{p} & \vec{a}\cdot\vec{q} & \vec{a}\cdot\vec{r}\\ \vec{b}\cdot\vec{p} & \vec{b}\cdot\vec{q} & \vec{b}\cdot\vec{r}\\ \vec{c}\cdot\vec{p} & \vec{c}\cdot\vec{q} & \vec{c}\cdot\vec{r}\end{vmatrix}$$

Thus,

$$[\vec{a}\,\vec{b}\,\vec{c}]^2=[\vec{a}\,\vec{b}\,\vec{c}][\vec{a}\,\vec{b}\,\vec{c}]=\begin{vmatrix}\vec{a}\cdot\vec{a} & \vec{a}\cdot\vec{b} & \vec{a}\cdot\vec{c}\\ \vec{b}\cdot\vec{a} & \vec{b}\cdot\vec{b} & \vec{b}\cdot\vec{c}\\ \vec{c}\cdot\vec{a} & \vec{c}\cdot\vec{b} & \vec{c}\cdot\vec{c}\end{vmatrix}$$

6. $[x_1\vec{a}+y_1\vec{b}+z_1\vec{c}\;\; x_2\vec{a}+y_2\vec{b}+z_2\vec{c}\;\; x_3\vec{a}+y_3\vec{b}+z_3\vec{c}]$

$$=\begin{vmatrix}x_1 & y_1 & z_1\\ x_2 & y_2 & z_2\\ x_3 & y_3 & z_3\end{vmatrix}[\vec{a}\,\vec{b}\,\vec{c}]$$

Proof:

$[x_1\vec{a}+y_1\vec{b}+z_1\vec{c}\;\; x_2\vec{a}+y_2\vec{b}+z_2\vec{c}\;\; x_3\vec{a}+y_3\vec{b}+z_3\vec{c}]$

$=(x_1\vec{a}+y_1\vec{b}+z_1\vec{c})\cdot(x_2\vec{a}+y_2\vec{b}+z_2\vec{c})\times(x_3\vec{a}+y_3\vec{b}+z_3\vec{c})$

$=(x_1\vec{a}+y_1\vec{b}+z_1\vec{c})\cdot[(x_2y_3-x_3y_2)\vec{a}\times\vec{b}+$
$(y_2z_3-y_3z_2)\vec{b}\times\vec{c}+(x_3z_2-x_2z_3)\vec{c}\times\vec{a}]$

$=[x_1(y_2z_3-y_3z_2)-y_1(x_2z_3-x_3z_2)+z_1(x_2y_3-x_3y_2)]\vec{a}\cdot(\vec{b}\times\vec{c})$

$$=\begin{vmatrix}x_1 & y_1 & z_1\\ x_2 & y_2 & z_2\\ x_3 & y_3 & z_3\end{vmatrix}[\vec{a}\,\vec{b}\,\vec{c}]$$

Illustration 3.65

Let vectors $\vec{a}=x\hat{i}+12\hat{j}-\hat{k}$, $\vec{b}=2\hat{i}+2x\hat{j}+\hat{k}$ and $\vec{c}=\hat{i}+\hat{k}$. If vectors $\vec{b}, \vec{c}, \vec{a}$ in order forms a left-handed system, then find values of x.

Sol. For a left-handed system $[\vec{b}\,\vec{c}\,\vec{a}]<0$.

$$\therefore \quad \begin{vmatrix}x & 12 & -1\\ 2 & 2x & 1\\ 1 & 0 & 1\end{vmatrix}<0$$

$\Rightarrow \quad x(2x-0)-12(2-1)-1(0-2x)<0$

$\Rightarrow \quad x^2+x-6<0$

$\Rightarrow \quad (x+3)(x-2)<0$

$\Rightarrow \quad -3<x<2$

Illustration 3.66

If $\vec{a}, \vec{b}$ and $\vec{c}$ are three non-coplanar vectors, then find the value of

$$\frac{\vec{a}\cdot(\vec{b}\times\vec{c})}{\vec{b}\cdot(\vec{c}\times\vec{a})}+\frac{\vec{b}\cdot(\vec{c}\times\vec{a})}{\vec{c}\cdot(\vec{a}\times\vec{b})}+\frac{\vec{c}\cdot(\vec{b}\times\vec{a})}{\vec{a}\cdot(\vec{b}\times\vec{c})}.$$

Sol. Since $[\vec{a}\;\vec{b}\;\vec{c}]\neq 0$, we have

$$\frac{\vec{a}\cdot(\vec{b}\times\vec{c})}{\vec{b}\cdot(\vec{c}\times\vec{a})}+\frac{\vec{b}\cdot(\vec{c}\times\vec{a})}{\vec{c}\cdot(\vec{a}\times\vec{b})}+\frac{\vec{c}\cdot(\vec{b}\times\vec{a})}{\vec{a}\cdot(\vec{b}\times\vec{c})}$$

$$=\frac{[\vec{a}\;\vec{b}\;\vec{c}]}{[\vec{b}\;\vec{c}\;\vec{a}]}+\frac{[\vec{b}\;\vec{c}\;\vec{a}]}{[\vec{c}\;\vec{a}\;\vec{b}]}+\frac{[\vec{c}\;\vec{b}\;\vec{a}]}{[\vec{a}\;\vec{b}\;\vec{c}]}$$

$$=\frac{[\vec{a}\;\vec{b}\;\vec{c}]}{[\vec{a}\;\vec{b}\;\vec{c}]}+\frac{[\vec{a}\;\vec{b}\;\vec{c}]}{[\vec{a}\;\vec{b}\;\vec{c}]}-\frac{[\vec{a}\;\vec{b}\;\vec{c}]}{[\vec{a}\;\vec{b}\;\vec{c}]}$$

$=1+1-1=1$

Illustration 3.67

Prove that $[\vec{a}+\vec{b}\;\; \vec{b}+\vec{c}\;\; \vec{c}+\vec{a}]=2\,[\vec{a}\;\;\vec{b}\;\;\vec{c}]$.

Sol. We have,

$[\vec{a}+\vec{b}\;\; \vec{b}+\vec{c}\;\; \vec{c}+\vec{a}]=(\vec{a}+\vec{b})\cdot((\vec{b}+\vec{c})\times(\vec{c}+\vec{a}))$

$=(\vec{a}+\vec{b})\cdot(\vec{b}\times\vec{c}+\vec{b}\times\vec{a}+\vec{c}\times\vec{a})$

$=[\vec{a}\;\vec{b}\;\vec{c}]+[\vec{b}\;\vec{c}\;\vec{a}]=2[\vec{a}\;\vec{b}\;\vec{c}]$

Illustration 3.68

If $\vec{u}, \vec{v}$ and $\vec{w}$ are three non-coplanar vectors, then prove that $(\vec{u}+\vec{v}-\vec{w})\cdot(\vec{u}-\vec{v})\times(\vec{v}-\vec{w}) = \vec{u}\cdot\vec{v}\times\vec{w}$.

Sol. We have, $(\vec{u}+\vec{v}-\vec{w})\cdot(\vec{u}-\vec{v})\times(\vec{v}-\vec{w})$

$$= (\vec{u}+\vec{v}-\vec{w})\cdot(\vec{u}\times\vec{v}-\vec{u}\times\vec{w}-\vec{v}\times\vec{v}+\vec{v}\times\vec{w})$$

$$= (\vec{u}+\vec{v}-\vec{w})\cdot(\vec{u}\times\vec{v}-\vec{u}\times\vec{w}+\vec{v}\times\vec{w})$$

$$= 0 - 0 + \vec{u}\cdot(\vec{v}\times\vec{w})+0-\vec{v}\cdot(\vec{u}\times\vec{w})+0-\vec{w}\cdot(\vec{u}\times\vec{v}) + 0 - 0$$

$$= [\vec{u}\ \vec{v}\ \vec{w}] + [\vec{v}\ \vec{w}\ \vec{u}] - [\vec{w}\ \vec{u}\ \vec{v}] = \vec{u}\cdot(\vec{v}\times\vec{w})$$

Illustration 3.69

If $[\vec{a}\ \vec{b}\ \vec{c}] = 2$, then find the value of $[(\vec{a}+2\vec{b}-\vec{c})\ (\vec{a}-\vec{b})\ (\vec{a}-\vec{b}-\vec{c})]$.

Sol. We have $[(\vec{a}+2\vec{b}-\vec{c})\ (\vec{a}-\vec{b})\ (\vec{a}-\vec{b}-\vec{c})]$

$$= \begin{vmatrix} 1 & 2 & -1 \\ 1 & -1 & 0 \\ 1 & -1 & -1 \end{vmatrix} [\vec{a}\ \vec{b}\ \vec{c}]$$

$$= 3[\vec{a}\ \vec{b}\ \vec{c}] = 6$$

Illustration 3.70

If $|\vec{a}| = 2, |\vec{b}| = 3$ and $|\vec{a}+\vec{b}| = \sqrt{19}$ then find the value of $[\vec{a}\ \vec{b}\ \vec{a}\times\vec{b}]$.

Sol. We have $|\vec{a}+\vec{b}| = \sqrt{19}$

$\therefore \quad |\vec{a}+\vec{b}|^2 = 19$

$\Rightarrow \quad |\vec{a}|^2 + |\vec{b}|^2 + 2\vec{a}\cdot\vec{b} = 19$

$\Rightarrow \quad 4+9+2\vec{a}\cdot\vec{b} = 19$

$\Rightarrow \quad \vec{a}\cdot\vec{b} = 3$

$\Rightarrow \quad 2\times 3\cos\theta = 3$

$\Rightarrow \quad \cos\theta = \frac{1}{2}$

$\Rightarrow \quad \theta = 60°$

Now, $[\vec{a}\ \vec{b}\ \vec{a}\times\vec{b}] = (\vec{a}\times\vec{b})\cdot(\vec{a}\times\vec{b})$

$$= |\vec{a}\times\vec{b}|^2$$

$$= (|\vec{a}||\vec{b}|\sin\theta)^2$$

$$= \left(2\times 3\times\frac{\sqrt{3}}{2}\right)^2 = 27$$

Illustration 3.71

Let $\vec{a} = \hat{i}+\hat{j}+\hat{k}, \vec{b}$ and $\vec{c} = \hat{j}-\hat{k}$ be three vectors such that $\vec{a}\times\vec{b} = \vec{c}$. Then find the length projection vector of the vector $\vec{b}$ on the vector $\vec{a}\times\vec{c}$.

Sol. Projection of $\vec{b}$ on the vector $\vec{a}\times\vec{c}$

$$= \frac{|\vec{b}\cdot(\vec{a}\times\vec{c})|}{|\vec{a}\times\vec{c}|}$$

$$= \frac{|\vec{c}\cdot(\vec{a}\times\vec{b})|}{|\vec{a}\times\vec{c}|}$$

$$= \frac{|\vec{c}\cdot\vec{c}|}{|\vec{a}\times\vec{c}|} = \frac{|\vec{c}|^2}{|\vec{a}\times\vec{c}|}$$

Now, $\vec{a}\times\vec{c} = \begin{vmatrix} \hat{i} & \hat{j} & \hat{k} \\ 1 & 1 & 1 \\ 0 & 1 & -1 \end{vmatrix} = -2\hat{i}+\hat{j}+\hat{k}$

Therefore, the required projection is: $\frac{2}{\sqrt{6}}$

Illustration 3.72

If $\vec{a}, \vec{b}$ and $\vec{c}$ are mutually perpendicular vectors and $\vec{a} = \alpha(\vec{a}\times\vec{b})+\beta(\vec{b}\times\vec{c})+\gamma(\vec{c}\times\vec{a})$ and $[\vec{a}\ \vec{b}\ \vec{c}] = 1$, then find the value of $\alpha+\beta+\gamma$.

Sol. We have, $\vec{a} = \alpha(\vec{a}\times\vec{b})+\beta(\vec{b}\times\vec{c})+\gamma(\vec{c}\times\vec{a})$

Taking dot product with $\vec{a}, \vec{b}$ and $\vec{c}$, respectively, we get

$$|\vec{a}|^2 = \beta\cdot[\vec{a}\ \vec{b}\ \vec{c}] = \beta \quad (\because [\vec{a}\ \vec{b}\ \vec{c}] = 1)$$

$$0 = \gamma\cdot[\vec{a}\ \vec{b}\ \vec{c}] = \gamma$$

and $\quad 0 = \alpha\cdot[\vec{a}\ \vec{b}\ \vec{c}] = \alpha$

$\therefore \quad \alpha+\beta+\gamma = |\vec{a}|^2$

Illustration 3.73

Let $\vec{a}, \vec{b}, \vec{c}$ be three unit vectors and $\vec{a}\cdot\vec{b} = \vec{a}\cdot\vec{c} = 0$. If the angle between $\vec{b}$ and $\vec{c}$ is $\frac{\pi}{3}$, then find the value of $|[\vec{a}\ \vec{b}\ \vec{c}]|$.

Sol. We have, $\vec{a}\cdot\vec{b} = \vec{a}\cdot\vec{c} = 0$.

Therefore, $\vec{a}$ is perpendicular to both $\vec{b}$ and $\vec{c}$.

Thus, $\vec{a} = \lambda(\vec{b}\times\vec{c})$

$\Rightarrow \quad |\vec{a}| = |\lambda(\vec{b}\times\vec{c})| = \left|\lambda\frac{\sqrt{3}}{2}\right| = 1 \Rightarrow |\lambda| = \frac{2}{\sqrt{3}}$.

Now,

$$|[\vec{a}\,\vec{b}\,\vec{c}]| = |\vec{a}\cdot(\vec{b}\times\vec{c})|$$
$$= |\lambda||(\vec{b}\times\vec{c})|^2$$
$$= \frac{2}{\sqrt{3}}|\vec{b}|^2|\vec{c}|^2\sin^2\left(\frac{\pi}{3}\right) = \frac{2}{\sqrt{3}}\times\left(\frac{\sqrt{3}}{2}\right)^2 = \frac{\sqrt{3}}{2}$$

Illustration 3.74

If $[\vec{a}\,\vec{b}\,\vec{c}] = 6$ then find the value of $\begin{vmatrix} \vec{a}\cdot\vec{a} & \vec{a}\cdot\vec{b} & \vec{a}\cdot\vec{c} \\ \vec{b}\cdot\vec{a} & \vec{b}\cdot\vec{b} & \vec{b}\cdot\vec{c} \\ \vec{c}\cdot\vec{a} & \vec{c}\cdot\vec{b} & \vec{c}\cdot\vec{c} \end{vmatrix}$.

Sol. $\begin{vmatrix} \vec{a}\cdot\vec{a} & \vec{a}\cdot\vec{b} & \vec{a}\cdot\vec{c} \\ \vec{b}\cdot\vec{a} & \vec{b}\cdot\vec{b} & \vec{b}\cdot\vec{c} \\ \vec{c}\cdot\vec{a} & \vec{c}\cdot\vec{b} & \vec{c}\cdot\vec{c} \end{vmatrix} = [\vec{a}\,\vec{b}\,\vec{c}]^2 = 6^2 = 36$

Illustration 3.75

If $\vec{e}_1, \vec{e}_2, \vec{e}_3$ and $\vec{E}_1, \vec{E}_2, \vec{E}_3$ are two sets of vectors such that $\vec{e}_i\cdot\vec{E}_j = 1$, if $i = j$ and $\vec{e}_i\cdot\vec{E}_j = 0$ and if $i \neq j$, then prove that $[\vec{e}_1\ \vec{e}_2\ \vec{e}_3][\vec{E}_1\ \vec{E}_2\ \vec{E}_3] = 1$.

Sol. We know that

$$[\vec{e}_1\ \vec{e}_2\ \vec{e}_3][\vec{E}_1\ \vec{E}_2\ \vec{E}_3] = \begin{vmatrix} \vec{e}_1\cdot\vec{E}_1 & \vec{e}_1\cdot\vec{E}_2 & \vec{e}_1\cdot\vec{E}_3 \\ \vec{e}_2\cdot\vec{E}_1 & \vec{e}_2\cdot\vec{E}_2 & \vec{e}_2\cdot\vec{E}_3 \\ \vec{e}_3\cdot\vec{E}_1 & \vec{e}_3\cdot\vec{E}_2 & \vec{e}_3\cdot\vec{E}_3 \end{vmatrix}$$
$$= \begin{vmatrix} 1 & 0 & 0 \\ 0 & 1 & 0 \\ 0 & 0 & 1 \end{vmatrix}$$
$$= 1$$

Illustration 3.76

If $\vec{a}$, $\vec{b}$ and $\vec{c}$ are three non-coplanar vectors and $\vec{r}$ is any arbitrary vector. Prove that

$[\vec{b}\ \vec{c}\ \vec{r}]\vec{a} + [\vec{c}\ \vec{a}\ \vec{r}]\vec{b} + [\vec{a}\ \vec{b}\ \vec{r}]\vec{c} = [\vec{a}\ \vec{b}\ \vec{c}]\vec{r}$.

Sol. Let $\vec{r} = x_1\vec{a} + x_2\vec{b} + x_3\vec{c} \Rightarrow \vec{r}\cdot(\vec{b}\times\vec{c}) = x_1\vec{a}\cdot(\vec{b}\times\vec{c})$

or $$x_1 = \frac{[\vec{r}\ \vec{b}\ \vec{c}]}{[\vec{a}\ \vec{b}\ \vec{c}]}$$

Also, $\vec{r}\cdot(\vec{c}\times\vec{a}) = x_2\vec{b}\cdot(\vec{c}\times\vec{a})$ or $x_2 = \dfrac{[\vec{r}\ \vec{c}\ \vec{a}]}{[\vec{a}\ \vec{b}\ \vec{c}]}$

and $\vec{r}\cdot(\vec{a}\times\vec{b}) = x_3\vec{c}\cdot(\vec{a}\times\vec{b})$ or $x_3 = \dfrac{[\vec{r}\ \vec{a}\ \vec{b}]}{[\vec{a}\ \vec{b}\ \vec{c}]}$

$$\Rightarrow \quad \vec{r} = \frac{[\vec{r}\ \vec{b}\ \vec{c}]}{[\vec{a}\ \vec{b}\ \vec{c}]}\vec{a} + \frac{[\vec{r}\ \vec{c}\ \vec{a}]}{[\vec{a}\ \vec{b}\ \vec{c}]}\vec{b} + \frac{[\vec{r}\ \vec{a}\ \vec{b}]}{[\vec{a}\ \vec{b}\ \vec{c}]}\vec{c}$$

or $$[\vec{b}\ \vec{c}\ \vec{r}]\vec{a} + [\vec{c}\ \vec{a}\ \vec{r}]\vec{b} + [\vec{a}\ \vec{b}\ \vec{r}]\vec{c} = [\vec{a}\ \vec{b}\ \vec{c}]\vec{r}$$

Illustration 3.77

If $\vec{r}\cdot\vec{a} = 0$, $\vec{r}\cdot\vec{b} = 1$ and $[\vec{r}\ \vec{a}\ \vec{b}] = 1$, $\vec{a}\cdot\vec{b} \neq 0$, $(\vec{a}\cdot\vec{b})^2 - |\vec{a}|^2|\vec{b}|^2 = 1$, then find $\vec{r}$ in terms of $\vec{a}$ and $\vec{b}$.

Sol. Writing $\vec{r}$ as a linear combination of $\vec{a}, \vec{b}$ and $\vec{a}\times\vec{b}$, we have

$$\vec{r} = x\vec{a} + y\vec{b} + z(\vec{a}\times\vec{b})$$

Taking dot product with $\vec{a}$, we get

$$0 = \vec{r}\cdot\vec{a} = x|\vec{a}|^2 + y\vec{a}\cdot\vec{b}$$

Taking dot product with $\vec{b}$, we get

$$1 = \vec{r}\cdot\vec{b} = x\vec{a}\cdot\vec{b} + y|\vec{b}|^2$$

Solving, we get $y = \dfrac{|\vec{a}|^2}{|\vec{a}|^2|\vec{b}|^2 - (\vec{a}\cdot\vec{b})^2} = -|\vec{a}|^2$

and $x = \dfrac{\vec{a}\cdot\vec{b}}{(\vec{a}\cdot\vec{b})^2 - |\vec{a}|^2|\vec{b}|^2} = \vec{a}\cdot\vec{b}$

Now, taking dot product with $\vec{a}\times\vec{b}$, we get

$$1 = [\vec{r}\ \vec{a}\ \vec{b}] = z|\vec{a}\times\vec{b}|^2$$

or $$z = \frac{1}{|a\times b|^2}$$

Thus, $$\vec{r} = ((\vec{a}\cdot\vec{b})\vec{a} - |\vec{a}|^2\vec{b}) + \frac{\vec{a}\times\vec{b}}{|\vec{a}\times\vec{b}|^2}$$

$$= \vec{a}\times(\vec{a}\times\vec{b}) + \frac{\vec{a}\times\vec{b}}{|\vec{a}\times\vec{b}|^2}$$

Illustration 3.78

Prove that

$$\vec{R}+\frac{\vec{R}\cdot(\vec{\beta}\times(\vec{\beta}\times\vec{\alpha}))}{|\vec{\alpha}\times\vec{\beta}|^2}\vec{\alpha}+\frac{\vec{R}\cdot(\vec{\alpha}\times(\vec{\alpha}\times\vec{\beta}))}{|\vec{\alpha}\times\vec{\beta}|^2}\vec{\beta}=\frac{\vec{R}\cdot(\vec{\alpha}\times\vec{\beta})}{|\vec{\alpha}\times\vec{\beta}|^2}(\vec{\alpha}\times\vec{\beta})$$

Sol. Clearly, here vectors $\vec{\alpha}$ and $\vec{\beta}$ are non-collinear.

Thus, three vectors $\vec{\alpha}, \vec{\beta}$ and $\vec{\alpha}\times\vec{\beta}$ are non–coplanar.

Therefore, vector $\vec{R}$ can be represented as a linear combination of these vectors as

$$\vec{R}=k_1\vec{\alpha}+k_2\vec{\beta}+k_3(\vec{\alpha}\times\vec{\beta}) \quad \ldots\text{(i)}$$

Taking dot product with $(\vec{\alpha}\times\vec{\beta})$, we get

$$\vec{R}\cdot(\vec{\alpha}\times\vec{\beta})=k_3(\vec{\alpha}\times\vec{\beta})\cdot(\vec{\alpha}\times\vec{\beta})=k_3|\vec{\alpha}\times\vec{\beta}|^2$$

or $$k_3=\frac{\vec{R}\cdot(\vec{\alpha}\times\vec{\beta})}{|\vec{\alpha}\times\vec{\beta}|^2}=\frac{[\vec{R}\,\vec{\alpha}\,\vec{\beta}]}{|\vec{\alpha}\times\vec{\beta}|^2}$$

In (i), taking dot product with $\vec{\alpha}\times(\vec{\alpha}\times\vec{\beta})$, we get

$$\Rightarrow \vec{R}\cdot(\vec{\alpha}\times(\vec{\alpha}\times\vec{\beta}))=k_2(\vec{\alpha}\times(\vec{\alpha}\times\vec{\beta}))\cdot\vec{\beta}$$

$$=-k_2((\vec{\alpha}\times\vec{\beta})\times\vec{\alpha})\cdot\vec{\beta}$$

$$=-k_2(\vec{\alpha}\times\vec{\beta})\cdot(\vec{\alpha}\times\vec{\beta}) \text{ (interchanging dot and cross)}$$

$$=-k_2|\vec{\alpha}\times\vec{\beta}|^2$$

$$\Rightarrow k_2=\frac{-[\vec{R}\cdot(\vec{\alpha}\times(\vec{\alpha}\times\vec{\beta}))]}{|\vec{\alpha}\times\vec{\beta}|^2}$$

Similarly, $$k_1=-\frac{[\vec{R}\cdot(\vec{\beta}\times(\vec{\beta}\times\vec{\alpha}))]}{|\vec{\alpha}\times\vec{\beta}|^2}$$

$$\therefore \vec{R}=-\frac{\vec{R}\cdot(\vec{\beta}\times(\vec{\beta}\times\vec{\alpha}))}{|\vec{\alpha}\times\vec{\beta}|^2}\vec{\alpha}-\frac{\vec{R}\cdot(\vec{\alpha}\times(\vec{\alpha}\times\vec{\beta}))}{|\vec{\alpha}\times\vec{\beta}|^2}\vec{\beta}+\frac{\vec{R}\cdot(\vec{\alpha}\times\vec{\beta})}{|\vec{\alpha}\times\vec{\beta}|^2}(\vec{\alpha}\times\vec{\beta})$$

$$\Rightarrow \vec{R}+\frac{\vec{R}\cdot(\vec{\beta}\times(\vec{\beta}\times\vec{\alpha}))}{|\vec{\alpha}\times\vec{\beta}|^2}\vec{\alpha}+\frac{\vec{R}\cdot(\vec{\alpha}\times(\vec{\alpha}\times\vec{\beta}))}{|\vec{\alpha}\times\vec{\beta}|^2}\vec{\beta}=\frac{\vec{R}\cdot(\vec{\alpha}\times\vec{\beta})}{|\vec{\alpha}\times\vec{\beta}|^2}(\vec{\alpha}\times\vec{\beta})$$

3.4.3 Volume of Parallelepiped

Let the three non-coplanar vectors $\vec{a}, \vec{b}, \vec{c}$ form a right-handed system. These vectors may be seen as three adjacent edges of the parallelepiped originating from a common vertex O, such that the base plane is determined by the vectors $\vec{b}$ and $\vec{c}$.

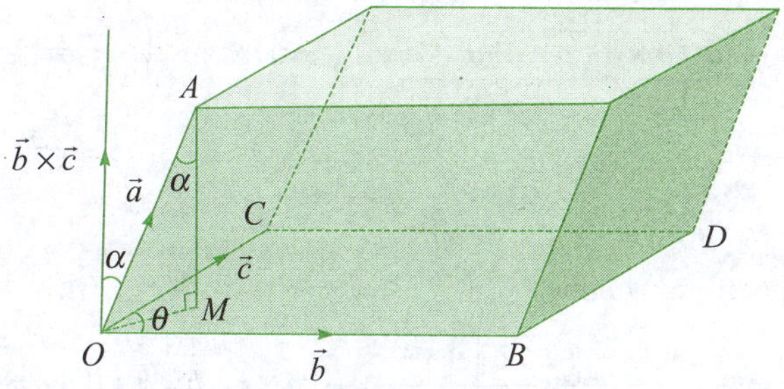

Now $[\vec{a}\,\vec{b}\,\vec{c}]=\vec{a}\cdot(\vec{b}\times\vec{c})=\vec{a}\cdot(|\vec{b}||\vec{c}|\sin\theta\hat{n})$

$=|\vec{b}||\vec{c}|\sin\theta(\vec{a}\cdot\hat{n})$

Here $\hat{n}$ is a vector perpendicular to the plane of the vectors $\vec{b}$ and $\vec{c}$.

Now, angle between vectors $\vec{a}$ and $\hat{n}$ is α.

$\therefore \quad [\vec{a}\,\vec{b}\,\vec{c}]=|\vec{b}||\vec{c}|\sin\theta(|\vec{a}||\hat{n}|\cos\alpha)$

$=(|\vec{a}|\cos\alpha)(|\vec{b}||\vec{c}|\sin\theta)$

$=$ Height $\times$ Area of the base

$=$ Volume of the parallelepiped

If vectors $\vec{a}, \vec{b}, \vec{c}$ forms a left-handed system, then volume is $-[\vec{a}\,\vec{b}\,\vec{c}]$.

Thus, volume of parallelepiped regardless of the handedness of the vectors is $|[\vec{a}\,\vec{b}\,\vec{c}]|$

3.4.4 Volume of Tetrahedron

Tetrahedron is a pyramid having a triangular base.

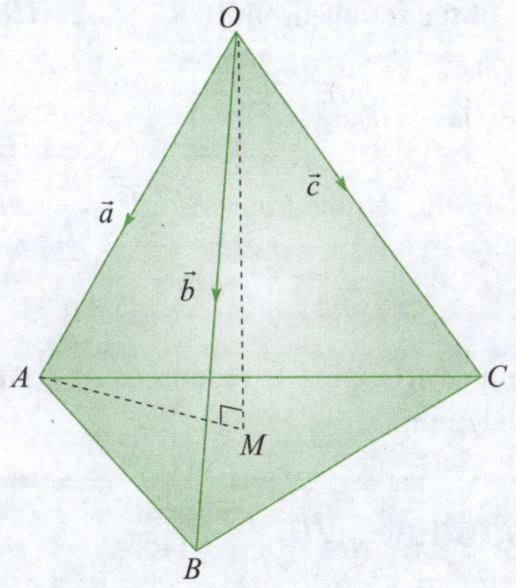

Consider tetrahedron $OABC$ formed by three vectors $\vec{a}, \vec{b}, \vec{c}$ emanating from vertex O.

So, the position vectors of A, B and C are $\vec{a}$, $\vec{b}$ and $\vec{c}$ respectively.

Now, the volume of the tetrahedron

$$=\frac{1}{3}\text{ Height}\times\text{Area of base}$$

$$=\frac{1}{3}OM\times\text{Area of }\Delta ABC$$

Area of $\Delta ABC=\frac{1}{2}|\vec{a}\times\vec{b}+\vec{b}\times\vec{c}+\vec{c}\times\vec{a}|$ and vector normal to the plane ABC is $(\vec{a}\times\vec{b}+\vec{b}\times\vec{c}+\vec{c}\times\vec{a})$

So, $OM=$ projection of vector $\overrightarrow{OA}$ on $(\vec{a}\times\vec{b}+\vec{b}\times\vec{c}+\vec{c}\times\vec{a})$

$$=\frac{|\vec{a}\cdot(\vec{a}\times\vec{b}+\vec{b}\times\vec{c}+\vec{c}\times\vec{a})|}{|\vec{a}\times\vec{b}+\vec{b}\times\vec{c}+\vec{c}\times\vec{a}|}$$

$$=\frac{|[\vec{a}\ \vec{b}\ \vec{c}]|}{|\vec{a}\times\vec{b}+\vec{b}\times\vec{c}+\vec{c}\times\vec{a}|}$$

Therefore, volume

$$=\frac{1}{3}\times\frac{|[\vec{a}\ \vec{b}\ \vec{c}]|}{|\vec{a}\times\vec{b}+\vec{b}\times\vec{c}+\vec{c}\times\vec{a}|}\times\frac{1}{2}|\vec{a}\times\vec{b}+\vec{b}\times\vec{c}+\vec{c}\times\vec{a}|$$

$$=\frac{1}{6}|[\vec{a}\ \vec{b}\ \vec{c}]|$$

Thus, the volume of a tetrahedron formed by three vectors is one sixth of the volume of the parallelepiped formed by the same three vectors.

This can be justified by noting that a parallelepiped can be exactly divided into six non-overlapping tetrahedrons of equal volume. Three of these, with vertices (O, A, B, C), (A, C, B, B') and (A, B, C', B'), are shown in the figure.

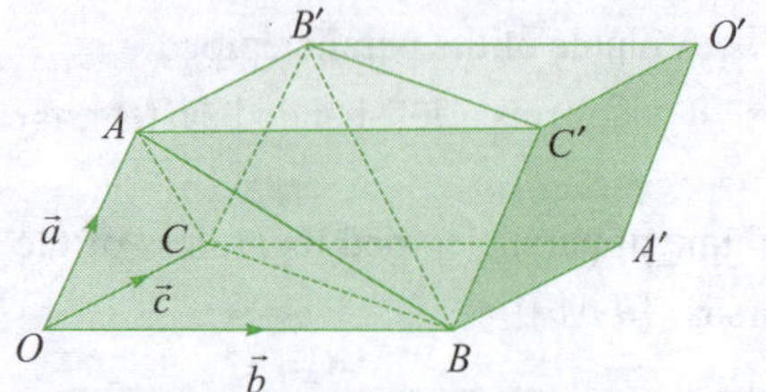

Clearly, the volume of the tetrahedron $OABC=\frac{1}{6}|[\vec{a}\ \vec{b}\ \vec{c}]|$.

The volume of the tetrahedron $ACBB'=\frac{1}{6}|[\overrightarrow{AB}\ \overrightarrow{AC}\ \overrightarrow{AB'}]|$

$$=\frac{1}{6}|[(\vec{b}-\vec{a})\ (\vec{c}-\vec{a})\ \vec{c}]|=\frac{1}{6}|[\vec{a}\ \vec{b}\ \vec{c}]|$$

and the volume of the tetrahedron $ABC'B'=\frac{1}{6}|[\overrightarrow{AB'}\ \overrightarrow{AC'}\ \overrightarrow{AB}]|$

$$=\frac{1}{6}|[\vec{c}\ \vec{b}\ (\vec{b}-\vec{a})]|=\frac{1}{6}|[\vec{a}\ \vec{b}\ \vec{c}]|$$

Thus, six non-overlapping tetrahedrons of equal volume can fit inside parallelepiped.

Illustration 3.79

If the vectors $2\hat{i}-3\hat{j}$, $\hat{i}+\hat{j}-\hat{k}$ and $3\hat{i}-\hat{k}$ form three concurrent edges of a parallelepiped, then find the volume of the parallelepiped.

Sol. Here, $\overrightarrow{OA}=2\hat{i}-3\hat{j}$

$\overrightarrow{OB}=\hat{i}+\hat{j}-\hat{k}$

and $\overrightarrow{OC}=3\hat{i}-\hat{k}$

Hence, Volume $=[\overrightarrow{OA}\ \overrightarrow{OB}\ \overrightarrow{OC}]$

$$=\begin{vmatrix}2 & -3 & 0\\ 1 & 1 & -1\\ 3 & 0 & -1\end{vmatrix}=4$$

Illustration 3.80

The position vectors of the four angular points of a tetrahedron are $A(\hat{j}+2\hat{k})$, $B(3\hat{i}+\hat{k})$, $C(4\hat{i}+3\hat{j}+6\hat{k})$ and $D(2\hat{i}+3\hat{j}+2\hat{k})$. Find the volume of the tetrahedron $ABCD$.

Sol. Vertices of the tetrahedron are

$A(\hat{j}+2\hat{k})$, $B(3\hat{i}+\hat{k})$, $C(4\hat{i}+3\hat{j}+6\hat{k})$ and $D(2\hat{i}+3\hat{j}+2\hat{k})$.

Therefore,

$$\overrightarrow{AB}=3\hat{i}-\hat{j}-\hat{k},\ \overrightarrow{AC}=4\hat{i}+2\hat{j}+4\hat{k} \text{ and } \overrightarrow{AD}=2\hat{i}+2\hat{j}$$

$\therefore$ Volume, $V=\frac{1}{6}[\overrightarrow{AB}\ \overrightarrow{AC}\ \overrightarrow{AD}]$

$$=\frac{1}{6}\begin{vmatrix}3 & -1 & -1\\ 4 & 2 & 4\\ 2 & 2 & 0\end{vmatrix}$$

$= 6$ cubic unit

Illustration 3.81

Find the value of a such that the volume of the parallelepiped formed by vectors $\hat{i}+a\hat{j}+\hat{k}$, $\hat{j}+a\hat{k}$ and $a\hat{i}+\hat{k}$ is minimum.

Sol. Volume, $V=\begin{vmatrix}1 & a & 1\\ 0 & 1 & a\\ a & 0 & 1\end{vmatrix}=1-a+a^3$

$\Rightarrow \quad \frac{dV}{da}=3a^2-1$

Now, $\frac{dV}{da}=0$ if $a=\frac{1}{\sqrt{3}}$

Hence, V is minimum at $a=\frac{1}{\sqrt{3}}$

Illustration 3.82

Find the altitude of a parallelepiped whose three coterminous edges are vectors $\vec{A}=\hat{i}+\hat{j}+\hat{k}$, $\vec{B}=2\hat{i}+4\hat{j}-\hat{k}$ and $\vec{C}=\hat{i}+\hat{j}+3\hat{k}$ with $\vec{A}$ and $\vec{B}$ as the sides of the base of the parallelepiped.

Sol. The height, $h=\frac{\text{Volume of the parallelepiped}}{\text{Area of the base}}$

$$=\frac{[\vec{A}\ \vec{B}\ \vec{C}]}{|\vec{A}\times\vec{B}|}=\frac{\begin{vmatrix}1 & 1 & 1\\ 2 & 4 & -1\\ 1 & 1 & 3\end{vmatrix}}{\left\|\begin{matrix}\hat{i} & \hat{j} & \hat{k}\\ 1 & 1 & 1\\ 2 & 4 & -1\end{matrix}\right\|}=\frac{4}{|-5\hat{i}+3\hat{j}+2\hat{k}|}=\frac{2\sqrt{38}}{19}$$

Illustration 3.83

Let G_1, G_2 and G_3 be the centroids of the trianglular faces *OBC*, *OCA* and *OAB*, respectively, of a tetrahedron *OABC*. If V_1 denotes the volume of the tetrahedron *OABC* and V_2 denotes the parallelepiped with OG_1, OG_2 and OG_3 as three concurrent edges, then prove that $4V_1 = 9V_2$.

Sol. Let *O* be the origin and let the position vectors of *A*, *B* and *C* be $\vec{a}$, $\vec{b}$ and $\vec{c}$, respectively. Then the position vectors G_1, G_2 and G_3 are $\frac{\vec{b}+\vec{c}}{3}$, $\frac{\vec{c}+\vec{a}}{3}$ and $\frac{\vec{a}+\vec{b}}{3}$, respectively. Therefore,

$$V_1 = \frac{1}{6}[\vec{a}\ \vec{b}\ \vec{c}] \text{ and } V_2 = [\overrightarrow{OG_1}\ \overrightarrow{OG_2}\ \overrightarrow{OG_3}]$$

Now, $V_2 = [\overrightarrow{OG_1}\ \overrightarrow{OG_2}\ \overrightarrow{OG_3}]$

$$= \frac{1}{27}[\vec{b}+\vec{c}\ \ \vec{c}+\vec{a}\ \ \vec{a}+\vec{b}]$$

$$= \frac{2}{27}[\vec{a}\ \vec{b}\ \vec{c}]$$

$$= \frac{2}{27}\times 6V_1 \text{ or } 9V_2 = 4V_1$$

Aim to complete within 40 minutes

Concept Application Exercise 3.3

1. Prove that $((\vec{a}\times\hat{j})\cdot\hat{k})\hat{i}+((\vec{a}\times\hat{k})\cdot\hat{i})\hat{j}+((\vec{a}\times\hat{i})\cdot\hat{j})\hat{k} = \vec{a}$.
2. Prove that $(\vec{a}\cdot(\vec{b}\times\hat{i}))\hat{i}+(\vec{a}\cdot(\vec{b}\times\hat{j}))\hat{j}+(\vec{a}\cdot\vec{b}\times\hat{k})\hat{k} = \vec{a}\times\vec{b}$.
3. If there does not exist real λ and μ such that $\vec{c}=\lambda\vec{a}+\mu\vec{b}$, then find the value of α if equation $[\vec{a}\ \vec{b}\ 2\vec{c}]x^2+[\vec{b}\ \alpha\vec{a}\ \vec{c}]x+[\alpha\vec{c}+\vec{a}\ \ \vec{a}+\vec{b}\ \ \vec{b}+2.5\vec{c}]=0$ has real roots.
4. If $\vec{x}\cdot\vec{a}=0$, $\vec{x}\cdot\vec{b}=0$ and $\vec{x}\cdot\vec{c}=0$ for some non-zero vector $\vec{x}$, then prove that $[\vec{a}\ \vec{b}\ \vec{c}]=0$.
5. If the volume of a parallelepiped whose adjacent edges are $\vec{a}=2\hat{i}+3\hat{j}+4\hat{k}$, $\vec{b}=\hat{i}+\alpha\hat{j}+2\hat{k}$, $\vec{c}=\hat{i}+2\hat{j}+\alpha\hat{k}$ is 15, then the value of α is
   (1) 2 (2) 3 (3) 4 (4) 4.5
6. Volume of tetrahedron whose coterminous edges are vectors $\vec{a}$, $\vec{b}$ and $\vec{c}$ is 12 cu. units. If the scalar projection of vector $\vec{a}$ on $\vec{b}\times\vec{c}$ is 4 units, then the value of $|\vec{b}\times\vec{c}|$ is
   (1) 9 (2) 18 (3) 24 (4) 27
7. If $\vec{a}$, $\vec{b}$, $\vec{c}$ are three vectors such each having magnitude 2 units and are equally inclined to each other at an angle 60°, then the value of $|[\vec{a}+\vec{b}\ \ \vec{b}+\vec{c}\ \ \vec{c}+\vec{a}]|$ is
   (1) $2\sqrt{2}$ (2) $4\sqrt{2}$ (3) 8 (4) $8\sqrt{2}$
8. Given unit vectors $\hat{m}$, $\hat{n}$ and $\hat{p}$ such that angle between $\hat{m}$ and $\hat{n}$ is α and angle between $\hat{p}$ and $(\hat{m}\times\hat{n})$ is also α, if $[\hat{n}\ \hat{p}\ \hat{m}] = 1/4$, then the value of α is
   (1) 15° (2) 30° (3) 45° (4) 60°
9. $\vec{a},\vec{b},\vec{c}$ are three unit vectors and every two are inclined to each other at an angle $\cos^{-1}(3/5)$. If $\vec{a}\times\vec{b}=p\vec{a}+q\vec{b}+r\vec{c}$, where *p*, *q*, *r* are scalars, then find the value of *q* is
   (1) $\frac{-3}{11}$ (2) $\frac{3}{55}$ (3) $\frac{-3}{\sqrt{55}}$ (4) $\frac{4}{\sqrt{55}}$
10. Prove that if $[\vec{l}, \vec{m}, \vec{n}]$ are three non-coplanar vectors, then $[\vec{l}\ \vec{m}\ \vec{n}](\vec{a}\times\vec{b}) = \begin{vmatrix}\vec{l}\cdot\vec{a} & \vec{l}\cdot\vec{b} & \vec{l}\\ \vec{m}\cdot\vec{a} & \vec{m}\cdot\vec{b} & \vec{m}\\ \vec{n}\cdot\vec{a} & \vec{n}\cdot\vec{b} & \vec{n}\end{vmatrix}$.
11. Let $\vec{a}=a_1\hat{i}+a_2\hat{j}+a_3\hat{k}$, $\vec{b}=b_1\hat{i}+b_2\hat{j}+b_3\hat{k}$ and $\vec{c}=c_1\hat{i}+c_2\hat{j}+c_3\hat{k}$ be three non-zero vectors such that $\vec{c}$ is a unit vector perpendicular to both $\vec{a}$ and $\vec{b}$. If the angle between $\vec{a}$ and $\vec{b}$ is $\frac{\pi}{6}$, then prove that $\begin{vmatrix}a_1 & a_2 & a_3\\ b_1 & b_2 & b_3\\ c_1 & c_2 & c_3\end{vmatrix}^2 = \frac{1}{4}(a_1^2+a_2^2+a_3^2)(b_1^2+b_2^2+b_3^2)$.
12. Let three vectors $\vec{a},\vec{b}$ and $\vec{c}$ be such that and $\vec{a}\times\vec{b}=\vec{c}$, $\vec{b}\times\vec{c}=\vec{a}$ and $|\vec{a}|=2$. Then find
    a. the value of $[\vec{a}\ \vec{b}\ \vec{c}]+[\vec{c}\ \vec{a}\ \vec{b}]$
    b. the scalar projection of $\vec{a}$ on $(\vec{b}\times\vec{c})$
    c. the value of $|3\vec{a}+\vec{b}-2\vec{c}|^2$

Answers

3. $\alpha\le -2$ or $\alpha\ge 10$ **5.** (4)
6. (2) **7.** (4)
8. (1) **9.** (3)
12. a. 8 **b.** 2 **c.** 53

3.5 VECTOR TRIPLE PRODUCT

3.5.1 Definition of the Vector Triple Product

The vector triple product of three vectors $\vec{a}, \vec{b}$ and $\vec{c}$ is the vector $\vec{a}\times(\vec{b}\times\vec{c}) = (\vec{a}\cdot\vec{c})\vec{b} - (\vec{a}\cdot\vec{b})\vec{c}$, which lies in the plane of vectors $\vec{b}$ and $\vec{c}$.

Also, $(\vec{a}\times\vec{b})\times\vec{c} = (\vec{a}\cdot\vec{c})\vec{b} - (\vec{b}\cdot\vec{c})\vec{a}$, which lies in the plane of vectors $\vec{a}$ and $\vec{b}$.

Let's derive the formula.

Consider three non-zero vectors $\vec{a}, \vec{b}$ and $\vec{c}$.

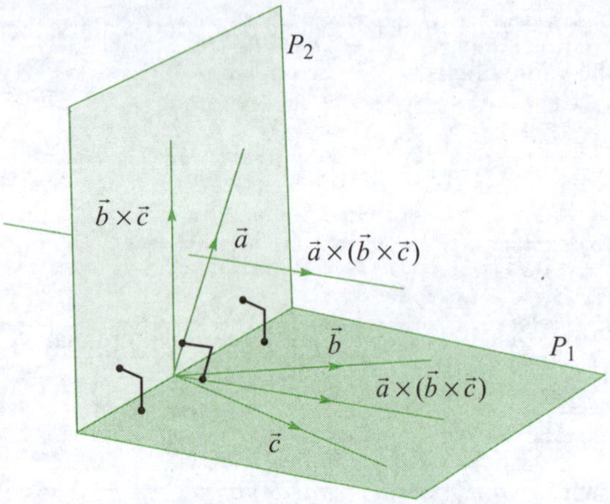

These vectors may be coplanar or non-coplanar.

Consider vectors to be non-coplanar.

We have plane P_1 determined by vectors and $\vec{b}$ and $\vec{c}$.

So, vector $\vec{b}\times\vec{c}$ is perpendicular to plane P_1.

Plane P_2 is determined by vectors $\vec{a}$ and $\vec{b}\times\vec{c}$.

For any position of vector $\vec{a}$, plane P_2 will be perpendicular to the plane P_1.

So, vector $\vec{a}\times(\vec{b}\times\vec{c})$ will be parallel to the plane P_1, i.e. it lies in the plane P_1.

$\therefore \quad \vec{a}\times(\vec{b}\times\vec{c}) = x\vec{b} + y\vec{c} \qquad \ldots(i)$

Here x and y are scalars.

Taking dot product with vector $\vec{a}$, we get

$\vec{a}\cdot\vec{a}\times(\vec{b}\times\vec{c}) = x\vec{a}\cdot\vec{b} + y\vec{a}\cdot\vec{c}$

$\therefore \quad 0 = x\vec{a}\cdot\vec{b} + y\vec{a}\cdot\vec{c} \qquad (\because \vec{a}\cdot\vec{a}\times(\vec{b}\times\vec{c}) = [\vec{a}\ \vec{a}\ \vec{b}\times\vec{c}] = 0)$

$\Rightarrow \quad \dfrac{x}{\vec{a}\cdot\vec{c}} = -\dfrac{y}{\vec{a}\cdot\vec{b}} = \lambda$

$\Rightarrow \quad \vec{a}\times(\vec{b}\times\vec{c}) = \lambda[(\vec{a}\cdot\vec{c})\vec{b} - (\vec{a}\cdot\vec{b})\vec{c}] \qquad \ldots(ii)$

Now, this is an identity which is true for any set of vectors $\vec{a}, \vec{b}$ and $\vec{c}$.

Let's choose $\vec{a} = \hat{i}, \vec{b} = \hat{i}+\hat{k}$ and $\vec{c} = \hat{k}$.

Thus, in (ii), L.H.S. = $\hat{i}\times((\hat{i}+\hat{k})\times\hat{k}) = \hat{i}\times(-\hat{j}) = -\hat{k}$

and R.H.S. = $\lambda[(\hat{i}\cdot\hat{k})(\hat{i}+\hat{k}) - (\hat{i}\cdot(\hat{i}+\hat{k}))\hat{k}] = -\lambda\hat{k}$

So, we have $-\hat{k} = -\lambda\hat{k}$ or $\lambda = 1$

Hence from (ii), we have

$\vec{a}\times(\vec{b}\times\vec{c}) = (\vec{a}\cdot\vec{c})\vec{b} - (\vec{a}\cdot\vec{b})\vec{c}$

NOTES

In general, $\vec{a}\times(\vec{b}\times\vec{c}) \ne (\vec{a}\times\vec{b})\times\vec{c}$.

However, if $\vec{a}\times(\vec{b}\times\vec{c}) = (\vec{a}\times\vec{b})\times\vec{c}$, then we have $(\vec{a}\cdot\vec{c})\vec{b} - (\vec{a}\cdot\vec{b})\vec{c} = (\vec{a}\cdot\vec{c})\vec{b} - (\vec{b}\cdot\vec{c})\vec{a}$ or $(\vec{a}\cdot\vec{b})\vec{c} = (\vec{b}\cdot\vec{c})\vec{a}$.

Therefore, vectors $\vec{a}$ and $\vec{c}$ are collinear.

Illustration 3.84

Prove that

a. $\vec{a}\times(\vec{b}\times\vec{c}) + \vec{b}\times(\vec{c}\times\vec{a}) + \vec{c}\times(\vec{a}\times\vec{b}) = 0$

b. $\hat{i}\times(\vec{a}\times\hat{i}) + \hat{j}\times(\vec{a}\times\hat{j}) + \hat{k}\times(\vec{a}\times\hat{k}) = 2\vec{a}$

Sol.

a. $\vec{a}\times(\vec{b}\times\vec{c}) = (\vec{a}\cdot\vec{c})\vec{b} - (\vec{a}\cdot\vec{b})\vec{c}$

$\vec{b}\times(\vec{c}\times\vec{a}) = (\vec{b}\cdot\vec{a})\vec{c} - (\vec{b}\cdot\vec{c})\vec{a}$

$\vec{c}\times(\vec{a}\times\vec{b}) = (\vec{c}\cdot\vec{b})\vec{a} - (\vec{c}\cdot\vec{a})\vec{b}$

Adding we get

$\vec{a}\times(\vec{b}\times\vec{c}) + \vec{b}\times(\vec{c}\times\vec{a}) + \vec{c}\times(\vec{a}\times\vec{b}) = 0$

b. $\hat{i}\times(\vec{a}\times\hat{i}) + \hat{j}\times(\vec{a}\times\hat{j}) + \hat{k}\times(\vec{a}\times\hat{k})$

$= (\hat{i}\cdot\hat{i})\vec{a} - (\hat{i}\cdot\vec{a})\hat{i} + (\hat{j}\cdot\hat{j})\vec{a} - (\hat{j}\cdot\vec{a})\hat{j} + (\hat{k}\cdot\hat{k})\vec{a} - (\hat{k}\cdot\vec{a})\hat{k}$

$= 3\vec{a} - [(\hat{i}\cdot\vec{a})\hat{i} + (\hat{j}\cdot\vec{a})\hat{j} + (\hat{k}\cdot\vec{a})\hat{k}]$

$= 3\vec{a} - \vec{a} = 2\vec{a}$

Illustration 3.85

If $\hat{i}\times[(\vec{a}-\hat{j})\times\hat{i}] + \hat{j}\times[(\vec{a}-\hat{k})\times\hat{j}] + \hat{k}\times[(\vec{a}-\hat{i})\times\hat{k}] = 0$, then find vector $\vec{a}$.

Sol. We have $\hat{i}\times[(\vec{a}-\hat{j})\times\hat{i}] = (\hat{i}\cdot\hat{i})(\vec{a}-\hat{j}) - (\hat{i}\cdot(\vec{a}-\hat{j}))\hat{i}$

$= \vec{a} - \hat{j} - (\hat{i}\cdot\vec{a})\hat{i}$

Similarly, $\hat{j}\times[(\vec{a}-\hat{k})\times\hat{j}] = \vec{a} - \hat{k} - (\hat{j}\cdot\vec{a})\hat{j}$

And $\hat{k}\times[(\vec{a}-\hat{i})\times\hat{k}] = \vec{a} - \hat{i} - (\hat{k}\cdot\vec{a})\hat{k}$

$\therefore \quad \hat{i}\times[(\vec{a}-\hat{j})\times\hat{i}] + \hat{j}\times[(\vec{a}-\hat{k})\times\hat{j}] + \hat{k}\times[(\vec{a}-\hat{i})\times\hat{k}]$

$= \vec{a} - \hat{j} - (\hat{i}\cdot\vec{a})\hat{i} + \vec{a} - \hat{k} - (\hat{j}\cdot\vec{a})\hat{j} + \vec{a} - \hat{i} - (\hat{k}\cdot\vec{a})\hat{k} = 0$

$\therefore \quad 3\vec{a} - (\hat{i}+\hat{j}+\hat{k}) - \vec{a} = 0$

or $\quad \vec{a} = \dfrac{1}{2}(\hat{i}+\hat{j}+\hat{k})$

Illustration 3.86

If $\vec{a} = 4\hat{i}+3\hat{j}+\hat{k}$ and $\vec{b} = \hat{i}-2\hat{j}+2\hat{k}$, then find the vector $\vec{a}\times(\vec{a}\times(\vec{a}\times(\vec{a}\times\vec{b})))$.

Sol. We have, $\vec{p}=\vec{a}\times(\vec{a}\times\underbrace{(\vec{a}\times(\vec{a}\times\vec{b}))}_{(\vec{a}\cdot\vec{b})\vec{a}-(\vec{a}\cdot\vec{a})\vec{b}})$

$= \vec{a}\times(\vec{a}\times((\vec{a}\cdot\vec{b})\vec{a}-(\vec{a}\cdot\vec{a})\vec{b}))$

$= \vec{a}\times(\vec{a}\times(-26\vec{b}))$ $\quad (\because\ \vec{a}\cdot\vec{b}=0 \text{ and } \vec{a}\cdot\vec{a}=|\vec{a}|^2=26)$

$= -26\vec{a}\times(\vec{a}\times\vec{b})$

$= -26((\vec{a}\cdot\vec{b})\vec{a}-(\vec{a}\cdot\vec{a})\vec{b})$

$= -26(-26\vec{b})$

$= 676\vec{b}$

$= 676(\hat{i}-2\hat{j}+2\hat{k})$

Illustration 3.87

If three vectors $\vec{v}_1, \vec{v}_2$ and $\vec{v}_3$ are such that $\vec{v}_1=\vec{v}_2-\vec{v}_3$ where $\vec{v}_1=(\vec{a}\times\hat{i})\times\hat{i}$, $\vec{v}_2=(\vec{a}\times\hat{j})\times\hat{j}$ and $\vec{v}_3=(\vec{a}\times\hat{k})\times\hat{k}$. ($\vec{a}$ is a non zero vector). Then prove that projection of vector $\vec{a}$ on *y*-axis is zero.

Sol. We have,

$\vec{v}_1=(\vec{a}\times\hat{i})\times\hat{i}=(\vec{a}\cdot\hat{i})\hat{i}-\vec{a}$,

$\vec{v}_2=(\vec{a}\times\hat{j})\times\hat{j}=(\vec{a}\cdot\hat{j})\hat{j}-\vec{a}$, and

$\vec{v}_3=(\vec{a}\times\hat{k})\times\hat{k}=(\vec{a}\cdot\hat{k})\hat{k}-\vec{a}$

Now, $\vec{v}_1=\vec{v}_2-\vec{v}_3$

$\Rightarrow \quad (\vec{a}\cdot\hat{i})\hat{i}-\vec{a}=(\vec{a}\cdot\hat{j})\hat{j}-\vec{a}-(\vec{a}\cdot\hat{k})\hat{k}+\vec{a}$

$\Rightarrow \quad (\vec{a}\cdot\hat{i})\hat{i}-\{(\vec{a}\cdot\hat{i})\hat{i}+(\vec{a}\cdot\hat{j})\hat{j}+(\vec{a}\cdot\hat{k})\hat{k}\}=(\vec{a}\cdot\hat{j})\hat{j}-(\vec{a}\cdot\hat{k})\hat{k}$

$\Rightarrow \quad 2(\vec{a}\cdot\hat{j})\hat{j}=0$

$\Rightarrow \quad \vec{a}\cdot\hat{j}=0$

Therefore, the projection of $\vec{a}$ on *y*-axis is zero.

Illustration 3.88

Let $\hat{a}$ and $\hat{c}$ be unit vectors inclined at an angle $\frac{\pi}{3}$ to each other. If $(\hat{a}\times(\vec{b}\times\hat{c})).(\hat{a}\times\hat{c})=5$ then find the value of $[\vec{a}\ \vec{b}\ \vec{c}]$.

Sol. We have,

$(\hat{a}\times(\vec{b}\times\hat{c})).(\hat{a}\times\hat{c})$

$= ((\hat{a}\cdot\hat{c})\vec{b}-(\hat{a}\cdot\vec{b})\hat{c}).(\hat{a}\times\hat{c})$

$= (\hat{a}\cdot\hat{c})[\vec{b}\ \hat{a}\ \hat{c}]$

Now, $\hat{a}\cdot\hat{c}=(1)(1)\cos\frac{\pi}{3}=\frac{1}{2}$

$\therefore \quad (\hat{a}\times(\vec{b}\times\hat{c})).(\hat{a}\times\hat{c})=\frac{1}{2}[\vec{b}\ \hat{a}\ \hat{c}]$

So, $\quad (\hat{a}\times(\vec{b}\times\hat{c})).(\hat{a}\times\hat{c})=5$

$\Rightarrow \quad \frac{1}{2}[\vec{b}\ \hat{a}\ \hat{c}]=5$

$\Rightarrow \quad [\vec{b}\ \hat{a}\ \hat{c}]=10$

$\Rightarrow \quad [\hat{a}\ \vec{b}\ \hat{c}]=-10$

Illustration 3.89

Let $\vec{a}=\hat{i}-\hat{j}, \vec{b}=\hat{j}-\hat{k}, \vec{c}=\hat{k}-\hat{i}$. Then find unit vector $\vec{d}$ such that $\vec{a}.\vec{d}=0=[\vec{b}\ \vec{c}\ \vec{d}]$.

Sol.

Given that $\vec{a}.\vec{d}=0=[\vec{b}\ \vec{c}\ \vec{d}]$

$\therefore$ $\quad \vec{d}$ is a vector perpendicular to the vector $\vec{a}$ and coplanar with the vectors $\vec{b}$ and $\vec{c}$.

Hence, $\vec{d}$ is a vector collinear with $\vec{a}\times(\vec{b}\times\vec{c})$.

$\Rightarrow \quad \vec{d}=\pm\frac{\vec{a}\times(\vec{b}\times\vec{c})}{|\vec{a}\times(\vec{b}\times\vec{c})|}$

Now, $\vec{a}\times(\vec{b}\times\vec{c})$

$=(\vec{a}.\vec{c})\vec{b}-(\vec{b}.\vec{a})\vec{c}$

$=-\vec{b}+\vec{c}$

$=-(\hat{i}+\hat{j}-2\hat{k})$

$\Rightarrow \quad \vec{d}=\pm\left(\frac{\hat{i}+\hat{j}-2\hat{k}}{\sqrt{6}}\right)$

Illustration 3.90

If $\vec{a}, \vec{b}$ and $\vec{c}$ are non-coplanar unit vectors such that $\vec{a}\times(\vec{b}\times\vec{c})=\frac{\vec{b}+\vec{c}}{\sqrt{2}}$ then prove that the angle between $\vec{a}$ and $\vec{b}$ is $3\pi/4$.

Sol. We have, $\vec{a}\times(\vec{b}\times\vec{c})=\frac{\vec{b}+\vec{c}}{\sqrt{2}}$

$\Rightarrow \quad (\vec{a}\cdot\vec{c})\vec{b}-(\vec{a}\cdot\vec{b})\vec{c}=\frac{1}{\sqrt{2}}\vec{b}+\frac{1}{\sqrt{2}}\vec{c}$ $\quad$ …(i)

Since $\vec{b}$ and $\vec{c}$ are non-collinear, comparing coefficients of $\vec{c}$ on both sides of (i), we get

$-\vec{a}\cdot\vec{b}=\frac{1}{\sqrt{2}}$ or $\vec{a}\cdot\vec{b}=-\frac{1}{\sqrt{2}}$

$\Rightarrow \quad (1)(1)\cos\theta=-\frac{1}{\sqrt{2}}$

where θ is the angle between $\vec{a}$ and $\vec{b}$. Therefore,

$$\cos\theta = -\frac{1}{\sqrt{2}} \quad \text{or} \quad \theta = 135°$$

Illustration 3.91

If $\vec{b}, \vec{c}$ are unit vectors such that $|\vec{a}| = 7$ and $\vec{a}\times(\vec{b}\times\vec{c})+\vec{b}\times(\vec{c}\times\vec{a})=\frac{1}{2}\vec{a}$, then find the angle between vectors $\vec{a}$ and $\vec{c}$ and the angle between vectors $\vec{b}$ and $\vec{c}$.

Sol. We have, $\vec{a}\times(\vec{b}\times\vec{c})+\vec{b}\times(\vec{c}\times\vec{a})=\frac{1}{2}\vec{a}$

$\Rightarrow \quad (\vec{a}\cdot\vec{c})\vec{b}-(\vec{a}\cdot\vec{b})\vec{c}+(\vec{a}\cdot\vec{b})\vec{c}-(\vec{b}\cdot\vec{c})\vec{a}=\frac{1}{2}\vec{a}$

$\Rightarrow \quad (\vec{a}\cdot\vec{c})\vec{b}-\left((\vec{b}\cdot\vec{c})+\frac{1}{2}\right)\vec{a}=\vec{0}$

$\Rightarrow \quad \vec{a}\cdot\vec{c}=0,\ \vec{b}\cdot\vec{c}=-\frac{1}{2}$

Therefore, the angle between vectors $\vec{a}$ and $\vec{c}$ is 90°.

Also, $\vec{b}\cdot\vec{c}=-\frac{1}{2}$

$\Rightarrow \quad (1)(1)\cos\theta = -\frac{1}{2}$

$\Rightarrow \quad \theta = 120°$

Therefore, the angle between vectors $\vec{b}$ and $\vec{c}$ is 120°.

Illustration 3.92

Let $\vec{a}\times(\vec{b}\times\vec{c})=\frac{\vec{b}}{3}+\frac{\vec{c}}{2}$ and $\vec{b}\times(\vec{c}\times\vec{a})=-\frac{\vec{c}}{2}$. If $\vec{a},\vec{b}$ and $\vec{c}$ are non-collinear pair wise unit vectors, then find the volume of a parallelepiped, whose coterminous edges are $\vec{a}$, $\vec{b}$ and $\vec{c}$.

Sol. We have, $\vec{a}\times(\vec{b}\times\vec{c})=\frac{\vec{b}}{3}+\frac{\vec{c}}{2}$

$\Rightarrow \quad (\vec{a}\cdot\vec{c})\vec{b}-(\vec{a}\cdot\vec{b})\vec{c}=\frac{\vec{b}}{3}+\frac{\vec{c}}{2}$

$\Rightarrow \quad \vec{a}\cdot\vec{c}=\frac{1}{3}$ and $\vec{a}\cdot\vec{b}=-\frac{1}{2}$

Also, $\vec{b}\times(\vec{c}\times\vec{a})=-\frac{\vec{c}}{2}$

$\Rightarrow \quad (\vec{b}\cdot\vec{a})\vec{c}-(\vec{b}\cdot\vec{c})\vec{a}=-\frac{\vec{c}}{2}$

$\Rightarrow \quad \vec{b}\cdot\vec{c}=0$

Volume of parallelepiped $=\left|[\vec{a}\ \vec{b}\ \vec{c}]\right|$

Now, $[\vec{a}\ \ \vec{b}\ \ \vec{c}]^2=\begin{vmatrix}\vec{a}\cdot\vec{a} & \vec{a}\cdot\vec{b} & \vec{a}\cdot\vec{c}\\ \vec{b}\cdot\vec{a} & \vec{b}\cdot\vec{b} & \vec{b}\cdot\vec{c}\\ \vec{c}\cdot\vec{a} & \vec{c}\cdot\vec{b} & \vec{c}\cdot\vec{c}\end{vmatrix}$

$$=\begin{vmatrix}1 & -\frac{1}{2} & \frac{1}{3}\\ -\frac{1}{2} & 1 & 0\\ \frac{1}{3} & 0 & 1\end{vmatrix}=\frac{23}{36}$$

$\Rightarrow \quad$ Volume $=\frac{\sqrt{23}}{6}$

Illustration 3.93

If vector $\vec{x}$ satisfying $\vec{x}\times\vec{a}+(\vec{x}\cdot\vec{b})\vec{c}=\vec{d}$ is given by $\vec{x}=\lambda\vec{a}+\vec{a}\times\frac{\vec{a}\times(\vec{d}\times\vec{c})}{(\vec{a}\cdot\vec{c})|\vec{a}|^2}$, then find the value of λ.

Sol. We have, $\vec{x}\times\vec{a}+(\vec{x}\cdot\vec{b})\vec{c}=\vec{d}$

$\therefore \quad \{\vec{x}\times\vec{a}+(\vec{x}\cdot\vec{b})\vec{c}\}\times\vec{c}=\vec{d}\times\vec{c}$

$\Rightarrow \quad (\vec{x}\times\vec{a})\times\vec{c}+(\vec{x}\cdot\vec{b})(\vec{c}\times\vec{c})=\vec{d}\times\vec{c}$

$\Rightarrow \quad (\vec{x}\cdot\vec{c})\vec{a}-(\vec{a}\cdot\vec{c})\vec{x}=(\vec{d}\times\vec{c})$

$\Rightarrow \quad \vec{a}\times\{(\vec{x}\cdot\vec{c})\vec{a}-(\vec{a}\cdot\vec{c})\vec{x}\}=\vec{a}\times(\vec{d}\times\vec{c})$

$\Rightarrow \quad -(\vec{a}\cdot\vec{c})(\vec{a}\times\vec{x})=\vec{a}\times(\vec{d}\times\vec{c}) \qquad (\because \vec{a}\times\vec{a}=0)$

$\Rightarrow \quad \vec{x}\times\vec{a}=\frac{\vec{a}\times(\vec{d}\times\vec{c})}{\vec{a}\cdot\vec{c}}$

$\Rightarrow \quad \vec{a}\times(\vec{x}\times\vec{a})=\vec{a}\times\frac{\vec{a}\times(\vec{d}\times\vec{c})}{\vec{a}\cdot\vec{c}}$

$\Rightarrow \quad (\vec{a}\cdot\vec{a})\vec{x}-(\vec{a}\cdot\vec{x})\vec{a}=\vec{a}\times\frac{\vec{a}\times(\vec{d}\times\vec{c})}{\vec{a}\cdot\vec{c}}$

$\Rightarrow \quad (\vec{a}\cdot\vec{a})\vec{x}=(\vec{a}\cdot\vec{x})\vec{a}+\vec{a}\times\frac{\vec{a}\times(\vec{d}\times\vec{c})}{\vec{a}\cdot\vec{c}}$

$\Rightarrow \quad \vec{x}=\frac{(\vec{a}\cdot\vec{x})\vec{a}}{|\vec{a}|^2}+\vec{a}\times\frac{\vec{a}\times(\vec{d}\times\vec{c})}{(\vec{a}\cdot\vec{c})|\vec{a}|^2}$, where $\lambda=\frac{\vec{a}\cdot\vec{x}}{|\vec{a}|^2}$

Illustration 3.94

If $\vec{a}$ and $\vec{b}$ are two given vectors and k is any scalar, then find the vector $\vec{r}$ satisfying $\vec{r}\times\vec{a}+k\vec{r}=\vec{b}$.

Sol. We have, $\vec{r}\times\vec{a}+k\vec{r}=\vec{b}$ ...(i)

$\Rightarrow \quad (\vec{r}\times\vec{a})\times\vec{a}+k\,\vec{r}\times\vec{a}=\vec{b}\times\vec{a}$

$\Rightarrow \quad (\vec{r}\cdot\vec{a})\,\vec{a}-(\vec{a}\cdot\vec{a})\,\vec{r}+k\,(\vec{b}-k\,\vec{r})=\vec{b}\times\vec{a}$

$\Rightarrow \quad (\vec{r}\cdot\vec{a})\,\vec{a}+k\,\vec{b}-\vec{b}\times\vec{a}=(|\vec{a}|^2+k^2)\,\vec{r}$

$\Rightarrow \quad \vec{r}=\dfrac{(\vec{r}\cdot\vec{a})\,\vec{a}+k\,\vec{b}-\vec{b}\times\vec{a}}{|\vec{a}|^2+k^2}$

Also in Eq. (i), taking dot product with $\vec{a}$, we have

$(\vec{r}\times\vec{a})\cdot\vec{a}+k\,\vec{r}\cdot\vec{a}=\vec{b}\cdot\vec{a}$

or $\quad \vec{r}\cdot\vec{a}=\dfrac{\vec{b}\cdot\vec{a}}{k}$

$\Rightarrow \quad \vec{r}=\dfrac{1}{k^2+|\vec{a}|^2}\left[\dfrac{(\vec{a}\cdot\vec{b})\,\vec{a}}{k}+k\,\vec{b}+(\vec{a}\times\vec{b})\right]$

3.5.2 Cross Product of Four Vectors

1. Consider four non-zero vectors $\vec{a},\vec{b},\vec{c}$ and $\vec{d}$.

   Let's find vector $(\vec{a}\times\vec{b})\times(\vec{c}\times\vec{d})$.

   Let $\vec{a}\times\vec{b}=\vec{p}$

   $\therefore \quad (\vec{a}\times\vec{b})\times(\vec{c}\times\vec{d})=\vec{p}\times(\vec{c}\times\vec{d})$

   $=(\vec{p}\cdot\vec{d})\vec{c}-(\vec{p}\cdot\vec{c})\vec{d}$

   $=((\vec{a}\times\vec{b})\cdot\vec{d})\vec{c}-((\vec{a}\times\vec{b})\cdot\vec{c})\vec{d}$

   So, $(\vec{a}\times\vec{b})\times(\vec{c}\times\vec{d})=[\vec{a}\,\vec{b}\,\vec{d}]\vec{c}-[\vec{a}\,\vec{b}\,\vec{c}]\vec{d}$, which lies in the plane of vectors $\vec{c}$ and $\vec{d}$ say plane P_2

   Also, we can prove that $(\vec{a}\times\vec{b})\times(\vec{c}\times\vec{d})=[\vec{a}\,\vec{c}\,\vec{d}]\vec{b}-[\vec{b}\,\vec{c}\,\vec{d}]\vec{a}$, which lies in the plane of vectors $\vec{a}$ and $\vec{b}$ say plane P_1.

   Thus, vector $(\vec{a}\times\vec{b})\times(\vec{c}\times\vec{d})$ is common to two planes P_1 and P_2. So, it is along the line of intersection of planes P_1 and P_2.

2. $[\vec{a}\times\vec{b}\;\vec{b}\times\vec{c}\;\vec{c}\times\vec{a}]=[\vec{a}\,\vec{b}\,\vec{c}]^2$

   Proof:

   $[\vec{a}\times\vec{b}\;\vec{b}\times\vec{c}\;\vec{c}\times\vec{a}]$

   $=\vec{a}\times\vec{b}\cdot\{(\vec{b}\times\vec{c})\times(\vec{c}\times\vec{a})\}$

   $=\vec{a}\times\vec{b}\cdot\{((\vec{b}\times\vec{c})\cdot\vec{a})\vec{c}-((\vec{b}\times\vec{c})\cdot\vec{c})\vec{a}\}$

   $=\vec{a}\times\vec{b}\cdot\{[\vec{b}\,\vec{c}\,\vec{a}]\vec{c}\}$

   $=[\vec{b}\,\vec{c}\,\vec{a}]\{\vec{a}\times\vec{b}\cdot\vec{c}\}$

   $=[\vec{a}\,\vec{b}\,\vec{c}][\vec{a}\,\vec{b}\,\vec{c}]$

   $=[\vec{a}\,\vec{b}\,\vec{c}]^2$

   So, $[\vec{a}\,\vec{b}\,\vec{c}]=0 \Leftrightarrow [\vec{a}\times\vec{b}\;\vec{b}\times\vec{c}\;\vec{c}\times\vec{a}]=0$. Thus non-zero vectors $\vec{a},\vec{b}$ and $\vec{c}$ are coplanar iff vectors $\vec{a}\times\vec{b},\vec{b}\times\vec{c}$ and $\vec{c}\times\vec{a}$ coplanar.

   Also $[\vec{a}\,\vec{b}\,\vec{c}]\neq 0 \Leftrightarrow [\vec{a}\times\vec{b}\;\vec{b}\times\vec{c}\;\vec{c}\times\vec{a}]\neq 0$. Thus non-zero vectors $\vec{a},\vec{b}$ and $\vec{c}$ are non-coplanar iff vectors $\vec{a}\times\vec{b},\vec{b}\times\vec{c}$ and $\vec{c}\times\vec{a}$ are non-coplanar.

Illustration 3.95

If vectors $\vec{b},\vec{c}$ and $\vec{d}$ are non coplanar, then prove that vector $(\vec{a}\times\vec{b})\times(\vec{c}\times\vec{d})+(\vec{a}\times\vec{c})\times(\vec{d}\times\vec{b})+(\vec{a}\times\vec{d})\times(\vec{b}\times\vec{c})$ is parallel to $\vec{a}$.

Sol. We have, $(\vec{a}\times\vec{b})\times(\vec{c}\times\vec{d})+(\vec{a}\times\vec{c})\times(\vec{d}\times\vec{b})+(\vec{a}\times\vec{d})\times(\vec{b}\times\vec{c})$

Here $(\vec{a}\times\vec{b})\times(\vec{c}\times\vec{d})$

$=-(\vec{c}\times\vec{d}\cdot\vec{b})\,\vec{a}+(\vec{c}\times\vec{d}\cdot\vec{a})\,\vec{b}=[\vec{a}\,\vec{c}\,\vec{d}]\,\vec{b}-[\vec{b}\,\vec{c}\,\vec{d}]\,\vec{a}$ ...(i)

$(\vec{a}\times\vec{c})\times(\vec{d}\times\vec{b})$

$=-(\vec{d}\times\vec{b}\cdot\vec{c})\,\vec{a}+(\vec{d}\times\vec{b}\cdot\vec{a})\,\vec{c}=[\vec{a}\,\vec{d}\,\vec{b}]\,\vec{c}-[\vec{c}\,\vec{d}\,\vec{b}]\,\vec{a}$ ...(ii)

and $(\vec{a}\times\vec{d})\times(\vec{b}\times\vec{c})$

$=(\vec{a}\times\vec{d}\cdot\vec{c})\,\vec{b}-(\vec{a}\times\vec{d}\cdot\vec{b})\,\vec{c}=-[\vec{a}\,\vec{c}\,\vec{d}]\,\vec{b}-[\vec{a}\,\vec{d}\,\vec{b}]\,\vec{c}$ ...(iii)

Adding (i), (ii) and (iii), we get

Given vector $=-2\,[\vec{b}\,\vec{c}\,\vec{d}]\,\vec{a}=k\,\vec{a}$

Hence, the given vector is parallel to $\vec{a}$.

Illustration 3.96

If $\vec{a},\vec{b}$ and $\vec{c}$ are three non-coplanar vectors, then prove that

$$\vec{d}=\frac{\vec{a}\cdot\vec{d}}{[\vec{a}\,\vec{b}\,\vec{c}]}(\vec{b}\times\vec{c})+\frac{\vec{b}\cdot\vec{d}}{[\vec{a}\,\vec{b}\,\vec{c}]}(\vec{c}\times\vec{a})+\frac{\vec{c}\cdot\vec{d}}{[\vec{a}\,\vec{b}\,\vec{c}]}(\vec{a}\times\vec{b})$$

Sol. Since $\vec{a},\vec{b}$ and $\vec{c}$ are non-coplanar, vectors $\vec{a}\times\vec{b},\ \vec{b}\times\vec{c}$ and $\vec{c}\times\vec{a}$ are also non-coplanar. Let

$\vec{d}=l(\vec{b}\times\vec{c})+m\,(\vec{c}\times\vec{a})+n\,(\vec{a}\times\vec{b})$ ...(i)

Now multiplying both sides of (i) scalarly by $\vec{a}$, we have

$\vec{a}\cdot\vec{d}=l\,\vec{a}\cdot(\vec{b}\times\vec{c})+m\,\vec{a}\cdot(\vec{c}\times\vec{a})+n\,\vec{a}\cdot(\vec{a}\times\vec{b})=l\,[\vec{a}\,\vec{b}\,\vec{c}]$

$(\because [\vec{a}\,\vec{c}\,\vec{a}]=0=[\vec{a}\,\vec{a}\,\vec{b}])$

$\Rightarrow \quad l=(\vec{a}\cdot\vec{d})/[\vec{a}\,\vec{b}\,\vec{c}]$

Similarly, multiplying (i) scalarly by $\vec{b}$ and $\vec{c}$, we get

$$m = (\vec{b}\cdot\vec{d})/[\vec{a}\ \vec{b}\ \vec{c}] \text{ and } n = (\vec{c}\cdot\vec{d})/[\vec{a}\ \vec{b}\ \vec{c}]$$

Putting these values of l, m and n in (i), we get the required result.

Illustration 3.97

If $\vec{a}, \vec{b}$ and $\vec{c}$ are three non-coplanar non-zero vectors, then prove that $(\vec{a}\cdot\vec{a})\ \vec{b}\times\vec{c} + (\vec{a}\cdot\vec{b})\ \vec{c}\times\vec{a} + (\vec{a}\cdot\vec{c})\ \vec{a}\times\vec{b} = [\vec{b}\ \vec{c}\ \vec{a}]\ \vec{a}$.

Sol. As $\vec{a}, \vec{b}$ and $\vec{c}$ are non-coplanar, $\vec{b}\times\vec{a}$, $\vec{c}\times\vec{a}$ and $\vec{a}\times\vec{b}$ are also non-coplanar.

So, any vector can be expressed as a linear combination of these vectors.

Let $\vec{a} = \lambda\ \vec{b}\times\vec{c} + \mu\ \vec{c}\times\vec{a} + \nu\ \vec{a}\times\vec{b}$

$\therefore\quad \vec{a}\cdot\vec{a} = \lambda[\vec{b}\ \vec{c}\ \vec{a}],\ \vec{a}\cdot\vec{b} = \mu[\vec{c}\ \vec{a}\ \vec{b}],\ \vec{a}\cdot\vec{c} = \nu\ [\vec{a}\ \vec{b}\ \vec{c}]$

Therefore, $\vec{a} = \dfrac{(\vec{a}\cdot\vec{a})\vec{b}\times\vec{c}}{[\vec{b}\ \vec{c}\ \vec{a}]} + \dfrac{(\vec{a}\cdot\vec{b})\vec{c}\times\vec{a}}{[\vec{c}\ \vec{a}\ \vec{b}]} + \dfrac{(\vec{a}\cdot\vec{c})\vec{a}\times\vec{b}}{[\vec{a}\ \vec{b}\ \vec{c}]}$

Illustration 3.98

If $\vec{a} = \hat{i}+\hat{j}+2\hat{k}$, $\vec{b} = \hat{i}+2\hat{j}+2\hat{k}$ and $|\vec{c}| = 1$ such that $[\vec{a}\times\vec{b}\ \ \vec{b}\times\vec{c}\ \ \vec{c}\times\vec{a}]$ has the maximum value then find the value of $|(\vec{a}\times\vec{b})\times\vec{c}|^2$.

Sol. We have, $[\vec{a}\times\vec{b}\ \ \vec{b}\times\vec{c}\ \ \vec{c}\times\vec{a}]$

$= [\vec{a}\ \vec{b}\ \vec{c}]^2$

$= |\vec{c}.(\vec{a}\times\vec{b})|^2$

$= |\vec{c}|^2\ |\vec{a}\times\vec{b}|^2\ \cos^2\theta$

The maximum value of $[\vec{a}\ \vec{b}\ \vec{c}]^2$ is possible only when $\vec{c}$ is parallel to $\vec{a}\times\vec{b}$.

$\Rightarrow\quad \theta = 0$ or π

Hence, $|(\vec{a}\times\vec{b})\times\vec{c}|^2 = |\vec{a}\times\vec{b}|^2\ |\vec{c}|^2\ \sin^2\theta = 0$.

3.5.3 Lagrange's Identity

Lagrange's identity is a result stated as

$$(\vec{a}\times\vec{b})\cdot(\vec{c}\times\vec{d}) = \begin{vmatrix}\vec{a}\cdot\vec{c} & \vec{a}\cdot\vec{d}\\ \vec{b}\cdot\vec{c} & \vec{b}\cdot\vec{d}\end{vmatrix}$$

Proof:

$(\vec{a}\times\vec{b})\cdot(\vec{c}\times\vec{d}) = \vec{p}\cdot(\vec{c}\times\vec{d})$, where $\vec{p} = \vec{a}\times\vec{b}$

$= (\vec{p}\times\vec{c})\cdot\vec{d}$

(interchanging dot and cross)

$= ((\vec{a}\times\vec{b})\times\vec{c})\cdot\vec{d}$

$= ((\vec{a}\cdot\vec{c})\vec{b} - (\vec{b}\cdot\vec{c})\vec{a})\cdot\vec{d}$

$= (\vec{a}\cdot\vec{c})(\vec{b}\cdot\vec{d}) - (\vec{b}\cdot\vec{c})(\vec{a}\cdot\vec{d})$

$= \begin{vmatrix}\vec{a}\cdot\vec{c} & \vec{a}\cdot\vec{d}\\ \vec{b}\cdot\vec{c} & \vec{b}\cdot\vec{d}\end{vmatrix}$

Illustration 3.99

For any four vectors, $\vec{a}, \vec{b}$, and $\vec{c}, \vec{d}$ prove that $(\vec{b}\times\vec{c})\cdot(\vec{a}\times\vec{d}) + (\vec{c}\times\vec{a})\cdot(\vec{b}\times\vec{d}) + (\vec{a}\times\vec{b})\cdot(\vec{c}\times\vec{d}) = 0$.

Sol. $(\vec{b}\times\vec{c})\cdot(\vec{a}\times\vec{d}) = (\vec{b}\cdot\vec{a})(\vec{c}\cdot\vec{d}) - (\vec{b}\cdot\vec{d})(\vec{c}\cdot\vec{a})$

$(\vec{c}\times\vec{a})\cdot(\vec{b}\times\vec{d}) = (\vec{c}\cdot\vec{b})(\vec{a}\cdot\vec{d}) - (\vec{c}\cdot\vec{d})(\vec{a}\cdot\vec{b})$

$(\vec{a}\times\vec{b})\cdot(\vec{c}\times\vec{d}) = (\vec{a}\cdot\vec{c})(\vec{b}\cdot\vec{d}) - (\vec{a}\cdot\vec{d})(\vec{b}\cdot\vec{c})$

Adding we get,

$(\vec{b}\times\vec{c})\cdot(\vec{a}\times\vec{d}) + (\vec{c}\times\vec{a})\cdot(\vec{b}\times\vec{d}) + (\vec{a}\times\vec{b})\cdot(\vec{c}\times\vec{d}) = 0$

Illustration 3.100

If $\vec{b}$ and $\vec{c}$ are two non-collinear vectors such that $\vec{a}\ \|\ (\vec{b}\times\vec{c})$, then prove that $(\vec{a}\times\vec{b})\cdot(\vec{a}\times\vec{c})$ is equal to $|\vec{a}|^2(\vec{b}\cdot\vec{c})$.

Sol. $\vec{a}\ \|\ (\vec{b}\times\vec{c})$

$\Rightarrow\quad \vec{a} = \lambda(\vec{b}\times\vec{c})$ and $\vec{a} = \vec{a}\perp\vec{b}$ and $\vec{a}\perp\vec{c}$

Now, $(\vec{a}\times\vec{b})\cdot(\vec{a}\times\vec{c}) = \begin{vmatrix}\vec{a}\cdot\vec{a} & \vec{a}\cdot\vec{c}\\ \vec{b}\cdot\vec{a} & \vec{b}\cdot\vec{c}\end{vmatrix}$

$= \begin{vmatrix}\vec{a}.\vec{a} & 0\\ 0 & \vec{b}.\vec{c}\end{vmatrix} = |\vec{a}|^2\ (\vec{b}\cdot\vec{c})$

3.5.4 Reciprocal System of Vectors

In the Cartesian coordinate system, the unit vectors $\hat{i}, \hat{j}, \hat{k}$ along the x-, y-, and z-axes satisfy:

$$\hat{i}\cdot\hat{i} = \hat{j}\cdot\hat{j} = \hat{k}\cdot\hat{k} = 1,\ \hat{i}\cdot\hat{j} = \hat{j}\cdot\hat{k} = \hat{k}\cdot\hat{i} = 0.$$

That is, each vector has scalar product equal to unity with itself and zero with the other two.

This orthogonality of $\hat{i}, \hat{j}, \hat{k}$ makes resolution of vectors along the coordinate axes very simple.

In many practical situations, however, the three reference directions we choose may not be mutually perpendicular. Suppose

$\vec{a}, \vec{b}, \vec{c}$ are three **non-coplanar vectors**, which may form an oblique system of axes. In such a case, it is convenient to introduce another set of vectors $\vec{a'}, \vec{b'}, \vec{c'}$, which play the same role for $\vec{a}, \vec{b}, \vec{c}$ as $\hat{i}, \hat{j}, \hat{k}$ do for one another in the Cartesian system.

The system $(\vec{a'}, \vec{b'}, \vec{c'})$ is called the reciprocal system of vectors corresponding to $(\vec{a}, \vec{b}, \vec{c})$, if the following conditions hold.

$$\vec{a}\cdot\vec{a'} = 1, \vec{b}\cdot\vec{b'} = 1, \vec{c}\cdot\vec{c'} = 1,$$

$$\vec{a}\cdot\vec{b'} = \vec{a}\cdot\vec{c'} = \vec{b}\cdot\vec{a'} = \vec{b}\cdot\vec{c'} = \vec{c}\cdot\vec{a'} = \vec{c}\cdot\vec{b'} = 0.$$

Thus, each vector of the system $(\vec{a}, \vec{b}, \vec{c})$ has scalar product equal to unity with its corresponding reciprocal vector, and zero with the other two.

The reciprocal system is a generalization of the familiar orthogonal system. In fact, in the Cartesian case, the unit vectors $(\hat{i}, \hat{j}, \hat{k})$ themselves form their own reciprocal system.

Now let's find the formulas for the reciprocal vectors.

Let $\vec{a}, \vec{b}, \vec{c}$ be three non-coplanar vectors. We want to construct their reciprocal system $\vec{a'}, \vec{b'}, \vec{c'}$ such that

$$\vec{a}\cdot\vec{a'} = 1, \vec{b}\cdot\vec{b'} = 1, \vec{c}\cdot\vec{c'} = 1,$$

and $\vec{a}\cdot\vec{b'} = \vec{a}\cdot\vec{c'} = \vec{b}\cdot\vec{a'} = \vec{b}\cdot\vec{c'} = \vec{c}\cdot\vec{a'} = \vec{c}\cdot\vec{b'} = 0.$

Since $\vec{b}\cdot\vec{a'} = \vec{c}\cdot\vec{a'} = 0$, $\vec{a'}$ is perpendicular to both $\vec{b}$ and $\vec{c}$.

Thus, $\vec{a'}$ is parallel to the vector $\vec{b}\times\vec{c}$.

Therefore, $\vec{a'} = \lambda(\vec{b}\times\vec{c})$.

Also, $\vec{a}\cdot\vec{a'} = 1$

Therefore, $\vec{a}\cdot(\lambda(\vec{b}\times\vec{c})) = 1$

$$\Rightarrow \quad \lambda = \frac{1}{[\vec{a}\,\vec{b}\,\vec{c}]}$$

Hence, $\vec{a'} = \dfrac{\vec{b}\times\vec{c}}{[\vec{a}\,\vec{b}\,\vec{c}]}$.

Similarly, $\vec{b'} = \dfrac{\vec{c}\times\vec{a}}{[\vec{a}\,\vec{b}\,\vec{c}]}$ and $\vec{c'} = \dfrac{\vec{a}\times\vec{b}}{[\vec{a}\,\vec{b}\,\vec{c}]}$.

Illustration 3.101

If $\vec{a}, \vec{b}, \vec{c}$ are three non-coplanar vectors and $\vec{a}', \vec{b}', \vec{c}'$ are the corresponding reciprocal vectors then prove that $[\vec{a}\ \vec{b}\ \vec{c}][\vec{a}'\ \vec{b}'\ \vec{c}'] = 1$.

Sol. We have $[\vec{a}'\ \vec{b}'\ \vec{c}'] = \left[\dfrac{\vec{b}\times\vec{c}}{[\vec{a}\ \vec{b}\ \vec{c}]}\ \dfrac{\vec{c}\times\vec{a}}{[\vec{a}\ \vec{b}\ \vec{c}]}\ \dfrac{\vec{a}\times\vec{b}}{[\vec{a}\ \vec{b}\ \vec{c}]}\right]$

$$= \frac{1}{[\vec{a}\,\vec{b}\,\vec{c}]^3}[\vec{b}\times\vec{c}\ \ \vec{c}\times\vec{a}\ \ \vec{a}\times\vec{b}]$$

$$= \frac{1}{[\vec{a}\,\vec{b}\,\vec{c}]^3}[\vec{a}\ \vec{b}\ \vec{c}]^2 = \frac{1}{[\vec{a}\ \vec{b}\ \vec{c}]}$$

$\therefore \quad [\vec{a}\ \vec{b}\ \vec{c}][\vec{a}'\ \vec{b}'\ \vec{c}'] = 1$

Illustration 3.102

Find a set of vectors reciprocal to the set $-\hat{i}+\hat{j}+\hat{k}, \hat{i}-\hat{j}+\hat{k}, \hat{i}+\hat{j}+\hat{k}$.

Sol. Let $\vec{a} = -\hat{i}+\hat{j}+\hat{k}$, $\vec{b} = \hat{i}-\hat{j}+\hat{k}$, $\vec{c} = \hat{i}+\hat{j}+\hat{k}$

Then $\vec{b}\times\vec{c} = \begin{vmatrix} \hat{i} & \hat{j} & \hat{k} \\ 1 & -1 & 1 \\ 1 & 1 & 1 \end{vmatrix} = -2\hat{i}+2\hat{k}$.

$$\vec{c}\times\vec{a} = \begin{vmatrix} \hat{i} & \hat{j} & \hat{k} \\ 1 & 1 & 1 \\ -1 & 1 & 1 \end{vmatrix} = -2\hat{j}+2\hat{k},$$

and $\vec{a}\times\vec{b} = \begin{vmatrix} \hat{i} & \hat{j} & \hat{k} \\ -1 & 1 & 1 \\ 1 & -1 & 1 \end{vmatrix} = 2\hat{i}+2\hat{j}$

Also, $[\vec{a}\,\vec{b}\,\vec{c}] = \begin{vmatrix} -1 & 1 & 1 \\ 1 & -1 & 1 \\ 1 & 1 & 1 \end{vmatrix} = 4$

If a', b', c' is the reciprocal system of vectors, then

$$\vec{a'} = (\vec{b}\times\vec{c})/[\vec{a}\,\vec{b}\,\vec{c}] = \frac{1}{2}(-\hat{i}+\hat{k}),$$

$$\vec{b'} = (\vec{c}\times\vec{a})/[\vec{a}\,\vec{b}\,\vec{c}] = \frac{1}{2}(-\hat{j}+\hat{k}),$$

$$\vec{c'} = (\vec{a}\times\vec{b})/[\vec{a}\,\vec{b}\,\vec{c}] = \frac{1}{2}(\hat{i}+\hat{j})$$

Illustration 3.103

Let $\vec{a}, \vec{b}$ and $\vec{c}$ be a set of non-coplanar vectors and $\vec{a}', \vec{b}'$ and $\vec{c}'$ be its reciprocal set. Prove that

a. $\vec{a} = \dfrac{b'\times c'}{[\vec{a}'\,\vec{b}'\,\vec{c}']}$, $\vec{b} = \dfrac{c'\times a'}{[\vec{a}'\,\vec{b}'\,\vec{c}']}$ and $\vec{c} = \dfrac{a'\times b'}{[\vec{a}'\,\vec{b}'\,\vec{c}']}$.

b. $\vec{a'}\times\vec{b'}+\vec{b'}\times\vec{c'}+\vec{c'}\times\vec{a'} = \dfrac{\vec{a}+\vec{b}+\vec{c}}{[\vec{a}\,\vec{b}\,\vec{c}]}$

Sol. We have, $\vec{b'}\times\vec{c'} = \dfrac{(\vec{c}\times\vec{a})\times(\vec{a}\times\vec{b})}{[\vec{a}\,\vec{b}\,\vec{c}]^2}$

$$= \frac{\{(\vec{c}\times\vec{a})\cdot\vec{b}\}\vec{a} - \{(\vec{c}\times\vec{a})\cdot\vec{a}\}\vec{b}}{[\vec{a}\,\vec{b}\,\vec{c}]^2}$$

$$= \frac{[\vec{c}\,\vec{a}\,\vec{b}]\vec{a}-[\vec{c}\,\vec{a}\,\vec{a}]\vec{b}}{[\vec{a}\,\vec{b}\,\vec{c}]^2} = \frac{[\vec{a}\,\vec{b}\,\vec{c}]\vec{a}-0}{[\vec{a}\,\vec{b}\,\vec{c}]^2} = \frac{\vec{a}}{[\vec{a}\,\vec{b}\,\vec{c}]}$$

Also, $[\vec{a}'\,\vec{b}'\,\vec{c}'] = \vec{a}'\cdot(\vec{b}'\times\vec{c}') = \frac{\vec{b}\times\vec{c}}{[\vec{a}\,\vec{b}\,\vec{c}]}\cdot\frac{\vec{a}}{[\vec{a}\,\vec{b}\,\vec{c}]}$

$$= \frac{[\vec{a}\,\vec{b}\,\vec{c}]}{[\vec{a}\,\vec{b}\,\vec{c}]^2} = \frac{1}{[\vec{a}\,\vec{b}\,\vec{c}]}$$

$$\Rightarrow \quad \frac{\vec{b}'\times\vec{c}'}{[\vec{a}'\,\vec{b}'\,\vec{c}']} = \vec{a}$$

Similarly, $\vec{b} = \frac{\vec{c}'\times\vec{a}'}{[\vec{a}'\,\vec{b}'\,\vec{c}']}$ and $\vec{c} = \frac{\vec{a}'\times\vec{b}'}{[\vec{a}'\,\vec{b}'\,\vec{c}']}$

Also, $\vec{a}'\times\vec{b}'+\vec{b}'\times\vec{c}'+\vec{c}'\times\vec{a}' = \frac{\vec{a}+\vec{b}+\vec{c}}{[\vec{a}\,\vec{b}\,\vec{c}]}$

Illustration 3.104

If $\vec{a}, \vec{b}$ and $\vec{c}$ be three non-coplanar vectors and a', b' and c' constitute the reciprocal system of vectors, then prove that

a. $\vec{r} = (\vec{r}\cdot\vec{a}')\vec{a} + (\vec{r}\cdot\vec{b}')\vec{b} + (\vec{r}\cdot\vec{c}')\vec{c}$

b. $\vec{r} = (\vec{r}\cdot\vec{a})\vec{a}' + (\vec{r}\cdot\vec{b})\vec{b}' + (\vec{r}\cdot\vec{c})\vec{c}'$

Sol.

a. Since a vector can be expressed as a linear combination of three non-coplanar vectors, let

$$\vec{r} = x\vec{a} + y\vec{b} + z\vec{c} \quad \ldots\text{(i)}$$

where x, y and z are scalars.

Multiplying both sides of (i) scalarly by $\vec{a}'$, we get

$$\vec{r}\cdot\vec{a}' = x\,\vec{a}\cdot\vec{a}' + y\,\vec{b}\cdot\vec{a}' + z\,\vec{c}\cdot\vec{a}' = x\cdot 1 = x$$

$$(\because \vec{a}\cdot\vec{a}' = 1,\ \vec{b}.\vec{a}' = 0 = \vec{c}\cdot\vec{a}')$$

Similarly, multiplying both sides of (i) scalarly by $\vec{b}'$ and $\vec{c}'$, successively, we get

$$y = \vec{r}\cdot\vec{b}' \text{ and } z = \vec{r}\cdot\vec{c}'$$

Putting in (i), we get $\vec{r} = (\vec{r}\cdot\vec{a}')\vec{a} + (\vec{r}\cdot\vec{b}')\vec{b} + (\vec{r}\cdot\vec{c}')\vec{c}$

b. Since $\vec{a}'$, $\vec{b}'$ and $\vec{c}'$ are three non-coplanar vectors, we can take $\vec{r} = x\vec{a}' + y\vec{b}' + z\vec{c}'$ ...(ii)

Multiplying both sides of (ii) scalarly by $\vec{a}$, we get

$$\vec{r}\cdot\vec{a} = x(\vec{a}'\cdot\vec{a}) + y(\vec{b}'\cdot\vec{a}) + z(\vec{c}'\cdot\vec{a}) = x$$

$$(\because \vec{a}'\cdot\vec{a} = 1,\ \vec{b}'\cdot\vec{a} = 0 = \vec{c}'\cdot\vec{a})$$

Similarly, multiplying both sides of (i) scalarly by $\vec{b}$ and $\vec{c}$ successively, we get

$$y = \vec{r}\cdot\vec{b} \text{ and } z = \vec{r}\cdot\vec{c}$$

Putting in (ii), we get $\vec{r} = (\vec{r}\cdot\vec{a})\vec{a}' + (\vec{r}\cdot\vec{b})\vec{b}' + (\vec{r}\cdot\vec{c})\vec{c}'$.

Aim to complete within 60 minutes

Concept Application Exercise 3.4

1. For any four vectors $\vec{a}, \vec{b}, \vec{c}$ and $\vec{d}$, prove that $\vec{d}\cdot(\vec{a}\times(\vec{b}\times(\vec{c}\times\vec{d}))) = (\vec{b}\cdot\vec{d})[\vec{a}\,\vec{c}\,\vec{d}]$.

2. If $\vec{a}$ and $\vec{b}$ be two non-collinear unit vectors such that $\vec{a}\times(\vec{a}\times\vec{b}) = \frac{1}{2}\vec{b}$, then find the angle between $\vec{a}$ and $\vec{b}$.

3. Show that $(\vec{a}\times\vec{b})\times\vec{c} = \vec{a}\times(\vec{b}\times\vec{c})$ if and only if $\vec{a}$ and $\vec{c}$ are collinear or $(\vec{a}\times\vec{c})\times\vec{b} = \vec{0}$.

4. Let $\vec{a}$, $\vec{b}$ and $\vec{c}$ be non-zero vectors such that no two are collinear and $(\vec{a}\times\vec{b})\times\vec{c} = \frac{1}{3}|\vec{b}||\vec{c}|\vec{a}$. If θ is the acute angle between vectors $\vec{b}$ and $\vec{c}$, then the value of $\sin\theta$ is

   (1) $\frac{2}{3}$ (2) $\frac{\sqrt{3}}{2}$

   (3) $\frac{2\sqrt{2}}{3}$ (4) $\frac{\sqrt{3}}{6}$

5. If $\vec{a} = \hat{i}+\hat{j}+\hat{k}$ and $\vec{b} = \hat{i}-2\hat{j}+\hat{k}$, then find vector $\vec{c}$ such that $\vec{a}\cdot\vec{c} = 2$ and $\vec{a}\times\vec{c} = \vec{b}$.

6. If $\vec{p}, \vec{q}, \vec{r}$ denote vectors $\vec{b}\times\vec{c}, \vec{c}\times\vec{a}, \vec{a}\times\vec{b}$, respectively, show that $\vec{a}$ is parallel to $\vec{q}\times\vec{r}$, $\vec{b}$ is parallel to $\vec{r}\times\vec{p}$, $\vec{c}$ is parallel to $\vec{p}\times\vec{q}$.

7. If $\vec{a} = \vec{p}+\vec{q}$, $\vec{p}\times\vec{b} = \vec{0}$ and $\vec{q}\cdot\vec{b} = 0$, then prove that $\frac{\vec{b}\times(\vec{a}\times\vec{b})}{\vec{b}\cdot\vec{b}} = \vec{q}$.

8. If $\vec{u}, \vec{v}, \vec{w}$ be three vectors of magnitude $\sqrt{3}$, 1, 2 such that $\vec{u}\times(\vec{u}\times\vec{w})+3\vec{v} = 0$, then the acute angle between vectors $\vec{u}$ and $\vec{w}$ is

   (1) 30° (2) 45° (3) 60° (4) 90°

9. If $[\vec{a}\ \vec{b}\ \vec{c}]=-3$, then the value of $[\vec{a}\times(\vec{b}+\vec{c}),\ \vec{b}\times(\vec{c}+2\vec{a}),\vec{c}\times(\vec{a}+3\vec{b})]$ is

(1) –15 (2) –18 (3) –27 (4) –45

10. If $\vec{a}\cdot\vec{b}=1$, $\vec{b}\cdot\vec{c}=2$ and $\vec{c}\cdot\vec{a}=3$, then the value of $[\vec{a}\times(\vec{b}\times\vec{c}),\ \vec{b}\times(\vec{c}\times\vec{a}),\ \vec{c}\times(\vec{b}\times\vec{a})]$ is

(1) 0 (2) 1 (3) 2 (4) 3

11. If $[\vec{a}\times\vec{b}\ \ \vec{b}\times\vec{c}\ \ \vec{c}\times\vec{a}]+[\vec{a}+\vec{b}\ \ \vec{b}+\vec{c}\ \ \vec{c}+\vec{a}]=8$, then the find the possible value(s) of area of tetrahedron *OABC* with three of its sides as vectors $\vec{a},\vec{b},\vec{c}$.

12. Find the volume of a parallelopiped having three co-terminus vectors of equal magnitude $|\vec{a}|$ and equal inclination θ with each other.

13. Let *OABC* be a tetrahedron whose edges are of unit length. If $\overline{OA}=\vec{a}, \overline{OB}=\vec{b}$, and $\overline{OC}=\alpha(\vec{a}+\vec{b})+\beta(\vec{a}\times\vec{b})$, then find the value $\alpha\beta$ is

(1) $\frac{1}{\sqrt{3}}$ (2) $\frac{2}{\sqrt{3}}$ (3) $\frac{4}{\sqrt{3}}$ (4) $\frac{\sqrt{2}}{3}$

14. Let $\vec{a},\vec{b},\vec{c}$ be non-coplanar vectors and let the equations $\vec{a}',\vec{b}',\vec{c}'$ are reciprocal system of vector $\vec{a},\vec{b},\vec{c}$, then prove that $\vec{a}\times\vec{a}'+\vec{b}\times\vec{b}'+\vec{c}\times\vec{c}'$ is a null vector.

15. If $\vec{a}\times(\vec{c}\times\vec{d})=\frac{\vec{c}}{2}+\frac{\vec{d}}{3}$ and $(\vec{c}\times\vec{d})\times\vec{b}=\frac{\vec{d}}{4}-\frac{\vec{c}}{5}$, where $\vec{c}$ and $\vec{d}$ are non-collinear vectors, then find the value of $(\vec{a}\times\vec{b})\cdot(\vec{c}\times\vec{d})$ is

(1) $-\frac{1}{8}$ (2) $\frac{-7}{60}$ (3) $\frac{-23}{120}$ (4) $\frac{-31}{61}$

16. Unit vectors $\vec{a},\vec{b},\vec{c}$ are coplanar. A unit vector $\vec{d}$ is perpendicular to them. If $(\vec{a}\times\vec{b})\times(\vec{c}\times\vec{d})=\frac{1}{6}\hat{i}-\frac{1}{3}\hat{j}+\frac{1}{3}\hat{k}$ and the angle between $\vec{a}$ and $\vec{b}$ is 30°, then find vector $\vec{c}$.

Answers

2. Information is inconsistent. **4.** (3)

5. $\vec{c}=\frac{1}{3}(-\hat{i}+2\hat{j}+5\hat{k})$ **8.** (1)

9. (4) **10.** (1)

11. $\frac{1}{3},\frac{2}{3}$

12. $|\vec{a}|^3\sqrt{1+2\cos\theta}\,(1-\cos\theta)$

13. (4) **15.** (3)

16. $\vec{c}=\frac{1}{3}(\hat{i}-2\hat{j}+2\hat{k})$

SOLVED EXAMPLES

Example 3.1

If vectors $\vec{a}, \vec{b}$ and $\vec{c}$ are coplanar, show that

$$\begin{vmatrix} \vec{a} & \vec{b} & \vec{c} \\ \vec{a}\cdot\vec{a} & \vec{a}\cdot\vec{b} & \vec{a}\cdot\vec{c} \\ \vec{b}\cdot\vec{a} & \vec{b}\cdot\vec{b} & \vec{b}\cdot\vec{c} \end{vmatrix} = \vec{0}.$$

Sol. Given that $\vec{a}, \vec{b}$ and $\vec{c}$ are three coplanar vectors. Therefore, there exist scalars x, y and z, not all zero, such that

$$x\vec{a} + y\vec{b} + z\vec{c} = \vec{0} \quad \text{(i)}$$

Taking dot product of $\vec{a}$ and (i), we get

$$x\vec{a}\cdot\vec{a} + y\vec{a}\cdot\vec{b} + z\vec{a}\cdot\vec{c} = 0 \quad \text{(ii)}$$

Taking dot product of $\vec{b}$ and (i), we get

$$x\vec{b}\cdot\vec{a} + y\vec{b}\cdot\vec{b} + z\vec{b}\cdot\vec{c} = 0 \quad \text{(iii)}$$

Now Eqs. (i), (ii) and (iii) form a homogeneous system of equations, where x, y and z are not all zero.

Therefore, the system must have a non-trivial solution,

Hence, $\begin{vmatrix} \vec{a} & \vec{b} & \vec{c} \\ \vec{a}\cdot\vec{a} & \vec{a}\cdot\vec{b} & \vec{a}\cdot\vec{c} \\ \vec{b}\cdot\vec{a} & \vec{b}\cdot\vec{b} & \vec{b}\cdot\vec{c} \end{vmatrix} = 0$

Example 3.2

If $OABC$ is a regular tetrahedron in which D is the circumcentre of ΔOAB and E is the midpoint of edge AC. Prove that DE is equal to half of the edge of the tetrahedron.

Sol. Let $\overline{OA} = \vec{a}$ and $\overline{OB} = \vec{b}$, where O is the origin.

Since ΔABC is equilateral, the circumcentres of triangle coincide with the centroid. Therefore,

$$\overline{OD} = \frac{\vec{a}+\vec{b}}{3}$$

Also $\overline{OE} = \dfrac{\vec{a}+\vec{c}}{2}$

$$\Rightarrow \quad \overline{DE} = \frac{\vec{a}+3\vec{c}-2\vec{b}}{6}$$

$$\Rightarrow \quad |\overline{DE}|^2 = \left|\frac{\vec{a}+3\vec{c}-2\vec{b}}{6}\right|^2$$

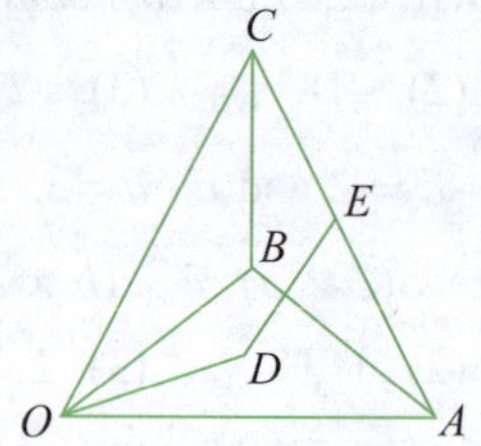

Since $|\vec{a}| = |\vec{b}| = |\vec{c}|$ and $\theta = 60°$, we have

$$|\overline{DE}|^2 = \frac{1}{36}\left(a^2 + 9a^2 + 4a^2 + \frac{6a^2}{2} - \frac{12a^2}{2} - \frac{4a^2}{2}\right)$$

$$= \frac{a^2}{4}$$

or $\quad |\overline{DE}| = \dfrac{a}{2}$

Example 3.3

If $A_1, A_2, \ldots, A_n$ are the vertices of a regular plane polygon with n sides and O as its centre. Then show that

$$\sum_{i=1}^{n-1} (\overrightarrow{OA}_i \times \overrightarrow{OA}_{i+1}) = (1-n)(\overrightarrow{OA}_2 \times \overrightarrow{OA}_1).$$

Sol. Vectors $\overrightarrow{OA}_1, \overrightarrow{OA}_2, \ldots, \overrightarrow{OA}_n$. All vectors are of same magnitude, say a. The angle between any two consecutive vectors is the same, that is, $2\pi/n$. Let $\hat{p}$ be the unit vector perpendicular to the plane of the polygon.

Let $\overrightarrow{OA}_1 \times \overrightarrow{OA}_2 = a^2 \sin\dfrac{2\pi}{n}\hat{p}$ …(i)

Therefore, $\displaystyle\sum_{i=1}^{n-1} \overrightarrow{OA}_i \times \overrightarrow{OA}_{i+1} = \sum_{i=1}^{n-1} a^2 \sin\frac{2\pi}{n}\hat{p}$

$$= (n-1)\, a^2 \sin\frac{2\pi}{n}\hat{p}$$

$$= (n-1)\,[-\overrightarrow{OA}_2 \times \overrightarrow{OA}_1]\ \text{[Using (i)]}$$

$$= (1-n)\,[\overrightarrow{OA}_2 \times \overrightarrow{OA}_1]$$

= R.H.S.

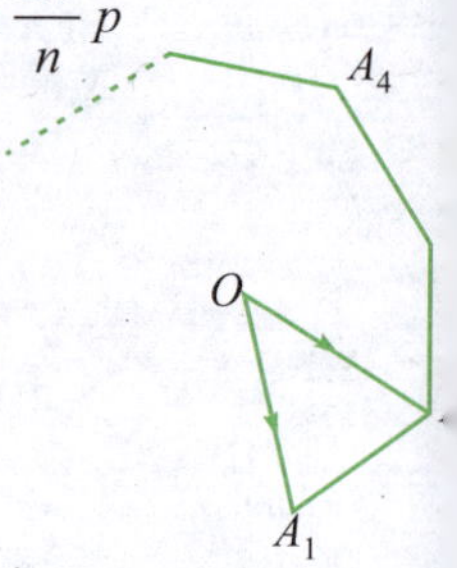

Example 3.4

Let O be an interior point of ΔABC such that $\overrightarrow{OA} + 2\,\overrightarrow{OB} + 3\,\overrightarrow{OC} = \vec{0}$. Then find the ratio of the area of ΔABC to the area of ΔAOC.

Sol. Let O be the origin and let $A(\vec{a})$, $B(\vec{b})$ and $C(\vec{c})$.

$$\frac{\text{Area of } \Delta ABC}{\text{Area of } \Delta AOC} = \frac{\frac{1}{2}|\vec{a}\times\vec{b}+\vec{b}\times\vec{c}+\vec{c}\times\vec{a}|}{\frac{1}{2}|\vec{a}\times\vec{c}|}$$

Now $\overrightarrow{OA}+2\overrightarrow{OB}+3\overrightarrow{OC}=\vec{0}$

or $\vec{a}+2\vec{b}+3\vec{c}=\vec{0}$

Taking cross product with $\vec{b}$, we get

$$\vec{a}\times\vec{b}+3\vec{c}\times\vec{b}=0$$

$\Rightarrow$ $\vec{a}\times\vec{b}=(\vec{b}\times\vec{c})$

Taking cross product with $\vec{a}$, we get

$$2\vec{a}\times\vec{b}+3\vec{a}\times\vec{c}=0$$

$\Rightarrow$ $\vec{a}\times\vec{b}=\frac{3}{2}(\vec{c}\times\vec{a})$

$\Rightarrow$ $\vec{a}\times\vec{b}=\frac{3}{2}(\vec{c}\times\vec{a})=3(\vec{b}\times\vec{c})$

Let $(\vec{c}\times\vec{a})=\vec{p}$ then $\vec{a}\times\vec{b}=\frac{3\vec{p}}{2}$ and $\vec{b}\times\vec{c}=\frac{\vec{p}}{2}$

Hence, $\frac{|\vec{a}\times\vec{b}+\vec{b}\times\vec{c}+\vec{c}\times\vec{a}|}{|\vec{c}\times\vec{a}|}=\frac{\left|\frac{3\vec{p}}{2}+\frac{\vec{p}}{2}+\vec{p}\right|}{|\vec{p}|}$

$$=\frac{3|\vec{p}|}{|\vec{p}|}=3$$

Example 3.5

The vectors $\vec{A}=a\hat{i}+b\hat{j}+c\hat{k}$, $\vec{B}=d\hat{i}+3\hat{j}+4\hat{k}$, $\vec{C}=3\hat{i}+\hat{j}-2\hat{k}$ represent the sides of a triangle taken in order.

If the area of the triangle is $5\sqrt{6}$, find the values of *a*, *b*, *c*, and *d*.

Sol. For the sides of a triangle taken in order

$$\vec{A}+\vec{B}+\vec{C}=\vec{0}$$

Substitute the given vectors, we get

$$(a+d+3)\hat{i}+(b+4)\hat{j}+(c+2)\hat{k}=0$$

$\Rightarrow$ $a=-(d+3), b=-4, c=-2.$

The area of the triangle is given by

$$\frac{1}{2}|\vec{B}\times\vec{C}|=5\sqrt{6} \text{ (given)}$$

$\therefore$ $|\vec{B}\times\vec{C}|=10\sqrt{6}$.

Now, $\vec{B}\times\vec{C}=\begin{vmatrix}\hat{a} & \hat{j} & k\\ d & 3 & 4\\ 3 & 1 & -2\end{vmatrix}$

$$=-10\hat{i}+(2d+12)\hat{j}+(d-9)\hat{k}$$

Therefore, $|\vec{B}\times\vec{C}|^2=600$

$\Rightarrow$ $d^2+6d-55=0.$

$\Rightarrow$ $(d-5)(d+11)=0$

$\Rightarrow$ $d=5,-11$

For $d=5, a=-8$

For $d=-11, a=8$

Thus, $(a, b, c, d)=(5, -8, -4, -2)$ or $(-11, 8, -4, -2)$

Example 3.6

If $\vec{a},\vec{b},\vec{c}$ are three given non-coplanar vectors and $\vec{r}$ is any arbitrary vector in space, where

$$\Delta_1=\begin{vmatrix}\vec{r}\cdot\vec{a} & \vec{b}\cdot\vec{a} & \vec{c}\cdot\vec{a}\\ \vec{r}\cdot\vec{b} & \vec{b}\cdot\vec{b} & \vec{c}\cdot\vec{b}\\ \vec{r}\cdot\vec{c} & \vec{b}\cdot\vec{c} & \vec{c}\cdot\vec{c}\end{vmatrix}, \Delta_2=\begin{vmatrix}\vec{a}\cdot\vec{a} & \vec{r}\cdot\vec{a} & \vec{c}\cdot\vec{a}\\ \vec{a}\cdot\vec{b} & \vec{r}\cdot\vec{b} & \vec{c}\cdot\vec{b}\\ \vec{a}\cdot\vec{c} & \vec{r}\cdot\vec{c} & \vec{c}\cdot\vec{c}\end{vmatrix}$$

$$\Delta_3=\begin{vmatrix}\vec{a}\cdot\vec{a} & \vec{b}\cdot\vec{a} & \vec{r}\cdot\vec{a}\\ \vec{a}\cdot\vec{b} & \vec{b}\cdot\vec{b} & \vec{r}\cdot\vec{b}\\ \vec{a}\cdot\vec{c} & \vec{b}\cdot\vec{c} & \vec{r}\cdot\vec{c}\end{vmatrix}, \Delta=\begin{vmatrix}\vec{a}\cdot\vec{a} & \vec{b}\cdot\vec{a} & \vec{c}\cdot\vec{a}\\ \vec{a}\cdot\vec{b} & \vec{b}\cdot\vec{b} & \vec{c}\cdot\vec{b}\\ \vec{a}\cdot\vec{c} & \vec{b}\cdot\vec{c} & \vec{c}\cdot\vec{c}\end{vmatrix},$$

then

prove that $\vec{r}=\frac{\Delta_1}{\Delta}\vec{a}+\frac{\Delta_2}{\Delta}\vec{b}+\frac{\Delta_3}{\Delta}\vec{c}$.

Sol. We have, $\Delta_1=\begin{vmatrix}\vec{r}\cdot\vec{a} & \vec{b}\cdot\vec{a} & \vec{c}\cdot\vec{a}\\ \vec{r}\cdot\vec{b} & \vec{b}\cdot\vec{b} & \vec{c}\cdot\vec{b}\\ \vec{r}\cdot\vec{c} & \vec{b}\cdot\vec{c} & \vec{c}\cdot\vec{c}\end{vmatrix}=[\vec{r}\,\vec{b}\,\vec{c}][\vec{a}\,\vec{b}\,\vec{c}]$,

$$\Delta_2=\begin{vmatrix}\vec{a}\cdot\vec{a} & \vec{r}\cdot\vec{a} & \vec{c}\cdot\vec{a}\\ \vec{a}\cdot\vec{b} & \vec{r}\cdot\vec{b} & \vec{c}\cdot\vec{b}\\ \vec{a}\cdot\vec{c} & \vec{r}\cdot\vec{c} & \vec{c}\cdot\vec{c}\end{vmatrix}=[\vec{r}\,\vec{c}\,\vec{a}][\vec{a}\,\vec{b}\,\vec{c}],$$

and $\Delta_3=\begin{vmatrix}\vec{a}\cdot\vec{a} & \vec{b}\cdot\vec{a} & \vec{r}\cdot\vec{a}\\ \vec{a}\cdot\vec{b} & \vec{b}\cdot\vec{b} & \vec{r}\cdot\vec{b}\\ \vec{a}\cdot\vec{c} & \vec{b}\cdot\vec{c} & \vec{r}\cdot\vec{c}\end{vmatrix}=[\vec{r}\,\vec{a}\,\vec{b}][\vec{a}\,\vec{b}\,\vec{c}]$

Also, $\Delta=\begin{vmatrix}\vec{a}\cdot\vec{a} & \vec{b}\cdot\vec{a} & \vec{c}\cdot\vec{a}\\ \vec{a}\cdot\vec{b} & \vec{b}\cdot\vec{b} & \vec{c}\cdot\vec{b}\\ \vec{a}\cdot\vec{c} & \vec{b}\cdot\vec{c} & \vec{c}\cdot\vec{c}\end{vmatrix}=[\vec{a}\,\vec{b}\,\vec{c}]^2$

Now $\vec{r}=x\vec{a}+y\vec{b}+z\vec{c}$

Taking dot product with $\vec{a}\times\vec{b}$, we have

$$\vec{r}\cdot(\vec{a}\times\vec{b})=z\vec{c}\cdot(\vec{a}\times\vec{b})$$

or $z = \dfrac{[\vec{r}\,\vec{a}\,\vec{b}]}{[\vec{a}\,\vec{b}\,\vec{c}]} = \dfrac{\Delta_3}{\Delta}$

Similarly $x = \dfrac{\Delta_1}{\Delta}$ and $y = \dfrac{\Delta_2}{\Delta}$

Hence, $\vec{r} = \dfrac{\Delta_1}{\Delta}\vec{a} + \dfrac{\Delta_2}{\Delta}\vec{b} + \dfrac{\Delta_3}{\Delta}\vec{c}$

Example 3.7

If c is a given non-zero scalar, and $\vec{A}$ and $\vec{B}$ are given non-zero vectors such that $\vec{A} \perp \vec{B}$, then find vector $\vec{X}$ which satisfies the equations $\vec{A}\cdot\vec{X} = c$ and $\vec{A}\times\vec{X} = \vec{B}$.

Sol. We have $\vec{A}\times\vec{X} = \vec{B}$

$\Rightarrow \quad (\vec{A}\times\vec{X})\times\vec{A} = \vec{B}\times\vec{A}$

$\Rightarrow \quad (\vec{A}\cdot\vec{A})\vec{X} - (\vec{X}\cdot\vec{A})\vec{A} = \vec{B}\times\vec{A}$

$\Rightarrow \quad (\vec{A}\cdot\vec{A})\,\vec{X} - c\,\vec{A} = \vec{B}\times\vec{A}$

$\Rightarrow \quad \vec{X} = \dfrac{\vec{B}\times\vec{A} + c\,\vec{A}}{(\vec{A}\cdot\vec{A})}$

Example 3.8

If vectors $\vec{b}$, $\vec{c}$ and $\vec{d}$ are not coplanar, then prove that vector $(\vec{a}\times\vec{b})\times(\vec{c}\times\vec{d}) + (\vec{a}\times\vec{c})\times(\vec{d}\times\vec{b}) + (\vec{a}\times\vec{d})\times(\vec{b}\times\vec{c})$ is parallel to $\vec{a}$.

Sol. We have,

$(\vec{a}\times\vec{b})\times(\vec{c}\times\vec{d}) = -(\vec{c}\times\vec{d}\cdot\vec{b})\,\vec{a} + (\vec{c}\times\vec{d}\cdot\vec{a})\,\vec{b}$

$= [\vec{a}\ \vec{c}\ \vec{d}]\,\vec{b} - [\vec{b}\ \vec{c}\ \vec{d}]\,\vec{a}$ …(i)

$(\vec{a}\times\vec{c})\times(\vec{d}\times\vec{b}) = -(\vec{d}\times\vec{b}\cdot\vec{c})\,\vec{a} + (\vec{d}\times\vec{b}\cdot\vec{a})\,\vec{c}$

$= [\vec{a}\ \vec{d}\ \vec{b}]\,\vec{c} - [\vec{c}\ \vec{d}\ \vec{b}]\,\vec{a}$ …(ii)

$(\vec{a}\times\vec{d})\times(\vec{b}\times\vec{c}) = (\vec{a}\times\vec{d}\cdot\vec{c})\,\vec{b} - (\vec{a}\times\vec{d}\cdot\vec{b})\,\vec{c}$

$= -[\vec{a}\ \vec{c}\ \vec{d}]\,\vec{b} - [\vec{a}\ \vec{d}\ \vec{b}]\,\vec{c}$ …(iii)

$(\vec{a}\times\vec{b})\times(\vec{c}\times\vec{d}) + (\vec{a}\times\vec{c})\times(\vec{d}\times\vec{b}) + (\vec{a}\times\vec{d})\times(\vec{b}\times\vec{c}) =$
$-2\,[\vec{b}\ \vec{c}\ \vec{d}]\,\vec{a} = k\,\vec{a}$

Hence, given vector is parallel to $\vec{a}$.

Example 3.9

Let $\vec{u}$ and $\vec{v}$ be unit vectors. If $\vec{w}$ is a vector such that $\vec{w} + (\vec{w}\times\vec{u}) = \vec{v}$, then prove that $|(\vec{u}\times\vec{v})\,.\,\vec{w}| \le \dfrac{1}{2}$ and that the equality holds if and only if $\vec{u}$ is perpendicular to $\vec{v}$.

Sol. We have $\vec{w} + (\vec{w}\times\vec{u}) = \vec{v}$ …(i)

Let's find $\vec{w}$ in the term of vectors $\vec{u}$ and $\vec{w}$

In (i), taking cross product with $\vec{u}$, we get

$$\vec{w}\times\vec{u} + (\vec{w}\times\vec{u})\times\vec{u} = \vec{v}\times\vec{u}$$

$\Rightarrow \quad \vec{v} - \vec{w} + (\vec{w}\cdot\vec{u})\,\vec{u} - |\vec{u}|^2\,\vec{w} = \vec{v}\times\vec{u}$ …(ii)

Now, in (i), taking dot product with $\vec{u}$, we get

$$\vec{w}\cdot\vec{u} + 0 = \vec{v}\cdot\vec{u}$$

Therefore, from (i), we have

$$\vec{v} - \vec{w} + (\vec{v}\cdot\vec{u})\vec{u} - \vec{w} = \vec{v}\times\vec{u}$$

$\Rightarrow \quad \vec{w} = \dfrac{1}{2}((\vec{v}\cdot\vec{u})\vec{u} + \vec{v} + \vec{u}\times\vec{v})$

$\Rightarrow \quad |(\vec{u}\times\vec{v}).\vec{w}|$

$= \dfrac{1}{2}|0 + 0 + \vec{u}\times\vec{v}|^2$

$= \dfrac{1}{2}|\vec{u}\times\vec{v}|^2$

$= \dfrac{1}{2}\sin^2\theta$, where θ is angle between $\vec{u}$ and $\vec{v}$

$\le \dfrac{1}{2}$

Equally holds when $\sin\theta = 1$ or $\theta = 90°$, that means $\vec{u} \perp \vec{v}$.

Example 3.10

The position vectors of the vertices A, B and C of a tetrahedron $ABCD$ are $\hat{i} + \hat{j} + \hat{k}$, $\hat{i}$ and $3\hat{i}$, respectively. The altitude from vertex D to the opposite face ABC meets the median line through A of triangle ABC at a point E. If the length of the side AD is 4 and the volume of the tetrahedron is $2\sqrt{2}/3$, find the position vectors of the point E for all its possible positions.

Sol. We are given $AD = 4$

Volume of tetrahedron $= \dfrac{2\sqrt{2}}{3}$

$\Rightarrow \quad \dfrac{1}{3}$ (Area of ΔABC) $p = \dfrac{2\sqrt{2}}{3}$

$\Rightarrow \quad \dfrac{1}{2}|\overrightarrow{BA}\times\overrightarrow{BC}|\,p = 2\sqrt{2}$

$\frac{1}{2}|(\hat{j}+\hat{k})\times 2\hat{i}|p = 2\sqrt{2}$

or $|\hat{j}-\hat{k}|p = 2\sqrt{2}$

or $\sqrt{2}p = 2\sqrt{2}$ or $p = 2$

We have to find the P.V. of point E. Let it divide median AF in the ratio $\lambda : 1$.

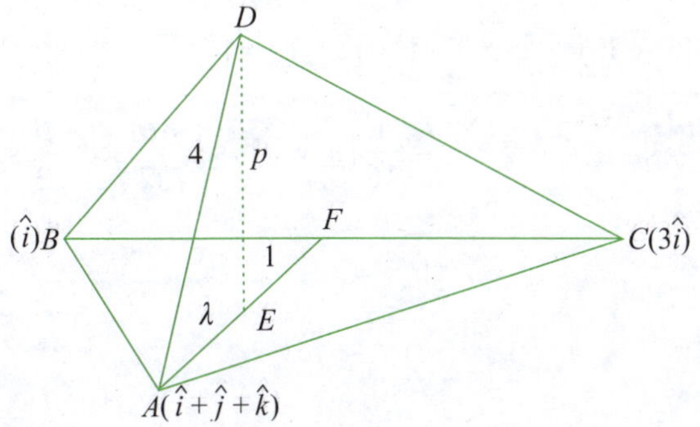

$$\text{P.V. of } E = \frac{\lambda \cdot 2\hat{i} + (\hat{i}+\hat{j}+\hat{k})}{\lambda+1} \quad \text{...(i)}$$

$$\Rightarrow \quad \overrightarrow{AE} = \text{P.V. or } E - \text{P.V. of } A = \frac{\lambda(\hat{i}-\hat{j}-\hat{k})}{\lambda+1}$$

$$|\overrightarrow{AE}|^2 = 3\left(\frac{\lambda}{\lambda+1}\right)^2 \quad \text{...(ii)}$$

$$\text{Now, } 4 + 3\left(\frac{\lambda}{\lambda+1}\right)^2 = 16$$

$$\left(\frac{\lambda}{\lambda+1}\right) = \pm 2$$

$$\lambda = -2 \text{ or } -2/3$$

Putting the value of λ in (i), we get the P.V. of possible positions of E as $-\hat{i}+3\hat{j}+3\hat{k}$ or $3\hat{i}-\hat{j}-\hat{k}$.

Example 3.11

If the incident ray on a surface is along the unit vector $\hat{v}$, the reflected ray is along the unit vector $\hat{w}$ and the normal is along the unit vector $\hat{a}$ outwards, express $\hat{w}$ in terms of $\hat{a}$ and $\hat{v}$.

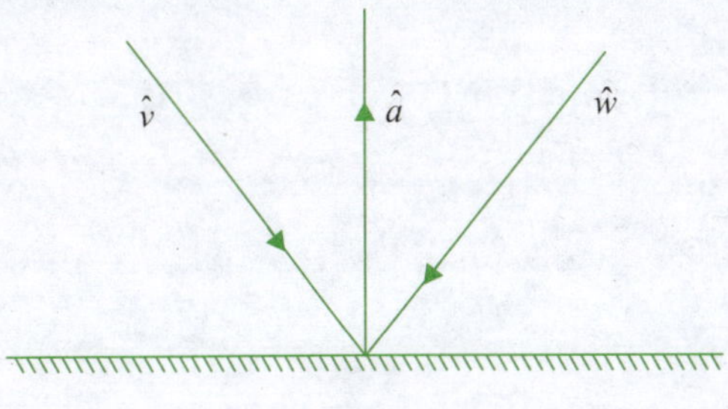

Sol. Given that the incident ray is along $\hat{v}$, the reflected ray is along $\hat{w}$ and the normal is along $\hat{a}$, outwards. The given figure can be redrawn as shown in the following figure.

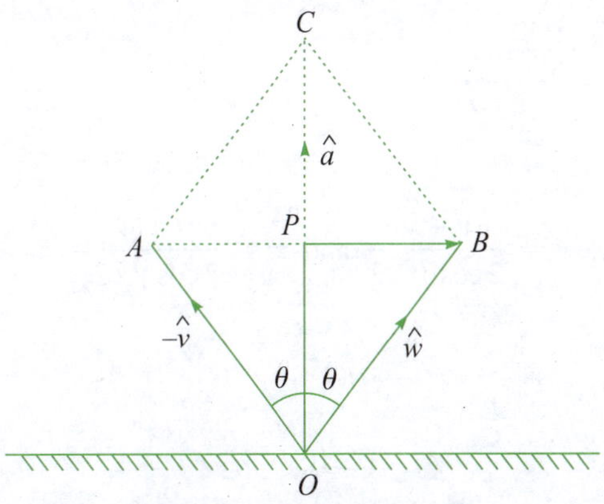

We know that the incident ray, the reflected ray, and the normal lie in a plane, and the angle of incidence is equal to the angle of reflection.

Therefore, $\hat{a}$ will be along the angle bisector of $\hat{w}$ and $-\hat{v}$, i.e.,

$$\hat{a} = \frac{\hat{w}+(-\hat{v})}{|\hat{w}-\hat{v}|} \quad \text{...(i)}$$

But $\hat{a}$ is a unit vector

where $\quad |\hat{w}-\hat{v}| = OC = 2OP = 2|\hat{w}|\cos\theta = 2\cos\theta$

Substituting this value in (i), we get

$$\hat{a} = \frac{\hat{w}-\hat{v}}{2\cos\theta}$$

or $\quad \hat{w} = \hat{v} + (2\cos\theta)\hat{a}$

or $\quad \hat{w} = \hat{v} - 2(\hat{a}\cdot\hat{v})\hat{a}$

Example 3.12

Let $\vec{u}, \vec{v}$ and $\vec{w}$ be three non-coplanar unit vectors and α, β and γ be the angles between $\vec{u}$ and $\vec{v}$, $\vec{v}$ and $\vec{w}$, and $\vec{w}$ and $\vec{u}$, respectively. If $\vec{x}, \vec{y}$ and $\vec{z}$ are unit vectors along the bisectors of the angles α, β and γ, respectively. Then prove that $[\vec{x}\times\vec{y} \ \ \vec{y}\times\vec{z} \ \ \vec{z}\times\vec{x}] = \frac{1}{16}[\vec{u}\ \vec{v}\ \vec{w}]^2 \sec^2\frac{\alpha}{2}\sec^2\frac{\beta}{2}\sec^2\frac{\gamma}{2}$.

Sol. We know that $[\vec{x}\times\vec{y} \ \ \vec{y}\times\vec{z} \ \ \vec{z}\times\vec{x}] = [\vec{x}\ \vec{y}\ \vec{z}]^2$

Also a vector along the bisector of given two unit vectors $\vec{u}, \vec{v}$ is $\vec{u}+\vec{v}$.

Therefore a unit vector along the bisector is $\dfrac{\vec{u}+\vec{v}}{\left|\vec{u}+\vec{v}\right|}$, where

$$|\vec{u}+\vec{v}|^2 = 1+1+2\vec{u}\cdot\vec{v} = 2+2\cos\alpha = 4\cos^2\frac{\alpha}{2}$$

$$\Rightarrow \quad \vec{x} = \frac{\vec{u}+\vec{v}}{2\cos\dfrac{\alpha}{2}}$$

Similarly, $\vec{y} = \dfrac{\vec{v}+\vec{w}}{2\cos\beta/2}$ and $\vec{z} = \dfrac{\vec{u}+\vec{w}}{2\cos\gamma/2}$

$$\Rightarrow \quad [\vec{x}\ \vec{y}\ \vec{z}] = \frac{1}{8}[\vec{u}+\vec{v}\ \ \vec{v}+\vec{w}\ \ \vec{u}+\vec{w}] \times \sec\frac{\alpha}{2}\sec\frac{\beta}{2}\sec\frac{\gamma}{2}$$

$$= \frac{1}{8}2[\vec{u}\ \vec{v}\ \vec{w}]\sec\frac{\alpha}{2}\sec\frac{\beta}{2}\sec\frac{\gamma}{2}$$

$$= \frac{1}{4}[\vec{u}\ \vec{v}\ \vec{w}]\sec\frac{\alpha}{2}\sec\frac{\beta}{2}\sec\frac{\gamma}{2}$$

$$\Rightarrow \quad [\vec{x}\times\vec{y}\ \ \vec{y}\times\vec{z}\ \ \vec{z}\times\vec{x}] = [\vec{x}\ \ \vec{y}\ \ \vec{z}]^2$$

$$= \frac{1}{16}[\vec{u}\ \vec{v}\ \vec{w}]^2\sec^2\frac{\alpha}{2}\sec^2\frac{\beta}{2}\sec^2\frac{\gamma}{2}$$

EXERCISES

[Note: Bulleted questions in Single Correct Answer Type, Numerical Value Type and Integer Answer Type sections are of high difficulty level.]

Single Correct Answer Type

Dot Product of Vectors

1. If $\hat{a}$, $\hat{b}$ and $\hat{c}$ are unit vectors satisfying $\hat{a}-\sqrt{3}\hat{b}+\hat{c}=\vec{0}$, then the angle between the vectors $\hat{a}$ and $\hat{c}$ is
(1) 30° (2) 60°
(3) 90° (4) 120°

2. Let $\vec{a}(x) = (\sin x)\hat{i} + (\cos x)\hat{j}$ and $\vec{b}(x) = (\cos 2x)\hat{i} + (\sin 2x)\hat{j}$ be two variable vectors ($x \in R$). Then $\vec{a}(x)$ and $\vec{b}(x)$ are
(1) collinear for unique value of x
(2) perpendicular for infinite values of x
(3) zero vectors for unique value of x
(4) parallel for infinity many values of x.

3. Let $\vec{f}(t)=[t]\hat{i}+(t-[t])\hat{j}+[t+1]\hat{k}$, where [.] denotes the greatest integer function. Then the vectors $\vec{f}\left(\frac{5}{4}\right)$ and $\vec{f}(t)$, $0<t<1$, are
(1) parallel to each other
(2) perpendicular to each other
(3) inclined at $\cos^{-1}\frac{2}{\sqrt{7(1-t^2)}}$
(4) inclined at $\cos^{-1}\frac{8+t}{9\sqrt{1+t^2}}$

4. The value of x for which the angle between $\vec{a}=2x^2\hat{i}+4x\hat{j}+\hat{k}$ and $\vec{b}=7\hat{i}-2\hat{j}+x\hat{k}$ is obtuse and the angle between $\vec{b}$ and the z-axis is acute and less than $\pi/6$ is
(1) $a<x<1/2$ (2) $1/2<x<15$
(3) $x>1/2$ or $x<0$ (4) no real value of x

5. If $\vec{a}$, $\vec{b}$ and $\vec{c}$ are three mutually perpendicular vectors, then the vector which is equally inclined to these vectors is
(1) $\vec{a}+\vec{b}+\vec{c}$ (2) $\frac{\vec{a}}{|\vec{a}|}+\frac{\vec{b}}{|\vec{b}|}+\frac{\vec{c}}{|\vec{c}|}$
(3) $\frac{\vec{a}}{|\vec{a}|^2}+\frac{\vec{b}}{|\vec{b}|^2}+\frac{\vec{c}}{|\vec{c}|^2}$ (4) $|\vec{a}|\vec{a}-|\vec{b}|\vec{b}+|\vec{c}|\vec{c}$

6. A parallelogram is constructed with adjacent edges as vectors $3\vec{a}+\vec{b}$ and $\vec{a}-4\vec{b}$, where $|\vec{a}|=6$ and $|\vec{b}|=8$, and $\vec{a}$ and $\vec{b}$ are anti-parallel. Then the length of the longer diagonal is
(1) 40 (2) 64 (3) 32 (4) 48

7. Let the vectors $\vec{u}, \vec{v}$ and $\vec{w}$ be such that $|\vec{u}|=1, |\vec{v}|=2$ and $|\vec{w}|=3$. If the projection of $\vec{v}$ on $\vec{u}$ is equal to the projection of $\vec{w}$ on $\vec{u}$ and the vectors $\vec{v}$ and $\vec{w}$ are perpendicular to each other, then $|\vec{u}-\vec{v}+\vec{w}|$ equals
(1) 2 (2) $\sqrt{7}$ (3) $\sqrt{14}$ (4) 14

8. Let $\vec{a}, \vec{b}$ and $\vec{c}$ be three vectors of equal magnitude. The angle between each pair of vectors is $\pi/3$ such that $|\vec{a}+\vec{b}+\vec{c}|=\sqrt{6}$. Then $|\vec{a}|$ is equal to
(1) 2 (2) –1 (3) 1 (4) $\sqrt{6}/3$

9. The unit vector orthogonal to vector $-\hat{i}+2\hat{j}+2\hat{k}$ and making equal angles with the x- and y-axes are
(1) $\pm\frac{1}{3}(2\hat{i}+2\hat{j}-\hat{k})$ (2) $\pm\frac{1}{3}(\hat{i}+\hat{j}-\hat{k})$
(3) $\pm\frac{1}{3}(2\hat{i}-2\hat{j}-\hat{k})$ (4) $\pm\frac{1}{3}(\hat{i}-\hat{j}+2\hat{k})$

10. Let $\vec{a}=2\hat{i}-\hat{j}+\hat{k}$, $\vec{b}=\hat{i}+2\hat{j}-\hat{k}$, $\vec{c}=\hat{i}+\hat{j}-2\hat{k}$. A vector coplanar with $\vec{b}$ and $\vec{c}$ whose projection on $\vec{a}$ has magnitude $\sqrt{\frac{2}{3}}$ is
(1) $2\hat{i}+3\hat{j}-3\hat{k}$ (2) $-2\hat{i}-\hat{j}+5\hat{k}$
(3) $2\hat{i}+3\hat{j}+3\hat{k}$ (4) $2\hat{i}+\hat{j}+5\hat{k}$

11. Position vector $\hat{k}$ is rotated about the origin by angle 135° in such a way that the plane made by it bisects the angle between $\hat{i}$ and $\hat{j}$. Then its new position is
(1) $\frac{\hat{i}}{\sqrt{2}}+\frac{\hat{j}}{\sqrt{2}}$ (2) $\frac{\hat{i}}{2}+\frac{\hat{j}}{2}-\frac{\hat{k}}{\sqrt{2}}$
(3) $\frac{\hat{i}}{\sqrt{2}}-\frac{\hat{k}}{\sqrt{2}}$ (4) $\frac{\hat{i}}{2}+\frac{\hat{j}}{2}+\frac{\hat{k}}{\sqrt{2}}$

12. A vector $\hat{a}$ in the plane of vectors $\vec{b}=2\hat{i}+\hat{j}$ and $\vec{c}=\hat{i}-\hat{j}+\hat{k}$ is such that it is equally inclined to $\vec{b}$ and $\vec{d}$ where $\vec{d}=\hat{j}+2\hat{k}$. Then the vector $\hat{a}$ is

(1) $\frac{\hat{i}+\hat{j}+\hat{k}}{\sqrt{3}}$ (2) $\frac{\hat{i}-\hat{j}+\hat{k}}{\sqrt{3}}$

(3) $\frac{2\hat{i}+\hat{j}}{\sqrt{5}}$ (4) $\frac{2\hat{i}-\hat{j}}{\sqrt{5}}$

13. Vectors $3\vec{a}-5\vec{b}$ and $2\vec{a}+\vec{b}$ are mutually perpendicular. If $\vec{a}+4\vec{b}$ and $\vec{b}-\vec{a}$ are also mutually perpendicular, then the cosine of the angle between $\vec{a}$ and $\vec{b}$ is

(1) $\frac{19}{5\sqrt{43}}$ (2) $\frac{19}{3\sqrt{43}}$

(3) $\frac{19}{2\sqrt{45}}$ (4) $\frac{19}{6\sqrt{43}}$

14. If $\hat{a}, \hat{b}$ and $\hat{c}$ are three unit vectors inclined to each other at an angle θ, then the maximum value of θ is

(1) $\frac{\pi}{3}$ (2) $\frac{\pi}{2}$

(3) $\frac{2\pi}{3}$ (4) $\frac{5\pi}{6}$

15. Let $\vec{a}$ and $\vec{c}$ be unit vectors and $|\vec{b}| = 4$. If the angle between $\vec{a}$ and $\vec{c}$ is $\cos^{-1}(1/4)$ and $\vec{b}-2\vec{c}=\lambda\vec{a}$. Then the value of λ is

(1) 3, –4 (2) 1/4, 3/4

(3) –3, 4 (4) –1/4, 3/4

16. If vectors $\vec{a}$ and $\vec{b}$ are two adjacent sides of a parallelogram, then the vector representing the altitude of the parallelogram which is perpendicular to $\vec{a}$ is

(1) $\vec{b}+\frac{\vec{b}\times\vec{a}}{|\vec{a}|^2}$ (2) $\frac{\vec{a}\cdot\vec{b}}{|\vec{b}|^2}$

(3) $\vec{b}-\frac{\vec{b}\cdot\vec{a}}{|\vec{a}|^2}\vec{a}$ (4) $\frac{\vec{a}\times(\vec{b}\times\vec{a})}{|\vec{b}|^2}$

17. If in a right-angled triangle ABC, the hypotenuse $AB = p$, then $\overrightarrow{AB}\cdot\overrightarrow{AC}+\overrightarrow{BC}\cdot\overrightarrow{BA}+\overrightarrow{CA}\cdot\overrightarrow{CB}$ is equal to

(1) $2p^2$ (2) $\frac{p^2}{2}$

(3) p^2 (4) $3p^2$

18. Given three vectors $\vec{a}$, $\vec{b}$ and $\vec{c}$, such that no two of them are collinear. If $(\vec{a}+\vec{b})$ is collinear with $\vec{c}$, $(\vec{b}+\vec{c})$ is collinear with $\vec{a}$ and $|\vec{a}|=|\vec{b}|=|\vec{c}|=\sqrt{2}$. Then the value of $\vec{a}\cdot\vec{b}+\vec{b}\cdot\vec{c}+\vec{c}\cdot\vec{a}$ is

(1) 3 (2) –3

(3) 0 (4) 1

19. Let $\vec{a}$ and $\vec{b}$ be two vectors such that $|2\vec{a}+3\vec{b}|=|3\vec{a}+\vec{b}|$ and the angle between $\vec{a}$ and $\vec{b}$ is 60°. If $\frac{1}{8}\vec{a}$ is a unit vector, then $|\vec{b}|$ is equal to

(1) 4 (2) 6 (3) 5 (4) 8

20. If P is any arbitrary point on the circumcircle of an equilateral triangle of side length l units, then $|\overrightarrow{PA}|^2+|\overrightarrow{PB}|^2+|\overrightarrow{PC}|^2$ is always equal to

(1) $2l^2$ (2) $2\sqrt{3}l^2$ (3) l^2 (4) $3l^2$

21. If l, m, n are the direction cosines of a vector such that $|3l+4m-5n|$ takes the maximum value then the value of $|lm+mn+nl|$ is

(1) 21/25 (2) 33/50 (3) 23/50 (4) 1/2

22. A non-zero vector $\vec{a}$ is such that its projections on vectors $\frac{\hat{i}+\hat{j}}{\sqrt{2}}, \frac{-\hat{i}+\hat{j}}{\sqrt{2}}$ and $\hat{k}$ are equal, then the unit vector along $\vec{a}$ is

(1) $\frac{\sqrt{2}\hat{j}-\hat{k}}{\sqrt{3}}$ (2) $\frac{\hat{j}-\sqrt{2}\hat{k}}{\sqrt{3}}$

(3) $\frac{\sqrt{2}}{\sqrt{3}}\hat{j}+\frac{\hat{k}}{\sqrt{3}}$ (4) $\frac{\hat{j}-\hat{k}}{\sqrt{2}}$

23. If the two adjacent sides of two rectangles are represented by vectors $\vec{p}=5\vec{a}-3\vec{b}$; $\vec{q}=-\vec{a}-2\vec{b}$ and $\vec{r}=-4\vec{a}-\vec{b}$; $\vec{s}=-\vec{a}+\vec{b}$, respectively, then the angle between the vectors $\vec{x}=\frac{1}{3}(\vec{p}+\vec{r}+\vec{s})$ and $\vec{y}=\frac{1}{5}(\vec{r}+\vec{s})$ is

(1) $-\cos^{-1}\left(\frac{19}{5\sqrt{43}}\right)$ (2) $\cos^{-1}\left(\frac{19}{5\sqrt{43}}\right)$

(3) $\pi+\cos^{-1}\left(\frac{19}{5\sqrt{43}}\right)$ (4) cannot be evaluated

24. The position vectors of the vertices A, B and C of a triangle are three unit vectors $\hat{a}, \hat{b}$ and $\hat{c}$, respectively. A vector $\vec{d}$ is such that $\vec{d}\cdot\hat{a}=\vec{d}\cdot\hat{b}=\vec{d}\cdot\hat{c}$ and $\vec{d}=\lambda(\hat{b}+\hat{c})$. Then the triangle ABC is

(1) acute angled (2) obtuse angled

(3) right angled (4) isosceles

25. In a quadrilateral $ABCD$, $\overrightarrow{AC}$ is the bisector of $\overrightarrow{AB}$ and $\overrightarrow{AD}$. If the angle between $\overrightarrow{AB}$ and $\overrightarrow{AD}$ is $2\pi/3$, and $15|\overrightarrow{AC}|=3|\overrightarrow{AB}|=5|\overrightarrow{AD}|$. Then the angle between $\overrightarrow{BA}$ and $\overrightarrow{CD}$ is

(1) $\cos^{-1}\frac{\sqrt{14}}{7\sqrt{2}}$ (2) $\cos^{-1}\frac{\sqrt{21}}{7\sqrt{3}}$

(3) $\cos^{-1}\frac{2}{\sqrt{7}}$ (4) $\cos^{-1}\frac{2\sqrt{7}}{14}$

Cross Product of Vectors

26. If $\vec{a}$ and $\vec{b}$ are two vectors, such that $\vec{a}\cdot\vec{b} < 0$ and $|\vec{a}\cdot\vec{b}| = |\vec{a}\times\vec{b}|$, then the angle between vectors $\vec{a}$ and $\vec{b}$ is

(1) π (2) $7\pi/4$ (3) $\pi/4$ (4) $3\pi/4$

27. Let $P(\vec{p})$ and $Q(\vec{q})$ be the position vectors of two fixed points and $R(\vec{r})$ be the position vector of a variable point. If R moves such that $(\vec{r}-\vec{p})\times(\vec{r}-\vec{q}) = \vec{0}$, then the locus of R is

(1) a plane containing the origin O and parallel to two non-collinear vectors $\overrightarrow{OP}$ and $\overrightarrow{OQ}$

(2) a surface of a sphere described on PQ as its diameter

(3) a line passing through points P and Q

(4) a set of lines parallel to line PQ

28. It two adjacent sides of a parallelogram $ABCD$ are $2\hat{i}+4\hat{j}-5\hat{k}$ and $\hat{i}+2\hat{j}+3\hat{k}$. Then the value of $|\overrightarrow{AC}\times\overrightarrow{BD}|$ is

(1) $20\sqrt{5}$ (2) $22\sqrt{5}$ (3) $24\sqrt{5}$ (4) $26\sqrt{5}$

29. If $\vec{a}, \vec{b}$ are non-parallel unit vectors and $\vec{c}$ is a vector which is perpendicular to both $\vec{a}$ and $\vec{b}$ such that $|\vec{c}| = \frac{5}{2}$ and $|\vec{a}+\vec{b}|^2 = |\vec{a}\times\vec{b}|$, then the value of $\vec{a}\cdot\vec{b}$ is

(1) 2/5 (2) 3/5 (3) –3/5 (4) –2/5

30. Let the pairs of vectors $\vec{a}, \vec{b}$ and $\vec{c}, \vec{d}$ each determine a plane. Then the planes will be parallel if

(1) $(\vec{a}\times\vec{c})\times(\vec{b}\times\vec{d}) = \vec{0}$

(2) $(\vec{a}\times\vec{c})\cdot(\vec{b}\times\vec{d}) = \vec{0}$

(3) $(\vec{a}\times\vec{b})\times(\vec{c}\times\vec{d}) = \vec{0}$

(4) $(\vec{a}\times\vec{b})\cdot(\vec{c}\times\vec{d}) = 0$

31. If $4\vec{a}+5\vec{b}+9\vec{c} = 0$, then $(\vec{a}\times\vec{b})\times[(\vec{b}\times\vec{c})\times(\vec{c}\times\vec{a})]$ is equal to

(1) a vector perpendicular to the plane of $\vec{a}, \vec{b}$ and $\vec{c}$

(2) a scalar quantity

(3) $\vec{0}$

(4) a vector lying in the plane of $\vec{a}, \vec{b}, \vec{c}$

32. If $\vec{a}$ satisfies $\vec{a}\times(\hat{i}+2\hat{j}+\hat{k}) = \hat{i}-\hat{k}$, then $\vec{a}$ is equal to

(1) $\lambda\hat{i}+(2\lambda-1)\hat{j}+\lambda\hat{k}, \lambda\in R$

(2) $\lambda\hat{i}+(1-2\lambda)\hat{j}+\lambda\hat{k}, \lambda\in R$

(3) $\lambda\hat{i}+(2\lambda+1)\hat{j}+\lambda\hat{k}, \lambda\in R$

(4) $\lambda\hat{i}-(1+2\lambda)\hat{j}+\lambda\hat{k}, \lambda\in R$

33. Let the position vectors of the points P and Q be $4\hat{i}+\hat{j}+\lambda\hat{k}$ and $2\hat{i}-\hat{j}+\lambda\hat{k}$, respectively. A vector $\hat{i}-\hat{j}+6\hat{k}$ is perpendicular to the plane containing the origin and the points P and Q. Then λ is equal to

(1) –1/2 (2) 1/2 (3) 1 (4) 3/2

34. If $\vec{a}$ and $\vec{b}$ are two vectors of magnitudes 2 and 3, respectively, such that $|2(\vec{a}\times\vec{b})|+|3(\vec{a}\cdot\vec{b})| = k$, then the maximum value of k is

(1) $\sqrt{13}$ (2) $2\sqrt{13}$ (3) $6\sqrt{13}$ (4) $10\sqrt{13}$

35. If $\vec{a}$ and $\vec{b}$ are two vectors of magnitudes 1 and 2, respectively, such that $(1-3\vec{a}\cdot\vec{b})^2 + |2\vec{a}+\vec{b}+3(\vec{a}\times\vec{b})|^2 = 47$, then the angle between $\vec{a}$ and $\vec{b}$ is

(1) $\pi/3$ (2) $\pi - \cos^{-1}(1/4)$

(3) $\frac{2\pi}{3}$ (4) $\cos^{-1}(1/4)$

36. Vector $\vec{c}$ is perpendicular to vectors $\vec{a} = (2, -3, 1)$ and $\vec{b} = (1, -2, 3)$ and satisfies the condition $\vec{c}\cdot(\hat{i}+2\hat{j}-7\hat{k}) = 10$. Then vector $\vec{c}$ is equal to

(1) (7, 5, 1) (2) (–7, –5, –1)

(3) (1, 1, –1) (4) (14, 10, 3)

37. If $\vec{a}$ and $\vec{b}$ are unit vectors such that $(\vec{a}+\vec{b})\cdot(2\vec{a}+3\vec{b})\times(3\vec{a}-2\vec{b}) = \vec{0}$, then angle between $\vec{a}$ and $\vec{b}$ is

(1) 0 (2) $\pi/2$

(3) π (4) indeterminate

38. Let $\vec{a}, \vec{b}$ and $\vec{c}$ be the three vectors having magnitudes 1, 5 and 3, respectively, such that the angle between $\vec{a}$ and $\vec{b}$ is θ and $\vec{a}\times(\vec{a}\times\vec{b}) = \vec{c}$. Then $\tan\theta$ is equal to

(1) 0 (2) 2/3 (3) 3/5 (4) 3/4

39. If $\vec{a}, \vec{b}, \vec{c}$ are unit vectors such that $\vec{a}\cdot\vec{b} = 0 = \vec{a}\cdot\vec{c}$ and the angle between $\vec{b}$ and $\vec{c}$ is $\pi/3$, then the value of $|\vec{a}\times\vec{b} - \vec{a}\times\vec{c}|$ is

(1) 1/2 (2) 1 (3) 2 (4) $\sqrt{3}$

40. The vertex A of triangle ABC is on the line $\vec{r} = \hat{i} + \hat{j} + \lambda\hat{k}$ and the vertices B and C have the position vectors $\hat{i}$ and $\hat{j}$ respectively. Let Δ be the area of the triangle such that $\Delta \in [3/2, \sqrt{33}/2]$. Then the range of the values of λ is

(1) $[-8, -4] \cup [4, 8]$ (2) $[-4, 4]$
(3) $[-2, 2]$ (4) $[-4, -2] \cup [2, 4]$

41. In the figure, AB, DE and GF are parallel to each other and AD, BG and EF are parallel to each other. If $CD : CE = CG : CB = 2 : 1$, then the ratio of the areas of ΔAEG and ΔABD is:

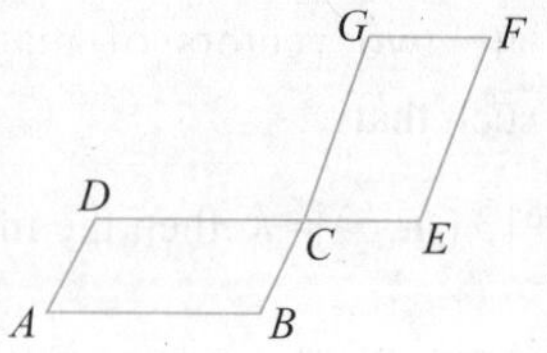

(1) 7/2 (2) 3 (3) 4 (4) 9/2

42. If the vector product of a fixed vector $\overrightarrow{OA}$ with a variable vector $\overrightarrow{OB}$ in a fixed plane OAB is a fixed vector, then the locus of B is

(1) a straight line perpendicular to $\overrightarrow{OA}$
(2) a circle with centre O and radius equal to $|\overrightarrow{OA}|$
(3) a straight line parallel to $\overrightarrow{OA}$
(4) a straight line passing through O and perpendicular to $\overrightarrow{OA}$

43. Let $\hat{a}$ and $\hat{b}$ be mutually perpendicular unit vectors. Then any arbitrary vector $\vec{r}$ is

(1) $\vec{r} = (\vec{r}\cdot\hat{a})\hat{a} + (\vec{r}\cdot\hat{b})\hat{b} + (\vec{r}\cdot(\hat{a}\times\hat{b}))(\hat{a}\times\hat{b})$
(2) $\vec{r} = (\vec{r}\cdot\hat{a}) - (\vec{r}\cdot\hat{b})\hat{b} - (\vec{r}\cdot(\hat{a}\times\hat{b}))(\hat{a}\times\hat{b})$
(3) $\vec{r} = (\vec{r}\cdot\hat{a})\hat{a} - (\vec{r}\cdot\hat{b})\hat{b} + (\vec{r}\cdot(\hat{a}\times\hat{b}))(\hat{a}\times\hat{b})$
(4) $\vec{r} = (\vec{r}\cdot\hat{a})\hat{a} + (\vec{r}\cdot\hat{b})\hat{b} - (\vec{r}\cdot(\hat{a}\times\hat{b}))(\hat{a}\times\hat{b})$

44. Let $\vec{a}$ and $\vec{b}$ be two vectors such that $|\vec{a}| = 1, |\vec{b}| = 4$ and $\vec{a}\cdot\vec{b} = 2$. If $\vec{c} = (2\vec{a}\times\vec{b}) - 3\vec{b}$, then the angle between $\vec{b}$ and $\vec{c}$ is

(1) $\frac{\pi}{3}$ (2) $\frac{\pi}{6}$
(3) $\frac{3\pi}{4}$ (4) $\frac{5\pi}{6}$

45. If $\vec{a} = \hat{i} + \hat{j}$ and $\vec{b} = 2\hat{i} - \hat{k}$. Then vector $\vec{r}$ satisfying the equations $\vec{r}\times\vec{a} = \vec{b}\times\vec{a}$ and $\vec{r}\times\vec{b} = \vec{a}\times\vec{b}$ is

(1) $\hat{i} - \hat{j} + \hat{k}$ (2) $3\hat{i} - \hat{j} + \hat{k}$
(3) $3\hat{i} + \hat{j} - \hat{k}$ (4) $\hat{i} - \hat{j} - \hat{k}$

46. Let $\vec{a}\cdot\vec{b} = 0$, where $\vec{a}$ and $\vec{b}$ are unit vectors. If the unit vector $\vec{c}$ is inclined at an angle θ to both $\vec{a}$ and $\vec{b}$ and $\vec{c} = m\vec{a} + n\vec{b} + p(\vec{a}\times\vec{b}), (m, n, p \in R)$, then

(1) $-\frac{\pi}{4} \le \theta \le \frac{\pi}{4}$ (2) $\frac{\pi}{4} \le \theta \le \frac{3\pi}{4}$
(3) $0 \le \theta \le \frac{\pi}{4}$ (4) $0 \le \theta \le \frac{3\pi}{4}$

47. Let $\vec{a} = 2\hat{i} + \hat{j} - 2\hat{k}$ and $\vec{b} = \hat{i} + \hat{j}$. Let $\vec{c}$ be a vector such that $|\vec{c} - \vec{a}| = 3$, $|(\vec{a}\times\vec{b})\times\vec{c}| = 3$ and the angle between $\vec{c}$ and $\vec{a}\times\vec{b}$ be 30°.
Then $\vec{a}\cdot\vec{c}$ is equal to

(a) $\frac{1}{8}$ (b) $\frac{25}{8}$ (c) 2 (d) 5

48. Let $\vec{a} = \hat{i} - 3\hat{j} + 7\hat{k}$, $\vec{b} = 2\hat{i} - \hat{j} + \hat{k}$ and $\vec{c}$ be a vector such that $(\vec{a} + 2\vec{b})\times\vec{c} = 3(\vec{c}\times\vec{a})$. If $\vec{a}.\vec{c} = 130$, then $\vec{b}.\vec{c}$ is equal to

(1) 15 (2) 20 (3) 30 (4) 40

49. Let $\vec{a} = \hat{i} + 2\hat{j} - 3\hat{k}$ and $\vec{b} = 2\hat{i} - 3\hat{j} + 5\hat{k}$.
If $\vec{r}\times\vec{a} = \vec{b}\times\vec{r}$, $\vec{r}.(\alpha\hat{i} + 2\hat{j} + \hat{k}) = 3$ and $\vec{r}.(2\hat{i} + 5\hat{j} - \alpha\hat{k}) = -1$, $\alpha \in R$, then the value of $\alpha + |\vec{r}|^2$ is equal to

(1) 9 (2) 15 (3) 13 (4) 11

50. Let $\vec{a} = 6\hat{i} + 9\hat{j} + 12\hat{k}$, $\vec{b} = \alpha\hat{i} + 11\hat{j} - 2\hat{k}$ and $\vec{c}$ be vectors such that $\vec{a}\times\vec{c} = \vec{a}\times\vec{b}$. If $\vec{a}\cdot\vec{c} = -12$, $\vec{c}\cdot(\hat{i} - 2\hat{j} + \hat{k}) = 5$, then $\vec{c}\cdot(\hat{i} + \hat{j} + \hat{k})$ is equal to

(1) 9 (2) 10 (3) 11 (4) 12

51. Let $\vec{a}$ and $\vec{b}$ be two unit vectors that are perpendicular. A unit vector that is equally inclined to $\vec{a}$, $\vec{b}$ and $\vec{a}\times\vec{b}$ is

(1) $\frac{1}{\sqrt{2}}(\vec{a} + \vec{b} + \vec{a}\times\vec{b})$ (2) $\frac{1}{2}(\vec{a}\times\vec{b} + \vec{a} + \vec{b})$
(3) $\frac{1}{\sqrt{3}}(\vec{a} + \vec{b} + \vec{a}\times\vec{b})$ (4) $\frac{1}{3}(\vec{a} + \vec{b} + \vec{a}\times\vec{b})$

52. The resolved part of vector $\vec{a}$ along vector $\vec{b}$ is $\vec{a}_1$, and the resolved part of $\vec{a}$ perpendicular to $\vec{b}$ is $\vec{a}_2$. Then, $\vec{a}_1\times\vec{a}_2$ is equal to:

(1) $\frac{(\vec{a}\times\vec{b})\cdot\vec{b}}{|\vec{b}|^2}$ (2) $\frac{(\vec{a}\cdot\vec{b})\,\vec{a}}{|\vec{a}|^2}$
(3) $\frac{(\vec{a}\cdot\vec{b})(\vec{b}\times\vec{a})}{|\vec{b}|^2}$ (4) $\frac{(\vec{a}\cdot\vec{b})(\vec{b}\times\vec{a})}{|\vec{b}\times\vec{a}|}$

53. Let $ABCD$ be a tetrahedron such that the edges AB, AC and AD are mutually perpendicular. Let the area of triangles

ABC, ACD and ADB be 3, 4 and 5 sq. units, respectively. Then the area (in sq. unit) of triangle BCD is

(1) $5\sqrt{2}$ (2) 5 (3) $\frac{\sqrt{5}}{2}$ (4) $\frac{5}{2}$

Scalar Triple Product

54. The volume of a tetrahedron formed by the coterminus edges $\vec{a}$, $\vec{b}$ and $\vec{c}$ is 3. Then the volume of the parallelepiped formed by the coterminus edges $\vec{a}+\vec{b}$, $\vec{b}+\vec{c}$ and $\vec{c}+\vec{a}$ is

(1) 6 (2) 18 (3) 36 (4) 9

55. If $\vec{a}$, $\vec{b}$ and $\vec{c}$ are three mutually perpendicular unit vectors, then the triple product $[\vec{a}+\vec{b}+\vec{c}\ \ \vec{a}+\vec{b}\ \ \vec{b}+\vec{c}]$ equals

(1) 0 (2) 1 or –1 (3) 1 (4) 3

56. Given $\vec{a}=x\hat{i}+y\hat{j}+2\hat{k}$, $\vec{b}=\hat{i}-\hat{j}+\hat{k}$, $\vec{c}=\hat{i}+2\hat{j}$; $\vec{a}\perp\vec{b}$, $\vec{a}\cdot\vec{c}=4$. Then

(1) $[\vec{a}\ \vec{b}\ \vec{c}]^2=|\vec{a}|$ (2) $[\vec{a}\ \vec{b}\ \vec{c}]=|\vec{a}|$

(3) $[\vec{a}\ \vec{b}\ \vec{c}]=0$ (4) $[\vec{a}\ \vec{b}\ \vec{c}]=|\vec{a}|^2$

57. Let $\vec{r}$, $\vec{a}$, $\vec{b}$ and $\vec{c}$ be four non-zero vectors such that $\vec{r}\cdot\vec{a}=0$, $|\vec{r}\times\vec{b}|=|\vec{r}||\vec{b}|$ and $|\vec{r}\times\vec{c}|=|\vec{r}||\vec{c}|$. Then $[\vec{a}\vec{b}\vec{c}]$ is equal to

(1) $|\vec{a}||\vec{b}||\vec{c}|$ (2) $-|\vec{a}||\vec{b}||\vec{c}|$

(3) 0 (4) $\frac{|\vec{a}||\vec{b}||\vec{c}|}{2}$

58. Three vectors $\hat{i}+\hat{j}$, $\hat{j}+\hat{k}$ and $\hat{k}+\hat{i}$ taken two at a time form three planes. The three unit vectors drawn perpendicular to these three planes form a parallelepiped of volume

(1) 1/3 (2) 4 (3) $(3\sqrt{3})/4$ (4) $\frac{4}{3\sqrt{3}}$

59. If $a(\vec{\alpha}\times\vec{\beta})+b(\vec{\beta}\times\vec{\gamma})+c(\vec{\gamma}\times\vec{\alpha})=0$ and at least one of a, b and c is non-zero, then vectors $\vec{\alpha}$, $\vec{\beta}$ and $\vec{\gamma}$ are

(1) parallel (2) coplanar

(3) mutually perpendicular (4) linearly independent

60. If $\vec{p}=\frac{\vec{b}\times\vec{c}}{[\vec{a}\ \vec{b}\ \vec{c}]}$, $\vec{q}=\frac{\vec{c}\times\vec{a}}{[\vec{a}\ \vec{b}\ \vec{c}]}$ and $\vec{r}=\frac{\vec{a}\times\vec{b}}{[\vec{a}\ \vec{b}\ \vec{c}]}$, where $\vec{a}$, $\vec{b}$ and $\vec{c}$ are three non-coplanar vectors, then the value of the expression $(\vec{a}+\vec{b}+\vec{c})\cdot(\vec{p}+\vec{q}+\vec{r})$ is

(1) 3 (2) 2 (3) 1 (4) 0

61. If V is the volume of a tetrahedron T and V' be the volume of another tetrahedran formed by the centroids of the faces of the tetrahedron T such that $V=KV'$, then K is equal to

(1) 9 (2) 12 (3) 27 (4) 81

62. If $\vec{r}=x_1(\vec{a}\times\vec{b})+x_2(\vec{b}\times\vec{c})+x_3(\vec{c}\times\vec{a})$ and $4[\vec{a}\ \vec{b}\ \vec{c}]=1$, then $x_1+x_2+x_3$ is equal to

(1) $\frac{1}{2}\vec{r}\cdot(\vec{a}+\vec{b}+\vec{c})$ (2) $\frac{1}{4}\vec{r}\cdot(\vec{a}+\vec{b}+\vec{c})$

(3) $2\vec{r}\cdot(\vec{a}+\vec{b}+\vec{c})$ (4) $4\vec{r}\cdot(\vec{a}+\vec{b}+\vec{c})$

63. If $\vec{a}$, $\vec{b}$ and $\vec{c}$ are such that $[\vec{a}\ \vec{b}\ \vec{c}]=1$, $\vec{c}=\lambda\vec{a}\times\vec{b}$, angle between $\vec{a}$ and $\vec{b}$ is $2\pi/3$, $|\vec{a}|=\sqrt{2}$, $|\vec{b}|=\sqrt{3}$ and $|\vec{c}|=\frac{1}{\sqrt{3}}$, then the angle between $\vec{a}$ and $\vec{b}$ is

(1) $\frac{\pi}{6}$ (2) $\frac{\pi}{4}$ (3) $\frac{\pi}{3}$ (4) $\frac{\pi}{2}$

• **64.** If $\vec{d}=\vec{a}\times\vec{b}+\vec{b}\times\vec{c}+\vec{c}\times\vec{a}$ is a non-zero vector and $|(\vec{d}\cdot\vec{c})(\vec{a}\times\vec{b})+(\vec{d}\cdot\vec{a})(\vec{b}\times\vec{c})+(\vec{d}\cdot\vec{b})(\vec{c}\times\vec{a})|=0$, then

(1) $|\vec{a}|=|\vec{b}|=|\vec{c}|$ (2) $|\vec{a}|+|\vec{b}|+|\vec{c}|=|\vec{d}|$

(3) $\vec{a}$, $\vec{b}$ and $\vec{c}$ are coplanar (4) $\vec{a}\perp\vec{b}$

• **65.** If the two diagonals of one of the faces of the parallelepiped are $6\hat{i}+6\hat{k}$ and $4\hat{j}+2\hat{k}$ and one of the edges not containing the given diagonals is $\vec{c}=4\hat{j}-8\hat{k}$, then the volume of a parallelepiped is

(1) 60 (2) 80 (3) 100 (4) 120

• **66.** If $\vec{r}\cdot\vec{a}=\vec{r}\cdot\vec{b}=\vec{r}\cdot\vec{c}=\frac{1}{2}$ for some non-zero vector $\vec{r}$, then the area of the triangle whose vertices are $A(\vec{a})$, $B(\vec{b})$ and $C(\vec{c})$ is ($\vec{a},\vec{b},\vec{c}$ are non-coplanar)

(1) $|[\vec{a}\ \vec{b}\ \vec{c}]|$ (2) $|\vec{r}|$

(3) $|[\vec{a}\ \vec{b}\ \vec{c}]\vec{r}|$ (4) $|[\vec{a}\vec{b}\vec{c}]|$

• **67.** If $A(\vec{a})$, $B(\vec{b})$ and $C(\vec{c})$ are the vertices of a triangle ABC and $R(\vec{r})$ is any point in the plane of triangle ABC, then $\vec{r}.(\vec{a}\times\vec{b}+\vec{b}\times\vec{c}+\vec{c}\times\vec{a})$ is always equal to

(1) zero (2) $[\vec{a}\ \vec{b}\ \vec{c}]$

(3) $-[\vec{a}\ \vec{b}\ \vec{c}]$ (4) $2[\vec{a}\vec{b}\vec{c}]$

• **68.** If $\vec{a}\perp\vec{b}$, then vector $\vec{v}$ in the form of $\vec{a}$ and $\vec{b}$ satisfying the equations $\vec{v}\cdot\vec{a}=0$ and $\vec{v}\cdot\vec{b}=1$ and $[\vec{v}\ \vec{a}\ \vec{b}]=1$ is

(1) $\dfrac{\vec{b}}{|\vec{b}|^2}+\dfrac{\vec{a}\times\vec{b}}{|\vec{a}\times\vec{b}|^2}$ (2) $\dfrac{\vec{b}}{|\vec{b}|}+\dfrac{\vec{a}\times\vec{b}}{|\vec{a}\times\vec{b}|^2}$

(3) $\dfrac{\vec{b}}{|\vec{b}|^2}+\dfrac{\vec{a}\times\vec{b}}{|\vec{a}\times\vec{b}|}$ (4) $\dfrac{\vec{b}}{|\vec{b}|}-\dfrac{\vec{a}\times\vec{b}}{|\vec{a}\times b^2|^2}$

Vector Triple Product

69. If $|\vec{a}| = 2$ and $|\vec{b}| = 3$ and $\vec{a}\cdot\vec{b} = 0$, then $(\vec{a}\times(\vec{a}\times(\vec{a}\times(\vec{a}\times\vec{b}))))$ is equal to

(1) $48\hat{b}$ (2) $-48\hat{b}$ (3) $48\hat{a}$ (4) $-48\hat{a}$

70. A vector of magnitude $\sqrt{2}$ units coplanar with the vectors $\vec{a}=\hat{i}+\hat{j}+2\hat{k}$ and $\vec{b}=\hat{i}+2\hat{j}+\hat{k}$, and which is perpendicular to the vector $\vec{c}=\hat{i}+\hat{j}+\hat{k}$, is

(1) $-\hat{j}+\hat{k}$ (2) $\hat{i}-\hat{k}$ (3) $\hat{i}-\hat{j}$ (4) $-\hat{j}$

71. If $\vec{\alpha}\parallel(\vec{\beta}\times\vec{\gamma})$, then $(\vec{\alpha}\times\vec{\beta})\cdot(\vec{\alpha}\times\vec{\gamma})$ is equal to

(1) $|\vec{\alpha}|^2(\vec{\beta}\cdot\vec{\gamma})$ (2) $|\vec{\beta}|^2(\vec{\gamma}\cdot\vec{\alpha})$

(3) $|\vec{\gamma}|^2(\vec{\alpha}\cdot\vec{\beta})$ (4) $|\vec{\alpha}||\vec{\beta}||\vec{\gamma}|$

72. If $\vec{a}$ is parallel to $\vec{b}\times\vec{c}$, then $(\vec{a}\times\vec{b})\cdot(\vec{a}\times\vec{c})$ is equal to

(1) $|\vec{a}|^2(\vec{b}\cdot\vec{c})$ (2) $|\vec{b}|^2(\vec{a}\cdot\vec{c})$

(3) $|\vec{c}|^2(\vec{a}\cdot\vec{b})$ (4) none of these

73. Value of $[\vec{a}\times\vec{b}\ \ \vec{a}\times\vec{c}\ \ \vec{d}]$ is equal to

(1) $(\vec{a}\cdot\vec{d})[\vec{a}\,\vec{b}\,\vec{c}]$ (2) $(\vec{a}\cdot\vec{c})[\vec{a}\,\vec{b}\,\vec{d}]$

(3) $(\vec{a}\cdot\vec{b})[\vec{a}\,\vec{b}\,\vec{d}]$ (4) $(\vec{b}\cdot\vec{d})[\vec{a}\,\vec{b}\,\vec{c}]$

74. Let $\vec{a}$ and $\vec{b}$ be unit vectors perpendicular to each other. Then $[\vec{a}+(\vec{a}\times\vec{b})\ \ \vec{b}+(\vec{a}\times\vec{b})\ \ \vec{a}\times\vec{b}]$ is equal to

(1) 1 (2) 0

(3) –1 (4) 1/2

75. Let $\vec{b}$ and $\vec{c}$ be two unit vectors. Then for any arbitrary vector $\vec{a}$, $(((\vec{a}\times\vec{b})+(\vec{a}\times\vec{c}))\times(\vec{b}\times\vec{c}))\cdot(\vec{b}-\vec{c})$ is equal to

(1) $|\vec{a}|$ (2) $\dfrac{1}{2}|\vec{a}|$

(3) $\dfrac{1}{3}|\vec{a}|$ (4) 0

76. If $\vec{a}.\vec{b}=\beta$ and $\vec{a}\times\vec{b}=\vec{c}$, then $\vec{b}$ is

(1) $\dfrac{(\beta\vec{a}-\vec{a}\times\vec{c})}{|\vec{a}|^2}$ (2) $\dfrac{(\beta\vec{a}+\vec{a}\times\vec{c})}{|\vec{a}|^2}$

(3) $\dfrac{(\beta\vec{c}-\vec{a}\times\vec{c})}{|\vec{a}|^2}$ (4) $\dfrac{(\beta\vec{c}+\vec{a}\times\vec{c})}{|\vec{a}|^2}$

77. If $(\vec{a}\times\vec{b})\times(\vec{b}\times\vec{c})=\vec{b}$, where $\vec{a},\vec{b}$ and $\vec{c}$ are non-zero vectors, then

(1) $\vec{a},\vec{b}$ and $\vec{c}$ can be coplanar

(2) $\vec{a},\vec{b}$ and $\vec{c}$ must be coplanar

(3) $\vec{a},\vec{b}$ and $\vec{c}$ cannot be coplanar

(4) none of these

78. If $\vec{a}$ and $\vec{b}$ are two unit vectors inclined at an angle $\pi/3$, then $\{\vec{a}\times(\vec{b}+\vec{a}\times\vec{b})\}\cdot\vec{b}$ is equal to

(1) $\dfrac{-3}{4}$ (2) $\dfrac{1}{4}$ (3) $\dfrac{3}{4}$ (4) $\dfrac{1}{2}$

79. If $\vec{a},\vec{b},\vec{c}$ are any three non-coplanar vectors, then the equation $[\vec{b}\times\vec{c}\ \ \vec{c}\times\vec{a}\ \ \vec{a}\times\vec{b}]x^2 + [\vec{a}+\vec{b}\ \ \vec{b}+\vec{c}\ \ \vec{c}+\vec{a}]x + 1 + [\vec{b}-\vec{c}\ \ \vec{c}-\vec{a}\ \ \vec{a}-\vec{b}]=0$ has roots

(1) real and distinct (2) real

(3) equal (4) imaginary

80. If $\vec{a}=2\hat{i}+\hat{j}+\hat{k}$, $\vec{b}=\hat{i}+2\hat{j}+2\hat{k}$, $\vec{c}=\hat{i}+\hat{j}+2\hat{k}$ and $(1+\alpha)\hat{i}+\beta(1+\alpha)\hat{j}+\gamma(1+\alpha)(1+\beta)\hat{k}=\vec{a}\times(\vec{b}\times\vec{c})$, then α,β and γ are respectively

(1) $-2, -4, -\dfrac{2}{3}$ (2) $2, -4, \dfrac{2}{3}$

(3) $-2, 4, \dfrac{2}{3}$ (4) $2, 4, -\dfrac{2}{3}$

81. If $\vec{a},\vec{b}$ and $\vec{c}$ are non-coplanar vectors and $\vec{a}\times\vec{c}$ is perpendicular to $\vec{a}\times(\vec{b}\times\vec{c})$, then the value of $[\vec{a}\times(\vec{b}\times\vec{c})]\times\vec{c}$ is equal to

(1) $[\vec{a}\,\vec{b}\,\vec{c}]\vec{c}$ (2) $[\vec{a}\,\vec{b}\,\vec{c}]\vec{b}$

(3) $\vec{0}$ (4) $[\vec{a}\,\vec{b}\,\vec{c}]\vec{a}$

82. The product $[(\vec{a}\times\vec{b})\times(\vec{b}\times\vec{c})\ \ (\vec{b}\times\vec{c})\times(\vec{c}\times\vec{a})\ \ (\vec{c}\times\vec{a})\times(\vec{a}\times\vec{b})]$ is equal to (where $\vec{a},\vec{b}$ and $\vec{c}$ are non-zero non-coplanar vectors)

(1) $[\vec{a}\,\vec{b}\,\vec{c}]^2$ (2) $[\vec{a}\,\vec{b}\,\vec{c}]^3$ (3) $[\vec{a}\,\vec{b}\,\vec{c}]^4$ (4) $[\vec{a}\,\vec{b}\,\vec{c}]$

83. If $\vec{a}=\hat{i}+\hat{j}$, $\vec{b}=\hat{j}+\hat{k}$, $\vec{c}=\hat{k}+\hat{i}$, then in the reciprocal system of vectors $\vec{a},\vec{b},\vec{c}$ the reciprocal vector $\vec{a}'$ corresponding to vector $\vec{a}$ is

(1) $\frac{\hat{i}+\hat{j}+\hat{k}}{2}$ (2) $\frac{\hat{i}-\hat{j}+\hat{k}}{2}$

(3) $\frac{-\hat{i}-\hat{j}+\hat{k}}{2}$ (4) $\frac{\hat{i}+\hat{j}-\hat{k}}{2}$

84. If $\vec{a'} = \hat{i} + \hat{j}$, $\vec{b'} = \hat{i} - \hat{j} + 2\hat{k}$ and $\vec{c'} = 2\hat{i} + \hat{j} - \hat{k}$, then the altitude of the parallelepiped formed by the vectors $\vec{a}$, $\vec{b}$ and $\vec{c}$ having base formed by $\vec{b}$ and $\vec{c}$ is (where $\vec{a'}$ is the reciprocal vector corresponding to vector $\vec{a}$, etc.)

(1) 1 (2) $3\sqrt{2}/2$ (3) $1/\sqrt{6}$ (4) $1/\sqrt{2}$

85. Let $\vec{a}, \vec{b}$ and $\vec{c}$ be three non-coplanar vectors and $\vec{r}$ be any arbitrary vector. Then $(\vec{a}\times\vec{b})\times(\vec{r}\times\vec{c})+(\vec{b}\times\vec{c})\times(\vec{r}\times\vec{a})+(\vec{c}\times\vec{a})\times(\vec{r}\times\vec{b})$ is equal to

(1) $[\vec{a}\,\vec{b}\,\vec{c}]\vec{r}$ (2) $2[\vec{a}\,\vec{b}\,\vec{c}]\vec{r}$

(3) $3[\vec{a}\,\vec{b}\,\vec{c}]\vec{r}$ (4) $-2[\vec{a}\,\vec{b}\,\vec{c}]\vec{r}$

• **86.** If $\vec{a}$ and $\vec{b}$ are orthogonal unit vectors, then for a vector $\vec{r}$ which is non-coplanar with $\vec{a}$ and $\vec{b}$, vector $\vec{r}\times\vec{a}$ is equal to

(1) $[\vec{r}\,\vec{a}\,\vec{b}]\vec{b} - (\vec{r}\cdot\vec{b})(\vec{b}\times\vec{a})$

(2) $[\vec{r}\,\vec{a}\,\vec{b}](\vec{a}+\vec{b})$

(3) $[\vec{r}\,\vec{a}\,\vec{b}]\vec{a} + (\vec{r}\cdot\vec{a})\vec{a}\times\vec{b}$

(4) $[\vec{r}\,\vec{a}\,\vec{b}]\vec{b} + (\vec{r}\cdot\vec{b})(\vec{b}\times\vec{a})$

• **87.** For any two vectors $\vec{a}$ and $\vec{b}$, $(\vec{a}\times\hat{i})\cdot(\vec{b}\times\hat{i})+(\vec{a}\times\hat{j})\cdot(\vec{b}\times\hat{j})+(\vec{a}\times\hat{k})\cdot(\vec{b}\times\hat{k})$ is equal to

(1) $\vec{a}\cdot\vec{b}$ (2) $2\,\vec{a}\cdot\vec{b}$

(3) zero (4) $3\,\vec{a}\cdot\vec{b}$

• **88.** If $\vec{a},\vec{b},\vec{p},\vec{q}$ are four vectors such that $\vec{a}+\vec{b}=\mu\vec{p}$, $\vec{b}\cdot\vec{q}=0$ and $(\vec{b})^2=1$, where μ is a scalar. Then $|(\vec{a}\cdot\vec{q})\vec{p}-(\vec{p}\cdot\vec{q})\vec{a}|$ is equal to

(1) $2|\vec{p}\cdot\vec{q}|$ (2) $(1/2)|\vec{p}\cdot\vec{q}|$

(3) $|\vec{p}\times\vec{q}|$ (4) $|\vec{p}\cdot\vec{q}|$

• **89.** The condition for the equations $\vec{r}\times\vec{a}=\vec{b}$ and $\vec{r}\times\vec{c}=\vec{d}$ to be consistent is

(1) $\vec{b}\cdot\vec{c}=\vec{a}\cdot\vec{d}$ (2) $\vec{a}\cdot\vec{b}=\vec{c}\cdot\vec{d}$

(3) $\vec{b}\cdot\vec{c}+\vec{a}\cdot\vec{d}=0$ (4) $\vec{a}\cdot\vec{b}+\vec{c}\cdot\vec{d}=0$

Numerical Value Type

1. Let $\vec{a}$ and $\vec{b}$ be unit vectors such that $|\vec{a}+\vec{b}|=\sqrt{3}$. Then the value of $(2\vec{a}+5\vec{b})\cdot(3\vec{a}+\vec{b}+\vec{a}\times\vec{b})$ is ________.

2. Let $\vec{a}=\alpha\hat{i}+\hat{j}+\beta\hat{k}$ and $\vec{b}=3\hat{i}-5\hat{j}+4\hat{k}$ be two vectors, such that $\vec{a}\times\vec{b}=-\hat{i}+9\hat{i}+12\hat{k}$. Then the projection of $\vec{b}-2\vec{a}$ on $\vec{b}+\vec{a}$ is equal to ________.

3. Let $\vec{a}=2\hat{i}-\hat{j}+2\hat{k}$ and $\vec{b}=\hat{i}+2\hat{j}-\hat{k}$. Let a vector $\vec{v}$ be in the plane containing $\vec{a}$ and $\vec{b}$. If $\vec{v}$ is perpendicular to the vector $3\hat{i}+2\hat{j}-\hat{k}$ and its projection on $\vec{a}$ is a 19 units, then $|\vec{v}|^2$ is equal to ________.

4. Let $\vec{a}=\hat{i}-\hat{j}$, $\vec{b}=\hat{i}+\hat{j}+\hat{k}$ and $\vec{c}$ be a vectors such that $\vec{a}\times\vec{c}+\vec{b}=\vec{0}$ and $\vec{a}.\vec{c}=4$, then $|\vec{c}|^2$ is equal to ________.

5. Let $\vec{a}=2\hat{i}+\hat{j}-2\hat{k}$ and $\vec{b}=\hat{i}+\hat{j}$. If $\vec{c}$ is a vector such that $\vec{a}.\vec{c}=|\vec{c}|, |\vec{c}-\vec{a}|=2\sqrt{2}$ and the angle between $(\vec{a}\times\vec{b})$ and $\vec{c}$ is $\frac{\pi}{6}$, then the value of $|(\vec{a}\times\vec{b})\times\vec{c}|$ is ________.

6. Let $\vec{a}, \vec{b}$ and $\vec{c}$ be three vectors such that $\vec{a}=\vec{b}\times(\vec{b}\times\vec{c})$. If magnitudes of the vectors $\vec{a}, \vec{b}$ and $\vec{c}$ are $\sqrt{2}$, 1 and 2 respectively and the angle between is $\vec{b}$ and $\vec{c}$ is $\theta\left(0<\theta<\frac{\pi}{2}\right)$, then the value of $\cos^2\theta$ is ________.

7. Let $\vec{a}=\hat{i}+2\hat{j}+3\hat{k}$ and $\vec{b}=\hat{i}+\hat{j}-\hat{k}$. If $\vec{c}$ is a vector such that $\vec{a}.\vec{c}=11$, $\vec{b}.(\vec{a}\times\vec{c})=27$ and $\vec{b}.\vec{c}=-\sqrt{3}|\vec{b}|$, then the square of the area of the triangle formed by vectors $\vec{a}$ and $\vec{c}$ is ________.

• **8.** If $|\vec{a}|=1, |\vec{b}|=2, |\vec{c}|=3$, $\vec{a},\vec{b}$ are perpendicular to $\vec{c}$ and $\vec{a}\cdot\vec{b}=1$ then the value of $\left[\frac{\vec{a}\times\vec{b}}{5}\quad \vec{b}\times\vec{c}\quad \frac{\vec{a}+\vec{b}+\vec{c}}{\sqrt{3}}\right]$ is ________.

• **9.** Let a unit vector $\vec{u}=x\hat{i}+y\hat{j}+z\hat{k}$ make angles $\frac{\pi}{2}, \frac{\pi}{3}$ and $\frac{2\pi}{3}$ with the vectors $\frac{1}{\sqrt{2}}\hat{i}+\frac{1}{\sqrt{2}}\hat{k}$, $\frac{1}{\sqrt{2}}\hat{j}+\frac{1}{\sqrt{2}}\hat{k}$ and $\frac{1}{\sqrt{2}}\hat{i}+\frac{1}{\sqrt{2}}\hat{j}$ respectively. If $\vec{v}=\frac{1}{\sqrt{2}}\hat{i}+\frac{1}{\sqrt{2}}\hat{j}+\frac{1}{\sqrt{2}}\hat{k}$, then $|\vec{u}-\vec{v}|^2$ is equal to ________.

Integer Answer Type

1. If $\vec{a}$ and $\vec{b}$ are any two unit vectors, then the greatest positive integer in the range of $\frac{3|\vec{a}+\vec{b}|}{2}+2|\vec{a}-\vec{b}|$ is ________.

2. Let $\vec{u}$ be a vector in a rectangular coordinate system with sloping angle 60°. Suppose that $|\vec{u}-\hat{i}|$ is geometric mean of $|\vec{u}|$ and $|\vec{u}-2\hat{i}|$, where $\hat{i}$ is the unit vector along the x-axis. Then the value of $(\sqrt{2}+1)|\vec{u}|$ is ________.

3. The absolute value of parameter t for which the area of the triangle whose vertices are $A(-1, 1, 2)$; $B(1, 2, 3)$ and $C(t, 1, 1)$ is minimum. is ________.

4. If $\vec{a} = a_1\hat{i} + a_2\hat{j} + a_3\hat{k}$; $\vec{b} = b_1\hat{i} + b_2\hat{j} + b_3\hat{k}$, $\vec{c} = c_1\hat{i} + c_2\hat{j} + c_3\hat{k}$ and $[3\vec{a} + \vec{b}\ 3\vec{b} + \vec{c}\ 3\vec{c} + \vec{a}]$

$$= \lambda \begin{vmatrix} \vec{a}\cdot\hat{i} & \vec{a}\cdot\hat{j} & \vec{a}\cdot\hat{k} \\ \vec{b}\cdot\hat{i} & \vec{b}\cdot\hat{j} & \vec{b}\cdot\hat{k} \\ \vec{c}\cdot\hat{i} & \vec{c}\cdot\hat{j} & \vec{c}\cdot\hat{k} \end{vmatrix}$$, then the value of λ is____.

5. Let $\vec{a} = \alpha\hat{i} + 2\hat{j} - 3\hat{k}$, $\vec{b} = \hat{i} + 2\alpha\hat{j} - 2\hat{k}$ and $\vec{c} = 2\hat{i} - \alpha\hat{j} + \hat{k}$. The value of 30α, such that $\{(\vec{a}\times\vec{b})\times(\vec{b}\times\vec{c})\}\times(\vec{c}\times\vec{a}) = 0$ is____.

6. The greatest value of λ if the volume of a tetrahedron whose vertices with position vectors $\hat{i} - 6\hat{j} + 10\hat{k}, -\hat{i} - 3\hat{j} + 7\hat{k}, 5\hat{i} - \hat{j} + \lambda\hat{k}$ and $7\hat{i} - 4\hat{j} + 7\hat{k}$ is 11 cubic unit is____.

7. Given that $\vec{u} = \hat{i} - 2\hat{j} + 3\hat{k}$; $\vec{v} = 2\hat{i} + \hat{j} + 4\hat{k}$; $\vec{w} = \hat{i} + 3\hat{j} + 3\hat{k}$ and $(\vec{u}\cdot\vec{R} - 15)\hat{i} + (\vec{v}\cdot\vec{R} - 30)\hat{j} + (\vec{w}\cdot\vec{R} - 20)\hat{k} = \vec{0}$. Then the value of $|\vec{R}|^2$ is____.

8. If $\vec{a}, \vec{b}, \vec{c}$ are unit vectors such that $\vec{a}\cdot\vec{b} = 0 = \vec{a}\cdot\vec{c}$ and the angle between $\vec{b}$ and $\vec{c}$ is $\frac{\pi}{3}$, then the value of $|\vec{a}\times\vec{b} - \vec{a}\times\vec{c}|$ is____.

9. Let $\vec{a} = 2\hat{i} - \hat{j} + 2\hat{k}$, $\vec{b} = \hat{i} + 2\hat{j}$ and $\vec{c} = x\hat{i} + y\hat{j} + z\hat{k}$ where $x, y, z \in I$; $|x| \le 10$. If $\vec{a}, \vec{b}$ and $\vec{c}$ are mutually perpendicular to each other, then minimum value of $|x + y + z|$ is ____.

10. A unit vector $\vec{r}$ has equal length of projections along the vectors $\frac{2\hat{i}+\hat{j}}{\sqrt{5}}, \frac{-2\hat{i}+\hat{j}}{\sqrt{5}}$ and $\hat{k}$. If $\vec{r}.(\hat{i}+\hat{j}+\hat{k})$ is positive, then number of such vectors $\vec{r}$ will be____.

11. If $\vec{a} = 5\hat{i} - \hat{j} + \hat{k}$, $\vec{b} = 2\hat{i} - 3\hat{j} - \hat{k}$, $\vec{c} = -3\hat{i} + \hat{j} + \hat{k}$ and $\vec{d} = 2\hat{j} + \hat{k}$, then the value of $\vec{d}\cdot(\vec{a}\times(\vec{b}\times(\vec{c}\times\vec{d})))$ is____.

12. The least value of the volume of parallelepiped formed by the vectors $\vec{V_1} = \hat{i} + \hat{j}$, $\vec{V_2} = \hat{i} + (2\operatorname{cosec}\alpha)\hat{j} + \hat{k}$ and $\vec{V_3} = \hat{j} + (2\operatorname{cosec}\alpha)\hat{k}$, where $\alpha \in (0, \pi)$, is____.

13. Let $\overline{OA} = \vec{a}, \overline{OB} = 10\vec{a} + 2\vec{b}$ and $\overline{OC} = \vec{b}$, where O, A and C are non-collinear points. Let p denote the area of quadrilateral $OACB$, and let q denote the area of parallelogram with OA and OC as adjacent sides. If $p = k\,q$, then the value of k is____.

• **14.** Let $\vec{a}, \vec{b}, \vec{c}$ be unit vectors such that $|\vec{a} + \vec{b} + 3\vec{c}| = \sqrt{14}$. i.e. Angle between vectors $\vec{a}$ and $\vec{b}$ is α, between $\vec{b}$ and $\vec{c}$ is β, and between $\vec{a}$ and $\vec{c}$ is γ. If $\alpha \in \left[\frac{\pi}{6}, \frac{2\pi}{3}\right]$ then maximum value of $\cos\alpha + 3\cos\beta$ is____.

• **15.** If $\vec{x}, \vec{y}$ are two non-zero and non-collinear vectors satisfying $[(a-2)\alpha^2 + (b-3)\alpha + c]\vec{x} + [(a-2)\beta^2 + (b-3)\beta + c]\vec{y} + [(a-2)\gamma^2 + (b-3)\gamma + c](\vec{x}\times\vec{y}) = 0$, where α, β, γ are three distinct real numbers, then the value of $(a^2 + b^2 + c^2)$ is____.

• **16.** Let $\vec{u}$ and $\vec{v}$ be unit vectors such that $\vec{u}\times\vec{v} + \vec{u} = \vec{w}$ and $\vec{w}\times\vec{u} = \vec{v}$. Then the value of $[\vec{u}\ \vec{v}\ \vec{w}]$ is____.

• **17.** Let a vector $\vec{V}$ satisfy the equation, $2\vec{V} + \vec{V}\times(\hat{i} + 2\hat{j}) = 2\hat{i} + \hat{k}$. If $3|\vec{V}| = \sqrt{m}$, then the value of m is____.

Multiple Correct Answers Type

1. If unit vectors $\vec{a}$ and $\vec{b}$ are inclined at an angle 2θ such that $|\vec{a} - \vec{b}| < 1$ $(0 \le \theta \le \pi)$, then θ can lie in the intervals

(1) $[0, \pi/6)$
(2) $(5\pi/6, \pi]$
(3) $[\pi/6, \pi/2)$
(4) $(\pi/2, 5\pi/6]$

2. $\vec{b}$ and $\vec{c}$ are non-collinear if $\vec{a}\times(\vec{b}\times\vec{c}) + (\vec{a}\cdot\vec{b})\vec{b} = (4 - 2x - \sin y)\vec{b} + (x^2 - 1)\vec{c}$ and $(\vec{c}\cdot\vec{c})\vec{a} = \vec{c}$. Then

(1) $x = 1$
(2) $x = -1$
(3) $y = (4n+1)\frac{\pi}{2}$, $n \in I$
(4) $y = (2n+1)\frac{\pi}{2}$, $n \in I$

3. Unit vectors $\vec{a}$ and $\vec{b}$ are perpendicular, and unit vector $\vec{c}$ is inclined at an angle θ to both $\vec{a}$ and $\vec{b}$. If $\vec{c} = \alpha\vec{a} + \beta\vec{b} + \gamma(\vec{a}\times\vec{b})$, then

(1) $a = \beta$
(2) $\gamma^2 = 1 - 2\alpha^2$
(3) $\gamma^2 = -\cos 2\theta$
(4) $\beta^2 = \frac{1+\cos 2\theta}{2}$

4. Let $\vec{a}$ and $\vec{b}$ be two given vectors. With these vectors as adjacent sides, a parallelogram is constructed. Which of the following vectors represents the altitude of the parallelogram that is perpendicular to $\vec{a}$?

(1) $\frac{(\vec{a}\cdot\vec{b})}{|\vec{a}|^2}\vec{a} - \vec{b}$

(2) $\frac{1}{|\vec{a}|^2}\{|\vec{a}|^2\vec{b} - (\vec{a}\cdot\vec{b})\vec{a}\}$

(3) $\frac{\vec{a}\times(\vec{a}\times\vec{b})}{|\vec{a}|^2}$

(4) $\frac{\vec{a}\times(\vec{b}\times\vec{a})}{|\vec{b}|^2}$

5. If $\vec{a} \times (\vec{b} \times \vec{c})$ is perpendicular to $(\vec{a} \times \vec{b}) \times \vec{c}$, then which of the following is (are) possible?

(1) $(\vec{a}\cdot\vec{c})|\vec{b}|^2 = (\vec{a}\cdot\vec{b})(\vec{b}\cdot\vec{c})$

(2) $\vec{a}\cdot\vec{b} = 0$

(3) $\vec{a}\cdot\vec{c} = 0$

(4) $\vec{b}\cdot\vec{c} = 0$

6. Let $\vec{a}$, $\vec{b}$ and $\vec{c}$ be vectors forming a right-hand triad. If $\vec{p} = \dfrac{\vec{b}\times\vec{c}}{[\vec{a}\ \vec{b}\ \vec{c}]}$, $\vec{q} = \dfrac{\vec{c}\times\vec{a}}{[\vec{a}\ \vec{b}\ \vec{c}]}$ and $\vec{r} = \dfrac{\vec{a}\times\vec{b}}{[\vec{a}\ \vec{b}\ \vec{c}]}$, $(x \in R^+)$ then

(1) $x[\vec{a}\ \vec{b}\ \vec{c}] + \dfrac{[\vec{p}\ \vec{q}\ \vec{r}]}{x}$ has least value 2

(2) $x^4[\vec{a}\ \vec{b}\ \vec{c}]^2 + \dfrac{[\vec{p}\ \vec{q}\ \vec{r}]}{x^2}$ has least value $(3/2^{2/3})$

(3) $[\vec{p}\ \vec{q}\ \vec{r}] > 0$

(4) none of these

7. Let $a_1, a_2, a_3 \in R - \{0\}$ and $a_1 + a_2 \cos 2x + a_3 \sin^2 x = 0$ for all $x \in R$, then

(1) vectors $\vec{a} = a_1\hat{i} + a_2\hat{j} + a_3\hat{k}$ and $\vec{b} = 4\hat{i} + 2\hat{j} + \hat{k}$ are perpendicular.

(2) vectors $\vec{a} = a_1\hat{i} + a_2\hat{j} + a_3\hat{k}$ and $\vec{b} = -\hat{i} + \hat{j} + 2\hat{k}$ are parallel.

(3) if vector $\vec{a} = a_1\hat{i} + a_2\hat{j} + a_3\hat{k}$ is of length $\sqrt{6}$ units, then one of the ordered triplet (a_1, a_2, a_3) is (1, –1,–2).

(4) if $2a_1 + 3a_2 + 6a_3 = 26$, then $|a_1\hat{i} + a_2\hat{j} + a_3\hat{k}| = 2\sqrt{6}$.

8. If $\vec{a}$ and $\vec{b}$ are two vectors having angle between them θ, then

(1) $|\vec{a}\times\vec{b}|^2 + (\vec{a}\cdot\vec{b})^2 = |\vec{a}|^2|\vec{b}|^2$

(2) $|\vec{a}\times\vec{b}| = (\vec{a}\cdot\vec{b})$, if $\theta = \pi/4$

(3) $\vec{a}\times\vec{b} = (\vec{a}\cdot\vec{b})\hat{n}$ (where $\hat{n}$ is a normal unit vector), if $\theta = \pi/4$

(4) $(\vec{a}\times\vec{b})\cdot(\vec{a}+\vec{b}) = 0$

9. Let $\vec{a}$ and $\vec{b}$ be two non-zero perpendicular vectors. A vector $\vec{r}$ satisfying the equation $\vec{r}\times\vec{b} = \vec{a}$ can be

(1) $\vec{b} - \dfrac{\vec{a}\times\vec{b}}{|\vec{b}|^2}$ (2) $2\vec{b} - \dfrac{\vec{a}\times\vec{b}}{|\vec{b}|^2}$

(3) $|\vec{a}|\vec{b} - \dfrac{\vec{a}\times\vec{b}}{|\vec{b}|^2}$ (4) $|\vec{b}|\vec{b} - \dfrac{\vec{a}\times\vec{b}}{|\vec{b}|^2}$

10. If $\vec{x} + \vec{c}\times\vec{y} = \vec{a}$ and $\vec{y} + \vec{c}\times\vec{x} = \vec{b}$, where $\vec{c}$ is a non-zero vector, then

(1) $\vec{x} = \dfrac{\vec{b}\times\vec{c} + \vec{a} + (\vec{c}\cdot\vec{a})\vec{c}}{1+\vec{c}\cdot\vec{c}}$

(2) $\vec{x} = \dfrac{\vec{c}\times\vec{b} + \vec{b} + (\vec{c}\cdot\vec{a})\vec{c}}{1+\vec{c}\cdot\vec{c}}$

(3) $\vec{y} = \dfrac{\vec{a}\times\vec{c} + \vec{b} + (\vec{c}\cdot\vec{b})\vec{c}}{1+\vec{c}\cdot\vec{c}}$

(4) $\vec{y} = \dfrac{\vec{c}\times\vec{a} + \vec{b} + (\vec{c}\cdot\vec{b})\vec{c}}{1+\vec{c}\cdot\vec{c}}$

11. Let $\vec{r}$ be a unit vector satisfying $\vec{r}\times\vec{a} = \vec{b}$, where $|\vec{a}| = \sqrt{3}$ and $|\vec{b}| = \sqrt{2}$. Then

(1) $\vec{r} = \frac{2}{3}(\vec{a} + \vec{a}\times\vec{b})$ (2) $\vec{r} = \frac{1}{3}(\vec{a} + \vec{a}\times\vec{b})$

(3) $\vec{r} = \frac{2}{3}(\vec{a} - \vec{a}\times\vec{b})$ (4) $\vec{r} = \frac{1}{3}(-\vec{a} + \vec{a}\times\vec{b})$

12. If $\vec{a}$ and $\vec{b}$ are unequal unit vectors such that $(\vec{a} - \vec{b})\times[(\vec{b} + \vec{a})\times(2\vec{a} + \vec{b})] = \vec{a} + \vec{b}$, then the angle θ between $\vec{a}$ and $\vec{b}$ is

(1) 0 (2) $\pi/2$ (3) $\pi/4$ (4) π

13. If $\vec{a}$ and $\vec{b}$ are two unit vectors perpendicular to each other and $\vec{c} = \lambda_1\vec{a} + \lambda_2\vec{b} + \lambda_3(\vec{a}\times\vec{b})$, then

(1) $\lambda_1 = \vec{a}\cdot\vec{c}$

(2) $\lambda_2 = |\vec{b}\times\vec{c}|$

(3) $\lambda_3 = |(\vec{a}\times\vec{b})\times\vec{c}|$

(4) $\lambda_1 + \lambda_2 + \lambda_3 = (\vec{a} + \vec{b} + \vec{a}\times\vec{b})\cdot\vec{c}$

14. If vectors $\vec{a}$ and $\vec{b}$ are non-collinear, then $\dfrac{\vec{a}}{|\vec{a}|} + \dfrac{\vec{b}}{|\vec{b}|}$ is

(1) a unit vector

(2) in the plane of $\vec{a}$ and $\vec{b}$

(3) equally inclined to $\vec{a}$ and $\vec{b}$

(4) perpendicular to $\vec{a}\times\vec{b}$

15. If $\vec{a}$ and $\vec{b}$ are non-zero vectors such that $|\vec{a} + \vec{b}| = |\vec{a} - 2\vec{b}|$, then

(1) $2\vec{a}\cdot\vec{b} = |\vec{b}|^2$

(2) $\vec{a}\cdot\vec{b} = |\vec{b}|^2$

(3) the least value of $\vec{a}\cdot\vec{b} + \dfrac{1}{|\vec{b}|^2 + 2}$ is $\sqrt{2}$

(4) the least value of $\vec{a}\cdot\vec{b} + \dfrac{1}{|\vec{b}| + 2}$ is $\sqrt{2} - 1$

16. Let $\vec{a}, \vec{b}$ and $\vec{c}$ be non-zero vectors and $\vec{V_1} = \vec{a} \times (\vec{b} \times \vec{c})$, $\vec{V_2} = (\vec{a} \times \vec{b}) \times \vec{c}$. If vectors $\vec{V_1}$ and $\vec{V_2}$ are equal. Then which of the following is/are true?

(1) $\vec{a}$ and $\vec{b}$ are orthogonal

(2) $\vec{a}$ and $\vec{c}$ are collinear

(3) $\vec{b}$ and $\vec{c}$ are orthogonal

(4) $\vec{b} = \lambda(\vec{a} \times \vec{c})$ when λ is a scalar

17. If vectors $\vec{A}$ and $\vec{B}$ satisfy the equations $\vec{A} + \vec{B} = \vec{a}$, $\vec{A} \times \vec{B} = \vec{b}$ and $\vec{A} \cdot \vec{a} = 1$, then

(1) $\vec{A} = \dfrac{(\vec{a} \times \vec{b}) - \vec{a}}{a^2}$

(2) $\vec{B} = \dfrac{(\vec{b} \times \vec{a}) + \vec{a}(a^2 - 1)}{a^2}$

(3) $\vec{A} = \dfrac{(\vec{a} \times \vec{b}) + \vec{a}}{a^2}$

(4) $\vec{B} = \dfrac{(\vec{b} \times \vec{a}) - \vec{a}(a^2 - 1)}{a^2}$

18. A vector $\vec{d}$ is equally inclined to the three vectors $\vec{a} = \hat{i} - \hat{j} + \hat{k}$, $\vec{b} = 2\hat{i} + \hat{j}$ and $\vec{c} = 3\hat{j} - 2\hat{k}$.

Let $\vec{x}, \vec{y}$ and $\vec{z}$ be three vectors in the plane of $\vec{a}, \vec{b}$; $\vec{b}, \vec{c}$; $\vec{c}, \vec{a}$, respectively. Then

(1) $\vec{x} \cdot \vec{d} = -1$

(2) $\vec{y} \cdot \vec{d} = 1$

(3) $\vec{z} \cdot \vec{d} = 0$

(4) $\vec{r} \cdot \vec{d} = 0$, where $\vec{r} = \lambda \vec{x} + \mu \vec{y} + \delta \vec{z}$

19. Vector(s) perpendicular to $\hat{i} - \hat{j} - \hat{k}$ and in the plane of $\hat{i} + \hat{j} + \hat{k}$ and $-\hat{i} + \hat{j} + \hat{k}$ is/are

(1) $\hat{i} + \hat{k}$ (2) $2\hat{i} + \hat{j} + \hat{k}$

(3) $3\hat{i} + 2\hat{j} + \hat{k}$ (4) $-4\hat{i} - 2\hat{j} - 2\hat{k}$

20. If side $\overrightarrow{AB}$ of an equilateral triangle ABC lying in the xy-plane is $3\hat{i}$, then side $\overrightarrow{CB}$ can be

(1) $-\dfrac{3}{2}(\hat{i} - \sqrt{3}\hat{j})$ (2) $\dfrac{3}{2}(\hat{i} - \sqrt{3}\hat{j})$

(3) $-\dfrac{3}{2}(\hat{i} + \sqrt{3}\hat{j})$ (4) $\dfrac{3}{2}(\hat{i} + \sqrt{3}\hat{j})$

21. The angles of a triangle, two of whose sides are represented by vectors $\sqrt{3}(\hat{a} \times \vec{b})$ and $\vec{b} - (\hat{a} \cdot \vec{b})\hat{a}$, where $\vec{b}$ is a non-zero vector and $\hat{a}$ is a unit vector in the direction of $\vec{a}$, are

(1) $\tan^{-1}(\sqrt{3})$ (2) $\tan^{-1}(1/\sqrt{3})$

(3) $\cot^{-1}(0)$ (4) $\tan^{-1}(1)$

22. Let the three unit vectors $\vec{a}, \vec{b}$ and $\vec{c}$ be coplanar. A unit vector $\vec{d}$ is perpendicular to them and $(\vec{a} \times \vec{b}) \times (\vec{c} \times \vec{d}) = \frac{1}{6}\hat{i} - \frac{1}{3}\hat{j} + \frac{1}{3}\hat{k}$. If the angle between $\vec{a}$ and $\vec{b}$ is 30°, then $\vec{c}$ can be

(1) $(\hat{i} - 2\hat{j} + 2\hat{k})/3$ (2) $(-\hat{i} + 2\hat{j} - 2\hat{k})/3$

(3) $(2\hat{i} + 2\hat{j} - \hat{k})/3$ (4) $(-2\hat{i} - 2\hat{j} + \hat{k})/3$

23. If $\vec{a} + 2\vec{b} + 3\vec{c} = \vec{0}$, then $\vec{a} \times \vec{b} + \vec{b} \times \vec{c} + \vec{c} \times \vec{a} =$

(1) $2(\vec{a} \times \vec{b})$ (2) $6(\vec{b} \times \vec{c})$

(3) $3(\vec{c} \times \vec{a})$ (4) $\vec{0}$

24. Let $\vec{a}$ and $\vec{b}$ be two non-collinear unit vectors. If $\vec{u} = \vec{a} - (\vec{a} \cdot \vec{b})\vec{b}$ and $\vec{v} = \vec{a} \times \vec{b}$. Then $|\vec{v}|$ is

(1) $|\vec{u}|$ (2) $|\vec{u}| + |\vec{u} \cdot \vec{b}|$

(3) $|\vec{u}| + |\vec{u} \cdot \vec{a}|$ (4) none of these

25. If $\vec{a} \times \vec{b} = \vec{c}$, $\vec{b} \times \vec{c} = \vec{a}$, where $\vec{c} \neq \vec{0}$, then

(1) $|\vec{a}| = |\vec{c}|$ (2) $|\vec{a}| = |\vec{b}|$

(3) $|\vec{b}| = 1$ (4) $|\vec{a}| = |\vec{b}| = |\vec{c}| = 1$

26. Let $\vec{a}, \vec{b}$ and $\vec{c}$ be three non-coplanar vectors and $\vec{d}$ be a non-zero vector, which is perpendicular to $(\vec{a} + \vec{b} + \vec{c})$. If $\vec{d} = (\vec{a} \times \vec{b}) \sin x + (\vec{b} \times \vec{c}) \cos y + 2(\vec{c} \times \vec{a})$. Then

(1) $\dfrac{\vec{d} \cdot (\vec{a} + \vec{c})}{[\vec{a}\ \vec{b}\ \vec{c}]} = 2$

(2) $\dfrac{\vec{d} \cdot (\vec{a} + \vec{c})}{[\vec{a}\ \vec{b}\ \vec{c}]} = -2$

(3) minimum value of $x^2 + y^2$ is $\pi^2/4$

(4) minimum value of $x^2 + y^2$ is $5\pi^2/4$

27. If $\vec{a}, \vec{b}$ and $\vec{c}$ are three unit vectors such that $\vec{a} \times (\vec{b} \times \vec{c}) = \frac{1}{2}\vec{b}$, then ($\vec{b}$ and $\vec{c}$ being non-parallel)

(1) the angle between $\vec{a}$ and $\vec{b}$ is $\pi/3$

(2) the angle between $\vec{a}$ and $\vec{c}$ is $\pi/3$

(3) the angle between $\vec{a}$ and $\vec{b}$ is $\pi/2$

(4) the angle between $\vec{a}$ and $\vec{c}$ is $\pi/2$

28. If in triangle ABC, $\overrightarrow{AB} = \frac{\vec{u}}{|\vec{u}|} - \frac{\vec{v}}{|\vec{v}|}$ and $\overrightarrow{AC} = \frac{2\vec{u}}{|\vec{u}|}$, where $|\vec{u}| \neq |\vec{v}|$, then

(1) $1 + \cos 2A + \cos 2B + \cos 2C = 0$

(2) $\sin A = \cos C$

(3) projection of AC on BC is equal to BC

(4) projection of AB on BC is equal to AB

29. $[\vec{a} \times \vec{b} \quad \vec{c} \times \vec{d} \quad \vec{e} \times \vec{f}]$ is equal to

(1) $[\vec{a}\,\vec{b}\,\vec{d}][\vec{c}\,\vec{e}\,\vec{f}] - [\vec{a}\,\vec{b}\,\vec{c}][\vec{d}\,\vec{e}\,\vec{f}]$

(2) $[\vec{a}\,\vec{b}\,\vec{e}][\vec{f}\,\vec{c}\,\vec{d}] - [\vec{a}\,\vec{b}\,\vec{f}][\vec{e}\,\vec{c}\,\vec{d}]$

(3) $[\vec{c}\,\vec{d}\,\vec{a}][\vec{b}\,\vec{e}\,\vec{f}] - [\vec{a}\,\vec{d}\,\vec{b}][\vec{a}\,\vec{e}\,\vec{f}]$

(4) $[\vec{a}\,\vec{c}\,\vec{e}][\vec{b}\,\vec{d}\,\vec{f}]$

30. The scalars l and m such that $l\vec{a} + m\vec{b} = \vec{c}$, where $\vec{a}, \vec{b}$ and $\vec{c}$ are given vectors, are equal to

(1) $l = \frac{(\vec{c} \times \vec{b}) \cdot (\vec{a} \times \vec{b})}{(\vec{a} \times \vec{b})^2}$ (2) $l = \frac{(\vec{c} \times \vec{a}) \cdot (\vec{b} \times \vec{a})}{(\vec{b} \times \vec{a})}$

(3) $m = \frac{(\vec{c} \times \vec{a}) \cdot (\vec{b} \times \vec{a})}{(\vec{b} \times \vec{a})^2}$ (4) $m = \frac{(\vec{c} \times \vec{a}) \cdot (\vec{b} \times \vec{a})}{(\vec{b} \times \vec{a})}$

31. If $(\vec{a} \times \vec{b}) \times (\vec{c} \times \vec{d}) \cdot (\vec{a} \times \vec{d}) = 0$, then which of the following may be true?

(1) $\vec{a}, \vec{b}, \vec{c}$ and $\vec{d}$ are necessarily coplanar

(2) $\vec{a}$ lies in the plane of $\vec{c}$ and $\vec{d}$

(3) $\vec{b}$ lies in the plane of $\vec{a}$ and $\vec{d}$

(4) $\vec{c}$ lies in the plane of $\vec{a}$ and $\vec{d}$

32. If vectors $\vec{a}, \vec{b}$ and $\vec{c}$ are non-coplanar and l, m and n are distinct scalars, then $[(l\vec{a} + m\vec{b} + n\vec{c})(l\vec{b} + m\vec{c} + n\vec{a})(l\vec{c} + m\vec{a} + n\vec{b})] = 0$ implies

(1) $l + m + n = 0$

(2) the roots of the equation $lx^2 + mx + n = 0$ are real

(3) $l^2 + m^2 + n^2 = 0$

(4) $l^3 + m^3 + n^3 = 3lmn$

33. If $\vec{\alpha} = a\hat{i} + b\hat{j} + c\hat{k}$, $\vec{\beta} = b\hat{i} + c\hat{j} + a\hat{k}$ and $\vec{\gamma} = c\hat{i} + a\hat{j} + b\hat{k}$ are three coplanar vectors with $a \neq b$, then the vector $\vec{v} = \hat{i} + \hat{j} + \hat{k}$, is perpendicular to

(1) $\vec{\alpha}$ (2) $\vec{\beta}$

(3) $\vec{\gamma}$ (4) none of these

34. If vectors $\vec{A} = 2\hat{i} + 3\hat{j} + 4\hat{k}$, $\vec{B} = \hat{i} + \hat{j} + 5\hat{k}$ and $\vec{C}$ form a left-handed system, then $\vec{C}$ can be

(1) $11\hat{i} - 6\hat{j} - \hat{k}$ (2) $-11\hat{i} + 6\hat{j} + \hat{k}$

(3) $11\hat{i} - 6\hat{j} + \hat{k}$ (4) $-11\hat{i} + 6\hat{j} - \hat{k}$

35. If $\vec{a} = x\hat{i} + y\hat{j} + z\hat{k}$, $\vec{b} = y\hat{i} + z\hat{j} + x\hat{k}$ and $\vec{c} = z\hat{i} + x\hat{j} + y\hat{k}$, then $\vec{a} \times (\vec{b} \times \vec{c})$ is

(1) parallel to $(y - z)\hat{i} + (z - x)\hat{j} + (x - y)\hat{k}$

(2) orthogonal to $\hat{i} + \hat{j} + \hat{k}$

(3) orthogonal to $(y + z)\hat{i} + (z + x)\hat{j} + (x + y)\hat{k}$

(4) orthogonal to $x\hat{i} + y\hat{j} + z\hat{k}$

36. A parallelogram is constructed with vectors $\vec{a} = 3\vec{\alpha} - \vec{\beta}$ and $\vec{b} = \vec{\alpha} + 3\vec{\beta}$ as its adjacent edges. If $|\vec{\alpha}| = |\vec{\beta}| = 2$ and angle between $\vec{\alpha}$ and $\vec{\beta}$ is $\frac{\pi}{3}$, then the length of diagonals of the parallelogram are

(1) $4\sqrt{5}$ (2) $4\sqrt{3}$

(3) $4\sqrt{7}$ (4) $4\sqrt{2}$

Linked Comprehension Type

For Questions 1–3

Let $\vec{a} = 2\hat{i} + 3\hat{j} - 6\hat{k}$, $\vec{b} = 2\hat{i} - 3\hat{j} + 6\hat{k}$ and $\vec{c} = -2\hat{i} + 3\hat{j} + 6\hat{k}$. Let $\vec{a}_1$ be the projection of $\vec{a}$ on $\vec{b}$ and $\vec{a}_2$ be the projection of $\vec{a}_1$ on $\vec{c}$. Then

1. $\vec{a}_2$ is equal to

(1) $\frac{943}{49}(2\hat{i} - 3\hat{j} - 6\hat{k})$ (2) $\frac{943}{49^2}(2\hat{i} - 3\hat{j} - 6\hat{k})$

(3) $\frac{943}{49}(-2\hat{i} + 3\hat{j} + 6\hat{k})$ (4) $\frac{943}{49^2}(-2\hat{i} + 3\hat{j} + 6\hat{k})$

2. $\vec{a}_1 \cdot \vec{b}$ is equal to

(1) –41 (2) –41/7 (3) 41 (4) 287

3. Which of the following is true?

(1) $\vec{a}$ and $\vec{a}_2$ are collinear

(2) $\vec{a}_1$ and $\vec{c}$ are collinear

(3) $\vec{a}, \vec{a}_1$ and $\vec{b}$ are coplanar

(4) $\vec{a}, \vec{a}_1$ and $\vec{a}_2$ are coplanar

For Questions 4–6

Consider a triangular pyramid $ABCD$ the position vectors of whose angular points are $A(3, 0, 1)$, $B(-1, 4, 1)$, $C(5, 2, 3)$ and $D(0, -5, 4)$. Let G be the point of intersection of the medians of triangle BCD. Then

4. $9|\overrightarrow{AG}|^2 =$

5. The square of area of triangle ABC is

6. If the length of the perpendicular from vertex D on the opposite face is h, then $\sqrt{6}h$ is equal to

For Questions 7–8

Vertices of a parallelogram taken in order are $A(2, -1, 4)$; $B(1, 0 -1)$; $C(1, 2, 3)$ and D.

7. The square of the distance between the parallel lines AB and CD is
8. The distance of the point $P(8, 2, -12)$ from the plane of the parallelogram is $\frac{\lambda}{\sqrt{13.5}}$ then $\lambda =$

For Questions 9–11

Let $\vec{r}$ be a position vector of a variable point in Cartesian OXY plane such that $\vec{r}\cdot(10\hat{j}-8\hat{i}-\vec{r}) = 40$ and $p_1 = \max\{|\vec{r}+2\hat{i}-3\hat{j}|^2\}$, $p_2 = \min\{|\vec{r}+2\hat{i}-3\hat{j}|^2\}$. A tangent line is drawn to the curve $y = 8/x^2$ at point A with abscissa 2. The drawn line cuts the x-axis at a point B.

9. Then $p_2 + 4\sqrt{2}$ is equal to
10. $p_1 + p_2$ is equal to
11. $\overrightarrow{AB}\cdot\overrightarrow{OB}$ is equal to

For Questions 12–14

Let AB, AC and AD be three adjacent edges of a parallelepiped. The diagonal of the parallelepiped passing through A and directed away from it is vector $\vec{a}$. The vector area of the faces containing vertices A, B, C and A, B, D are $\vec{b}$ and $\vec{c}$, respectively, i.e., $\overrightarrow{AB}\times\overrightarrow{AC} = \vec{b}$ and $\overrightarrow{AD}\times\overrightarrow{AB} = \vec{c}$. The projection of each of the edges AB and AC on the diagonal vector $\vec{a}$ is $\frac{|\vec{a}|}{3}$.

12. Vector $\overrightarrow{AB}$ is
    (1) $\frac{1}{3}\vec{a} + \frac{\vec{a}\times(\vec{b}-\vec{c})}{|\vec{a}|^2}$
    (2) $\frac{1}{3}\vec{a} + \frac{\vec{a}\times(\vec{b}-\vec{c})}{|\vec{a}|^2} + \frac{3(\vec{b}\times\vec{a})}{|\vec{a}|^2}$
    (3) $\frac{1}{3}\vec{a} + \frac{\vec{a}\times(\vec{b}-\vec{c})}{|\vec{a}|^2} - \frac{3(\vec{c}\times\vec{a})}{|\vec{a}|^2}$
    (4) none of these
13. Vector $\overrightarrow{AC}$ is
    (1) $\frac{1}{3}\vec{a} + \frac{\vec{a}\times(\vec{b}-\vec{c})}{|\vec{a}|^2}$
    (2) $\frac{1}{3}\vec{a} + \frac{\vec{a}\times(\vec{b}-\vec{c})}{|\vec{a}|^2} + \frac{3(\vec{b}\times\vec{a})}{|\vec{a}|^2}$
    (3) $\frac{1}{3}\vec{a} + \frac{\vec{a}\times(\vec{b}-\vec{c})}{|\vec{a}|^2} - \frac{3(\vec{c}\times\vec{a})}{|\vec{a}|^2}$
    (4) none of these
14. Vector $\overrightarrow{AD}$ is
    (1) $\frac{1}{3}\vec{a} + \frac{\vec{a}\times(\vec{b}-\vec{c})}{|\vec{a}|^2}$
    (2) $\frac{1}{3}\vec{a} + \frac{\vec{a}\times(\vec{b}-\vec{c})}{|\vec{a}|^2} + \frac{3(\vec{b}\times\vec{a})}{|\vec{a}|^2}$
    (3) $\frac{1}{3}\vec{a} + \frac{\vec{a}\times(\vec{b}-\vec{c})}{|\vec{a}|^2} - \frac{3(\vec{c}\times\vec{a})}{|\vec{a}|^2}$
    (4) none of these

For Questions 15–17

Let $\vec{u}, \vec{v}$ and $\vec{w}$ be three unit vectors such that $\vec{u} + \vec{v} + \vec{w} = \vec{a}$, $\vec{u}\times(\vec{v}\times\vec{w}) = \vec{b}$, $(\vec{u}\times\vec{v})\times\vec{w} = \vec{c}$, $\vec{a}\cdot\vec{u} = 3/2$, $\vec{a}\cdot\vec{v} = 7/4$ and $|\vec{a}| = 2$. Then

15. Vector $\vec{u}$ is
    (1) $\vec{a} - \frac{2}{3}\vec{b} + \vec{c}$
    (2) $\vec{a} + \frac{4}{3}\vec{b} + \frac{8}{3}\vec{c}$
    (3) $2\vec{a} - \vec{b} + \frac{1}{3}\vec{c}$
    (4) $\frac{4}{3}\vec{a} - \vec{b} + \frac{2}{3}\vec{c}$
16. Vector $\vec{v}$ is
    (1) $2\vec{a} - 3\vec{c}$
    (2) $3\vec{b} - 4c$
    (3) $-4\vec{c}$
    (4) $\vec{a} + \vec{b} + 2\vec{c}$
17. Vector $\vec{w}$ is
    (1) $\frac{2}{3}(2\vec{c} - \vec{b})$
    (2) $\frac{1}{3}(\vec{a} - \vec{b} - \vec{c})$
    (3) $\frac{1}{3}\vec{a} - \frac{2}{3}\vec{b} - 2\vec{c}$
    (4) $\frac{4}{3}(\vec{c} - \vec{b})$

For Questions 18–20

Vectors $\vec{x}, \vec{y}$ and $\vec{z}$, each of magnitude $\sqrt{2}$, make an angle 60° with each other. If $\vec{x}\times(\vec{y}\times\vec{z}) = \vec{a}$, $\vec{y}\times(\vec{z}\times\vec{x}) = \vec{b}$ and $\vec{x}\times\vec{y} = \vec{c}$. Then

18. Vector $\vec{x}$ is
    (1) $\frac{1}{2}[(\vec{a} - \vec{b})\times\vec{c} + (\vec{a} + \vec{b})]$
    (2) $\frac{1}{2}[(\vec{a} + \vec{b})\times\vec{c} + (\vec{a} - \vec{b})]$
    (3) $\frac{1}{2}[-(\vec{a} + \vec{b})\times\vec{c} + (\vec{a} + \vec{b})]$
    (4) $\frac{1}{2}[(\vec{a} + \vec{b})\times\vec{c} - (\vec{a} + \vec{b})]$

19. Vector $\vec{y}$ is

(1) $\frac{1}{2}[(\vec{a}+\vec{c})\times\vec{b}-\vec{b}-\vec{a}]$

(2) $\frac{1}{2}[(\vec{a}-\vec{c})\times\vec{c}+\vec{b}+\vec{a}]$

(3) $\frac{1}{2}[(\vec{a}+\vec{b})\times\vec{c}+\vec{b}+\vec{a}]$

(4) $\frac{1}{2}[(\vec{a}-\vec{c})\times\vec{a}+\vec{b}-\vec{a}]$

20. Vector $\vec{z}$ is

(1) $\frac{1}{2}[(\vec{a}-\vec{c})\times\vec{c}-\vec{b}+\vec{a}]$

(2) $\frac{1}{2}[(\vec{a}+\vec{b})\times\vec{c}+\vec{b}-\vec{a}]$

(3) $\frac{1}{2}[\vec{c}\times(\vec{a}-\vec{b})+\vec{b}+\vec{a}]$

(4) none of these

For Questions 21–23

If $\vec{x}\times\vec{y}=\vec{a}$, $\vec{y}\times\vec{z}=\vec{b}$, $\vec{x}\cdot\vec{b}=\gamma$, $\vec{x}\cdot\vec{y}=1$ and $\vec{y}\cdot\vec{z}=1$. Then

21. Vector $\vec{x}$ is

(1) $\frac{1}{|\vec{a}\times\vec{b}|^2}[\vec{a}\times(\vec{a}\times\vec{b})]$

(2) $\frac{\gamma}{|\vec{a}\times\vec{b}|^2}[\vec{a}\times\vec{b}-\vec{a}\times(\vec{a}\times\vec{b})]$

(3) $\frac{\gamma}{|\vec{a}\times\vec{b}|^2}[\vec{a}\times\vec{b}+\vec{b}\times(\vec{a}\times\vec{b})]$

(4) none of these

22. Vector $\vec{y}$ is

(1) $\frac{\vec{a}\times\vec{b}}{\gamma}$ (2) $\vec{a}+\frac{\vec{a}\times\vec{b}}{\gamma}$

(3) $\vec{a}+\vec{b}+\frac{\vec{a}\times\vec{b}}{\gamma}$ (4) none of these

23. Vector $\vec{z}$ is

(1) $\frac{\gamma}{|\vec{a}\times\vec{b}|^2}[\vec{a}+\vec{b}\times(\vec{a}\times\vec{b})]$

(2) $\frac{\gamma}{|\vec{a}\times\vec{b}|^2}[\vec{a}+\vec{b}-\vec{a}\times(\vec{a}\times\vec{b})]$

(3) $\frac{\gamma}{|\vec{a}\times\vec{b}|^2}[\vec{a}\times\vec{b}+\vec{b}\times(\vec{a}\times\vec{b})]$

(4) none of these

For Questions 24–26

Given two orthogonal vectors $\vec{A}$ and $\vec{B}$ each of length unity. Let $\vec{P}$ be the vector satisfying the equation $\vec{P}\times\vec{B}=\vec{A}-\vec{P}$. Then

24. $(\vec{P}\times\vec{B})\times\vec{B}$ is equal to

(1) $\vec{P}$ (2) $-\vec{P}$ (3) $2\vec{B}$ (4) $\vec{A}$

25. $\vec{P}$ is equal to

(1) $\frac{\vec{A}}{2}+\frac{\vec{A}\times\vec{B}}{2}$ (2) $\frac{\vec{A}}{2}+\frac{\vec{B}\times\vec{A}}{2}$

(3) $\frac{\vec{A}\times\vec{B}}{2}-\frac{\vec{A}}{2}$ (4) $\vec{A}\times\vec{B}$

26. Which of the following statements is false?

(1) vectors $\vec{P}$, $\vec{A}$ and $\vec{P}\times\vec{B}$ are linearly dependent.

(2) vectors $\vec{P}$, $\vec{B}$ and $\vec{P}\times\vec{B}$ are linearly independent.

(3) $\vec{P}$ is orthogonal to $\vec{B}$ and has length $1/\sqrt{2}$.

(4) none of the above.

For Questions 27–29

Let $\vec{a},\vec{b},\vec{c}$ be non-coplanar unit vectors, equally inclined to each other at an angle θ. If $\vec{a}\times\vec{b}+\vec{b}\times\vec{c}=p\vec{a}+q\vec{b}+r\vec{c}$, then

27. The value of p is

(1) $\frac{1}{\sqrt{1+2\cos\theta}}$ (2) $\frac{1}{\sqrt{1-2\cos\theta}}$

(3) $\frac{2\cos\theta}{\sqrt{1+2\cos\theta}}$ (4) $\frac{-2\cos\theta}{\sqrt{1+2\cos\theta}}$

28. The value of q is

(1) $\frac{1}{\sqrt{1+2\cos\theta}}$ (2) $\frac{1}{\sqrt{1-2\cos\theta}}$

(3) $\frac{2\cos\theta}{\sqrt{1+2\cos\theta}}$ (4) $\frac{-2\cos\theta}{\sqrt{1+2\cos\theta}}$

29. The value of r is

(1) $\frac{1}{\sqrt{1+2\cos\theta}}$ (2) $\frac{1}{\sqrt{1-2\cos\theta}}$

(3) $\frac{2\cos\theta}{\sqrt{1+2\cos\theta}}$ (4) $\frac{-2\cos\theta}{\sqrt{1+2\cos\theta}}$

Matrix Match Type

1. Match the following lists:

List I	List II
(P) If $\|\vec{a}\| = \|\vec{b}\| = \|\vec{c}\|$, angle between each pair of vectors is $\frac{\pi}{3}$ and $\|\vec{a}+\vec{b}+\vec{c}\| = \sqrt{6}$, then $2\|\vec{a}\|$ is equal to	(I) 1
(Q) If $\vec{a}$ is perpendicular to $\vec{b}+\vec{c}$, $\vec{b}$ is perpendicular to $\vec{c}+\vec{a}$, $\vec{c}$ is perpendicular to $\vec{a}+\vec{b}$, $\|\vec{a}\| = 2$, $\|\vec{b}\| = 3$ and $\|\vec{c}\| = 6$, then $\|\vec{a}+\vec{b}+\vec{c}\| - 2$ is equal to	(II) 2
(R) $\vec{a} = 2\hat{i} + 3\hat{j} - \hat{k}$, $\vec{b} = -\hat{i} + 2\hat{j} - 4\hat{k}$, $\vec{c} = \hat{i} + \hat{j} + \hat{k}$ and $\vec{d} = 3\hat{i} + 2\hat{j} +$, then $\frac{1}{7}(\vec{a}\times\vec{b})\cdot(\vec{c}\times\vec{d})$ is equal to	(III) 3
(S) If $\|\vec{a}\| = \|\vec{b}\| = \|\vec{c}\| = 2$ and $\vec{a}\cdot\vec{b} = \vec{b}\cdot\vec{c} = \vec{c}\cdot\vec{a} = 2$, then $[\vec{a}\ \vec{b}\ \vec{c}]\cos 45°$ is equal to	(IV) 4
	(V) 5

(1) (P) → (I); (Q) → (II); (R) → (III); (S) → (IV)
(2) (P) → (III); (Q) → (V); (R) → (I); (S) → (IV)
(3) (P) → (I); (Q) → (V); (R) → (III); (S) → (IV)
(4) (P) → (I); (Q) → (IV); (R) → (III); (S) → (V)

2. Given two vectors $\vec{a} = -\hat{i} + \hat{j} + 2\hat{k}$ and $\vec{b} = -\hat{i} - 2\hat{j} - \hat{k}$. Now, match the following lists:

List I	List II
(P) Area of triangle formed by $\vec{a}$ and $\vec{b}$	(I) 3
(Q) Area of parallelogram having sides $\vec{a}$ and $\vec{b}$	(II) $12\sqrt{3}$
(R) Area of parallelogram having diagonals $2\vec{a}$ and $4\vec{b}$	(III) $3\sqrt{3}$
(S) Volume of parallelepiped formed by $\vec{a}$, $\vec{b}$ and $\vec{c} = \hat{i} + \hat{j} + \hat{k}$	(IV) $\frac{3\sqrt{3}}{2}$
	(V) $6\sqrt{3}$

(1) (P) → (IV); (Q) → (III); (R) → (I); (S) → (V)
(2) (P) → (IV); (Q) → (III); (R) → (II); (S) → (I)
(3) (P) → (V); (Q) → (II); (R) → (I); (S) → (III)
(4) (P) → (V); (Q) → (II); (R) → (IV); (S) → (I)

3. Match the following lists:

List I	List II
(P) If $\|\vec{a}+\vec{b}\| = \|\vec{a}+2\vec{b}\|$, then angle between $\vec{a}$ and $\vec{b}$ is	(I) 0°
(Q) If $\|\vec{a}+\vec{b}\| = \|\vec{a}-2\vec{b}\|$, then angle between $\vec{a}$ and $\vec{b}$ is	(II) acute
(R) If $\|\vec{a}+\vec{b}\| = \|\vec{a}-\vec{b}\|$, then angle between $\vec{a}$ and $\vec{b}$ is	(III) 90°
(S) Angle between $\vec{a}\times\vec{b}$ and a vector perpendicular to the vector $\vec{c}\times(\vec{a}\times\vec{b})$ is	(IV) obtuse
	(V) 180°

(1) (P) → (IV); (Q) → (III); (R) → (II); (S) → (I)
(2) (P) → (III); (Q) → (IV); (R) → (I); (S) → (II)
(3) (P) → (V); (Q) → (II); (R) → (I); (S) → (III)
(4) (P) → (IV); (Q) → (II); (R) → (III); (S) → (I)

4. Volume of parallelepiped formed by vectors $\vec{a}\times\vec{b}$, $\vec{b}\times\vec{c}$ and $\vec{c}\times\vec{a}$ is 36 sq. units. Now, match the following lists:

List I	List II
(P) Volume of parallelepiped formed by vectors $\vec{a}$, $\vec{b}$ and $\vec{c}$ is	(I) 0
(Q) Volume of tetrahedron formed by vectors $\vec{a}$, $\vec{b}$ and $\vec{c}$ is	(II) 1 sq. units
(R) Volume of parallelepiped formed by vectors $\vec{a}+\vec{b}$, $\vec{b}+\vec{c}$ and $\vec{c}+\vec{a}$ is	(III) 2 sq. units
(S) Volume of parallelepiped formed by vectors $\vec{a}-\vec{b}$, $\vec{b}-\vec{c}$ and $\vec{c}-\vec{a}$ is	(IV) 6 sq. units
	(V) 12 sq. units

(1) (P) → (IV); (Q) → (II); (R) → (V); (S) → (I)
(2) (P) → (V); (Q) → (III); (R) → (I); (S) → (II)
(3) (P) → (IV); (Q) → (III); (R) → (I); (S) → (I)
(4) (P) → (V); (Q) → (III); (R) → (IV); (S) → (I)

ARCHIVES

Single Correct Answer Type

1. Let $\overrightarrow{PR} = 3\hat{i} + \hat{j} - 2\hat{k}$ and $\overrightarrow{SQ} = \hat{i} - 3\hat{j} - 4\hat{k}$ determine diagonals of a parallelogram *PQRS*, and $\overrightarrow{PT} = \hat{i} + 2\hat{j} + 3\hat{k}$ be another vector. Then the volume of the parallelepiped determined by the vectors $\overrightarrow{PT}, \overrightarrow{PQ}$ and $\overrightarrow{PS}$ is

(1) 5 (2) 20 (3) 10 (4) 30

(JEE Advanced 2013)

2. Let *O* be the origin and let *PQR* be an arbitrary triangle. The point *S* is such that

$\overrightarrow{OP}\cdot\overrightarrow{OQ}+\overrightarrow{OR}\cdot\overrightarrow{OS}=\overrightarrow{OR}\cdot\overrightarrow{OP}+\overrightarrow{OQ}\cdot\overrightarrow{OS}=\overrightarrow{OQ}\cdot\overrightarrow{OR}+\overrightarrow{OP}\cdot\overrightarrow{OS}$

$\overrightarrow{OP}\cdot\overrightarrow{OQ}+\overrightarrow{OR}\cdot\overrightarrow{OS}=\overrightarrow{OR}\cdot\overrightarrow{OP}+\overrightarrow{OQ}\cdot\overrightarrow{OS}=\overrightarrow{OQ}\cdot\overrightarrow{OR}+\overrightarrow{OP}\cdot\overrightarrow{OS}$

then the triangle *PQR* has *S* as its

(1) centroid (2) circumcentre
(3) incentre (4) orthocenter

(JEE Advanced 2017)

3. Let the position vectors of the points *P*, *Q*, *R* and *S* be $\vec{a}=\hat{i}+2\hat{j}-5\hat{k}, \vec{b}=3\hat{i}+6\hat{j}+3\hat{k}$, $\vec{c}=\frac{17}{5}\hat{i}+\frac{16}{5}\hat{j}+7\hat{k}$ and $\vec{d}=2\hat{i}+\hat{j}+\hat{k}$, respectively. Then which of the following statements is true?

(1) The points *P*, *Q*, *R* and *S* are **NOT** coplanar

(2) $\frac{\vec{b}+2\vec{d}}{3}$ is the position vector of a point which divides *PR* internally in the ratio 5 : 4

(3) $\frac{\vec{b}+2\vec{d}}{3}$ is the position vector of a point which divides *PR* externally in the ratio 5 : 4

(4) The square of the magnitude of the vector $\vec{b}\times\vec{d}$ is 95

(JEE Advanced 2023)

Multiple Correct Answers Type

1. Let $\vec{x}, \vec{y}$ and $\vec{z}$ be three vectors each of magnitude $\sqrt{2}$ and the angle between each pair of them is $\frac{\pi}{3}$. If $\vec{a}$ is a non-zero vector perpendicular to $\vec{x}$ and $\vec{y}\times\vec{z}$ and $\vec{b}$ is a non-zero vector perpendicular to $\vec{y}$ and $\vec{z}\times\vec{x}$, then

(1) $\vec{b}=(\vec{b}\cdot\vec{z})(\vec{z}-\vec{x})$ (2) $\vec{a}=(\vec{a}\cdot\vec{y})(\vec{y}-\vec{z})$

(3) $\vec{a}\cdot\vec{b}=-(\vec{a}\cdot\vec{y})(\vec{b}\cdot\vec{z})$ (4) $\vec{a}=(\vec{a}\cdot\vec{y})(\vec{z}-\vec{y})$

(JEE Advanced 2014)

2. Let ΔPQR be a triangle. Let $\vec{a}=\overrightarrow{QR}$, $\vec{b}=\overrightarrow{RP}$ and $\vec{c}=\overrightarrow{PQ}$. If $|\vec{a}|=12, |\vec{b}|=4\sqrt{3}$ and $\vec{b}\cdot\vec{c}=24$, then which of the following is (are) true?

(1) $\frac{|\vec{c}|^2}{2}-|\vec{a}|=12$

(2) $\frac{|\vec{c}|^2}{2}+|\vec{a}|=30$

(3) $|\vec{a}\times\vec{b}+\vec{c}\times\vec{a}|=48\sqrt{3}$

(4) $\vec{a}\cdot\vec{b}=-72$

(JEE Advanced 2015)

3. Let $\hat{u}=u_1\hat{i}+u_2\hat{j}+u_3\hat{k}$ be a unit vector in R^3 and $\hat{\omega}=\frac{1}{\sqrt{6}}(\hat{i}+\hat{j}+2\hat{k})$. Given that there exists a vector $\vec{v}$ in R^3 such that $|\hat{u}\times\vec{v}|=1$ and $\hat{\omega}\cdot(\hat{u}\times\vec{v})=1$. Which of the following statement(s) is (are) correct ?

(1) There is exactly one choice for such $\vec{v}$

(2) There are infinitely many choices for such $\vec{v}$

(3) If $\hat{u}$ lies in the *xy*-plane then $|u_1|=|u_2|$

(4) If $\hat{u}$ lies in the *xz*-plane then $2|u_1|=|u_3|$

(JEE Advanced 2016)

4. Let *a* and *b* be positive real numbers. Suppose $\overrightarrow{PQ}=a\hat{i}+b\hat{j}$ and $\overrightarrow{PS}=a\hat{i}-b\hat{j}$ are adjacent sides of a parallelogram *PQRS*. Let $\vec{u}$ and $\vec{v}$ be the projection vectors of $\vec{w}=\hat{i}+\hat{j}$ along $\overrightarrow{PQ}$ and $\overrightarrow{PS}$, respectively. If $|\vec{u}|+|\vec{v}|=|\vec{w}|$ and if the area of the parallelogram *PQRS* is 8, then which of the following statements is/are TRUE?

(1) $a+b=4$

(2) $a-b=2$

(3) The length of the diagonal *PR* of the parallelogram *PQRS* is 4.

(4) $\vec{w}$ is an angle bisector of the vectors $\overrightarrow{PQ}$ and $\overrightarrow{PS}$.

(JEE Advanced 2020)

5. Let *O* be the origin and $\overrightarrow{OA}=2\hat{i}+2\hat{j}+\hat{k}$, $\overrightarrow{OB}=\hat{i}-2\hat{j}+2\hat{k}$ and $\overrightarrow{OC}=\frac{1}{2}(\overrightarrow{OB}-\lambda\overrightarrow{OA})$ for some $\lambda>0$. If $|\overrightarrow{OB}\times\overrightarrow{OC}|=\frac{9}{2}$, then which of the following statements is (are) TRUE?

(1) Projection of $\overrightarrow{OC}$ on $\overrightarrow{OA}$ is $-\frac{3}{2}$

(2) Area of the triangle OAB is $\frac{9}{2}$

(3) Area of the triangle ABC is $\frac{9}{2}$

(4) The acute angle between the diagonals of the parallelogram with adjacent sides $\overrightarrow{OA}$ and $\overrightarrow{OC}$ is $\frac{\pi}{3}$

(JEE Advanced 2021)

6. Let $\hat{i}, \hat{j}$ and $\hat{k}$ be the unit vectors along the three positive coordinate axes. Let

$\vec{a} = 3\hat{i} + \hat{j} - \hat{k}$,

$\vec{b} = \hat{i} + b_2\hat{j} + b_3\hat{k}, b_2, b_3 \in R$

$\vec{c} = c_1\hat{i} + c_2\hat{j} + c_3\hat{k}, c_1, c_2, c_3 \in R$

be three vectors such that $b_2b_3 > 0$, $\vec{a} \cdot \vec{b} = 0$

$$\begin{pmatrix} 0 & -c_3 & c_2 \\ c_3 & 0 & -c_1 \\ -c_2 & c_1 & 0 \end{pmatrix}\begin{pmatrix} 1 \\ b_2 \\ b_3 \end{pmatrix} = \begin{pmatrix} 3-c_1 \\ 1-c_2 \\ -1-c_3 \end{pmatrix}$$

Then, which of the following is/are TRUE?

(1) $\vec{a} \cdot \vec{c} = 0$ (2) $\vec{b} \cdot \vec{c} = 0$

(3) $|\vec{b}| > \sqrt{10}$ (4) $|\vec{c}| \le \sqrt{11}$

(JEE Advanced 2022)

Linked Comprehension Type

For Questions 1 and 2

Let O be the origin, and $\overline{OX}, \overline{OY}, \overline{OZ}$ be three unit vectors in the directions of the sides $\overline{QR}, \overline{RP}, \overline{PQ}$, respectively, of a triangle PQR. **(JEE Advanced 2017)**

1. $|\overline{OX} \times \overline{OY}| =$

(1) $\sin(P+Q)$ (2) $\sin 2R$

(3) $\sin(P+R)$ (4) $\sin(Q+R)$

2. If the triangle PQR varies, then the minimum value of $\cos(P+Q) + \cos(Q+R) + \cos(R+P)$ is

(1) $-\frac{5}{3}$ (2) $-\frac{3}{2}$ (3) $\frac{3}{2}$ (4) $\frac{5}{3}$

Matrix Match Type

1. Match the following lists:

List I	List II
(P) Volume of parallelepiped determined by vectors $\vec{a}, \vec{b}$ and $\vec{c}$ is 2. Then the volume of the parallelepiped determined by vectors $2(\vec{a} \times \vec{b}), 3(\vec{b} \times \vec{c})$ and $(\vec{c} \times \vec{a})$ is	**(I)** 100
(Q) Volume of parallelepiped determined by vectors $\vec{a}, \vec{b}$ and $\vec{c}$ is 5. Then the volume of the parallelepiped determined by vectors $3(\vec{a}+\vec{b}), (\vec{b}+\vec{c})$ and $2(\vec{c}+\vec{a})$ is	**(II)** 30
(R) Area of a triangle with adjacent sides determined by vectors $\vec{a}$ and $\vec{b}$ is 20. Then the area of the triangle with adjacent sides determined by vectors $(2\vec{a}+3\vec{b})$ and $(\vec{a}-\vec{b})$ is	**(III)** 24
(S) Area of a parallelogram with adjacent sides determined by vectors $\vec{a}$ and $\vec{b}$ is 30. Then the area of the parallelogram with adjacent sides determined by vectors $(\vec{a}+\vec{b})$ and $\vec{a}$ is	**(IV)** 60

(JEE Advanced 2013)

2. Match the statements given in List I with the values given in List II.

List I	List II
(P) Let $y(x) = \cos(3\cos^{-1}x), x \in [-1, 1]$, $x \neq \pm\frac{\sqrt{3}}{2}$. Then $\frac{1}{y(x)}\left\{(x^2-1)\frac{d^2y(x)}{dx^2} + x\frac{dy(x)}{dx}\right\}$ equals	**(I)** 1
(Q) Let $A_1, A_2, \ldots, A_n$ $(n > 2)$ be the vertices of a regular polygon of n sides with its centre at the origin. Let $\vec{a_k}$ be the position vector of the point A_k, $k = 1, 2, \ldots n$. If $\left|\sum_{k=1}^{n-1}(\vec{a_k} \times \vec{a_{k+1}})\right| = \left|\sum_{k=1}^{n-1}(\vec{a_k} \cdot \vec{a_{k+1}})\right|$, then the minimum value of n is	**(II)** 2
(R) If the normal from the point $P(h, 1)$ on the ellipse $\frac{x^2}{6} + \frac{y^2}{3} = 1$ is perpendicular to the line $x + y = 8$, then the value of h is	**(III)** 8
(S) Number of positive solutions satisfying the equation $\tan^{-1}\left(\frac{1}{2x+1}\right) + \tan^{-1}\left(\frac{1}{4x+1}\right) = \tan^{-1}\left(\frac{2}{x^2}\right)$ is	**(IV)** 9

Codes:

	(P)	(Q)	(R)	(S)
(1)	IV	III	II	I
(2)	II	IV	III	I
(3)	IV	III	I	II
(4)	II	IV	I	III

(JEE Advanced 2014)

3. Match the following lists:

List I	List II
(P) In R^2, if the magnitude of the projection vector of the vector $\alpha\hat{i}+\beta\hat{j}$ on $\sqrt{3}\hat{i}+\hat{j}$ is $\sqrt{3}$ and if $\alpha = 2 + \sqrt{3}\beta$, then possible value(s) of $\|\alpha\|$ is (are)	(I) 1
(Q) Let a and b be real numbers such that the function $f(x) = \begin{cases} -3ax^2-2, & x<1 \\ bx+a^2, & x\geq 1 \end{cases}$ is differentiable for all $x \in R$. Then possible value(s) of α is (are)	(II) 2
(R) Let $\omega \neq 1$ be a complex cube root of unity If $(3-3\omega+2\omega^2)^{4n+3}+(2+3\omega-3\omega^2)^{4n+3}+(-3+2\omega+3\omega^2)^{4n+3}=0$, then possible value(s) of n is (are)	(III) 3
(S) Let the harmonic mean of two positive real numbers a and b be 4. If q is a positive real number such that $a, 5, q, b$ is an arithmetic progression, then the value(s) of $\|q-a\|$ is (are)	(IV) 4
	(V) 5

(JEE Advanced 2015)

4. Let $\vec{w} = \hat{i}+\hat{j}-2\hat{k}$ and $\vec{u}$ and $\vec{v}$ be two vectors, such that $\vec{u}\times\vec{v}=\vec{w}$ and $\vec{v}\times\vec{w}=\vec{u}$. Let α, β, γ, and t be real numbers such that
$\vec{u}=\alpha\hat{i}+\beta\hat{j}+\gamma\hat{k}$, $-t\alpha+\beta+\gamma=0$, $\alpha-t\beta+\gamma=0$, and $\alpha+\beta-t\gamma=0$.
Match each entry in List – I to the correct entry in List – II and choose the correct option.

List I	List II
(P) $\|\vec{v}\|^2$ is equal to	(I) 0
(Q) If $\alpha=\sqrt{3}$, then γ^2 is equal to	(II) 1
(R) If $\alpha=\sqrt{3}$, then $(\beta+\gamma)^2$ is equal to	(III) 2
(S) If $\alpha=\sqrt{2}$, then t + 3 is equal to	(IV) 3
	(V) 5

(1) (P) → (II); (Q) → (I); (R) → (IV); (S) → (V)
(2) (P) → (II); (Q) → (IV); (R) → (III); (S) → (V)
(3) (P) → (II); (Q) → (I); (R) → (IV); (S) → (III)
(4) (P) → (IV); (Q) → (IV); (R) → (I); (S) → (III)

(JEE Advanced 2025)

Numerical Value Type

1. Consider the cube in the first octant with sides OP, OQ and OR of length 1, along the x-axis, y-axis and z-axis, respectively, where $O(0, 0, 0)$ is the origin. Let $S\left(\frac{1}{2},\frac{1}{2},\frac{1}{2}\right)$ be the centre of the cube and T be the vertex of the cube opposite to the origin O such that S lies on the diagonal OT. If $\vec{p}=\overline{SP}$, $\vec{q}=\overline{SQ}$, $\vec{r}=\overline{SR}$ and $\vec{t}=\overline{ST}$, then the value of $|(\vec{p}\times\vec{q})\times(\vec{r}\times\vec{t})|$ is ________.

(JEE Advanced 2018)

Integer Answer Type

1. Let $\vec{a}, \vec{b}$, and $\vec{c}$ be three non-coplanar unit vectors such that the angle between every pair of them is $\frac{\pi}{3}$. If $\vec{a}\times\vec{b}+\vec{b}\times\vec{c}=p\vec{a}+q\vec{b}+r\vec{c}$, where p, q and r are scalars, then the value of $\frac{p^2+2q^2+r^2}{q^2}$ is ______.

(JEE Advanced 2014)

2. Let $\vec{a}$ and $\vec{b}$ be two unit vectors such that $\vec{a}\cdot\vec{b}=0$. For some $x, y \in R$, let $\vec{c}=x\vec{a}+y\vec{b}+(\vec{a}\times\vec{b})$. If $|\vec{c}|=2$ and the vector $\vec{c}$ is inclined at the same angle α to both $\vec{a}$ and $\vec{b}$, then the value of $8\cos^2\alpha$ is ________.

(JEE Advanced 2018)

3. Let $\vec{a}=2\hat{i}+\hat{j}-\hat{k}$ and $\vec{b}=\hat{i}+2\hat{j}+\hat{k}$ be two vectors. Consider a vector $\vec{c}=\alpha\vec{a}+\beta\vec{b}$, $\alpha, \beta, \in R$. If the projection of $\vec{c}$ on the vector $(\vec{a}+\vec{b})$ is $3\sqrt{2}$, then the minimum value of $(\vec{c}-(\vec{a}\times\vec{b}))\cdot\vec{c}$ equals________.

(JEE Advanced 2019)

4. In a triangle PQR, let $\vec{a}=\overrightarrow{QR}$, $\vec{b}=\overrightarrow{RP}$ and $\vec{c}=\overrightarrow{PQ}$.
If $|\vec{a}|=3$, $|\vec{b}|=4$ and $\frac{\vec{a}\cdot(\vec{c}-\vec{b})}{\vec{c}\cdot(\vec{a}-\vec{b})}=\frac{|\vec{a}|}{|\vec{a}|+|\vec{b}|}$, then the value of $|\vec{a}\times\vec{b}|^2$ is ________. **(JEE Advanced 2020)**

5. Let $\vec{u}$, $\vec{v}$ and $\vec{w}$ be vectors in three-dimensional space, where $\vec{u}$ and $\vec{v}$ are unit vectors which are not perpendicular to each other and
$$\vec{u}\cdot\vec{w}=1, \vec{v}\cdot\vec{w}=1, \vec{w}\cdot\vec{w}=4$$
If the volume of the parallelepiped, whose adjacent sides are represented by the vectors $\vec{u}$, $\vec{v}$ and $\vec{w}$ is $\sqrt{2}$, then the value of $|3\vec{u}+5\vec{v}|$ is ________. **(JEE Advanced 2021)**

6. Let $\overrightarrow{OP}=\frac{\alpha-1}{\alpha}\hat{i}+\hat{j}+\hat{k}$, $\overrightarrow{OQ}=\hat{i}+\frac{\beta-1}{\beta}\hat{j}+\hat{k}$ and $\overrightarrow{OR}=\hat{i}+\hat{j}+\frac{1}{2}\hat{k}$ be three vectors, where $\alpha, \beta \in R-\{0\}$ and O denotes the origin. If $(\overrightarrow{OP}\times\overrightarrow{OQ})\cdot\overrightarrow{OR}=0$ and the point $(\alpha, \beta, 2)$ lies on the plane $3x+3y-z+l=0$, then the value of l is ________. **(JEE Advanced 2024)**

7. Let $\vec{p}=2\hat{i}+\hat{j}+3\hat{k}$ and $\vec{q}=\hat{i}-\hat{j}+\hat{k}$. If for some real numbers α, β, and γ, we have $15\hat{i}+10\hat{j}+6\hat{k}=\alpha(2\vec{p}+\vec{q})+\beta(\vec{p}-2\vec{q})+\gamma(\vec{p}\times\vec{q})$, then the value of γ is ______.

(JEE Advanced 2024)

ANSWER KEY

EXERCISES

Single Correct Answer Type

1. (2) **2.** (2) **3.** (4) **4.** (4) **5.** (2)
6. (4) **7.** (3) **8.** (3) **9.** (1) **10.** (2)
11. (2) **12.** (2) **13.** (1) **14.** (3) **15.** (1)
16. (3) **17.** (3) **18.** (2) **19.** (3) **20.** (1)
21. (3) **22.** (3) **23.** (2) **24.** (3) **25.** (3)
26. (4) **27.** (3) **28.** (2) **29.** (3) **30.** (3)
31. (3) **32.** (3) **33.** (1) **34.** (3) **35.** (3)
36. (1) **37.** (4) **38.** (4) **39.** (2) **40.** (4)
41. (1) **42.** (3) **43.** (1) **44.** (4) **45.** (3)
46. (2) **47.** (3) **48.** (3) **49.** (2) **50.** (3)
51. (3) **52.** (3) **53.** (1) **54.** (3) **55.** (2)
56. (4) **57.** (3) **58.** (4) **59.** (2) **60.** (1)
61. (3) **62.** (4) **63.** (2) **64.** (3) **65.** (4)
66. (3) **67.** (2) **68.** (1) **69.** (1) **70.** (1)
71. (1) **72.** (1) **73.** (1) **74.** (1) **75.** (4)
76. (1) **77.** (3) **78.** (1) **79.** (3) **80.** (1)
81. (3) **82.** (3) **83.** (4) **84.** (4) **85.** (2)
86. (1) **87.** (2) **88.** (4) **89.** (3)

Numerical Value Type

1. (19.5) **2.** (9.2) **3.** (237.5) **4.** (9.5) **5.** (1.5)
6. (0.5) **7.** (71.25) **8.** (6.6) **9.** (2.5)

Integer Answer Type

1. (5) **2.** (1) **3.** (2) **4.** (28) **5.** (20)
6. (7) **7.** (45) **8.** (1) **9.** (6) **10.** (4)
11. (98) **12.** (1) **13.** (6) **14.** (3) **15.** (13)
16. (1) **17.** (6)

Multiple Correct Answers Type

1. (1), (2) **2.** (1), (3)
3. (1), (2), (3), (4) **4.** (1), (2), (3)
5. (1), (3) **6.** (1), (3)
7. (1), (2), (3), (4) **8.** (1), (2), (3), (4)
9. (1), (2), (3), (4) **10.** (1), (3)
11. (2), (4) **12.** (2), (4)
13. (1), (4) **14.** (2), (3), (4)
15. (1), (4) **16.** (2), (4)
17. (2), (3) **18.** (3), (4)
19. (2), (4) **20.** (2), (4)
21. (1), (2), (3) **22.** (1), (2)
23. (1), (2), (3) **24.** (1), (2)
25. (1), (3) **26.** (2), (4)
27. (2), (3) **28.** (1), (2), (3)
29. (1), (2), (3) **30.** (1), (3)
31. (2), (3), (4) **32.** (1), (2), (4)
33. (1), (2), (3) **34.** (2), (4)
35. (1), (2), (3), (4) **36.** (2), (3)

Linked Comprehension Type

1. (2) **2.** (1) **3.** (3) **4.** (51) **5.** (96)
6. (14) **7.** (8) **8.** (32) **9.** (9) **10.** (18)
11. (3) **12.** (1) **13.** (2) **14.** (3) **15.** (2)
16. (3) **17.** (4) **18.** (4) **19.** (3) **20.** (2)
21. (2) **22.** (1) **23.** (3) **24.** (2) **25.** (2)
26. (4) **27.** (1) **28.** (4) **29.** (1)

Matrix Match Type

1. (3) **2.** (2) **3.** (4) **4.** (1)

ARCHIVES

Single Correct Answer Type

1. (3) **2.** (4) **3.** (2)

Multiple Correct Answers Type

1. (1), (2), (3) **2.** (1), (3), (4)
3. (2), (3) **4.** (1), (3)
5. (1), (2), (3) **6.** (2), (3), (4)

Linked Comprehension Type

1. (1) **2.** (2)

Matrix Match Type

1. (P) → (III); (Q) → (IV); (R) → (I); (S) → (II)
2. (1)
3. (P) → (I), (II)
4. (1)

Numerical Value Type

1. (0.5)

Integer Answer Type

1. (4) **2.** (3) **3.** (18.00) **4.** (108)
5. (7) **6.** (5) **7.** (2)

Chapter

4 Line and Plane

Chapter Prerequisites

Before starting this chapter, ensure that you understand the following concepts thoroughly:

1. **Coordinate Geometry in Two and Three Dimensions:** Understanding of coordinate axes, plotting points, the distance and section formulas in two dimensions, and basic familiarity with representing points in three dimensions using ordered triples.
2. **Vector Algebra in Three Dimensions:** Comfort with representing vectors in component form, computing magnitudes, using direction ratios, and interpreting directions and positions in space.
3. **Vector Products:** Understanding of dot product, cross product, scalar triple product, and vector triple product as tools for measuring angles, determining directions, checking coplanarity, and evaluating geometric quantities.
4. **Algebraic and Trigonometric Skills:** Ability to simplify expressions, manipulate equations, and use trigonometric relationships involving angles and direction.
5. **Spatial Visualization:** Ability to visualise lines and planes and their interactions in space, such as parallelism, perpendicularity, intersection, and relative position.

Topic List

Learning Objectives

Line in Space: Direction, Equation, Angle, & Distance

Understand direction ratios of a line, write equations of lines in space, find the angle between lines, and determine the distance of a point from a line.

Line–Plane Relationships

Determine when a line lies in a plane, is parallel to a plane, is perpendicular to a plane, or intersects a plane obliquely.

Coplanar and Non-Coplanar Lines

Understand coplanar and non-coplanar lines in space, compute the distance between parallel lines, and interpret skew lines geometrically.

Division of Space & Angle Bisector Planes

Understand how planes divide space, identify regions containing a given point, find the angle between planes, and interpret bisector planes.

Plane in Space & Family of Planes

Learn different forms of the equation of a plane and represent the family of planes containing the line of intersection of two planes.

Distance from a Plane

Learn to compute the perpendicular distance of a point from a plane and the distance between two parallel planes.

4.1 LINE IN SPACE

4.1.1 Direction Ratios of a Line

In the earlier chapter, we discussed the direction cosines of a vector, which are the cosines of the angles that a directed vector makes with the coordinate axes.

The same concept applies to a directed line in space.

If a line L makes angles α, β, and γ with the x-, y-, and z-axes respectively, then $\cos\alpha$, $\cos\beta$, and $\cos\gamma$ are called the direction cosines of the line.

Since a line in space can be extended infinitely in both directions, it has two possible directions, and hence two sets of direction cosines that differ only in sign.

To have a unique set of direction cosines for a given line, we assign one of its two directions as the positive or reference direction and consider the line as a directed line.

A directed line in space is therefore a line for which one of the two possible directions has been chosen as positive. Although the line itself extends infinitely in both directions, by fixing one direction as the reference, we can define a unique set of direction cosines corresponding to that direction.

If this chosen direction is reversed, the direction angles α, β, γ are replaced by their supplements $\pi-\alpha, \pi-\beta, \pi-\gamma$ and therefore, the signs of the direction cosines are reversed.

If a given line does not pass through the origin, a line parallel to it and passing through the origin can be considered. Both lines will have the same direction cosines.

Any three numbers that are proportional to the direction cosines of a line are called its **direction ratios** (or **direction numbers**).

If l, m, n are the direction cosines and a, b, c are the direction ratios of a line, then

$$l=\lambda a,\ m=\lambda b,\ n=\lambda c,\ \text{where } \lambda \neq 0$$

Since $l^2+m^2+n^2=1$, $\lambda=\pm\dfrac{1}{\sqrt{a^2+b^2+c^2}}$,

and hence,

$$l=\pm\frac{a}{\sqrt{a^2+b^2+c^2}},\ m=\pm\frac{b}{\sqrt{a^2+b^2+c^2}},\ n=\pm\frac{c}{\sqrt{a^2+b^2+c^2}}$$

The sign is chosen according to the selected direction of the line.

If a line is parallel to vector $a\hat{i}+b\hat{j}+c\hat{k}$, then it is also parallel to any non-zero scalar multiple of this vector, $\lambda(a\hat{i}+b\hat{j}+c\hat{k})$, where $\lambda \neq 0$.

When $\lambda>0$, both vectors represent the same direction, whereas for $\lambda<0$, they represent opposite directions.

In either case, they lie along the same line in space.

Therefore, a line can have infinitely many sets of direction ratios, all proportional to one another.

Illustration 4.1

A line passes through the points (6, −7, −1) and (2, −3, 1). Find the direction cosines of the line.

Sol. Let l, m, n be the DCs of the given line.

The line passes through $A(6,-7,-1)$ and $B(2,-3,1)$, therefore its DRs are $6-2, -7+3, -1-1 \equiv 4, -4, -2$.

Also, $AB = 6$

Hence the DCs of the given line are $\pm\dfrac{4}{6}, \mp\dfrac{4}{6}, \mp\dfrac{2}{6}$

or $\quad \pm\dfrac{2}{3}, \mp\dfrac{2}{3}, \mp\dfrac{1}{3}$.

Illustration 4.2

Let ABC be a triangle having vertices $A \equiv (2, 3, 5)$, $B \equiv (-1, 3, 2)$ and $C \equiv (\lambda, 5, \mu)$. If the median through A is equally inclined to the axes, then find the values of λ and μ.

Sol. Mid-point of BC is $\left(\dfrac{\lambda-1}{2}, 4, \dfrac{2+\mu}{2}\right)$.

Direction ratios of the median through A are

$$\frac{\lambda-1}{2}-2,\ 4-3,\ \frac{2+\mu}{2}-5$$

or $\quad \dfrac{\lambda-5}{2}, 1, \dfrac{\mu-8}{2}$.

Since the median is equally inclined to axes, the direction ratios must be equal.

So, $\quad \dfrac{\lambda-5}{2}=1=\dfrac{\mu-8}{2}$

$\Rightarrow \quad \lambda=7, \mu=10$.

Illustration 4.3

If l_1, m_1, n_1 and l_2, m_2, n_2 are the DCs of the two lines inclined to each other at an angle θ, then prove that the DCs of the

a. internal bisector of the angle between these lines are

$$\frac{l_1+l_2}{2\cos\theta/2}, \frac{m_1+m_2}{2\cos\theta/2}, \frac{n_1+n_2}{2\cos\theta/2} \quad \text{and}$$

b. external bisector of the angle between the lines are

$$\frac{l_1-l_2}{2\sin\theta/2}, \frac{m_1-m_2}{2\sin\theta/2}, \frac{n_1-n_2}{2\sin\theta/2}$$

Sol.

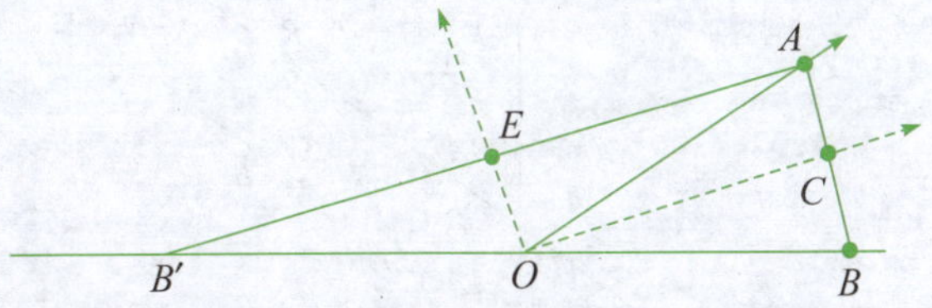

Let OA and OB be two lines with DCs l_1, m_1, n_1 and l_2, m_2, n_2; where O is origin.

Let $OA = OB = 1$.

Then the coordinates of A and B are (l_1, m_1, n_1) and (l_2, m_2, n_2) respectively.

Let OC be the bisector of $\angle AOB$.

Then C is the mid-point of AB and its coordinates are $\left(\frac{l_1+l_2}{2}, \frac{m_1+m_2}{2}, \frac{n_1+n_2}{2}\right)$

$\therefore$ DRs of OC are $\frac{l_1+l_2}{2}, \frac{m_1+m_2}{2}, \frac{n_1+n_2}{2}$

Also, we have $OC = \sqrt{\left(\frac{l_1+l_2}{2}\right)^2+\left(\frac{m_1+m_2}{2}\right)^2+\left(\frac{n_1+n_2}{2}\right)^2}$

$$= \frac{1}{2}\sqrt{(l_1^2+m_1^2+n_1^2)+(l_2^2+m_2^2+n_2^2)+2(l_1l_2+m_1m_2+n_1n_2)}$$

$$= \frac{1}{2}\sqrt{2+2\cos\theta} \qquad [\because \cos\theta = l_1l_2+m_1m_2+n_1n_2]$$

$$= \frac{1}{2}\sqrt{2(1+\cos\theta)} = \frac{1}{2}\sqrt{2.2\cos^2\frac{\theta}{2}} = \cos\left(\frac{\theta}{2}\right)$$

$\therefore$ DCs of $\overline{OC}$ are $\frac{l_1+l_2}{2(OC)}, \frac{m_1+m_2}{2(OC)}, \frac{n_1+n_2}{2(OC)}$

or $\frac{l_1+l_2}{2\cos\theta/2}, \frac{m_1+m_2}{2\cos\theta/2}, \frac{n_1+n_2}{2\cos\theta/2}$

In figure, OE is the external bisector.

The coordinates of E are $\left(\frac{l_1-l_2}{2}, \frac{m_1-m_2}{2}, \frac{n_1-n_2}{2}\right)$.

Therefore, DRs of OE are $\frac{l_1-l_2}{2}, \frac{m_1-m_2}{2}, \frac{n_1-n_2}{2}$.

Also, $OE = \frac{1}{2}\sqrt{2-2\cos\theta} = \frac{1}{2}\sqrt{2\cdot 2\sin^2\frac{\theta}{2}} = \sin\left(\frac{\theta}{2}\right)$

$\therefore$ DCs of $\overline{OE}$ are $\frac{l_1-l_2}{2\sin\theta/2}, \frac{m_1-m_2}{2\sin\theta/2}, \frac{n_1-n_2}{2\sin\theta/2}$

4.1.2 Equation of a Line through a Given Point Having a Given Direction

Vector Form

In a two-dimensional coordinate system, we learned that for a fixed line, one point and its direction(slope), or two points, are sufficient to determine the line.

In a three-dimensional coordinate system, to specify the direction of a line we require one vector to which the line is parallel.

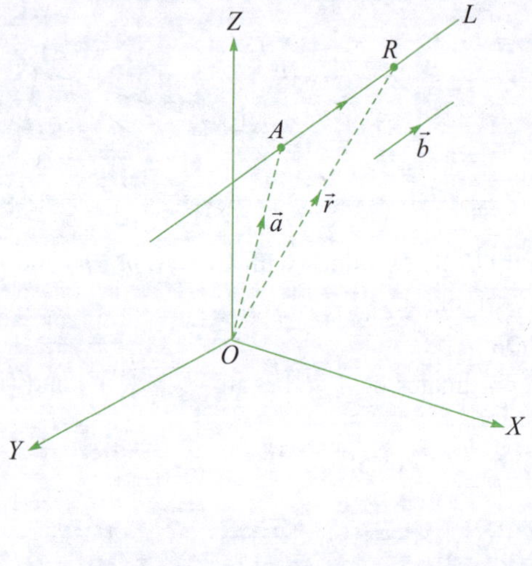

Let's find the equation of a line L that passes through a fixed point $A(\vec{a})$ and is parallel to a given vector $\vec{b}$ as shown in the following figure.

For any variable point $R(\vec{r})$ on the line, vector $\overrightarrow{AR}$ is parallel to the vector $\vec{b}$

$\therefore \quad \overrightarrow{AR} = \lambda\vec{b}$, where λ is a scalar.

$\Rightarrow \quad \overrightarrow{OR} - \overrightarrow{OA} = \lambda\vec{b}$

$\Rightarrow \quad \vec{r} - \vec{a} = \lambda\vec{b}$

$\Rightarrow \quad \vec{r} = \vec{a} + \lambda\vec{b},\ \lambda \in R$ ...(i)

This is the vector equation of a line, where λ is a real parameter. For different values of λ, we obtain different points on the line.

Now, vectors $\vec{r}-\vec{a}$ and $\vec{b}$ are collinear.

So, $(\vec{r}-\vec{a})\times\vec{b} = 0$ ...(ii)

This is the equation of a line in cross product form.

Cartesian Form

Let $\vec{r} = x\hat{i}+y\hat{j}+z\hat{k}$, $\vec{a} = x_1\hat{i}+y_1\hat{j}+z_1\hat{k}$ and $\vec{b} = b_1\hat{i}+b_2\hat{j}+b_3\hat{k}$. Then from equation (1), we have

$$(x\hat{i}+y\hat{j}+z\hat{k})-(x_1\hat{i}+y_1\hat{j}+z_1\hat{k}) = \lambda(b_1\hat{i}+b_2\hat{j}+b_3\hat{k})$$

or $\quad (x-x_1)\hat{i}+(y-y_1)\hat{j}+(z-z_1)\hat{k} = \lambda(b_1\hat{i}+b_2\hat{j}+b_3\hat{k})$

Hence, $\frac{x-x_1}{b_1} = \frac{y-y_1}{b_2} = \frac{z-z_1}{b_3} = \lambda,\ \lambda \in R$. ...(iii)

This is the equation of line in cartesian form and more specifically, in its symmetric form.

The line passes through the point (x_1, y_1, z_1) and has direction ratios b_1, b_2, b_3.

The coordinates of any point on the line are given by $(x_1+\lambda b_1,\ y_1+\lambda b_2,\ z_1+\lambda b_3)$, $\lambda \in R$.

4.1.2.1 Equation of a Line Parallel to One of the Coordinate Axes

Let the line pass through the point $P(x_1, y_1, z_1)$.

1. A line passing through point P and parallel to the x-axis is the line of intersection of the planes

   $y = y_1$ ...(i)

   and $z = z_1$ ...(ii)

   Equations (i) and (ii) together represent the line passing through the point P and parallel to the x-axis.

   The equation of the line in symmetric form can be written as

   $$\frac{x-x_1}{t} = \frac{y-y_1}{0} = \frac{z-z_1}{0} \text{ or } \frac{x-x_1}{1} = \frac{y-y_1}{0} = \frac{z-z_1}{0}$$

   Direction ratios of the line are 1, 0, 0 and the coordinates of any point on the line are $(x_1+\lambda, y_1, z_1)$.

   In particular case equation of x-axis given by $y = 0, z = 0$ or $\frac{x}{1} = \frac{y}{0} = \frac{z}{0}$

2. The equation of the line passing through the point P and parallel to the y-axis is given by $x = x_1$, $z = z_1$ or $\frac{x-x_1}{0} = \frac{y-y_1}{1} = \frac{z-z_1}{0}$.

The equation of the y-axis is given by $x = 0, z = 0$ or $\frac{x}{0} = \frac{y}{1} = \frac{z}{0}$.

3. The equation of the line passing through the point P and parallel to the z-axis is given by $x = x_1$, $y = y_1$ or $\frac{x - x_1}{0} = \frac{y - y_1}{0} = \frac{z - z_1}{1}$.

The equation of the z-axis is given by $x = 0, y = 0$ or $\frac{x}{0} = \frac{y}{0} = \frac{z}{1}$.

4.1.3 Equation of a Line Passing through Two Points

Vector Form

Let the line pass through two fixed points $A(\vec{a})$ and $B(\vec{b})$.

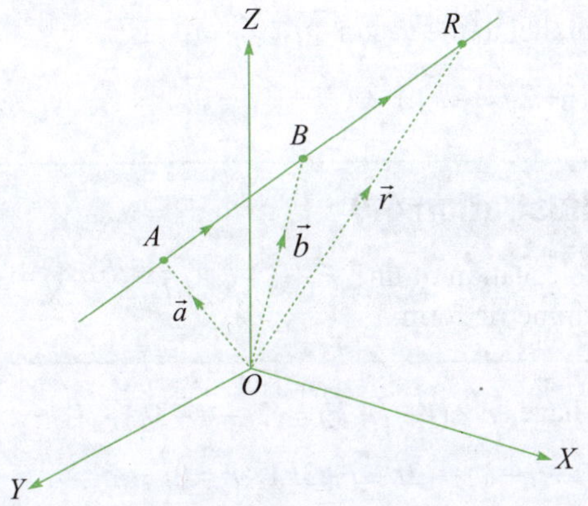

Consider a variable point $R(\vec{r})$ on the line AB.

Since points A, B and R are collinear, vectors $\overrightarrow{AR}$ and $\overrightarrow{AB}$ are collinear.

$\therefore \quad \overrightarrow{AR} = \lambda \overrightarrow{AB}$, where λ is a scalar.

$\Rightarrow \quad \vec{r} - \vec{a} = \lambda(\vec{b} - \vec{a})$

$\Rightarrow \quad \vec{r} = \vec{a} + \lambda(\vec{b} - \vec{a}), \lambda \in R \qquad \ldots\text{(i)}$

Also $\overrightarrow{BR} = \mu \overrightarrow{AB}$

$\Rightarrow \quad \vec{r} - \vec{b} = \mu(\vec{b} - \vec{a})$

$\Rightarrow \quad \vec{r} = \vec{b} + \mu(\vec{b} - \vec{a}), \mu \in R \qquad \ldots\text{(ii)}$

Thus, both equations (i) and (ii) represent the line passing through points $A(\vec{a})$ and $B(\vec{b})$.

Now, from (i), we have

$$\vec{r} = \frac{(1-\lambda)\vec{a} + \lambda\vec{b}}{(1-\lambda)+1} = \frac{\vec{a} + \frac{\lambda}{1-\lambda}\vec{b}}{1 + \frac{\lambda}{1-\lambda}}$$

$\therefore \quad \vec{r} = \frac{\vec{a} + t\vec{b}}{1+t}$, where $t = \frac{\lambda}{1-\lambda}$

This is the position vector of the variable point R on the line.

If $t > 0$ or $\frac{\lambda}{1-\lambda} > 0$ or $0 < \lambda < 1$, point $R(\vec{r})$ lies between the points A and B, and hence divides AB internally.

If $t < 0$ or $\frac{\lambda}{1-\lambda} < 0$ or $\lambda \in (-\infty, 0) \cup (1, \infty)$, point R does not lie between the points A and B and hence divides AB externally.

Cartesian Form

If $\vec{r} = x\hat{i} + y\hat{j} + z\hat{k}$, $\vec{a} = x_1\hat{i} + y_1\hat{j} + z_1\hat{k}$ and $\vec{b} = x_2\hat{i} + y_2\hat{j} + z_2\hat{k}$, then from equation (i),

$$(x\hat{i} + y\hat{j} + z\hat{k}) - (x_1\hat{i} + y_1\hat{j} + z_1\hat{k})$$
$$= \lambda[(x_2\hat{i} + y_2\hat{j} + z_2\hat{k}) - (x_1\hat{i} + y_1\hat{j} + z_1\hat{k})]$$

$\therefore$
$$(x - x_1)\hat{i} + (y - y_1)\hat{j} + (z - z_1)\hat{k}$$
$$= \lambda[(x_2 - x_1)\hat{i} + (y_2 - y_1)\hat{j} + (z_2 - z_1)\hat{k}]$$

$$\Rightarrow \quad \frac{x - x_1}{x_2 - x_1} = \frac{y - y_1}{y_2 - y_1} = \frac{z - z_1}{z_2 - z_1} = \lambda, \lambda \in R \qquad \ldots\text{(iii)}$$

Similarly, from equation (ii) we get

$$\frac{x - x_2}{x_2 - x_1} = \frac{y - y_2}{y_2 - y_1} = \frac{z - z_2}{z_2 - z_1} = \mu, \mu \in R \qquad \ldots\text{(iv)}$$

These are equations of the line in Cartesian form or symmetric form.

Illustration 4.4

A line passes through the point having position vector $2\hat{i} - 3\hat{j} + 4\hat{k}$ and is parallel to the vector $3\hat{i} + 4\hat{j} - 5\hat{k}$. Find the equation of the line in vector form, cartesian form and cross product form.

Sol. The line is passing through the point $2\hat{i} - 3\hat{j} + 4\hat{k}$ and is parallel to the vector $3\hat{i} + 4\hat{j} - 5\hat{k}$. Therefore, its vector equation is:

$$\vec{r} = 2\hat{i} - 3\hat{j} + 4\hat{k} + \lambda(3\hat{i} + 4\hat{j} - 5\hat{k}) \qquad \ldots\text{(i)}$$

Equivalently, line passes through the point (2, –3, 4) and has direction ratios 3, 4, –5.

So, the cartesian form of equation of line is $\frac{x-2}{3} = \frac{y+3}{4} = \frac{z-4}{-5}$.

From (i), $\vec{r} - (2\hat{i} - 3\hat{j} + 4\hat{k}) = \lambda(3\hat{i} + 4\hat{j} - 5\hat{k})$

So, $\quad [\vec{r} - (2\hat{i} - 3\hat{j} + 4\hat{k})] \parallel (3\hat{i} + 4\hat{j} - 5\hat{k})$.

$\therefore \quad [\vec{r} - (2\hat{i} - 3\hat{j} + 4\hat{k})] \times (3\hat{i} + 4\hat{j} - 5\hat{k}) = 0$

This is the equation of line in cross-product form.

Illustration 4.5

The cartesian equation of a line is $6x - 2 = 3y + 1 = 2z - 2$. Then

a. Write the equation of the line in the symmetric form.

b. Find the direction ratios of the line.

c. Find the equation of the line in vector form.

Sol. The equation of line is $6x - 2 = 3y + 1 = 2z - 2$.

or $6\left(x-\frac{1}{3}\right)=3\left(y+\frac{1}{3}\right)=2(z-1)$

or $\frac{x-1/3}{1}=\frac{y+1/3}{2}=\frac{z-1}{3}$.

This is the standard symmetric form.

Line passes through the point (1/3, –1/3, 1) and has direction ratios 1, 2, 3.

The equation of line in vector form is

$$\vec{r}=\frac{1}{3}\hat{i}-\frac{1}{3}\hat{j}+\hat{k}+\lambda(\hat{i}+2\hat{j}+3\hat{k})$$

Illustration 4.6

Find the vector equation of a line passing through the points $A(3, 4, -7)$ and $B(1, -1, 6)$. Also find its cartesian equations.

Sol. Since the line passes through the points $A(3\hat{i}+4\hat{j}-7\hat{k})$ and $B(\hat{i}-\hat{j}+6\hat{k})$, its vector equation is

$$\vec{r}=3\hat{i}+4\hat{j}-7\hat{k}+\lambda[(\hat{i}-\hat{j}+6\hat{k})-(3\hat{i}+4\hat{j}-7\hat{k})]$$

or $\vec{r}=3\hat{i}+4\hat{j}-7\hat{k}-\lambda(2\hat{i}+5\hat{j}-13\hat{k})$, ...(i)

where λ is a parameter.

The Cartesian equivalent of (i) is $\frac{x-3}{2}=\frac{y-4}{5}=\frac{z+7}{-13}$.

Illustration 4.7

Find the Cartesian and vector equations of the line which passes through the point (–2, 4, –5) and is parallel to the line $\frac{x+3}{3}=\frac{y-4}{5}=\frac{z+8}{6}$.

Sol. Equation of given line is $\frac{x+3}{3}=\frac{y-4}{5}=\frac{z+8}{6}$ ...(i)

Direction ratios of the line are 3, 5, 6

Let the required line be L. Since L is parallel to the given line, its direction ratios are 3, 5, 6.

Also, L passes through the point (–2, 4, –5).

Hence, the equation of line L is $\frac{x+2}{3}=\frac{y-4}{5}=\frac{z+5}{6}$

The equation of line L in vector form is

$$\vec{r}=(-2\hat{i}+4\hat{j}-5\hat{k})+\lambda(3\hat{i}+5\hat{j}+6\hat{k})$$

Illustration 4.8

Find the vector equations of the line passing through the point (1, 2, –4) and perpendicular to the lines whose equations are $\frac{x-8}{3}=\frac{y+19}{-16}=\frac{z-10}{7}$ and $\frac{x-15}{3}=\frac{y-29}{8}=\frac{z-5}{-5}$

Sol. Since the line to be determined is perpendicular to the given two straight lines, it is parallel to the

Vector $(3\hat{i}-16\hat{j}+7\hat{k})\times(3\hat{i}+8\hat{j}-5\hat{k})$

$$=\begin{vmatrix}\hat{i} & \hat{j} & \hat{k}\\ 3 & -16 & 7\\ 3 & 8 & -5\end{vmatrix}=24\hat{i}+36\hat{j}+72\hat{k}$$

Therefore, the direction ratios of the line are 24, 36, 72 or 2, 3, 6

Hence, the equation of line passing through the point (1, 2, –4) which is parallel to the vector $2\hat{i}+3\hat{j}+6\hat{k}$ is

$$\vec{r}=(\hat{i}+2\hat{j}-4\hat{k})+\lambda(2\hat{i}+3\hat{j}+6\hat{k})$$

Illustration 4.9

Write the equation of line $\vec{r}\times(\hat{i}+\hat{j}+\hat{k})=(2\hat{i}-\hat{j}-3\hat{k})\times\vec{r}$ in the symmetric form.

Sol. We have, $\vec{r}\times(\hat{i}+\hat{j}+\hat{k})=(2\hat{i}-\hat{j}-3\hat{k})\times\vec{r}$

$\therefore\quad \vec{r}\times(\hat{i}+\hat{j}+\hat{k})-(2\hat{i}-\hat{j}-3\hat{k})\times\vec{r}=0$

$\Rightarrow\quad \vec{r}\times(\hat{i}+\hat{j}+\hat{k})+\vec{r}\times(2\hat{i}-\hat{j}-3\hat{k})=0$

$\Rightarrow\quad \vec{r}\times[(\hat{i}+\hat{j}+\hat{k})+(2\hat{i}-\hat{j}-3\hat{k})]=0$

$\Rightarrow\quad \vec{r}\times(3\hat{i}+0\hat{j}-2\hat{k})=0$

$\Rightarrow\quad \vec{r}=\lambda(3\hat{i}+0\hat{j}-2\hat{k}),\ \lambda\in R$

$\Rightarrow\quad \frac{x}{3}=\frac{y}{0}=\frac{z}{-2}$

Line passes through the origin and has direction ratios 3, 0, –2.

Illustration 4.10

Equation of line is given by $x = 2, y = 3$. Find its equation in symmetric from.

Sol. We know that $x = 2$ is the equation of plane parallel to the yz-plane and $y = 3$ is the equation of plane parallel to the xz-plane.

So, given equations together represent line of intersection of planes $x = 2$ and $y = 3$.

This line passes through the point $(2, 3, \lambda)$, $\lambda\in R$ and is parallel to the z-axis (or perpendicular to the xy-plane).

So, the equation of line in the symmetric from is $\frac{x-2}{0}=\frac{y-3}{0}=\frac{z}{1}$

Illustration 4.11

Find the equation of a diameter of the sphere $x^2 + y^2 + z^2 - 2x + 4y - 6y - 7 = 0$, which is parallel to the x-axis.

Sol. The centre of the sphere is (1, –2, 3).

he diameter of the sphere is the line passing through its centre.
herefore, the equation of the diagonal which is parallel to the

-axis is $\frac{x-1}{0} = \frac{y+2}{0} = \frac{z-3}{1}$.

Illustration 4.12

Which of the following point(s) lie on the line $\frac{x-2}{-1} = \frac{y+1}{3} = \frac{z-4}{5}$?

a. (1, 2, 9) **b.** (3, –4, 1)

Sol. We have line, $\frac{x-2}{-1} = \frac{y+1}{3} = \frac{z-4}{5}$

utting point (1, 2, 9) in the equation, we get

$\frac{-2}{-1} = \frac{2+1}{3} = \frac{9-4}{5}$ or 1 = 1 = 1, which is true

herefore, point (1, 2, 9) lies on the line.

utting point (3, –4, 1) in the equation, we get $\frac{3-2}{-1} = \frac{-4+1}{3} = \frac{1-4}{5}$

r $-1 = -1 = -\frac{3}{5}$, which is not true.

ence, the point (3, –4, 1) does not lie on the line.

Illustration 4.13

Find the points on the line $\frac{x-1}{2} = \frac{y-2}{3} = \frac{z+3}{1}$ which are at distance $\sqrt{18}$ units from the origin?

Sol. Given line is $\frac{x-1}{2} = \frac{y-2}{3} = \frac{z+3}{1} = \lambda$ (say)

he variable point on the line is $(2\lambda + 1, 3\lambda + 2, \lambda - 3)$

he distance of variable point from the origin is $\sqrt{18}$ units.

$\sqrt{(2\lambda+1)^2 + (3\lambda+2)^2 + (\lambda-3)^2} = \sqrt{18}$

$\Rightarrow 14\lambda^2 + 10\lambda - 4 = 0$

$\Rightarrow 7\lambda^2 + 5\lambda - 2 = 0$

$\Rightarrow (7\lambda - 2)(\lambda + 1) = 0$

$\Rightarrow \lambda = \frac{2}{7}, -1$

utting the values of λ, we get two points $\left(\frac{11}{7}, \frac{20}{7}, \frac{-19}{7}\right)$ and
-1, –1, –4).

Illustration 4.14

Show that the line $\frac{x-3}{0} = \frac{y+6}{-11} = \frac{z-7}{5}$ is perpendicular to the x-axis but does not intersect the x-axis.

Sol. We have line $\frac{x-3}{0} = \frac{y+6}{-11} = \frac{z-7}{5} = \lambda$ (say)

The variable point on the line is $(3, -11\lambda - 6, 5\lambda + 7)$.

If the line intersects the x-axis, then we must have, $-11\lambda - 6 = 0$ and $5\lambda + 7 = 0$, which is not possible as we get different values of λ. Therefore, line does not intersect the x-axis.

Now, line is parallel to the vector $0\hat{i} - 11\hat{j} + 4\hat{k}$.

Vector along x-axis is $\hat{i}$.

Clearly $\hat{i} \cdot (0\hat{i} - 11\hat{j} + 4\hat{k}) = 0$

Therefore, line is perpendicular to x-axis.

Illustration 4.15

Find the points where the line $\frac{x-1}{2} = \frac{y+2}{-1} = \frac{z}{1}$ intersects xy, yz and zx planes.

Sol. The line meets the xy-plane where $z = 0$.

Putting $z = 0$ in the given equation of line, we have $\frac{x-1}{2} = \frac{y+2}{-1} = \frac{0}{1}$.

Thus, $x = 1$ and $y = -2$.

Therefore, line meets the xy-plane at (1, – 2, 0).

The line meets the yz-plane, where $x = 0$.

Putting $x = 0$ in the given equation of line, we have $\frac{0-1}{2} = \frac{y+2}{-1} = \frac{z}{1}$

$\Rightarrow \quad z = \frac{-1}{2}$ and $y = -\frac{3}{2}$

Therefore, the line meets the yz-plane at $\left(0, -\frac{3}{2}, \frac{-1}{2}\right)$.

The line meets the zx-plane, where $y = 0$.

Putting $y = 0$ in the given equation of line, we have $\frac{x-1}{2} = \frac{y+2}{-1} = \frac{z}{1}$

$\Rightarrow \quad z = -2, x = -3$

Therefore, the line meets zx-plane at (– 3, 0, –2).

Illustration 4.16

Line L_1 is parallel to the vector $\frac{x-1}{2} = \frac{y+2}{-1} = \frac{z}{1}$ and passes through the point $A(7, 6, 2)$ and line L_2 is parallel to the vector $\frac{x-1}{2} = \frac{y+2}{-1} = \frac{z}{1}$ and passes through the point $B(5, 3, 4)$. If line L_3 parallel to a vector $\vec{r} = 2\hat{i} - 2\hat{j} - \hat{k}$ intersects the lines L_1 and L_2 at points C and D, respectively, then find $\vec{r} = 2\hat{i} - 2\hat{j} - \hat{k}$.

Sol. Equation of line L_1 is $\vec{r} = 7\hat{i} + 6\hat{j} + 2\hat{k} + \lambda(-3\hat{i} + 2\hat{j} + 4\hat{k})$.

Equation of line L_2 is $\vec{r} = 5\hat{i} + 3\hat{j} + 4\hat{k} + \mu(2\hat{i} + \hat{j} + 3\hat{k})$.

Point C on the line L_1 can be taken as $7\hat{i} + 6\hat{j} + 2\hat{k} + \lambda(-3\hat{i} + 2\hat{j} + 4\hat{k})$ and point D on line L_2 can be taken as $5\hat{i} + 3\hat{j} + 4\hat{k} + \mu(2\hat{i} + \hat{j} + 3\hat{k})$.

$\therefore \quad \overrightarrow{CD} = 5\hat{i}+3\hat{j}+4\hat{k}+\mu(2\hat{i}+\hat{j}+3\hat{k})$
$\quad -[7\hat{i}+6\hat{j}+2\hat{k}+\lambda(-3\hat{i}+2\hat{j}+4\hat{k})]$

or $\quad \overrightarrow{CD} = (-2+2\mu+3\lambda)\hat{i}+(-3+\mu-2\lambda)\hat{j}+(2+3\mu-4\lambda)\hat{k}$

Given that $\overrightarrow{CD}$ is parallel to the vector $2\hat{i}-2\hat{j}-\hat{k}$.

$$\therefore \quad \frac{-2+2\mu+3\lambda}{2} = \frac{-3+\mu-2\lambda}{-2} = \frac{2+3\mu-4\lambda}{-1}$$

Solving, we get $\lambda = 2$ and $\mu = 1$.

Therefore, $\overrightarrow{CD} = 6\hat{i}-6\hat{j}-3\hat{k}$

Hence, $|\overrightarrow{CD}| = \sqrt{6^2+(-6)^2+(-3)^2} = \sqrt{81} = 9$.

Illustration 4.17

ABCD be a tetrahedron where, $A \equiv (2, 0, 0)$, $B \equiv (0, 4, 0)$ and $CD = \sqrt{14}$. If the edge *CD* lies on the line $\frac{x-1}{1} = \frac{y-2}{2} = \frac{z-3}{3}$, then find the locus of the centroid of tetrahedron.

Sol. The edge *CD* lies on the line $\frac{x-1}{1} = \frac{y-2}{2} = \frac{z-3}{3}$.

Let $\quad C \equiv (\lambda_1+1, 2\lambda_1+2, 3\lambda_1+3)$

and $\quad D \equiv (\lambda_2+1, 2\lambda_2+2, 3\lambda_2+3)$

Now, $CD = \sqrt{14}$

$\Rightarrow \quad (\lambda_1-\lambda_2)^2+4(\lambda_1-\lambda_2)^2+9(\lambda_1-\lambda_2)^2 = 14$

$\Rightarrow \quad \lambda_2-\lambda_1 = \pm 1$ or $\lambda_2 = \lambda_1 \pm 1$

Let centroid of tetrahedron *ABCD* is $G \equiv (\alpha, \beta, \gamma)$.

Then for $\lambda_2 = \lambda_1 + 1$,

$4\alpha = 5 + 2\lambda_1$

$4\beta = 10 + 4\lambda_1$

$4\gamma = 9 + 6\lambda_1$

Therefore, the locus of *G* is $\frac{4x-5}{2} = \frac{4y-10}{4} = \frac{4z-9}{6}$.

Similarly we find the locus for $\lambda_2 = \lambda_1 - 1$.

Aim to complete within 35 minutes

Concept Application Exercise 4.1

1. The line joining the points (–2, 1, –8) and (*a*, *b*, *c*) is parallel to the line whose direction ratios are 6, 2 and 3. Find the values of *a*, *b* and *c*.
2. Find the equation of a line which passes through the point (2, 3, 4) and which is equally inclined to axes.
3. Find the equation of the line which passes through the point (1, 2, 3) and makes angles $\frac{\pi}{4}, \frac{\pi}{4}, \frac{\pi}{2}$ with *x*, *y* and *z*-axes respectively.
4. If equation of line is $\vec{r} = (3\lambda-2)\hat{i}+(\lambda+4)\hat{j}+(5-7\lambda)\hat{k}$. Then, find the equation of line in vector form, cross product form and symmetric form. Also find the direction ratios of the line.
5. Find the equation of common diameter of the spheres $x^2+y^2+z^2 = 4$ and $x^2+y^2+z^2-2x+4y-6z-3 = 0$.
6. Find the equation of the line passing through the point (2, –1, –1) and parallel to the line $-6x-2 = 3y+1 = 2z-2$.
7. The straight line $\frac{x-3}{3} = \frac{y-2}{1} = \frac{z-1}{0}$ is
   (1) parallel to the *x*-axis
   (2) parallel to the *y*-axis
   (3) parallel to the *z*-axis
   (4) perpendicular to the *z*-axis
8. Find the point where line $\frac{x-2}{1} = \frac{y-4}{2} = \frac{z-5}{3}$ meets the *z*-axis.
9. Find the equation of line which has intercepts 4 units and 3 units on positive *x*-axis and positive *y*-axis respectively.
10. In triangle *ABC* find the equation of median through *A*; where $A \equiv (1, 2, -1)$, $B \equiv (0, 1, -4)$ and $C \equiv (6, 5, 0)$. Also verify that the centroid lies on the this median.
11. The equation of the line through the point (0, 1, 2) and perpendicular to the line $\frac{x-1}{2} = \frac{y+1}{3} = \frac{z-1}{-2}$ is
   (1) $\frac{x}{3} = \frac{y-1}{4} = \frac{z-2}{3}$
   (2) $\frac{x}{3} = \frac{y-1}{-4} = \frac{z-2}{3}$
   (3) $\frac{x}{3} = \frac{y-1}{4} = \frac{z-2}{-3}$
   (4) $\frac{x}{-3} = \frac{y-1}{4} = \frac{z-2}{3}$
12. Find the points where line $\frac{x-3}{2} = \frac{y-3}{1} = \frac{z+2}{-2}$ meets coordinates planes.
13. Consider triangle *ABC* formed by *A*(1, –1, 0), *B*(3, 5, 3) and *C*(–11, –5, 6). Find the equation of internal angle bisector of angle *A*.
14. If the line $L: \frac{x-1}{2} = \frac{y}{1}$; $z = 2$ meets the planes $y = 1$ at *A* and $x = 1$ at *B*, the distance *AB* is
   (1) 2 (2) $\sqrt{5}$
   (3) 3 (4) $\sqrt{6}$
15. The line passing through the point (–1, 2, 3) and perpendicular to the lines $\frac{x}{2} = \frac{y-1}{-3} = \frac{z+2}{-2}$ and $\frac{x+3}{-1} = \frac{y+3}{2} = \frac{z-1}{3}$ does not pass through the point
   (1) (4, 6, 2) (2) (–6, –2, 4)
   (3) (3, –1, 5) (4) (9, 10, 1)
16. The equations of motion of a rocket are $x = 2t$, $y = -4t$ and $z = 4t$, where time *t* is given in seconds, and the coordinates of a moving point in kilometres. What is the path of the rocket? At what distance will be the rocket from the starting point *O*(0, 0, 0) in 10 s?

Answers

1. $a = 6\lambda - 2, b = 2\lambda + 1, c = 3\lambda - 8; \lambda \in R$

2. $\frac{x-2}{1} = \frac{y-3}{1} = \frac{z-4}{1}$ 3. $\frac{x-1}{1} = \frac{y-2}{1} = \frac{z-3}{0}$

4. Vector form: $\vec{r} = -2\hat{i} + 4\hat{j} + 5\hat{k} + \lambda(3\hat{i} + \hat{j} - 7\hat{k})$

Cross-product form:

$(\vec{r} - (-2\hat{i} + 4\hat{j} + 5\hat{k})) \times (3\hat{i} + \hat{j} - 7\hat{k}) = 0$

Symmetric form: $\frac{x+2}{3} = \frac{y-4}{1} = \frac{z-5}{-7}$

5. $\frac{x}{1} = \frac{y}{-2} = \frac{z}{3}$ 6. $\frac{x-2}{-1} = \frac{y+1}{2} = \frac{z+1}{3}$

7. (4) 8. (0, 0, –1)

9. $\frac{x-4}{4} = \frac{y}{-3} = \frac{z}{0}$ 10. $\frac{x-1}{-2} = \frac{y-2}{-1} = \frac{z+1}{1}$

11. (4)

12. *xy*-plane: (1, 2, 0); *yz*-plane: $\left(0, \frac{3}{2}, 1\right)$; *xz*-plane: (–3, 0, 4).

13. $\frac{x-1}{-8} = \frac{y+1}{8} = \frac{z}{9}$

14. (2) 15. (3)

16. Path of the rocket is a straight line passing through origin; 60 km

4.2 ANGLE BETWEEN LINES AND DISTANCE OF A POINT FROM A LINE

4.2.1 Angle Between Lines

The angle between two lines in space is defined as the angle between the vectors to which the lines are parallel.

Let the given lines are $L_1 \equiv \vec{r} = \vec{a} + \lambda\vec{b}$ ...(i)

and $L_2 \equiv \vec{r} = \vec{c} + \mu\vec{d}$ ...(ii)

or $L_1 \equiv \frac{x-a_1}{b_1} = \frac{y-a_2}{b_2} = \frac{z-a_3}{b_3} = \lambda$ ...(iii)

and $L_2 \equiv \frac{x-c_1}{d_1} = \frac{y-c_2}{d_2} = \frac{z-c_3}{d_3} = \mu$ ...(iv)

where $\vec{a} = a_1\hat{i} + a_2\hat{j} + a_3\hat{k}$, $\vec{b} = b_1\hat{i} + b_2\hat{j} + b_3\hat{k}$,

$\vec{c} = c_1\hat{i} + c_2\hat{j} + c_3\hat{k}$, $\vec{d} = d_1\hat{i} + d_2\hat{j} + d_3\hat{k}$

The angle between the lines (i) and (ii) (or (iii) and (iv)) is the angle between the vectors $\vec{b}$ and $\vec{d}$ to which lines are parallel.

The acute angle θ, between the lines is given by,

$$\cos\theta = \frac{|\vec{b}\cdot\vec{d}|}{|\vec{b}||\vec{d}|} = \frac{|b_1d_1 + b_2d_2 + b_3d_3|}{\sqrt{b_1^2 + b_2^2 + b_3^2}\sqrt{d_1^2 + d_2^2 + d_3^2}}$$

- Lines are perpendicular if $\vec{b}\cdot\vec{d} = 0$ or $b_1d_1 + b_2d_2 + b_3d_3 = 0$.
- Lines are parallel if $\vec{b}\times\vec{d} = 0$ or $\vec{b} = \alpha\vec{d}$, where α is a scalar or $\frac{b_1}{d_1} = \frac{b_2}{d_2} = \frac{b_3}{d_3}$

Illustration 4.18

Find the angle between the following pair of lines:

a. $\vec{r} = 2\hat{i} - 5\hat{j} + \hat{k} + \lambda(3\hat{i} + 2\hat{j} + 6\hat{k})$

and $\vec{r} = 7\hat{i} - 6\hat{k} + \mu(\hat{i} + 2\hat{j} + 2\hat{k})$

b. $\frac{x}{2} = \frac{y}{2} = \frac{z}{1}$ and $\frac{x-5}{4} = \frac{y-2}{1} = \frac{z-3}{8}$

Sol.

a. The given lines are parallel to the vectors $\vec{b}_1 = 3\hat{i} + 2\hat{j} + 6\hat{k}$ and $\vec{b}_2 = \hat{i} + 2\hat{j} + 2\hat{k}$. If θ is the angle between the lines, then

$$\cos\theta = \frac{\vec{b}_1 \cdot \vec{b}_2}{|\vec{b}_1||\vec{b}_2|} = \frac{(3)(1)+(2)(2)+(6)(2)}{\sqrt{3^2+2^2+6^2}\sqrt{1^2+2^2+2^2}} = \frac{19}{7\times 3}$$

$$\therefore \quad \theta = \cos^{-1}\left(\frac{19}{21}\right)$$

b. The given lines are parallel to the vectors $\vec{b}_1 = 2\hat{i} + 2\hat{j} + \hat{k}$ and $\vec{b}_2 = 4\hat{i} + \hat{j} + 8\hat{k}$. If θ is the angle between the lines, then

$$\cos\theta = \frac{\vec{b}_1 \cdot \vec{b}_2}{|\vec{b}_1||\vec{b}_2|} = \frac{(2)(4)+(2)(1)+(1)(8)}{\sqrt{2^2+2^2+1^2}\sqrt{4^2+1^2+8^2}}$$

$$= \frac{18}{3\times 9} = \frac{2}{3}$$

$$\therefore \quad \theta = \cos^{-1}\left(\frac{2}{3}\right)$$

Illustration 4.19

Find the acute angle between the lines $\frac{x-1}{l} = \frac{y+1}{m} = \frac{z}{n}$ and $\frac{x+1}{m} = \frac{y-3}{n} = \frac{z-1}{l}$, where $l > m > n$, and l, m, n are roots of the equation $x^3 + x^2 - 4x = 4$.

Sol. Given lines are parallel to the vectors $l\hat{i} + m\hat{j} + n\hat{k}$ and $m\hat{i} + n\hat{j} + l\hat{k}$.

Angle θ between lines is given by

$$\cos\theta = \frac{lm + mn + nl}{l^2 + m^2 + n^2} \quad ...(i)$$

Now l, m, n are roots of the equation $x^3 + x^2 - 4x - 4 = 0$. Thus,

$$l + m + n = -1 \text{ and } lm + mn + nl = -4$$

Now, $(l + m + n)^2 = l^2 + m^2 + n^2 + 2(lm + mn + nl)$

$\Rightarrow \quad (-1)^2 = l^2 + m^2 + n^2 + 2(-4)$

$\Rightarrow \quad l^2 + m^2 + n^2 = 9$

$\therefore \quad \cos\theta = -\dfrac{4}{9}$

Therefore, the acute angle between the lines is $\cos^{-1}\dfrac{4}{9}$.

Illustration 4.20

Find the values of p for which lines $\dfrac{1-x}{3} = \dfrac{7y-14}{2p} = \dfrac{z-3}{2}$ and $\dfrac{7-7x}{3p} = \dfrac{y-5}{1} = \dfrac{6-z}{5}$ are perpendicular.

Sol. The given equations can be written in the standard symmetric form as

$$\frac{x-1}{-3} = \frac{y-2}{2p/7} = \frac{z-3}{2} \text{ and } \frac{x-1}{-3p/7} = \frac{y-5}{1} = \frac{z-6}{-5}$$

The direction ratios of the lines are $-3, \dfrac{2p}{7}, 2$ and $\dfrac{-3p}{7}, 1, -5$, respectively.

Now, lines are perpendicular. Therefore,

$$(-3)\cdot\left(\frac{-3p}{7}\right) + \left(\frac{2p}{7}\right)\cdot(1) + 2\cdot(-5) = 0$$

$$\Rightarrow \quad \frac{9p}{7} + \frac{2p}{7} = 10 \Rightarrow 11p = 70 \Rightarrow p = \frac{70}{11}$$

Illustration 4.21

Find the angle between the lines whose direction cosines satisfy

$$\ell + m + n = 0 \text{ and } \ell^2 + m^2 - 3n^2 = 0$$

Sol. We have, $\ell + m + n = 0$

or $\quad -(\ell + n) = m$

Putting the value of m in $\ell^2 + m^2 - 3n^2 = 0$, we get

$$\ell^2 - 3n^2 + (\ell + n)^2 = 0$$

or $\quad \ell^2 + \ell n - n^2 = 0$

$$\Rightarrow \quad \ell = \left(\frac{-1 \pm \sqrt{5}}{2}\right)n$$

If $\quad \ell = \left(\dfrac{-1-\sqrt{5}}{2}\right)n$, then

$$\ell : m : n = \ell : (\ell + m) : n$$

$$= \frac{-1-\sqrt{5}}{2} : \frac{\sqrt{5}-1}{2} : 1$$

If $\ell = \left(\dfrac{-1+\sqrt{5}}{2}\right)n$, then

$$\ell : m : n = \ell : (\ell + m) : n$$

$$= \frac{-1+\sqrt{5}}{2} : -\frac{1+\sqrt{5}}{2} : 1$$

Thus, the angle between lines is given by

$$\cos\theta = \frac{\left(\frac{-1-\sqrt{5}}{2}\right)\left(\frac{-1+\sqrt{5}}{2}\right) + \left(\frac{-1+\sqrt{5}}{2}\right)\left(-\frac{1+\sqrt{5}}{2}\right) + (1)(1)}{\sqrt{\left(\frac{-1+\sqrt{5}}{2}\right)^2 + \left(\frac{1+\sqrt{5}}{2}\right)^2 + 1}\sqrt{\left(\frac{-1-\sqrt{5}}{2}\right)^2 + \left(\frac{1-\sqrt{5}}{2}\right)^2 + 1}}$$

$$= \frac{\frac{1-5}{4} + \frac{1-5}{4} + 1}{2 \times 2} = \frac{-1}{4}$$

$$\therefore \quad \theta = \cos^{-1}\left(-\frac{1}{4}\right) = \pi - \cos^{-1}\frac{1}{4}.$$

Therefore, the acute angle between the lines is $\cos^{-1}\dfrac{1}{4}$.

Illustration 4.22

If two straight lines whose direction cosines are given by the relations $\ell + m - n = 0$, $3\ell^2 + m^2 + cn\ell = 0$ are parallel, then find the positive value(s) of c.

Sol. We have, $\ell + m - n = 0$ and $3\ell^2 + m^2 + cn\ell = 0$

Eliminating n, we get $3\ell^2 + m^2 + c\ell(\ell + m) = 0$

or $\quad (3 + c)\ell^2 + c\ell m + m^2 = 0$

$$\Rightarrow \quad (3+c)\left(\frac{\ell}{m}\right)^2 + c\left(\frac{\ell}{m}\right) + 1 = 0$$

Since lines are parallel, roots of the above equation must be equal.

$\Rightarrow \quad$ Discriminant $= 0$

$\Rightarrow \quad c^2 - 4(3 + c) = 0$

$\Rightarrow \quad c^2 - 4c - 12 = 0$

$\Rightarrow \quad (c - 6)(c + 2) = 0$

$\Rightarrow \quad c = -2, 6$

Since c is positive, we take $c = 6$.

Illustration 4.23

If $\vec{r} = (\hat{i} + 2\hat{j} + 3\hat{k}) + \lambda(\hat{i} - \hat{j} + \hat{k})$

and $\vec{r} = (\hat{i} + 2\hat{j} + 3\hat{k}) + \mu(\hat{i} + \hat{j} - \hat{k})$

are two lines, then find the equation of the bisector of the acute angle between them.

Sol. Given lines are $\vec{r} = (\hat{i} + 2\hat{j} + 3\hat{k}) + \lambda(\hat{i} - \hat{j} + \hat{k})$

and $\quad \vec{r} = (\hat{i} + 2\hat{j} + 3\hat{k}) + \mu(\hat{i} + \hat{j} - \hat{k})$

Both the lines are passing through the point (1, 2, 3).

The angle bisector also passes through the this point.

Lines are along the vector $(\hat{i}-\hat{j}+\hat{k})$ and $(\hat{i}+\hat{j}-\hat{k})$.

The angle between these vectors is

$$\cos^{-1}\frac{(1)\times(1)+(-1)(1)+(1)(-1)}{\sqrt{3}\,\sqrt{3}}=\cos^{-1}\left(-\frac{1}{3}\right), \text{ which is obtuse.}$$

Therefore, the acute angle bisector is parallel to the vector

$$\frac{\hat{i}-\hat{j}+\hat{k}}{\sqrt{3}}-\frac{\hat{i}+\hat{j}-\hat{k}}{\sqrt{3}}=-\frac{2\hat{j}}{\sqrt{3}}+\frac{2\hat{k}}{\sqrt{3}}$$

Therefore, the equation of the acute angle bisector is $\vec{r}=(\hat{i}+2\hat{j}+3\hat{k})+t(\hat{j}-\hat{k})$, $t\in R$.

4.2.2 Distance of a Point from a Line

Vector Form

Let the given line is $L\equiv\vec{r}=\vec{a}+\lambda\vec{b}$, $\lambda\in R$ ...(i)

Let us find the position vector of the foot of perpendicular M from the point $P(\vec{p})$ on this line.

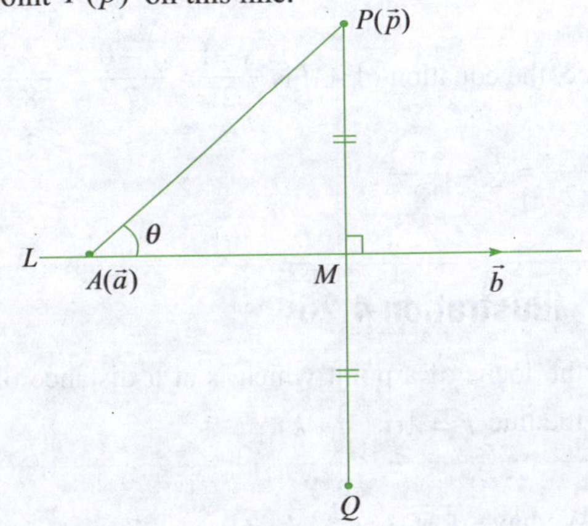

In equation (i), $\vec{r}$ is the position vector of any point on the line L.

Therefore, the position vector of point M can be taken as $\vec{a}+\lambda\vec{b}$.

Now, there will be a unique value of the parameter λ for point M.

To get this value of λ, we use the fact that PM is perpendicular to the line L.

Now, $\overrightarrow{PM}=\vec{a}+\lambda\vec{b}-\vec{p}$

Since, $\overrightarrow{PM}\perp\vec{b}$, we have $(\vec{a}+\lambda\vec{b}-\vec{p})\cdot\vec{b}=0$

$$\therefore\quad \lambda=\frac{\vec{b}\cdot(\vec{p}-\vec{a})}{|\vec{b}|^2}$$

Putting this value of λ in equation (i), we get the position vector of point M as $\vec{a}+\left(\frac{\vec{b}\cdot(\vec{p}-\vec{a})}{|\vec{b}|^2}\right)\vec{b}$.

Now, the distance of the point P from the line, is

$$|\overrightarrow{PM}|=|\vec{a}+\lambda\vec{b}-\vec{p}|$$

$$=\left|\vec{a}+\left(\frac{\vec{b}\cdot(\vec{p}-\vec{a})}{|\vec{b}|^2}\right)\vec{b}-\vec{p}\right| \quad \text{...(ii)}$$

Alternatively, in triangle AMP,

$$|\overrightarrow{PM}|=|\overrightarrow{AP}|\times\sin\theta$$

$$=|\overrightarrow{AP}|\times\frac{|\overrightarrow{AP}\times\vec{b}|}{|\overrightarrow{AP}||\vec{b}|}$$

$$=\frac{|(\vec{p}-\vec{a})\times\vec{b}|}{|\vec{b}|} \quad \text{...(iii)}$$

Also, from (ii), we have $\left|\vec{a}+\left(\frac{\vec{b}\cdot(\vec{p}-\vec{a})}{|\vec{b}|^2}\right)\vec{b}-\vec{p}\right|$

$$=\left|\frac{(\vec{b}\cdot(\vec{p}-\vec{a}))\vec{b}-|\vec{b}|^2(\vec{p}-\vec{a})}{|\vec{b}|^2}\right|$$

$$=\left|\frac{\vec{b}\times(\vec{b}\times(\vec{p}-\vec{a}))}{|\vec{b}|^2}\right|$$

$$=\left|\frac{|\vec{b}|\times|\vec{b}\times(\vec{p}-\vec{a})|\sin 90^\circ}{|\vec{b}|^2}\right| \quad (\because (\vec{b}\times(\vec{p}-\vec{a}))\perp\vec{b})$$

$$=\left|\frac{\vec{b}\times(\vec{p}-\vec{a})}{|\vec{b}|}\right|$$

$$=\frac{|(\vec{p}-\vec{a})\times\vec{b}|}{|\vec{b}|}$$

Cartesian Form

Let $\vec{r}=x\hat{i}+y\hat{j}+z\hat{k}$, $\vec{a}=x_1\hat{i}+y_1\hat{j}+z_1\hat{k}$ and $\vec{b}=b_1\hat{i}+b_2\hat{j}+b_3\hat{k}$.

Hence, the equation of the line L in cartesian form is $\frac{x-x_1}{b_1}=\frac{y-y_1}{b_2}=\frac{z-z_1}{b_3}=\lambda$.

Let $P(x_2, y_2, z_2)$.

The Coordinates of point M can be taken as $(x_1+\lambda b_1, y_1+\lambda b_2, z_1+\lambda b_3)$.

Since PM is perpendicular to the line, we have

$$b_1(x_1+\lambda b_1-x_2)+b_2(y_1+\lambda b_2-y_2)+b_3(z_1+\lambda b_3-z_2)=0$$

$$\Rightarrow\quad \lambda=\frac{b_1(x_2-x_1)+b_2(y_2-y_1)+b_3(z_2-z_1)}{b_1^{\,2}+b_2^{\,2}+b_3^{\,2}}$$

Putting this value of λ in $(x_1+\lambda b_1, y_1+\lambda b_2, z_1+\lambda b_3)$, we get the coordinates of the foot of perpendicular M.

Now, using the distance formula, we can find the distance PM.

Also, if the image (reflection) of the point P in the line is Q, then M is the midpoint of PQ.

Hence, point Q can be obtained using the section formula.

Illustration 4.24

If the foot of the perpendicular from a point $P(1, 0, 2)$ on the line $\frac{x+1}{3}=\frac{y-2}{-2}=\frac{z+1}{-1}$ is M, then find the coordinates of M. Find the image of point P in line. Also find the distance PM and the equation of PM.

Sol. Given line L is $\frac{x+1}{3}=\frac{y-2}{-2}=\frac{z+1}{-1}=\lambda$ ...(i)

Given point is $P(1, 0, 2)$

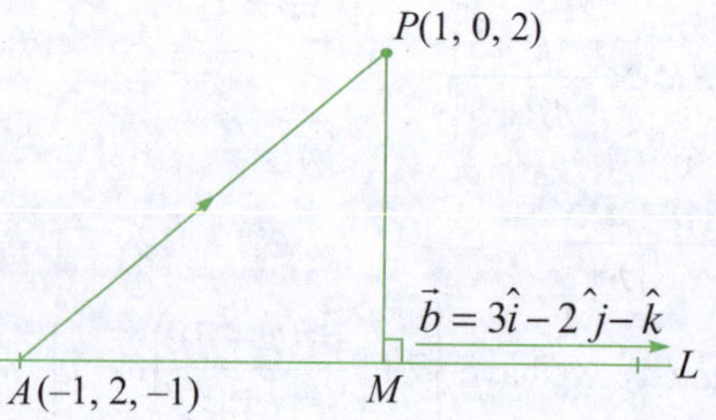

Let the foot of the perpendicular from the point P on the line L be M.

$\therefore \quad M \equiv (3\lambda - 1, -2\lambda + 2, -\lambda - 1)$

Therefore, direction ratios of PM are $3\lambda - 2, -2\lambda + 2, -\lambda - 3$.

Direction ratios of line L are $3, -2, -1$

Since $PM \perp L$, we have

$$3(3\lambda - 2) - 2(-2\lambda + 2) - 1(-\lambda - 3) = 0$$

$$\Rightarrow \quad 14\lambda = 7 \Rightarrow \lambda = \frac{1}{2} \quad \therefore M \equiv \left(\frac{1}{2}, 1, -\frac{3}{2}\right)$$

$$\Rightarrow \quad PM = \sqrt{\left(1 - \frac{1}{2}\right)^2 + (0-1)^2 + \left(2 + \frac{3}{2}\right)^2} = \frac{3\sqrt{3}}{\sqrt{2}}$$

Alternatively, $PM = |\overrightarrow{AP}| \sin\theta$

$$= \frac{|\overrightarrow{AP}||\vec{b}|\sin\theta}{|\vec{b}|}$$

$$= \frac{|\overrightarrow{AP} \times \vec{b}|}{|\vec{b}|}$$

$$= \frac{|8\hat{i} + 11\hat{j} + 2\hat{k}|}{\sqrt{14}} = \frac{\sqrt{189}}{\sqrt{14}} = \frac{3\sqrt{3}}{\sqrt{2}}$$

Also, the equation of line PM is $\dfrac{x-1}{1} = \dfrac{y}{-2} = \dfrac{z-2}{7}$

If Q is the image of P in the line, then M will be the mid point of PQ. Hence, using the section formula we get

$$Q \equiv (0, 2, -5)$$

Illustration 4.25

Find the co-ordinates of the foot of the perpendicular drawn from the point $A(1, 0, 3)$ on the line joining points $B(4, 7, 1)$ and $C(3, 5, 3)$. Also find the equation of the altitude of triangle ABC from the vertex A.

Sol. Let point M be the foot of perpendicular from vertex A on the side BC, which divides BC in the ratio $\lambda : 1$.

Then the co-ordinates of M are $\left(\dfrac{3\lambda+4}{\lambda+1}, \dfrac{5\lambda+7}{\lambda+1}, \dfrac{3\lambda+1}{\lambda+1}\right)$

$$\therefore \quad \overrightarrow{AM} = \left(\frac{3\lambda+4}{\lambda+1} - 1\right)\hat{i} + \left(\frac{5\lambda+7}{\lambda+1}\right)\hat{j} + \left(\frac{3\lambda+1}{\lambda+1} - 3\right)\hat{k}$$

$$= \left(\frac{2\lambda+3}{\lambda+1}\right)\hat{i} + \left(\frac{5\lambda+7}{\lambda+1}\right)\hat{j} + \left(\frac{-2}{\lambda+1}\right)\hat{k}$$

Now, $\overrightarrow{BC} = -\hat{i} - 2\hat{j} + 2\hat{k}$.

Since $\overrightarrow{AM} \perp \overrightarrow{BC}$, we have $\overrightarrow{AM} \cdot \overrightarrow{BC} = 0$

$$\Rightarrow \quad (-1)(2\lambda + 3) + (-2)(5\lambda + 7) + (2)(-2) = 0$$

$$\Rightarrow \quad \lambda = -\frac{7}{4}$$

Hence, point $M \equiv \left(\dfrac{5}{3}, \dfrac{7}{3}, \dfrac{17}{3}\right)$

Now in triangle ABC, AM is the altitude.

Direction ratios of AM are $\dfrac{5}{3} - 1, \dfrac{7}{3} - 0, \dfrac{17}{3} - 3$ or $\dfrac{2}{3}, \dfrac{7}{3}, \dfrac{8}{3}$.

Hence, the equation of AM is $\dfrac{x-1}{\frac{2}{3}} = \dfrac{y-0}{\frac{7}{3}} = \dfrac{z-3}{\frac{8}{3}}$

or $\dfrac{x-1}{2} = \dfrac{y}{7} = \dfrac{z-3}{8}$.

Illustration 4.26

Find the locus of a point which is at a distance of 3 units from the line $\vec{r} = \lambda(\hat{i} + \hat{j} + \hat{k})$.

Sol. We have $\overrightarrow{PM} \cdot (\hat{i} + \hat{j} + \hat{k}) = 0$

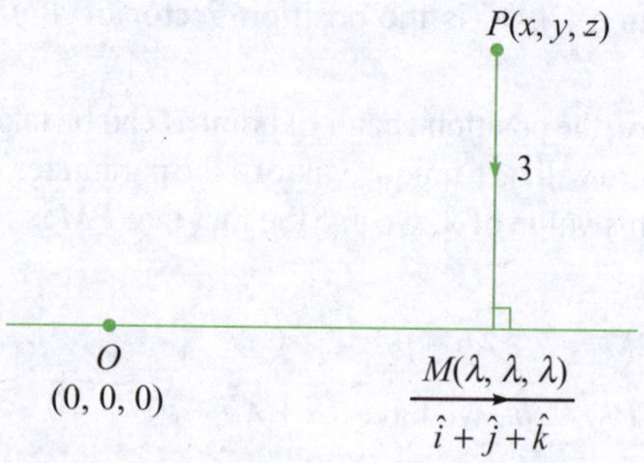

$$\Rightarrow \quad x - \lambda + y - \lambda + z - \lambda = 0$$

$$\Rightarrow \quad \lambda = \frac{x+y+z}{3}$$

Now, $PM^2 = 9$

$$\Rightarrow \quad (x - \lambda)^2 + (y - \lambda)^2 + (z - \lambda)^2 = 9$$

Putting $\lambda = \dfrac{x+y+z}{3}$, we get

$x^2 + y^2 + z^2 - xy - yz - zx = 27/2$, which is the required locus.

Aim to complete within 15 minutes

Concept Application Exercise 4.2

1. The angle between the lines $2x = 3y = -z$ and $6x = -y = -4z$ is
   (1) 30° (2) 45° (3) 60° (4) 90°

2. If the lines $\frac{x-1}{-3} = \frac{y-2}{2k} = \frac{z-3}{2}$ and $\frac{x-1}{3k} = \frac{y-5}{1} = \frac{z-6}{-5}$ are at right angle, then the value of k is
   (1) $\frac{-10}{7}$ (2) $\frac{-5}{7}$ (3) $\frac{6}{5}$ (4) $\frac{3}{4}$

3. Find the angle between the lines whose direction cosines satisfy $\ell + m - 2n = 0$ and $\ell m + mn + 3n\ell = 0$.

4. The vertices B and C of a ΔABC lie on the line, $\frac{x+2}{3} = \frac{y-1}{0} = \frac{z}{4}$ such that $BC = 5$ units. Then the area (in sq. units) of this triangle, given that the point $A(1, -1, 2)$, is
   (1) $5\sqrt{17}$ (2) 6
   (3) $\sqrt{34}$ (4) $2\sqrt{34}$

5. Find the equation of the smallest sphere which has centre at the origin and touches the line $2(x+1) = 2 - y = z + 3$.

6. If the image of point (1, 2, 3) in the line $\frac{x-6}{3} = \frac{y-7}{2} = \frac{z-7}{-2}$ is (α, β, γ) then $\alpha + \gamma - \beta$ is equal to
   (1) 10 (2) 11 (3) 12 (4) 15

Answers

1. (4) 2. (1)
3. $\theta = \cos^{-1}\left(\frac{-7}{\sqrt{193}}\right)$ 4. (3)
5. $9(x^2 + y^2 + z^2) = 5$ 6. (3)

4.3 COPLANAR AND NON-COPLANAR LINES

Two distinct lines drawn in a plane are either parallel or intersecting. However, two lines drawn in space may be parallel, intersecting, or non-coplanar (that is, neither parallel nor intersecting).

It should be noted that parallel or intersecting lines are always coplanar.

Consider two lines

$$L_1 \equiv \vec{r} = \vec{a} + \lambda\vec{b},\ \lambda \in R \quad \ldots\text{(i)}$$

and $$L_2 \equiv \vec{r} = \vec{c} + \mu\vec{d},\ \mu \in R \quad \ldots\text{(ii)}$$

Let $\vec{a} = a_1\hat{i} + a_2\hat{j} + a_3\hat{k}$, $\vec{b} = b_1\hat{i} + b_2\hat{j} + b_3\hat{k}$,

$\vec{c} = c_1\hat{i} + c_2\hat{j} + c_3\hat{k}$, $\vec{d} = d_1\hat{i} + d_2\hat{j} + d_3\hat{k}$

Then, the equations of two lines in Cartesian form are:

$$L_1 \equiv \frac{x-a_1}{b_1} = \frac{y-a_2}{b_2} = \frac{z-a_3}{b_3} \quad \ldots\text{(iii)}$$

and $$L_2 \equiv \frac{x-c_1}{d_1} = \frac{y-c_2}{d_2} = \frac{z-c_3}{d_3} \quad \ldots\text{(iv)}$$

Now, we shall study the conditions and properties of coplanar and non-coplanar lines.

4.3.1 Coplanar Lines

If the lines are parallel, then the vectors $\vec{b}$ are $\vec{d}$ are parallel.

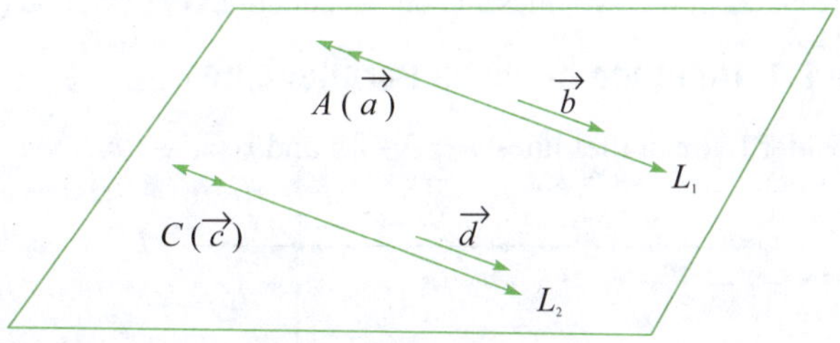

If the lines are intersecting, then the vectors $\vec{c} - \vec{a}, \vec{b}$ and $\vec{d}$ are coplanar.

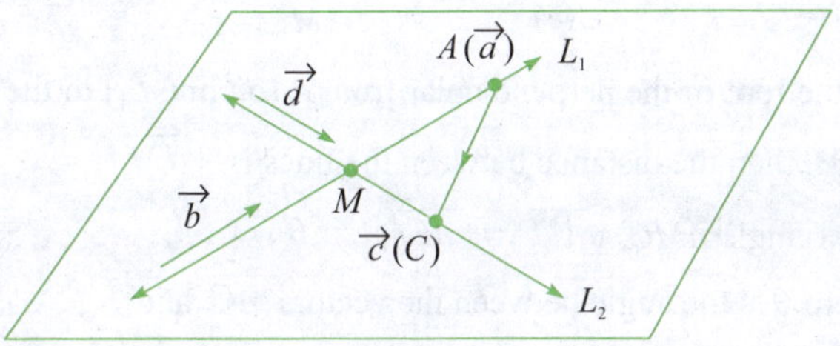

Thus, in either case, $\begin{vmatrix} c_1 - a_1 & c_2 - a_2 & c_3 - a_3 \\ b_1 & b_2 & b_3 \\ d_1 & d_2 & d_3 \end{vmatrix} = 0$.

This is the necessary and sufficient condition for two lines to be coplanar.

Alternatively, the position vector of any point on the line L_1 is $\vec{a} + \lambda\vec{b}$ and that on line L_2 is $\vec{c} + \mu\vec{d}$.

At the point of intersection, we have

$$\vec{a} + \lambda\vec{b} = \vec{c} + \mu\vec{d}$$

or $$(a_1 + \lambda b_1)\hat{i} + (a_2 + \lambda b_2)\hat{j} + (a_3 + \lambda b_3)\hat{k} = (c_1 + \mu d_1)\hat{i} + (c_2 + \mu d_2)\hat{j} + (c_3 + \mu d_3)\hat{k}$$

Thus, we get

$$a_1 + \lambda b_1 = c_1 + \mu d_1 \quad \ldots\text{(v)}$$

$$a_2 + \lambda b_2 = c_2 + \mu d_2 \quad \ldots\text{(vi)}$$

$$a_3 + \lambda b_3 = c_3 + \mu d_3 \quad \ldots\text{(vii)}$$

Since the lines intersect, the values of λ and μ obtained from any two of these equations will satisfy the third.

Therefore, the system of equations (v), (vi) and (vii) is consistent.

Hence, $\begin{vmatrix} c_1 - a_1 & c_2 - a_2 & c_3 - a_3 \\ b_1 & b_2 & b_3 \\ d_1 & d_2 & d_3 \end{vmatrix} = 0$

(**Note:** If the lines do not intersect, the system of equation will be inconsistent.)

If solution of the above system of equations is (λ_1, μ_1), then position vector of point of intersection M will be $\vec{a}+\lambda_1\vec{b}$ or $\vec{c}+\mu_1\vec{d}$.

In cartesian form, the coordinates of any point on line L_1 are $(a_1+\lambda b_1, a_2+\lambda b_2, a_3+\lambda b_3)$ and those on L_2 are $(c_1+\mu d_1, c_2+\mu d_2, c_3+\mu d_3)$.

At the point of intersection, we equate the corresponding coordinates of the two lines and obtain equations (v), (vi) and (vii).

4.3.1.1 Distance between Parallel Lines

Consider two parallel lines $\vec{r}=\vec{a}+\lambda\vec{b}$ and $\vec{r}=\vec{c}+\mu\vec{b}$.

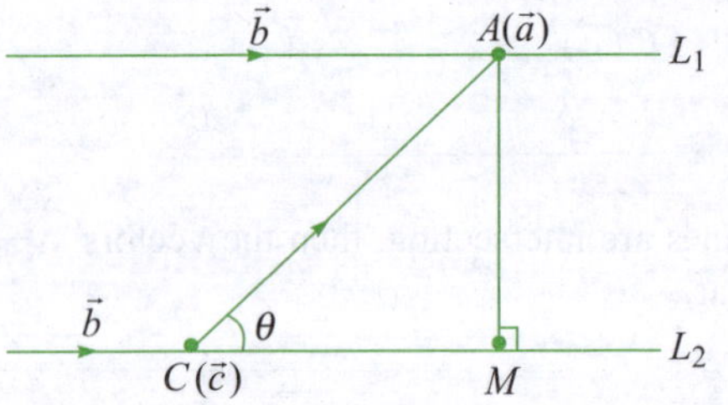

If the foot of the perpendicular from A (on line L_1) to the line L_2 is M, then the distance between the lines is $|\overrightarrow{AM}|$.

In triangle AMC, $|\overrightarrow{AM}|=|\overrightarrow{AC}|\times\sin\theta$ ...(viii)

Here θ is the angle between the vectors $\overrightarrow{AC}$ and $\vec{b}$.

$$\therefore \quad \sin\theta=\frac{|\overrightarrow{AC}\times\vec{b}|}{|\overrightarrow{AC}||\vec{b}|}$$

So, from (viii),

$$|\overrightarrow{AM}|=|\overrightarrow{AC}|\times\sin\theta$$

$$=|\overrightarrow{AC}|\times\frac{|\overrightarrow{AC}\times\vec{b}|}{|\overrightarrow{AC}||\vec{b}|}$$

$$=\frac{|(\vec{c}-\vec{a})\times\vec{b}|}{|\vec{b}|}$$

This is the shortest distance between parallel lines L_1 and L_2.

Illustration 4.27

Determine whether each of the following pairs of lines intersect. If they do find the point of intersection.

a. $\vec{r}=\hat{i}-\hat{j}+\lambda(2\hat{i}+\hat{k}),\ \vec{r}=2\hat{i}-\hat{j}+\mu(\hat{i}+\hat{j}-\hat{k})$

b. $\vec{r}=\hat{i}+\hat{j}-\hat{k}+\lambda(3\hat{i}-\hat{j}),\ \vec{r}=4\hat{i}-\hat{k}+\mu(2\hat{i}+3\hat{k})$

Sol.

a. We have lines

$$\vec{r}=\hat{i}-\hat{j}+\lambda(2\hat{i}+\hat{k}) \quad ...(i)$$

$$\vec{r}=2\hat{i}-\hat{j}+\mu(\hat{i}+\hat{j}-\hat{k}) \quad ...(ii)$$

If lines intersect then for the common point on the lines, we have

$$\hat{i}-\hat{j}+\lambda(2\hat{i}+\hat{k})=2\hat{i}-\hat{j}+\mu(\hat{i}+\hat{j}-\hat{k})$$

$$\Rightarrow \quad -\hat{i}+\lambda(2\hat{i}+\hat{k})=\mu(\hat{i}+\hat{j}-\hat{k})$$

$$\Rightarrow \quad -1+2\lambda=\mu,\ 0=\mu \text{ and } \lambda=-\mu$$

Putting $\mu=0$ in $-1+2\lambda=\mu$ and $\lambda=-\mu$, we get $\lambda=1/2$ and $\lambda=0$.

This is not possible as we cannot have different values of λ for the same point.

So, lines do not intersect.

b. We have lines

$$\vec{r}=\hat{i}+\hat{j}-\hat{k}+\lambda(3\hat{i}-\hat{j}) \quad ...(i)$$

and $\vec{r}=4\hat{i}-\hat{k}+\mu(2\hat{i}+3\hat{k})$ ...(ii)

If lines intersect then for the common point on the lines, we have

$$\hat{i}+\hat{j}-\hat{k}+\lambda(3\hat{i}-\hat{j})=4\hat{i}-\hat{k}+\mu(2\hat{i}+3\hat{k})$$

$$\Rightarrow \quad -3\hat{i}+\hat{j}+\lambda(3\hat{i}-\hat{j})=\mu(2\hat{i}+3\hat{k})$$

$$\Rightarrow \quad -3+3\lambda=2\mu,\ 1-\lambda=0 \text{ and } 0=3\mu$$

$$\Rightarrow \quad \mu=0 \text{ and } \lambda=1$$

These values satisfy equation $-3+3\lambda=2\mu$.

So, lines intersect.

Putting $\mu=0$ in (ii), we get $\vec{r}=4\hat{i}-\hat{k}$, which is the point of intersection.

Alternatively

The equations of the lines are

$$\frac{x-1}{3}=\frac{y-1}{-1}=\frac{z+1}{0} \text{ and } \frac{x-4}{2}=\frac{y}{0}=\frac{z+1}{3}$$

Any point on line (i) can be written as $(3\lambda+1, -\lambda+1, -1)$. Substituting this point into the equation of the second line, we get

$$\frac{3\lambda-3}{2}=\frac{-\lambda+1}{0}=\frac{-1+1}{3}$$

From the comparison of the first two ratios, we get $0=-\lambda+1$ or $\lambda=1$.

From the comparison of last two ratios, we get $3(3\lambda-3)=0$ or $\lambda=1$.

Thus, we obtain the same value of λ from two comparisons. Therefore, the lines intersect.

(**Note:** If we do not obtain the same value of λ from the two comparisons, the lines do not intersect).

Also, for $\lambda=1$, we get the point of intersection as (4, 0, –1).

Illustration 4.28

Find the equation of a line which passes through the point (1, 1, 1) and intersects the lines $\frac{x-1}{2}=\frac{y-2}{3}=\frac{z-3}{4}$ and $\frac{x+2}{1}=\frac{y-3}{2}=\frac{z+1}{4}$.

Sol. Equation of line passing through the point (1, 1, 1) can be taken as $\frac{x-1}{a}=\frac{y-1}{b}=\frac{z-1}{c}$. ...(i)

This line intersects the line $\frac{x-1}{2}=\frac{y-2}{3}=\frac{z-3}{4}$,

if $a:b:c \neq 2:3:4$ and $\begin{vmatrix} 1-1 & 2-1 & 3-1 \\ a & b & c \\ 2 & 3 & 4 \end{vmatrix}=0$

$\Rightarrow \quad a-2b+c=0 \qquad \ldots$(ii)

Also, line (i) intersects the line $\frac{x-(-2)}{1}=\frac{y-3}{2}=\frac{z-(-1)}{4}$,

if $a:b:c \neq 1:2:4$ and $\begin{vmatrix} -2-1 & 3-1 & -1-1 \\ a & b & c \\ 1 & 2 & 4 \end{vmatrix}=0$

$\Rightarrow \quad 6a+5b-4c=0 \qquad \ldots$(iii)

From (ii) and (iii) by cross multiplication, we have

$$\frac{a}{8-5}=\frac{b}{6+4}=\frac{c}{5+12}$$

or $\frac{a}{3}=\frac{b}{10}=\frac{c}{17}$

So, the required equation of the line is $\frac{x-1}{3}=\frac{y-1}{10}=\frac{z-1}{17}$.

Illustration 4.29

Find the shortest distance between the lines

$\vec{r}=(\hat{i}+2\hat{j}+\hat{k})+\lambda(2\hat{i}+\hat{j}+2\hat{k})$ and

$\vec{r}=2\hat{i}-\hat{j}-\hat{k}+\mu(2\hat{i}+\hat{j}+2\hat{k})$.

Sol. We have lines

$$\vec{r}=\underbrace{(\hat{i}+2\hat{j}+\hat{k})}_{a}+\lambda\underbrace{(2\hat{i}+\hat{j}+2\hat{k})}_{b} \qquad \ldots\text{(i)}$$

and $$\vec{r}=\underbrace{(2\hat{i}-\hat{j}-\hat{k})}_{c}+\lambda\underbrace{(2\hat{i}+\hat{j}+2\hat{k})}_{b} \qquad \ldots\text{(ii)}$$

Both the lines are parallel to the vector $2\hat{i}+\hat{j}+2\hat{k}$.

Therefore, the distance between lines is $\frac{|(\vec{c}-\vec{a})\times\vec{b}|}{|\vec{b}|}$

$$=\frac{\left\|\begin{matrix} \hat{i} & \hat{j} & \hat{k} \\ 1 & -3 & -2 \\ 2 & 1 & 2 \end{matrix}\right\|}{\sqrt{2^2+1^2+2^2}}$$

$$=\frac{|-4\hat{i}-6\hat{j}+7\hat{k}|}{3}=\frac{\sqrt{16+36+49}}{3}=\frac{\sqrt{101}}{3}$$

4.3.2 Non-Coplanar Lines

Non-coplanar lines are non-parallel and non-intersecting. Such lines are called skew lines.

Skew lines L_1 and L_2 are as shown in the following figure.

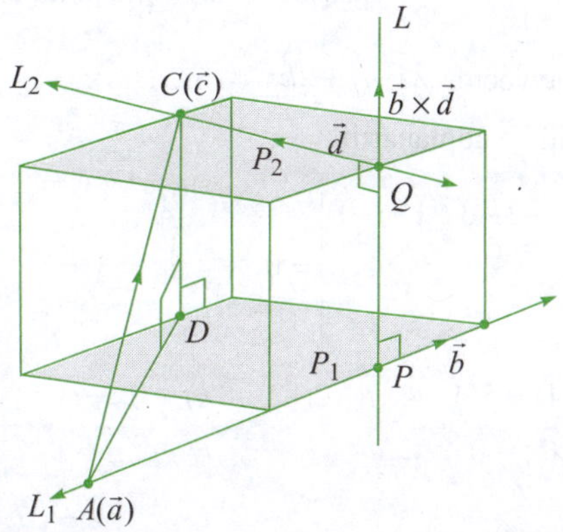

Line L_1 lies in the plane P_1, the lower face of the cuboid and line L_2 lies in the plane P_2, the upper face of the cuboid.

Thus, lines L_1 and L_2 are non-coplanar lines lying in two parallel planes.

It is always possible to draw parallel planes through non-coplanar lines.

Now, we can observe that the vectors $\vec{c}-\vec{a}, \vec{b}$ and $\vec{d}$ are non-coplanar.

Therefore, $[\vec{c}-\vec{a}\ \vec{b}\ \vec{d}]\neq 0$

As shown in the figure, the shortest distance between the lines is PQ on line L which is perpendicular to lines L_1 and L_2.

Since line L is perpendicular to both the lines L_1 and L_2, it is parallel to the vector $\vec{b}\times\vec{d}$.

Hence, the shortest distance between the lines = PQ = CD (height of the cuboid)

= projection of $\overrightarrow{AC}$ on $\vec{b}\times\vec{d}$

$$=\frac{|(\vec{c}-\vec{a})\cdot(\vec{b}\times\vec{d})|}{|\vec{b}\times\vec{d}|}$$

$$=\frac{\left\|\begin{matrix} c_1-a_1 & c_2-a_2 & c_3-a_3 \\ b_1 & b_2 & b_3 \\ d_1 & d_2 & d_3 \end{matrix}\right\|}{\left\|\begin{matrix} \hat{i} & \hat{j} & \hat{k} \\ b_1 & b_2 & b_3 \\ d_1 & d_2 & d_3 \end{matrix}\right\|}$$

If the lines intersect then $[\vec{c}-\vec{a}\ \vec{b}\ \vec{d}]=0$ and the shortest distance becomes zero.

Illustration 4.30

The least positive integral value of λ for which lines $\frac{x-1}{2}=\frac{y+2}{3}=\frac{z-1}{4}$ and $\frac{x}{\lambda}=\frac{y-3}{-1}=\frac{z+1}{2}$ are non-coplanar is

Sol. Line $\frac{x-1}{2}=\frac{y+2}{3}=\frac{z-1}{4}$ passes through the point (1, –2, 1) and parallel to the vector $2\hat{i}+3\hat{j}+4\hat{k}$.

Line $\frac{x}{\lambda} = \frac{y-3}{-1} = \frac{z+1}{2}$ passes through the point (0, 3, –1) and parallel to the vector $\lambda\hat{i} - \hat{j} + 2\hat{k}$.

The lines will be coplanar if

$$\begin{vmatrix} 1-0 & -2-3 & 1-(-1) \\ 2 & 3 & 4 \\ \lambda & -1 & 2 \end{vmatrix} = 0$$

$\Rightarrow \quad 1(6+4) + 5(4-4\lambda) + 2(-2-3\lambda) = 0$

$\Rightarrow \quad 26 - 26\lambda = 0$

$\Rightarrow \quad \lambda = 1$

Hence, the least value of λ for which lines are non-coplanar is 2.

Illustration 4.31

If the foot of the perpendicular from point $P(4, 3, 8)$ on the line $L_1 : \frac{x-a}{\ell} = \frac{y-2}{3} = \frac{z-b}{4}$, $\ell \neq 0$ is (3, 5, 7), then find the shortest distance between the line L_1 and line $L_2 : \frac{x-2}{3} = \frac{y-4}{4} = \frac{z-5}{5}$.

Sol. Clearly, point $M(3, 5, 7)$ satisfies the line

$$L_1 : \frac{3-a}{\ell} = \frac{5-2}{3} = \frac{7-b}{4}$$

$\Rightarrow \quad \frac{3-a}{\ell} = 1$ and $\frac{7-b}{4} = 1$

$\Rightarrow \quad b = 3$ and $a + \ell = 3$ ...(i)

Also, $\overrightarrow{PM} = -\hat{i} + 2\hat{j} - \hat{k}$ and the vector along the line L_1 is $\vec{v_1} = \ell\hat{i} + 3\hat{j} + 4\hat{k}$.

Now, $\overrightarrow{PM} \cdot \vec{v} = 0$

$\Rightarrow \quad -\ell + 6 - 4 = 0$ or $\ell = 2$

Thus, from (i), $a = 1$.

So, $\quad L_1 : \frac{x-1}{2} = \frac{y-2}{3} = \frac{z-3}{4}$

Also, $L_2 : \frac{x-2}{3} = \frac{y-4}{4} = \frac{z-5}{5}$

Line L_1 passes through the point $A(1, 2, 3)$ and parallel to the vector $\vec{v_1} = 2\hat{i} + 3\hat{j} + 4\hat{k}$.

Line L_2 passes through the point $B(2, 4, 5)$ and parallel to the vector $\vec{v_2} = 3\hat{i} + 4\hat{j} + 5\hat{k}$.

$$\therefore \quad \overrightarrow{AB} = \hat{i} + 2\hat{j} + 2\hat{k} \text{ and } \vec{v_1} \times \vec{v_2} = \begin{vmatrix} \hat{i} & \hat{j} & \hat{k} \\ 2 & 3 & 4 \\ 3 & 4 & 5 \end{vmatrix} = -\hat{i} + 2\hat{j} - \hat{k}$$

Hence, the shortest distance between the lines is

$$= \frac{|\overrightarrow{AB} \cdot (\vec{v_1} \times \vec{v_2})|}{|\vec{v_1} \times \vec{v_2}|} = \frac{|-1+4-2|}{\sqrt{1+4+1}} = \frac{1}{\sqrt{6}}$$

Illustration 4.32

Find the shortest distance between the lines $\frac{x-1}{2} = \frac{y-2}{3} = \frac{z-3}{4}$ and $\frac{x-2}{3} = \frac{y-4}{4} = \frac{z-5}{5}$. Also obtain the equation of the line of the shortest distance, or find the equation of the line which intersects the given lines perpendicularly.

Sol.

The given lines are $\frac{x-1}{2} = \frac{y-2}{3} = \frac{z-3}{4} = r_1$ (say) ...(i)

and $\frac{x-2}{3} = \frac{y-4}{4} = \frac{z-5}{5} = r_2$ (say) ...(ii)

Any point on (i) is given by $P\,(2r_1 + 1, 3r_1 + 2, 4r_1 + 3)$

And any point on (ii) is given by $Q\,(3r_2 + 2, 4r_2 + 4, 5r_2 + 5)$

Direction ratios of PQ are given by $3r_2 - 2r_1 + 1$, $4r_2 - 3r_1 + 2$ and $5r_2 - 4r_1 + 2$

Since PQ is perpendicular to (i), we have

$2(3r_2 - 2r_1 + 1) + 3(4r_2 - 3r_1 + 2) + 4(5r_2 - 4r_1 + 2) = 0$

or $\quad 38r_2 - 29r_1 + 16 = 0$ ...(iii)

Also PQ is perpendicular to (ii). So, we have

$3(3r_2 - 2r_1 + 1) + 4(4r_2 - 3r_1 + 2) + 5(5r_2 - 4r_1 + 2) = 0$

or $\quad 50r_2 - 38r_1 + 21 = 0$ ...(iv)

Solving (iii) and (iv), we obtain $r_2 = -(1/6)$, $r_1 = (1/3)$.

Therefore, coordinates of P and Q are $\left(\frac{5}{3}, 3, \frac{13}{3}\right)$ and $\left(\frac{3}{2}, \frac{10}{3}, \frac{25}{6}\right)$, respectively. Thus,

$$PQ^2 = \left(\frac{3}{2} - \frac{5}{3}\right)^2 + \left(\frac{10}{3} - 3\right)^2 + \left(\frac{25}{6} - \frac{13}{3}\right)^2$$

$$= \left(-\frac{1}{6}\right)^2 + \left(\frac{1}{3}\right)^2 + \left(-\frac{1}{6}\right)^2$$

$$= \frac{1}{6}$$

or $\quad PQ = \frac{1}{\sqrt{6}}$

The equation of the line of the shortest distance (line PQ) is given by

$$\frac{x-(5/3)}{(3/2)-(5/3)}=\frac{y-3}{(10/3)-3}=\frac{z-(13/3)}{(25/6)-(13/3)}$$

or $$\frac{x-(5/3)}{-(1/6)}=\frac{y-3}{(1/3)}=\frac{z-(13/3)}{-(1/6)}$$

or $$\frac{x-(5/3)}{1}=\frac{y-3}{-2}=\frac{z-(13/3)}{1}$$

Aim to complete within 20 minutes

Concept Application Exercise 4.3

1. Find the equation of the line passing through the intersection of $\frac{x-1}{2}=\frac{y-2}{-3}=\frac{z-3}{4}$ and $\frac{x-4}{5}=\frac{y-1}{2}=z$ and also through the point (2, 1, – 2)
2. If the straight lines $x=-1+s$, $y=3-\lambda s$, $z=1+\lambda s$ and $x=\frac{t}{2}$, $y=1+t$, $z=2-t$, with parameters s and t, respectively, are coplanar, then λ is equal to
   (1) –2 (2) –1
   (3) 1 (4) 2
3. Find the area of the square whose two sides have equations $\frac{x-1}{2}=\frac{y+2}{3}=\frac{z-3}{4}$ and $\frac{x}{2}=\frac{y-1}{3}=\frac{z+1}{4}$.
4. The shortest distance between the lines $\vec{r}=(1-\lambda)\hat{i}+(\lambda-2)\hat{j}+(3-2\lambda)\hat{k}$ and $\vec{r}=(\mu+1)\hat{i}+(2\mu-1)\hat{j}-(2\mu+1)\hat{k}$ is
   (1) $\frac{1}{5}$ (2) $\frac{1}{\sqrt{5}}$
   (3) $\frac{2}{\sqrt{3}}$ (4) $\frac{5}{\sqrt{2}}$
5. If for some $\alpha\in R$, the lines
   $L_1:\frac{x+1}{2}=\frac{y-2}{-1}=\frac{z-1}{1}$ and
   $L_2:\frac{x+2}{\alpha}=\frac{y+1}{5-\alpha}=\frac{z+1}{1}$ are coplanar, then the line L_2 passes through the point
   (1) (–2, 10, 2) (2) (10, 2, 2)
   (3) (10, –2, 2) (4) (2, –10, –2)
6. If the shortest distance between the straight lines $3(x-1)=6(y-2)=2(z-1)$ and $4(x-2)=2(y-\lambda)=(z-3)$, $\lambda\in R$ is $\frac{1}{\sqrt{38}}$, then the integral value of λ is equal to
   (1) 3 (2) 5
   (3) 2 (4) –1

Answers

1. $\frac{x+1}{3}=\frac{y+1}{2}=\frac{z+1}{-1}$
2. (1)
3. $\frac{673}{29}$
4. (2)
5. (4)
6. (1)

4.4 PLANE IN SPACE

A plane is a flat, two-dimensional surface extending infinitely in three-dimensional space.

Let us determine the **equations of a plane** under different conditions.

4.4.1 Equation of Plane in Point–Normal Form

To determine a plane uniquely, we must specify its position and orientation.

The orientation of a plane is defined by a normal vector, which is perpendicular to the plane.

Infinitely many parallel planes can be drawn perpendicular to a given vector $\vec{n}$.

Among this family of planes, only one passes through the given point $A(\vec{a})$.

Let's find the equation of this plane.

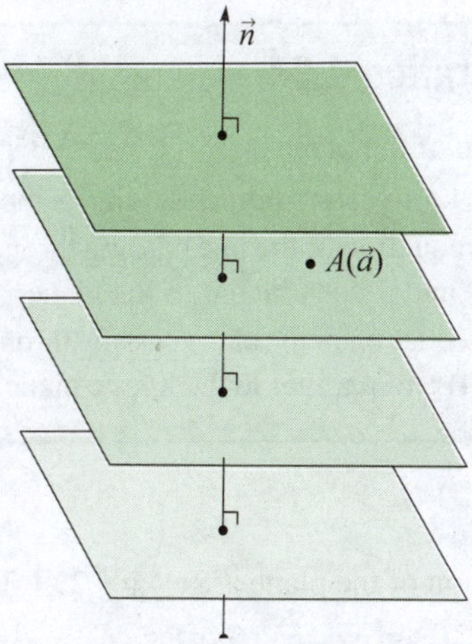

The equation of a plane represents the **locus** of a variable point on the plane.

Let the variable point on the plane be R,having the position vector $\vec{r}$.

For all position of point R on the plane, $\overrightarrow{AR} \perp \vec{n}$.

Hence, $(\vec{r}-\vec{a})\cdot\vec{n}=0$ ...(i)

This equation is known as the vector equation of a plane in **point–normal** form.

From (i), $\vec{r}\cdot\vec{n}=\vec{a}\cdot\vec{n}$

or $\vec{r}\cdot\vec{n}=D$, where $D=\vec{a}\cdot\vec{n}$ ...(ii)

If $\vec{a}=x_1\hat{i}+y_1\hat{j}+z_1\hat{k}$, $\vec{n}=a\hat{i}+b\hat{j}+c\hat{k}$ and $\vec{r}=x\hat{i}+y\hat{j}+z\hat{k}$, then from (i), we have

$$a(x-x_1)+b(y-y_1)+c(z-z_1)=0 \quad \text{...(iii)}$$

This is the cartesian equation of a plane passing through a given point (x_1, y_1, z_1) and perpendicular to a given vector $a\hat{i}+b\hat{j}+c\hat{k}$.

Equation (i) can be written as,

$$ax+by+cz+d=0, \text{ where } d=-(ax_1+by_1+cz_1)$$

This is the general equation of a plane in Cartesian form, where $a\hat{i}+b\hat{j}+c\hat{k}$ is a vector that is normal to the plane.

Since any plane parallel to the above plane is perpendicular to the vector $a\hat{i}+b\hat{j}+c\hat{k}$, the equation of a plane parallel to the above plane is given by $ax+by+cz+d'=0$.

Illustration 4.33

Find the equation of the plane that passes through the point (2, 3, 4) and is perpendicular to the vector $-\hat{i}+2\hat{j}-3\hat{k}$.

Sol. Equation of the plane is

$$-1(x-2)+2(y-3)-3(z-4)=0$$

or $x-2y+3z-8=0$

Illustration 4.34

If the equation of the plane is $2x-y+2z+3=0$, then

a. find the unit vectors perpendicular to the plane

b. find the equation of the line passing through the point (1, 2, 3) and perpendicular to the plane.

c. find the equation of the plane passing through the point (–2, 3, 4) and parallel to the above plane.

Sol.

a. The equation of the plane is $2x-y+2z+3=0$

So, the vector that is normal to the plane is $\vec{n}=2\hat{i}-\hat{j}+2\hat{k}$.

$$|\vec{n}|=\sqrt{2^2+(-1)^2+2^2}=\sqrt{9}=3$$

Hence, unit vectors perpendicular to the plane are $\pm\frac{1}{3}(2\hat{i}-\hat{j}+2\hat{k})$.

b. Line perpendicular to the plane will be parallel to the vector that normal to the plane.

Line passes through the point (1, 2, 3).

Hence, the equation of the line is $\dfrac{x-1}{2}=\dfrac{y-2}{-1}=\dfrac{z-3}{2}$.

c. The equation of the plane parallel to the given plane is $2x-y+2z+\lambda=0$

If plane passes through the point (–2, 3, 4) then

$$2(-2)-3+2(4)+\lambda=0$$

or $\lambda=-1$

Hence, the equation of the plane is $2x-y+2z-1=0$.

Illustration 4.35

Find the equation of the plane such that image of the point (1, 2, 3) in it is (–1, 0, 1).

Sol. Since the image of the point A (1, 2, 3) in the plane is $B(-1,0,1)$, the plane passes through the midpoint of AB that is (0, 1, 2).

Also, the vector that is normal to the plane is $\overrightarrow{AB}\equiv-2\hat{i}-2\hat{j}-2\hat{k}$.

Hence, the equation of the plane is $-2(x-0)-2(y-1)-2(z-2)=0$

or $x+y+z-3=0$.

Illustration 4.36

Find the equation of the perpendicular bisector of the line segment joining points (2, 3, 2) and (– 4, 1, 4) which passes through the point (– 3, 6, 1).

Also find the equation of the plane on which all the perpendicular bisectors of the line segments joining the points (2, 3, 2) and (– 4, 1, 4) lie.

Sol. Given points are $A(2, 3, 2)$ and $B(-4, 1, 4)$.

The midpoint of AB is $M\equiv\left(\dfrac{2-4}{2},\dfrac{3+1}{2},\dfrac{2+4}{2}\right)\equiv(-1, 2, 3)$.

The perpendicular bisector passes through the point $C(-3, 6, 1)$

So, the Direction ratios of CM are $<-1-(-3), 2-6, 3-1>$ or $<2, -4, 2>$.

Therefore, the equation of the line CM is $\dfrac{x+3}{1}=\dfrac{y-6}{-2}=\dfrac{z-1}{1}$

All the perpendicular bisectors of AB lie on the plane that passes through the point M and perpendicular to AB.

Direction ratios of AB are $<-4-2, 1-3, 4-2>$ or $<-6, -2, 2>$

Therefore, the equation of plane is $6(x+1)+2(y-2)-2(z-3)=0$ or $3x+y-z+4=0$

Illustration 4.37

Find the equation of the plane passing through the point (0, 1, –1) which is parallel to the lines $\vec{r} = \hat{i} + \hat{j} + \lambda(2\hat{i} + \hat{j} + 4\hat{k})$ and $\frac{x+1}{-3} = \frac{y-3}{2} = \frac{z+2}{1}$.

Sol. The plane is parallel to the given lines, which are parallel to the vectors $\vec{a} = 2\hat{i} + \hat{j} + 4\hat{k}$ and $\vec{b} = -3\hat{i} + 2\hat{j} + 1\hat{k}$.

So, the vector that is normal to the plane is

$$\vec{a} \times \vec{b} \equiv \begin{vmatrix} \hat{i} & \hat{j} & \hat{k} \\ 2 & 1 & 4 \\ -3 & 2 & 1 \end{vmatrix} = -7\hat{i} - 14\hat{j} + 7\hat{k}$$

The plane passes through the point (0, 1, –1).

Therefore, the equation of the plane is

$$-7(x - 0) - 14(y - 1) + 7(z + 1) = 0$$

or $7x + 14y - 7z = 21$

Illustration 4.38

Find the equation of the plane which passes through the point (2, 1, 0) and contains the line $\frac{x-1}{2} = \frac{y+2}{-3} = \frac{z-4}{1}$.

Sol. Plane contains the line $\frac{x-1}{2} = \frac{y+2}{-3} = \frac{z\ 4}{1}$

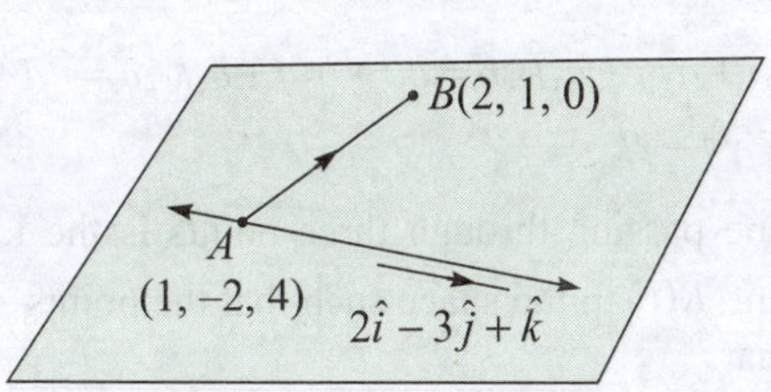

Thus, the plane contains the point A(1, –2, 4) and is parallel to the vector $\vec{p} = 2\hat{i} - 3\hat{j} + \hat{k}$

Also, the plane passes through the point B(2, 1, 0).

So, the plane is parallel to the vector $\overrightarrow{AB} = \hat{i} + 3\hat{j} - 4\hat{k}$

Therefore, the vector that is normal to the plane is

$$\vec{p} \times \overrightarrow{AB} = \begin{vmatrix} \hat{i} & \hat{j} & \hat{k} \\ 2 & -3 & 1 \\ 1 & 3 & -4 \end{vmatrix}$$

$$= 9\hat{i} + 9\hat{j} + 9\hat{k}$$

Hence, the equation of the plane is $9(x - 2) + 9(y - 1) + 9(z - 0) = 0$ or $x + y + z - 3 = 0$

Illustration 4.39

Find the equation of the plane which passes through the points (–1, 2, –4) and (0, 1, 3) and parallel to the straight line $\frac{x-1}{2} = \frac{y+2}{-3} = \frac{z-4}{1}$.

Sol. Plane passes through the points A(–1, 2, –4) and B(0, 1, 3).

So, the plane is parallel to the vector $\overrightarrow{AB} = \hat{i} - \hat{j} + 7\hat{k}$.

Also, the plane is parallel to the line $\frac{x-1}{2} = \frac{y+2}{-3} = \frac{z-4}{1}$.

So, it is parallel to the vector $\vec{p} = 2\hat{i} - 3\hat{j} + \hat{k}$.

Thus, the vector that is normal to the plane is $\overrightarrow{AB} \times \vec{p} = \begin{vmatrix} \hat{i} & \hat{j} & \hat{k} \\ 1 & -1 & 7 \\ 2 & -3 & 1 \end{vmatrix}$

$= 20\hat{i} + 13\hat{j} - \hat{k}$.

Therefore, equation of plane is $20(x + 1) + 13(y - 2) - (z + 4) = 0$ or $20x + 13y - z - 10 = 0$.

Illustration 4.40

Find the equation of the plane which contains lines $\frac{x}{2} = \frac{3-y}{4} = \frac{z+7}{-5}$ and $\vec{r} = (1+2\lambda)\hat{i} + (2-4\lambda)\hat{j} - (1+5\lambda)\hat{k}$

Sol. Given lines are $\frac{x}{2} = \frac{y-3}{-4} = \frac{z+7}{-5}$

and $\vec{r} = \hat{i} + 2\hat{j} - \hat{k} + \lambda(2\hat{i} - 4\hat{j} - 5\hat{k})$

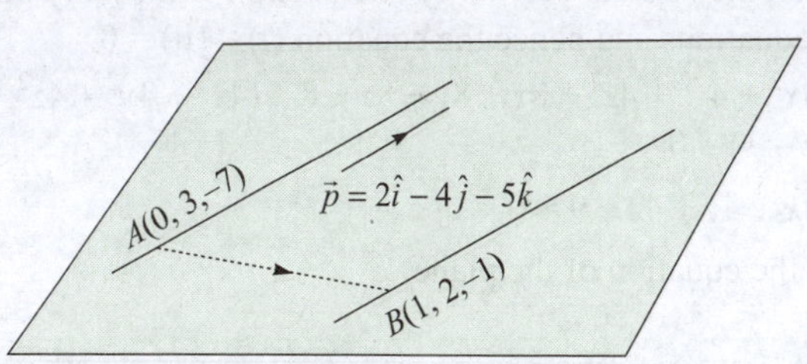

Thus, both the lines are parallel to the vector $\vec{p} = 2\hat{i} - 4\hat{j} - 5\hat{k}$

Plane passes through these lines.

So, the plane passes through the points A(0, 3, –7) and B(1, 2, –1)

Now, $\overrightarrow{AB} = \hat{i} - \hat{j} + 6\hat{k}$

Therefore, the plane is perpendicular to the vector

$$\vec{p} \times \overrightarrow{AB} = \begin{vmatrix} \hat{i} & \hat{j} & \hat{k} \\ 2 & -4 & -5 \\ 1 & -1 & 6 \end{vmatrix} = -29\hat{i} - 17\hat{j} + 2\hat{k}$$

Hence, the equation of the plane is $-29(x - 0) - 17(y - 3) + 2(z + 7) = 0$ (using point A) or $29x + 17y - 2z - 65 = 0$

Illustration 4.41

Find the equation of the plane, $\vec{r} = \hat{i} - \hat{j} + \lambda(\hat{i} + \hat{j} + \hat{k}) + \mu(\hat{i} - 2\hat{j} + 3\hat{k})$ in the cartesian form.

Sol. The equation of the plane is

$$\vec{r} = \hat{i} - \hat{j} + \lambda(\hat{i} + \hat{j} + \hat{k}) + \mu(\hat{i} - 2\hat{j} + 3\hat{k}).$$

Let $\vec{r} = x\hat{i} + y\hat{j} + z\hat{k}$

Hence, the equation is

$$(x\hat{i} + y\hat{j} + z\hat{k}) - (\hat{i} - \hat{j}) = \lambda(\hat{i} + \hat{j} + \hat{k}) + \mu(\hat{i} - 2\hat{j} + 3\hat{k})$$

Thus, vectors $(x\hat{i} + y\hat{j} + z\hat{k}) - (\hat{i} - \hat{j}), \hat{i} + \hat{j} + \hat{k}, \hat{i} - 2\hat{j} + 3\hat{k}$ are coplanar.

Therefore, the equation of the plane is $\begin{vmatrix} x-1 & y-(-1) & z-0 \\ 1 & 1 & 1 \\ 1 & -2 & 3 \end{vmatrix} = 0$

or $\quad 5x - 2y - 3z - 7 = 0$

Illustration 4.42

It is given that the spheres $x^2 + y^2 + z^2 + 2x + 2y + 2z + 2 = 0$ and $4x^2 + 4y^2 + 4z^2 + 4x + 4y + 4z - 1 = 0$ intersect. Then prove that the circular section of the intersection lies on the plane. Find the equation of that plane.

Sol. Given spheres are $x^2 + y^2 + z^2 + 2x + 2y + 2z + 2 = 0$

or $\quad 4x^2 + 4y^2 + 4z^2 + 8x + 8y + 8z + 8 = 0 \qquad \ldots$(i)

and $\quad 4x^2 + 4y^2 + 4z^2 + 4x + 4y + 4z - 1 = 0 \qquad \ldots$(ii)

Points on the circular section of intersection will satisfy both the above equations and hence the equation (i) – (ii) = 0.

or $\quad 4x^2 + 4y^2 + 4z^2 + 8x + 8y + 8z + 8 - (4x^2 + 4y^2 + 4z^2 + 4x + 4y + 4z - 1) = 0$

or $\quad 4x + 4y + 4z + 9 = 0.$

This is the equation of the plane.

Illustration 4.43

Prove that the lines $\dfrac{x-a+d}{\alpha-\delta} = \dfrac{y-a}{\alpha} = \dfrac{z-a-d}{\alpha+\delta}$, $\dfrac{x-b+c}{\beta-\gamma} = \dfrac{y-b}{\beta} = \dfrac{z-b-c}{\beta+\gamma}$ are coplanar. Also find the equation of the plane in which they lie. $(\alpha\gamma \neq \beta\delta)$.

Sol. Given lines are $\dfrac{x-(a-d)}{\alpha-\delta} = \dfrac{y-a}{\alpha} = \dfrac{z-(a+d)}{\alpha+\delta}$,

$$\frac{x-(b-c)}{\beta-\gamma} = \frac{y-b}{\beta} = \frac{z-(b+c)}{\beta+\gamma}$$

Now,

$$\begin{vmatrix} a-d-b+c & a-b & a+d-b-c \\ \alpha-\delta & \alpha & \alpha+\delta \\ \beta-\gamma & \beta & \beta+\gamma \end{vmatrix}$$

$$= \begin{vmatrix} -d+c & a-b & d-c \\ -\delta & \alpha & \delta \\ -\gamma & \beta & \gamma \end{vmatrix} = 0$$

(Applying $C_1 \to C_1 - C_2$ and $C_3 \to C_3 - C_2$)

Hence, lines are coplanar.

Equation of the plane containing these lines is

$$\begin{vmatrix} x \quad a+d & y \quad a & z \quad a \quad d \\ \alpha \quad \delta & \alpha & \alpha+\delta \\ \beta \quad \gamma & \beta & \beta+\gamma \end{vmatrix} = 0$$

or $\begin{vmatrix} x-2y+z & y-a & z-a-d \\ 0 & \alpha & \alpha+\delta \\ 0 & \beta & \beta+\gamma \end{vmatrix} = 0$

(Applying $C_1 \to C_1 + C_3 - 2C_2$)

or $\quad (x - 2y + z)(\alpha\gamma - \beta\delta) = 0$

or $\quad x - 2y + z = 0 \qquad [\because (\alpha\gamma \neq \beta\delta)]$

4.4.2 Equation of Plane Passing Through Three Given Points

We know that three non–collinear points in the space form a unique plane or surface.

Four points are said to be coplanar if the fourth point lies in the plane formed by other three points.

Consider three non-collinear points $A(\vec{a})$, $B(\vec{b})$ and $C(\vec{c})$.

Let $\vec{a} = a_1\hat{i} + a_2\hat{j} + a_3\hat{k}$, $\vec{b} = b_1\hat{i} + b_2\hat{j} + b_3\hat{k}$, $\vec{c} = c_1\hat{i} + c_2\hat{j} + c_3\hat{k}$ and $\vec{r} = x\hat{i} + y\hat{j} + z\hat{k}$

Now plane passing through three points is the locus of the variable point $R(\vec{r})$ in the space such that the points A, B, C and R are coplanar.

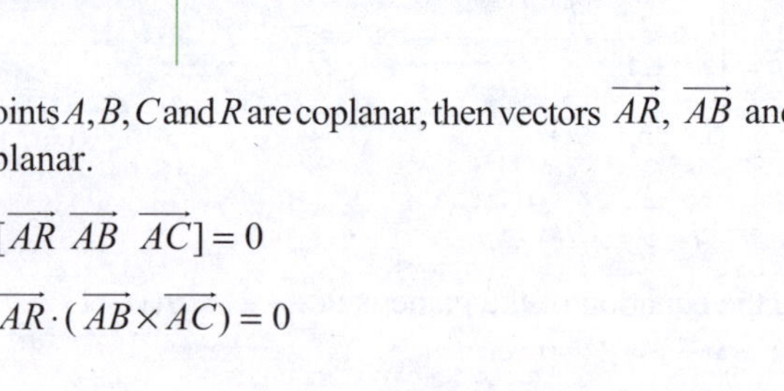

If points A, B, C and R are coplanar, then vectors $\vec{AR}$, $\vec{AB}$ and $\vec{AC}$ are coplanar.

$\therefore \quad [\vec{AR}\ \vec{AB}\ \vec{AC}] = 0$

or $\quad \vec{AR} \cdot (\vec{AB} \times \vec{AC}) = 0$

Thus, $\begin{vmatrix} x-x_1 & y-y_1 & z-z_1 \\ x_2-x_1 & y_2-y_1 & z_2-z_1 \\ x_3-x_1 & y_3-y_1 & z_3-z_1 \end{vmatrix} = 0.$

This is equation of plane passing through the point $A \equiv (x_1, y_1, z_1)$, $B \equiv (x_2, y_2, z_2)$ and $C \equiv (x_3, y_3, z_3)$.

Here $\overrightarrow{AB} \times \overrightarrow{AC}$ is a vector normal to the plane.

Illustration 4.44

Find the equation of the plane passing through the points A (2, 2, –1), B (3, 4, 2) and C (7, 0, 6). Also find the unit vector perpendicular to this plane.

Sol. Here $(x_1, y_1, z_1) \equiv (2, 2, -1)$, $(x_2, y_2, z_2) \equiv (3, 4, 2)$ and $(x_3, y_3, z_3) \equiv (7, 0, 6)$.

Then the equation of the plane is

$$\begin{vmatrix} x-x_1 & y-y_1 & z-z_1 \\ x_2-x_1 & y_2-y_1 & z_2-z_1 \\ x_3-x_1 & y_3-y_1 & z_3-z_1 \end{vmatrix} = 0$$

or $\begin{vmatrix} x-2 & y-2 & z-(-1) \\ 3-2 & 4-2 & 2-(-1) \\ 7-2 & 0-2 & 6-(-1) \end{vmatrix} = 0$

or $5x + 2y - 3z = 17$

A vector normal to this plane is $\vec{d} = 5\hat{i} + 2\hat{j} - 3\hat{k}$ ...(i)

Therefore, a unit vector normal to the plane is

$$\hat{n} = \frac{\vec{d}}{|\vec{d}|} = \frac{5\hat{i} + 2\hat{j} - 3\hat{k}}{\sqrt{25+4+9}} = \frac{1}{\sqrt{38}}(5\hat{i} + 2\hat{j} - 3\hat{k})$$

4.4.3 Equation of Plane in Intercepts Form

Let the plane P intersect x, y and z-axes at $A(a, 0, 0)$, $B(0, b, 0)$ and $C(0, 0, c)$.

Thus, the x, y and z -intercepts of the plane are a, b and c respectively.

Then, the equation of the plane is passing through points A, B and C is $\begin{vmatrix} x-a & y-0 & z-0 \\ 0-a & b-0 & 0-0 \\ 0-a & 0-0 & c-0 \end{vmatrix} = 0$

or $(x - a)(bc) - y(-ac) + z(ab) = 0$

or $bcx + acy + abz = abc$

Dividing throughout by abc, we get

$$\frac{x}{a} + \frac{y}{b} + \frac{z}{c} = 1$$

This is the equation of the plane in intercept form.

Illustration 4.45

If a plane meets the coordinate axes at A, B and C such that the centroid of the triangle ABC is (1, 2, 4) then find the equation of the plane.

Sol. Let the plane meets the co-ordinates axes at $A(a, 0, 0)$, $B(0, b, 0)$ $C(0, 0, c)$

The centroid of the triangle ABC is (1, 2, 4).

$$\therefore \quad \left(\frac{a+0+0}{3}, \frac{0+b+0}{3}, \frac{0+0+c}{3}\right) \equiv (1, 2, 4)$$

$\Rightarrow \quad a = 3, b = 6, c = 12$

Hence, using the intercept form, the equation of the plane is

$\frac{x}{3} + \frac{y}{6} + \frac{z}{12} = 1$ or $4x + 2y + z = 12$

4.4.4 Equation of Plane in Normal Form

The equation of a plane in normal form is expressed in terms of:

- the perpendicular distance of the plane from the origin, and
- a unit normal vector drawn from the origin perpendicular to the plane.

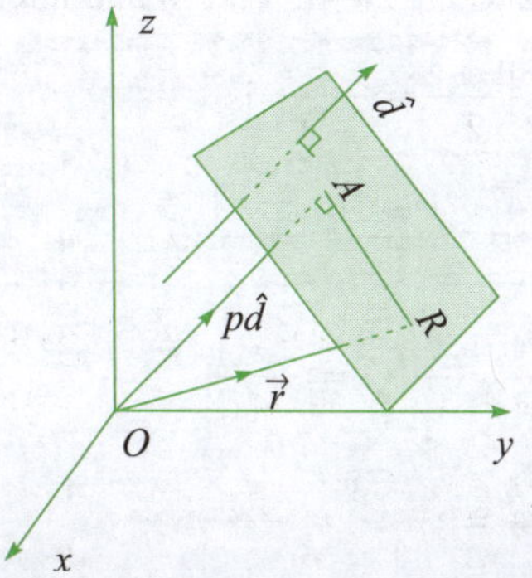

Let A be the foot of the perpendicular from the origin to the plane, and let $OA = d$.

Let $\hat{n}$ be the unit normal vector in the direction of $\overline{OA}$.

Then $\overline{OA} = d\hat{n}$.

If $\vec{r} = x\hat{i} + y\hat{j} + z\hat{k}$ is the position vector of an arbitrary point R on the plane, then $\overrightarrow{AR}$ is perpendicular to $\overrightarrow{OA}$.

Therefore, $\overrightarrow{AR} \cdot \overrightarrow{OA} = 0$

$\Rightarrow \quad (\vec{r} - d\hat{n}) \cdot (d\hat{n}) = 0$

$\Rightarrow \quad (\vec{r} - d\hat{n}) \cdot \hat{n} = 0$

$\Rightarrow \quad \vec{r} \cdot \hat{n} = d$

This is the vector equation of the plane in normal form.

If $\hat{n} = \ell\hat{i} + m\hat{j} + n\hat{k}$, then the equation of the plane is

$$(x\hat{i} + y\hat{j} + z\hat{k}) \cdot (\ell\hat{i} + m\hat{j} + n\hat{k}) = d$$

or $\quad \ell x + my + nz = d$

This is cartesian form of the equation of plane in normal form.

Illustration 4.46

Write the equation of the plane $2x - y - 2z - 9 = 0$ in the standard normal form and find its distance from the origin.

Sol. The vector equation of the plane is

$\vec{r}\cdot(2\hat{i}-\hat{j}-2\hat{k})=9$

or $\quad \vec{r}\cdot\vec{n}=9$ ...(i)

Here $\vec{n}=2\hat{i}+\hat{j}+2\hat{k}$

$\therefore \quad \hat{n}=\dfrac{\vec{n}}{|\vec{n}|}=\dfrac{2\hat{i}-\hat{j}-2\hat{k}}{3}$

So, from (i), $\vec{r}\cdot\dfrac{\vec{n}}{|\vec{n}|}=\dfrac{9}{|\vec{n}|}$ or $\vec{r}\cdot\hat{n}=3$, which is the standard normal form of the equation of the plane.

Hence, the distance of the plane from the origin is 3 units.

Aim to complete within 50 minutes

Concept Application Exercise 4.4

1. The equation of the line passing through (–4, 3, 1), parallel to the plane $x + 2y - z - 5 = 0$ and intersecting the line $\dfrac{x+1}{-3}=\dfrac{y-3}{2}=\dfrac{z-2}{-1}$ is

(1) $\dfrac{x+4}{-1}=\dfrac{y-3}{1}=\dfrac{z-1}{1}$

(2) $\dfrac{x+4}{3}=\dfrac{y-3}{-1}=\dfrac{z-1}{1}$

(3) $\dfrac{x+4}{1}=\dfrac{y-3}{1}=\dfrac{z-1}{3}$

(4) $\dfrac{x-4}{2}=\dfrac{y+3}{1}=\dfrac{z+1}{4}$

2. The equation of the plane containing the straight line $\dfrac{x}{2}=\dfrac{y}{3}=\dfrac{z}{4}$ and perpendicular to the plane containing the straight lines $\dfrac{x}{3}=\dfrac{y}{4}=\dfrac{z}{2}$ and $\dfrac{x}{4}=\dfrac{y}{2}=\dfrac{z}{3}$ is

(1) $x + 2y - 2z = 0$
(2) $x - 2y + z = 0$
(3) $5x + 2y - 4z = 0$
(4) $3x + 2y - 3z = 0$

3. Find the unit vector perpendicular to the plane $\vec{r}\cdot(2\hat{i}+\hat{j}+2\hat{k})=5$.

4. Find the equation of line passing through the point $3\hat{i}-5\hat{j}+7\hat{k}$ and perpendicular to the plane $3x - 4y + 5z = 8$.

5. The foot of the perpendicular drawn from the origin to a plane is (1, 2, –3). Find the equation of the plane.

6. Find the equation of the plane which passes through the point (1, 2, –3) and parallel to the vectors $\hat{i}-2\hat{j}+2\hat{k}$ and $3\hat{j}+4\hat{k}$.

7. Find the equation of the plane perpendicular to the line $\dfrac{x-1}{2}=\dfrac{y-3}{-1}=\dfrac{z-4}{2}$ and passing through the origin.

8. A plane P meets the coordinate axes at A, B and C respectively. The centroid of $DABC$ is given to be (1, 1, 2). Then the equation of the line through this centroid and perpendicular to the plane P is

(1) $\dfrac{x-1}{1}=\dfrac{y-1}{2}=\dfrac{z-2}{2}$

(2) $\dfrac{x-1}{2}=\dfrac{y-1}{2}=\dfrac{z-2}{1}$

(3) $\dfrac{x-1}{2}=\dfrac{y-1}{1}=\dfrac{z-2}{1}$

(4) $\dfrac{x-1}{1}=\dfrac{y-1}{1}=\dfrac{z-2}{2}$

9. Find the equation of the plane passing through the points (1, 0, –1) and (3, 2, 2) and parallel to the line $x-1=\dfrac{1-y}{2}=\dfrac{z-2}{3}$.

10. The plane through the points (2, 3, 1) and (4, –5, 3) and parallel to the x-axis does not pass through the point

(1) (0, 3, 1) (2) (3, –1, 2)
(3) (1, 2, –3) (4) (4, –5, 3)

11. Find the equation of a plane containing the lines $\dfrac{x-5}{4}=\dfrac{y-7}{4}=\dfrac{z+3}{-5}$ and $\dfrac{x-8}{7}=\dfrac{y-4}{1}=\dfrac{z-5}{3}$.

12. Find the equation of the plane passing through the line $\dfrac{x-1}{5}=\dfrac{y+2}{6}=\dfrac{z-3}{4}$ and point (4, 3, 7).

13. Find the equation of a plane which passes through the point (1, 2, 3) and which is at the maximum distance from the point (–1, 0, 2).

14. How many lines can be drawn through the point (2, 3, –1) and perpendicular to the line $\dfrac{x-1}{2}=\dfrac{y-2}{-1}=\dfrac{z}{3}$?

Find the equation of plane in which all such lines lie.

15. Find the equation of the plane through the points (2, 1, 0), (3, –2, –2) and (3, 1, 7).

16. Write the equation of plane $\vec{r}\cdot(2\hat{i}+\hat{j}-2\hat{k})=-4$ in standard normal form. Also find the distance of plane from the origin.

17. Find the equation of the locus of all those points which are equidistant from A(1, 2, 3), B(2, 3, 1) and C(3, 1, 2).

Answers

1. (2)
2. (2)
3. $\frac{1}{3}(2\hat{i}+\hat{j}+2\hat{k})$
4. $\frac{x-3}{3}=\frac{y+5}{-4}=\frac{z-7}{5}$
5. $x+2y-3z-14=0$
6. $14x+4y-3z-31=0$
7. $2x-y+2z=0$
9. $4x-y-2z-9=0$
10. (3)
11. $17x-47y-24z+172=0$
12. $4x-8y+7z=41$
13. $2x+2y+z=9$
14. Infinite lines; Plane: $2x-y+3z+2=0$
15. $7x+3y-z=17$
16. $\vec{r}\cdot\left(\frac{-2\hat{i}-\hat{j}+2\hat{k}}{3}\right)=\frac{4}{3}$, Distance $=\frac{4}{3}$
17. $x=y=z$

4.5 FAMILY OF PLANES CONTAINING THE LINE OF INTERSECTION OF TWO PLANES

Let P_1 and P_2 be two planes with equations $\vec{r}\cdot\vec{n}_1=d_1$ and $\vec{r}\cdot\vec{n}_2=d_2$ respectively.

These planes intersect in line L as shown in the following figure.

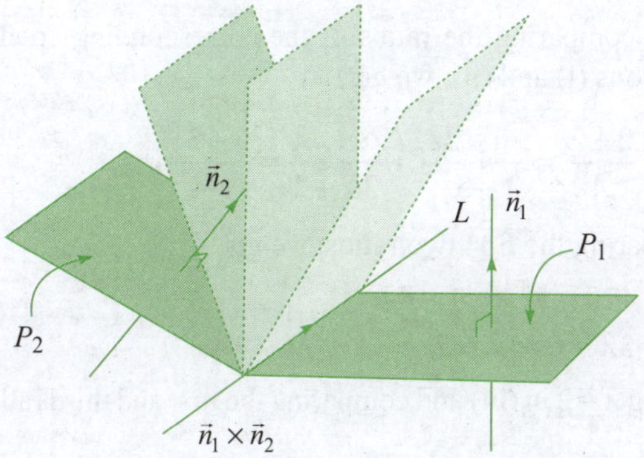

We can draw as many planes as we want passing through this line.

Points on the line L satisfy both the equations of the planes. Hence, these points also satisfy the equation

$$\vec{r}\cdot\hat{n}_1-d_1+\lambda(\vec{r}\cdot\hat{n}_2-d_2)=0,\ \lambda\in R. \quad \text{...(i)}$$

The above equation is an identity for all points on the line L and is therefore the general equation of the family of planes passing through the line L.

If $\vec{n}_1=A_1\hat{i}+B_1\hat{j}+C_1\hat{k}$, $\vec{n}_2=A_2\hat{i}+B_2\hat{j}+C_2\hat{k}$ and $\vec{r}=x\hat{i}+y\hat{j}+z\hat{k}$, then equation (i) becomes

$$A_1x+B_1y+C_1z-d_1)+\lambda(A_2x+B_2y+C_2z-d_2)=0,\ \lambda\in R \quad \text{...(ii)}$$

This is the general equation of the family of planes passing through the line of intersection of the planes

$$A_1x+B_1y+C_1z-d_1=0 \quad \text{...(iii)}$$

and $$A_2x+B_2y+C_2z-d_2=0 \quad \text{...(iv)}$$

Line of Intersection of Two Planes

Since points on the line of intersection of the planes satisfy both the equations of the planes, equations (iii) and (iv) together represent the line of intersection of the given planes.

Also, the line L is perpendicular to the normal vectors $\vec{n}_1$ and $\vec{n}_2$. Hence, it is parallel to the vector $\vec{n}_1\times\vec{n}_2$.

Illustration 4.47

Find the equation of the plane which contains the line of intersection of the planes $\vec{r}\cdot(\hat{i}+2\hat{j}+3\hat{k})-4=0$ and $\vec{r}\cdot(2\hat{i}+\hat{j}-\hat{k})+5=0$ and which is perpendicular to the plane $\vec{r}\cdot(5\hat{i}+3\hat{j}-6\hat{k})+8=0$.

Sol. The equations of the given planes are

$$\vec{r}\cdot(\hat{i}+2\hat{j}+3\hat{k})-4=0 \quad \text{...(i)}$$

$$\vec{r}\cdot(2\hat{i}+\hat{j}-\hat{k})+5=0 \quad \text{...(ii)}$$

The equation of the family of planes passing through the line of intersection of the above planes is

$$[\vec{r}\cdot(\hat{i}+2\hat{j}+3\hat{k})-4]+\lambda[\vec{r}\cdot(2\hat{i}+\hat{j}-\hat{k})+5]=0$$

or $$\vec{r}\cdot[(2\lambda+1)\hat{i}+(\lambda+2)\hat{j}+(3-\lambda)\hat{k}]+(5\lambda-4)=0 \quad \text{...(iii)}$$

The plane in (iii) is perpendicular to the plane, $\vec{r}\cdot(5\hat{i}+3\hat{j}-6\hat{k})+8=0$. Thus,

$$5(2\lambda+1)+3(\lambda+2)-6(3-\lambda)=0$$

$\Rightarrow$ $19\lambda-7=0$

$\Rightarrow$ $\lambda=\frac{7}{19}$

Substituting $\lambda=\frac{7}{19}$ in (iii), we get

$$\vec{r}\cdot\left[\frac{33}{19}\hat{i}+\frac{45}{19}\hat{j}+\frac{50}{19}\hat{k}\right]-\frac{41}{19}=0$$

or $\vec{r}\cdot(33\hat{i}+45\hat{j}+50\hat{k})-41=0$

or $33x+45y+50z-41=0$, which is the required equation of the plane.

Illustration 4.48

Find the equation of the plane containing the line of intersection of the planes $x+y+z-6=0$ and $2x+3y+4z+5=0$ and passing through the point (1, 1, 1).

Sol. The equation of the family of planes passing through the line of intersection of the given planes is

$$(x+y+z-6)+\lambda(2x+3y+4z+5)=0 \quad \text{...(i)}$$

If plane passes through the point (1, 1, 1), then $(1+1+1-6)+\lambda(2+3+4+5)=0$,

$\Rightarrow$ $\lambda=\frac{3}{14}$

Putting $\lambda = 3/14$ in (i), we get the equation of plane as

$$(x+y+z-6)+\frac{3}{14}(2x+3y+4z+5)=0$$

or $20x+23y+26z-69=0$

Illustration 4.49

Find the equation of the plane passing through the line of intersection of the planes $\vec{r}\cdot(\hat{i}+3\hat{j})-6=0$ and $\vec{r}\cdot(3\hat{i}-\hat{j}-4\hat{k})=0$, such that its perpendicular distance from the origin is unity.

Sol. Equation of the family of planes passing through the line intersection of the given planes is

$$\vec{r}\cdot(\hat{i}+3\hat{j})-6+\lambda[\vec{r}\cdot(3\hat{i}-\hat{j}-4\hat{k})]=0,\ \lambda\in R.$$

or $\vec{r}\times[(1+3\lambda)\hat{i}+(3-\lambda)\hat{j}+(-4\lambda)\hat{k}]=6$ ...(i)

Using the standard normal form of the equation of the plane, we have

$$\frac{\vec{r}.[(1+3\lambda)\hat{i}+(3-\lambda)\hat{j}+(-4\lambda)\hat{k}]}{\sqrt{(1+3\lambda)^2+(3-\lambda)^2+(-4\lambda)^2}}=\frac{6}{\sqrt{(1+3\lambda)^2+(3-\lambda)^2+(-4\lambda)^2}}$$

Since, the perpendicular distance of the plane from the origin is unity, we have

$$\frac{6}{\sqrt{(1+3\lambda)^2+(3-\lambda)^2+(-4\lambda)^2}}=1$$

$\Rightarrow$ $(1+3\lambda)^2+(3-\lambda)^2+(-4\lambda)^2=36$

$\Rightarrow$ $1+9\lambda^2+6\lambda+9+\lambda^2-6\lambda+16\lambda^2=36$

$\Rightarrow$ $\lambda^2=1$

$\Rightarrow$ $\lambda=\pm 1$

Putting these values of λ in equation (i), we get

$$\vec{r}.[(1\pm 3)\hat{i}+(3\mp 1)\hat{j}+(\mp 4)\hat{k}]=6$$

$\Rightarrow$ $\vec{r}.(4\hat{i}+2\hat{j}-4\hat{k})=6$ and $\vec{r}.(-2\hat{i}+4\hat{j}+4\hat{k})=6$

or $4x+2y-4z-6=0$ and $-2x+4y+4z-6=0$

These are the required equations of planes.

Illustration 4.50

Find the equation of plane containing the line $x-2y+3z-4=0=5x-6y+z-3$ and parallel to the line $x+y-z-3=0=2x-y+4z+6$.

Sol. Equation of the family of planes through the line $x-2y+3z-4=0=5x-6y+z-3$ is

$$x-2y+3z-4+\lambda(5x-6y+z-3)=0,\ \lambda\in R.$$

or $(1+5\lambda)x+(-2-6\lambda)y+(3+\lambda)z-4-3\lambda=0$ ...(i)

Now, line $x+y-z-3=0=2x-y+4z+6$ is parallel to the vector

$$\begin{vmatrix}\hat{i} & \hat{j} & \hat{k}\\ 1 & 1 & -1\\ 2 & -1 & 4\end{vmatrix}=3\hat{i}-6\hat{j}-3\hat{k}$$

Plane (i) is parallel to the above vector.

$\therefore$ $3(1+5\lambda)-6(-2-6\lambda)-3(3+\lambda)=0$

$\Rightarrow$ $16\lambda+2=0$

$\Rightarrow$ $\lambda=-\frac{1}{8}$

Putting this value in equation (i), we get the equation of plane as $3x-10y+23z-29=0$

Illustration 4.51

Let the equation of the plane passing through the line of intersection of the planes $2x-y+\alpha z+6=0$ and $x-2y+4z-5=0$ is $4x-5y+(3\alpha+2)z-4=0$. Then find the value of α.

Sol. Equation of the family of planes passing through the line of intersection of first two planes is

$$2x-y+\alpha z+6+\lambda(x-2y+4z-5)=0$$

or $(2+\lambda)x+(-1-2\lambda)y+(\alpha+4\lambda)+6-5\lambda=0$ ...(i)

One of the members of this family is $4x-5y+(3\alpha+2)z-4=0$ ...(ii)

Then, comparing the ratios of the corresponding coefficients of equations (i) and (ii), we get

$$\frac{2+\lambda}{4}=\frac{-1-2\lambda}{-5}=\frac{\alpha+4\lambda}{3\alpha+2}=\frac{6-5\lambda}{-4}\quad ...(iii)$$

Comparing the first two ratios, we get

$$-10-5\lambda=-4-8\lambda$$

$\Rightarrow$ $3\lambda=6$ or $\lambda=2$

Putting $\lambda=2$ in (iii) and comparing the first and third ratios, we get

$$1=\frac{\alpha+8}{3\alpha+2}$$

$\Rightarrow$ $3\alpha+2=\alpha+8$

$\Rightarrow$ $2\alpha=6$

$\Rightarrow$ $\alpha=3$

Illustration 4.52

Find the equation of the image of the plane $x-2y+2z-3=0$ in the plane $x+y+z-1=0$.

Sol. The image of the plane $x-2y+2z-3=0$ ...(i)
in the plane $x+y+z-1=0$ ...(ii)
passes through the line of intersection of the given planes. Therefore, the image plane is one of the members of the family of planes having equation

$(x - 2y + 2z - 3) + t\,(x + y + z - 1) = 0, \quad t \in R$

or $(1 + t)\,x + (-2 + t)\,y + (2 + t)\,z - 3 - t = 0$ ...(iii)

Now, plane (ii) makes the same angle with plane (i) and image plane (iii). Thus,

$$\frac{1-2+2}{3\sqrt{3}} = \pm \frac{1+t-2+t+2+t}{\sqrt{3}\sqrt{(t+1)^2+(t-2)^2+(2+t)^2}}$$

$$\Rightarrow \quad \frac{1}{3} = \pm \frac{3t+1}{\sqrt{3t^2+2t+9}}$$

$\Rightarrow \quad 3t^2 + 2t + 9 = 9\,(9t^2 + 6t + 1)$

$\Rightarrow \quad 78t^2 + 52t = 0$

$\Rightarrow \quad t = 0 \text{ or } t = -\frac{2}{3}$

For $t = 0$, we get plane (i), hence for the image plane, $t = -\frac{2}{3}$.

Thus, the equation of the image plane is

$3\,(x - 2y + 2z - 3) - 2\,(x + y + z - 1) = 0$

or $x - 8y + 4z - 7 = 0$

Illustration 4.53

Write the equation of the line $x - y + 2z = 5$, $3x + y + z = 6$ in symmetric form.

or

Find the line of intersection of planes $x - y + 2z = 5$ and $3x + y + z = 6$.

Sol. We have planes

$x - y + 2z = 5$ ...(i)

and $3x + y + z = 6$ ...(ii)

These equations together represent the line of intersection of the planes.

From (i) + (ii), we get

$4x + 3z = 11$ ...(iii)

From 2(2) – (1), we get

$5x + 3y = 7$ ...(iv)

Comparing x values from (iii) and (iv), we get the equation of the line as

$$x = \frac{7-3y}{5} = \frac{11-3z}{4}$$

$$\text{or} \quad x = \frac{y-\frac{7}{3}}{-\frac{5}{3}} = \frac{z-\frac{11}{3}}{-\frac{4}{3}}$$

$$\text{or} \quad \frac{x}{-3} = \frac{y-\frac{7}{3}}{5} = \frac{z-\frac{11}{3}}{4}$$

Alternatively, putting $x = 0$ in first two equations, we get

$-y + 2z = 5$

and $y + z = 6$.

Solving, we get $y = \frac{11}{3}$ and $z = \frac{7}{3}$

Since the line is perpendicular to the vectors normal to the planes, it is along the vector

$$(\hat{i} - \hat{j} + 2\hat{k}) \times (3\hat{i} + \hat{j} + \hat{k})$$

$$= \begin{vmatrix} \hat{i} & \hat{j} & \hat{k} \\ 1 & -1 & 2 \\ 3 & 1 & 1 \end{vmatrix} = -3\hat{i} + 5\hat{j} + 4\hat{k}$$

Illustration 4.54

Find the equations of the line passing through the point (3, 0, 1) and parallel to the planes $x + 2y = 0$ and $3y - z = 0$.

Sol. Equation of the two planes are $x + 2y = 0$ and $3y - z = 0$

Vectors normal to the planes are $\vec{n}_1 = \hat{i} + 2\hat{j}$ and $\vec{n}_2 = 3\hat{j} - \hat{k}$.

Since required the line is parallel to the given two planes, it is perpendicular to $\vec{n}_1$ and $\vec{n}_2$

Therefore, the line is parallel to the vector

$$\vec{n}_1 \times \vec{n}_2 = \begin{vmatrix} \hat{i} & \hat{j} & \hat{k} \\ 1 & 2 & 0 \\ 0 & 3 & -1 \end{vmatrix} = -2\hat{i} + \hat{j} + 3\hat{k}$$

Hence, the equation of the line through the point (3, 0, 1) and parallel to the given two planes is

$$\frac{x-3}{-2} = \frac{y}{1} = \frac{z-1}{3}$$

Illustration 4.55

Find the angle between the lines $x - 3y - 4 = 0$, $4y - z + 5 = 0$ and $x + 3y - 11 = 0$, $2y - z + 6 = 0$.

Sol. Line $x - 3y - 4 = 0 = 4y - z + 5$ is the line of intersection of planes $x - 3y - 4 = 0$ and $4y - z + 5 = 0$.

Vector along the line of intersection is

$$\begin{vmatrix} \hat{i} & \hat{j} & \hat{k} \\ 1 & -3 & 0 \\ 0 & 4 & -1 \end{vmatrix} = 3\hat{i} + \hat{j} + 4\hat{k} = \vec{a}$$

Line $x + 3y - 11 = 0 = 2y - z + 6$ is the line of intersection of planes $x + 3y - 11 = 0$ and $2y - z + 6 = 0$.

Vector along line of intersection is $\begin{vmatrix} \hat{i} & \hat{j} & \hat{k} \\ 1 & 3 & 0 \\ 0 & 2 & -1 \end{vmatrix} = -3\hat{i} + \hat{j} + 2\hat{k} = \vec{b}$

Now, $\vec{a}\cdot\vec{b} = (3)(-3) + (1)(1) + (4)(2) = 0$

Therefore, the lines are perpendicular.

Illustration 4.56

If the line $x = y = z$ intersects the line $x \sin A + y \sin B + z \sin C = 2d^2$, $x \sin 2A + y \sin 2B + z \sin 2C = d^2$, then find the value of $\sin\frac{A}{2}\cdot\sin\frac{B}{2}\cdot\sin\frac{C}{2}$, where A, B, C are the angles of a triangle.

Sol. Any point on the line $x = y = z$ is $(\lambda, \lambda, \lambda)$, $\lambda \in R$.

We have

$$x \sin A + y \sin B + z \sin C = 2d^2 \quad \ldots\text{(i)}$$

and $$x \sin 2A + y \sin 2B + z \sin 2C = d^2 \quad \ldots\text{(ii)}$$

Above two equations of planes together represent the line of intersection of planes.

If the line $x = y = z$ intersects this line of intersection of planes then point $(\lambda, \lambda, \lambda)$ will satisfy equations (i) and (ii).

$\therefore \quad \lambda \sin A + \lambda \sin B + \lambda \sin C = 2d^2$

and $\quad \lambda \sin 2A + \lambda \sin 2B + \lambda \sin 2C = d^2$

Eliminating d^2 we get

$$\sin A + \sin B + \sin C = 2(\sin 2A + \sin 2B + \sin 2C).$$

$$\Rightarrow \quad 4\cos\frac{A}{2}\cos\frac{B}{2}\cos\frac{C}{2} = 8\sin A\sin B\sin C$$

$$\Rightarrow \quad \sin\frac{A}{2}\sin\frac{B}{2}\sin\frac{C}{2} = \frac{1}{16}$$

Illustration 4.57

Find the point of intersection of the line passing through the point (0, 0, 1) and intersecting the lines $x + 2y + z = 1$, $-x + y - 2z = 2$ and $x + y = 2$, $x + z = 2$ with the xy plane.

Sol. Equation of lines are

$$x + 2y + z - 1 + \lambda(-x + y - 2z - 2) = 0$$

and $$x + y - 2 + \mu(x + z - 2) = 0$$

This line passes through the point (0, 0, 1).

Substituting $x = 0$, $y = 0$ and $z = 1$, we get

$$\lambda = 0, \mu = -2$$

Hence, equation of the line is $x + 2y + z - 1 = 0 = -x + y - 2z + 2$

For the point of intersection with xy plane, we put $z = 0$.

$\therefore \quad x + 2y = 1, -x + y + 2 = 0$

Solving, we get the point of intersection as $\left(\frac{5}{3}, \frac{-1}{3}, 0\right)$.

Illustration 4.58

A horizontal plane $4x - 3y + 7z = 0$ is given. Find the equation of the line of greatest slope in the plane $2x + y - 5z = 0$, which passes through the point (2, 1, 1).

Sol.

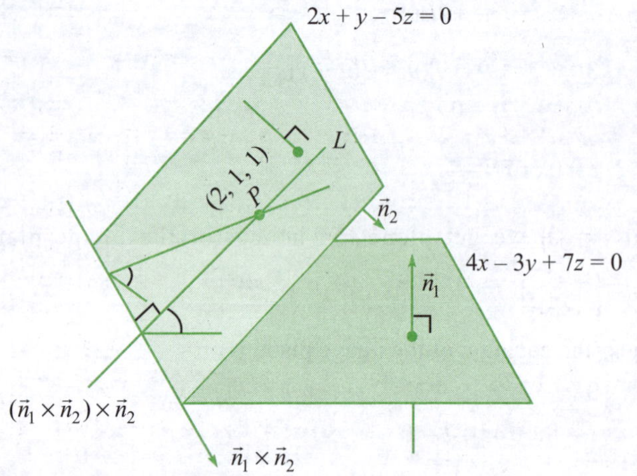

From the figure clearly, the required line L through the point $P(2, 1, 1)$ in the plane $2x + y - 5z = 0$ having greatest slope is the one which is perpendicular to the line of intersection of the planes. $2x + y - 5z = 0$ and $4x - 3y + 7z = 0$

Vector along the line of intersection of planes is $\vec{n}_1\times\vec{n}_2$

$\therefore$ Vector along the required line L is

$$(\vec{n_1}\times\vec{n_2})\times\vec{n_2}$$
$$= (\vec{n_1}\cdot\vec{n_2})\vec{n_2} - |\vec{n_2}|^2\,\vec{n_1}$$
$$= (8 - 3 - 35)(2\hat{i} + \hat{j} - 5\hat{k}) - (4 + 1 + 25)(4\hat{i} - 3\hat{j} + 7\hat{k})$$
$$= -30(2\hat{i} + \hat{j} - 5\hat{k}) - 30(4\hat{i} - 3\hat{j} + 7\hat{k})$$
$$= -30(6\hat{i} - 2\hat{j} + 2\hat{k})$$

So, equation of required line L is $\frac{x-2}{3} = \frac{y-1}{-1} = \frac{z-1}{1}$.

Aim to complete within 25 minutes

Concept Application Exercise 4.5

1. Find the vector equation of the line passing through (1, 2, 3) and parallel to the planes $\vec{r}\cdot(\hat{i} - \hat{j} + 2\hat{k}) = 5$ and $\vec{r}\cdot(3\hat{i} + \hat{j} + \hat{k}) = 6$.
2. Find the equation of plane containing the line $x - y = 2$, $z = 2$ and parallel to the line of intersection of planes $\frac{x}{2} - \frac{z}{3} = 1$ and $y = 2$.
3. Find the equation of the plane which contains the x-axis and passes through the point $(-2, 3, 5)$.

4. Plane P_3 containing the line of intersection of the planes $P_1: x + \alpha y - 3z + 4 = 0$ and $P_2: 2x - y + 4z - 5 = 0$ passes through the points $(0, -11, 0)$ and $(1, 5, 1)$. Find the value of α and equation of plane P_3.

5. If the equation of the plane containing the line $x + 2y + 3z - 4 = 0 = 2x + y - z + 5$ and perpendicular to the plane $\vec{r} = (\hat{i} - \hat{j}) + \lambda(\hat{i} + \hat{j} + \hat{k}) + \mu(\hat{i} - 2\hat{j} + 3\hat{k})$ is $ax + by + cz = 4$, then $(a - b + c)$ is equal to
(1) 24 (2) 22
(3) 20 (4) 18

6. Let the line $\frac{x-3}{7} = \frac{y-2}{-1} = \frac{z-3}{-4}$ intersect the plane containing the lines $\frac{x-4}{1} = \frac{y+1}{-2} = \frac{z}{1}$ and $4ax - y + 5z - 7a = 0 = 2x - 5y - z - 3$, $a \in R$ at the point $P(\alpha, \beta, \gamma)$. Then the value of $\alpha + \beta + \gamma$ equals
(1) 10 (2) 12
(3) 14 (4) 22

Answers

1. $\vec{r} = (\hat{i} + 2\hat{j} + 3\hat{k}) + \lambda(-3\hat{i} + 5\hat{j} + 4\hat{k})$
2. $3x - 5y - 2z - 2 = 0$
3. $5y - 3z = 0$
4. $\alpha = 2$, $7x - y + 9z - 11 = 0$
5. (2)
6. (2)

4.6 DISTANCE OF A POINT FROM A PLANE

Let the equation of the given plane be $\vec{r} \cdot \vec{n} = D$, and let the given point be $P(\vec{p})$, which does not lie on the plane.

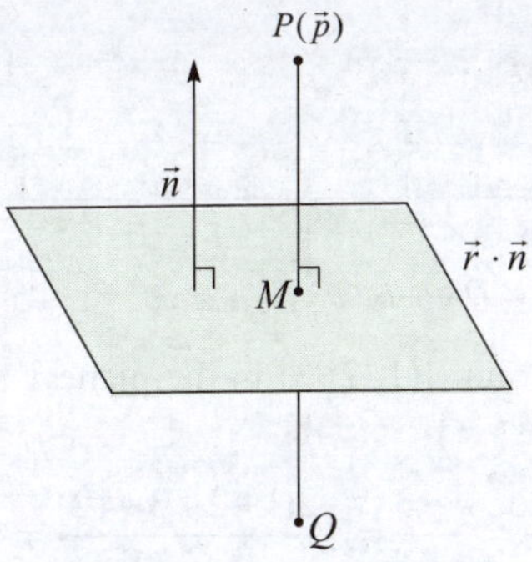

We want to find the distance of the point P from the plane.

Let M be the foot of perpendicular drawn from point P to the plane.

The Line PM passes through P and is parallel to the vector $\vec{n}$, which is normal to the plane.

Therefore, the equation of the line PM is $\vec{r} = \vec{p} + \lambda\vec{n}$, $\lambda \in R$ ...(i)

For point M we substitute $\vec{r} = \vec{p} + \lambda\vec{n}$ in the equation of the plane.

So, we have

$$(\vec{p} + \lambda\vec{n}) \cdot \vec{n} = D$$

$$\Rightarrow \quad \vec{p} \cdot \vec{n} + \lambda\vec{n} \cdot \vec{n} = D$$

$$\Rightarrow \quad \lambda = \frac{D - \vec{p} \cdot \vec{n}}{|\vec{n}|^2}$$

This is the value of λ for point M.

Putting the value of λ in (i), we get the position vector of M as

$$\vec{r} = \vec{p} + \left(\frac{D - \vec{p} \cdot \vec{n}}{|\vec{n}|^2}\right)\vec{n}$$

Now, $\overrightarrow{PM} = P.V.$ of $M - P.V.$ of P

$$= \vec{p} + \left(\frac{D - \vec{p} \cdot \vec{n}}{|\vec{n}|^2}\right)\vec{n} - \vec{p}$$

$$= \left(\frac{D - \vec{p} \cdot \vec{n}}{|\vec{n}|^2}\right)\vec{n}$$

$$\therefore \quad \text{Distance } PM = |\overrightarrow{PM}| = \left|\left(\frac{D - \vec{p} \cdot \vec{n}}{|\vec{n}|^2}\right)\vec{n}\right| = \frac{|D - \vec{p} \cdot \vec{n}|}{|\vec{n}|} \quad \text{...(ii)}$$

If $\vec{p} = x_1\hat{i} + y_1\hat{j} + z_1\hat{k}$ and $\vec{n} = a\hat{i} + b\hat{j} + c\hat{k}$, then the equation of plane becomes $ax + by + cz - D = 0$ or $ax + by + cz + d = 0$, where $d = -D$.

From (ii), we get $PM = \frac{|-d - ax_1 - by_1 - cz_1|}{\sqrt{a^2 + b^2 + c^2}}$

$$= \frac{|ax_1 + by_1 + cz_1 + d|}{\sqrt{a^2 + b^2 + c^2}}$$

Alternatively, let $P \equiv (x_1, y_1, z_1)$ and let the equation of the plane be $ax + by + cz + d = 0$.

Then the equation of line PM is

$$\frac{x - x_1}{a} = \frac{y - y_1}{b} = \frac{z - z_1}{c} = \lambda \quad \text{...(iii)}$$

The coordinates of any point on this line are $(x_1 + \lambda a, y_1 + \lambda b, z_1 + \lambda c)$, where $\lambda \in R$.

For point M, we substitute these coordinates in the equation of the plane.

$$\therefore \quad a(x_1 + \lambda a) + b(y_1 + \lambda b) + c(z_1 + \lambda c) + d = 0$$

$$\Rightarrow \quad \lambda = -\frac{ax_1 + by_1 + cz_1 + d}{a^2 + b^2 + c^2} \quad \text{...(iv)}$$

$$\therefore PM = \sqrt{(x_1 + \lambda a - x_1)^2 + (y_1 + \lambda b - y_1)^2 + (z_1 + \lambda c - z_1)^2}$$

$$= \sqrt{(a^2 + b^2 + c^2)\lambda^2}$$

$$= |\lambda|\sqrt{a^2 + b^2 + c^2}$$

$$= \left|\frac{-(ax_1 + by_1 + cz_1 + d)}{a^2 + b^2 + c^2}\right|\sqrt{a^2 + b^2 + c^2}$$

$$= \frac{|ax_1 + by_1 + cz_1 + d|}{\sqrt{a^2 + b^2 + c^2}}$$

If $M \equiv (x_2, y_2, z_2)$, then using (iii), we get

$$\frac{x_2 - x_1}{a} = \frac{y_2 - y_1}{b} = \frac{z_2 - z_1}{c} = -\frac{ax_1 + by_1 + cz_1 + d}{a^2 + b^2 + c^2} \quad \ldots(v)$$

Using this formula, we can find the coordinates of the foot of perpendicular drawn from a given point the given plane.

Also, if $Q(x_3, y_3, z_3)$ is the image (reflection) of the point P in the plane, then point M will be the midpoint of PQ.

$$\therefore \quad M \equiv \left(\frac{x_1 + x_3}{2}, \frac{y_1 + y_3}{2}, \frac{z_1 + z_3}{2}\right)$$

Putting these coordinates in (v), we get

$$\frac{\frac{x_1 + x_3}{2} - x_1}{a} = \frac{\frac{y_1 + y_3}{2} - y_1}{b} = \frac{\frac{z_1 + z_3}{2} - z_1}{c}$$

$$= -\frac{ax_1 + by_1 + cz_1 + d}{a^2 + b^2 + c^2}$$

or $$\frac{x_3 - x_1}{a} = \frac{y_3 - y_1}{b} = \frac{z_3 - z_1}{c} = -\frac{2(ax_1 + by_1 + cz_1 + d)}{a^2 + b^2 + c^2}$$

Using this formula, we can find the coordinates of the image of the given point in the given plane.

Illustration 4.59

Foot of the perpendicular from the point $P(7, 14, 5)$ on the plane $2x + 4y - z = 2$ is M. Find the length PM and the coordinates of point M. Also, find the coordinates of the image of the point P in the plane.

Sol. Point M is the foot of perpendicular from point $P(7, 14, 5)$ on the plane $2x + 4y - z = 2$.

$$\therefore \quad PM = \frac{|2(7) + 4(14) - (5) - 2|}{\sqrt{2^2 + 4^2 + (-1)^2}}$$

$$= \frac{|14 + 56 - 5 - 2|}{\sqrt{4 + 16 + 1}} = \frac{63}{\sqrt{21}}$$

Let $M \equiv (\alpha, \beta, \gamma)$

$$\therefore \quad \frac{\alpha - 7}{2} = \frac{\beta - 14}{4} = \frac{\gamma - 5}{-1} = -\frac{(2(7) + 4(14) - (5) - 2)}{2^2 + 4^2 + (-1)^2}$$

$$\Rightarrow \quad \frac{\alpha - 7}{2} = \frac{\beta - 14}{4} = \frac{\gamma - 5}{-1} = -3$$

So, $\alpha = 1, \beta = 2, \gamma = 8$

Hence $M \equiv (1, 2, 8)$

Let Q be the image of point P in the plane.

Now, Point M is the mid-point of PQ.

Then, using the section formula, we get $Q \equiv (-5, -10, 11)$.

Illustration 4.60

Find the locus of a point, whose sum of the squares of the perpendicular distances from the planes $x - z = 0$, $x - 2y + z = 0$ and $x + y + z = 0$ is 36.

Sol. Given planes are

$$x - z = 0, x - 2y + z = 0, x + y + z = 0$$

Let the point whose locus is required be $P(\alpha, \beta, \gamma)$.

According to the question,

$$\frac{|\alpha - \gamma|^2}{2} + \frac{|\alpha - 2\beta + \gamma|^2}{6} + \frac{|\alpha + \beta + \gamma|^2}{3} = 36$$

$$\Rightarrow \quad 3(\alpha^2 + \gamma^2 - 2\alpha\gamma) + \alpha^2 + 4\beta^2 + \gamma^2 - 4\alpha\beta - 4\beta\gamma + 2\alpha\gamma$$
$$+ 2(\alpha^2 + \beta^2 + \gamma^2 + 2\alpha\beta + 2\beta\gamma + 2\alpha\gamma) = 36 \times 6$$

$$\Rightarrow \quad 6\alpha^2 + 6\beta^2 + 6\gamma^2 = 36 \times 6$$

$$\Rightarrow \quad \alpha^2 + \beta^2 + \gamma^2 = 36$$

Hence, the required equation of the locus is $x^2 + y^2 + z^2 = 36$.

Illustration 4.61

A ray of light passing through the point $A(1, 2, 3)$, strikes the plane $x + y + z = 12$ at B and after reflection, passes through the point $C(3, 5, 9)$. Find the coordinates of the point B.

Sol.

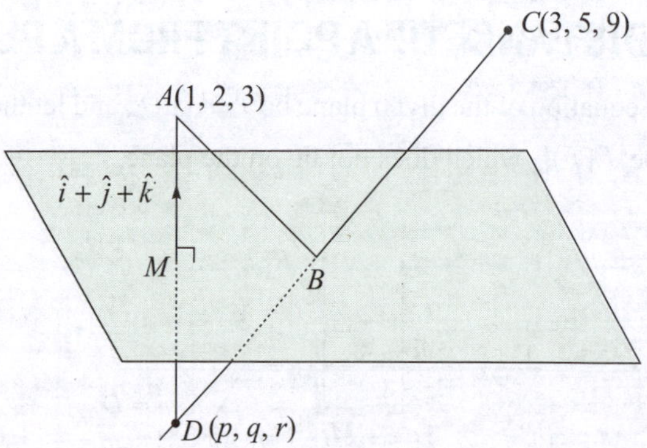

Let the image of point $A(1, 2, 3)$ in the plane $x + y + z = 12$ is $D(p, q, r)$.

$$\therefore \quad \frac{p - 1}{1} = \frac{q - 2}{1} = \frac{r - 3}{1} = \frac{-2(1 + 2 + 3 - 12)}{1^2 + 1^2 + 1^2} = 4$$

Thus, $D \equiv (p, q, r) \equiv (5, 6, 7)$

Now, the equation of line CD is $\frac{x - 5}{3 - 5} = \frac{y - 6}{5 - 6} = \frac{z - 7}{9 - 7}$

or $$\frac{x - 5}{-2} = \frac{y - 6}{-1} = \frac{z - 7}{2}$$

Point B is the point of intersection of CD with the plane.

Let $B \equiv (-2\lambda + 5, -\lambda + 6, 2\lambda + 7)$

This point lies on the plane.

$\therefore \quad -2\lambda + 5 - \lambda + 6 + 2\lambda + 7 = 12$ or $\lambda = 6$

Therefore, $B \equiv (-7, 0, 19)$.

Illustration 4.62

Show that the plane $2x - 2y + z + 12 = 0$ touches the sphere $x^2 + y^2 + z^2 - 2x - 4y + 2z - 3 = 0$.

Sol. The given plane will touch the given sphere if the perpendicular distance from the centre of the sphere to the plane is equal to the radius of the sphere. The centre of the given sphere is (1, 2, –1) and radius is $\sqrt{1^2 + 2^2 + (-1)^2 - (-3)} = 3$.

Length of the perpendicular from (1, 2, –1) to the plane $2x - 2y + z + 12 = 0$ is

$$\left|\frac{2(1) - 2(2) + (-1) + 12}{\sqrt{2^2 + (-2)^2 + 1^2}}\right| = \frac{9}{3} = 3$$

Thus, the given plane touches the given sphere.

Illustration 4.63

If the intercepts of a plane in two systems of rectangular axes having the same origin are a, b, c and a', b', c', then prove that $\frac{1}{a^2} + \frac{1}{b^2} + \frac{1}{c^2} = \frac{1}{a'^2} + \frac{1}{b'^2} + \frac{1}{c'^2}$.

Sol. Equations of the plane in two systems of rectangular axes are

$$\frac{x}{a} + \frac{y}{b} + \frac{z}{c} = 1 \quad \ldots\text{(i)}$$

and $$\frac{x}{a'} + \frac{y}{b'} + \frac{z}{c'} = 1 \quad \ldots\text{(ii)}$$

Clearly, in both the cases the length of the perpendicular from the origin to the plane will be the same.

$$\therefore \quad \frac{\left|\frac{0}{a} + \frac{0}{b} + \frac{0}{c} - 1\right|}{\sqrt{\frac{1}{a^2} + \frac{1}{b^2} + \frac{1}{c^2}}} = \frac{\left|\frac{0}{a'} + \frac{0}{b'} + \frac{0}{c'} - 1\right|}{\sqrt{\frac{1}{a'^2} + \frac{1}{b'^2} + \frac{1}{c'^2}}}$$

$$\Rightarrow \quad \frac{1}{a^2} + \frac{1}{b^2} + \frac{1}{c^2} = \frac{1}{a'^2} + \frac{1}{b'^2} + \frac{1}{c'^2}$$

Illustration 4.64

Find the radius of the circular section of the sphere $|\vec{r}| = 5$ cut by the plane $\vec{r} \cdot (\hat{i} + \hat{j} + \hat{k}) = 4\sqrt{3}$.

Sol.

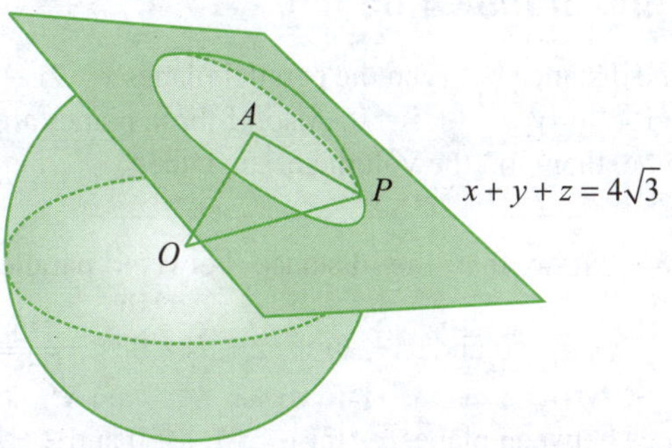

Given sphere is $|\vec{r}| = 5$.

Centre of the sphere is $O(0, 0, 0)$ and the radius is 5 units.

Equation of the plane is $\vec{r} \cdot (\hat{i} + \hat{j} + \hat{k}) = 4\sqrt{3}$

or $\quad x + y + z - 4\sqrt{3} = 0$

Plane cuts the sphere in a circular section.

The foot of perpendicular from O on this plane is A

$$\therefore \quad OA = \frac{|0 + 0 + 0 - 4\sqrt{3}|}{\sqrt{1^2 + 1^2 + 1^2}} = 4$$

Also, OP= radius of sphere = 5

So, in the right angled triangle OAP, $AP^2 = OP^2 - OA^2 = 25 - 16 = 9$

$\therefore \quad AP = 3$ units, which is the radius of the circular section.

4.6.1 Distance Between Two Parallel Planes

Let equations of two parallel planes be

$$ax + by + cz + d = 0 \quad \ldots\text{(i)}$$

and $$ax + by + cz + d' = 0 \quad \ldots\text{(ii)}$$

Distance between planes is the distance of any point on plane (i) from plane (ii).

Consider point $P(x_1, y_1, z_1)$ on the plane (i).

So, $\quad ax_1 + by_1 + cz_1 + d = 0 \quad \ldots\text{(iii)}$

$\therefore$ Distance of point P from the plane (ii)

$$= \frac{|ax_1 + by_1 + cz_1 + d'|}{\sqrt{a^2 + b^2 + c^2}}$$

$$= \frac{|-d + d'|}{\sqrt{a^2 + b^2 + c^2}} \quad \text{(Using (iii))}$$

$$= \frac{|d - d'|}{\sqrt{a^2 + b^2 + c^2}}$$

The distance between two parallel planes $\vec{r} \cdot \vec{n} = d_1$ and $\vec{r} \cdot \vec{n} = d_2$ is given by $\frac{|d_1 - d_2|}{|\vec{n}|}$.

Illustration 4.65

Find the distance between the parallel planes $x + 2y - 2z + 1 = 0$ and $2x + 4y - 4z + 5 = 0$. Also, if these planes are faces of a cube, then find the volume of the cube.

Sol. We know that the distance between parallel planes $ax + by + cz + d_1 = 0$ and $ax + by + cz + d_2 = 0$ is $\frac{|d_1 - d_2|}{\sqrt{a^2 + b^2 + c^2}}$

$\therefore$ Distance between planes $x + 2y - 2z + 1 = 0$ and $x + 2y - 2z + \frac{5}{2} = 0$ is $\frac{|5/2 - 1|}{\sqrt{1+4+4}} = \frac{1}{2}$.

If the given plane represent faces of a cube, then the length of each side of the cube is $\frac{1}{2}$.

Therefore, the volume of the cube is $\left(\frac{1}{2}\right)^3 = \frac{1}{8}$ cu. units .

Aim to complete within 20 minutes

Concept Application Exercise 4.6

1. Find the equations of plane which are parallel to the plane $x - 2y + 2z = 5$ and whose distance from the point (1, 2, 3) is 1.
2. The radius of the circular section of the sphere $|\vec{r}| = 5$ by the plane $\vec{r} \cdot (\hat{i} + 2\hat{j} - \hat{k}) = 4\sqrt{3}$ is
   (1) 2 units (2) 3 units
   (3) 4 units (4) 6 units
3. If the radius of the right cylinder is 3 units and equations of two closed ends are $x + y - z = 4$ and $x + y - z = 10$. Then the volume of the cylinder in units is
   (1) 18π (2) $9\sqrt{2}\pi$
   (3) $18\sqrt{3}\pi$ (4) $12\sqrt{3}\pi$
4. Find the equation of the locus of all those points in space which are equidistant from planes $x + y + z = 3$ and $x + y + z = 9$.
5. A perpendicular drawn from point P on the line $\frac{x+1}{-1} = \frac{y-3}{2} = \frac{z-2}{1}$ to the plane $x + 2y + z - 5 = 0$ such that foot the foot of the perpendicular Q also lies on the plane $x - y + z - 2 = 0$. Then find points P and Q.
6. If the distance between the plane, $23x - 10y - 2z + 48 = 0$ and the plane containing the lines $\frac{x+1}{2} = \frac{y-3}{4} = \frac{z+1}{3}$ and $\frac{x+3}{2} = \frac{y+2}{6} = \frac{z-1}{\lambda}$ $(\lambda \in R)$ is equal to $\frac{k}{\sqrt{633}}$, then $k + \lambda$ is equal to
   (1) −1 (2) −2
   (3) −3 (4) −4

Answers

1. $x - 2y + 2z - 6 = 0$ or $x - 2y + 2z = 0$
2. (2)
3. (3)
4. $x + y + z - 6 = 0$
5. $P \equiv (1, -1, 0), Q \equiv (2, 1, 1)$
6. (4)

4.7 DIVISION OF SPACE BY PLANES AND ANGLE BISECTOR PLANES

4.7.1 Division of Space by One Plane

Regions of one plane

A plane divides the space into two half-regions, the positive region and the negative region.

The positive region of the plane is the region of space toward which the normal vector points.

For example, consider the plane $x + y + z - 1 = 0$.

This plane cuts the coordinate axes at the points (1, 0, 0), (0, 1, 0) and (0, 0, 1) and divides the whole space into two parts (or half- spaces).

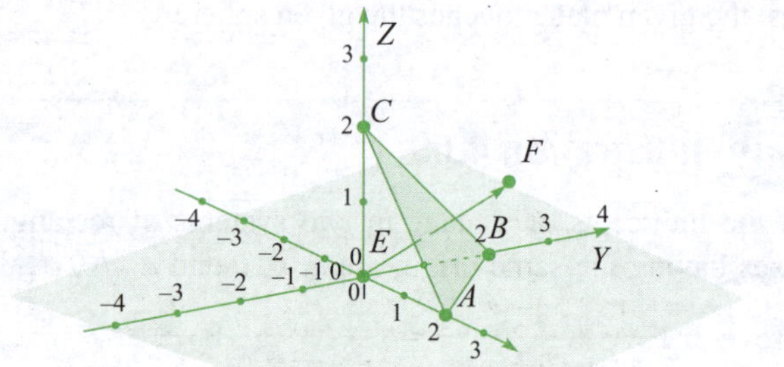

Let $f(x, y, z) = x + y + z - 1$

For the origin, $f(0, 0, 0) = 0 + 0 + 0 - 1 < 0$.

Thus, the origin lies in the negative region of the plane.

The vector normal to the plane is $\vec{n} = \hat{i} + \hat{j} + \hat{k}$, which points from origin towards the plane.

Hence, the positive region of the plane is the region above the plane (i.e., on the side opposite to the origin), where, where $x + y + z - 1 > 0$.

Therefore, the normal vector emerges from the plane into the positive region, indicating the direction in which the value of $x + y + z - 1$ increases.

Since function $f(x, y, z)$ is continuous, there is no sudden jump in sign of $f(x, y, z)$.

A change in the sign of $f(x, y, z)$ can occur only when the point crosses the plane, where $f(x, y, z) = 0$.

Hence, the sign of $f(x, y, z)$ for one point in a region remains the same for all points in that region.

Therefore, any two points (x_1, y_1, z_1) and (x_2, y_2, z_2):

- The points lie on the same side of the plane if $f(x_1, y_1, z_1)$ and $f(x_2, y_2, z_2)$ have the same sign, and
- The points lie on the opposite sides of the plane if $f(x_1, y_1, z_1)$ and $f(x_2, y_2, z_2)$ have opposite signs.

Alternatively, this result can be proved using the section formula which shows that if two points lie on opposite sides of a plane, then the plane divides the line joining them internally in the ratio $|f(x_1, y_1, z_1)|:|f(x_2, y_2, z_2)|$.

For a general plane $ax+by+cz+d=0$, the vector $\vec{n} = a\hat{i} + b\hat{j} + c\hat{k}$ is the normal to the plane and points towards the positive region, where $ax + by + cz + d > 0$.

The origin lies in the positive region if $d > 0$ and in the negative region if $d < 0$.

Illustration 4.66

If points (1, 2, λ) and (–2, λ, 3) lie on the same side of the plane $x + 2y - 3z - 4 = 0$ then find the integral values of λ.

Sol. Let $f(x, y, z) = x + 2y - 3z - 4$

Then $f(1, 2, \lambda) = 1 + 4 - 3\lambda - 4 = 1 - 3\lambda$,

and $f(-2, \lambda, 3) = -2 + 2\lambda - 9 - 4 = 2\lambda - 15$

Since points lie on the same side of the plane,

$f(1, 2, \lambda) \cdot f(-2, \lambda, 3) > 0$

$\Rightarrow \quad (1 - 3\lambda)(2\lambda - 15) > 0$

$\Rightarrow \quad (3\lambda - 1)(2\lambda - 15) < 0$

$\Rightarrow \quad \dfrac{1}{3} < \lambda < \dfrac{15}{2}$

$\Rightarrow \quad \lambda = 1, 2, 3, 4, 5, 6, 7$

4.7.2 Division of Space by Two Intersecting Planes

Consider two intersecting planes

$P_1 : a_1x + b_1y + c_1z + d_1 = 0$

and

$P_2 : a_2x + b_2y + c_2z + d_2 = 0.$

The two planes intersect along a line and divide space into four wedge–shaped regions, denoted by S_1, S_2, S_3, and S_4.

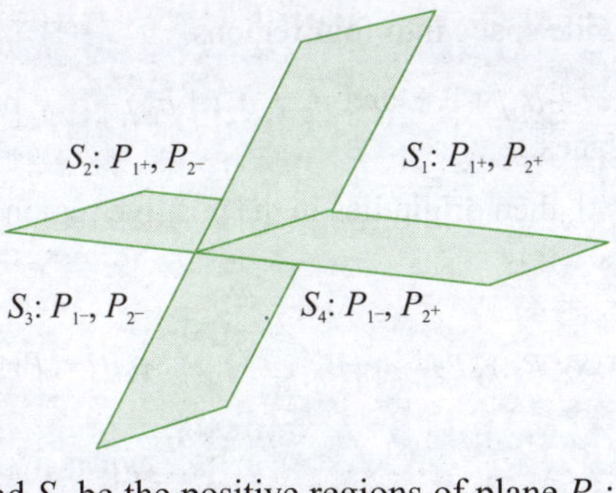

Let S_1 and S_2 be the positive regions of plane P_1; then S_3 and S_4 are its negative regions.

Similarly, if S_1 and S_4 are the positive regions of plane P_2, then S_2 and S_3 will be its negative regions.

Thus, in the vertically opposite regions, the functions $f_1(x, y, z) = a_1x + b_1y + c_1z + d_1$ and $f_2(x, y, z) = a_2x + b_2y + c_2z + d_2$ have either the same sign or opposite signs, depending on the orientation of the planes.

4.7.2.1 Angle (Dihedral Angle) Between Planes

Consider two planes

$P_1 : \vec{r} \cdot \vec{n_1} = d_1$ ... (i)

and $P_2 : \vec{r} \cdot \vec{n_2} = d_2$ ...(ii)

Vector normal to the plane P_1 is $\vec{n_1} = a_1\hat{i} + b_1\hat{j} + c_1\hat{k}$ and vector normal to the plane P_2 is $\vec{n_2} = a_2\hat{i} + b_2\hat{j} + c_2\hat{k}$.

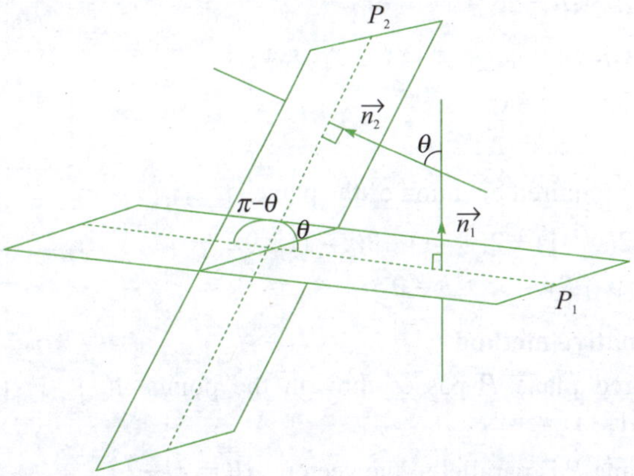

The angle θ between the two planes is defined as the acute (internal) angle between their normal vectors.

Thus, the angle θ between the planes is given by

$$\cos\theta = \frac{|\vec{n_1} \cdot \vec{n_2}|}{|\vec{n_1}||\vec{n_2}|} \text{ or } \theta = \cos^{-1}\frac{|\vec{n_1} \cdot \vec{n_2}|}{|\vec{n_1}||\vec{n_2}|}.$$

The obtuse (external) angle between the planes is

$$\pi - \cos^{-1}\frac{|\vec{n_1} \cdot \vec{n_2}|}{|\vec{n_1}||\vec{n_2}|}.$$

If the planes are perpendicular then vectors normal to the plane are also perpendicular giving $\vec{n_1} \cdot \vec{n_2} = 0$.

Illustration 4.67

Find the acute and obtuse angles between the planes $2x + y - 2z + 3 = 0$ and $\vec{r} \cdot (6\hat{i} + 3\hat{j} + 2\hat{k}) = 5$.

Sol. Vectors normal to the planes are

$2\hat{i} + \hat{j} - 2\hat{k}$ and $6\hat{i} + 3\hat{j} + 2\hat{k}$

Then one of the angles between the planes is

$$\theta = \frac{(2\hat{i} + \hat{j} - 2\hat{k}) \cdot (6\hat{i} + 3\hat{j} + 2\hat{k})}{\sqrt{(2)^2 + (1)^2 + (-2)^2}\sqrt{(6)^2 + (3)^2 + (2)^2}} = \cos^{-1}\frac{11}{21}.$$

Clearly, $\cos^{-1}\dfrac{11}{21}$ is an acute angle.

Hence, the obtuse angle between the planes is $\pi - \cos^{-1}\dfrac{11}{21}$.

Illustration 4.68

Find the equation of plane passing through the points (–1, 1, 1) and (1, –1, 1) and perpendicular to the plane $x + 2y + 2z = 5$.

Sol. Equation of plane passing through the point $A(-1, 1, 1)$ is

$a(x + 1) + b(y - 1) + c(z - 1) = 0$

Also, Plane passes through the point $B(1, -1, 1)$

$\therefore \quad a(1 + 1) + b(-1 - 1) + c(1 - 1) = 0$

or $\quad a - b = 0 \qquad \ldots(i)$

Now, plane is perpendicular to the given plane $x + 2y + 2z = 5$

$\therefore \quad a + 2b + 2c = 0 \qquad \ldots(ii)$

From (i), putting $a = b$ in (ii), we get

$$3a + 2c = 0$$

Thus, $a : b : c = 2 : 2 : -3$

So, the required equation of the plane is

$$2(x + 1) + 2(y - 1) - 3(z - 1) = 0$$

or $\quad 2x + 2y - 3z + 3 = 0$

Alternative method

Required plane P passes through the points $A(-1, 1, 1)$ and $B(1, -1, 1)$.

So, plane P is parallel to the vector, $\overrightarrow{AB} = 2\hat{i} - 2\hat{j}$.

Also, the required plane is perpendicular to the plane $x + 2y + 2z = 5$.

So, vector normal to this plane, which is $\vec{p} = \hat{i} + 2\hat{j} + 2\hat{k}$ will be parallel to the required plane P.

Thus, plane P is parallel to vectors $\overrightarrow{AB}$ and $\vec{p}$.

So, the vector that is normal to the plane P is

$$\overrightarrow{AB} \times \vec{p} = \begin{vmatrix} \hat{i} & \hat{j} & \hat{k} \\ 2 & -2 & 0 \\ 1 & 2 & 2 \end{vmatrix} = -4\hat{i} - 4\hat{j} + 6\hat{k}$$

Therefore, the equation of the plane is $-4(x + 1) - 4(y - 1) + 6(z - 1) = 0$ or $2x + 2y - 3z + 3 = 0$.

Illustration 4.69

Find the equations of planes passing through (1, 0, 0) and (0, 1, 0) and making an angle 45° with the plane $x + y - 3 = 0$.

Sol. Equation of the plane passing through the point (1, 0, 0) is

$$a(x - 1) + b(y - 0) + c(z - 0) = 0$$

Plane also passes through the point (0, 1, 0).

$\therefore \quad -a + b = 0$ or $a = b$.

So, equation of plane becomes

$ax + ay + cz - a = 0 \qquad \ldots(i)$

This plane makes an angle 45° with the plane $x + y - 3 = 0$.

$$\therefore \quad \cos 45° = \frac{|a + a + 0|}{\sqrt{a^2 + a^2 + c^2}\sqrt{1^2 + 1^2}}$$

$$\Rightarrow \quad |2a| = \sqrt{2a^2 + c^2}$$

$$\Rightarrow \quad 4a^2 = 2a^2 + c^2$$

$$\Rightarrow \quad 2a^2 = c^2$$

$$\Rightarrow \quad c = \pm\sqrt{2}a$$

So, the equations of the planes are $ax + ay \pm \sqrt{2}az - a = 0$ or $x + y \pm \sqrt{2}z - 1 = 0$

Illustration 4.70

The plane $ax + by = 0$ is rotated through an angle α about its line of intersection with the plane $z = 0$. Show that the equation of the plane in the new position is $ax + by \pm z\sqrt{a^2 + b^2}\tan\alpha = 0$.

Sol. Given planes are $ax + by = 0 \qquad \ldots(i)$

and $z = 0 \qquad \ldots(ii)$

Then, the General equation of planes passing through the line of intersection of planes (i) and (ii) is

$ax + by + \lambda z = 0 \qquad \ldots(iii)$

Given that the angle between planes (ii) and (iii) is α, then

$$\cos\alpha = \frac{a \cdot a + b \cdot b + \lambda \cdot 0}{\sqrt{a^2 + b^2 + \lambda^2}\sqrt{a^2 + b^2}} = \sqrt{\frac{a^2 + b^2}{a^2 + b^2 + \lambda^2}}$$

$$\Rightarrow \quad \frac{\lambda^2}{a^2 + b^2} + 1 = \sec^2\alpha$$

$$\Rightarrow \quad \lambda^2 = (a^2 + b^2)(\sec^2\alpha - 1) = (a^2 + b^2)\tan^2\alpha$$

$$\Rightarrow \quad \lambda = \pm\sqrt{a^2 + b^2}\tan\alpha$$

So, equations of the required planes are

$$ax + by \pm z\sqrt{a^2 + b^2}\tan\alpha = 0$$

4.7.2.1 Region Containing Given Point

Consider two intersecting and non-perpendicular planes,

$$P_1 : f_1(x, y, z) = a_1x + b_1y + c_1z + d_1 = 0$$

$$P_2 : f_2(x, y, z) = a_2x + b_2y + c_2z + d_2 = 0$$

Planes divide space into four regions.

Let $\vec{n_1} = a_1\hat{i} + b_1\hat{j} + c_1\hat{k}$ and $\vec{n_2} = a_2\hat{i} + b_2\hat{j} + c_2\hat{k}$ be the normal, to the two planes.

If $d_1, d_2 > 0$, then origin lies in the positive region (S_1) of both planes.

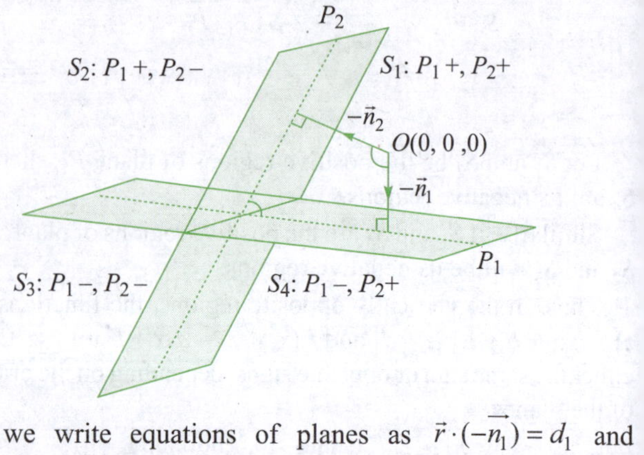

So, we write equations of planes as $\vec{r} \cdot (-\vec{n_1}) = d_1$ and $\vec{r} \cdot (-\vec{n_2}) = d_2$, where $-\vec{n_1}$ and $-\vec{n_2}$ are the vectors normal to the plane from the origin.

Now if the angle between planes in the region where the origin lies is acute then the angle between the vectors $-\vec{n_1}$ and $-\vec{n_2}$ will be obtuse.

$\therefore \quad (-\vec{n_1})\cdot(-\vec{n_2}) < 0$

or $\quad \vec{n_1}\cdot\vec{n_2} < 0$

$\Rightarrow \quad a_1a_2 + b_1b_2 + c_1c_2 < 0$

Similarly, if $d_1, d_2 < 0$, then origin lies in the negative region (S_3) of both the planes.

So, we write equations of planes as $\vec{r}\cdot\vec{n_1} = -d_1$ and $\vec{r}\cdot\vec{n_2} = -d_2$, where $\vec{n_1}$ and $\vec{n_2}$ are the vectors normal to the plane from the origin.

Now, if the angle between the planes in the region where the origin lies is acute then the angle between the vectors $\vec{n_1}$ and $\vec{n_2}$ will be obtuse.

$\therefore \quad \vec{n_1}\cdot\vec{n_2} < 0$

$\Rightarrow \quad a_1a_2 + b_1b_2 + c_1c_2 < 0$

Thus, if $d_1 \cdot d_2 > 0$ and $a_1a_2 + b_1b_2 + c_1c_2 < 0$, then origin lies in the acute angle region.

Similarly, it can be shown that if $d_1 \cdot d_2 > 0$ and $a_1a_2 + b_1b_2 + c_1c_2 > 0$ then origin lies in the obtuse angle region.

Further if $d_1 > 0$ and $d_2 < 0$, then the origin lies in the positive region plane of P_1 and negative region of plane P_2.

So, we write the equations of planes as $\vec{r}\cdot(-\vec{n_1}) = d_1$ and $\vec{r}\cdot\vec{n_2} = -d_2$, where $-\vec{n_1}$ and $\vec{n_2}$ are vectors normal to the plane from the origin.

Now, if the angle between the planes in the region where origin lies is acute then the angle between vectors $-\vec{n_1}$ and $\vec{n_2}$ will be obtuse.

$\therefore \quad (-\vec{n_1})\cdot\vec{n_2} < 0$

$\Rightarrow \quad \vec{n_1}\cdot\vec{n_2} > 0$

$\Rightarrow \quad a_1a_2 + b_1b_2 + c_1c_2 > 0$

Similarly, if $d_1 < 0$, $d_2 > 0$ and the origin lies in the acute angle region then also we get $a_1a_2 + b_1b_2 + c_1c_2 > 0$.

Finally, if d_1 and d_2 have opposite signs then the origin will lie in the obtuse angle region if $a_1a_2 + b_1b_2 + c_1c_2 < 0$.

The above conclusions are also true for any point $P(x_1, y_1, z_1)$. i.e. if $f_1 = f_1(x_1, y_1, z_1)$ and $f_2 = f_2(x_1, y_1, z_1)$, then

Point P lies in the acute angle region if $f_1 \cdot f_2 > 0$ and $a_1a_2 + b_1b_2 + c_1c_2 < 0$ or $f_1 \cdot f_2 < 0$ and $a_1a_2 + b_1b_2 + c_1c_2 > 0$.

Point P lies in the obtuse angle region if $f_1 \cdot f_2 > 0$ and $a_1a_2 + b_1b_2 + c_1c_2 > 0$ or $f_1 \cdot f_2 < 0$ and $a_1a_2 + b_1b_2 + c_1c_2 < 0$.

Illustration 4.71

Consider the planes $2x - y + 2z - 3 = 0$, $x + 2y - 2z + 4 = 0$. In which regions of the planes origin and point (1, 2, 3) lie?

Sol. We have the planes

$2x - y + 2z - 3 = 0 \quad$ ...(i)

and $\quad x + 2y - 2z + 4 = 0 \quad$ ...(ii)

Let $f_1(x, y, z) = 2x - y + 2z - 3$ and $f_2(x, y, z) = x + 2y - 2z + 4$

We have $a_1a_2 + b_1b_2 + c_1c_2 = (2)(1) + (-1)(2) + (2)(-2) < 0$

For the origin, $f_1(0, 0, 0) = -3$ and $f_2(0, 0, 0) = 4$

Thus, $f_1 \times f_2 < 0$ and $a_1a_2 + b_1b_2 + c_1c_2 < 0$

So, origin lies in the obtuse angle region between the two planes. We can also say that in obtuse angle regions, f_1 and f_2 have opposite signs.

For the point (1, 2, 3), $f_1(1, 2, 3) = 3$ and $f_2(1, 2, 3) = 3$

Thus, $f_1f_2 > 0$. So, point (1, 2, 3) lies in the acute angle region between the planes. We can also say that in acute angle regions, f_1 and f_2 have the same sign.

Illustration 4.72

Find the values of 'α' if origin lies in the obtuse angle region of the planes $x - 2y + 3z + 5 = 0$ and $\alpha x + 3y - z + 7 = 0$.

For which values of 'α' origin will lie in the acute angle region?

Sol. We have $f_1(x, y, z) = x - 2y + 3z + 5$ and $f_2(x, y, z) = \alpha x + 3y - z + 7$.

Therefore, $f_1(0, 0, 0) = 5$ and $f_2(0, 0, 0) = 7$.

So, origin lies in the obtuse angle region if $a_1a_2 + b_1b_2 + c_1c_2 > 0$.

$\therefore (1)(\alpha) + (-2)(3) + (3)(-1) > 0 \Rightarrow \alpha > 9$

Also, origin will lie in the acute angle region if $a_1a_2 + b_1b_2 + c_1c_2 < 0$

$\Rightarrow \quad \alpha < 9$

4.7.3 Bisector Planes

Consider two intersecting and non-perpendicular planes

$$P_1 : f_1(x, y, z) = a_1x + b_1y + c_1z + d_1 = 0$$

and $\quad P_2 : f_2(x, y, z) = a_2x + b_2y + c_2z + d_2 = 0$

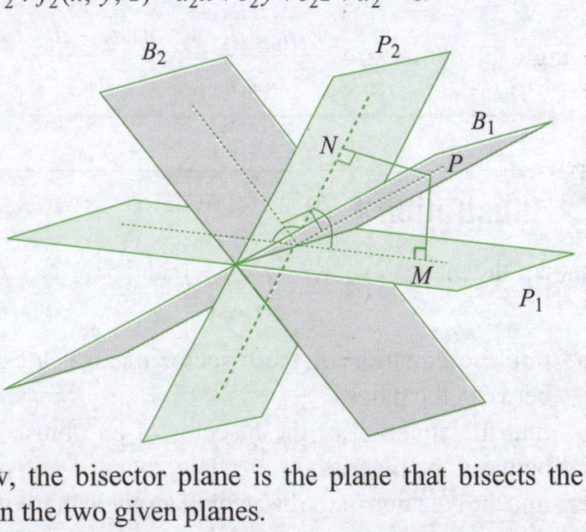

Now, the bisector plane is the plane that bisects the angle between the two given planes.

There are two bisector planes, one corresponding to the acute angle and the other to the obtuse angle.

To obtain the equations of the bisector planes, we use the condition that any point on a bisector plane is equidistant from the given planes.

So, for a variable point P on one of the bisector planes, say B_1, we have

$PM = PN$

$$\therefore \quad \frac{|a_1x+b_1y+c_1z+d_1|}{\sqrt{a_1^2+b_1^2+c_1^2}} = \frac{|a_2x+b_2y+c_2z+d_2|}{\sqrt{a_2^2+b_2^2+c_2^2}}$$

$$\text{or} \quad B_1: \frac{a_1x+b_1y+c_1z+d_1}{\sqrt{a_1^2+b_1^2+c_1^2}} = \frac{a_2x+b_2y+c_2z+d_2}{\sqrt{a_2^2+b_2^2+c_2^2}} \quad \ldots\text{(i)}$$

$$\text{and} \quad B_2: \frac{a_1x+b_1y+c_1z+d_1}{\sqrt{a_1^2+b_1^2+c_1^2}} = -\frac{a_2x+b_2y+c_2z+d_2}{\sqrt{a_2^2+b_2^2+c_2^2}} \quad \ldots\text{(ii)}$$

These are the equations of planes bisecting the angle between the planes P_1 and P_2.

Clearly, plane B_1 bisects the angle between the planes in the region where signs of f_1 and f_2 are same and plane B_2 bisects the angle between the planes in the region where signs of f_1 and f_2 are different.

If $d_1 \times d_2 > 0$ and $a_1a_2 + b_1b_2 + c_1c_2 < 0$ then origin lies in the acute angle region and acute angle bisector is B_1.

If $d_1 \times d_2 < 0$ and $a_1a_2 + b_1b_2 + c_1c_2 > 0$, then origin lies in the acute angle region and acute angle bisector is B_2.

If $d_1 \times d_2 > 0$ and $a_1a_2 + b_1b_2 + c_1c_2 > 0$ then origin lies in the obtuse angle region and obtuse angle bisector is B_1.

If $d_1 \times d_2 < 0$ and $a_1a_2 + b_1b_2 + c_1c_2 < 0$, then origin lies in the obtuse angle region and obtuse angle bisector is B_2.

If $f_1(x_1, y_1, z_1)$ and $f_2(x_1, y_1, z_1)$ have the same sign then the bisector of the angle containing point $P(x_1, y_1, z_1)$ is B_1.

If $f_1(x_1, y_1, z_1)$ and $f_2(x_1, y_1, z_1)$ have opposite signs then the bisector of the angle containing point $P(x_1, y_1, z_1)$ is B_2.

NOTES

If we consider the equations of planes in the vector form i.e. $\vec{r}\cdot\vec{n_1} = d_1$ and $\vec{r}\cdot\vec{n_2} = d_2$, then the equations of the angle bisectors are $\dfrac{\vec{r}\cdot\vec{n_1} - d_1}{|\vec{n_1}|} = \pm\dfrac{\vec{r}\cdot\vec{n_2} - d_2}{|\vec{n_2}|}$

Illustration 4.73

Consider the planes $2x - y + 2z + 3 = 0$ and $3x - 2y - 6z - 8 = 0$. Then

a. find the equation of the bisector of an acute angle between the planes.

b. find the equation of the bisector of an obtuse angle between the planes.

c. find the equation of the bisector of an angle that contains the origin.

d. find the equation of the bisector of an angle that contains the point $(2, -1, -3)$.

e. find the equations of the planes that pass through the point $(1, 2, 3)$ and equally inclined to the given planes.

Sol. Given planes are

$$2x - y + 2z + 3 = 0 \quad \ldots\text{(i)}$$

$$\text{and} \quad 3x - 2y - 6z - 8 = 0 \quad \ldots\text{(ii)}$$

Compare equations with $a_1x + b_1y + c_1z + d_1 = 0$ and $a_2x + b_2y + c_2z + d_2 = 0$.

We find that d_1 and d_2 have opposite signs and $a_1a_2 + b_1b_2 + c_1c_2 = (2)(3) + (-1)(-2) + (2)(-6) < 0$

So, equation of the bisector of the obtuse angle is

$$\frac{2x - y + 2z + 3}{\sqrt{2^2 + (-1)^2 + 2^2}} = -\frac{3x - 2y - 6z - 8}{\sqrt{3^2 + (-2)^2 + (-6)^2}}$$

$$\text{or} \quad \frac{2x - y + 2z + 3}{3} = -\frac{3x - 2y - 6z - 8}{7}$$

$$\text{or} \quad 23x - 13y - 4z - 3 = 0 \quad \ldots\text{(iii)}$$

and the equation of the bisector of the acute angle is

$$\frac{2x - y + 2z + 3}{3} = \frac{3x - 2y - 6z - 8}{7}$$

$$\text{or} \quad 5x - y + 32z + 45 = 0 \quad \ldots\text{(iv)}$$

Alternative method

In this method, we first find the equations of the two bisectors planes.

Consider one of the bisectors, say $23x - 13y - 4z - 3 = 0$.

Here we do not know which angle this plane bisects.

Therefore, we find the angle between this bisector plane and one of the given planes, say $2x - y + 2z + 3 = 0$.

If this angle is $\dfrac{\theta}{2}$, then

$$\cos\frac{\theta}{2} = \frac{|46 + 13 - 8|}{\sqrt{529 + 169 + 16}\sqrt{4 + 1 + 4}} < \frac{1}{\sqrt{2}}$$

$$\therefore \quad \frac{\theta}{2} > 45^\circ$$

(as cosine is decreasing function in the first quadrant)

$$\Rightarrow \quad \theta > 90^\circ$$

So, plane $23x - 13y - 4z - 3 = 0$ bisects the obtuse angle.

c. Since d_1 and d_2 have opposite signs, the bisector that contains the origin is the one given by equation (iii).

d. Let $f_1(x, y, z) = 2x - y + 2z + 3$ and

$f_2(x, y, z) = 3x - 2y - 6z - 8$

$\therefore \quad f_1(2, -1, -3) = 2(2) - (-1) + 2(-3) + 3 > 0$

and $\quad f_2(2, -1, -3) = 3(2) - 2(-1) - 6(-3) - 8 > 0$

So, the bisector that contains the point $(2, -1, -3)$ is the one given by equation (iv).

e. Planes which are equally inclined to the given planes are parallel to the angle bisector planes.

So, the required equations of the planes are $23x - 13y - 4z + \lambda_1 = 0$ and $5x - y + 32z + \lambda_2 = 0$

These planes are passing through the point $(1, 2, 3)$.

$\therefore 23(1) - 13(2) - 4(3) + \lambda_1 = 0$ and $5(1) - (2) + 32(3) + \lambda_2 = 0$

$\Rightarrow \lambda_1 = 15$ and $\lambda_2 = 99$

So, the equations of the planes are $23x - 13y - 4z + 15 = 0$ and $5x - y + 32z - 99 = 0$

Aim to complete within 25 minutes

Concept Application Exercise 4.7

1. If acute angle between planes $x + y - z - 4 = 0$ and $\alpha x - y + 2z - 5 = 0$ is $\cos^{-1}\dfrac{1}{3\sqrt{3}}$, then the sum of the value(s) of α is
   (1) 4.5 (2) 5.5
   (3) 6 (4) 6.75
2. The plane passing through the point (–1, 3, 2) and perpendicular to the planes $x + 2y + 3z = 5$ and $3x + 3y + z = 0$ is $\alpha x + \beta y + 3z + \gamma = 0$ then the value of $\alpha + \beta + \gamma$ is
   (1) 22 (2) 23
   (3) 24 (4) 28
3. The equation of the plane passing through the straight line $\dfrac{x-1}{2} = \dfrac{y+2}{-3} = \dfrac{z}{5}$ and perpendicular to the plane $x - y + z + 2 = 0$ is $ax + by + cz + 4 = 0$, then the value of $a + b + c$ is
   (1) 4 (2) 5
   (3) 6 (4) 7
4. Find the equation of a plane which passes through the point (1, 2, 3) and which is equally inclined to the planes $x - 2y + 2z - 3 = 0$ and $8x - 4y + z - 7 = 0$.
5. In which region points (1, 2, –3) and (–10, 3, –5) lie w.r.t. planes $x - 2y + 2z - 3 = 0$ and $2x + y - 2z + 4 = 0$ (acute or obtuse angle region)? Also find the equations of bisectors of those regions.
6. Plane $x - y + z - 5 = 0$ is rotated by an angle 90° about its line of intersection with the plane $x + y - z + 4 = 0$. Find the equation of plane in new position.
7. A plane containing the line L: $\alpha x + y - 5z + 6 = 0$ and $3x - (\alpha + 3)y + z - 7 = 0$ and perpendicular to the plane $x + y - z - 3 = 0$ is $13x - 14y - z - 22 = 0$ then the value of α is
   (1) –1 (2) 1 (3) 2 (4) 4

Answers

1. (4)
2. (3)
3. (3)
4. $11x - 10y + 7z + 12 = 0$, $5x + 2y - 5z + 6 = 0$
5. (1, 2, –3) lies in the acute angle region and (–10, 3, –3) lies in the obtuse angle region.
   $3x - y + 1 = 0$ is obtuse angle bisector
   $x + 3y - 4z + 7 = 0$ is acute angle bisector.
6. $-2x - 4y + 4z - 17 = 0$
7. (2)

4.8 LINE AND PLANE

4.8.1 Line Lies in the Plane

Consider line $L \equiv \vec{r} = \vec{a} + \lambda\vec{b}$ and plane $\pi \equiv \vec{r}\cdot\vec{n} = d$

The line is parallel to the vector $\vec{b}$ and the plane is perpendicular to the vector $\vec{n}$.

In the following figure, line L is lying in the plane π.

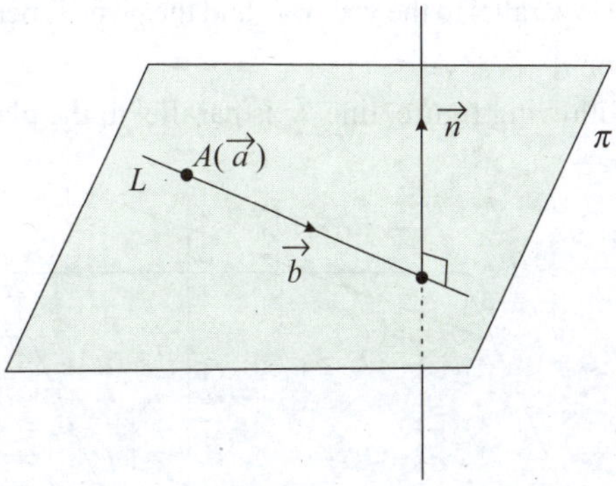

Therefore, the point A on the line having position vector $\vec{a}$ lies in the plane, which gives

$$\vec{a}\cdot\vec{n} = d \quad \text{...(i)}$$

Also, the direction vector $\vec{b}$ of the line is perpendicular to the normal vector $\vec{n}$ of the plane.

Hence, $\vec{b}\cdot\vec{n} = 0$ ...(ii)

Alternatively, putting $\vec{r} = \vec{a} + \lambda\vec{b}$ in the equation of the plane, we have

$$(\vec{a} + \lambda\vec{b})\cdot\vec{n} = 0$$

or $\vec{a}\cdot\vec{n} + \lambda\vec{b}\cdot\vec{n} = 0$

Now, the above equation is an identity which is true for all real values of λ.

Hence, $\vec{a}\cdot\vec{n} = d$ and $\vec{b}\cdot\vec{n} = 0$.

Illustration 4.74

If line $\dfrac{x-1}{2} = \dfrac{y+2}{\alpha} = \dfrac{z-3}{5}$ lies in the plane $2x - y + 3z + \beta = 0$, then find the values of α and β.

Sol. Coordinates of any point on the given line are $(2\lambda + 1, \alpha\lambda - 2, 5\lambda + 3)$.

Since the line lies on the plane, these coordinates satisfy the equation of the plane for all real values of λ.

Putting the coordinates in the equation of the plane, we get

$$2(2\lambda + 1) - (\alpha\lambda - 2) + 3(5\lambda + 3) + \beta = 0$$

or $\lambda(19 - \alpha) + 13 + \beta = 0$

This is true for all real values of λ.

$\therefore$ $19 - \alpha = 0$ and $13 + \beta = 0$

$\Rightarrow$ $\alpha = 19$ and $\beta = -13$

Alternatively, the vector normal to the plane and the vector to which line is parallel are perpendicular.

$\therefore\quad (2)(2) + (\alpha)(-1) + (5)(3) = 0$

$\Rightarrow\quad \alpha = 19.$

Also, the point (1, –2, 3) on the line lies in the plane

$\therefore\quad 2(1) - (-2) + 3(3) + \beta = 0 \Rightarrow \beta = -13$

4.8.2 Line Parallel to the Plane

Consider line $L \equiv \vec{r} = \vec{a} + \lambda\vec{b}$ and plane $\pi \equiv \vec{r}\cdot\vec{n} = d$

The line is parallel to the vector $\vec{b}$ and the plane is perpendicular to the vector $\vec{n}$.

In the following figure, line L is parallel to the plane π.

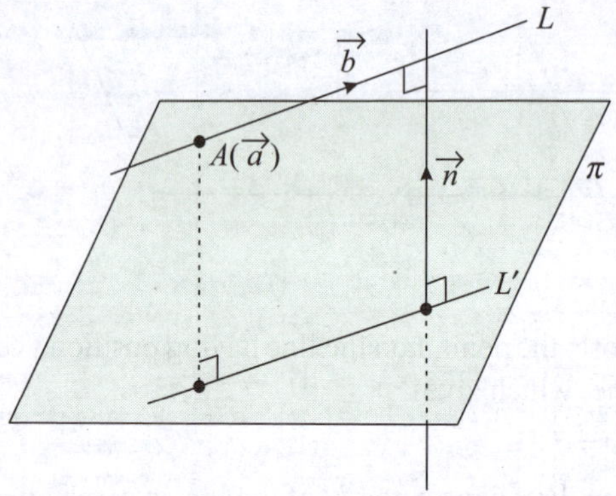

Clearly, vector $\vec{b}$ is perpendicular to the vector $\vec{n}$, but point $A(\vec{a})$ does not lie in the plane.

Therefore,

$$\vec{b}\cdot\vec{n} = 0 \text{ and } \vec{a}\cdot\vec{n} \neq d.$$

Illustration 4.75

Find the equation of the line passing through the point (1, 2, 3), parallel to the plane $x - y - 2z - 3 = 0$ and intersecting the line $\frac{x+1}{3} = \frac{y-4}{2} = \frac{z+2}{-1}$.

Sol. Let the direction ratios of required line are a, b, c.

So, the required equation of the line is $\frac{x-1}{a} = \frac{y-2}{b} = \frac{z-3}{c}$

This line is parallel to the plane $x - y - 2z - 3 = 0$.

So, $a - b - 2c = 0$ …(i)

The required line intersects the line $\frac{x+1}{3} = \frac{y-4}{2} = \frac{z+2}{-1}$.

$$\therefore\quad \begin{vmatrix} 1-(-1) & 2-4 & 3-(-2) \\ 3 & 2 & -1 \\ a & b & c \end{vmatrix} = 0$$

$$\Rightarrow\quad \begin{vmatrix} 2 & -2 & 5 \\ 3 & 2 & -1 \\ a & b & c \end{vmatrix} = 0$$

$\Rightarrow\quad -8a + 17b + 10c = 0$ …(ii)

Solving (i) and (ii), we get $a : b : c = 8 : 2 : 3$

So, the required equation of the line is $\frac{x-1}{8} = \frac{y-2}{2} = \frac{z-3}{3}$

Illustration 4.76

Find the equation of the projection and the image of the line $\frac{x-1}{9} = \frac{y-2}{-1} = \frac{z+3}{-3}$ in the plane $3x - 3y + 10z - 26 = 0$.

Sol. Given line is parallel to the vector $\vec{b} = 9\hat{i} - \hat{j} - 3\hat{k}$. Vector normal to the given plane is $3\hat{i} - 3\hat{j} + 10\hat{k}$.

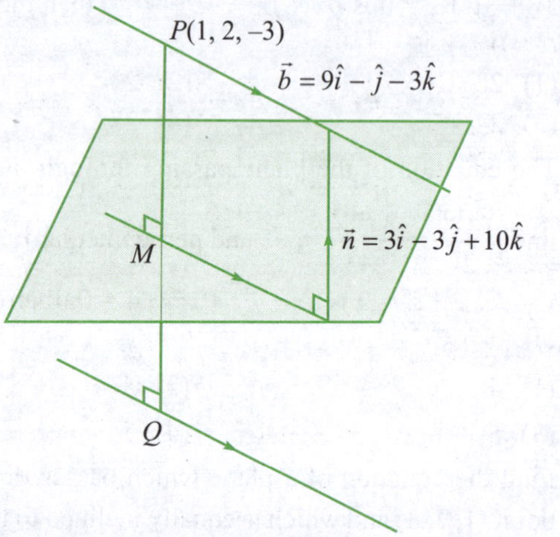

Now, $\vec{b}\cdot\vec{n} = (9)(3) + (-1)(-3) + (-3)(10) = 27 + 3 - 30 = 0.$

Also, line is passing through the point $P(1, 2, -3)$ which does not satisfy the equation of the given plane.

So, the line is parallel to the plane.

Hence the projection of the line on the plane and image of line in the plane are parallel to the given line and have direction ratios 9, –1, –3.

Now, if $M(x_1, y_1, z_1)$ is the foot of the perpendicular from the point P on the plane, then

$$\frac{x_1-1}{3} = \frac{y_1-2}{-3} = \frac{z_1+3}{10} = -\frac{3(1)-3(2)+10(-3)-26}{9+9+100} = \frac{1}{2}$$

$\therefore\quad M \equiv (5/2, 1/2, 2)$

Therefore, the equation of the projection of the line on the plane is $\frac{x-5/2}{9} = \frac{y-1/2}{-1} = \frac{z-2}{-3}$

Also, if the image of point P in the plane is $Q(x_2, y_2, z_2)$, then M is mid-point of PQ.

$\therefore\quad Q \equiv (4, -1, 7)$

Hence, the equation of the image line is $\frac{x-4}{9} = \frac{y+1}{-1} = \frac{z-7}{-3}$

4.8.3 Line Perpendicular to the Plane

Consider line $L \equiv \vec{r} = \vec{a} + \lambda\vec{b}$ and plane $\pi \equiv \vec{r}\cdot\vec{n} = d$

The line is parallel to the vector $\vec{b}$ and the plane is perpendicular to the vector $\vec{n}$.

In the following figure, line L is perpendicular to the plane π.

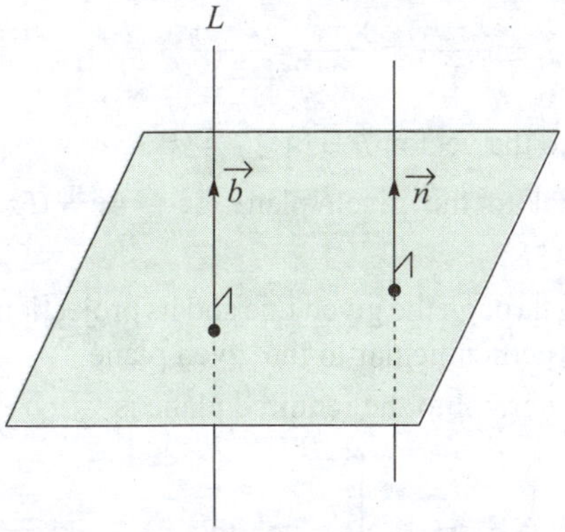

Therefore, vector $\vec{b}$ is parallel to the vector $\vec{n}$.

Hence, $\vec{b} = \lambda\vec{n}$ or $\vec{b} \times \vec{n} = 0$.

Point A may or may not lie on the plane.

Illustration 4.77

If the projection of the line $\frac{x-1}{2} = \frac{y+2}{-3} = \frac{z-3}{4}$ on the plane $3x + \alpha y - \beta z - 15 = 0$ is a point, then find the values of α and β.

Sol. Normal vector the plane $3x + \alpha y - \beta z - 15 = 0$ is $3\hat{i} + \alpha\hat{j} - \beta\hat{k}$.

Line $\frac{x-1}{2} = \frac{y+2}{-3} = \frac{z-3}{4}$ is parallel to the vector $2\hat{i} - 3\hat{j} + 4\hat{k}$.

Since, the projection of line on the plane is a point, line is perpendicular to the plane.

Therefore, vectors $3\hat{i} + \alpha\hat{j} - \beta\hat{k}$ and $2\hat{i} - 3\hat{j} + 4\hat{k}$ are parallel.

$$\therefore \quad \frac{3}{2} = \frac{\alpha}{-3} = \frac{-\beta}{4}$$

$$\Rightarrow \quad \alpha = -\frac{9}{2}, \beta = -6$$

4.8.4 Line Intersecting the Plane Obliquely

Consider line $L \equiv \vec{r} = \vec{a} + \lambda\vec{b}$ and plane $\pi \equiv \vec{r} \cdot \vec{n} = d$

The line is parallel to the vector $\vec{b}$ and the plane is perpendicular to the vector $\vec{n}$.

In the following figure, line L is intersecting the plane π obliquely at an angle θ.

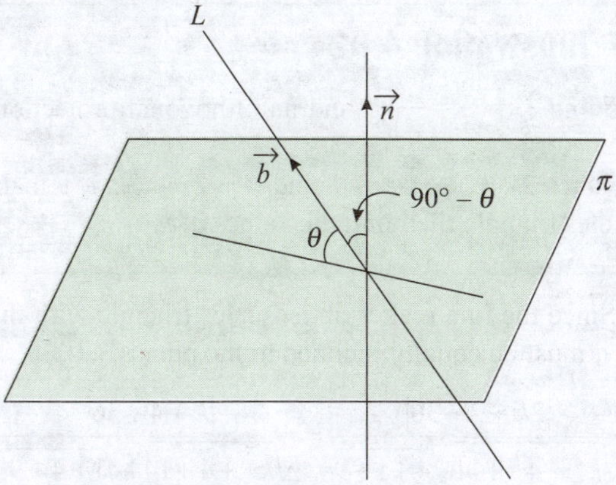

- Since line is not parallel to the plane, we have $\vec{b} \cdot \vec{n} \neq 0$
- Also, line is not perpendicular to the plane.

  So, vector $\vec{b}$ is not parallel to vector $\vec{n}$.

  $$\therefore \quad \vec{b} \times \vec{n} \neq 0$$

- Point A may or may not lie on the plane.
- **Angle between line and plane.**

  From the figure,

  $$\cos(90° - \theta) = \frac{|\vec{b} \cdot \vec{n}|}{|\vec{b}||\vec{n}|}$$

  $$\Rightarrow \quad \sin\theta = \frac{|\vec{b} \cdot \vec{n}|}{|\vec{b}||\vec{n}|}$$

  $$\Rightarrow \quad \theta = \sin^{-1}\frac{|\vec{b} \cdot \vec{n}|}{|\vec{b}||\vec{n}|}$$

Illustration 4.78

Find the distance of the point $P(3, 8, 2)$ from the line $\frac{x-1}{2} = \frac{y-3}{4} = \frac{z-2}{3}$ measured parallel to the plane $3x + 2y - 2z + 15 = 0$.

Sol. Line $\frac{x-1}{2} = \frac{y-3}{4} = \frac{z-2}{3}$ is parallel to the vector $\vec{b} = 2\hat{i} + 4\hat{j} + 3\hat{k}$

Vector normal to the plane $3x + 2y - 2z + 15 = 0$ is $\vec{n} = 3\hat{i} + 2\hat{j} - 2\hat{k}$.

Clearly line is not perpendicular to the plane.

Also $\vec{b} \cdot \vec{n} \neq 0$

So, line is not parallel to the plane.

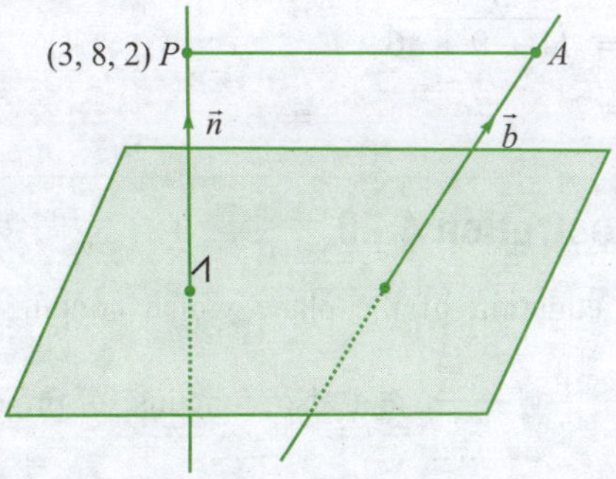

We have to find the distance of the point $P(3, 8, 2)$ from the line measured parallel to the plane.

So, we draw a line through the point P parallel to plane intersecting the given line at A.

Let $A \equiv (2\lambda + 1, 4\lambda + 3, 3\lambda + 2)$.

Now, vector $\overrightarrow{PA}$ is perpendicular to the vector $\vec{n} = 3\hat{i} + 2\hat{j} - 2\hat{k}$

$$\therefore \quad ((2\lambda - 2)\hat{i} + (4\lambda - 5)\hat{j} + (3\lambda)\hat{k}) \cdot (3\hat{i} + 2\hat{j} - 2\hat{k}) = 0$$

$$\Rightarrow \quad 6\lambda - 6 + 8\lambda - 10 - 6\lambda = 0$$

$$\Rightarrow \quad \lambda = 2$$

$$\therefore \quad A \equiv (5, 11, 8)$$

So, $PA = \sqrt{(5-3)^2 + (11-8)^2 + (8-2)^2}$

$= \sqrt{4 + 9 + 36} = 7$

Illustration 4.79

Find the distance of the point (1, 0, – 3) from the plane $x - y - z = 9$ measured parallel to the line, $\frac{x-2}{2} = \frac{y+2}{3} = \frac{z-6}{-6}$.

Sol. Given line is neither parallel nor perpendicular to the plane.

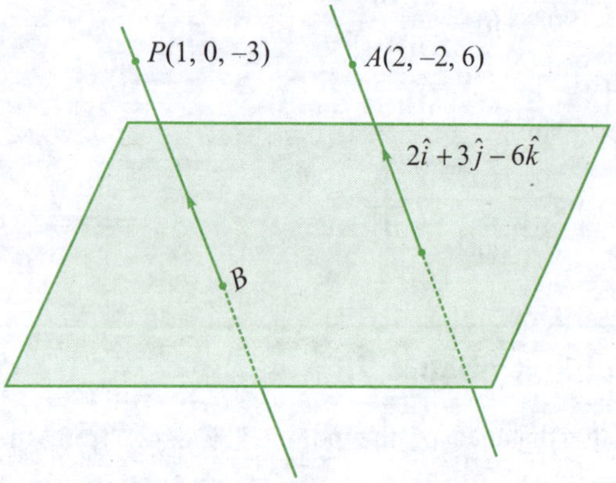

We have to find the distance of the point $P(1, 0, -3)$ from the plane measured parallel to the given line.
Equation of line through point P parallel to the given line is

$$\frac{x-1}{2} = \frac{y}{3} = \frac{z+3}{-6} = \lambda$$

This line intersects the plane at $B(2\lambda + 1, 3\lambda, -6\lambda - 3)$

$\therefore \quad 2\lambda + 1 - 3\lambda + 6\lambda + 3 = 9$

$\Rightarrow \quad \lambda = 1$

$\Rightarrow \quad A \equiv (3, 3, -9)$

Hence, $PB = \sqrt{(3-1)^2 + (3-0)^2 + (-9+3)^2}$

$= \sqrt{4 + 9 + 36}$

$= 7$

Illustration 4.80

Find the equation of the plane which contains the line $\frac{x-2}{1} = \frac{y+3}{2} = \frac{z-4}{-3}$ and also contains its projection on the plane $2x - 4y + 5z + 7 = 0$

Sol.

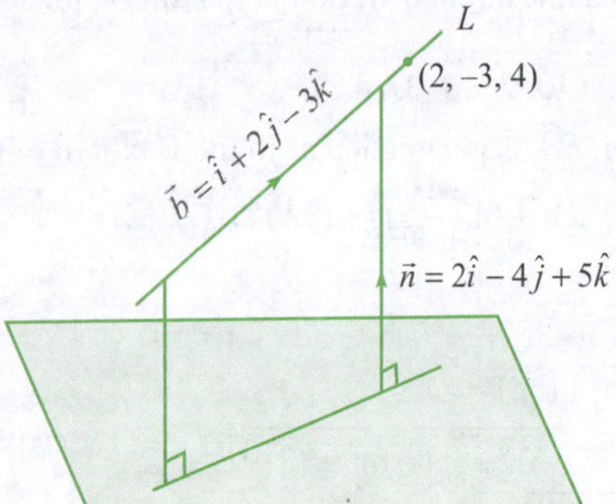

Given line is $L \equiv \frac{x-2}{1} = \frac{y+3}{2} = \frac{z-4}{-3}$.

It is parallel to the vector $\vec{b} = \hat{i} + 2\hat{j} - 3\hat{k}$.

Vector normal to the given plane $2x - 4y + 5z + 7 = 0$ is $\vec{n} = 2\hat{i} - 4\hat{j} + 5\hat{k}$

Plane passing through the given line and its projection on the given plane will be perpendicular to the given plane.

$\therefore$ Vector normal to the required plane is

$$\vec{b} \times \vec{n} = \begin{vmatrix} \hat{i} & \hat{j} & \hat{k} \\ 1 & 2 & -3 \\ 2 & -4 & 5 \end{vmatrix} = -2\hat{i} - 11\hat{j} - 8\hat{k}$$

Hence, the equation of the required plane is $-2(x - 2) - 11(y + 3) - 8(z - 4) = 0$ or $2x + 11y + 8z - 3 = 0$.

Illustration 4.81

Find the angle between the line $\vec{r} = \hat{i} + 2\hat{j} - \hat{k} + \lambda(\hat{i} - \hat{j} + \hat{k})$ and the plane $\vec{r} \cdot (2\hat{i} - \hat{j} + \hat{k}) = 4$.

Sol. We know that if θ is the angle between the line $\vec{r} = \vec{a} + \lambda\vec{b}$ and the plane $\vec{r} \cdot \vec{n} = p$, then $\sin\theta = \left|\frac{\vec{b} \cdot \vec{n}}{|\vec{b}||\vec{n}|}\right|$.

Therefore, if θ is the angle between the line $\vec{r} = \hat{i} + 2\hat{j} - \hat{k} + \lambda(\hat{i} - \hat{j} + \hat{k})$ and the plane $\vec{r} \cdot (2\hat{i} - \hat{j} + \hat{k}) = 4$, then

$$\sin\theta = \left|\frac{(\hat{i} - \hat{j} + \hat{k}) \cdot (2\hat{i} - \hat{j} + \hat{k})}{|\hat{i} - \hat{j} + \hat{k}||2\hat{i} - \hat{j} + \hat{k}|}\right|$$

$$= \frac{2+1+1}{\sqrt{1+1+1}\sqrt{4+1+1}}$$

$$= \frac{4}{\sqrt{3}\sqrt{6}} = \frac{4}{3\sqrt{2}}$$

or $\quad \theta = \sin^{-1}\left(\frac{4}{3\sqrt{2}}\right)$

Illustration 4.82

If the line $\frac{x}{k} = \frac{y}{2} = \frac{z}{-12}$ and its orthogonal projections on the planes $2x + y + 3z = 1$ and $x + 2y - 3z = 1$ make an isosceles triangle then find the values of k.

Sol. Since the line makes an isosceles triangle with the given planes, it must be equally inclined to the planes.

$$\therefore \quad \frac{|2k + 2 - 36|}{\sqrt{k^2 + 4 + 144}\sqrt{4+1+9}} = \frac{|k + 4 + 36|}{\sqrt{k^2 + 4 + 144}\sqrt{1+4+9}}$$

$\Rightarrow \quad 2k-34=\pm(k+40)$

$\Rightarrow \quad 2k-34=-k-40$ or $2k-34=k+40$

$\Rightarrow \quad 3k=-6$ or $k=74$

$\Rightarrow \quad k=-2$ or $k=74$

Aim to complete within 40 minutes

Concept Application Exercise 4.8

1. The distance of the point (–1, –5, –10) from the point of intersection of the line $\frac{x-2}{3}=\frac{y+1}{4}=\frac{z-2}{12}$ and the plane $x-y+z=5$ is
   (1) 10 (2) 11 (3) 12 (4) 13
2. If the distance between the line $\frac{x+1}{-3}=\frac{y-3}{2}=\frac{z-2}{1}$ and the plane $x+y+z+3=0$ is $\sqrt{\frac{m}{n}}$, H.C.F $(m, n)=1$, then the value of $(m-16n)$ is
   (1) –1 (2) 0 (3) 1 (4) 2
3. Find the distance of the point $\vec{a}$ from the plane $\vec{r}\cdot\hat{n}=d$ measured parallel to the line $\vec{r}=\vec{b}+t\vec{c}$.
4. The value of m for which the straight line $3x-2y+z+3=0=4x-3y+4z+1$ is parallel to the plane $2x-y+mz-2=0$ is
   (1) –3 (2) –2 (3) 2 (4) 3
5. Find the angle between the line $\frac{x+1}{3}=\frac{y-1}{2}=\frac{z-1}{4}$ and the plane $2x+y-3z+4=0$.
6. Find the angle between the line $\vec{r}=(\hat{i}+2\hat{j}-\hat{k})+\lambda(\hat{i}-\hat{j}+\hat{k})$ and the normal to the plane $\vec{r}\cdot(2\hat{i}-\hat{j}+\hat{k})=4$.
7. Find the direction ratios of orthogonal projection of line $\frac{x-1}{1}=\frac{y+1}{-2}=\frac{z-2}{3}$ in the plane $x-y+2z-3=0$. Also find the direction ratios of the image of the line in the plane.
8. The length of projection of line segment joining (2, 3, 4) and (5, 6, 7) on plane $2x+y+z=1$ is
   (1) $\sqrt{2}$ (2) $\sqrt{3}$ (3) 2 (4) $2\sqrt{2}$
9. If line of the shortest distance between the lines $\frac{x-1}{2}=\frac{y+1}{3}=\frac{z}{-1}$ and $\frac{x}{1}=\frac{y-3}{4}=\frac{z+2}{2}$ meets the plane $\alpha x+3y+6z-7=0$ at an angle $\cos^{-1}\frac{\sqrt{5}}{\sqrt{6}}$, find the value of α.

Answers

1. (4)
2. (3)
3. $\frac{|d-\vec{a}\cdot\hat{n}|}{|\vec{c}\cdot\hat{n}|}|\vec{c}|$
4. (2)
5. $\sin^{-1}\frac{4}{\sqrt{406}}$
6. $\cos^{-1}\left(\frac{2\sqrt{2}}{3}\right)$
7. D.R.s if orthogonal projection are 1, 1, 0; D.R.s of image are 2, –1, 3.
8. (2)
9. $\alpha=2, -6$

SOLVED EXAMPLES

Example 4.1

If variable lines in two adjacent positions have direction cosines l, m and n and $(l + \delta l)$, $(m + \delta m)$, $(n + \delta n)$, show that the small angle $\delta\theta$ between the two positions is given by $(\delta\theta)^2 = (\delta l)^2 + (\delta m)^2 + (\delta n)^2$.

Sol. Since l, m and n, and $(l + \delta m)$, $(m + \delta m)$, $(n + \delta n)$ are the direction cosines, we have

$$l^2 + m^2 + n^2 = 1 \quad \text{...(i)}$$

$$(l + \delta l)^2 + (m + \delta m)^2 + (n + \delta n)^2 = 1$$

or $l^2 + m^2 + n^2 + 2l\,\delta l + 2m\,\delta m + 2n\,\delta n + (\delta l)^2 + (\delta m)^2 + (\delta n)^2 = 1$

or $2(l\,\delta l + m\,\delta m + n\,\delta n) = -(\{\delta l)^2 + (\delta m)^2 + (\delta n)^2\}$ ...(ii)

Now it is given that $\delta\theta$ is the angle between two adjacent positions of the line. Therefore

$$\cos\delta\theta = l(l + \delta l) + m(m + \delta m) + n(n + \delta n) \quad \text{...(iii)}$$

Now $\cos\delta\theta = 1 - \frac{(\delta\theta)^2}{2!} + \frac{(\delta\theta)^2}{4!} - \cdots$

If $\delta\theta$ is small, then $\cos\delta\theta = 1 - \frac{(\delta\theta)^2}{2}$

Then from (iii), we have

$$1 - \frac{(\delta\theta)^2}{2} = (l^2 + m^2 + n^2) + (l\,\delta l + m\,\delta m + n\,\delta n)$$

or $1 - \frac{(\delta\theta)^2}{2} = 1 - \frac{1}{2}\{(\delta l)^2 + (\delta m)^2 + (\delta n)^2\}$

[using (i) and (ii)]

or $(\delta\theta)^2 = (\delta l)^2 + (\delta m)^2 + (\delta n)^2$

Example 4.2

Find the equation of the plane containing the line $\frac{y}{b} + \frac{z}{c} = 1$, $x = 0$, and parallel to the line $\frac{x}{a} - \frac{z}{c} = 1$, $y = 0$.

Sol. The equation of the first line can be written as

$$\frac{y}{b} - \frac{1}{2} = \frac{1}{2} - \frac{z}{c},\ x = 0$$

or $\frac{x}{0} = \frac{y - \frac{1}{2}b}{b} = \frac{z - \frac{1}{2}c}{-c}$ ...(i)

Similarly, the equation of the second line can be written as

$$\frac{x - \frac{1}{2}a}{a} = \frac{y}{0} = \frac{z + \frac{1}{2}c}{c} \quad \text{...(ii)}$$

The equation of any plane passing through the line (i) is

$$A(x) + B\left(y - \frac{1}{2}b\right) + C\left(z - \frac{1}{2}c\right) = 0 \quad \text{...(iii)}$$

where $A\cdot 0 + B\cdot b - C\cdot c = 0$ ...(iv)

Now plane (iii) will be parallel to line (ii) if

$A\cdot a + B\cdot 0 - C\cdot c = 0$ ...(v)

Solving (iv) and (v), we set $\frac{A}{bc} = \frac{B}{-ca} = \frac{C}{-ab}$

Putting these values of A, B and C in (iii), we set the equation of the required plane as

$$bcx - ca\left(y - \frac{1}{2}b\right) - ab\left(z - \frac{1}{2}c\right) = 0$$

or $\frac{x}{a} - \frac{y}{b} - \frac{z}{c} + 1 = 0$

Example 4.3

A variable plane passes through a fixed point (α, β, γ) and meets the axes at A, B and C. Show that the locus of the point of intersection of the planes through A, B and C parallel to the coordinate planes is $\alpha x^{-1} + \beta y^{-1} + \gamma z^{-1} = 1$.

Sol. Let the equation of the variable plane be

$$\frac{x}{a} + \frac{y}{b} + \frac{z}{c} = 1, \quad \text{...(i)}$$

Plane (i) passes through the point (α, β, γ). Therefore,

$$\frac{\alpha}{a} + \frac{\beta}{b} + \frac{\gamma}{c} = 1 \quad \text{...(ii)}$$

Plane (i) meets the coordinate axes at points A, B and C. The equations of the planes passing through A, B and C and parallel to the coordinate planes are, respectively,

$$x = a,\ y = b,\ z = c \quad \text{...(iii)}$$

The locus of the point of intersection of these planes is obtained by eliminating the variables a, b and c between (ii) and (iii). Putting the values of a, b and c from (iii) in (ii), we get the required locus as $\frac{\alpha}{x} + \frac{\beta}{y} + \frac{\gamma}{z} = 1$ or $\alpha x^{-1} + \beta y^{-1} + \gamma z^{-1} = 1$

Example 4.4

Show that the straight lines whose direction cosines are given by the equations $al + bm + cn = 0$ and $ul^2 + vm^2 + wn^2 = 0$ are parallel or perpendicular as $\frac{a^2}{u} + \frac{b^2}{v} + \frac{c^2}{w} = 0$ or $a^2(v + w) + b^2(w + u) + c^2(u + v) = 0$.

Sol. Here, $l = -\dfrac{(bm+cn)}{a}$ and $ul^2 + m^2v + wn^2 = 0$.

Eliminating l, we get

$$\frac{u(bm+cn)^2}{a^2} + vm^2 + wn^2 = 0$$

$$u(b^2m^2 + 2bcmn + c^2n^2) + va^2m^2 + wa^2n^2 = 0$$

$$(b^2u + a^2v)m^2 + (2bcu)mn + (c^2u + a^2w)n^2 = 0$$

or $$(b^2u+a^2v)\left(\frac{m}{n}\right)^2 + (2bcu)\left(\frac{m}{n}\right) + (c^2u+a^2w) = 0$$

which is quadratic in (m/n) having roots m_1/n_1 and m_2/n_2.

a. If the straight lines are parallel, the quadratic in m/n has equal roots, i.e., discriminant = 0

$\Rightarrow$ $(2bcu)^2 - 4(b^2u + a^2v)(c^2u + a^2w) = 0$

or $b^2c^2u^2 = (b^2u + a^2v)(c^2u + a^2w)$

or $a^2vw + b^2uw + c^2uv = 0$

or $\dfrac{a^2}{u} + \dfrac{b^2}{v} + \dfrac{c^2}{w} = 0$

b. If the straight lines are perpendicular, then

$$\frac{m_1}{n_1}\frac{m_2}{n_2} = \frac{c^2u+a^2w}{b^2u+a^2v} \quad \text{(product of roots)}$$

or $$\frac{m_1m_2}{c^2u+a^2w} = \frac{n_1n_2}{b^2u+a^2v} \quad \text{...(i)}$$

Similarly, by eliminating n, we get

$$\frac{l_1l_2}{b^2w+c^2v} = \frac{m_1m_2}{c^2u+a^2w} \quad \text{...(ii)}$$

From (i) and (ii),

$$\frac{l_1l_2}{b^2w+c^2v} = \frac{m_1m_2}{c^2u+a^2w} = \frac{n_1n_2}{b^2u+a^2v} = \lambda$$

Since they are perpendicular,

$$l_1l_2 + m_1m_2 + n_1n_2 = 0$$

or $\lambda(b^2w+c^2v) + \lambda(c^2u+a^2w) + \lambda(b^2u+a^2v) = 0$

or $a^2(v+w) + b^2(w+u) + c^2(u+v) = 0$

Example 4.5

If the planes $x - cy - bz = 0$, $cx - y + az = 0$ and $bx + ay - z = 0$ pass through a straight line, then find the value of $a^2 + b^2 + c^2 + 2abc$.

Sol. The given planes are

$$x - cy - bz = 0 \quad \text{...(i)}$$

$$cx - y + az = 0 \quad \text{...(ii)}$$

$$bx + ay - z = 0 \quad \text{...(iii)}$$

The equation of the planes passing through the line of intersection of planes (i) and (ii) may be taken as

$$(x - cy - bz) + \lambda(cx - y + az) = 0$$

or $x(1 + \lambda c) - y(c + \lambda) + z(-b + a\lambda) = 0$ ...(iv)

If planes (iii) and (iv) are the same, then Eqs. (iii) and (iv) will be identical.

$$\frac{1+c\lambda}{b} = \frac{-(c+\lambda)}{a} = \frac{-b+a\lambda}{-1}$$

$$\lambda = -\frac{(a+bc)}{(ac+b)} \text{ and } \lambda = -\frac{(ab+c)}{(1-a^2)}$$

$$\therefore \quad \frac{-(a+bc)}{(ac+b)} = -\frac{(ab+c)}{(1-a^2)}$$

$$a - a^3 + bc - a^2bc = a^2bc + ac^2 + ab^2 + bc$$

or $2a^2bc + ac^2 + ab^2 + a^3 - a = 0$

or $a(2abc + c^2 + b^2 + a^2 - 1) = 0$

or $a^2 + b^2 + c^2 + 2abc = 1$

Alternate method:

Since the planes pass through the origin, the given planes have a common line of intersection if given system of equations has a non-trivial solution. Hence,

$$\begin{vmatrix} 1 & -c & -b \\ c & -1 & a \\ b & a & -1 \end{vmatrix} = 0$$

or $a^2 + b^2 + c^2 + 2abc = 1$

Example 4.6

Let a plane $ax + by + cz + 1 = 0$, where a, b and c are parameters, make an angle 60° with the line $x = y = z$, 45° with the line $x = y - z = 0$ and θ with the plane $x = 0$. The distance of the plane from point (2, 1, 1) is 3 units. Find the value of θ and the equation of the plane.

Sol. We have, plane $ax + by + cz + 1 = 0$ ...(i)

It makes an angle 60° with the line $x = y = z$.

Therefore, $\sin 60° = \dfrac{a+b+c}{\sqrt{3}\sqrt{\Sigma a^2}}$

or $3\sqrt{\Sigma a^2} = 2(a+b+c)$ ...(ii)

Plane (i) makes an angle of 45° with the line

$$x = y - z = 0 \left(\text{or } \frac{x}{0} = \frac{y}{1} = \frac{z}{1}\right)$$

Therefore, $\sin 45° = \dfrac{b+c}{\sqrt{2}\sqrt{\Sigma a^2}}$ or $\sqrt{\Sigma a^2} = b + c$ ...(iii)

Plane (i) makes an angle θ with the plane $x = 0$.

Therefore, $\cos\theta = \dfrac{a}{\sqrt{\Sigma a^2}}$ ...(iv)

From (ii) and (iii), we get

$$(\sqrt{\Sigma a^2}) = 2a$$

or $\dfrac{a}{\sqrt{\Sigma a^2}} = \dfrac{1}{2}$

Hence, from (iv), $\cos\theta = 1/2$ or $\theta = 60°$

Now, the distance of the plane (i) from the point (2, 1, 1) is 3 units.

Therefore, $\dfrac{2a+b+c+1}{\sqrt{\Sigma a^2}} = \pm 3$

or $\quad \pm 3\sqrt{\Sigma a^2} = 2a+b+c+1$

Case I:

$$3\sqrt{\Sigma a^2} = 2a+b+c+1 \quad \text{...(v)}$$

From (ii) and (v), we get

$$b+c-1=0 \quad \text{...(vi)}$$

and from (iii) and (iv), we get

$$2a+b+c+1 = 3(b+c) \quad \text{...(vii)}$$

From (vi) and (vii), we get

$$a = \frac{1}{2},\ b = \frac{(2\mp\sqrt{2})}{4} \text{ and } c = \frac{2\pm\sqrt{2}}{4}.$$

Hence, the equations of such planes are

$$2x + (2\pm\sqrt{2})y + (2\pm\sqrt{2})z + 4 = 0.$$

Case II:

$$-3\sqrt{\Sigma a^2} = 2a+b+c+1$$

$$a = \frac{-1}{10},\ b = \frac{-(2\pm\sqrt{2})}{20} \text{ and } c = \frac{-(2\mp\sqrt{2})}{20}$$

Hence, the equations of the planes are

$$2x + (2\pm\sqrt{2})y + (2\mp\sqrt{2})z - 20 = 0.$$

Example 4.7

Prove that the volume of tetrahedron bounded by the planes $\vec{r}\cdot(m\hat{j}+n\hat{k})=0$, $\vec{r}\cdot(n\hat{k}+\ell\hat{i})=0$, $\vec{r}\cdot(\ell\hat{i}+m\hat{j})=0$, $\vec{r}\cdot(\ell\hat{i}+m\hat{j}+n\hat{k})=p$ is $\dfrac{2p^3}{3\ell mn}$.

Sol. Planes $\vec{r}\cdot(m\hat{j}+n\hat{k})=0 \quad$ ...(i)

$$\vec{r}\cdot(n\hat{k}+\ell\hat{i})=0 \quad \text{...(ii)}$$

$$\vec{r}\cdot(\ell\hat{i}+m\hat{j})=0 \quad \text{...(iii)}$$

are passing through the origin O.

Let the planes (i) and (ii), (ii) and (iii), (i) and (iii) be intersect along edges OA, OB and OC, where A, B, C are the points where these vectors are tereminated by the plane

$$\vec{r}\cdot(\ell\hat{i}+m\hat{j}+n\hat{k})=p \quad \text{...(iv)}$$

Vectors along the edges OA, OB, OC are, respectively,

$$(m\hat{j}+n\hat{k})\times(\ell\hat{i}+n\hat{k}) = mn\hat{i}+n\ell\hat{j}-\ell m\hat{k}$$

$$(n\hat{k}+\ell\hat{i})\times(\ell\hat{i}+m\hat{j}) = -mn\hat{i}+n\ell\hat{j}+\ell m\hat{k}$$

$$(\ell\hat{i}+m\hat{j})\times(m\hat{j}+n\hat{k}) = mn\hat{i}-n\ell\hat{j}+\ell m\hat{k}$$

Equation of the line OA is $\vec{r}=\lambda(mn\hat{i}+n\ell\hat{j}-\ell m\hat{k})$

For its point of intersection with the plane $\vec{r}\cdot(\ell\hat{i}+m\hat{j}+n\hat{k})=p$, we have

$$\lambda(mn\hat{i}+n\ell\hat{j}-\ell m\hat{k})\cdot(\ell\hat{i}+m\hat{j}+n\hat{k})=p$$

$$\therefore \quad \lambda = \frac{p}{\ell mn}$$

Thus, $\overrightarrow{OA} = \dfrac{p}{\ell mn}(mn\hat{i}+n\ell\hat{i}-\ell m\hat{k})$

Similarly, $\overrightarrow{OB} = \dfrac{p}{\ell mn}(-mn\hat{i}+n\ell\hat{j}+\ell m\hat{k})$ and

$$\overrightarrow{OC} = \frac{p}{\ell mn}(mn\hat{i}-n\ell\hat{j}+\ell m\hat{k})$$

Therefore, the volume of the tetrahedron formed by the vectors $\overrightarrow{OA}, \overrightarrow{OB}, \overrightarrow{OC}$ is

$$\frac{1}{6}\left(\frac{p}{\ell mn}\right)^3 \begin{vmatrix} mn & n\ell & -\ell m \\ -mn & n\ell & \ell m \\ mn & -n\ell & \ell m \end{vmatrix}$$

$$= \frac{1}{6}\left(\frac{p}{\ell mn}\right)^3 \ell^2m^2n^2 \begin{vmatrix} 1 & 1 & -1 \\ -1 & 1 & 1 \\ 1 & -1 & 1 \end{vmatrix} = \frac{2}{3}\frac{p^3}{\ell mn}$$

Example 4.8

Let the foot of perpendicular from the point $A(4, 3, 1)$ on the plane $P: x-y+2z+3=0$ be N. If $B(5, \alpha, \beta)$, $\alpha, \beta \in Z$ is a point on plane P such that the area of the triangle ABN in $3\sqrt{2}$, then find the values of α and β.

Sol.

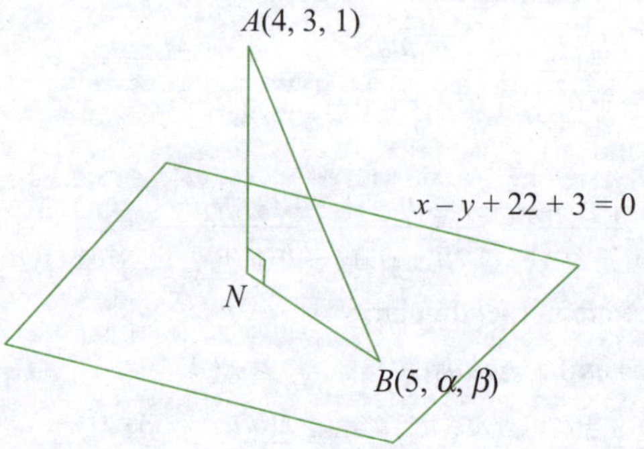

$$AN = \sqrt{6}$$

Point $(5, \alpha, \beta)$ lies on the plane $x-y+2z+3=0$

$$\therefore \quad 5-\alpha+2\beta+3=0$$

$$\Rightarrow \quad \alpha = 8+2\beta \quad \text{...(i)}$$

Point N is given by

$$\frac{x-4}{1} = \frac{y-3}{-1} = \frac{z-1}{2} = \frac{-(4-3+2+3)}{1+1+4}$$

$$\Rightarrow \quad x=3,\ y=4,\ z=-1$$

$$\Rightarrow \quad N \equiv (3, 4, -1)$$

$$\Rightarrow \quad BN = \sqrt{4+(\alpha-4)^2+(\beta+1)^2}$$

$$= \sqrt{4+(2\beta+4)^2+(\beta+1)^2}$$

Now, Area of $\Delta ABN = \dfrac{1}{2}\,AN \times BN = 3\sqrt{2}$ (given)

$\Rightarrow \quad \frac{1}{2} \times \sqrt{6} \times BN = 3\sqrt{2}$

$\Rightarrow \quad BN = 2\sqrt{3}$

$\Rightarrow \quad 4 + (2\beta + 4)^2 + (\beta + 1)^2 = 12$

$\Rightarrow \quad 5\beta^2 + 18\beta + 9 = 0$

$\Rightarrow \quad (5\beta + 3)(\beta + 3) = 0$

$\Rightarrow \quad \beta = -3$

$\Rightarrow \quad \alpha = 2$

Example 4.9

Prove that for all values of λ and μ, the planes

$\frac{2x}{a} + \frac{y}{b} + \frac{2z}{c} - 1 + \lambda\left(\frac{x}{a} - \frac{2y}{b} - \frac{z}{c} - 2\right) = 0$ and

$\frac{4x}{a} - \frac{3y}{b} - 5 + \mu\left(\frac{5y}{b} + \frac{4z}{c} + 3\right) = 0$ intersect on the same line.

Sol. We have $\frac{2x}{a} + \frac{y}{b} + \frac{2z}{c} - 1 + \lambda\left(\frac{x}{a} - \frac{2y}{b} - \frac{z}{c} - 2\right) = 0$ ...(i)

This is the family of planes passing through the line of intersection of the planes

$$\frac{2x}{a} + \frac{y}{b} + \frac{2z}{c} - 1 = 0 \quad \text{...(ii)}$$

and $$\frac{x}{a} - \frac{2y}{b} - \frac{z}{c} - 2 = 0 \quad \text{...(iii)}$$

Also, $$\frac{4x}{a} - \frac{3y}{b} - 5 + \mu\left(\frac{5y}{b} + \frac{4z}{c} + 3\right) = 0 \quad \text{...(iv)}$$

This the is the family of planes passing through the line of intersection of the planes

$$\frac{4x}{a} - \frac{3y}{b} - 5 = 0 \quad \text{...(v)}$$

and $$\frac{5y}{b} + \frac{4z}{c} + 3 = 0 \quad \text{...(vi)}$$

Now, eliminating z from (ii) and (iii), we get the equation (v), i.e. (ii) + 2(iii) = (v).

Also, eliminating x from (ii) and (iii), we get the equation (vi), i.e. (ii) – 2(iii) = (vi).

Thus, equations (i) and (iv) represent the same family of planes.

Hence, variable planes given by equations (iii) and (iv), intersect on the same line.

Example 4.10

Find equations of lines passing through the origin which intersect the line $\frac{x-3}{2} = \frac{y-3}{1} = \frac{z}{1}$ at an angle 60°.

Sol. Equation of the given line is $\frac{x-3}{2} = \frac{y-3}{1} = \frac{z}{1} = \lambda$ (say)

So, direction ratios of the line are $(2, 1, 1) \equiv (a_1, b_1, c_1)$

Any point on the given line is $P(2\lambda + 3, \lambda + 3, \lambda)$

$\therefore$ Direction ratios of OP are $(2\lambda + 3, \lambda + 3, \lambda) \equiv (a_2, b_2, c_2)$

Now angle between given line and OP is $\frac{\pi}{3}$

$$\therefore \quad \cos\frac{\pi}{3} = \frac{a_1a_2 + b_1b_2 + c_1c_2}{\sqrt{a_1^2 + b_1^2 + c_1^2}\sqrt{a_2^2 + b_2^2 + c_2^2}}$$

$$\therefore \quad \frac{1}{2} = \frac{(4\lambda + 6) + (\lambda + 3) + (\lambda)}{\sqrt{6}\sqrt{(2\lambda + 3)^2 + (\lambda + 3)^2 + \lambda^2}}$$

$$\Rightarrow \quad \frac{\sqrt{6}}{2} = \frac{6\lambda + 9}{\sqrt{6\lambda^2 + 18\lambda + 18}}$$

$\Rightarrow \quad \sqrt{\lambda^2 + 3\lambda + 3} = 2\lambda + 3$

$\Rightarrow \quad \lambda^2 + 3\lambda + 3 = 4\lambda^2 + 12\lambda + 9$

$\Rightarrow \quad \lambda^2 + 3\lambda + 2 = 0$

$\Rightarrow \quad (\lambda + 1)(\lambda + 2) = 0$

$\therefore \quad \lambda = -1, -2$

So, the directions of line are 1, 2, –1 and –1, 1, –2.

Also, both the required lines pass through origin.

So, the required equations of lines are $\frac{x}{1} = \frac{y}{2} = \frac{z}{-1}$ and $\frac{x}{-1} = \frac{y}{1} = \frac{z}{-2}$.

Example 4.11

Prove that the projection of the line $\frac{x-1}{2} = \frac{y+1}{-1} = \frac{z-3}{4}$ on the plane $x + 2y + z = 9$ is the line of intersection of plane $9x - 2y - 3z + 4 = 0$ and the given plane. Find the direction ratio of the line of projection. Also find the direction ratios of the image of line in the plane.

Sol.

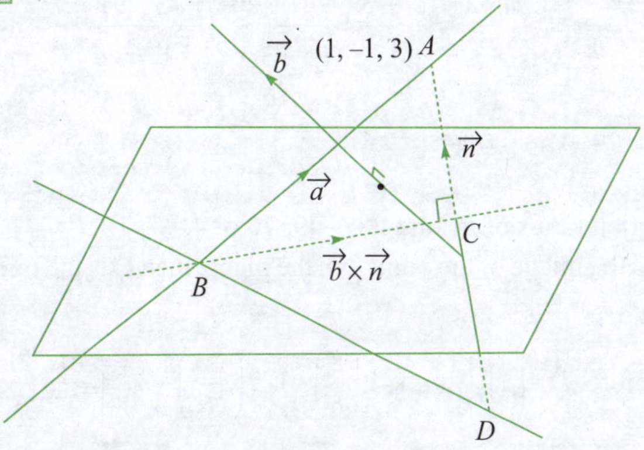

Given line is $\frac{x-1}{2} = \frac{y+1}{-1} = \frac{z-3}{4}$ ...(i)

Line is along the vector $\vec{a} = 2\hat{i} - \hat{j} + 4\hat{k}$

Given plane is $x + 2y + z = 9$

Vector normal to the plane is $\vec{n} = \hat{i} + 2\hat{j} + \hat{k}$.

Clearly, line is neither parallel nor perpendicular to the plane.

Line passes through point $A(1, -1, 3)$ and intersects the plane at B.

The foot of the perpendicular from the point A on the plane is C.

So, the plane ABC is perpendicular to the given plane.

Normal to the plane ABC is

$$\vec{b} = \vec{a} \times \vec{n}$$

$$= \begin{vmatrix} \hat{i} & \hat{j} & \hat{k} \\ 2 & -1 & 4 \\ 1 & 2 & 1 \end{vmatrix} = -9\hat{i} + 2\hat{j} + 5\hat{k}$$

So, the equation of the plane ABC is

$$(-9)(x-1) + 2(y+1) + 5(z-3) = 0$$

or $\quad 9x - 2y - 5z + 4 = 0$

Thus, the projection BC is the line of intersection of the given plane and plane ABC.

Its equation is,

$$x + 2y + z - 9 = 0 = 9x - 2y - 5z + 4.$$

Vector along the projection line is

$$\vec{b} \times \vec{n}$$

$$= \begin{vmatrix} \hat{i} & \hat{j} & \hat{k} \\ -9 & 2 & 5 \\ 1 & 2 & 1 \end{vmatrix} = -8\hat{i} + 14\hat{j} - 20\hat{k}$$

Thus, the direction ratios of projection line are 4, –7, 10

Alternative method

Let's first find the point of intersection B of the line and the plane.

Let $B \equiv (2\lambda + 1, -\lambda - 1, 4\lambda + 3)$

This point lies on the plane.

$\therefore \quad (2\lambda + 1) + 2(-\lambda - 1) + (4\lambda + 3) - 9 = 0$

$\Rightarrow \quad 4\lambda = 7$ or $\lambda = 7/4$

$\therefore \quad B \equiv \left(\frac{18}{4}, -\frac{11}{4}, \frac{40}{4}\right)$

Also, the foot of the perpendicular drawn from C to A is given by

$$\frac{x_1 - 1}{1} = \frac{y_1 + 1}{2} = \frac{z_1 - 3}{1} = \frac{-(1 + 2(-1) + 3 - 9)}{1^2 + 2^2 + 1^2} = \frac{7}{6}$$

$\therefore \quad C \equiv (x_1, y_1, z_1) \equiv \left(\frac{13}{6}, \frac{8}{6}, \frac{25}{6}\right)$

Direction ratios of BC are 28, – 49, 70 or 4, –7, 10.

If D is the image of the point A in the plane, then C is the midpoint of AD.

Therefore, $D \equiv \left(2\left(\frac{13}{6}\right) - 1,\ 2\left(\frac{8}{6}\right) + 1,\ 2\left(\frac{25}{6}\right) - 3\right) \equiv \left(\frac{10}{3}, \frac{11}{3}, \frac{16}{3}\right)$

Hence, the direction ratios of BD are –14, 77, –56 or –2, 11, –8.

Example 4.12

A ray of light emanating from a point $A(1, 2, 3)$ and strikes a line mirror $L : \frac{x+2}{1} = \frac{y-3}{2} = \frac{z-1}{3}$ at point P and after the reflection it passes through the point $B(3, 6, 9)$. Then find point P.

Sol.

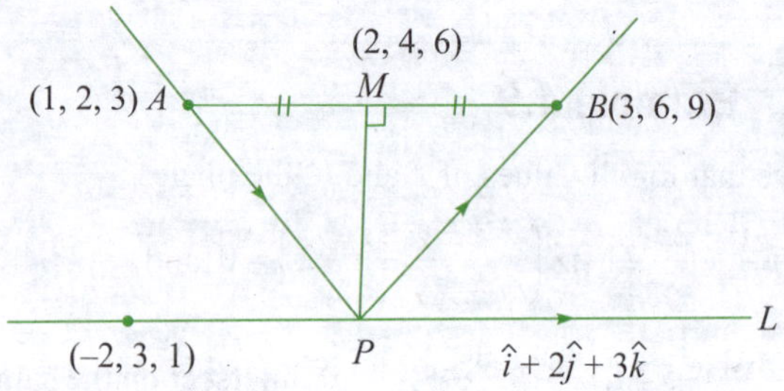

We have, $\overrightarrow{AB} = 2\hat{i} + 4\hat{j} + 6\hat{k}$, which is parallel to the line L.

So, triangle APB is an isosceles triangle.

Point M is the mid-point of AB, which is the foot of perpendicular from point A on the line L.

Thus, $M \equiv (2, 4, 6)$

Let $P \equiv (\lambda - 2, 2\lambda + 3, 3\lambda + 1)$

Now, $\overrightarrow{PM} \perp \overrightarrow{AB}$

$\therefore \quad (\lambda - 4) + (2\lambda - 1)2 + (3\lambda - 5)3 = 0$

$\Rightarrow \quad 14\lambda = 21$

$\Rightarrow \quad \lambda = \frac{3}{2}$

$\Rightarrow \quad P \equiv \left(-\frac{1}{2}, 6, \frac{11}{2}\right)$

Example 4.13

Find the minimum distance of the point (1, 1, 1) from the plane $x + y + z = 1$ measured perpendicular to the line $\frac{x - x_1}{1} = \frac{y - y_1}{2} = \frac{z - z_1}{3}$

Sol. Given plane is $P_1: x + y + z = 1$ ...(i)

Given line is L: $\frac{x - x_1}{1} = \frac{y - y_1}{2} = \frac{z - z_1}{3}$ ...(ii)

This is a variable line having direction ratios 1, 2, 3.

Given point is $P(1, 1, 1)$

Let d = minimum distance of point P from the plane P_1 measured perpendicular to the line L.

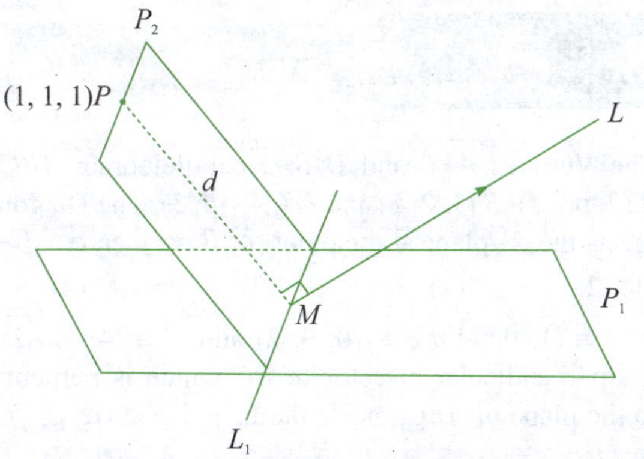

From the figure 'd' is distance of point P from the line of intersection of the given plane and the plane P_2 which is passing through the point P and perpendicular to the variable line L.

Thus, the equation of the plane P_2 is $x + 2y + 3z - 6 = 0$ …(iii)

From (iii) – (i), we get

$$y + 2z - 5 = 0 \quad \text{…(iv)}$$

From (iii) – 2(ii), we get

$$-x + z - 4 = 0 \quad \text{…(v)}$$

Then, from (iv) and (v), we get

$$L_1: \frac{x+4}{1} = \frac{y-5}{-2} = \frac{z-0}{1} = \lambda$$

Now, let the foot of the perpendicular from the point P on the line L_1 is M.

Therefore, $M \equiv (\lambda - 4, -2\lambda + 5, \lambda)$

Since, $\overrightarrow{PM} \perp (\hat{i} - 2\hat{j} + \hat{k})$, we have

$$(\lambda - 5) - 2(-2\lambda + 4) + (\lambda - 1) = 0$$

$$\Rightarrow \quad 6\lambda - 14 = 0$$

$$\Rightarrow \quad \lambda = \frac{7}{3}$$

Therefore, $\overrightarrow{PM} = -\frac{8}{3}\hat{i} - \frac{2}{3}\hat{j} + \frac{4}{3}\hat{k}$

Hence, $|\overrightarrow{PM}| = \frac{1}{3}\sqrt{64+4+16} = \frac{2}{3}\sqrt{21}$ units.

Example 4.14

Find the value of λ for which the planes $x - y + z + 1 = 0, \lambda x + 3y + 2z - 3 = 0, 3x + \lambda y + z - 2 = 0$ form a triangular prism.

Sol. The condition for the three planes to form a triangular prism are:

(a) lines of intersection of planes are parallel

(b) no two planes are parallel

The given planes are parallel to the vectors $\overrightarrow{n_1} = \hat{i} - \hat{j} + \hat{k}$, $\overrightarrow{n_2} = \lambda\hat{i} + 3\hat{j} + 2\hat{k}$ and $\overrightarrow{n_3} = 3\hat{i} + \lambda\hat{j} + \hat{k}$

The lines of intersection of planes are along the vectors $\overrightarrow{n_1} \times \overrightarrow{n_2}$, $\overrightarrow{n_2} \times \overrightarrow{n_3}$ and $\overrightarrow{n_3} \times \overrightarrow{n_1}$.

Since lines of intersection are parallel, these vectors must be parallel.

Therefore, vectors $\overrightarrow{n_1}$, $\overrightarrow{n_2}$ and $\overrightarrow{n_3}$ are coplanar, but not all the vectors are parallel.

$$\text{Thus, } \Delta = \begin{vmatrix} 1 & -1 & 1 \\ \lambda & 3 & 2 \\ 3 & \lambda & 1 \end{vmatrix} = 0$$

$$\therefore \quad 1(3 - 2\lambda) + 1(\lambda - 6) + 1(\lambda^2 - 9) = 0$$

$$\Rightarrow \quad \lambda^2 - \lambda - 12 = 0$$

$$\Rightarrow \quad (\lambda - 4)(\lambda + 3) = 0$$

$$\Rightarrow \quad \lambda = 4, -3.$$

For both these values of λ, no two of the three vectors $\overrightarrow{n_1}$, $\overrightarrow{n_2}$ and $\overrightarrow{n_3}$ are parallel.

Therefore, $\lambda = 4, -3$.

EXERCISES

[Note: Bulleted questions in Single Correct Answer Type, Numerical Value Type and Integer Answer Type sections are of high difficulty level.]

Single Correct Answer Type

Line in Space

1. If the distance of any point on the line L passing through point (1,2,3) from the line $\vec{r} = (3\lambda - 1)\hat{i} + (-2\lambda + 3)\hat{j} + (4 + \lambda)\hat{k}$ is constant then the line L does not pass through the point
(1) (4, 0, 4) (2) (–2, 4, 2)
(3) (7, –2, 5) (4) (–5, 6, 2)

2. The line passing through the point $\hat{i} + 3\hat{j} + 2\hat{k}$ and perpendicular to the lines $\vec{r} = (\hat{i} + 2\hat{j} - \hat{k}) + \lambda(2\hat{i} + \hat{j} + \hat{k})$ and $\vec{r} = (2\hat{i} + 6\hat{j} + \hat{k}) + \mu(\hat{i} + 2\hat{j} + 3\hat{k})$ is
(1) $\vec{r} = (\hat{i} + 2\hat{j} - \hat{k}) + \lambda(-\hat{i} + 5\hat{j} - 3\hat{k})$
(2) $\vec{r} = \hat{i} + 3\hat{j} + 2\hat{k} + \lambda(\hat{i} - 5\hat{j} + 3\hat{k})$
(3) $\vec{r} = \hat{i} + 3\hat{j} + 2\hat{k} + \lambda(\hat{i} - 5\hat{j} + 3\hat{k})$
(4) $\vec{r} = \hat{i} + 3\hat{j} + 2\hat{k} + \lambda(-\hat{i} - 5\hat{j} - 3\hat{k})$

3. The line L through the point (–1, 0, 2) is parallel to the line $\vec{r} = (3 + \lambda)\hat{i} + (-1 + \alpha\lambda)\hat{i} + (2 - \alpha^2\lambda)\hat{k}, \alpha > 0$ and perpendicular to the line $(\vec{r} - (\hat{i} - \hat{j} + \hat{k})) \times (\hat{i} + 2\hat{j} + 3\hat{k}) = 0$, does not pass through the point
(1) (0, 1, 1) (2) (1, 2, 0)
(3) (5, 6, –4) (4) (–3, –2, 3)

4. If the vertices of a triangle are $A(1, 2, -1)$, $B(2, 4, 1)$ and $C(-7, -2, 0)$, then the internal angle bisector of $\angle A$ does not pass through the point
(1) (–4, 4, 6)
(2) (6, 0, –8)
(3) (–9, 6, 6)
(4) (16, –4, –22)

5. An angle between the lines whose direction cosines are given by the equations, $l + 3m + 5n = 0$ and $5lm - 2mn + 6nl = 0$, is:
(1) $\cos^{-1}\left(\frac{1}{3}\right)$ (2) $\cos^{-1}\left(\frac{1}{4}\right)$
(3) $\cos^{-1}\left(\frac{1}{6}\right)$ (4) $\cos^{-1}\left(\frac{1}{8}\right)$

6. If three coinitial vectors $\hat{i} - 2\hat{j} + \hat{k}$, $2\hat{i} - \hat{j} + \hat{k}$ and $\hat{i} + \hat{j} + \hat{k}$ at point $A(1, -1, 0)$ are sides of a parallelepiped, then the diagonal of the parallelepiped through point A passes through the point
(1) (5, –3, 6) (2) (9, 3, –6)
(3) (13, –7, 9) (4) (7, –4, 1)

7. The Vertices A, B and D of a parallelogram $ABCD$ are A (2, 0, –3), B (1, 2, 5) and D (–1, –4, 3). The Diagonal AC meets the yz–plane at the point (α, β, γ), then $\alpha + \beta + \gamma =$
(1) 2 (2) 3 (3) 4 (4) 5

8. If $A \equiv (1, 0, -1)$, $B \equiv (0, 3, 2)$ and $C \equiv (4, 1, -2)$, then the perpendicular bisector of BC which is perpendicular to the plane of ABC, meets the xz–plane at (α, β, γ). Then $\alpha + \beta + \gamma =$
(1) 2 (2) 3 (3) 4 (4) 6

9. The point on the line $\frac{x-2}{1} = \frac{y+3}{-2} = \frac{z+5}{-2}$ at a distance 6 units from the point (2, – 3, – 5) is
(1) (3, – 5, – 3) (2) (4, – 7, –9)
(3) (0, 2, – 1) (4) (–3, 5, 3)

10. If the lines $x = y = z$ and $x = \frac{y}{2} = \frac{z}{3}$ and a third line passing through the point (1, 1, 1), form a triangle of area $\sqrt{6}$ units, then the point of intersection of the third line with the second line will be
(1) (1, 2, 3) (2) (2, 4, 6)
(3) $\left(\frac{4}{3}, \frac{8}{3}, \frac{12}{3}\right)$ (4) (1, 6, 4)

11. The line $\frac{x-2}{3} = \frac{y+1}{2} = \frac{z-1}{-1}$ intersects the curve $xy = c^2$, $z = 0$ if c is equal to
(1) ± 1 (2) ± 1/3
(3) $\pm\sqrt{5}$ (4) ± 1/2

12. The line $\frac{x-1}{-1} = \frac{y}{-2} = \frac{z-2}{1}$ meets the yz, zx and xy-plane at points A, B, C respectively, then which of the following is NOT true?
(1) point B divides AC internally in the ratio 1 : 2
(2) point C divides AB externally in the ratio 3 : 2
(3) point A divides BC externally in the ratio 4 : 1
(4) point A, B, C are collinear

13. Points C and D lie on the lines $L_1 : \frac{x-1}{2} = \frac{y-3}{1} = \frac{z-2}{2}$ and $L_2 = \frac{x-2}{1} = \frac{y-2}{2} = \frac{z-3}{3}$ respectively such that CD is parallel to x–axis. Then $|\overrightarrow{CD}| =$
(1) 4 (2) 5 (3) 6 (4) 8

•**14.** Line L is perpendicular to lines L_1 and L_2. The equation of line L_1 is $\frac{2x-1}{3} = \frac{3-y}{1} = \frac{3z-4}{-1}$ and line L_2 is perpendicular to the lines $\vec{r} = (3\hat{i} - \hat{j} + \hat{k}) + \lambda(\hat{i} + \hat{j} - \hat{k})$ and $\frac{x-1}{-2} = \frac{y}{1} = \frac{z-4}{5}$. If line L passes through the point (1, 1, 1), then it also passes through the point

(1) (2, 3, –4) (2) (9, 14, –2)
(3) (0, 2, –3) (4) (13, –12, 14)

•**15.** If the line $\frac{x-\alpha}{2} = \frac{y-\alpha^2}{4} = \frac{z+2}{\alpha-3}$ does not meet the xy–plane then distance between points where the line meets the other two planes is

(1) $\sqrt{5}$ (2) $3\sqrt{3}$
(3) $\frac{3}{2}\sqrt{5}$ (4) $4\sqrt{2}$

•**16.** Consider the triangle AOB in the x–y plane, where $A \equiv (1, 0, 0)$, $B \equiv (0, 2, 0)$, and $O \equiv (0, 0, 0)$. The new position of O, when the triangle is rotated about the side AB by 90°, can be

(1) $\left(\frac{4}{5}, \frac{3}{5}, \frac{2}{\sqrt{5}}\right)$ (2) $\left(\frac{-3}{5}, \frac{\sqrt{2}}{5}, \frac{2}{\sqrt{5}}\right)$
(3) $\left(\frac{4}{5}, \frac{2}{5}, \frac{2}{\sqrt{5}}\right)$ (4) $\left(\frac{4}{5}, \frac{2}{5}, \frac{1}{\sqrt{5}}\right)$

Angle Between Lines and Distance of a Point From a Line

17. The coordinates of the foot of the perpendicular drawn from the origin to the line joining the points (–9, 4, 5) and (10, 0, –1) is

(1) (–3, 2, 1) (2) (1, 2, 2)
(3) (4, 5, 3) (4) $\left(\frac{58}{59}, \frac{112}{59}, \frac{109}{59}\right)$

18. The distance of the point A (–2, 3, 1) from the line PQ passing through P (–3, 5, 2), which makes equal angles with the axes, is

(1) $2/\sqrt{3}$ (2) $\sqrt{14/3}$ (3) $16/\sqrt{3}$ (4) $5/\sqrt{3}$

19. The length of the perpendicular drawn from (1, 2, 3) to the line $\frac{x-6}{3} = \frac{y-7}{2} = \frac{z-7}{-2}$ is

(1) 4 (2) 5 (3) 6 (4) 7

20. The line $\frac{x+6}{5} = \frac{y+10}{3} = \frac{z+14}{8}$ is the hypotenuse of an isosceles right-angled triangle whose opposite vertex is (7, 2, 4). Then which of the following is not the side of the triangle?

(1) $\frac{x-7}{2} = \frac{y-2}{-3} = \frac{z-4}{6}$

(2) $\frac{x-7}{3} = \frac{y-2}{6} = \frac{z-4}{2}$

(3) $\frac{x-4}{5} = \frac{y+4}{3} = \frac{z-2}{8}$

(4) none of these

21. The coordinates of the point P on the line $\vec{r} = (\hat{i} + \hat{j} + \hat{k}) + \lambda(-\hat{i} + \hat{j} - \hat{k})$ which is nearest to the origin are

(1) $\left(\frac{2}{3}, \frac{4}{3}, \frac{2}{3}\right)$ (2) $\left(-\frac{2}{3}, -\frac{4}{3}, \frac{2}{3}\right)$
(3) $\left(\frac{2}{3}, \frac{4}{3}, -\frac{2}{3}\right)$ (4) None of these

22. If the length of the perpendicular form the point $(\beta, 0, \beta)$ $(\beta \neq 0)$ to the line, $\frac{x}{1} = \frac{y-1}{0} = \frac{z+1}{-1}$ is $\sqrt{\frac{3}{2}}$, then β is equal

(1) 1 (2) –1 (3) 2 (4) – 2

•**23.** L_1 and L_2 are two lines whose vector equations are
L_1: $\vec{r} = \lambda((\cos\theta + \sqrt{3})\hat{i} + (\sqrt{2}\sin\theta)\hat{j} + (\cos\theta - \sqrt{3})\hat{k})$
L_2: $\vec{r} = \mu(a\hat{i} + b\hat{j} + c\hat{k})$, where λ and μ are scalars. Let α be the acute angle between L_1 and L_2. If the angle α is independent of θ, then the value of α is

(1) $\frac{\pi}{6}$ (2) $\frac{\pi}{4}$ (3) $\frac{\pi}{3}$ (4) $\frac{\pi}{2}$

•**24.** A straight line passes through $\vec{a}$ point P whose position vector is $\vec{p}$, is parallel to the vector which is equally inclined to two unit vectors $\vec{a}$ and $\vec{b}$, in the plane of $\vec{a}$ and $\vec{b}$. Then the distance of the point Q with the position vector $\vec{q}$ from the line is

(1) $|(q-p)\times(\vec{a}+\vec{b})|$ (2) $|(\vec{q}-\vec{p}).(\vec{a}+\vec{b})|$
(3) $\frac{|(\vec{q}-\vec{p})\times(\vec{a}+\vec{b})|}{|\vec{a}+\vec{b}|}$ (4) $\frac{|(\vec{q}-\vec{p}).(\vec{a}+\vec{b})|}{|\vec{a}+\vec{b}|}$

Coplanar and Non-Coplanar Lines

25. The lines $\frac{x-2}{1} = \frac{y-3}{1} = \frac{z-4}{-k}$ and $\frac{x-1}{k} = \frac{y-4}{2} = \frac{z-5}{1}$ are coplanar if

(1) $k = 1$ or -1 (2) $k = 0$ or -3
(3) $k = 3$ or -3 (4) $k = 0$ or -1

26. If the point of intersection of the lines $\frac{x-5}{3} = \frac{y-7}{-1} = \frac{z+2}{1}$ and $\frac{x+3}{-36} = \frac{y-3}{2} = \frac{z-6}{4}$ is (α, β, γ), then $\alpha - \beta - \gamma =$

(1) 16 (2) 18 (3) 21 (4) 27

27. The shortest distance between the lines $\frac{x-1}{1} = \frac{y-1}{1} = \frac{z-1}{1}$ and $\frac{x-2}{1} = \frac{y-3}{1} = \frac{z-4}{1}$ is equal to

(1) $\sqrt{14}$ (2) $\sqrt{7}$
(3) $\sqrt{2}$ (4) $\sqrt{3}$

28. The shortest distance between the lines $\frac{x-1}{0}=\frac{y+1}{-1}=\frac{z}{1}$ and $x+y+z+1=0,\ 2x-y+z+3=0$ is

(1) $\frac{1}{2}$ (2) 1

(3) $\frac{1}{\sqrt{2}}$ (4) $\frac{1}{\sqrt{3}}$

•**29.** If the lines $\vec{r}=\vec{a}+\lambda(\vec{b}\times\vec{c})$ and $\vec{r}=\vec{b}+\mu(\vec{c}\times\vec{a})$ intersect, then which of the following is possible?

(1) $\vec{a}\times\vec{c}=\vec{b}\times\vec{c}$ (2) $\vec{a}\cdot\vec{c}=\vec{b}\cdot\vec{c}$

(3) $\vec{b}\times\vec{a}=\vec{c}\times\vec{a}$ (4) $\vec{a}\cdot(\vec{b}\times\vec{c})=0$

•**30.** Let $A(\vec{a})$ and $B(\vec{b})$ be points on two skew lines $\vec{r}=\vec{a}+\lambda\vec{p}$ and $\vec{r}=\vec{b}+u\vec{q}$ and let the shortest distance between the skew lines be 1, where $\vec{p}$ and $\vec{q}$ are unit vectors forming adjacent sides of a parallelogram enclosing an area of $\frac{1}{2}$ 59 units. If the angle between AB and the line of shortest distance is 60°, then $AB=$

(1) $\frac{1}{2}$ (2) 2

(3) 1 (4) $\lambda\in R-\{0\}$

•**31.** If $\vec{a}=\hat{i}+\hat{j}$ and $\vec{b}=2\hat{i}-\hat{k}$, then the point of intersection of the lines $\vec{r}\times\vec{a}=\vec{b}\times\vec{a}$ and $\vec{r}\times\vec{b}=\vec{a}\times\vec{b}$ is

(1) (3, –1, 1) (2) (3, 1, –1)

(3) (–3, 1, 1) (4) (–3, –1, –1)

Plane in Space

32. If the foot of the perpendicular drawn from the origin to a plane is $P(a, b, c)$, then the equation of the plane is

(1) $\frac{x}{a}+\frac{y}{b}+\frac{z}{c}=3$

(2) $ax+by+cz=3$

(3) $ax+by+cz=a^2+b^2+c^2$

(4) $ax+by+cz=a+b+c$

33. The Cartesian equation of the plane $\vec{r}=(1+\lambda-\mu)\hat{i}+(2-\lambda)\hat{j}+(3-2\lambda+2\mu)\hat{k}$ is

(1) $2x+y=5$ (2) $2x-y=5$

(3) $2x+z=5$ (4) $2x-z=5$

34. The equation of the plane passing through the point (3, 2, 1) and containing the y-axis is

(1) $x=3z$ (2) $x+y-2z=3$

(3) $x-y-z=0$ (4) $x-y+2z=3$

35. A plane P meets the coordinate axes at A, B and C respectively. The centroid of $DABC$ is given to be (1, 1, 2). Then the equation of the line through this centroid and perpendicular to the plane P is

(1) $\frac{x-1}{1}=\frac{y-1}{2}=\frac{z-2}{2}$

(2) $\frac{x-1}{2}=\frac{y-1}{2}=\frac{z-2}{1}$

(3) $\frac{x-1}{2}=\frac{y-1}{1}=\frac{z-2}{1}$

(4) $\frac{x-1}{1}=\frac{y-1}{1}=\frac{z-2}{2}$

36. If the intercepts made by the plane $\vec{r}.\vec{n}=q$ on the x, y, z-axis are a_1, a_2 and a_3 respectively then

(1) $\vec{n}=(qa_1)\hat{i}+(qa_2)\hat{j}+(qa_3)\hat{k}$

(2) $\vec{n}=\frac{a_1}{q}\hat{i}+\frac{a_2}{q}\hat{j}+\frac{a_3}{q}\hat{k}$

(3) $\vec{n}=\frac{q}{a_1}\hat{i}+\frac{q}{a_2}\hat{j}+\frac{q}{a_3}\hat{k}$

(4) $\vec{n}=\left(\frac{1}{a_1q}\right)\hat{i}+\left(\frac{1}{a_2q}\right)\hat{j}+\left(\frac{1}{a_3q}\right)\hat{k}$

37. If the plane $\frac{x}{2}+\frac{y}{3}+\frac{z}{6}=1$ cuts the coordinate axes at points A, B and C, then the area of the triangle ABC.

(1) 18 sq unit (2) 36 sq unit

(3) $3\sqrt{14}$ sq unit (4) $2\sqrt{14}$ sq unit

38. The equation of the plane which passes through the z-axis and perpendicular to the line $\frac{x-a}{\cos\theta}=\frac{y+2}{\sin\theta}=\frac{z-3}{0}$ is

(1) $x+y\tan\theta=0$ (2) $y+x\tan\theta=0$

(3) $x\cos\theta-y\sin\theta=0$ (4) $x\sin\theta-y\cos\theta=0$

39. Which of the following is NOT the equation of the plane passing through the points $P(1, 1, -1)$, $Q(3, 0, 2)$ and $R(-2, 1, 0)$?

(1) $x+11y+3z-9=0$

(2) $((x-9)\hat{i}+y\hat{j}+z\hat{k})\cdot(\hat{i}+11\hat{j}+3\hat{k})=0$

(3) $((2\hat{i}-\hat{j}+3\hat{k})\times(3\hat{i}-\hat{k}))\cdot(x\hat{i}+y\hat{j}+z\hat{k})=9$

(4) None of these

40. The ratio in which the plane $\vec{r}\cdot(\hat{i}-2\hat{j}+3\hat{k})=17$ divides the line joining the points $-2\hat{i}+4\hat{j}+7\hat{k}$ and $3\hat{i}-5\hat{j}+8\hat{k}$ is

(1) 1 : 5 (2) 1 : 10 (3) 3 : 5 (4) 3 :10

41. The equation of the plane passing through the points (2, 2, 1) and (9, 3, 6), and perpendicular to the plane $2x+6y+6z-1=0$ is

(1) $3x+4y+5z=9$ (2) $3x+4y-5z=9$

(3) $3x-4y-5z=9$ (4) $3x-4y+6z=9$

42. The intersection of the spheres $x^2+y^2+z^2+7x-2y-z=13$ and $x^2+y^2+z^2-3x+3y+4z=8$ is the same as the intersection of one of the spheres and the plane

(1) $x-y-z=1$ (2) $x-2y-z=1$

(3) $x-y-2z=1$ (4) $2x-y-z=1$

43. From the point $P(a, b, c)$, let perpendiculars PL and PM be drawn to YOZ and ZOX planes, respectively. Then the equation of the plane OLM, where O is origin is

(1) $\frac{x}{a}+\frac{y}{b}+\frac{z}{c}=0$ (2) $\frac{x}{a}+\frac{y}{b}-\frac{z}{c}=0$

(3) $\frac{x}{a}-\frac{y}{b}-\frac{z}{c}=0$ (4) $\frac{x}{a}-\frac{y}{b}+\frac{z}{c}=0$

44. The equation of the plane passing through the lines $\frac{x-4}{1}=\frac{y-3}{1}=\frac{z-2}{2}$ and $\frac{x-3}{1}=\frac{y-2}{-4}=\frac{z}{5}$ is

(1) $11x-y-3z=35$ (2) $11x+y-3z=35$

(3) $11x-y+3z=35$ (4) $-13x+3y+5z+33=0$

45. The intercepts made on the axes by the plane which bisects the line segment joining the points (1, 2, 3) and (–3, 4, 5) at right angle are

(1) $\left(-\frac{9}{2}, 9, 9\right)$ (2) $\left(\frac{9}{2}, 9, 9\right)$

(3) $\left(9, -\frac{9}{2}, 9\right)$ (4) $\left(9, \frac{9}{2}, 9\right)$

46. The plane which passes through the point (3, 2, 0) and the line $\frac{x-3}{1}=\frac{y-6}{5}=\frac{z-4}{4}$ is

(1) $x-y+z=1$ (2) $x+y+z=5$

(3) $x+2y-z=1$ (4) $2x-y+z=5$

•**47.** The centre of the circle given by $\vec{r}\cdot(\hat{i}+2\hat{j}+2\hat{k})=15$ and $|\vec{r}-(\hat{j}+2\hat{k})|=4$ is

(1) (0, 1, 2) (2) (1, 3, 4)

(3) (–1, 3, 4) (4) (0, –1, –2)

•**48.** The equation of the plane which passes through the point of intersection of lines $\frac{x-1}{3}=\frac{y-2}{1}=\frac{z-3}{2}$, and $\frac{x-3}{1}=\frac{y-1}{2}=\frac{z-2}{3}$ and is at the greatest distance from the point (0, 0, 0) is

(1) $4x+3y+5z=25$ (2) $4x+3y=5z=50$

(3) $3x+4y+5z=49$ (4) $x+7y-5z=2$

•**49.** A plane passing through the point (1, 1, 1) cuts the positive directions of the coordinate axes at A, B and C, then the volume V of the tetrahedron $OABC$ satisfies

(1) $V\le\frac{9}{2}$ (2) $V\ge\frac{9}{2}$

(3) $V=\frac{9}{2}$ (4) $V=\frac{-9}{2}$

•**50.** The plane containing the line $\frac{x-3}{2}=\frac{y-1}{4}=\frac{z-2}{5}$ is rotated through an angle 90° about this line to contain the point (4, 3, 7). The equation of plane before rotation is

(1) $3x-4y+2z=9$ (2) $x-2y-2z=1$

(3) $x-3y+2z=8$ (4) $x+2y-2z=1$

•**51.** A plane passes through a fixed point (a, b, c). The locus of the foot of the perpendicular on it drawn from the origin is the sphere of radius

(1) $\frac{1}{2}\sqrt{a^2+b^2+c^2}$ (2) $\sqrt{a^2+b^2+c^2}$

(3) $a^2+b^2+c^2$ (4) $\frac{1}{2}(a^2+b^2+c^2)$

Family of Planes Containing the Line of Intersection of Two Planes

52. Consider three planes $x+ay+(b+c)z+d=0$, $x+by+(c+a)z+d=0$ and $x+cy+(a+b)z+d=0$; where $a\ne b\ne c$. Then

(1) They meet at a point

(2) They form a triangular prism

(3) They pass through a line

(4) They are at equal distance from the origin

53. The equation of the plane passing through the line of intersection of the planes $x+2y+3z-4=0$ and $4x+3y+2z+1=0$ and the origin is

(1) $17x+14y+11z=0$ (2) $7x+4y+z=0$

(3) $x+14y+11z=0$ (4) $17x+y+z=0$

54. The three planes $4y+6z=5$, $2x+3y+5z=5$ and $6x+5y+9z=10$

(1) meet in a point (2) have a line in common

(3) form a triangular prism (4) none of these

55. The variable plane $(2\lambda+1)x+(3-\lambda)y+z=4$, $(\lambda\in R)$ contains the line

(1) $\frac{x}{0}=\frac{y}{0}=\frac{z+4}{1}$ (2) $\frac{x}{1}=\frac{y}{2}=\frac{z}{-3}$

(3) $\frac{x}{1}=\frac{y}{2}=\frac{z-4}{-7}$ (4) $\frac{x}{2}=\frac{y}{-1}=\frac{z}{0}$

56. The vector equation of the plane passing through the origin and the line of intersection of the planes $\vec{r}\cdot\vec{a}=\lambda$ and $\vec{r}\cdot\vec{b}=\mu$ is

(1) $\vec{r}\cdot(\lambda\vec{a}-\mu\vec{b})=0$ (2) $\vec{r}\cdot(\lambda\vec{b}-\mu\vec{a})=0$

(3) $\vec{r}\cdot(\lambda\vec{a}+\mu\vec{b})=0$ (4) $\vec{r}\cdot(\lambda\vec{b}+\mu\vec{a})=0$

57. A unit vector parallel to the line of intersection of the planes $\vec{r}\cdot\left(\hat{i}-\hat{j}+\hat{k}\right)=5$ and $\vec{r}\cdot\left(2\hat{i}+\hat{j}-3\hat{k}\right)=4$ is

(1) $\frac{2\hat{i}+5\hat{j}-3\hat{k}}{\sqrt{38}}$ (2) $\frac{2\hat{i}-5\hat{j}+3\hat{k}}{\sqrt{38}}$

(3) $\frac{-2\hat{i}-5\hat{j}-3\hat{k}}{\sqrt{38}}$ (4) $\frac{-2\hat{i}+5\hat{j}-3\hat{k}}{\sqrt{38}}$

58. Let L be the line of intersection of the planes $2x+3y+z=1$ and $x+3y+2z=2$.

If L makes an angle α with the positive x-axis, then $\cos\alpha$ equals

(1) $\frac{1}{2}$ (2) 1 (3) $\frac{1}{\sqrt{2}}$ (4) $\frac{1}{\sqrt{3}}$

59. If $P_1: \vec{r}\cdot\vec{n_1} - d_1 = 0, P_2: \vec{r}\cdot\vec{n_2} - d_2 = 0$ and $P_3: \vec{r}\cdot\vec{n_3} - d_3 = 0$ are three planes and $\vec{n_1}, \vec{n_2}$ and $\vec{n_3}$ are three non-coplanar vectors, then three lines $P_1 = 0, P_2 = 0; P_2 = 0, P_3 = 0$ and $P_3 = 0, P_1 = 0$ are

(1) parallel lines (2) coplanar lines
(3) coincident lines (4) concurrent lines

•**60.** Direction ratios of line L on which all the points lie at the same distance from the points $A(1, 2, -1)$, $B(0, 1, -3)$ and $C(-1, -2, 3)$ are

(1) 2, 3, –1 (2) 6, –3, 1
(3) 6, –4, –1 (4) 2, 3, –4

•**61.** The equation of the plane through the line of intersection of the planes $ax + by + cz + d = 0$ and $a'x + b'y + c'z + d' = 0$ and parallel to the line $y = 0, z = 0$ is

(1) $(ab' - a'b)x + (bc' - b'c)y + (ad' - a'd) = 0$
(2) $(ab' - a'b)x + (bc' - b'c)y + (ad' - a'd)z = 0$
(3) $(ab' - a'b)y + (ac' - a'c)z + (ad' - a'd) = 0$
(4) none of these

•**62.** The equation of the plane which passes through the line of intersection of the planes $\vec{r}\cdot\vec{n_1} = q_1, \vec{r}\cdot\vec{n_2} = q_2$ and parallel to the line of intersection of planes $\vec{r}\cdot\vec{n_3} = q_3$ and $\vec{r}\cdot\vec{n_4} = q_4$ is

(1) $[\vec{n_2}\,\vec{n_3}\,\vec{n_4}](\vec{r}\cdot\vec{n_1} - q_1) = [\vec{n_1}\,\vec{n_3}\,\vec{n_4}](\vec{r}\cdot\vec{n_2} - q_2)$
(2) $[\vec{n_1}\,\vec{n_2}\,\vec{n_3}](\vec{r}\cdot\vec{n_4} - q_4) = [\vec{n_4}\,\vec{n_3}\,\vec{n_1}](\vec{r}\cdot\vec{n_2} - q_2)$
(3) $[\vec{n_4}\,\vec{n_3}\,\vec{n_1}](\vec{r}\cdot\vec{n_4} - q_4) = [\vec{n_1}\,\vec{n_2}\,\vec{n_3}](\vec{r}\cdot\vec{n_2} - q_2)$
(4) none of these

Distance of a Point From a Plane

63. If a point P whose position vector is $\vec{p}$, does not lie on the plane ABC, where $A(2, 3, 0)$ $B(3, -2, 0)$ $C(2, -1, 0)$, then the distance of P from the plane ABC is given by

(1) $|\vec{p}\cdot\hat{k}|$
(2) $|(\vec{p} - (2\hat{i} + 3\hat{j}))\cdot(\hat{i} - 5\hat{j})|$
(3) $|(\vec{p} - (2\hat{i} + 3\hat{j}))\cdot(\hat{j})|$
(4) $|(\vec{p}\cdot(\hat{i} - 5\hat{j}))|$

64. The shortest distance from the plane $12x + y + 3z = 327$ to the sphere $x^2 + y^2 + z^2 + 4x - 2y - 6z = 155$ is

(1) 39 (2) 26
(3) $41\frac{4}{13}$ (4) 13

65. Let $\vec{\alpha} = 3\hat{i} + \hat{j} + 2\hat{k}$ and $\vec{\beta} = \hat{i} - 2\hat{j} - 4\hat{k}$ be the position vectors of the points A and B. Then the distance of the point $-\hat{i} + \hat{j} + \hat{k}$ from the plane passing through B and perpendicular to AB is

(1) 5 (2) 10 (3) 15 (4) 20

66. Let Q be the foot of perpendicular drawn from the point P (1, 2, 3) to the plane $x + 2y + z = 14$. If R is a point on the plane such that $\angle PRQ = 60°$, then the area of ΔPQR is equal to

(1) $\frac{\sqrt{3}}{2}$ (2) $\sqrt{3}$ (3) $2\sqrt{3}$ (4) 3

67. Let A (1, 1, 1), B (2, 3, 5) and C (–1, 0, 2) be three points. Then the equation of a plane parallel to the plane ABC and at a distance of 2 units from it is

(1) $2x - 3y + z + 2\sqrt{14} = 0$
(2) $2x - 3y + z - \sqrt{14} = 0$
(3) $2x - 3y + z + 2 = 0$
(4) $2x - 3y + z - 2 = 0$

68. If the shortest distance between the y-axis and the line $2x + 3y + 5z + 1 = 3x + 4y + 6z + 2 = 0$ is $\frac{2}{\sqrt{k}}$, then the value of 'k' is

(1) 2 (2) 3 (3) 5 (4) 6

69. The image of the point (–1, 3, 4) in the plane $x - 2y = 0$ is

(1) $\left(-\frac{17}{3}, -\frac{19}{3}, 4\right)$ (2) (15, 11, 4)
(3) $\left(-\frac{17}{3}, -\frac{19}{3}, 1\right)$ (4) $\left(\frac{9}{5}, -\frac{13}{5}, 4\right)$

70. The radius of the circle in which the sphere $x^2 + y^2 + z^2 + 2z - 2y - 4z - 19 = 0$ is cut by the plane $x + 2y + 2z + 7 = 0$ is

(1) 2 (2) 3 (3) 4 (4) 1

71. The length of the perpendicular from the origin to the plane passing through the point $\vec{a}$ and containing the line $\vec{r} = \vec{b} + \lambda\vec{c}$ is

(1) $\frac{|[\vec{a}\,\vec{b}\,\vec{c}]|}{|\vec{a}\times\vec{b} + \vec{b}\times\vec{c} + \vec{c}\times\vec{a}|}$ (2) $\frac{|[\vec{a}\,\vec{b}\,\vec{c}]|}{|\vec{a}\times\vec{b} + \vec{b}\times\vec{c}|}$
(3) $\frac{|[\vec{a}\,\vec{b}\,\vec{c}]|}{|\vec{b}\times\vec{c} + \vec{c}\times\vec{a}|}$ (4) $\frac{|[\vec{a}\,\vec{b}\,\vec{c}]|}{|\vec{c}\times\vec{a} + \vec{a}\times\vec{b}|}$

•**72.** The distance of the point P $(\vec{p})$ from the plane $\vec{r}\cdot\vec{n} = 0$ is

(1) $|\vec{p}\cdot\vec{n}|$ (2) $\frac{|\vec{p}\times\vec{n}|}{|\vec{n}|}$
(3) $\frac{|\vec{p}\cdot\vec{n}|}{|\vec{n}|}$ (4) $|\vec{p}\times\vec{n}|$

•**73.** The reflection of the point $\vec{a}$ in the plane $\vec{r}\cdot\vec{n} = q$ is

(1) $\vec{a} + \frac{(\vec{q} - \vec{a}\cdot\vec{n})}{|\vec{n}|}$

(2) $\vec{a}+2\left(\frac{(\vec{q}-\vec{a}\cdot\vec{n})}{|\vec{n}|^2}\right)\vec{n}$

(3) $\vec{a}+\frac{2(\vec{q}+\vec{a}\cdot\vec{n})}{|\vec{n}|}\vec{n}$

(4) $\vec{a}+\frac{2(\vec{q}+\vec{a}\cdot\vec{n})}{|\vec{n}|}$

•**74.** The projection of the point $P(\vec{p})$ on the plane $\vec{r}\cdot\vec{n}=q$ is $(\vec{s})$, then

(1) $\vec{s}=\frac{(q-\vec{p}\cdot\vec{n})\vec{n}}{|\vec{n}|^2}$ (2) $\vec{s}=\vec{p}+\frac{(q-\vec{p}\cdot\vec{n})\vec{n}}{|\vec{n}|^2}$

(3) $\vec{s}=\vec{p}-\frac{(\vec{p}\cdot\vec{n})\vec{n}}{|\vec{n}|^2}$ (4) $\vec{s}=\vec{p}-\frac{(q-\vec{p}\cdot\vec{n})\vec{n}}{|\vec{n}|^2}$

•**75.** The equation of a straight line lying in the plane $\vec{r}\cdot\vec{n}=d$ and parallel to the line $\vec{r}=\vec{a}+\lambda\vec{b}$, which passes through the foot of the perpendicular drawn from the point $P(\vec{a})$ to the plane $\vec{r}\cdot\vec{n}=d$ (where $\vec{n}\cdot\vec{b}=0$) is

(1) $\vec{r}=\vec{a}+\left(\frac{d-\vec{a}\cdot\vec{n}}{n^2}\right)\vec{n}+\lambda\vec{b}$

(2) $\vec{r}=\vec{a}+\left(\frac{d-\vec{a}\cdot\vec{n}}{n}\right)\vec{n}+\lambda\vec{b}$

(3) $\vec{r}=\vec{a}+\left(\frac{\vec{a}\cdot\vec{n}-d}{n^2}\right)\vec{n}+\lambda\vec{b}$

(4) $\vec{r}=\vec{a}+\left(\frac{\vec{a}\cdot\vec{n}-d}{n}\right)\vec{n}+\lambda\vec{b}$

•**76.** If the distance between the planes $ax-2y+z=b$ and the plane containing the lines $\frac{x-1}{2}=\frac{y-2}{3}=\frac{z-3}{4}$ and $\frac{x-2}{3}=\frac{y-3}{4}=\frac{z-4}{5}$ is $\sqrt{6}$, then $|b|$ is

(1) 3 (2) 4 (3) 5 (4) 6

Division of Space by Planes and Angle Bisector Planes

77. Acute angle between planes

$P_1:\vec{r}=(\hat{i}-\hat{j})+\lambda(2\hat{i}+\hat{j}-\hat{k})+\mu(\hat{i}-\hat{j}+\hat{k})$ and

$P_2:\vec{r}\cdot(2\hat{i}-\hat{j}+2\hat{k})=3$ is

(1) $\cos^{-1}\frac{1}{3}$ (2) $\cos^{-1}\frac{1}{3\sqrt{2}}$

(3) $\cos^{-1}\frac{1}{2\sqrt{3}}$ (4) $\cos^{-1}\frac{1}{2\sqrt{2}}$

78. A tetrahedron has vertices $O(0, 0, 0)$, $A(1, 2, 1)$, $B(2, 1, 3)$ and $C(-1, 1, 2)$, then the angle between faces OAB and ABC is

(1) $\cos^{-1}\left(\frac{17}{31}\right)$ (2) 30°

(3) 90° (4) $\cos^{-1}\left(\frac{19}{35}\right)$

79. If angle between the planes $x-2y+3z-5=0$ and $x+\alpha y+2z+7=0$ is $\cos^{-1}\frac{1}{14}$, then the difference between the values of α is

(1) $\frac{12}{11}$ (2) $\frac{62}{55}$ (3) $\frac{31}{11}$ (4) $\frac{8}{5}$

80. The plane $4x+7y+4z+81=0$ is rotated through a right angle about its line of intersection with the plane $5x+3y+10z=25$. The equation of the plane in its new position is

(1) $x-4y+6z=106$

(2) $x-8y+13z=103$

(3) $x-4y+6z=110$

(4) $x-8y+13z=105$

•**81.** The direction ratios of the normal to the plane passing through the points (1, 0, 0) and (0, 1, 0), making an angle of $\frac{\pi}{4}$ with the plane $x+y=3$, are

(1) $<1, \sqrt{2}, 1>$ (2) $<1, 1, \sqrt{2}>$

(3) $<1, 1, 2>$ (4) $<\sqrt{2}, 1, 1>$

•**82.** The number of integral values α for which the point $(2, 3, \alpha)$ lies in the acute angle region of the planes $x+2y+3z-5=0$ and $x-y+z-7=0$ is

(1) 6 (2) 7 (3) 8 (4) 9

•**83.** If the plane $x+2y+z+18=0$ bisects the angle between the planes $x-2y+2z-5=0$ and $ax+by+cz+d=0$, then the value of $a+b+c+d$ is

(1) 7 (2) 8 (3) 9 (4) 10

Line and Plane

84. Let the equations of a line and a plane be $\frac{x+3}{2}=\frac{y-4}{3}=\frac{z+5}{2}$ and $4x-2y-z=1$, respectively, then

(1) the line is parallel to the plane

(2) the line is perpendicular to the plane

(3) the line lies in the plane

(4) none of these

85. The value of m for which the straight line $3x-2y+z+3=0=4x-3y+4z+1$ is parallel to the plane $2x-y+mz-2=0$ is

(1) −2 (2) 8 (3) −18 (4) 11

86. The value of λ such that the line $\frac{x-1}{2}=\frac{y-1}{3}=\frac{z-1}{\lambda}$ is perpendicular to the normal to the plane $\vec{r}\cdot(2\hat{i}+3\hat{j}+4\hat{k})=0$ is

(1) $-\frac{13}{4}$ (2) $-\frac{17}{4}$

(3) 4 (4) 3

87. The distance between the line $\vec{r} = 2\hat{i} - 2\hat{j} + 3\hat{k} + \lambda(\hat{i} - \hat{j} + 4\hat{k})$ and the plane $\vec{r} \cdot (\hat{i} + 5\hat{j} + \hat{k}) = 5$ is

(1) $\frac{10}{3\sqrt{3}}$ (2) $\frac{10}{9}$

(3) $\frac{10}{3}$ (4) $\frac{3}{10}$

88. If the angle θ between the line $\frac{x+1}{1} = \frac{y-1}{2} = \frac{z-2}{2}$ and the plane $2x - y + \sqrt{\lambda}\, z + 4 = 0$ is such that $\sin\theta = \frac{1}{3}$, then the value of λ is

(1) $\frac{-3}{5}$ (2) $\frac{5}{3}$

(3) $\frac{-4}{3}$ (4) $\frac{3}{4}$

89. Let θ be the angle between the planes $P_1 : \vec{r} \cdot (\hat{i} + \hat{j} + 2\hat{k}) = 9$ and $P_2 : \vec{r} \cdot (2\hat{i} - \hat{j} + \hat{k}) = 15$. Let L be the line that meets P_2 at the point $(4, -2, 5)$ and makes an angle θ with the normal of P_2. If α is the angle between L and P_2 then $(\tan^2\theta)(\cot^2\alpha)$ is equal to

(1) 3 (2) 6 (3) 9 (4) 12

•**90.** Equation of a line lying in the plane $\pi \equiv 2x - y + z - 4 = 0$ and perpendicular to the line l, whose equation is $\frac{x-2}{1} = \frac{y-2}{-1} = \frac{z-3}{-2}$ and which passes through the point of intersection of l and π is

(1) $\frac{x-2}{1} = \frac{y-1}{5} = \frac{z-1}{-1}$ (2) $\frac{x-1}{3} = \frac{y-3}{5} = \frac{z-5}{-1}$

(3) $\frac{x+2}{2} = \frac{y+1}{-1} = \frac{z+1}{1}$ (4) $\frac{x-2}{2} = \frac{y-1}{-1} = \frac{z-1}{1}$

•**91.** The projection of the line $\frac{x+1}{-1} = \frac{y}{2} = \frac{z-1}{3}$ on the plane $x - 2y + z = 6$ is the line of intersection of this plane with the plane

(1) $2x + y + 2 = 0$ (2) $3x + y - z = 2$

(3) $2x - 3y + 8z = 3$ (4) none of these

•**92.** If for $a > 0$, the feet of perpendiculars from the points $A(a, -2a, 3)$ and $B(0, 4, 5)$ on the plane $lx + my + nz = 0$ are points $C(0, -a, -1)$ and D respectively, then the length of line segment CD is equal to

(1) $\sqrt{31}$ (2) $\sqrt{41}$ (3) $\sqrt{55}$ (4) $\sqrt{66}$

•**93.** If the distance of the point $(1, -2, 3)$ from the plane $x + 2y - 3z + 10 = 0$ measured parallel to the line, $\frac{x-1}{3} = \frac{2-y}{m} = \frac{z+3}{1}$ is $\sqrt{\frac{7}{2}}$, then the value of $|m|$ is equal to

(1) 1 (2) 2 (3) 3 (4) 4

Numerical Value Type

1. If the line $\frac{2x-3}{4} = \frac{y-\alpha}{3} = \frac{z-1}{2}$ meets the z-axis, then the value of α is ________.

2. If the image of the point $P(1, -2, 3)$ in the plane, $2x + 3y - 4z + 22 = 0$ measured parallel to line, $\frac{x}{1} = \frac{y}{4} = \frac{z}{5}$ is $Q(\alpha, \beta, \gamma)$ then the value of $\frac{2\beta}{\gamma - \alpha}$ is ______.

3. Let A be a point on the line $\vec{r} = (1 - 3\mu)\hat{i} + (\mu - 1)\hat{j} + (2 + 5\mu)\hat{k}$ and $B(3, 2, 6)$ be a point in the space. Then the value of μ for which the vector $\overline{AB}$ is parallel to the plane $x - 4y + 3z = 1$ is ______.

4. Let P be the plane, which contains the line of intersection of the planes, $x + y + z - 6 = 0$ and $2x + 3y + z + 5 = 0$ and it is perpendicular to the xy–plane. Then the square of the distance of the point $(0, 0, 256)$ from P is equal to ________.

5. Let $a, b \in R$. If the mirror image of the point $P(a, 6, 9)$ with respect to the line $\frac{x-3}{7} = \frac{y-2}{5} = \frac{z-1}{-9}$ is $(20, b, -a-9)$, then $\frac{a}{b}$ is equal to ________.

•**6.** The distance of the point $(-1, 2, 3)$ from the plane $\vec{r} \cdot (\hat{i} - 2\hat{j} + 3\hat{k}) = 10$ parallel to the line of the shortest distance between the lines $\vec{r} = (\hat{i} - \hat{j}) + \lambda(2\hat{i} + \hat{k})$ and $\vec{r} = (2\hat{i} - \hat{j}) + \mu(\hat{i} - \hat{j} + \hat{k})$ is α, then the value of $\frac{132}{\alpha^2}$ is ________.

•**7.** One vertex of a rectangular parallelopiped is at the origin O and the lengths of its edges along x, y and z axes are 3, 4 and 5 units respectively. Let P be the vertex $(3, 4, 5)$. Then the shortest distance between the diagonal OP and an edge parallel to z axis, not passing through O or P is ______.

•**8.** Let the equation of plane passing through the line of intersection of the planes $x + 2y + az = 2$ and $x - y + z = 3$ be $5x - 11y + bz = 6a - 1$. For $c \in Z$, if the distance of this plane from the point $(a, -c, c)$ is $\frac{2}{\sqrt{a}}$, then the value of $\left|\frac{c}{a+b}\right|$ is ________.

•**9.** If the equation of plane passing through the mirror image of a point $(2, 3, 1)$ with respect to line $\frac{x+1}{2} = \frac{y-3}{1} = \frac{z+2}{-1}$ and containing the line $\frac{x-2}{3} = \frac{1-y}{2} = \frac{z+1}{1}$ is $\alpha x + \beta y + \gamma z = 24$, then the value of $\frac{3\alpha}{\beta - \gamma}$ is ________.

Integer Answer Type

1. The distance of the z-axis from the image of the point M (2, –3, 3) in the plane $x - 2y - z + 1 = 0$, is_____.

2. If the angle between the plane $x - 3y + 2z = 1$ and the line $\frac{x-1}{2} = \frac{y-1}{1} = \frac{z-1}{-3}$ is θ, then the value of cosec θ. is _____.

3. Let $P(a, b, c)$ be any point on the plane $3x + 2y + z = 7$, then the least value of $2(a^2 + b^2 + c^2)$, is _____.

4. The plane $P_1 : 4x + 7y + 4z + 81 = 0$ is rotated through a right angle about its line of intersection with the plane P_2 : $5x + 3y + 10z = 25$. If the plane in its new position is P, then the distance of this plane from the origin is $\frac{106}{\sqrt{\lambda}}$, where λ= _____.

5. The value of k such that line $\frac{x-4}{1} = \frac{y-2}{1} = \frac{z-k}{2}$ lies in the plane $2x - 4y + z = 7$ is_____.

6. A plane which is perpendicular to two planes $2x - 2y + z = 0$ and $x - y + 2z = 4$ passes through the point (1, –2, 1). The square of distance of the plane from the point (1, 2, 2) is_____.

7. If the line contained by the planes $2x + y + z - 3 = 0$ and $x - y + 2z + 5 = 0$ is parallel to the plane $\alpha x + 2y + 3z + 1 = 0$, then the value of α is_____.

8. A plane P is perpendicular to the vector $2\hat{i} + 3\hat{j} + 6\hat{k}$ contains the terminal point of the vector $\hat{i} + 5\hat{j} + 3\hat{k}$ through the origin. Then the distance of the plane P from the origin is_____.

9. Line L whose direction ratios are 2, 2 and 1 intersects the lines $L_1 : \frac{x-7}{3} = \frac{y-5}{2} = \frac{z-3}{1}$ and $L_2 : \frac{x-1}{2} = \frac{y+1}{4} = \frac{z+1}{3}$ at A and B, then the distance AB is _____.

10. Lines passing through the point (2, –2, 6) are parallel to the plane $x - 2y + 2z - 3 = 0$. Then all such lines lie in the plane whose distance from the origin is _____.

•11. If the length of the projection of the line segment joining points (1, 0, – 1) and (–1, 2, 2) on the plane $x + 3y - 5z = 6$ is $\sqrt{\frac{\lambda}{35}}$, then λ = _____.

•12. Let the equation of the plane containing the line $x - y - z - 4 = 0 = x + y + 2z - 4$ and parallel to the line of intersection of the planes $2x + 3y + z = 1$ and $x + 3y + 2z = 2$ be $x + Ay + Bz + C = 0$. Then the value of $|A + B + C|$ is_____.

•13. Twice the distance of the point $P(-2, 3, -4)$ from the line $\frac{x+2}{3} = \frac{2y+3}{4} = \frac{3z+4}{5}$ measured parallel to the plane $4x + 12y - 3z + 1 = 0$ is_____.

•14. The position vectors of the four angular points of a tetrahedron $OABC$ are (0, 0, 0), (0, 0, 2), (0, 4, 0) and (6, 0, 0), respectively. A point P inside the tetrahedron is at the same distance r from the four plane faces of the tetrahedron. Then the value of $9r$ is _____.

•15. A variable plane at a distance of 1 unit from the origin cuts the coordinate axes at A, B and C. If the centroid $D(x, y, z)$ of triangle ABC satisfies the relation $\frac{1}{x^2} + \frac{1}{y^2} + \frac{1}{z^2} = k$, then the value of k is _____.

Multiple Correct Answers Type

1. Let PM be the perpendicular from the point $P(1, 2, 3)$ to the xy plane. If $\overrightarrow{OP}$ makes an angle θ with the positive direction of the z-axis and $\overrightarrow{OM}$ makes an angle ϕ with the positive direction of x-axis, where O is the origin and θ and ϕ are acute angles, then

(1) $\cos\theta\cos\phi = 1/\sqrt{14}$ (2) $\sin\theta\sin\phi = 2/\sqrt{14}$
(3) $\tan\phi = 2$ (4) $\tan\theta = \sqrt{5}/3$

2. A line with direction cosines proportional to 1, –5 and –2 meets lines $x = y + 5 = z + 1$ and $x + 5 = 3y = 2z$. The coordinates of points of the intersection are given by

(1) (2, –3, 1) (2) (1, 2, 3)
(3) (0, 5/3, 5/2) (4) (3, –2, 2)

3. The equation of the lines $x + y + z - 1 = 0$ and $4x + y - 2z + 2 = 0$ written in the symmetrical form is

(1) $\frac{x-1}{2} = \frac{y+2}{-1} = \frac{z-2}{2}$

(2) $\frac{x+(1/2)}{1} = \frac{y-1}{-2} = \frac{z-(1/2)}{1}$

(3) $\frac{x}{1} = \frac{y}{-2} = \frac{z-1}{1}$

(4) $\frac{x+1}{1} = \frac{y-2}{-2} = \frac{z-0}{1}$

4. If the lines $\frac{x-2}{1} = \frac{y-3}{1} = \frac{z-4}{\lambda}$ and $\frac{x-1}{\lambda} = \frac{y-4}{2} = \frac{z-5}{1}$ intersect, then

(1) $\lambda = -1$ (2) $\lambda = 2$
(3) $\lambda = -3$ (4) $\lambda = 0$

5. L_1, L_2 and L_3 are three lines in 3-dimensional space. If all three lines are pair wise skew and the shortest distance between any two of them is same, then which of the following statements can be TRUE

(1) lines are edges of a cube
(2) lines are all parallel
(3) lines are edges of any parallelepiped
(4) lines are perpend

6. For the line $\frac{x-1}{1} = \frac{y-2}{2} = \frac{z-3}{3}$ which of the following statements is/are true?

(1) It lies in the plane $x - 2y + z = 0$
(2) It is same as $\frac{x}{1} = \frac{y}{2} = \frac{z}{3}$
(3) It passe through the point (2, 3, 5)
(4) It is parallel to the plane $x - 2y + z - 6 = 0$

7. Let P_1 be plane to which the vector $(\hat{i}+\hat{j})$ is normal and which contains the line L, whose equation is $\vec{r}=\hat{i}+\hat{j}+\hat{k}+\lambda(\hat{i}-\hat{j}-\hat{k})$. Let P_2 be the plane containing the line L and the point whose position vector is $\hat{j}$. Which of the following statements is (are) correct?
(1) The equation of P_1 is $x+y=2$.
(2) The equation of P_2 is $\vec{r}\cdot(\hat{i}-2\hat{j}+\hat{k})=2$.
(3) The acute angle between P_1 and P_2 is $\cot^{-1}(\sqrt{3})$.
(4) The angle between the plane P_2 and the line L is $\tan^{-1}\sqrt{3}$.

8. Let $P=0$ be the equation of the plane passing through the line of intersection of the planes $2x-y=0$ and $3z-y=0$ and perpendicular to the plane $4x+5y-3z=8$. Then the points which lie on the plane $P=0$ is/are
(1) (0, 9, 17) (2) (1/7, 2, 1/9)
(3) (1, 3, – 4) (4) (1/2, 1, 1/3)

9. Consider the planes $3x-6y+2z+5=0$ and $4x-12y+3z=3$. The plane $67x-162y+47z+44=0$ bisects the angle between the given planes which
(1) contains origin
(2) is acute
(3) is obtuse
(4) contains the point (1, 1, 1)

10. The equation(s) of the plane which passes through (0, 0, 0) and which is equally inclined to the planes $x-y+z-3=0$ and $x+y+z+4=0$ is/are
(1) $y=0$ (2) $x=0$
(3) $x+y=0$ (4) $x+z=0$

11. The *xy*-plane is rotated about its line of intersection with the *yz*-plane by 45°, then the equation(s) of the new plane(s) may be
(1) $z+x=0$ (2) $z-y=0$
(3) $x+y+z=0$ (4) $z-x=0$

12. The equation(s) of the plane passing through the origin and equally inclined to the lines $\frac{x-1}{2}=\frac{y}{-2}=\frac{z+2}{-1}$ and $\frac{x+3}{8}=\frac{y-4}{1}=\frac{z}{-4}$ and is/are
(1) $14x-5y-7z=0$ (2) $2x+7y-z=0$
(3) $3x-4y-z=0$ (4) $x+2y-5z=0$

13. Which of the following lines lie on the plane $x+2y-z+4=0$?
(1) $\frac{x-1}{1}=\frac{y}{-1}=\frac{z-5}{-1}$
(2) $x-y+z=2x+y-z=0$
(3) $\vec{r}=2\hat{i}-\hat{j}+4\hat{k}+\lambda(3\hat{i}+\hat{j}+5\hat{k})$
(4) none of these

14. A rod of length 2 units has one end at (1, 0, –1). If its other end touches the plane $x-2y+2z+4=0$. Then
(1) the rod sweeps out the figure whose volume is π cubic units.
(2) the area of the region which the rod traces out on the plane is 2π.
(3) the length of projection of the rod on the plane is $\sqrt{3}$ units.
(4) the centre of the region which the rod traces out on the plane is $\left(\frac{2}{3},\frac{2}{3},\frac{-5}{3}\right)$.

15. Consider a set of points R in the space which are at a distance of 2 units from the line $\frac{x}{1}=\frac{y-1}{-1}=\frac{z+2}{2}$ between the planes $x-y+2z+3=0$ and $x-y+2z-2=0$. Then
(1) The volume of the bounded figure by the set of points R and the planes is $(10/3\sqrt{3})\pi$ cube units.
(2) The area of the curved surface formed by the set of points R is $(20\pi/\sqrt{6})$ sq. units.
(3) The volume of the bounded figure by the set of point R and the planes is $(20\pi/\sqrt{6})$ cubic units.
(4) The area of the curved surface formed by the set of points R is $(10/\sqrt{3})\pi$ sq. units.

16. The equation of the line passing through the point $\vec{a}$ parallel to the plane $\vec{r}.\vec{n}=q$ and perpendicular to the line $\vec{r}=\vec{b}+t\vec{c}$ is
(1) $\vec{r}=\vec{a}+\lambda(\vec{n}\times\vec{c})$ (2) $(\vec{r}-\vec{a})\times(\vec{n}\times\vec{c})=0$
(3) $\vec{r}=\vec{b}+\lambda(\vec{n}\times\vec{c})$ (4) $\vec{r}-\vec{b}-\lambda(\vec{n}\times\vec{c})$

17. If the volume of the tetrahedron $ABCD$ is 1 cubic units where $A(0, 1, 2)$, $B(-1, 2, 1)$ and $C(1, 2, 1)$, then the locus of point D is
(1) $x+y-z=3$ (2) $y+z=6$
(3) $y+z=0$ (4) $y+z=-3$

Linked Comprehension Type

For Questions 1–3

Given four points $A(2, 1, 0)$, $B(1, 0, 1)$, $C(3, 0, 1)$ and $D(0, 0, 2)$. Point D lies on the line L which is orthogonal to the plane determined by the points A, B and C. Then

1. The equation of the plane ABC is
(1) $x+y+z-3=0$ (2) $y+z-1=0$
(3) $x+z-1=0$ (4) $2y+z-1=0$

2. The equation of the line L is
(1) $\vec{r}=2\hat{k}+\lambda(\hat{i}+\hat{k})$ (2) $\vec{r}=2\hat{k}+\lambda(2\hat{j}+\hat{k})$
(3) $\vec{r}=2\hat{k}+\lambda(2\hat{j}+\hat{k})$ (4) none

3. The perpendicular distance of D from the plane ABC is
(1) $\sqrt{2}$ (2) 1/2
(3) 2 (4) $1/\sqrt{2}$

For Questions 4–6

A ray of light comes along the line L and strikes the plane mirror kept along the plane P at B. Let $A(2, 1, 6)$ be the point on the line L whose image about plane P is A'. If the line L is $\frac{x-2}{3}=\frac{y-1}{4}=\frac{z-6}{5}$ and the plane P is $x + y - 2z = 3$. Then answer the following questions.

4. The coordinates of A' are
(1) (6, 5, 2) (2) (6, 5, –2)
(3) (6, –5, 2) (4) (–6, 5, –2)

5. The coordinates of B are
(1) (5, 10, 6) (2) (10, 15, 11)
(3) (–10, –15, –14) (4) (–5, 10, –6)

6. If L_1 is the reflected ray, then its equation is
(1) $\frac{x+10}{4}=\frac{y-5}{4}=\frac{z+2}{3}$
(2) $\frac{x+10}{3}=\frac{y+15}{5}=\frac{z+14}{5}$
(3) $\frac{x+10}{4}=\frac{y+15}{5}=\frac{z+14}{3}$
(4) $\frac{x+10}{2}=\frac{y-5}{3}=\frac{z-2}{5}$

For Questions 7–9

Consider a plane $x + y - z = 1$ and a point $A(1, 2, -3)$. A line L has the equation $x = 1 + 3r$, $y = 2 - r$ and $z = 3 + 4r$. Then

7. The coordinate of a point B on the line L such that AB is parallel to the plane is
(1) (10, –1, 15) (2) (–5, 4, –5)
(3) (4, 1, 7) (4) (–8, 5, –9)

8. The equation of the plane containing the line L and the point A is
(1) $x - 3y + 5 = 0$ (2) $x + 3y - 7 = 0$
(3) $3x - y - 1 = 0$ (4) $3x + y - 5 = 0$

9. The distance between the points on the line which are at a distance of $4/\sqrt{3}$ units from the plane is
(1) $4\sqrt{26}$ (2) 20 (3) $10\sqrt{13}$ (4) $2\sqrt{104}$

For Questions 10–11

Plane P contains the lines $3x - 2y + z - 6 = 0$, $x + y - 4z + 5 = 0$ and parallel to the lines $x - y + z - 3 = 0$, $2x + y - z - 7 = 0$.

10. The distance of P from the origin is
(1) $\frac{9}{23\sqrt{2}}$ (2) $\frac{23}{9\sqrt{2}}$
(3) $\frac{23}{6\sqrt{2}}$ (4) $\frac{\sqrt{23}}{9\sqrt{2}}$

11. If plane P is perpendicular to the plane $\vec{r}=\hat{i}+\hat{j}-\hat{k}+\lambda(2\hat{i}-\hat{j}+3\hat{k})+\mu(-\hat{i}+2\hat{j}+\alpha\hat{k})$, then the value of α is ________
(1) 1 (2) 0 (3) –1 (4) 2

For Questions 12-13

A plane P passes through the lines $[\vec{r}-(-\hat{i}+3\hat{j}-2\hat{k})]\times(-3\hat{i}+2\hat{j}+\hat{k})=0$ and $\frac{x}{1}=\frac{7-y}{3}=\frac{z+7}{\alpha}$.

12. The value of α is ________

13. If the points $(2, -9, \lambda)$ and $(\lambda, -1, -3)$ lie on the opposite sides of a plane P then number of integral values of λ is ________

For Questions 14–15

Consider a plane $P_1: x - 2y + z = 0$ and the plane P_2 which passes through the point (2, 3, 5). Line of intersection of planes P_1 and P_2 is $L: \frac{x-\alpha}{3}=\frac{y-2}{5}=\frac{z-3}{\beta}$.

14. The square of distance of point (1, 4, –3) from the plane P_2 is $\frac{\sqrt{\lambda}}{3}$, then the value of λ is ________

15. If (p, q, r) is the nearest point on the line L from the origin, then $-83(p + q + r) =$ ________

Matrix Match Type

1. Match the following lists:

List I	List II
(P) A vector perpendicular to the line $x = 2t + 1, y = t + 2$ and $z = -t - 3$ is	(I) $7\hat{i}+3\hat{j}+5\hat{k}$
(Q) A vector parallel to the planes $x + y + z - 3 = 0$ and $2x - y + 3z = 0$ is	(II) $4\hat{i}-\hat{j}-3\hat{k}$
(R) A vector along which the distance between the lines $\frac{x}{2}=\frac{y}{-3}=\frac{z}{-1}$ and $\vec{r}=(3\hat{i}-\hat{j}+\hat{k})+t(\hat{i}+\hat{j}-2\hat{k})$ is the shortest is	(III) $-11\hat{i}+7\hat{j}+5\hat{k}$
(S) A vector normal to the plane $\vec{r}=-\hat{i}+4\hat{j}-6\hat{k}+\lambda(\hat{i}+3\hat{j}-2\hat{k})+\mu(-\hat{i}+2\hat{j}-5\hat{k})$ is	(IV) $\hat{i}+3\hat{j}+5\hat{k}$
	(V) $\hat{i}+\hat{j}-3\hat{k}$

(1) (P) → (IV); (Q) → (II); (R) → (I); (S) → (III)
(2) (P) → (IV); (Q) → (III); (R) → (I); (S) → (II)
(3) (P) → (I); (Q) → (II); (R) → (IV); (S) → (III)
(4) (P) → (IV); (Q) → (II); (R) → (I); (S) → (V)

2. Match the following lists:

List I	List II
(P) The point on the line $x = 4y + 5$, $z = 3y - 6$ at a distance 3 from the point (5, 3, – 6) is	(I) (–1, – 2, 0)
(Q) The plane containing the lines $\frac{x-2}{3}=\frac{y+3}{5}=\frac{z+5}{7}$ and parallel to $\hat{i}+4\hat{j}+7\hat{k}$ passes through the point	(II) (5, 0, –6)

(R) A line passes through two points $A(2, -3, -1)$ and $B(8, -1, 2)$. The point on this line nearer to the origin and at a distance of 14 units from A is	**(III)** (2, 5, 7)
(S) The foot of the perpendicular from the point (3, –1, 11) on the line $\frac{x}{2}=\frac{y-2}{3}=\frac{z-3}{4}$ is	**(IV)** (1, 2, –3)
	(V) (–10, –7, –7)

(1) (P) → (II); (Q) → (I); (R) → (IV); (S) → (V)
(2) (P) → (IV); (Q) → (I); (R) → (V); (S) → (II)
(3) (P) → (II); (Q) → (I); (R) → (V); (S) → (III)
(4) (P) → (III); (Q) → (II); (R) → (V); (S) → (IV)

3. Match the following lists:

List I	List II
(P) The distance between the line $\vec{r}=(2\hat{i}-2\hat{j}+3\hat{k})+\lambda(\hat{i}-\hat{j}+4\hat{k})$ and plane $\vec{r}\cdot(\hat{i}+5\hat{j}+\hat{k})=5$ is	**(I)** $\frac{25}{3\sqrt{14}}$
(Q) The distance between planes $\vec{r}\cdot(2\hat{i}-\hat{j}+3\hat{k})=4$ and $\vec{r}\cdot(6\hat{i}-3\hat{j}+9\hat{k})+13=0$ is	**(II)** 13/7
(R) The distance of a point (2, 5, – 3) from the plane $\vec{r}\cdot(6\hat{i}-3\hat{j}+2\hat{k})=4$ is	**(III)** $\frac{10}{3\sqrt{3}}$
(S) The distance of the point (1, 0, – 3) from the plane $x-y-z-9=0$ measured parallel to line $\frac{x-2}{2}=\frac{y+2}{3}=\frac{z-6}{-6}$	**(IV)** 6
	(V) 7

(1) (P) → (III); (Q) → (IV); (R) → (II); (S) → (I)
(2) (P) → (IV); (Q) → (I); (R) → (II); (S) → (III)
(3) (P) → (II); (Q) → (I); (R) → (V); (S) → (III)
(4) (P) → (III); (Q) → (I); (R) → (II); (S) → (V)

4. Match the following lists:

List I	List II
(P) Image of the point (3, 5, 7) in the plane $2x+y+z=-18$ is	**(I)** (– 1, – 1, –1)
(Q) The point of intersection of the line $\frac{x-2}{-3}=\frac{y-1}{-2}=\frac{z-3}{2}$ and the plane $2x+y-z=3$ is	**(II)** (–21, –7, –5)
(R) The foot of the perpendicular from the point (1, 1, 2) to the plane $2x-2y+4z+5=0$ is	**(III)** $\left(\frac{5}{2},\frac{2}{3},\frac{8}{3}\right)$
(S) The intersection point of the lines $\frac{x-1}{2}=\frac{y-2}{3}=\frac{z-3}{4}$ and $\frac{x-4}{5}=\frac{y-1}{2}=z$ is	**(IV)** $\left(-\frac{1}{12},\frac{25}{12},\frac{-2}{12}\right)$
	(V) (1, –1, 1)

(1) (P) → (III); (Q) → (IV); (R) → (II); (S) → (I)
(2) (P) → (II); (Q) → (III); (R) → (IV); (S) → (I)
(3) (P) → (II); (Q) → (I); (R) → (V); (S) → (III)
(4) (P) → (I); (Q) → (IV); (R) → (III); (S) → (V)

5. Match the following columns

List I	List II
(P) Line $\frac{x-3}{2}=\frac{y-6}{4}=\frac{z+3}{1}$	**(I)** Meets one of the coordinate axes
(Q) Line $\frac{2x+5}{3}=\frac{y+1}{2}=\frac{z-3}{6}$	**(II)** Meets two of the coordinate axes
(R) Line $\frac{x}{0}=\frac{y-3}{3}=\frac{z}{-4}$	**(III)** Meets all the three coordinate axes
(S) Line $\frac{x}{-3}=\frac{y}{5}=\frac{z}{6}$	**(IV)** Is perpendicular to one of the coordinate axes
	(V) Does not meet any coordinate axes

(1) (P) → (I); (Q) → (II); (R) → (III); (S) → (V)
(2) (P) → (II); (Q) → (III); (R) → (IV); (S) → (I)
(3) (P) → (I); (Q) → (V); (R) → (III); (S) → (IV)
(4) (P) → (I); (Q) → (V); (R) → (II); (S) → (III)

ARCHIVES

Single Correct Answer Type

1. Perpendiculars are drawn from points on the line $\frac{x+2}{2}=\frac{y+1}{-1}=\frac{z}{3}$ to the plane $x+y+z=3$. The feet of perpendiculars lie on the line

(1) $\frac{x}{5}=\frac{y-1}{8}=\frac{z-2}{-13}$ (2) $\frac{x}{2}=\frac{y-1}{3}=\frac{z-2}{-5}$

(3) $\frac{x}{4}=\frac{y-1}{3}=\frac{z-2}{-7}$ (4) $\frac{x}{2}=\frac{y-1}{-7}=\frac{z-2}{5}$

(JEE Advanced 2013)

2. Let P be the image of the point (3, 1, 7) with respect to the plane $x-y+z=3$. Then the equation of the plane passing through P and containing the straight line $\frac{x}{1}=\frac{y}{2}=\frac{z}{1}$

(1) $x+y-3z=0$ (2) $3x+z=0$
(3) $x-4y+7z=0$ (4) $2x-y=0$

(JEE Advanced 2016)

3. The equation of the plane passing through the point (1, 1, 1) and perpendicular to the planes $2x+y-2z=5$ and $3x-6y-2z=7$, is
(1) $14x+2y+15z=31$
(2) $14x+2y-15z=1$
(3) $-14x+2y+15z=3$
(4) $14x-2y+15z=27$ **(JEE Advanced 2017)**

4. Let Q be the cube with the set of vertices $\{(x_1, x_2, x_3)\in\mathbb{R}^3 : x_1, x_2, x_3 \in \{0, 1\}\}$. Let F be the set of all twelve lines containing the diagonals of the six faces of the cube Q. Let S be the set of all four lines containing the main diagonals of the cube Q; for instance, the line passing through the vertices (0, 0, 0) and (1, 1, 1) is in S. For lines l_1 and l_2, let $d(l_1, l_2)$ denote the shortest distance between them. Then the maximum value of $d(l_1, l_2)$, as l_1 varies over F and l_2 varies over S, is

(1) $\frac{1}{\sqrt{6}}$ (2) $\frac{1}{\sqrt{8}}$

(3) $\frac{1}{\sqrt{3}}$ (4) $\frac{1}{\sqrt{12}}$

(JEE Advanced 2023)

Multiple Correct Answers Type

1. Two lines $L_1 : x=5, \frac{y}{3-\alpha}=\frac{z}{-2}$ and $L_2 : x=\alpha, \frac{y}{-1}=\frac{z}{2-\alpha}$ are coplanar. Then α can take value(s)

(1) 1 (2) 2
(3) 3 (4) 4

(JEE Advanced 2013)

2. A line l passing through the origin is perpendicular to the lines

$l_1 : (3+t)\hat{i} + (-1+2t)\hat{j} + (4+2t)\hat{k}, -\infty < t < \infty,$
$l_2 : (3+2s)\hat{i} + (3+2s)\hat{j} + (2+s)\hat{k}, -\infty < s < \infty$

Then, the coordinate(s) of the point(s) on l_2 at a distance of $\sqrt{17}$ from the point of intersection of l and l_1 is(are)

(1) $\left(\frac{7}{3}, \frac{7}{3}, \frac{5}{3}\right)$ (2) (−1, −1, 0)

(3) (1, 1, 1) (4) $\left(\frac{7}{9}, \frac{7}{9}, \frac{8}{9}\right)$

(JEE Advanced 2013)

3. In R^3, let L be a straight line passing through the origin. Suppose that all the points on L are at a constant distance from the two planes $P_1 : x+2y-z+1=0$ and $P_2 : 2x-y+z-1=0$. Let M be the locus of the feet of the perpendiculars drawn from the points on L to the plane P_1. Which of the following points lie(s) on M?

(1) $\left(0, -\frac{5}{6}, -\frac{2}{3}\right)$ (2) $\left(-\frac{1}{6}, -\frac{1}{3}, \frac{1}{6}\right)$

(3) $\left(-\frac{5}{6}, 0, \frac{1}{6}\right)$ (4) $\left(-\frac{1}{3}, 0, \frac{2}{3}\right)$

(JEE Advanced 2015)

4. In R^3, consider the planes $P_1 : y=0$ and $P_2 : x+z=1$. Let P_3 be a plane, different from P_1 and P_2, which passes through the intersection of P_1 and P_2. If the distance of the point (0, 1, 0) from P_3 is 1 and the distance of a point (α, β, γ) from P_3 is 2, then which of the following relations is (are) true?
(1) $2\alpha+\beta+2\gamma+2=0$ (2) $2\alpha-\beta+2\gamma+4=0$
(3) $2\alpha+\beta-2\gamma-10=0$ (4) $2\alpha-\beta+2\gamma-8=0$
(JEE Advanced 2015)

5. Consider a pyramid $OPQRS$ located in the first octant ($x \geq 0, y \geq 0, z \geq 0$) with O as origin, and OP and OR along the x-axis and the y-axis, respectively. The base $OPQR$ of the pyramid is a square with $OP = 3$. The point S is directly above the midpoint T of diagonal OQ such that $TS = 3$. Then
(1) the acute angle between OQ and OS is $\pi/3$
(2) the equation of the plane containing the triangle OQS is $x-y=0$
(3) the length of the perpendicular from P to the plane containing the triangle OQS is $\frac{3}{\sqrt{2}}$
(4) the perpendicular distance from O to the straight line containing RS is $\sqrt{\frac{15}{2}}$

(JEE Advanced 2016)

6. Let $P_1 : 2x + y - z = 3$ and $P_2 : x + 2y + z = 2$ be two planes. Then, which of the following statement(s) is (are) **TRUE?**

(1) The line of intersection of P_1 and P_2 has direction ratios 1, 2, –1

(2) The line $\frac{3x-4}{9} = \frac{1-3y}{9} = \frac{z}{3}$ is perpendicular to the line of intersection of P_1 an P_2

(3) The acute angle between P_1 and P_2 is 60º

(4) If P_3 is the plane passing through the point (4, 2, –2) and perpendicular to the line of intersection of P_1 and P_2, then the distance of the point (2, 1, 1) from the plane P_3 is $\frac{2}{\sqrt{3}}$

(JEE Advanced 2018)

7. Let L_1 and L_2 denote the lines $\vec{r} = \hat{i} + \lambda(-\hat{i} + 2\hat{j} + 2\hat{k}), \lambda \in R$ and $\vec{r} = \mu(2\hat{i} - \hat{j} + 2\hat{k}), \mu \in R$, respectively. If L_3 is a line which is perpendicular to both L_1 and L_2 and cuts both of them, then which of the following options describe(s) L_3?

(1) $\vec{r} = \frac{1}{3}(2\hat{i} + \hat{k}) + t(2\hat{i} + 2\hat{j} - \hat{k}), t \in R$

(2) $\vec{r} = \frac{2}{9}(4\hat{i} + \hat{j} + \hat{k}) + t(2\hat{i} + 2\hat{j} - \hat{k}), t \in R$

(3) $\vec{r} = \frac{2}{9}(2\hat{i} - \hat{j} + 2\hat{k}) + t(2\hat{i} + 2\hat{j} - \hat{k}), t \in R$

(4) $\vec{r} = t(2\hat{i} + 2\hat{j} - \hat{k}), t \in R$ **(JEE Advanced 2019)**

8. Three lines

$L_1 : \vec{r} = \lambda\hat{i}, \lambda \in R$

$L_2 : \vec{r} = \hat{k} + \mu\hat{j}, \mu \in R$ and

$L_3 : \vec{r} = \hat{i} + \hat{j} + v\hat{k}, v \in R$

are given. For which point(s) Q on L_2 can we find a point P on L_1 and a point R on L_3 so that P, Q and R are collinear?

(1) $\hat{k} + \frac{1}{2}\hat{j}$ (2) $\hat{k} + \hat{j}$

(3) $\hat{k}$ (4) $\hat{k} - \frac{1}{2}\hat{j}$

(JEE Advanced 2019)

9. Let L_1 and L_2 be the following straight lines.

$L_1 : \frac{x-1}{1} = \frac{y}{-1} = \frac{z-1}{3}$ and

$L_2 : \frac{x-1}{-3} = \frac{y}{-1} = \frac{z-1}{1}$

Suppose the straight line $L : \frac{x-\alpha}{l} = \frac{y-1}{m} = \frac{z-\gamma}{-2}$ lies in the plane containing L_1 and L_2, and passes through the point of intersection of L_1 and L_2. If the line L bisects the acute angle between the lines L_1 and L_2, then which of the following statements is/are TRUE?

(1) $\alpha - \gamma = 3$

(2) $l + m = 2$

(3) $\alpha - \gamma = 1$

(4) $l + m = 0$ **(JEE Advanced 2020)**

10. Let $\alpha, \beta, \gamma, \delta$ be real numbers such that $\alpha^2 + \beta^2 + \gamma^2 \neq 0$ and $\alpha + \gamma = 1$. Suppose the point (3, 2, – 1) is the mirror image of the point (1, 0, –1) with respect to the plane $\alpha x + \beta y + \gamma z = \delta$. Then which of the following statements is/are TRUE?

(1) $\alpha + \beta = 2$

(2) $\delta - \gamma = 3$

(3) $\delta + \beta = 4$

(4) $\alpha + \beta + \gamma = \delta$ **(JEE Advanced 2020)**

11. Let P_1 and P_2 be two planes given by

P_1: $10x + 15y + 12z - 60 = 0$,

P_2 : $-2x + 5y + 4z - 20 = 0$.

Which of the following straight lines can be an edge of some tetrahedron whose two faces lie on P_1 and P_2?

(1) $\frac{x-1}{0} = \frac{y-1}{0} = \frac{z-1}{5}$ (2) $\frac{x-6}{-5} = \frac{y}{2} = \frac{z}{3}$

(3) $\frac{x}{-2} = \frac{y-4}{5} = \frac{z}{4}$ (4) $\frac{x}{1} = \frac{y-4}{-2} = \frac{z}{3}$

(JEE Advanced 2022)

12. Let S be the reflection of a point Q with respect to the plane given by

$\vec{r} = -(t + p)\hat{i} + t\hat{j} + (1 + p)\hat{k}$

where t, p are real parameters and $\hat{i}, \hat{j}, \hat{k}$ are the unit vectors along the three positive coordinate axes. If the position vectors of Q and S are $10\hat{i} + 15\hat{j} + 20\hat{k}$ and $\alpha\hat{i} + \beta\hat{j} + \gamma\hat{k}$ respectively, then which of the following is/are TRUE ?

(1) $3(\alpha + \beta) = -101$ (2) $3(\beta + \gamma) = -71$

(3) $3(\gamma + \alpha) = -86$ (4) $3(\alpha + \beta + \gamma) = -121$

(JEE Advanced 2022)

13. Let R^3 denote the three-dimensional space. Take two points $P = (1, 2, 3)$ and $Q = (4, 2, 7)$. Let dist (X, Y) denote the distance between two points X and Y in R^3. Let

$S = \{X \in R^3 : (\text{dist}(X, P))^2 - (\text{dist}(X, Q))^2 = 50\}$ and

$T = \{Y \in R^3 : (\text{dist}(Y, Q))^2 - (\text{dist}(Y, P))^2 = 50\}$

Then which of the following statements is (are) TRUE?

(1) There is a triangle whose area is 1 and all of whose vertices are from S.

(2) There are two distinct points L and M in T such that each point on the line segment LM is also in T.

(3) There are infinitely many rectangles of perimeter 48, two of whose vertices are from S and the other two vertices are from T.

(4) There is a square of perimeter 48, two of whose vertices are from S and the other two vertices are from T.

(JEE Advanced 2024)

14. A straight line drawn from the point $P(1, 3, 2)$, parallel to the line $\frac{x-2}{1} = \frac{y-4}{2} = \frac{z-6}{1}$, intersects the plane $L_1 : x - y + 3z = 6$ at the point Q. Another straight line which passes through Q and is perpendicular to the plane L_1 intersects the plane $L_2 : 2x - y + z = -4$ at the point R.

Then which of the following statements is (are) TRUE?

(1) The length of the line segment PQ is $\sqrt{6}$.

(2) The coordinates of R are (1, 6, 3).

(3) The centroid of the triangle PQR is $\left(\frac{4}{3}, \frac{14}{3}, \frac{5}{3}\right)$.

(4) The perimeter of the triangle PQR is $\sqrt{2} + \sqrt{6} + \sqrt{11}$

(JEE Advanced 2024)

15. Let L_1 be the line of intersection of the planes given by the equations

$2x + 3y + z = 4$ and $x + 2y + z = 5$.

Let L_2 be the line passing through the point $P(2, -1, 3)$ and parallel to L_1. Let M denote the plane given by the equation $2x + y - 2z = 6$.

Suppose that the line L_2 meets the plane M at the point Q. Let R be the foot of the perpendicular drawn from P to the plane M.

Then which of the following statements is (are) TRUE?

(1) The length of the line segment of PQ is $9\sqrt{3}$

(2) The length of the line segment QR is 15

(3) The area of ΔPQR is $\frac{3}{2}\sqrt{234}$

(4) The acute angle between the line segments PO and PR is $\cos^{-1}\left(\frac{1}{2\sqrt{3}}\right)$

(JEE Advanced 2025)

Matrix Match Type

1. Consider the lines

$$L_1 : \frac{x-1}{2} = \frac{y}{-1} = \frac{z+3}{1}, L_2 : \frac{x-4}{1} = \frac{y+3}{1} = \frac{z+3}{2}$$

and the planes $P_1 : 7x + y + 2z = 3$, $P_2 : 3x + 5y - 6z = 4$. Let $ax + by + cz = d$ be the equation of the plane passing through the point of intersection of lines L_1 and L_2, and perpendicular to planes P_1 and P_2.

Match List I with List II.

List I	List II
(P) $a =$	**(I)** 13
(Q) $b =$	**(II)** −3
(R) $c =$	**(III)** 1
(S) $d =$	**(IV)** −2

The correct option is

(1) (P) → (III); (Q) → (IV); (R) → (I); (S) → (II)

(2) (P) → (III); (Q) → (IV); (R) → (II); (S) → (I)

(3) (P) → (III); (Q) → (II); (R) → (IV); (S) → (I)

(4) (P) → (II); (Q) → (III); (R) → (IV); (S) → (I)

(JEE Advanced 2013)

2. Let l_1 and l_2 be the lines $\vec{r_1} = \lambda(\vec{i} + \vec{j} + \vec{k})$ and $\vec{r_2} = (\vec{j} - \vec{k}) + \mu(\vec{i} + \vec{k})$, respectively. Let X be the set of all the planes H that contain the line l_1. For a plane H, let $d(H)$ denote the smallest possible distance between the points of l_2 and H. Let H_0 be plane in X for which $d(H_0)$ is the maximum value of $d(H)$ as H varies over all planes in X.

Match each entry in **List I** to the correct entries in **List II.**

List I	List II
(P) The value of $d(H_0)$ is	**(I)** $\sqrt{3}$
(Q) The distance of the point (0,1,2) from H_0 is	**(II)** $\frac{1}{\sqrt{3}}$
(R) The distance of origin from H_0 is	**(III)** 0
(S) The distance of origin from the point of intersection of planes $y = z$, $x = 1$ and H_0 is	**(IV)** $\sqrt{2}$
	(V) $\frac{1}{\sqrt{2}}$

The correct option is:

(1) (P) → (II); (Q) → (IV); (R) → (V); (S) → (I)

(2) (P) → (V); (Q) → (IV); (R) → (III); (S) → (I)

(3) (P) → (II); (Q) → (I); (R) → (III); (S) → (II)

(4) (P) → (V); (Q) → (I); (R) → (IV); (S) → (II)

(JEE Advanced 2023)

3. Let $\gamma \in R$ be such that the lines $L_1 : \frac{x+11}{1} = \frac{y+21}{2} = \frac{z+29}{3}$ and $L_2 : \frac{x+16}{3} = \frac{y+11}{2} = \frac{z+4}{\gamma}$ intersect. Let R_1 be the point of intersection of L_1 and L_2. Let $O = (0, 0, 0)$, and $\hat{n}$ denote a unit normal vector to the plane containing both the lines L_1 and L_2.

Match each entry in **List I** to the correct entry in **List II.**

List I	List II
(P) γ equals	**(I)** $-\hat{i} - \hat{j} + \hat{k}$
(Q) A possible choice for $\hat{n}$ is	**(II)** $\sqrt{\frac{3}{2}}$
(R) $\overrightarrow{OR_1}$ equals	**(III)** 1
(S) A possible value of $\overrightarrow{OR_1} \cdot \hat{n}$ is	**(IV)** $\frac{1}{\sqrt{6}}\hat{i} - \frac{2}{\sqrt{6}}\hat{j} + \frac{1}{\sqrt{6}}\hat{k}$
	(V) $\sqrt{\frac{2}{3}}$

The correct option is

(1) (P) → (III); (Q) → (IV); (R) → (I); (S) → (II)

(2) (P) → (V); (Q) → (IV); (R) → (I); (S) → (II)

(3) (P) → (III); (Q) → (IV); (R) → (I); (S) → (V)

(4) (P) → (III); (Q) → (I); (R) → (IV); (S) → (V)

(JEE Advanced 2024)

Numerical Value Type

1. Three lines are given by $\vec{r} = \lambda\hat{i}, \lambda \in R$, $\vec{r} = \mu(\hat{i}+\hat{j}), \mu \in R$ and $\vec{r} = \nu(\hat{i}+\hat{j}+\hat{k}), \nu \in R$. Let the lines cut the plane $x+y+z=1$ at the points A, B and C, respectively. If the area of the triangle ABC is Δ then the value of $(6\Delta)^2$ equals ________. **(JEE Advanced 2019)**

Integer Answer Type

1. Let P be a point in the first octant, whose image Q in the plane $x+y=3$ (that is, the line segment PQ is perpendicular to the plane $x+y=3$ and the mid-point of PQ lies in the plane $x+y=3$) lies on the z-axis. Let the distance of P from the x-axis be 5. If R is the image of P in the xy-plane, then the length of PR is ________. **(JEE Advanced 2018)**

2. Let P be the plane $\sqrt{3}x+2y+3z=16$ and let $S=\left\{\alpha\hat{i}+\beta\hat{j}+\gamma\hat{k}: \alpha^2+\beta^2+\gamma^2=1 \text{ and the distance of } (\alpha, \beta, \gamma) \text{ from the plane } P \text{ is } \frac{7}{2}\right\}$.

   Let $\vec{u}, \vec{v}$ and $\vec{w}$ be three distinct vectors in S such that $|\vec{u}-\vec{v}|=|\vec{v}-\vec{w}|=|\vec{w}-\vec{u}|$. Let V be the volume of the parallelepiped determined by vectors $\vec{u}, \vec{v}$ and $\vec{w}$. Then the value of $\frac{80}{\sqrt{3}}V$ is ________. **(JEE Advanced 2023)**

3. Let $\overrightarrow{OP} = \frac{\alpha-1}{\alpha}\hat{i}+\hat{j}+\hat{k}$, $\overrightarrow{OQ} = \hat{i}+\frac{\beta-1}{\beta}\hat{j}+\hat{k}$ and $\overrightarrow{OR} = \hat{i}+\hat{j}+\frac{1}{2}\hat{k}$ be three vectors, where $\alpha, \beta \in R - \{0\}$ and O denotes the origin. If $(\overrightarrow{OP}\times\overrightarrow{OQ})\cdot\overrightarrow{OR} = 0$ and the point $(\alpha, \beta, 2)$ lies on the plane $3x+3y-z+l=0$, then the value of l is ________. **(JEE Advanced 2024)**

4. Let $\vec{p} = 2\hat{i}+\hat{j}+3\hat{k}$ and $\vec{q} = \hat{i}-\hat{j}+\hat{k}$. If for some real numbers α, β, and γ, we have $15\hat{i}+10\hat{j}+6\hat{k} = \alpha(2\vec{p}+\vec{q})+\beta(\vec{p}-2\vec{q})+\gamma(\vec{p}\times\vec{q})$, then the value of γ is ________. **(JEE Advanced 2024)**

ANSWER KEY

EXERCISES

Single Correct Answer Type

1. (4) 2. (2) 3. (4) 4. (3) 5. (3)
6. (3) 7. (2) 8. (4) 9. (2) 10. (2)
11. (3) 12. (3) 13. (3) 14. (2) 15. (3)
16. (3) 17. (4) 18. (2) 19. (4) 20. (4)
21. (1) 22. (2) 23. (1) 24. (3) 25. (2)
26. (1) 27. (3) 28. (4) 29. (2) 30. (2)
31. (2) 32. (3) 33. (3) 34. (1) 35. (2)
36. (3) 37. (3) 38. (1) 39. (4) 40. (4)
41. (2) 42. (4) 43. (2) 44. (4) 45. (1)
46. (1) 47. (2) 48. (2) 49. (2) 50. (4)
51. (1) 52. (3) 53. (1) 54. (2) 55. (3)
56. (2) 57. (3) 58. (4) 59. (4) 60. (3)
61. (3) 62. (1) 63. (1) 64. (4) 65. (1)
66. (2) 67. (1) 68. (3) 69. (4) 70. (2)
71. (3) 72. (3) 73. (2) 74. (2) 75. (1)
76. (4) 77. (2) 78. (4) 79. (2) 80. (1)
81. (2) 82. (3) 83. (4) 84. (1) 85. (1)
86. (1) 87. (1) 88. (2) 89. (3) 90. (2)
91. (1) 92. (4) 93. (2)

Numerical Value Type

1. (2.25) 2. (1.2) 3. (0.25) 4. (24.2) 5. (1.75)
6. (5.5) 7. (2.4) 8. (0.25) 9. (2.1)

Integer Answer Type

1. (1) 2. (2) 3. (7) 4. (53) 5. (7)
6. (8) 7. (5) 8. (5) 9. (3) 10. (6)
11. (474) 12. (6) 13. (17) 14. (6) 15. (9)

Multiple Correct Answers Type

1. (2), (3), (4) 2. (1), (2) 3. (2), (3), (4)
4. (1), (4) 5. (1), (3) 6. (1), (2), (4)
7. (1), (3) 8. (1), (4) 9. (1), (2), (4)
10. (1), (3) 11. (1), (4) 12. (1), (2)
13. (1), (3) 14. (1), (3), (4) 15. (2), (3)
16. (1), (2) 17. (2), (3)

Linked Comprehension Type

1. (2) 2. (3) 3. (4) 4. (2) 5. (3)
6. (3) 7. (4) 8. (2) 9. (4) 10. (2)
11. (1) 12. (2) 13. (2) 14. (126) 15. (12)

Matrix Match Type

1. (1) 2. (3) 3. (4) 4. (2) 5. (4)

ARCHIVES

Single Correct Answer Type

1. (4) 2. (3) 3. (1) 4. (1)

Multiple Correct Answers Type

1. (1), (4) 2. (2), (4) 3. (1), (2)
4. (2), (4) 5. (2), (3), (4) 6. (3), (4)
7. (1), (2), (3) 8. (1), (4) 9. (1), (2)
10. (1), (2), (3) 11. (1), (2), (4) 12. (1), (2), (3)
13. (1), (2), (3) 14. (1), (3) 15. (1), (3)

Matrix Match Type

1. (3) 2. (2) 3. (3)

Numerical Value Type

1. (0.75)

Integer Answer Type

1. (8) 2. (45) 3. (5) 4. (2)

SOLUTIONS

Introduction to Three-dimensional Geometry

Concept Application Exercises

Exercise 1.1

1. Feet of perpendiculars from point $P(-4, 5, 9)$ on coordinate axes are $(-4, 0, 0)$, $(0, 5, 0)$ and $C(0, 0, 9)$.

   Feet of perpendiculars from point P on coordinate planes are $(-4, 5, 0)$, $(0, 5, 9)$ and $(-4, 0, 9)$.

2. Distance of point $P(3, -5, 6)$ from x-axis

   $$= \sqrt{y^2 + z^2} = \sqrt{(-5)^2 + 6^2} = \sqrt{61}$$

   Distance of point $P(3, -5, 6)$ from y-axis

   $$= \sqrt{x^2 + z^2} = \sqrt{3^2 + 6^2} = \sqrt{45}$$

   Distance of point $P(3, -5, 6)$ from z-axis

   $$= \sqrt{x^2 + y^2} = \sqrt{3^2 + (-5)^2} = \sqrt{34}$$

   Distance of point $P(3, -5, 6)$ from xy-plane $= |z| = 6$

   Distance of point $P(3, -5, 6)$ from yz-plane $= |x| = 3$

   Distance of point $P(3, -5, 6)$ from zx-plane $= |y| = |-5| = 5$

3. Given point is $(-2, 3, -5)$.

   a. equation of plane through given point and parallel to xy-plane is $z = -5$

   b. equation of plane through given point and parallel to yz-plane is $x = -2$

   c. equation of plane through given point and parallel to zx-plane is $y = 3$

4. Given point is $(2, 3, -4)$

   a. equations of line through point which is parallel to x-axis is $y = 3, z = -4$

   b. equations of line through point which is parallel to y-axis is $x = 2, z = -4$

   c. equations of line through point which is parallel to z-axis is $x = 2, y = 3$

5. We have $x^2 - 2x - 3 = 0$ ...(i)

   $\Rightarrow \quad (x - 3)(x + 1) = 0$

   $\Rightarrow \quad x = 3$ and $x = -1$

   Thus, equation (i) is the combined equation of two planes $x = -1$ and $x = 3$ which are parallel to yz-plane.

   Distance between two planes is 4 units.

6. **a.** Distance of plane $z = 6$ from xy-plane is 6 units.

   Distance of point $P(3, 4, -5)$ from xy-plane is 5 units.

   But point P and plane $z = 6$ lie on the opposite sides of xy-plane.

   So, distance of point P from plane $z = 6$ is $5 + 6 = 11$ units.

   b. Distance of plane $x = 7$ from yz-plane is 7 units.

   Distance of point $P(3, 4, -5)$ from yz-plane is 3 units.

   Since point P and plane $x = 7$ lies on the same side of yz-plane, required distance is $7 - 3 = 4$ units.

7. Point $P(-3, 2, 4)$ lies in second octant.

   Image of P in xy-plane is $P'(-3, 2, -4)$ which lies in the sixth octant.

   Image of P in yz-plane is $P''(3, 2, 4)$ which lies in the first octant.

   Image of P in xz-plane is $P'''(-3, -2, 4)$ which lies in the third octant.

Exercise 1.2

1. Lengths of the edges of the cuboid are $x_2 - x_1, y_2 - y_1, z_2 - z_1$ or $6 - 3, 8 - 4, 10 - 8$ or 3, 4, 2

   $\therefore$ The length of the diagonal is $\sqrt{9 + 16 + 4} = \sqrt{29}$

2. Let $P \equiv (x_1, y_1, z_1)$ and $Q \equiv (x_2, y_2, z_2)$.

   Given that

   $$PQ^2 = (x_1 - x_2)^2 + (y_1 - y_2)^2 + (z_1 - z_2)^2 = 100, (x_1 - x_2)^2 = 9$$

   and $(y_1 - y_2)^2 = 16$.

   $\therefore \quad (z_1 - z_2)^2 = 100 - 9 - 16 = 75$ or $|z_1 - z_2| = 5\sqrt{3}$

3. **a.** Let points $(0, 7, -10)$, $(1, 6, -6)$, and $(4, 9, -6)$ be denoted by A, B, and C, respectively.

   $$AB = \sqrt{(1 - 0)^2 + (6 - 7)^2 + (-6 + 10)^2}$$
   $$= \sqrt{1 + 1 + 16} = \sqrt{18} = 3\sqrt{2}$$
   $$BC = \sqrt{(4 - 1)^2 + (9 - 6)^2 + (-6 + 6)^2}$$
   $$= \sqrt{3^2 + 3^2} = \sqrt{18} = 3\sqrt{2}$$
   $$CA = \sqrt{(0 - 4)^2 + (7 - 9)^2 + (-10 + 6)^2}$$
   $$= \sqrt{(-4)^2 + (-2)^2 + (-4)^2} = \sqrt{36} = 6$$

   Here, $AB = BC \neq CA$

   Thus, the given points are the vertices of an isosceles triangle.

   b. Let $(-1, 2, 1)$, $(1, -2, 5)$, $(4, -7, 8)$, and $(2, -3, 4)$ be denoted by A, B, C, and D, respectively.

   $$AB = \sqrt{(1 + 1)^2 + (-2 - 2)^2 + (5 - 1)^2}$$
   $$= \sqrt{4 + 16 + 16} = \sqrt{36} = 6$$
   $$BC = \sqrt{(4 - 1)^2 + (-7 + 2)^2 + (8 - 5)^2}$$
   $$= \sqrt{9 + 25 + 9} = \sqrt{43}$$
   $$CD = \sqrt{(2 - 4)^2 + (-3 + 7)^2 + (4 - 8)^2}$$
   $$= \sqrt{4 + 16 + 16} = \sqrt{36} = 6$$
   $$DA = \sqrt{(-1 - 2)^2 + (2 + 3)^2 + (1 - 4)^2}$$
   $$= \sqrt{9 + 25 + 9} = \sqrt{43}$$

   Here, $AB = CD = 6$, $BC = AD = \sqrt{43}$

   Hence, the opposite sides of quadrilateral $ABCD$, whose vertices are taken in order, are equal.

   Therefore, $ABCD$ is a parallelogram.

   Hence, the given points are the vertices of a parallelogram.

4. Let point P be (x, y, z)

   Now $x^2 + y^2 + z^2 = 49$ ...(i)

   $y^2 + z^2 = 13$ ...(ii)

   and $x^2 + z^2 = 40$ ...(iii)

   $\therefore \quad x^2 = 36, y^2 = 9$ and $z^2 = 4$

   $\therefore \quad (x, y, z) \equiv (\pm 6, \pm 3, \pm 2)$

   Thus, there are eight points.

5.

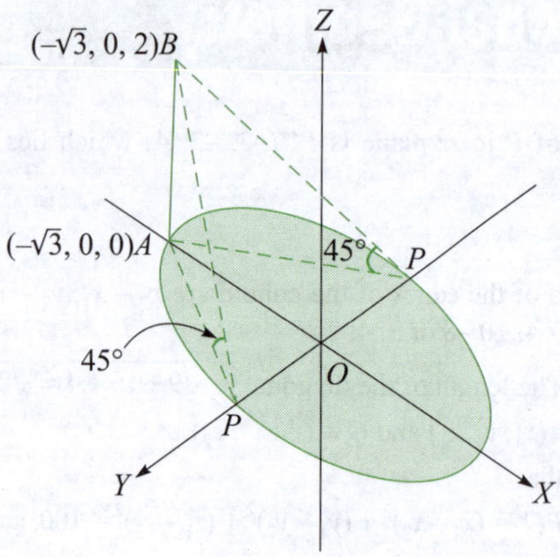

Point P lies on

$$x^2 + 3y^2 = 3 \quad \text{...(i)}$$

Now, from diagram according to given conditions $AP = AB$

or $\quad (x + \sqrt{3})^2 + (y - 0)^2 = 4$

or $\quad (x + \sqrt{3})^2 + y^2 = 4 \quad \text{...(ii)}$

Solving (i) and (ii), we get $x = 0$ and $y = \pm 1$.

So, point P has coordinates $(0, \pm 1, 0)$.

Exercise 1.3

1. Let C divide AB in the ratio $\lambda : 1$. Then

$$C \equiv \left(\frac{5\lambda + 3}{\lambda + 1}, \frac{4\lambda + 2}{\lambda + 1}, \frac{-6\lambda - 4}{\lambda + 1}\right) \equiv (9, 8\ -10)$$

Comparing x-coordinate, we get

$$5\lambda + 3 = 9\lambda + 9 \text{ or } 4\lambda = -6$$

$\therefore \quad \lambda = -3/2$

Also, comparing y and z-coordinates, we get $4\lambda + 2 = 8\lambda + 8$ and $-6\lambda - 4 = -10\lambda - 10$. These give the same value of λ.

Thus, points are collinear.

2. Let A and B be the points that trisect the line segment joining points $P(4, 2, -6)$ and $Q(10, -16, 6)$.

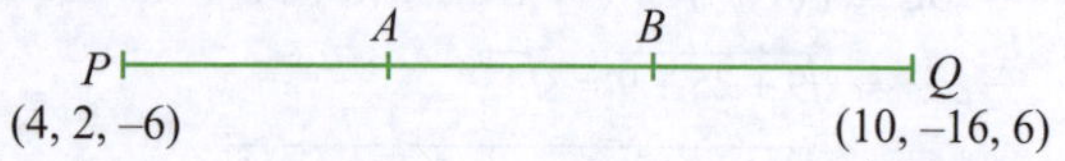

Point A divides PQ in the ratio 1 : 2. Therefore, by section formula, the coordinates of point A are given by

$$\left(\frac{1(10) + 2(4)}{1 + 2}, \frac{1(-16) + 2(2)}{1 + 2}, \frac{1(6) + 2(-6)}{1 + 2},\right)$$

$$= (6, -4, -2)$$

Point B divides PQ in the ratio 2 : 1. Therefore, by section formula, the coordinates of point B are given by

$$\left(\frac{2(10) + 1(4)}{2 + 1}, \frac{2(-16) + 1(2)}{2 + 1}, \frac{2(6) - 1(-6)}{2 + 1}\right)$$

$$= (8, -10, 6)$$

Thus, $(6, -4, -2)$ and $(8, -10, 6)$ are the points that trisect the line segment joining points $P\ (4, 2, -6)$ and $Q(10, -16, 6)$.

3. Let given plane divides the join of A and B at point P in the ratio $\lambda : 1$.

$$\therefore \quad P \equiv \left(\frac{7\lambda + 2}{\lambda + 1}, \frac{2\lambda + 3}{\lambda + 1}, \frac{-4}{\lambda + 1}\right)$$

Point P lies on the given plane.

So, $\quad \dfrac{7\lambda + 2}{\lambda + 1} - \dfrac{2\lambda + 3}{\lambda + 1} - \dfrac{8}{\lambda + 1} - 7 = 0$

or $\quad 5\lambda - 9 - 7\lambda - 7 = 0$

or $\quad \lambda = -8$

Thus point P divides AB externally in the ratio 8 : 1.

So, points A and B lie on the same side of the plane.

4. Consider tetrahedron $A_1A_2A_3A_4$ with $A_i \equiv (x_i, y_i, z_i)$; $i = 1, 2, 3, 4$.

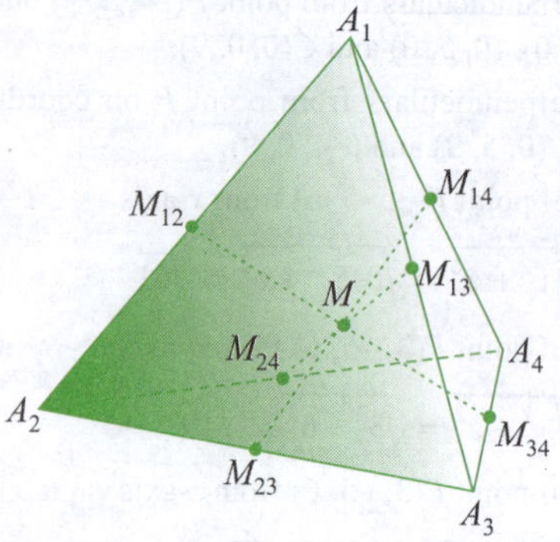

We have three pairs of opposite edges, (A_1A_2, A_3A_4), (A_1A_3, A_2A_4) and (A_1A_4, A_2A_3).

Midpoint of A_1A_2 is

$$M_{12}\left(\frac{x_1 + x_2}{2}, \frac{y_1 + y_2}{2}, \frac{z_1 + z_2}{2}\right).$$

Midpoint of A_3A_4 is

$$M_{34}\left(\frac{x_3 + x_4}{2}, \frac{y_3 + y_4}{2}, \frac{z_3 + z_4}{2}\right).$$

Now, midpoint of $M_{12}M_{34}$ is

$$M\left(\frac{\frac{x_1 + x_2}{2} + \frac{x_3 + x_4}{2}}{2}, \frac{\frac{y_1 + y_2}{2} + \frac{y_3 + y_4}{2}}{2}, \frac{\frac{z_1 + z_2}{2} + \frac{z_3 + z_4}{2}}{2}\right)$$

$$\equiv M\left(\frac{\sum_{i=1}^{4} x_i}{4}, \frac{\sum_{i=1}^{4} y_i}{4}, \frac{\sum_{i=1}^{4} z_i}{4}\right)$$

We can show that point M also bisect the line segment joining midpoints of other two pairs of opposite edges.

Exercise 1.4

1. Equation of the sphere whose centre is (3, 1, 2) and radius is 5 is

$$(x - 3)^2 + (y - 1)^2 + (z - 2)^2 = 5^2$$

or $\quad x^2 + y^2 + z^2 - 6x - 2y - 4z - 11 = 0$

2. Let the equation of the sphere passing through (0, 0, 0), (1, 0, 0), (0, 1, 0) and (0, 0,1) be

$$x^2 + y^2 + z^2 + 2ux + 2vy + 2wz + d = 0 \quad \text{...(i)}$$

As (i) passes through (0, 0, 0), (1, 0, 0), (0, 1, 0) and (0, 0, 1) we must have d = 0 ; $1 + 2u + d = 0$; $1 + 2v + d = 0$; $1 + 2w + d = 0$

Since $d = 0$, we get $2u = 2v = 2w = -1$.

Thus, the equation of the required sphere is $x^2 + y^2 + z^2 - x - y - z = 0$.

3. Let $P(x, y, z)$ be any point on the locus. Then

$PA^2 + PB^2 + PC^2 = 120$

$\Rightarrow$ $(x-1)^2 + (y-2)^2 + (z-3)^2 + (x-2)^2 + (y+3)^2 + (z-5)^2 + (x-0)^2 + (y-7)^2 + (z-4)^2 = 120$

or $3x^2 + 3y^2 + 3z^2 - 6x - 12y - 24z + 117 = 120$

or $x^2 + y^2 + z^2 - 2x - 4y - 8z - 1 = 0$

This represents a sphere with centre at (1, 2, 4) and radius as $\sqrt{1^2 + 2^2 + 4^2 + 1} = \sqrt{22}$.

4. Since $3PA = 2PB$, we get

$9PA^2 = 4PB^2$

$\Rightarrow$ $9[(x-1)^2 + (y-3)^2 + (z-4)^2] = 4[(x-1)^2 + (y+2)^2 + (z+1)^2]$

$\Rightarrow$ $9[x^2 + y^2 + z^2 - 2x - 6y - 8z + 26] = 4[x^2 + y^2 + z^2 - 2x + 4y + 2z + 6]$

$\Rightarrow$ $5x^2 + 5y^2 - 10x - 70y - 80z + 210 = 0$

$\Rightarrow$ $x^2 + y^2 + z^2 - 2x - 14y - 16z + 42 = 0$

This represents a sphere with centre at (1, 7, 8) and radius equal to $\sqrt{1^2 + 7^2 + 8^2 - 42} = \sqrt{72} = 6\sqrt{2}$.

5. Required equation of sphere is $(x-1)(x+2) + (y-2)(y-4) + (z-3)(z+1) = 0$

or $x^2 + y^2 + z^2 + x - 6y - 2z + 3 = 0$

Exercises

Single Correct Answer Type

1. (3) Given, point is $P(3, 4, 5)$.

Distance of P from xy-plane

$= |z \text{ coordinate of } P| = 5$

2. (2) We know that, on the y-axis, $x = 0$ and $z = 0$

Thus, foot of perpendicular from P on y-axis is $A(0, 4, 0)$.

$\therefore$ $PA = \sqrt{(0-3)^2 + (4-4)^2 + (0-5)^2}$

$= \sqrt{9 + 0 + 25} = \sqrt{34}$

3. (1) We know that, on the xy-plane $z = 0$.

Hence, the coordinates of point M are $(-1, 7, 0)$.

4. (1) Clearly, $A \equiv (1, 0, 0)$, $B \equiv (0, 2, 0)$ and $C \equiv (0, 0, 3)$.

$\therefore$ Centroid of triangle ABC is

$\therefore$ $G \equiv \left(\frac{1+0+0}{3}, \frac{0+2+0}{3}, \frac{0+0+3}{3}\right)$

$\equiv \left(\frac{1}{3}, \frac{2}{3}, 1\right)$

$OG \equiv \sqrt{\left(\frac{1}{3}\right)^2 + \left(\frac{2}{3}\right)^2 + (1)^2} \equiv \frac{\sqrt{14}}{3}$

5. (2) Clearly $A \equiv (1, 2, 0)$, $B \equiv (0, 2, 3)$ and $C \equiv (1, 0, 3)$.

$\therefore$ Centroid of triangle ABC is

$G \equiv \left(\frac{1+0+1}{3}, \frac{2+2+0}{3}, \frac{0+3+3}{3}\right)$

$\equiv \left(\frac{2}{3}, \frac{4}{3}, 2\right)$

$\therefore$ $OG = \sqrt{\left(\frac{2}{3}\right)^2 + \left(\frac{4}{3}\right)^2 + (2)^2} = \frac{2\sqrt{14}}{3}$

6. (3) Suppose R divides PQ in the ratio $k : 1$.

$\therefore$ $R \equiv \left(\frac{8k+2}{k+1}, \frac{-3}{k+1}, \frac{10k+4}{k+1}\right)$

It is given that the x–coordinate of point R is 6.

$\therefore$ $\frac{8k+2}{k+1} = 6 \Rightarrow 8k + 2 = 6k + 6$

$\Rightarrow$ $2k = 4 \Rightarrow k = 2$

Therefore, the coordinates of point R are given by

$\left(6, \frac{-3}{2+1}, \frac{10 \times 2 + 4}{2+1}\right) \equiv (6, -1, 8)$

7. (2) We have $x^2 - 3x - 4 = 0$

$\Rightarrow$ $(x-4)(x+1) = 0$

$\Rightarrow$ $x = 4, x = -1$

Distance between these two parallel planes (parallel to yz-plane) is 5 units.

We have $y^2 - 5y + 6 = 0$

$\Rightarrow$ $(y-2)(y-3) = 0$

$\Rightarrow$ $y = 2, y = 3$

Distance between these two parallel planes (parallel to xz-plane) is 1 unit.

We have $z^2 + 6z + 8 = 0$

$\Rightarrow$ $(z+2)(z+4) = 0$

$\Rightarrow$ $z = -2, z = -4$

Distance between these two parallel planes (parallel to xy-plane) is 2 unit.

So, volume of cuboid is $5 \times 1 \times 2 = 10$ cu. units.

8. (4) Clearly, there will be eight spheres, one in each octant.

9. (2) Let $P(x, y, z)$. Now under given condition, we get

$\left[\sqrt{x^2+y^2}\right]^2 + \left[\sqrt{y^2+z^2}\right]^2 + \left[\sqrt{z^2+x^2}\right]^2 = 36$

$\Rightarrow$ $x^2 + y^2 + z^2 = 18$

Then distance from origin to the point (x, y, z) is

$\sqrt{x^2+y^2+z^2} = \sqrt{18} = 3\sqrt{2}$

10. (2) Let xy-plane divides the line joining the points $A(-1, 3, 4)$ and $B(2, -5, 6)$ in the ratio $\lambda : 1$ at point P then $P \equiv \left(\frac{2\lambda-1}{\lambda+1}, \frac{-5\lambda+3}{\lambda+1}, \frac{6\lambda+4}{\lambda+1}\right)$.

Point P lies on xy-plane.

So, $\frac{6\lambda+4}{\lambda+1} = 0$ or $\lambda = -\frac{2}{3}$

Thus, xy plane divides AB externally in the ratio 2 : 3.

11. (1) Let $(5, 5, \lambda)$ divide the segment joining $(-1, 3, 2)$ and $(-4, 2, -2)$ in the ratio $k : 1$.

$\therefore$ $\left(\frac{-4k-1}{k+1}, \frac{2k+3}{k+1}, \frac{-2k+2}{k+1}\right) \equiv (5, 5, \lambda)$

So, $-4k - 1 = 5k + 5$ and $2k + 3 = 5k + 5$

$\therefore$ $k = -2/3$

So, comparing z-coordinate with $k = -2/3$, we get

$4/3 + 2 = \lambda/3$

$\therefore$ $\lambda = 10$

12. (3) Distance from origin $= \sqrt{1+4+9} = \sqrt{14}$

and distance from y-axis $= \sqrt{1+9} = \sqrt{10}$

13. (4) Distance from x-axis $= \sqrt{y^2+z^2} = \sqrt{4+9} = \sqrt{13}$

Distance from y-axis = $\sqrt{1+9} = \sqrt{10}$

Distance from z-axis = $\sqrt{1+4} = \sqrt{5}$

14. (3) $0 = \dfrac{a-2+4}{3} \Rightarrow a = -2$

$0 = \dfrac{1+b+7}{3} \Rightarrow b = -8$

and $0 = \dfrac{3-5+c}{3} \Rightarrow c = 2$

15. (2) For xy-plane, $z = 0$.

So, $\dfrac{6\lambda+1}{\lambda+1} = 0 \Rightarrow \lambda = -\dfrac{1}{6}$

$\therefore \quad x = \dfrac{-5+18}{5} = \dfrac{13}{5}, y = \dfrac{-1+24}{5} = \dfrac{23}{5}$

16. (1) Let point be (x, y, z). Then

$x^2 + y^2 + z^2 = (x-a)^2 + y^2 + z^2$
$= x^2 + (y-b)^2 + z^2$
$= x^2 + y^2 + (z-c)^2$

Therefore $x = \dfrac{a}{2}, y = \dfrac{b}{2}$ and $z = \dfrac{c}{2}$

17. (2) Here $AB = BC = CD = DA = 7$.

Also, $AC = \sqrt{16+25+81} = \sqrt{122}$

and $BD = \sqrt{64+1+9} = \sqrt{74}$

Thus, points are vertices of rhombus.

18. (4) Centre of the sphere is $C(2, 3, -4)$ and radius is $r = \sqrt{4+9+16-(-7)} = 6$.

For point $P(1, 2, 0)$, $CP = \sqrt{1+1+16} = \sqrt{18} < 6$

So, point $P(1, 2, 0)$ lies inside the sphere.

For other points, we have $CP > 6$.

19. (2) For sphere $x^2 + y^2 + z^2 = 36$, centre is $C_1(0, 0, 0)$ and radius is $r_1 = 6$.

For sphere $x^2 + y^2 + z^2 - 20x + 84 = 0$, centre is $C_2(10, 0, 0)$ and radius is $r_2 = 4$.

$C_1C_2 = 10$ and $r_1 + r_2 = 6 + 4 = 10$

Thus, two spheres touch externally at point $(6, 0, 0)$.

20. (2) Given sphere is $x^2 + y^2 + z^2 - 2x - 4y + 6z - 4 = 0$

Since it cuts the xy-plane where $z = 0$, equation of circular section is $x^2 + y^2 - 2x - 4y - 4 = 0$.

The radius of the circle is $\sqrt{1+4-(-4)} = \sqrt{9} = 3$.

21. (1) We have plane $6x + 3y - 2z = 18$.

Coordinates of A, B, C are, respectively, $(3, 0, 0)$, $(0, 6, 0)$ and $(0, 0, -9)$.

Thus, centroid of ΔABC is $\left(\dfrac{3}{3}, \dfrac{6}{3}, \dfrac{-9}{3}\right)$ or $(1, 2, -3)$.

22. (1) Let the given point is $P \equiv (2, 3, -4)$.

Then, the reflection of P in the x-axis is $Q \equiv (-2, -3, -4)$

The reflection of P in the xy-plane is $R \equiv (2, 3, 4)$.

Then $QR = 2\sqrt{29}$

23. (3) The given point $P(2, -3, -5)$ lies in the 8th octant.

The reflection of P in the x-axis is $(2, 3, 5)$, which lies in the 1st octant.

The reflection of P in the xy-plane is $(2, -3, 5)$, which lies in the 4th octant.

The reflection of P in the y-axis is $(-2, 3, 5)$, which lies in the 3rd octant.

The reflection of P in the yz-plane is $(-2, -3, -5)$, which lies in the 7th octant.

Introduction to Vectors

Concept Application Exercises

Exercise 2.1

1. (2) We have $|\vec{a}|=|\vec{b}|=|\vec{a}-\vec{b}|$.

So, we construct equilateral triangle as shown in the figure.

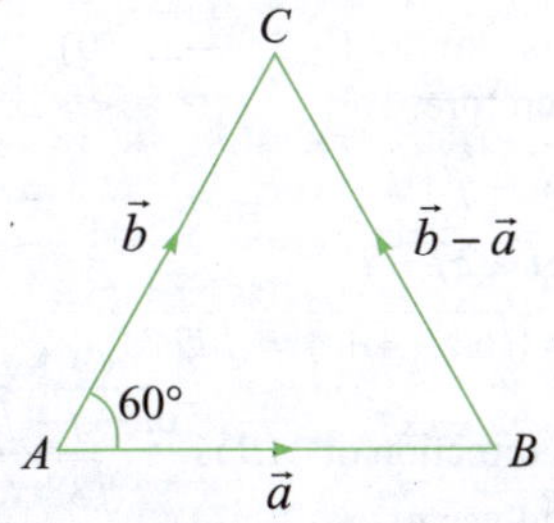

From the figure angle between vectors $\vec{a}$ and $\vec{b}$ is 60°.

2. (3) We have $|\vec{a}|=|\vec{b}|$.

So, parallelogram formed by vectors $\vec{a}$ and $\vec{b}$ is rhombus.

Therefore, angle between diagonal vectors $\vec{a}+\vec{b}$ and $\vec{a}-\vec{b}$ is 90°.

3. Since the diagonals of a rhombus bisect each other, $\overrightarrow{OA} = -\overrightarrow{OC}$

and $\overrightarrow{OB} = -\overrightarrow{OD}$ and so $\overrightarrow{OA} + \overrightarrow{OB} + \overrightarrow{OC} + \overrightarrow{OD} = \vec{0}$.

4. (3) Since the diagonals of a parallelogram bisect each other, P is the middle point of AC and BD both. Therefore,

$$\overrightarrow{OA} + \overrightarrow{OC} = 2\overrightarrow{OP} \text{ and } \overrightarrow{OB} + \overrightarrow{OD} = 2\overrightarrow{OP}$$

5.

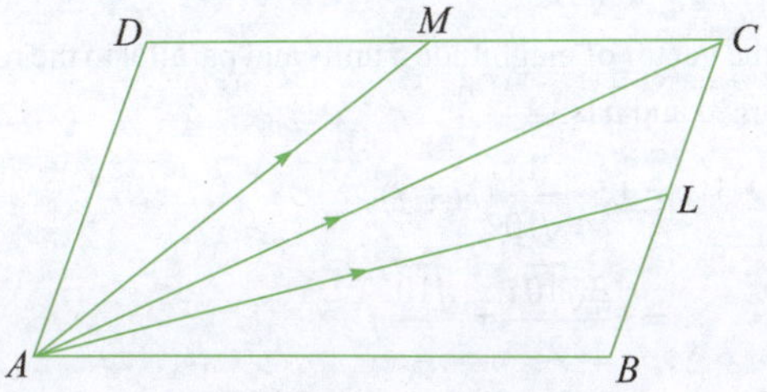

$$\overrightarrow{AL} = \overrightarrow{AB} + \overrightarrow{BL} = \overrightarrow{AB} + \frac{1}{2}\overrightarrow{BC}$$

$$= \overrightarrow{AB} + \frac{1}{2}\overrightarrow{AD}$$

$$\overrightarrow{AM} = \overrightarrow{AD} + \overrightarrow{DM} = \overrightarrow{AD} + \frac{1}{2}\overrightarrow{DC}$$

$$= \overrightarrow{AD} + \frac{1}{2}\overrightarrow{AB}$$

Adding, $\overrightarrow{AL} + \overrightarrow{AM} = \frac{3}{2}(\overrightarrow{AB} + \overrightarrow{AD})$

$$= \frac{3}{2}(\overrightarrow{AB} + \overrightarrow{BC}) = \frac{3}{2}\overrightarrow{AC}$$

6. We know that the figure formed by the lines joining the midpoints of the sides of a quadrilateral is a parallelogram. Hence, $MPNQ$ is a parallelogram, whose diagonals are MN and PQ intersecting at E, which is the midpoint of both MN and PQ. For any origin O, we have $\overrightarrow{OA} + \overrightarrow{OB} = 2(\overrightarrow{OM})$ (as M is the midpoint of AB).

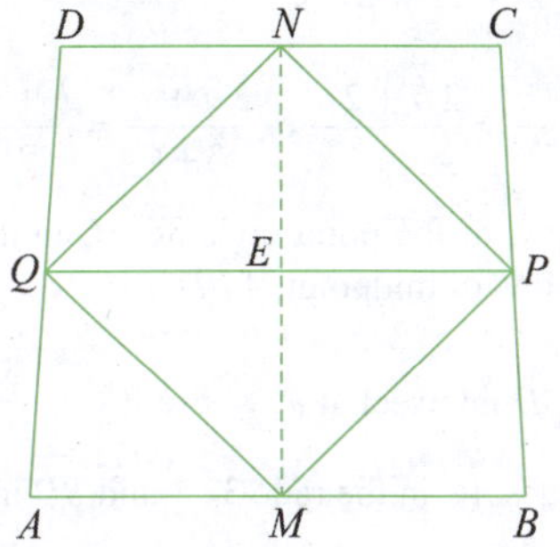

$$\overrightarrow{OC} + \overrightarrow{OD} = 2(\overrightarrow{ON})$$

(as N is the midpoint of CD)

$$\Rightarrow \overrightarrow{OA} + \overrightarrow{OB} + \overrightarrow{OC} + \overrightarrow{OD} = 2(\overrightarrow{OM} + \overrightarrow{ON})$$

$$= 2(2\overrightarrow{OE}) = 4\overrightarrow{OE}$$

where E is the midpoint of MN as it is the intersection of the diagonals of a parallelogram.

7. Since, $\vec{a}, \vec{b}$ and $\vec{c}$ are mutually perpendicular vectors or equal magnitudes, they can be considered as three coterminous edges of a cube.

Then the vector $\vec{a}+\vec{b}+\vec{c}$ is the body diagonal of the cube.

Let $\vec{a}, \vec{b}$ and $\vec{c}$ lie along the positive x-, y- and z-axis respectively.

Also, let $|\vec{a}|=|\vec{b}|=|\vec{c}|=\lambda$

Then, the length of the diagonal vector of the cuboid is

$$|\vec{a}+\vec{b}+\vec{c}| = \sqrt{|\vec{a}|^2+|\vec{b}|^2+|\vec{c}|^2}$$

$$= \sqrt{3\lambda^2} = \sqrt{3}\lambda$$

Now, the angle of the vector $\vec{a}+\vec{b}+\vec{c}$ and the positive x-axis or vector $\vec{a}$ is given by,

$$\cos\theta = \frac{\lambda}{\sqrt{3}\lambda} = \frac{1}{\sqrt{3}}$$

Hence, $\theta = \cos^{-1}\frac{1}{\sqrt{3}}$

8. (1) $|\vec{a}|=|\vec{b}|=1$

Hence, vectors $\vec{a}$ and $\vec{b}$ are adjacent sides of rhombus $OACB$

Angle between $\vec{a}$ and $\vec{b}$ is 120°.

Diagonals of rhombus are OC and AB. Thus,

$$\overrightarrow{OC} = \vec{a}+\vec{b} \text{ and } \overrightarrow{AB} = \vec{b}-\vec{a}$$

Clearly from the figure, angle between $\vec{b}$ and $\vec{b}-\vec{a}$ is 30°.

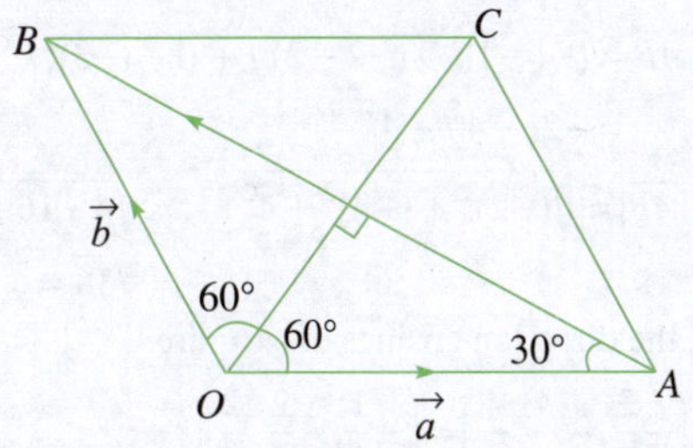

9. Since $3\vec{a}-2\vec{b}+\vec{c}-2\vec{d}=\vec{0}$

$$3\vec{a}+\vec{c}=2\vec{b}+2\vec{d}$$

$$\Rightarrow \quad \frac{3\vec{a}+\vec{c}}{4}=\frac{2\vec{b}+2\vec{d}}{4} \text{ or } \frac{3\vec{a}+\vec{c}}{3+1}=\frac{\vec{b}+\vec{d}}{2}$$

Therefore, P.V. of the point dividing AC in the ratio 1 : 3 is the same as the P.V. of midpoint of BD.

So AC and BD intersect at P, whose P.V. is $\frac{3\vec{a}+\vec{c}}{4}$ or $\frac{\vec{b}+\vec{d}}{2}$.

Point P divides AC in the ratio 3 : 1 and BD in the ratio 1 : 1.

Exercise 2.2

1. The unit vector $\hat{a}$ in the direction of vector $\vec{a}=\hat{i}+\hat{j}+2\hat{k}$ is given by $\hat{a}=\frac{\vec{a}}{|a|}$

$$|\vec{a}|=\sqrt{1^2+1^2+2^2}=\sqrt{1+1+4}=\sqrt{6}$$

$$\therefore \quad \hat{a}=\frac{\vec{a}}{|a|}=\frac{\hat{i}+\hat{j}+2\hat{k}}{\sqrt{6}}=\frac{1}{\sqrt{6}}\hat{i}+\frac{1}{\sqrt{6}}\hat{j}+\frac{2}{\sqrt{6}}\hat{k}$$

2. Let $\vec{a}=\hat{i}+2\hat{j}+3\hat{k}$

$$\therefore \quad |\vec{a}|=\sqrt{1^2+2^2+3^2}=\sqrt{1+4+9}=\sqrt{14}$$

Hence, the direction cosines of $\vec{a}$ are

$$\left(\frac{1}{\sqrt{14}}, \frac{2}{\sqrt{14}}, \frac{3}{\sqrt{14}}\right).$$

3. (1) We have vector $\overline{AB}=-\hat{i}-2\hat{j}+3\hat{k}$, where with $A(2, 1, -4)$.

Thus, $\overline{OA}=2\hat{i}-\hat{j}+4\hat{k}$.

Now, $\overline{AB}=\overline{OB}-\overline{OA}=\hat{i}-2\hat{j}+3\hat{k}$

$$\therefore \quad \overline{OB}=-4\hat{i}-2\hat{j}+3\hat{k}+\overline{OA}=-4\hat{i}-2\hat{j}+3\hat{k}+2\hat{i}-\hat{j}+4\hat{k}$$

$$=-2\hat{i}-3\hat{j}+7\hat{k}$$

Thus, the terminal point is $B(-2, -3, 7)$, which lies in the 3rd octant.

4. (3) Since, the angle between vectors is obtuse,

$$(a)(a)+(-2)(2)+(-4)(3)<0$$

or $\quad a^2-16<0$

$\Rightarrow \quad -4<a<4$

Therefore, the integral values of 'a' are $-3, -2, -1, 0, 1, 2, 3$.

5. (4) Let $P \equiv (x, y, z)$.

Since, $\overline{OP}$ makes an acute angle with the x-axis, $x<0$.

Therefore, point P may lie in 2nd, 3rd, 6th or 7th octant.

6. The given points are A (1, 2, –3) and B (–1, –2, 1). Therefore,

$$\overline{AB}=(-1-1)\hat{i}+(-2-2)\hat{j}+\{1-(-3)\}\hat{k}$$

$$=-2\hat{i}-4\hat{j}+4\hat{k}$$

$$\therefore \quad |\overline{AB}|=\sqrt{(-2)^2+(-4)^2+4^2}=\sqrt{4+16+16}$$

$$=\sqrt{36}=6$$

Hence, the direction cosines of $\overline{AB}$ are

$$\left(-\frac{2}{6}, -\frac{4}{6}, \frac{4}{6}\right)=\left(-\frac{1}{3}, -\frac{2}{3}, \frac{2}{3}\right).$$

7. Given points are $P(5\hat{i}+4\hat{j}+a\hat{k})$ and $Q(-\hat{i}+2\hat{j}-2\hat{k})$

Now $\overline{PQ}=(-\hat{i}+2\hat{j}-2\hat{k})-(5\hat{i}+4\hat{j}+a\hat{k})$

$$=(-6\hat{i}-2\hat{j}-(2+a)\hat{k})$$

Given $|\overline{PQ}|=7$

$$\therefore \quad \sqrt{36+4+(a+2)^2}=7$$

or $\quad (a+2)^2=9$

or $\quad a+2=\pm 3$

or $\quad a=-5, 1.$

8. Given points are $A(-3, -2, 0)$, $B(3, -3, 1)$ and $C(5, 0, 2)$. Therefore,

$$\overline{AB}=6\hat{i}-\hat{j}+\hat{k}$$

and $\quad \overline{AC}=8\hat{i}+2\hat{j}+2\hat{k}$

$$|\overline{AC}|=\sqrt{64+4+4}=\sqrt{72}$$

Unit vector in direction of $\overline{AB}$ is $\frac{6\hat{i}-\hat{j}+\hat{k}}{\sqrt{38}}$

$\therefore \quad$ Required vector

$$=\frac{\sqrt{72}}{\sqrt{38}}(6\hat{i}-\hat{j}+\hat{k})=\frac{6}{\sqrt{19}}(6\hat{i}-\hat{j}+\hat{k})$$

9. We have,

$$\vec{a}=2\hat{i}+3\hat{j}-\hat{k} \text{ and } \vec{b}=\hat{i}-2\hat{j}+\hat{k}$$

Let $\vec{c}$ be the resultant of $\vec{a}$ and $\vec{b}$. Then

$$\vec{c}=\vec{a}+\vec{b}=(2+1)\hat{i}+(3-2)\hat{j}+(-1+1)\hat{k}$$

$$=3\hat{i}+\hat{j}$$

$$\therefore \quad |\vec{c}|=\sqrt{3^2+1^2}=\sqrt{9+1}=\sqrt{10}$$

$$\therefore \quad \hat{c}=\frac{\vec{c}}{|\vec{c}|}=\frac{(3\hat{i}+\hat{j})}{\sqrt{10}}$$

Hence, the vector of magnitude 5 units and parallel to the resultant of vectors $\vec{a}$ and $\vec{b}$ is

$$\pm 5\cdot\hat{c}=\pm 5\cdot\frac{1}{\sqrt{10}}(3\hat{i}+\hat{j})$$

$$=\pm\frac{3\sqrt{10}\,\hat{i}}{2}\pm\frac{\sqrt{10}}{2}\hat{j}.$$

10. We have $\vec{a}=3\hat{i}+4\hat{j}-2\hat{k}$. Therefore,

$$|\vec{a}|=\sqrt{9+16+4}=\sqrt{29}$$

Therefore, the unit vector parallel to $\vec{a}$ is

$$\frac{\vec{a}}{|\vec{a}|}=\frac{1}{\sqrt{(29)}}(3\hat{i}+4\hat{j}-2\hat{k}).$$

Now suppose $\vec{b}$ is the vector which when added to $\vec{a}$ gives the resultant $\hat{i}$. Then

$$\vec{a}+\vec{b}=\hat{i} \text{ or } \vec{b}=\hat{i}-\vec{a}=\hat{i}-(3\hat{i}+4\hat{j}-2\hat{k})$$

$$\therefore \quad \vec{b}=-2\hat{i}-4\hat{j}+2\hat{k}$$

11. (4) $|\overrightarrow{OA}|=|\overrightarrow{OB}|=\sqrt{14}$

ΔAOB is isosceles. Hence, the bisector of angle AOB will bisect the base AB.

Hence, P is the midpoint $(2, 2, -2)$ of AB. Therefore,

$$\overline{OP}=2(\hat{i}+\hat{j}-\hat{k})$$

12. A vector bisecting the angle between $\vec{a}$ and $\vec{b}$ is $\frac{\vec{a}}{|\vec{a}|}+\frac{\vec{b}}{|\vec{b}|}$.

Here $\frac{2\hat{i}+\hat{j}-\hat{k}}{\sqrt{6}}+\frac{\hat{i}-2\hat{j}+\hat{k}}{\sqrt{6}}$

i.e., $\frac{3\hat{i}-\hat{j}}{\sqrt{6}}$

A vector of magnitude 3 along this vector is

$$\frac{3(3\hat{i}-\hat{j})}{\sqrt{10}}$$

13. (2) If mid-point of BC is D, then median vector

$$\overrightarrow{AD}=\frac{1}{2}(\overrightarrow{AB}+\overrightarrow{AC})$$
$$=\frac{1}{2}((2\hat{i}+3\hat{j}-\hat{k})+(-4\hat{i}-\hat{j}+3\hat{k}))$$
$$=-\hat{i}+\hat{j}+\hat{k}$$
$$\therefore\quad |\overrightarrow{AD}|=|-\hat{i}+\hat{j}+\hat{k}|=\sqrt{3}$$

14. (4) Here, $\cos^2\alpha+\cos^2(90°-\alpha)+\cos^2\gamma=1$

$\Rightarrow\quad \cos^2\alpha+\sin^2\alpha+\cos^2\gamma=1$

$\Rightarrow\quad \cos^2\gamma+1=1 \Rightarrow \gamma=90°$

15. (2) $\cos 2\alpha+\cos 2\beta+\cos 2\gamma$

$=2\cos^2\alpha-1+2\cos^2\beta-1+2\cos^2\gamma-1$

$=2(\cos^2\alpha+\cos^2\beta+\cos^2\gamma)-3=2-3=-1$

16. We have point $P(2, 3, 6)$.

$\therefore\quad \cos\alpha=\frac{2}{\sqrt{2^2+3^2+6^2}}=\frac{2}{7},$

$\cos\beta=\frac{3}{7}$ and $\cos\gamma=\frac{6}{7}$

Hence, angle α is greatest as cosine is decreasing function for angles lying in $(0, \pi)$.

17. Since $|\overrightarrow{AB}|=|\overrightarrow{BC}|=|\overrightarrow{CA}|$, the incentre is same as the circumcentre, and hence IA is perpendicular to BC. Therefore, $\vec{r}$ is parallel to BC.

$$\vec{r}=\lambda(\hat{i}-\hat{j})$$

Hence, unit vector $\vec{r}=\pm\frac{1}{\sqrt{2}}(\hat{i}-\hat{j})$

Exercise 2.3

1. Consider $2\vec{a}-\vec{b}+3\vec{c}=x(\vec{a}+\vec{b}-2\vec{c})+y(\vec{a}+\vec{b}-3\vec{c})$

or $2\vec{a}-\vec{b}+3\vec{c}=(x+y)\vec{a}+(x+y)\vec{b}+(-2x-3y)\vec{c}$

$x+y=2$ ...(i)

$x+y=-1$ ...(ii)

$-2x-3y=3$ ...(iii)

Clearly this is not possible

Therefore, the given vectors are not coplanar.

Alternate method:

We have determinant of co-efficients as

$$\begin{vmatrix}2 & -1 & 3\\ 1 & 1 & -2\\ 1 & 1 & -3\end{vmatrix}=-3\neq 0$$

Hence, vectors are non-coplanar.

2. a. Let $\vec{a}=\hat{i}+\hat{j}+\hat{k},\ \vec{b}=2\hat{i}+3\hat{j}-\hat{k},\ \vec{c}=-\hat{i}-2\hat{j}+3\hat{k}$

$$\begin{vmatrix}1 & 1 & 1\\ 2 & 3 & -1\\ -1 & -2 & 3\end{vmatrix}=-1$$

Hence, vectors are non-coplanar and linearly independent.

b. Let $\vec{a}=3\hat{i}+\hat{j}-\hat{k},\ \vec{b}=2\hat{i}-\hat{j}+7\hat{k},\ \vec{c}=7\hat{i}-\hat{j}+13\hat{k}$

$$\begin{vmatrix}3 & 1 & -1\\ 2 & -1 & 7\\ 7 & -1 & 13\end{vmatrix}=0$$

Hence, vectors are coplanar and linearly dependent.

3. (3) Putting the values of $\vec{A}$ and $\vec{B}$, and then equating the coefficients of $\vec{a}$ and $\vec{b}$ on both sides, we get

$3(p+4q)=2(-2p+q+2)$

$3(2p+q+1)=2(2p-3q-1)$

$7p+10q=4$ and $2p+9q=-5$

Solving, we get $p=2$ and $q=-1$

4. Points $A(l_1\vec{a}+m_1\vec{b}+n_1\vec{c}), B(l_2\vec{a}+m_2\vec{b}+n_2\vec{c}), C(l_3\vec{a}+m_3\vec{b}+n_3\vec{c}), D(l_4\vec{a}+m_4\vec{b}+n_4\vec{c})$ are coplanar.

$\Rightarrow$ Vectors

$$\overrightarrow{AB}=(l_2-l_1)\vec{a}+(m_2-m_1)\vec{b}+(n_2-n_1)\vec{c},$$
$$\overrightarrow{AC}=(l_3-l_1)\vec{a}+(m_3-m_1)\vec{b}+(n_3-n_1)\vec{c},$$
$$\overrightarrow{AD}=(l_4-l_1)\vec{a}+(m_4-m_1)\vec{b}+(n_4-n_1)\vec{c}$$

are coplanar

$$\Rightarrow\quad \begin{vmatrix}l_2-l_1 & m_2-m_1 & n_2-n_1\\ l_3-l_1 & m_3-m_1 & n_3-n_1\\ l_4-l_1 & m_4-m_1 & n_4-n_1\end{vmatrix}=0$$

Now if $\begin{vmatrix}l_1 & l_2 & l_3 & l_4\\ m_1 & m_2 & m_3 & m_4\\ n_1 & n_2 & n_3 & n_4\\ 1 & 1 & 1 & 1\end{vmatrix}=0$

Then applying $C_2\to C_2-C_1$, $C_3\to C_3-C_1$, $C_4\to C_4-C_1$, we get

$$\begin{vmatrix}l_1 & l_2-l_1 & l_3-l_1 & l_4-l_1\\ m_1 & m_2-m_1 & m_3-m_1 & m_4-m_1\\ n_1 & n_2-n_1 & n_3-n_1 & n_4-n_1\\ 1 & 0 & 0 & 0\end{vmatrix}=0$$

or $\begin{vmatrix}l_2-l_1 & m_2-m_1 & n_2-n_1\\ l_3-l_1 & m_3-m_1 & n_3-n_1\\ l_4-l_1 & m_4-m_1 & n_4-n_1\end{vmatrix}=0$

5. Any vector $\vec{r}$ can be uniquely expressed as a linear combination of three non-coplanar vectors.

Let us choose that

$$7\vec{a}-11\vec{b}+15\vec{c}=x(\vec{a}-2\vec{b}+3\vec{c})$$
$$+y(2\vec{a}-3\vec{b}+4\vec{c})+z(3\vec{a}-4\vec{b}+5\vec{c})$$

Comparing the coefficients of $\vec{a}$, $\vec{b}$ and $\vec{c}$ on both sides, we get

$x+2y+3z=7, -2x-3y-4z=-11,$

$3x+4y+5z=15$

Eliminating x and then solving for y and z, we get $x=1, y=3, z=0$

6. (1) Vectors $a\hat{i}+a\hat{j}+c\hat{k}, \hat{i}+\hat{k}, c\hat{i}+c\hat{j}+b\hat{k}$ are coplanar. Thus,

$$\begin{vmatrix} a & a & c \\ 1 & 0 & 1 \\ c & c & b \end{vmatrix}=0$$

or $c^2-ab=0$

For equation $ax^2+2cx+b=0$

$D=4c^2-4ab=4(c^2-ab)=0$

Hence, roots are real and equal

7. (3) Let $QM:MT=\lambda:1$ and $PM:MS=\mu:1$

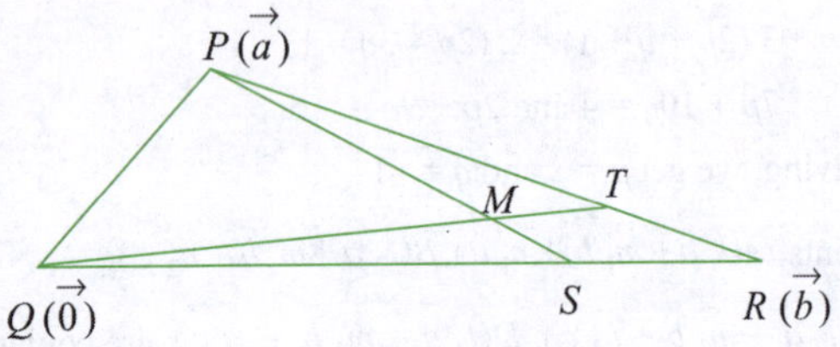

and $\overrightarrow{QP}=\vec{a}, \overrightarrow{QR}=\vec{b}$

$\Rightarrow \overrightarrow{QT}=\dfrac{4\vec{b}+\vec{a}}{5}$

and $\overrightarrow{QM}=\dfrac{\lambda}{\lambda+1}\left(\dfrac{4\vec{b}+\vec{a}}{5}\right)$ ...(i)

$\overrightarrow{QS}=\dfrac{3}{4}\vec{b}, \overrightarrow{QM}=\dfrac{\mu\left(\frac{3}{4}\vec{b}\right)+\vec{a}}{\mu+1}$ ...(ii)

From (i) and (ii), we have

$$\frac{1}{\mu+1}=\frac{\lambda}{5(\lambda+1)} \text{ and } \frac{4\lambda}{5(\lambda+1)}=\frac{3\mu}{4(\mu+1)}$$

$\Rightarrow \lambda=15/4$ and $\mu=16/3$

$\therefore QM:MT=15:4$

8. We have, $\begin{vmatrix} al+a_1l_1 & am+a_1m_1 & an+a_1n_1 \\ bl+b_1l_1 & bm+b_1m_1 & bn+b_1n_1 \\ cl+c_1l_1 & cm+c_1m_1 & cn+c_1n_1 \end{vmatrix}$

$$=\begin{vmatrix} a & a_1 & 0 \\ b & b_1 & 0 \\ c & c_1 & 0 \end{vmatrix}\begin{vmatrix} l & l_1 & 0 \\ m & m_1 & 0 \\ n & n_1 & 0 \end{vmatrix}=0$$

Therefore, the given vectors are coplanar.

Exercises

Single Correct Answer Type

1. (1) Since, $\vec{a}$, $\vec{b}$ and $\vec{c}$ are pairwise mutually perpendicular, these vectors can be taken as coterminous edges of the cuboid.

Then the vector $\vec{a}+\vec{b}+\vec{c}$ is one of the diagonals.

Therefore, $|\vec{a}+\vec{b}+\vec{c}|=\sqrt{1+4+4}=3$

2. (4) $3\vec{a}+4\vec{b}+5\vec{c}=0$

Hence, $\vec{a}, \vec{b}$ and $\vec{c}$ are coplanar.

No other conclusion can be derived from it.

3. (1) We have $\overrightarrow{GB}+\overrightarrow{GC}=2\overrightarrow{GD}$, where D is the midpoint of BC.

Therefore,

$$\overrightarrow{GA}+\overrightarrow{GB}+\overrightarrow{GC}=\overrightarrow{GA}+2\overrightarrow{GD}$$
$$=\overrightarrow{GA}-\overrightarrow{GA}=0$$

($\because G$ divides AC in the ratio 2 : 1, $\therefore 2\overrightarrow{GD}=-\overrightarrow{GA}$)

4. (3) $m\vec{a}$ is a unit vector if and only if

$|m\vec{a}|=1 \Rightarrow |m|\ |\vec{a}|=1 \Rightarrow |m|\ a=1 \Rightarrow a=\dfrac{1}{|m|}$

5. (4) $\vec{a}-\vec{b}=2(\vec{d}-\vec{c})$

$\therefore \dfrac{\vec{a}+2\vec{c}}{2+1}=\dfrac{\vec{b}+2\vec{d}}{2+1}$

Hence, AC and BD trisect each other as L.H.S. is the position vector of a point trisecting A an C, and R.H.S. that of B and D.

6. (4) $\vec{c}-\vec{b}=\alpha\vec{d}$ and $\vec{p}=\overrightarrow{AC}+\overrightarrow{BD}=\mu\overrightarrow{AD}$

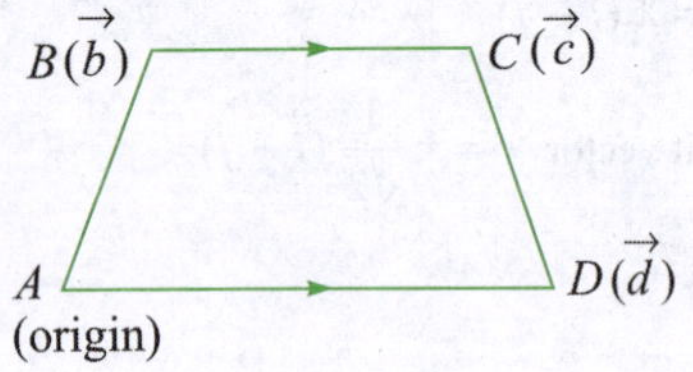

Hence, $\vec{p}=\vec{c}+\vec{d}-\vec{b}=\mu\vec{d}$

$\Rightarrow \alpha\vec{d}+\vec{d}=\mu\vec{d}$ (using $\vec{c}-\vec{b}=\alpha\vec{d}$)

or $\alpha+1=\mu$

7. (3)

8. (3) $|\vec{a}+\vec{b}|<|\vec{a}-\vec{b}|$

So, we have following parallelogram formed by vector $\vec{a}$ and $\vec{b}$

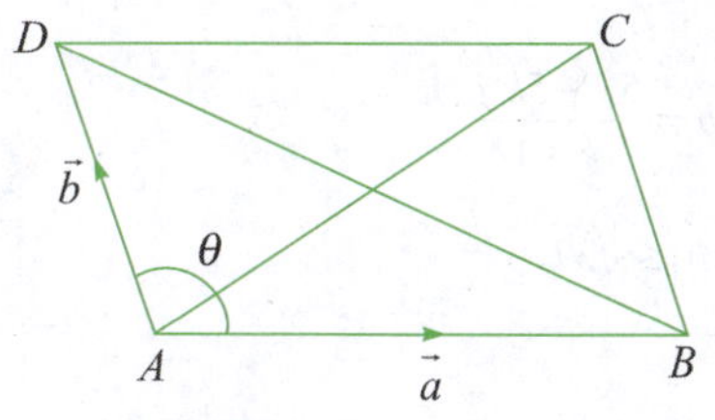

We have $|\vec{a}+\vec{b}| < |\vec{a}-\vec{b}|$

$\Rightarrow \quad AC < BD$

$\Rightarrow \quad \theta \in (\pi/2, \pi)$

9. (3) The position vector of the point O with respect to itself is

$$\frac{\overrightarrow{OA}\sin 2A + \overrightarrow{OB}\sin 2B + \overrightarrow{OC}\sin 2C}{\sin 2A + \sin 2B + \sin 2C}$$

$$\Rightarrow \quad \frac{\overrightarrow{OA}\sin 2A + \overrightarrow{OB}\sin 2B + \overrightarrow{OC}\sin 2C}{\sin 2A + \sin 2B + \sin 2C} = \vec{0}$$

or $\quad \overrightarrow{OA}\sin 2A + \overrightarrow{OB}\sin 2B + \overrightarrow{OC}\sin 2C = \vec{0}$

10. (2) Vector in the direction of angle bisector of $\vec{a}$ and $\vec{b}$ is $\vec{a}+\vec{b}$.

Unit vector in this direction is $\dfrac{\vec{a}+\vec{b}}{|\vec{a}+\vec{b}|}$.

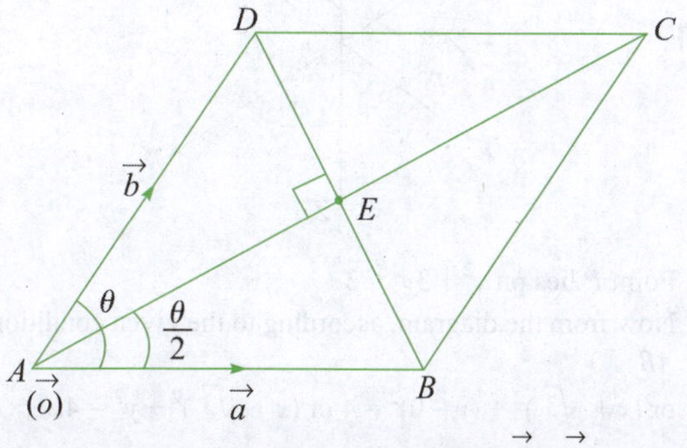

From the figure, position vector of E is $\dfrac{\vec{a}+\vec{b}}{2}$

Now in triangle AEB, $AE = AB\cos\dfrac{\theta}{2}$

$$\Rightarrow \quad \left|\frac{\vec{a}+\vec{b}}{2}\right| = \cos\frac{\theta}{2}$$

Hence, unit vector along the bisector is $\dfrac{\vec{a}+\vec{b}}{2\cos(\theta/2)}$

11. (2) From the diagram, it is obvious that locus is a cone having axis as the positive x-axis, vertex at the origin and the slant height equal to the magnitude of the vector.

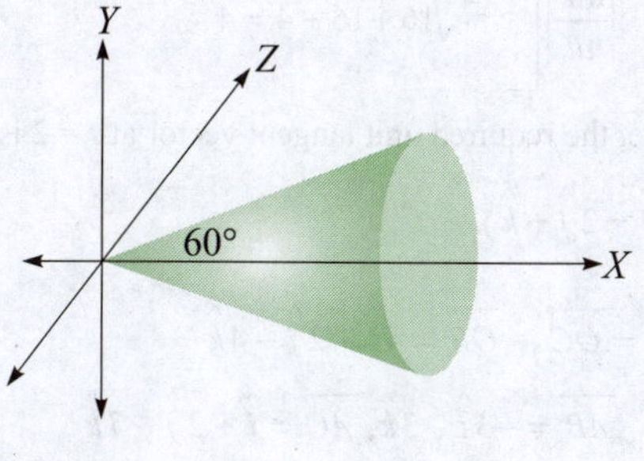

12. (3)

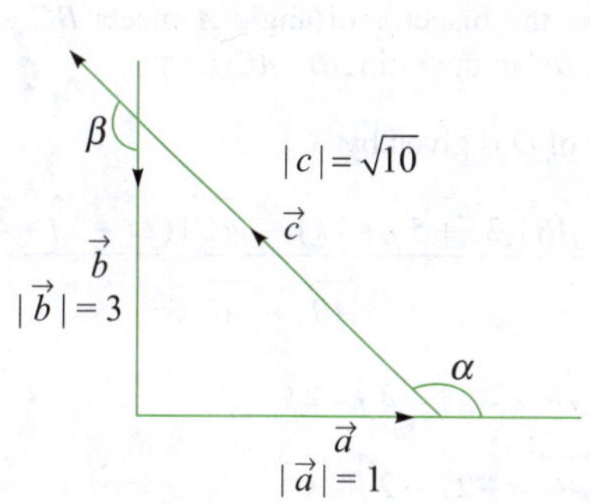

From the figure, $\pi - \alpha + \pi - \beta = \dfrac{\pi}{2}$

or $\quad \alpha + \beta = \dfrac{3\pi}{2}$

13. (2)

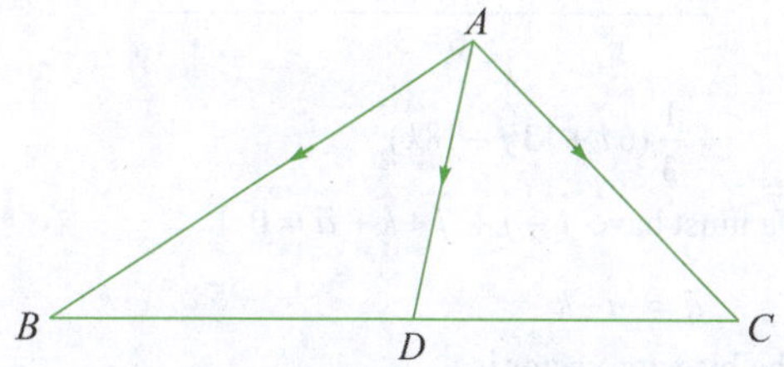

$$\overrightarrow{AB} + \overrightarrow{AC} = 2\,\overrightarrow{AD}$$

$$\therefore \quad \overrightarrow{AD} = \frac{1}{2}\{(-3\hat{i}+4\hat{k}) + (5\hat{i}-2\hat{j}+4\hat{k})\}$$

$$= \hat{i} - \hat{j} + 4\hat{k}$$

Length of $AD = \sqrt{1+1+16} = \sqrt{18}$

14. (2) $|a| + |b| + |c| = \sqrt{a^2+b^2+c^2}$

$\Leftrightarrow \quad 2\,|ab| + 2\,|bc| + 2\,|ca| = 0$

$\Leftrightarrow \quad ab = bc = ca = 0 \Leftrightarrow$ any two of a, b and c are zero

15. (3) $\vec{\alpha} = \vec{a} + \vec{b} + \vec{c} = 6\hat{i} + 12\hat{j}$

Let $\quad \vec{\alpha} = x\,\vec{a} + y\,\vec{b} \Rightarrow 6x + 2y = 6$

and $\quad -3x - 6y = 12$

$\therefore \quad x = 2, y = -3$

$\therefore \quad \vec{\alpha} = 2\,\vec{a} - 3\,\vec{b}$

16. (2) Note that $\vec{a} + \vec{b} = \vec{c}$

17. (1) $\hat{a} = \dfrac{-4i+3\hat{k}}{5}$; $\hat{b} = \dfrac{14\hat{i}+2\hat{j}-5\hat{k}}{15}$

A vector $\vec{V}$ bisecting the angle between $\vec{a}$ and $\vec{b}$ is

$$\vec{V} = \hat{a} + \hat{b}$$

$$= \frac{-12\hat{i}+9\hat{k}+14\hat{i}+2\hat{j}-5\hat{k}}{15}$$

$$= \frac{2\hat{i}+2\hat{j}+4\hat{k}}{15}$$

Required vector $\vec{d} = \sqrt{6}\,\hat{V} = \hat{i} + \hat{j} + 2\hat{k}$

18. (3) Suppose the bisector of angle A meets BC at D. Then AD divides BC in the ratio $AB : AC$.

So, P.V. of D is given by

$$\frac{|\overrightarrow{AB}|(2\hat{i}+5\hat{j}+7\hat{k})+|\overrightarrow{AC}|(2\hat{i}+3\hat{j}+4\hat{k})}{|\overrightarrow{AB}|+|\overrightarrow{AC}|}$$

But $\overrightarrow{AB}=-2\hat{i}-4\hat{j}-4\hat{k}$

and $\overrightarrow{AC}=-2\hat{i}-2\hat{j}-\hat{k}$

$\Rightarrow \quad |\overrightarrow{AB}|=6$ and $|\overrightarrow{AC}|=3$

Therefore, P.V. of D is given by

$$\frac{6(2\hat{i}+5\hat{j}+7\hat{k})+3(2\hat{i}+3\hat{j}+4\hat{k})}{6+3}$$

$$=\frac{1}{3}(6\hat{i}+13\hat{j}+18\hat{k})$$

19. (1) We must have $\hat{i}-\hat{j}+\hat{j}+\hat{k}+\vec{a}=\vec{0}$

$\therefore \quad \vec{a}=-\hat{i}-\hat{k}$

20. (1) The bisector vector is

$$\frac{2\hat{i}+2\hat{j}-\hat{k}}{3}+\frac{-2\hat{i}-3\hat{j}+6\hat{k}}{7}=\frac{8\hat{i}+5\hat{j}+11\hat{k}}{21}$$

Therefore, the angle of this vector with the x-axis is given by

$$\cos\alpha=\frac{8}{\sqrt{64+25+121}}=\frac{8}{\sqrt{210}}$$

21. (2) We have,

$$\cos 60^\circ=\frac{(2)(3)+(-3)(2)+(a)(-4)}{\sqrt{4+9+a^2}\sqrt{9+4+16}}$$

$$\Rightarrow \quad \frac{1}{2}=\frac{-4a}{\sqrt{13+a^2}\sqrt{29}}$$

$$\Rightarrow \quad 29(13+a^2)=64a^2$$

$$\Rightarrow \quad a^2=\frac{377}{35} \text{ or } a=-\sqrt{\frac{377}{35}}$$

22. (4) We must have $\lambda(\hat{i}-3\hat{j}+5\hat{k})=a+\frac{2\hat{k}+2\hat{j}-\hat{i}}{3}$. Therefore,

$$3a=3\lambda(\hat{i}-3\hat{j}+5\hat{k})-(2\hat{k}+2\hat{j}-\hat{i})$$

$$=\hat{i}(3\lambda+1)-\hat{j}(2+9\lambda)+\hat{k}(15\lambda-2)$$

or $\quad 3|\hat{a}|=\sqrt{(3\lambda+1)^2+(2+9\lambda)^2+(15\lambda-2)^2}$

or $\quad 9=(3\lambda+1)^2+(2+9\lambda)^2+(15\lambda-2)^2$

or $\quad 315\lambda^2-18\lambda=0 \Rightarrow \lambda=0, \frac{2}{35}$

If $\lambda=0$, $\vec{a}=\hat{i}-2\hat{j}-2\hat{k}$ (not acceptable)

For $\lambda=\frac{2}{35}$, $\vec{a}=\frac{41}{105}\hat{i}-\frac{88}{105}\hat{j}-\frac{40}{105}\hat{k}$

23. (2) The point that divides $5\hat{i}$ and $5\hat{j}$ in the ratio of $k:1$ is $\frac{(5\hat{j})k+(5\hat{i})1}{k+1}$

$$\therefore \quad \vec{b}=\frac{5\hat{i}+5k\hat{j}}{k+1}$$

Also, $|\vec{b}|\le\sqrt{37}$

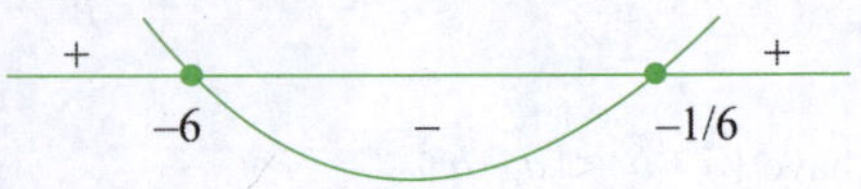

$$\Rightarrow \quad \frac{1}{k+1}\sqrt{25+25k^2}\le\sqrt{37}$$

or $\quad 5\sqrt{1+k^2}\le\sqrt{37}(k+1)$

Squaring both sides, we get

$$25(1+k^2)\le 37(k^2+2k+1)$$

or $\quad 6k^2+37k+6\ge 0$ or $(6k+1)(k+6)\ge 0$

$$k\in(-\infty,-6]\cup\left[-\frac{1}{6},\infty\right)$$

24. (1)

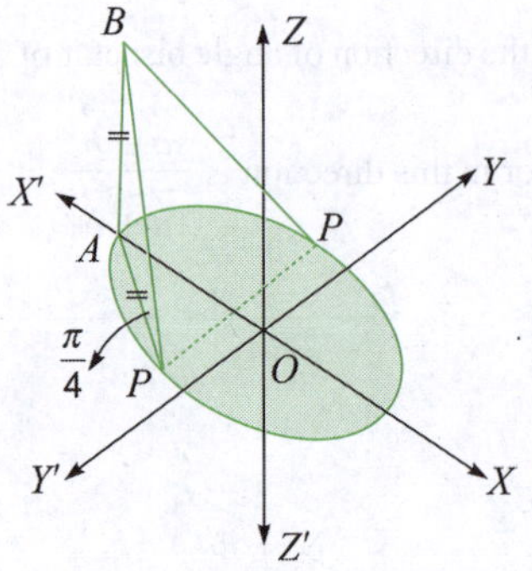

Point P lies on $x^2+3y^2=3$ ...(

Now from the diagram, according to the given conditions, AP AB

or $(x+\sqrt{3})^2+(y-0)^2=4$ or $(x+\sqrt{3})^2+y^2=4$ ...(i

Solving (i) and (ii), we get $x=0$ and $y=\pm 1$

Hence, point P has position vector $\pm\hat{j}$.

25. (1) The position vector of any point on the curve is

$$\vec{r}=(2+t^2)\hat{i}+(4t-5)\hat{j}+(2t^2-6t)\hat{k}$$

$$\Rightarrow \quad \frac{d\vec{r}}{dt}=2t\,\hat{i}+4\hat{j}+(4t-6)\hat{k}$$

$$\Rightarrow \quad \left.\frac{d\vec{r}}{dt}\right|_{t=2}=4\hat{i}+4\hat{j}+2\hat{k}$$

and $\quad \left.\left|\frac{d\vec{r}}{dt}\right|\right|_{t=2}=\sqrt{16+16+4}=4$

Hence, the required unit tangent vector at $t=2$ is $\frac{1}{3}(2\hat{i}+2\hat{j}+\hat{k})$.

26. (3) $\overrightarrow{BC}=\overrightarrow{OC}-\overrightarrow{OB}=4\hat{i}+2\hat{j}-4\hat{k}$

$$\overrightarrow{AB}=-3\hat{i}-3\hat{k},\ \overrightarrow{AC}=\hat{i}+2\hat{j}-7\hat{k}$$

$BC^2 = 36, AB^2 = 18, AC^2 = 54$

Clearly, $AC^2 = BC^2 + AB^2$

$\therefore \quad \angle B = 90°$

27. (3) Given $\vec{\alpha} + \vec{\beta} + \vec{\gamma} = a\vec{\delta}$ ...(i)

$\vec{\beta} + \vec{\gamma} + \vec{\delta} = b\vec{\alpha}$ ...(ii)

From (i), $\vec{\alpha} + \vec{\beta} + \vec{\gamma} + \vec{\delta} = (a+1)\vec{\delta}$ ...(iii)

From (ii), $\vec{\alpha} + \vec{\beta} + \vec{\gamma} + \vec{\delta} = (b+1)\vec{\alpha}$ ...(iv)

From (iii) and (iv), we get

$(a+1)\vec{\delta} = (b+1)\vec{\alpha}$ ...(v)

Since $\vec{\alpha}$ is not parallel to $\vec{\delta}$,

From (v), $a + 1 = 0$ and $b + 1 = 0$

From (iii), $\vec{\alpha} + \vec{\beta} + \vec{\gamma} + \vec{\delta} = 0$

28. (4) $\vec{r_1} + 2\vec{r_2} = (p\vec{a} + q\vec{b} + \vec{c}) + 2(\vec{a} + p\vec{b} + q\vec{c})$

$= (p+2)\vec{a} + (q+2p)\vec{b} + (1+2q)\vec{c}$

$2\vec{r_1} + \vec{r_2} = (2p+1)\vec{a} + (2q+p)\vec{b} + (2+q)\vec{c}$

$$\frac{p+2}{2p+1} = \frac{q+2p}{2q+p} = \frac{1+2q}{2+q}$$

$$= \frac{p+q+2p+2q+3}{p+q+2p+2q+3} = 1$$

$\Rightarrow \quad p = 1$ and $q = 1$

29. (4) $\sqrt{3}\tan\theta + 1 = 0$ and $\sqrt{3}\sec\theta - 2 = 0$

$\Rightarrow \quad \tan\theta = -\frac{1}{\sqrt{3}}$ and $\cos\theta = \frac{\sqrt{3}}{2}$

$\Rightarrow \quad \theta = \frac{11\pi}{6}$

$\Rightarrow \quad \theta = 2n\pi + \frac{11\pi}{6}, \; n \in Z$

30. (3) We have $A(\hat{i} - \hat{j} + \hat{k})$, $B(2\hat{i} + \hat{j} - \hat{k})$, $C(3\hat{i} + 4\hat{j} + \hat{k})$ and $D(\alpha\hat{i} + 2\hat{j} + 3\hat{k})$

Points A, B, C, D are coplanar

$\therefore$ Vectors $\overrightarrow{AB}, \overrightarrow{AC}, \overrightarrow{AD}$ are coplanar

So, $\begin{vmatrix} 1 & 2 & -2 \\ 2 & 5 & 0 \\ \alpha-1 & 3 & 2 \end{vmatrix} = 0$

$\Rightarrow \quad \alpha = 2$

31. (3) Given vectors are coplanar

$\therefore \quad \begin{vmatrix} \alpha & 1 & -\beta \\ 1 & -2 & 3 \\ 2 & 1 & -4 \end{vmatrix} = 0$

$\Rightarrow \quad \alpha - \beta + 2 = 0$

$\Rightarrow$ Point (α, β) lies on the line $x - y + 2 = 0$

32. (3) Given $a_1\vec{r_1} + a_2\vec{r_2} + \cdots + a_n\vec{r_n} = 0$

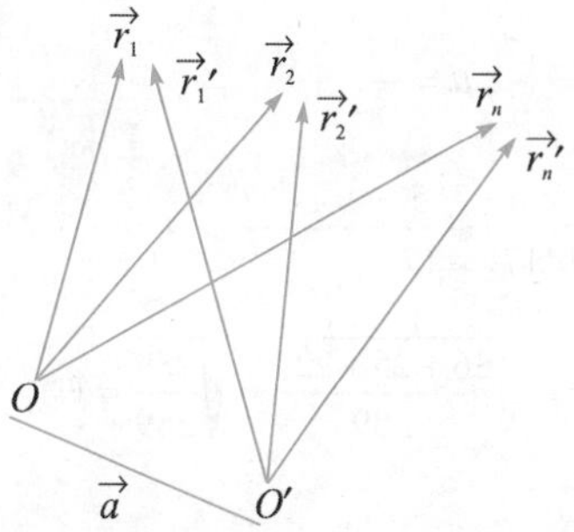

Now $\vec{a} + \vec{r_1'} = \vec{r_1}$ and so on

Hence,

$a_1(\vec{a} + \vec{r_1'}) + a_2(\vec{a} + \vec{r_2'}) + \cdots + a_n(\vec{a} + \vec{r_n'}) = 0$

$a_1\vec{r_1'} + a_2\vec{r_2'} + \cdots + a_n\vec{r_n'} + \vec{a}(a_1 + a_2 + \cdots + a_n) = 0$

Hence, $a_1\vec{r_1'} + a_2\vec{r_2'} + \cdots + a_n\vec{r_n'} = 0$

if $\quad a_1 + a_2 + \cdots + a_n = 0.$

33. (2) Let $\overrightarrow{OA} = \vec{a}$, $\overrightarrow{OB} = \vec{b}$ and $\overrightarrow{OC} = \vec{c}$,

$\overrightarrow{AB} = \vec{b} - \vec{a}$ and $\overrightarrow{OP} = \frac{1}{3}\vec{a}$,

$\overrightarrow{OQ} = \frac{1}{2}\vec{b}$, $\overrightarrow{OR} = \frac{1}{3}\vec{c}$.

then

Since P, Q, R and S are coplanar, then

$\overrightarrow{PS} = \alpha\,\overrightarrow{PQ} + \beta\,\overrightarrow{PR}$

($\overrightarrow{PS}$ can be written as a linear combination of $\overrightarrow{PQ}$ and $\overrightarrow{PR}$)

$= \alpha(\overrightarrow{OQ} - \overrightarrow{OP}) + \beta(\overrightarrow{OR} - \overrightarrow{OP})$

i.e., $\overrightarrow{OS} - \overrightarrow{OP} = -(\alpha + \beta)\frac{\vec{a}}{3} + \frac{\alpha}{2}\vec{b} + \frac{\beta}{3}\vec{c}$

$\Rightarrow \quad \overrightarrow{OS} = (1 - \alpha - \beta)\frac{\vec{a}}{3} + \frac{\alpha}{2}\vec{b} + \frac{\beta}{3}\vec{c}$ ...(i)

Given $\overrightarrow{OS} = \lambda\,\overrightarrow{AB} = \lambda(\vec{b} - \vec{a})$ ...(ii)

From (i) and (ii), $\beta = 0, \frac{1-\alpha}{3} = -\lambda$ and $\frac{\alpha}{2} = \lambda$

$\Rightarrow \quad 2\lambda = 1 + 3\lambda$

or $\quad \lambda = -1$

34. (1) We have

$A(\lambda\hat{i} - \hat{j} + 2\hat{k})$, $B(2\hat{i} + 4\hat{j} - 5\hat{k})$ and $C(\hat{i} - 2\hat{j} + \mu\hat{k})$

$\therefore \quad \overrightarrow{AB} = (2 - \lambda)\,\hat{i} + 5\hat{j} - 7\hat{k}$ and $\overrightarrow{BC} = \hat{i} - 6\hat{j} + (\mu + 5)\hat{k}$

Points, A, B, C are collinear.

$\therefore$ Vectors $\overrightarrow{AB}$ and $\overrightarrow{BC}$ are collinear

So, $\dfrac{2-\lambda}{1}=\dfrac{5}{-6}=\dfrac{-7}{m+5}$

$\Rightarrow \quad \lambda=\dfrac{17}{6}, \mu=\dfrac{17}{5}$

$\Rightarrow \quad \vec{p}=\dfrac{6}{17}\hat{i}+\dfrac{5}{17}\hat{j}+\dfrac{2\sqrt{57}}{17}\hat{k}$

$\Rightarrow \quad |\vec{p}|=\sqrt{\dfrac{36+25+228}{289}}=\sqrt{\dfrac{289}{289}}=1$

35. (1) We have

$x\hat{i}+y\hat{j}+z\hat{k}-(\hat{i}+\hat{j}+\hat{k})=\lambda(\hat{i}-\hat{j}+\hat{k})+\mu(2\hat{i}+\hat{j}+\hat{k});\ \lambda,\mu\in R$

or $(x-1)\ \hat{i}+(y-1)y\hat{j}+(z-1)\hat{k}=\lambda(\hat{i}-\hat{j}+\hat{k})+\mu(2\hat{i}+\hat{j}+\hat{k})$

This means vectors $(x-1)\hat{i}+(y-1)y\hat{j}+(z-1)\hat{k}, \hat{i}-\hat{j}+\hat{k}$ and $2\hat{i}+\hat{j}+\hat{k}$ are coplanar.

Therefore, $\begin{vmatrix} x-1 & y-1 & z-1 \\ 1 & -1 & 1 \\ 2 & 1 & 1 \end{vmatrix}=0$

$\Rightarrow \quad -2(x-1)+(y-1)+3(z-1)=0$

$\Rightarrow \quad 2x-y-3z+2=0$

Numerical Value Type

1. (1.25)

We have $\vec{a}_{\text{Old}}=3p\hat{i}+\hat{j}$

and $\quad \vec{a}_{\text{New}}=(p+1)\hat{i}+\sqrt{10}\hat{j}$

$\Rightarrow \quad |\vec{a}_{\text{Old}}|=|\vec{a}_{\text{New}}|$

$\Rightarrow \quad 9p^2+1=(p+1)^2+10$

$\Rightarrow \quad 4p^2-p-5=0$

$\Rightarrow \quad (4p-5)(p+1)=0$

$\Rightarrow \quad p=\dfrac{5}{4}, -1$

2. (7.25)

We must have $\begin{vmatrix} 1 & 2 & 3 \\ 3 & -1 & m \\ 1 & -2 & 2 \end{vmatrix}=0$

$\Rightarrow \quad 3(-5)-m(-4)+2(-7)=0$

$\Rightarrow \quad 4m=29$

$\Rightarrow \quad m=7.25$

3. (8.8)

Given points are $A(1, 5, 35)$, $B(7, 5, 5)$, $C(1, \lambda, 7)$, $D(2\lambda, 1, 2)$

$\therefore \quad \overrightarrow{AB}=6\hat{i}-30\hat{k}, \overrightarrow{BC}=-6\hat{i}+(\lambda-5)\hat{j}+2\hat{k},$

$\overrightarrow{CD}=(2\lambda-1)\hat{i}+(1-\lambda)\hat{j}-5\hat{k}$

Points are coplanar

$\therefore \quad \begin{vmatrix} 6 & 0 & -30 \\ -6 & \lambda-5 & 2 \\ 2\lambda-1 & 1-\lambda & -5 \end{vmatrix}=0$

$\Rightarrow \quad 5\lambda^2-44\lambda+39=0$

$\Rightarrow \quad$ Sum of values of $\lambda=\dfrac{44}{5}$.

4. (1.6)

We must have, $\overrightarrow{AB}+\overrightarrow{BC}+\overrightarrow{CA}=\vec{0}$

$\Rightarrow \quad (-2\hat{i}+\hat{j}+3\hat{k})+(-\alpha\hat{i}-\beta\hat{j}-8\hat{k})+(4\hat{i}+3\hat{j}+\gamma\hat{k})=\vec{0}$

$\Rightarrow \quad (-\alpha+2)\hat{i}+(4-\beta)\hat{j}+(-5+\gamma)\hat{k}=\vec{0}$

$\Rightarrow \quad \alpha=2, \beta=4, \gamma=5$

5. (1.5)

Since, $|\vec{a}+\vec{b}|=|\vec{a}|+|\vec{b}|$, $\vec{a}$ and $\vec{b}$ are parallel.

Therefore, $\dfrac{1}{2\mu}=\dfrac{\lambda}{1}=\dfrac{1}{1}$

$\Rightarrow \quad \lambda=1, \mu=\dfrac{1}{2}$

$\Rightarrow \quad \lambda+\mu=1+\dfrac{1}{2}=1.5$

6. (5.5)

$|\overrightarrow{BC}|=a=3, |\overrightarrow{CA}|=b=5$ and $|\overrightarrow{BA}|=c=7,$

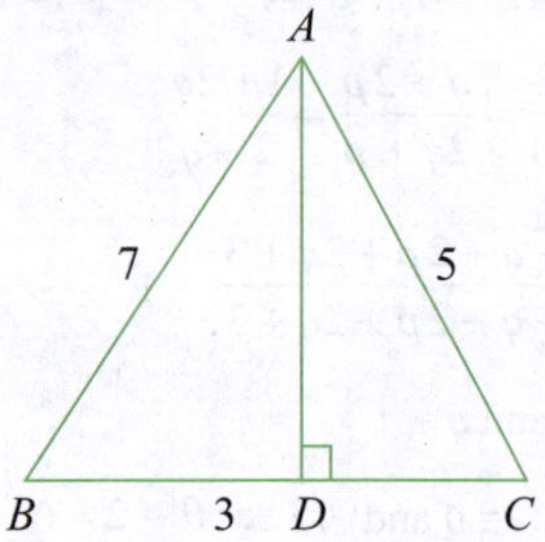

Projection of $\overrightarrow{BA}$ on $\overrightarrow{BC}$

$BD=|c\cos B|$

$=7\left|\dfrac{a^2+c^2-b^2}{2\,ac}\right|$

$=7\left|\dfrac{7^2+3^2-5^2}{2\times7\times3}\right|=\dfrac{11}{2}$

7. (0.75) Given, $\vec{a}+\vec{b}=\vec{c}$

Now vector $\vec{c}$ is along the diagonal of the parallelogra which has adjacent side vectors $\vec{a}$ and $\vec{b}$.

Since $\vec{c}$ is also a unit vector, triangle formed by vecto $\vec{a}$ and $\vec{c}$ is an equilateral triangle. Then

Area of triangle $=\dfrac{\sqrt{3}}{4}$

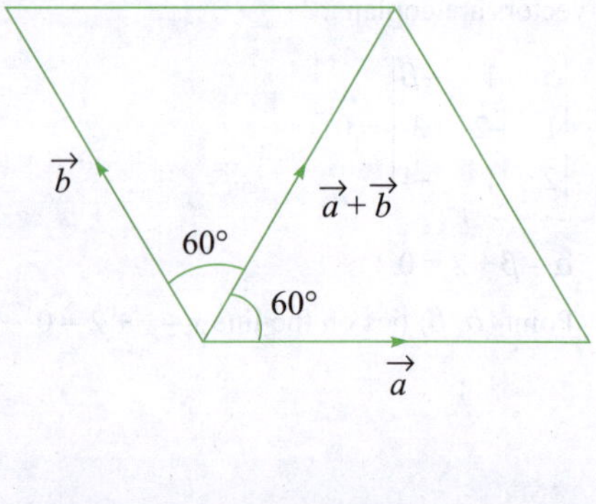

Integer Answer Type

1. (2) L.H.S. $= \vec{d} - \vec{a} + \vec{d} - \vec{b} + \vec{h} - \vec{c} + 3(\vec{g} - \vec{h})$

$$= 2\vec{d} - (\vec{a} + \vec{b} + \vec{c}) = 3\frac{(\vec{a} + \vec{b} + \vec{c})}{3} - 2\vec{h}$$

$$= 2\vec{d} - 2\vec{h} = 2(\vec{d} - \vec{h}) = 2\overrightarrow{HD}$$

$\Rightarrow \quad \lambda = 2$

2. (6) Let $\vec{R}$ be the resultant. Then

$$\vec{R} = \vec{F_1} + \vec{F_2} + \vec{F_3} = (p+1)\hat{i} + 4\hat{j}$$

Given, $|\vec{R}| = 5$, therefore $R^2 = 25$

$\because \quad (p+1)^2 + 16 = 25$ or $p + 1 = \pm 3$

or $\quad p = 2, -4$

3. (2) Let $\vec{a} = x\hat{i} - 3\hat{j} - \hat{k}$ and $\vec{b} = 2x\hat{i} + x\hat{j} - \hat{k}$ be the adjacent sides of the parallelogram.

Now angle between $\vec{a}$ and $\vec{b}$ is acute, i.e.,

$$|\vec{a} + \vec{b}| > |\vec{a} - \vec{b}|$$

$\Rightarrow \quad \left|3x\hat{i} + (x-3)\hat{j} - 2\hat{k}\right|^2 > \left|-x\hat{i} - (x+3)\hat{j}\right|^2$

or $\quad 9x^2 + (x-3)^2 + 4 > x^2 + (x+3)^2$

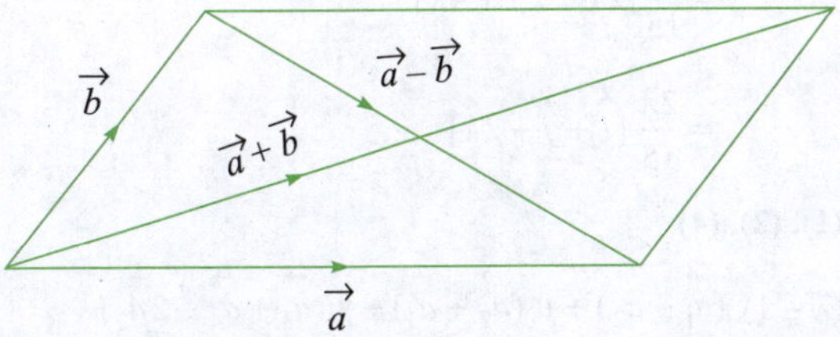

or $\quad 8x^2 - 12x + 4 > 0$

or $\quad 2x^2 - 3x + 1 > 0$

or $\quad (2x-1)(x-1) > 0$

$\Rightarrow \quad x < 1/2$ or $x > 1$

Hence, the least positive integral value is 2.

4. (7) Vectors along the sides are $\vec{a} = \hat{i} + 2\hat{j} + \hat{k}$ and $\vec{b} = 2\hat{i} + 4\hat{j} + \hat{k}$

Clearly the vector along the longer diagonal is

$$\vec{a} + \vec{b} = 3\hat{i} + 6\hat{j} + 2\hat{k}$$

Hence, length of the longer diagonal is

$$|\vec{a} + \vec{b}| = |3\hat{i} + 6\hat{j} + 2\hat{k}| = 7$$

5. (13) Vector $\vec{a} = \hat{i} + 2\hat{j} - \hat{k}, \vec{b} = 2\hat{i} - \hat{j} + \hat{k}, \quad \vec{c} = \lambda\hat{i} + \hat{j} + 2\hat{k}$ are coplanar

$$\Rightarrow \begin{vmatrix} 1 & 2 & -1 \\ 2 & -1 & 1 \\ \lambda & 1 & 2 \end{vmatrix} = 0$$

or $\quad \lambda - 3 + 2(-5) = 0$

or $\quad \lambda = 13$

6. (0) We have

$$\alpha(\hat{i} + 2\hat{j} + 3\hat{k}) + \beta(2\hat{i} + 3\hat{j} + \hat{k}) + \gamma(3\hat{i} + \hat{j} + 2\hat{k}) = -3\hat{i} + 3\hat{k}$$

Comparing the coefficients of $\hat{i}, \hat{j}$ and $\hat{k}$, we get

$$\alpha + 2\beta + 3\gamma = -3$$

$$2\alpha + 3\beta + \gamma = 0$$

$$3\gamma + \beta + 2\gamma = 3$$

Solving, we get

$$\alpha = 2, \beta = -1, \gamma = -1$$

$$\alpha + \beta + \gamma = 0$$

Multiple Correct Answers Type

1. (1), (2), (3), (4)

$x\hat{i} + (x+1)\hat{j} + (x+2)\hat{k}, (x+3)\hat{i} + (x+4)\hat{j} + (x+5)\hat{k}$ and $(x+6)\hat{i} + (x+7)\hat{j} + (x+8)\hat{k}$ are coplanar. We have determinant of their coefficients as $\begin{vmatrix} x & x+1 & x+2 \\ x+3 & x+4 & x+5 \\ x+6 & x+7 & x+8 \end{vmatrix}$

Applying $C_2 \to C_2 - C_1$ and $C_3 \to C_3 - C_1$, we have

$$\begin{vmatrix} x & 1 & 2 \\ x+3 & 1 & 2 \\ x+6 & 1 & 2 \end{vmatrix} = 0$$

Hence, $x \in R$.

2. (1), (4)

Let $\vec{a} = 2\hat{i} + 4\hat{j} - 5\hat{k}$ and $\vec{b} = \hat{i} + 2\hat{j} + 3\hat{k}$.

Then the diagonals of the parallelogram are $\vec{p} = \vec{a} + \vec{b}$ and $\vec{q} = \vec{b} - \vec{a}$,

i.e., $\quad \vec{p} = 3\hat{i} + 6\hat{j} - 2\hat{k}, \ \vec{q} = -\hat{i} - 2\hat{j} + 8\hat{k}$

So, unit vectors along the diagonals are

$$\frac{1}{7}(3\hat{i} + 6\hat{j} - 2\hat{k}) \text{ and } \frac{1}{\sqrt{69}}(-\hat{i} - 2\hat{j} + 8\hat{k}).$$

3. (2), (3)

We have, $\vec{a} = 2p\hat{i} + \hat{j}$

On rotation, let $\vec{b}$ be the vector with components $(p+1)$ and 1 so that $\vec{b} = (p+1)\hat{i} + \hat{j}$.

Now, $|\vec{a}| = |\vec{b}|$ or $a^2 = b^2$

$\Rightarrow \quad 4p^2 + 1 = (p+1)^2 + 1$

or $\quad 4p^2 = (p+1)^2$

or $\quad 2p = \pm(p+1)$

$\Rightarrow \quad 3p = -1$ or $p = 1$

$\therefore \quad p = -1/3$ or $p = 1$

4. (1), (2), (4)

Points $A(\hat{i} + \hat{j}), B(\hat{i} - \hat{j})$ and $C(p\hat{i} + q\hat{j} + r\hat{k})$ are collinear

Now $\overrightarrow{AB} = -2\hat{j}$ and $\overrightarrow{BC} = (p-1)\hat{i} + (q+1)\hat{j} + r\hat{k}$

Vectors $\overrightarrow{AB}$ and $\overrightarrow{BC}$ must be collinear

$\Rightarrow \quad p = 1, r = 0$ and $q \neq -1$

Similarly, if we consider $\overrightarrow{AB}$ and $\overrightarrow{AC}$ then $q \neq 1$

5. (1), (2), (3)

For coplanar vectors, $\begin{vmatrix} 1 & 2 & 3 \\ 0 & \lambda & \mu \\ 0 & 0 & 2\lambda-1 \end{vmatrix} = 0$

or $\quad (2\lambda - 1)\,\lambda = 0$ or $\lambda = 0, \dfrac{1}{2}$

6. (2), (3)

Let $\vec{R}$ be the resultant. Then

$$\vec{R} = \vec{F}_1 + \vec{F}_2 + \vec{F}_3 = (p+1)\,\hat{i} + 4\hat{j}$$

Given, $|\vec{R}| = 5$. Therefore,

$$(p+1)^2 + 16 = 25$$

or $\quad p + 1 = \pm 3$

or $\quad p = 2, -4$

7. (2), (3)

Let $\vec{\alpha} = \hat{i} + x\hat{j} + 3\hat{k},\ \vec{\beta} = 4\hat{i} + (4x-2)\hat{j} + 2\hat{k}$

Given, $2\,|\vec{\alpha}| = |\vec{\beta}|$

or $\quad 2\sqrt{10 + x^2} = \sqrt{20 + 4\,(2x-1)^2}$

or $\quad 10 + x^2 = 5 + (4x^2 - 4x + 1)$

or $\quad 3x^2 - 4x - 4 = 0$

or $\quad x = 2, -\dfrac{2}{3}$

8. (3), (4)

Let $\vec{a}, \vec{b}$ and $\vec{c}$ lie in the x–y plane.

Let $\vec{a} = \hat{i},\ \vec{b} = -\dfrac{1}{2}\hat{i} + \dfrac{\sqrt{3}}{2}\hat{i}$ and $\vec{c} = -\dfrac{1}{2}\hat{i} - \dfrac{\sqrt{3}}{2}\hat{j}$. Therefore,

$$|\vec{p} + \vec{q} + \vec{r}| = |\lambda\vec{a} + \mu\vec{b} + \nu\vec{c}|$$

$$= \left|\lambda\hat{i} + \mu\left(-\frac{1}{2}\hat{i} + \frac{\sqrt{3}}{2}\hat{j}\right) + \nu\left(-\frac{1}{2}\hat{i} - \frac{\sqrt{3}}{2}\hat{j}\right)\right|$$

$$= \left|\left(\lambda - \frac{\mu}{2} - \frac{\nu}{2}\right)\hat{i} + \frac{\sqrt{3}}{2}(\mu - \nu)\hat{j}\right|$$

$$= \sqrt{\left(\lambda - \frac{\mu}{2} - \frac{\nu}{2}\right)^2 + \frac{3}{4}(\mu-\nu)^2}$$

$$= \sqrt{\lambda^2 + \mu^2 + \nu^2 - \lambda\mu - \lambda\nu \quad \mu\nu}$$

$$= \frac{1}{\sqrt{2}}\sqrt{(\lambda-\mu)^2 + (\mu-\nu)^2 + (\nu-\lambda)^2}$$

$$\ge \frac{1}{\sqrt{2}}\sqrt{1+1+4} = \sqrt{3}$$

Hence, $|\vec{p} + \vec{q} + \vec{r}|$ can take a value equal to $\sqrt{3}$ and 2.

9. (2), (4)

Since $\vec{a}$ and $\vec{b}$ are equally inclined to $\vec{c}$, $\vec{c}$ must be of the form

$$t\left(\frac{\vec{a}}{|\vec{a}|} + \frac{\vec{b}}{|\vec{b}|}\right).$$

Now $\quad \dfrac{|\vec{b}|}{|\vec{a}| + |\vec{b}|}\vec{a} + \dfrac{|\vec{a}|}{|\vec{a}| + |\vec{b}|}\vec{b}$

$$= \frac{|\vec{a}||\vec{b}|}{|\vec{a}| + |\vec{b}|}\left(\frac{\vec{a}}{|\vec{a}|} + \frac{\vec{b}}{|\vec{b}|}\right)$$

Also, $\quad \dfrac{|\vec{b}|}{2|\vec{a}| + |\vec{b}|}\vec{a} + \dfrac{|\vec{a}|}{2|\vec{a}| + |\vec{b}|}\vec{b}$

$$= \frac{|\vec{a}||\vec{b}|}{2|\vec{a}| + |\vec{b}|}\left(\frac{\vec{a}}{|\vec{a}|} + \frac{\vec{b}}{|\vec{b}|}\right)$$

Other two vectors cannot be written in the form

$$t\left(\frac{\vec{a}}{|\vec{a}|} + \frac{\vec{b}}{|\vec{b}|}\right).$$

10. (1), (3), (4)

$$\overrightarrow{OA} = -4\hat{i} + 3\hat{k};\ \overrightarrow{OB} = 14\hat{i} + 2\hat{j} - 5\hat{k}$$

$$\hat{a} = \frac{-4\hat{i} + 3\hat{k}}{5};\ \hat{b} = \frac{14\hat{i} + 2\hat{j} - 5\hat{k}}{15}$$

So, bisector, $\vec{r} = \dfrac{\lambda}{15}[-12\hat{i} + 9\hat{j} + 14\hat{i} + 2\hat{j} - 5\hat{k}]$

$$= \frac{\lambda}{15}[2\hat{i} + 2\hat{j} + 4\hat{k}]$$

$$= \frac{2\lambda}{15}[\hat{i} + \hat{j} + 2\hat{k}]$$

11. (1), (2), (4)

$$(\lambda - 1)\,(\vec{a}_1 - \vec{a}_2) + \mu\,(\vec{a}_2 + \vec{a}_3) + \gamma\,(\vec{a}_3 + \vec{a}_4 - 2\vec{a}_2) + \vec{a}_3 + \delta\,\vec{a}_4 = \vec{0}$$

i.e., $(\lambda - 1)\,\vec{a}_1 + (1 - \lambda + \mu - 2\gamma)\vec{a}_2 + (\mu + \gamma + 1)\,\vec{a}_3 + (\gamma + \delta)\,\vec{a}_4 = \vec{0}$

Since $\vec{a}_1, \vec{a}_2, \vec{a}_3$ and $\vec{a}_4$ are linearly independent, we have $\lambda - 1 = 0,\ 1 - \lambda + \mu - 2\gamma = 0,\ \mu + \gamma + 1 = 0$ and $\gamma + \delta = 0$

i.e., $\quad \lambda = 1, \mu = 2\gamma, \mu + \gamma + 1 = 0, \gamma + \delta = 0$

Hence, $\lambda = 1, \mu = -\dfrac{2}{3}, \gamma = -\dfrac{1}{3}, \delta = \dfrac{1}{3}$

12. (1), (3)

We have, $\overrightarrow{AB} = -\hat{i} - \hat{j} - 4\hat{k}$, $\overrightarrow{BC} = -3\hat{i} + 3\hat{j}$ and $\overrightarrow{CA} = 4\hat{i} - 2\hat{j} + 4\hat{k}$. Therefore,

$$|\overrightarrow{AB}| = |\overrightarrow{BC}| = 3\sqrt{2} \text{ and } |\overrightarrow{CA}| = 6$$

Clearly, $|\overrightarrow{AB}|^2 + |\overrightarrow{BC}|^2 = |\overrightarrow{AC}|^2$

Hence, the triangle is right-angled isosceles triangle.

Linked Comprehension Type

1. (3), 2. (2), 3. (3)

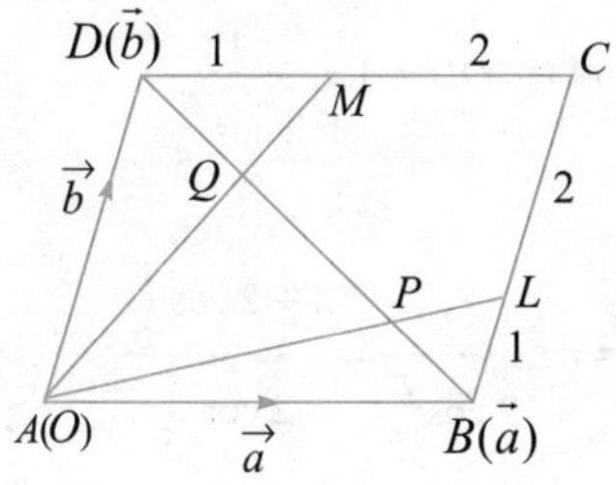

$$\overrightarrow{BL} = \frac{1}{3}\vec{b}$$

$\therefore \quad \overrightarrow{AL} = \vec{a} + \frac{1}{3}\vec{b}$

Let $\overrightarrow{AP} = \lambda\, \overrightarrow{AL}$ and P divides DB in the ratio $\mu : 1 - \mu$. Then

$$\overrightarrow{AP} = \lambda\, \vec{a} + \frac{\lambda}{3}\vec{b} \qquad \text{...(i)}$$

Also $\quad \overrightarrow{AP} = \mu\, \vec{a} + (1-\mu)\,\vec{b} \qquad$ ...(ii)

From (i) and (ii), $\lambda\, \vec{a} + \frac{\lambda}{3}\vec{b} = \mu\, \vec{a} + (1-\mu)\,\vec{b}$

$\therefore \quad \lambda = \mu$ and $\frac{\lambda}{3} = 1 - \mu$

$\therefore \quad \lambda = \frac{3}{4}$

Hence, P divides AL in the ratio 3 : 1 and P divides DB in the ratio 3 : 1.

Similarly, Q divides DB in the ratio 1 : 3.

Thus, $\quad DQ = \frac{1}{4}DB$ and $PB = \frac{1}{4}DB$

$\therefore \quad PQ = \frac{1}{2}DB$, i.e., $PQ : DB = 1 : 2$

4. (0.4), 5. (6)

Let the position vectors of A, B, C and D be $\vec{a}, \vec{b}, \vec{c}$ and $\vec{d}$, respectively. Then,

$OA : CB = 2 : 1$

$\Rightarrow \quad \overrightarrow{OA} = 2\, \overrightarrow{CB}$

$\Rightarrow \quad \vec{a} = 2(\vec{b} - \vec{c}) \qquad$ ...(i)

and $\quad OD : AB = 1 : 3$

$\Rightarrow \quad 3\,\overrightarrow{OD} = \overrightarrow{AB}$

$\Rightarrow \quad 3\vec{d} = (\vec{b} - \vec{a}) = \vec{b} - 2(\vec{b} - \vec{c}) \qquad$ [using (i)]

$\quad = -\vec{b} + 2\vec{c} \qquad$ ...(ii)

Let $OX : XC = \lambda : 1$ and $AX : XD = \mu : 1$

Now, X divides OC in the ratio $\lambda : 1$. Therefore,

$$\text{P.V. of } X = \frac{\lambda\vec{c}}{\lambda+1} \qquad \text{...(iii)}$$

X divides AD in the ratio $\mu : 1$. Therefore,

$$\text{P.V. of } X = \frac{\mu\vec{d} + \vec{a}}{\mu+1} \qquad \text{...(iv)}$$

From (iii) and (iv), we get

$$\frac{\lambda\vec{c}}{\lambda+1} = \frac{\mu\vec{d} + \vec{a}}{\mu+1}$$

or $\quad \left(\frac{\lambda}{\lambda+1}\right)\vec{c} = \left(\frac{\mu}{\mu+1}\right)\vec{d} + \left(\frac{1}{\mu+1}\right)\vec{a}$

or $\quad \left(\frac{\lambda}{\lambda+1}\right)\vec{c} = \left(\frac{\mu}{\mu+1}\right)\left(\frac{-\vec{b}+2\vec{c}}{3}\right) + \left(\frac{1}{\mu+1}\right)2\left(\vec{b}-\vec{c}\right)$

[using (i) and (ii)]

or $\quad \left(\frac{\lambda}{\lambda+1}\right)\vec{c} = \left(\frac{6-\mu}{3(\mu+1)}\right)\vec{b} + \left(\frac{2\mu-6}{3(\mu+1)}\right)\vec{c}$

or $\quad \left(\frac{6-\mu}{3(\mu+1)}\right)\vec{b} + \left(\frac{2\mu-6}{3(\mu+1)} - \frac{\lambda}{\lambda+1}\right)\vec{c} = \vec{0}$

or $\quad \frac{6-\mu}{3(\mu+1)} = 0$ and $\frac{2\mu-6}{3(\mu+1)} - \frac{\lambda}{\lambda+1} = 0$

(as $\vec{b}$ and $\vec{c}$ are non-collinear)

or $\quad \mu = 6, \lambda = \frac{2}{5}$

Hence, $OX : XC = 2 : 5$

6. (4), 7. (6)

Consider the regular hexagon $ABCDEF$ with centre at O (origin).

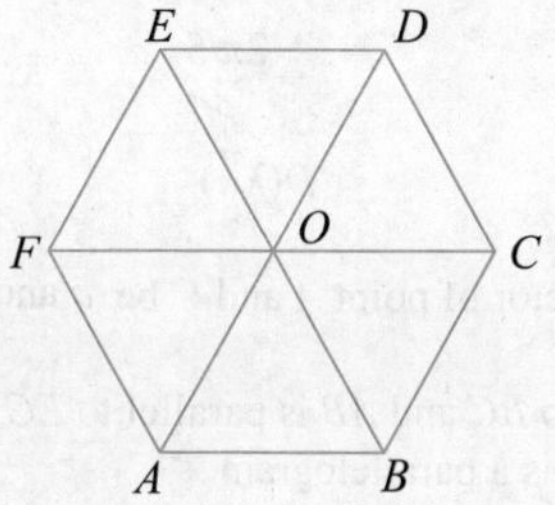

$$\overrightarrow{AD} + \overrightarrow{EB} + \overrightarrow{FC} = 2\,\overrightarrow{AO} + 2\,\overrightarrow{OB} + 2\,\overrightarrow{OC}$$

$$= 2(\overrightarrow{AO} + \overrightarrow{OB}) + 2\,\overrightarrow{OC}$$

$$= 2\,\overrightarrow{AB} + 2\,\overrightarrow{AB}$$

$(\because \overrightarrow{OC} = \overrightarrow{AB})$

$$= 4\,\overrightarrow{AB}$$

$$\vec{R} = \overrightarrow{AB} + \overrightarrow{AC} + \overrightarrow{AD} + \overrightarrow{AE} + \overrightarrow{AF}$$

$$= \overrightarrow{ED} + \overrightarrow{AC} + \overrightarrow{AD} + \overrightarrow{AE} + \overrightarrow{CD}$$

$(\because \overrightarrow{AB} = \overrightarrow{ED}$ and $\overrightarrow{AF} = \overrightarrow{CD})$

$$= (\overrightarrow{AC} + \overrightarrow{CD}) + (\overrightarrow{AE} + \overrightarrow{ED}) + \overrightarrow{AD}$$

$$= \overrightarrow{AD} + \overrightarrow{AD} + \overrightarrow{AD} = 3\,\overrightarrow{AD} = 6\,\overrightarrow{AO}$$

8. (13), 9. (37)

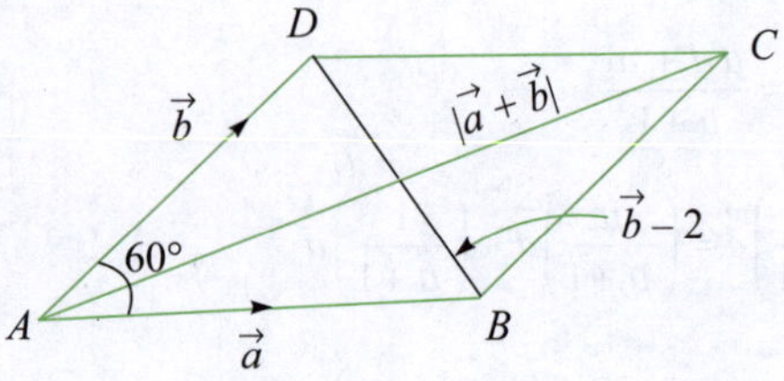

In ΔABD, using the cosine rule, we have

$$BD^2 = AD^2 + AB^2 - 2AD\,AB \cdot \cos 60°$$

$$= 16 + 9 - 2(4)(3) \cdot \frac{1}{2}$$

$$= 13$$

$\therefore \quad BD = |\vec{a} - \vec{b}| = \sqrt{13}$

Also, in ABC, using the cosine rule, we have

$$AC^2 = AB^2 + BC^2 - 2AB \cdot AC \cdot \cos 120°$$

$$= 16 + 9 - 2(4)(3)\left(-\frac{1}{2}\right)$$

$$= 37$$

10. (3), 11. (3)

Given $ABCDE$ is a regular pentagon

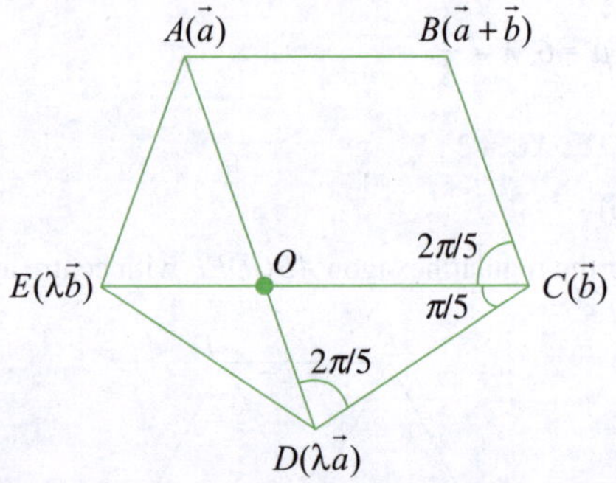

Let position vector of point A and C be $\vec{a}$ and $\vec{b}$, respectively.

AD is parallel to BC and AB is parallel to EC. Therefore,

$AOCB$ is a parallelogram

and position vector of B is $\vec{a} + \vec{b}$

The position vectors of E and D are $\lambda\vec{b}$ and $\lambda\vec{a}$ respectively.

Also $OA = BC = AB = OC = 1$ (let)

Therefore, $AOCB$ is rhombus.

$$\angle ABC = \angle AOC = \frac{3\pi}{5}$$

and $\angle OAB = \angle BCO = \pi - \dfrac{3\pi}{5} = \dfrac{2\pi}{5}$

Further $OA = AE = 1$ and $OC = CD = 1$

Thus, ΔEAO and ΔOCD are isosceles.

In ΔOCD, using sine rule we get $\dfrac{OC}{\sin\frac{2\pi}{5}} = \dfrac{OD}{\sin\frac{\pi}{5}}$

$$\Rightarrow \quad OD = \frac{1}{2\cos\frac{\pi}{5}} = OE$$

$$\Rightarrow \quad AD = OA + OD = 1 + \frac{1}{2\cos\frac{\pi}{5}}$$

$$\Rightarrow \quad \frac{AD}{BC} = 1 + \frac{1}{2\cos\frac{\pi}{5}} = \frac{1 + 2\cos\frac{\pi}{5}}{2\cos\frac{\pi}{5}}$$

And $\dfrac{OE}{OC} = \dfrac{1}{2\cos\frac{\pi}{5}}$

12. (2), 13. (4)

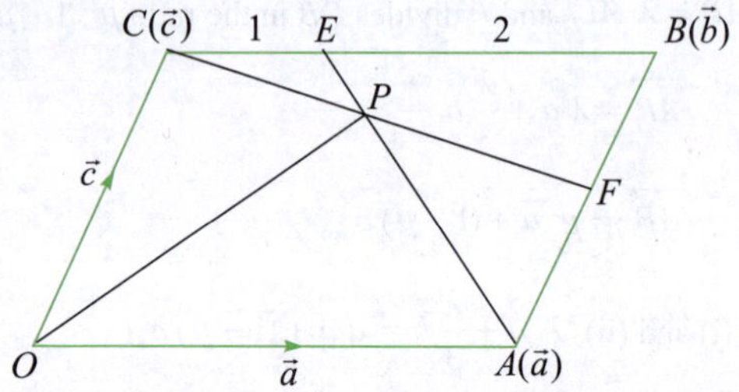

Let the position vector of A and C be $\vec{a}$ and $\vec{c}$ respectively

Therefore,

Position vector of $B = \vec{b} = \vec{a} + \vec{c}$ ...(i

Also Position vector of $E = \dfrac{\vec{b} + 2\vec{c}}{3} = \dfrac{\vec{a} + 3\vec{c}}{3}$ ...(ii

Now point P lies on angle bisector of $\angle AOC$. Thus,

Position vector of point $P = \lambda\left(\dfrac{\vec{a}}{|\vec{a}|} + \dfrac{\vec{b}}{|\vec{b}|}\right)$ ...(iii

Also let P divides EA in ration $\mu : 1$. Therefore,
Position vector of P

$$= \frac{\mu\vec{a} + \frac{\vec{a} + 3\vec{c}}{3}}{\mu + 1} = \frac{(3\mu + 1)\vec{a} + 3\vec{c}}{3(\mu + 1)} \quad ...(iv$$

Comparing (iii) and (iv), we get

$$\lambda\left(\frac{\vec{a}}{|\vec{a}|} + \frac{\vec{c}}{|\vec{c}|}\right) = \frac{(3\mu + 1)\vec{a} + 3\vec{c}}{3(\mu + 1)}$$

$$\Rightarrow \quad \frac{\lambda}{|\vec{a}|} = \frac{3\mu + 1}{3(\mu + 1)} \text{ and } \frac{\lambda}{|\vec{c}|} = \frac{1}{\mu + 1}$$

$$\Rightarrow \quad \frac{3|\vec{c}| - |\vec{a}|}{3|\vec{a}|} = \mu$$

$$\Rightarrow \quad \frac{\lambda}{|\vec{c}|} = \frac{1}{\frac{3|\vec{c}| - |\vec{a}|}{3|\vec{a}|} + 1}$$

$$\Rightarrow \quad \lambda = \frac{3|\vec{a}||\vec{c}|}{3|\vec{c}| + 2|\vec{a}|}$$

Hence, position vector of P is $\dfrac{3|\vec{a}||\vec{c}|}{3|\vec{c}|+2|\vec{a}|}\left(\dfrac{\vec{a}}{|\vec{a}|}+\dfrac{\vec{c}}{|\vec{c}|}\right)$

Let F divides AB in ratio $t:1$, then position vector of F is $\dfrac{t\vec{b}+\vec{a}}{t+1}$.

Now points C, P, F are collinear. Then,

$$\overrightarrow{CF} = m\overrightarrow{CP}$$

$$\Rightarrow \quad \frac{t(\vec{a}+\vec{c})}{t+1} - \vec{c} = m\left\{\frac{3|\vec{a}||\vec{c}|}{3|\vec{c}|+2|\vec{a}|}\left(\frac{\vec{a}}{|\vec{a}|}+\frac{\vec{c}}{|\vec{c}|}\right) - \vec{c}\right\}$$

Comparing coefficients we get

$$\frac{t}{t+1} = m\frac{3|\vec{c}|}{3|\vec{c}|+2|\vec{a}|} \text{ and } \frac{-1}{t+1} = m\frac{|\vec{a}|-3|\vec{c}|}{3|\vec{c}|+2|\vec{a}|}$$

$$t = \frac{3|\vec{c}|}{3|\vec{c}|-|\vec{a}|}$$

Matrix Match Type

1. (4)

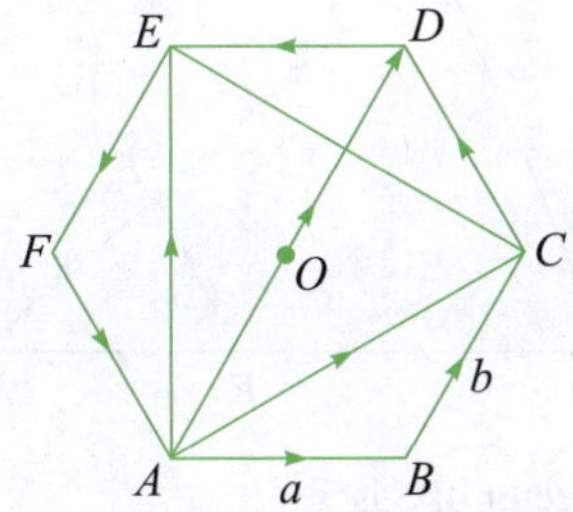

$$\overrightarrow{AB} = \vec{a},\ \overrightarrow{BC} = \vec{b}$$

$$\therefore \quad \overrightarrow{AC} = \overrightarrow{AB} + \overrightarrow{BC} = \vec{a} + \vec{b} \quad \ldots\text{(i)}$$

$$\overrightarrow{AD} = 2\,\overrightarrow{BC} = 2\vec{b} \quad \ldots\text{(ii)}$$

(because AD is parallel to BC and twice its length).

$$\overrightarrow{CD} = \overrightarrow{AD} - \overrightarrow{AC} = 2\vec{b} - (\vec{a}+\vec{b})$$

$$= \vec{b} - \vec{a}$$

$$\overrightarrow{FA} = -\overrightarrow{CD} = \vec{a} - \vec{b} \quad \ldots\text{(iii)}$$

$$\overrightarrow{DE} = -\overrightarrow{AB} = -\vec{a} \quad \ldots\text{(iv)}$$

$$\overrightarrow{EF} = -\overrightarrow{BC} = -\vec{b} \quad \ldots\text{(v)}$$

$$\overrightarrow{AE} = \overrightarrow{AD} + \overrightarrow{DE} = 2\vec{b} - \vec{a} \quad \ldots\text{(vi)}$$

$$\overrightarrow{CE} = \overrightarrow{CD} + \overrightarrow{DE} = \vec{b} - \vec{a} - \vec{a} = \vec{b} - 2\vec{a} \quad \ldots\text{(vii)}$$

2. (2)

Let $A \equiv (0, 0, 0)$. Then $B \equiv (2, 2, 1)$ and $C \equiv (2, 4, 4)$.

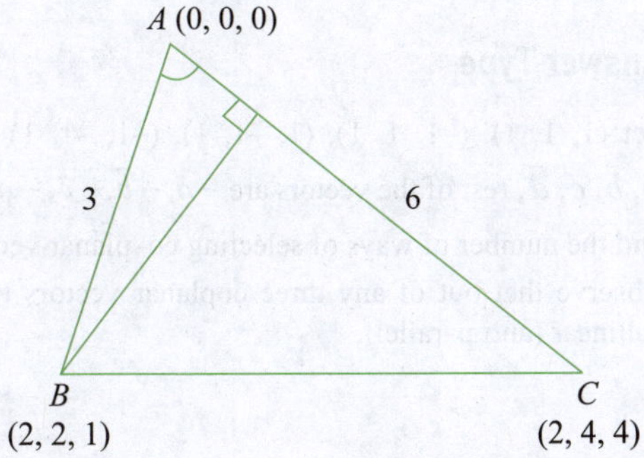

Therefore, $BC = \sqrt{(2-2)^2+(2-4)^2+(1-4)^2} = \sqrt{13}$

Mid-point of BC is (2, 3, 5/2) (say D).

Then the length of the median through A is:

$$AM = \sqrt{4+9+\frac{25}{4}} = \frac{\sqrt{77}}{2}$$

Now, $AB = 3$ and $AC = 6$.

Let, the angle bisector meets BC at M.

Then, using the angle bisector theorem, $\dfrac{BM}{CM} = \dfrac{AB}{AC} = \dfrac{1}{2}$

Therefore, using section formula,

$$M \equiv \left(\frac{2(2)+1(2)}{2+1}, \frac{2(2)+1(4)}{2+1}, \frac{2(1)+1(4)}{2+1}\right) \equiv \left(2, \frac{8}{3}, 2\right)$$

Hence, the length of the angle bisector through A is:

$$AM \equiv \sqrt{4+\frac{64}{9}+4} = \frac{2\sqrt{34}}{3}$$

Now, using cosine rule, $\cos(\angle BAC) = \dfrac{3^2+6^2-13}{2(3)(6)} = \dfrac{8}{9}$

Therefore, $\sin(\angle BAC) = \dfrac{\sqrt{17}}{9}$

Hence, the length of the altitude through B is:

$$BN = AB\sin(\angle BAC) = 3\times\frac{\sqrt{17}}{9} = \frac{\sqrt{17}}{3}.$$

3. (3)

(P) Since vectors are collinear.

$$\therefore \quad \frac{\lambda-2}{4\lambda-9} = \frac{1}{3}$$

$$\Rightarrow \quad 3\lambda - 6 = 4\lambda - 9$$

$$\Rightarrow \quad \lambda = 3$$

(Q) Since vectors are coplanar,

$$\begin{vmatrix} \mu & 1 & 1 \\ 1 & \mu & 1 \\ 1 & 1 & \mu \end{vmatrix} = 0$$

$$\Rightarrow \quad \begin{vmatrix} \mu-1 & 1-\mu & 0 \\ 0 & \mu-1 & 1-\mu \\ 1 & 1 & \mu \end{vmatrix} = 0 \quad (\text{Applying } R_1 \to R_1 - R_2 \text{ and } R_2 \to R_2 - R_3)$$

$$\Rightarrow (\mu-1)^2\begin{vmatrix} 1 & -1 & 0 \\ 0 & 1 & -1 \\ 1 & 1 & \mu \end{vmatrix} = 0$$

$$\Rightarrow (\mu-1)^2(\mu+2) = 0$$

$$\Rightarrow \quad \mu = 1, \mu = -2$$

(R) A, B, C will be collinear if

$$\overrightarrow{AB} \parallel \overrightarrow{AC}$$

$\therefore \quad \frac{1}{2}=\frac{\alpha-4}{-6} \quad \Rightarrow \alpha=1$

So, the smallest positive integral value of α for which A, B, C will be non-collinear is 2.

(S)

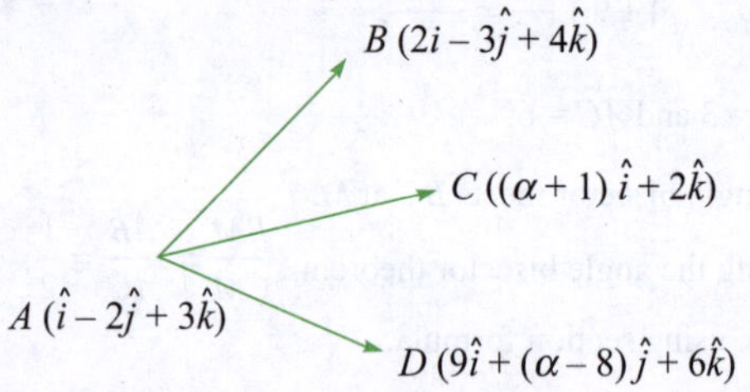

Points A, B, C and D are coplanar

$$\therefore \quad \begin{vmatrix} 1 & -1 & 1 \\ \alpha & 2 & -1 \\ 8 & \alpha-6 & 3 \end{vmatrix}=0$$

$\Rightarrow \quad \alpha^2-2\alpha-8=0$

$\Rightarrow \quad \alpha=-2, 4$

4. (3)

(P) If vectors $\vec{a}, \vec{b}$ and $\vec{c}$ from a triangle, then

$\vec{a}+\vec{b}+\vec{c}=\vec{0} \quad \Rightarrow \vec{c}=-(\vec{a}+\vec{b}) \quad = -4\hat{i}+2\hat{j}-3\hat{k}$

(Q) Vector bisecting the angle between $\vec{a}$ and $\vec{b}$ is

$$\frac{\vec{a}}{|\vec{a}|}+\frac{\vec{b}}{|\vec{b}|}=\frac{2\hat{i}+2\hat{j}-\hat{k}}{3}+\frac{2\hat{i}-4\hat{j}+4\hat{k}}{6}$$

$$=\frac{3\hat{i}+\hat{k}}{3}$$

(R) Resultant of $\vec{a}$ and $\vec{b}$ is:

$\vec{a}+\vec{b}=4\hat{i}-2\hat{j}+3\hat{k}$

(S) Let vector $p\hat{i}+a\hat{j}+r\hat{k}$ is equally inclined to $\vec{a}$ and $\vec{b}$ then

$$\frac{2p+2q-r}{3\sqrt{p^2+q^2+r^2}}=\frac{2p-4q+4r}{6\sqrt{p^2+q^2+r^2}}$$

$\Rightarrow \quad p+4q-3r=0$

Among the options $3\hat{j}+4\hat{k}$ satisfies this.

Archives

Matrix Match Type

1. R → I, II

We have $\overrightarrow{OX}=\sqrt{3}\hat{i}+\hat{j}$ and $\overrightarrow{OY}=\hat{i}+\sqrt{3}\hat{j}$

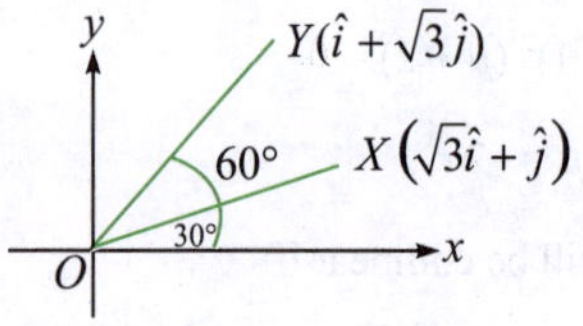

Hence, equation of acute angle bisector of $\overrightarrow{OX}$ and $\overrightarrow{OY}$ is

$y=x$

or $\quad x-y=0$

Now, distance of $\beta\hat{i}+(1-\beta)\hat{i}\equiv Z$ or $(\beta, 1-\beta)$ from $x-y=0$, is

$$\left|\frac{\beta-(1-\beta)}{\sqrt{2}}\right|=\frac{3}{\sqrt{2}} \text{ (given)}$$

$\Rightarrow \quad |2\beta-1|=3$

$\Rightarrow \quad 2\beta-1=\pm 3$

$\Rightarrow \quad 2\beta=4, -2$

$\Rightarrow \quad \beta=2, -1$

$\Rightarrow \quad |\beta|=2, 1$

Note: Solutions of the remaining parts are given in their respective chapters.

Numerical Value Type

1. (1.2)

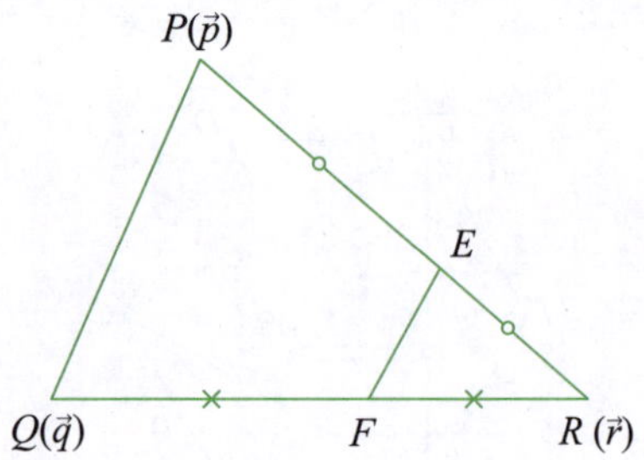

Let position vector of S is $\vec{s}$.

Given that $\overrightarrow{SP}+5\overrightarrow{QS}+6\overrightarrow{SR}=\vec{0}$

$\therefore \quad (\vec{p}-\vec{s})+5(\vec{q}-\vec{s})+6(\vec{r}-\vec{s})=\vec{0}$

So, $\vec{s}=\dfrac{\vec{p}+5\vec{q}+6\vec{r}}{12}$

$$\therefore \quad \frac{|\overrightarrow{EF}|}{|\overrightarrow{ES}|}$$

$$=\frac{\frac{1}{2}|\overrightarrow{PQ}|}{|\overrightarrow{ES}|}$$

$$=\frac{\frac{1}{2}|\vec{q}-\vec{p}|}{\left|\frac{\vec{p}+5\vec{q}+6\vec{r}}{12}-\frac{\vec{p}+\vec{r}}{2}\right|}$$

$$=\frac{\frac{1}{2}|\vec{q}-\vec{p}|}{\frac{5}{12}|\vec{q}-\vec{p}|}=1.2$$

Integer Answer Type

1. (5) Let (1, 1, 1), (–1, 1, 1), (1, –1, 1), (–1, –1, 1) be vectors $\vec{a}, \vec{b}, \vec{c}, \vec{d}$, rest of the vectors are $-\vec{a}, -\vec{b}, -\vec{c}, -\vec{d}$ and let us find the number of ways of selecting co–planar vectors.

Observe that out of any three coplanar vectors two will be collinear (anti parallel).

Number of ways of selecting the anti-parallel pair = 4

Number of ways of selecting the third vector = 6

Total = 24

Number of non-coplanar selections

$= {}^8C_3 - 24 = 32 = 2^5, p = 5$

$\therefore \quad p = 5$

2. (9) According to question $\vec{s} = 4\vec{p} + 3\vec{q} + 5\vec{r}$ and

$$\vec{s} = x(-\vec{p} + \vec{q} + \vec{r}) + y(\vec{p} - \vec{q} + \vec{r}) + z(-\vec{p} - \vec{q} + \vec{r})$$

$\therefore \quad -x + y - z = 4 \quad \ldots(i)$

$x - y - z = 3 \quad \ldots(ii)$

$x + y + z = 5 \quad \ldots(iii)$

Adding (i) and (ii), we get

$$z = -\frac{7}{2}$$

Adding (ii) and (iii), we get

$$x = 4$$

Adding (i) and (iii), we get

$$y = 9/2$$

3. (–2)

Vector $\vec{X}, \vec{Y}$ and $\vec{Z}$ are coplanar

$$\therefore \quad \begin{vmatrix} \alpha & \beta & -1 \\ -1 & \alpha & \beta \\ \beta & -1 & \alpha \end{vmatrix} = 0$$

$\Rightarrow \quad \alpha^3 + \beta^3 - 1 + 3\alpha\beta = 0$

$\Rightarrow \quad \alpha^3 + \beta^3 + (-1)3 = 3\alpha\beta(-1)$

$\Rightarrow \quad \alpha^3 + \beta + (-1) = 0$ (as α and β are distinct)

$\Rightarrow \quad \alpha^3 + \beta - 3 = -2$

Products of Vectors

Concept Application Exercises

Exercise 3.1

1. (3) Resultant force

$$\vec{F} = \vec{P_1} + \vec{P_2} + \vec{P_3} = 2\hat{j} - \hat{k}$$

And displacement $= \overrightarrow{AB}$

$= \text{P.V. of } B - \text{P.V. of } A$

$= (6\hat{i} + \hat{j} - 3\hat{k}) - (4\hat{i} - 3\hat{j} - 2\hat{k})$

$= 2\hat{i} + 4\hat{j} - \hat{k}$

$\therefore$ Work done $= \vec{F} \cdot \overrightarrow{AB}$

$= (2\hat{j} - \hat{k}) \cdot (2\hat{i} + 4\hat{j} - \hat{k})$

$= 9$ units

2. $(\vec{a} + \vec{b}) \cdot (\vec{a} - \vec{b}) = 8$

$\Rightarrow \quad \vec{a} \cdot \vec{a} - \vec{a} \cdot \vec{b} + \vec{b} \cdot \vec{a} - \vec{b} \cdot \vec{b} = 8$

$\Rightarrow \quad |\vec{a}|^2 - |\vec{b}|^2 = 8$

$\Rightarrow \quad (8|\vec{b}|)^2 - |\vec{b}|^2 = 8 \qquad [\because |\vec{a}| = 8|\vec{b}|]$

$\Rightarrow \quad |\vec{b}| = \sqrt{\dfrac{8}{63}}$

$\Rightarrow \quad |\vec{a}| = 8|\vec{b}| = \dfrac{8 \times 2\sqrt{2}}{3\sqrt{7}} = \dfrac{16\sqrt{2}}{3\sqrt{7}}$

3. $(|\vec{a}|\vec{b} + |\vec{b}|\vec{a}) \cdot (|\vec{a}|\vec{b} - |\vec{b}|\vec{a})$

$= |\vec{a}|^2 \vec{b} \cdot \vec{b} - |\vec{a}||\vec{b}|\vec{b} \cdot \vec{a} + |\vec{b}||\vec{a}|\vec{a} \cdot \vec{b} - |\vec{b}|^2 \vec{a} \cdot \vec{a}$

$= |\vec{a}|^2|\vec{b}|^2 - |\vec{b}|^2|\vec{a}|^2$

$= 0$

Hence, $|\vec{a}|\vec{b} + |\vec{b}|\vec{a}$ and $|\vec{a}|\vec{b} - |\vec{b}|\vec{a}$ are perpendicular to each other.

4. (1) $\hat{n} = a_1\hat{i} + a_2\hat{j} + a_3\hat{k}$, where $a_1^2 + a_2^2 + a_3^2 = 1$

Given that $\vec{u} \cdot \hat{n} = 0 \Rightarrow a_1 + a_2 = 0$

Also, $\vec{v} \cdot \hat{n} = 0 \Rightarrow a_1 - a_2 = 0$

$\Rightarrow \quad a_1 = a_2 = 0$

and $\quad a_3 = 1$ or -1

$\therefore \quad \hat{n} = \hat{k}$ or $-\hat{k}$

Hence, $|\vec{w}\hat{n}| = 3$

5. $\overrightarrow{AD} = \overrightarrow{AB} + \overrightarrow{BC} + \overrightarrow{CD} = \vec{a} + \vec{b} + \vec{c}$

$\overrightarrow{AC} = \overrightarrow{AB} + \overrightarrow{BC} = \vec{a} + \vec{b}$ or $\overrightarrow{CA} = -(\vec{a} + \vec{b})$

$\overrightarrow{BD} = \overrightarrow{BC} + \overrightarrow{CD} = \vec{b} + \vec{c}$

$\therefore \quad \overrightarrow{AB} \cdot \overrightarrow{CD} + \overrightarrow{BC} \cdot \overrightarrow{AD} + \overrightarrow{CA} \cdot \overrightarrow{BD}$

$= \vec{a} \cdot \vec{c} + \vec{b} \cdot (\vec{a} + \vec{b} + \vec{c}) + (-\vec{a} - \vec{b}) \cdot (\vec{b} + \vec{c})$

$= \vec{a} \cdot \vec{c} + \vec{b} \cdot \vec{a} + \vec{b} \cdot \vec{b} + \vec{b} \cdot \vec{c} - \vec{a} \cdot \vec{b} - \vec{a} \cdot \vec{c}$

$- \vec{b} \cdot \vec{b} - \vec{b} \cdot \vec{c} = 0$

6. (3) $3\vec{p} + \vec{q}$ and $5\vec{p} - 3\vec{q}$ are perpendicular. Therefore,

$(3\vec{p} + \vec{q}) \cdot (5\vec{p} - 3\vec{q}) = 0$

$\Rightarrow \quad 15\vec{p}^2 - 3\vec{q}^2 = 4\vec{p} \cdot \vec{q} \qquad \ldots\text{(i)}$

$2\vec{p} + \vec{q}$ and $4\vec{p} - 2\vec{q}$ are perpendicular. Therefore,

$(2\vec{p} + \vec{q}) \cdot (4\vec{p} - 2\vec{q}) = 0$

$\Rightarrow \quad 8\vec{p}^2 = 2\vec{q}^2$

$\Rightarrow \quad \vec{q}^2 = 4\vec{p}^2 \qquad \ldots\text{(ii)}$

Now, $\cos\theta = \dfrac{\vec{p} \cdot \vec{q}}{|\vec{p}||\vec{q}|}$

Substituting $\vec{q}^2 = 4\vec{p}^2$ in (i), $3\vec{p}^2 = 4\vec{p} \cdot \vec{q}$, we get

$\cos\theta = \dfrac{3}{4} \cdot \dfrac{\vec{p}^2}{|\vec{p}|2|\vec{p}|} = \dfrac{3}{8}$

or $\quad \theta = \cos^{-1}\dfrac{3}{8}$

7. $\vec{A} \cdot (\alpha\vec{A} + \vec{B}) = \vec{B} \cdot (\alpha\vec{A} + \vec{B})$

or $\quad \alpha + \vec{A} \cdot \vec{B} = \alpha\vec{A} \cdot \vec{B} + 1$

or $\quad (\vec{A} \cdot \vec{B})(1 - \alpha) = (1 - \alpha)$

Since $\vec{A} \cdot \vec{B} \neq 0 \Rightarrow \alpha = 1$

8. (1) Let $\vec{a} = \hat{i} + 3\hat{j} + 7\hat{k}$ and $\vec{b} = 7\hat{i} - \hat{j} + 8\hat{k}$.

Now, projection of vector $\vec{a}$ on $\vec{b}$ is given by

$\dfrac{1}{|\vec{b}|}(\vec{a} \cdot \vec{b}) = \dfrac{1}{\sqrt{7^2 + (-1)^2 + 8^2}}\{1(7) + 3(-1) + 7(8)\}$

$= \dfrac{7 - 3 + 56}{\sqrt{49 + 1 + 64}} = \dfrac{60}{\sqrt{114}}$

9. We have $\vec{a} + \vec{b} + \vec{c} = 0$

$\Rightarrow \quad (\vec{a} + \vec{b} + \vec{c})^2 = 0$

$\Rightarrow \quad |\vec{a}|^2 + |\vec{b}|^2 + |\vec{c}|^2 + 2(\vec{a} \cdot \vec{b} + \vec{b} \cdot \vec{c} + \vec{c} \cdot \vec{a}) = 0$

$\Rightarrow \quad \vec{c} \cdot \vec{b} + \vec{b} \cdot \vec{c} + \vec{c} \cdot \vec{a} = -\dfrac{3}{2} \qquad (\because |\vec{a}| = |\vec{b}| = |\vec{c}| = 1)$

10. $|4\vec{a} + 3\vec{b}| = \sqrt{(4\vec{a} + 3\vec{b}) \cdot (4\vec{a} + 3\vec{b})}$

$= \sqrt{16|\vec{a}|^2 + 9|\vec{b}|^2 + 24\vec{a} \cdot \vec{b}}$

$= \sqrt{144 + 144 + 24 \times 3 \times 4 \times \left(\dfrac{-1}{2}\right)}$

$= 12$

11. Given, $\vec{a}+\vec{b}+\vec{c}=\vec{0}$

$\vec{a}+\vec{b}=-\vec{c}$

$(\vec{a}+\vec{b})\cdot(\vec{a}+\vec{b})=(-\vec{c})\cdot(-\vec{c})$

$\Rightarrow \quad a^2+b^2+2\vec{a}\cdot\vec{b}=c^2$

$\Rightarrow \quad 9+25+2(\vec{a}\cdot\vec{b})=49$

$\Rightarrow \quad \vec{a}\cdot\vec{b}=15/2$

$\Rightarrow \quad ab\cos\theta=15/2$ or $3\cdot 5\cos\theta=15/2$

$\Rightarrow \quad \cos\theta=1/2 \Rightarrow \theta=\pi/3$

12. (2) $\vec{a}+\vec{b}+\vec{c}=\vec{x}$

Taking dot with $\vec{x}$ on both sides, we get

$\vec{x}\cdot\vec{a}+\vec{x}\cdot\vec{b}+\vec{x}\cdot\vec{c} = \vec{x}\cdot\vec{x}=|x|^2=4$

or $\quad 1+\frac{3}{2}+\vec{x}\cdot\vec{c}=4$ or $\vec{x}\cdot\vec{c}=\frac{3}{2}$

If θ is the angle between $\vec{c}$ and $\vec{x}$, then

$|\vec{x}||\vec{c}|\cos\theta=3/2$

$\Rightarrow \quad \cos\theta=3/4$ or $\theta=\cos^{-1}(3/4)$

13. (3) Let θ be an angle between unit vectors $\vec{a}+\vec{b}$. Then

$\vec{a}\cdot\vec{b}=\cos\theta$

$\therefore \quad |\vec{a}+\vec{b}|^2=|\vec{a}|^2+|\vec{b}|^2+2\vec{a}\cdot\vec{b}$

$=2+2\cos\theta=4\cos^2\theta/2$

or $\quad |\vec{a}+\vec{b}|=2\cos\frac{\theta}{2}$

Similarly, $|\vec{a}-\vec{b}|=2\sin\frac{\theta}{2}$

$\Rightarrow \quad |\vec{a}+\vec{b}|+|\vec{a}-\vec{b}|=2\left(\cos\frac{\theta}{2}+\sin\frac{\theta}{2}\right)\le 2\sqrt{2}$

14. $|\vec{a}+\vec{b}|^2+|\vec{a}-\vec{b}|^2=2(|\vec{a}|^2+|\vec{b}|^2)$

$\Rightarrow \quad 100+64=2(25+|\vec{b}|^2)$

or $\quad 82=25+|\vec{b}|^2$

or $\quad |\vec{b}|^2=57$

15. (2) Let A be the origin, and the position vectors of B, C and D be $\vec{b},\vec{c},\vec{d}$.

$\overline{AB}\cdot\overline{CD}=k(|\overline{AD}|^2+|\overline{BC}|^2-|\overline{AC}|^2-|\overline{BD}|^2)$

or $\quad (\vec{b})\cdot(\vec{d}-\vec{c})$

$=k[(\vec{d})^2+(\vec{c}-\vec{b})^2-(\vec{c})^2-(\vec{d}-\vec{b})^2]$

or $\quad \vec{b}\cdot\vec{d}-\vec{b}\cdot\vec{c}=k(-2\vec{b}\cdot\vec{c}+2\vec{b}\cdot\vec{d})$

or $\quad k=1/2$

16. (2) $|\vec{a}-\vec{b}|^2+|\vec{a}-\vec{c}|^2=8$

$\Rightarrow \quad |\vec{a}|^2+|\vec{b}|^2-2\vec{a}\cdot\vec{b}+|\vec{a}|^2+|\vec{c}|^2-2\vec{a}\cdot\vec{c}=8$

$\Rightarrow \quad 4-2(\vec{a}\cdot\vec{b}+\vec{a}\cdot\vec{c})=8$

$\Rightarrow \quad \vec{a}\cdot\vec{b}+\vec{a}\cdot\vec{c}=-2$

$|\vec{a}+2\vec{b}|^2+|\vec{a}+2\vec{c}|^2$

$=|a^2|+4|\vec{b}|^2+4\vec{a}\cdot\vec{b}+|\vec{a}|^2+4|\vec{c}|^2+4\vec{a}\cdot\vec{c}$

$=10+4(\vec{a}.\vec{b}+\vec{a}.\vec{c})$

$=10-8$

$=2$

17. (3) Let $\vec{\beta}_1=\lambda\vec{\alpha}$

Now, $\vec{\beta}_2=\vec{\beta}-\vec{\beta}_1$

$=(\hat{i}+2\hat{j}-4\hat{k})-\lambda(4\hat{i}+3\hat{j}+5\hat{k})$

$=(1-4\lambda)\hat{i}+(2-3\lambda)\hat{j}-(5\lambda+4)\hat{k}$

Also, $\vec{\beta}_2.\vec{\alpha}=0$

$\Rightarrow \quad 4(1-4\lambda)+3(2-3\lambda)-5(5\lambda+4)=0$

$\Rightarrow \quad 50\lambda=-10$

$\Rightarrow \quad \lambda=\frac{-1}{5}$

$\therefore \quad \vec{\beta}_2=\left(1+\frac{4}{5}\right)\hat{i}+\left(1+\frac{3}{5}\right)\hat{j}-(-1+4)\hat{k}$

or $\quad \vec{\beta}_2=\frac{9}{5}\hat{i}+\frac{13}{5}\hat{j}-3\hat{k}$

So, $\quad 5\vec{\beta}_2.(\hat{i}+\hat{j}+\hat{k})=9+13-15=7$

18. $\vec{x}=\lambda\vec{a}+\mu\vec{b}$ (λ and μ are scalars)

$\vec{x}=\hat{i}(2\lambda+\mu)+\hat{j}(2\mu-\lambda)+\hat{k}(\lambda-\mu)$

Since $\vec{x}\cdot(3\hat{i}+2\hat{j}-\hat{k})=0$

$\Rightarrow \quad 3\lambda+8\mu=0 \quad$ ...(i)

Also, Projection of $\vec{x}$ on $\vec{a}$ is $\frac{17\sqrt{6}}{2}$

$\therefore \quad \frac{\vec{x}\cdot\vec{a}}{|\vec{a}|}=\frac{17\sqrt{6}}{2}$

$\Rightarrow \quad 6\lambda-\mu=51 \quad$ ...(ii)

Solving (1) and (2), we get

$\lambda=8,\ \mu=-3$

$\therefore \quad \vec{x}=13\hat{i}-14\hat{j}+11\hat{k}$

$\Rightarrow \quad |\vec{x}|^2=486$

Exercise 3.2

1.

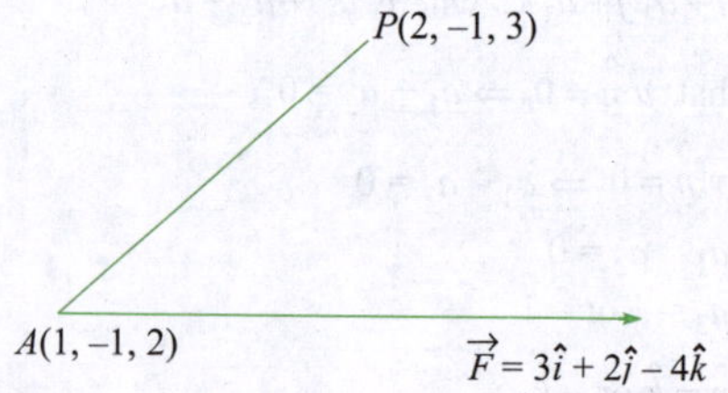

$\vec{F}=3\hat{i}+2\hat{j}-4\hat{k}$

A is (1, –1, 2), P is (2, –1, 3)

$\therefore \quad \vec{PA}$ = P.V. of A – P.V. of P

$=(\hat{i}-\hat{j}+2\hat{k})-(2\hat{i}-\hat{j}+3\hat{k})$

$=-\hat{i}-\hat{k}$

Required vector moment $=\vec{PA}\times\vec{F}$

$=(-\hat{i}-\hat{k})\times(3\hat{i}+2\hat{j}-4\hat{k})$

$=\begin{vmatrix}\hat{i} & \hat{j} & \hat{k}\\ -1 & 0 & -1\\ 3 & 2 & -4\end{vmatrix}$

$=2\hat{i}-7\hat{j}-2\hat{k}$

2. (2) $\vec{a}\times\vec{b}=\begin{vmatrix}\hat{i} & \hat{j} & \hat{k}\\ 2 & 3 & -5\\ m & n & 12\end{vmatrix}$

$=(36+5n)\hat{i}-(24+5m)\hat{j}+(2n-3m)\hat{k}$

$=\vec{0}$

$\therefore\quad m=\frac{-24}{5},\ n=\frac{-36}{5}$

3. $\vec{a}\times\vec{b}=\begin{vmatrix}\hat{i} & \hat{j} & \hat{k}\\ 3 & 2 & x\\ 1 & -1 & 1\end{vmatrix}$

$=(2+x)\hat{i}-(3-x)\hat{j}-5\hat{k}$

$\therefore\quad |\vec{a}\times\vec{b}|=\sqrt{(2+x)^2+(3-x)^2+25}$

$=\sqrt{2x^2-2x+38}$

$=\sqrt{2}\sqrt{x^2-x+19}$

$=\sqrt{2}\sqrt{\left(x-\frac{1}{2}\right)^2+\frac{75}{4}}$

$\geq\sqrt{2}\sqrt{\frac{75}{4}}=\frac{5\sqrt{3}}{\sqrt{2}}$

$\therefore\quad r\geq 5\sqrt{\frac{3}{2}}$

4. (1) We have $\vec{v}\times\vec{w}=\vec{u}+\lambda\vec{v}$ ...(i)

First, we find the value of λ.

In (i), taking dot product with $\vec{v}$, we get

$0=(2-1+2)+\lambda(6)$

$\Rightarrow\quad \lambda=-\frac{1}{2}$

Thus, (i) becomes,

$\vec{v}\times\vec{w}=\vec{u}-\frac{1}{2}\vec{v}$ ...(ii)

Now, in above equation, taking dot product with $\vec{w}$, we get

$\Rightarrow\quad 0=\vec{u}.\vec{w}-\frac{1}{2}(2)$

$\Rightarrow\quad \vec{u}.\vec{w}=1$

5. Since $\vec{a}\times\vec{b}=\vec{b}\times\vec{c}\neq 0$, we have

$\vec{a}\times\vec{b}-\vec{b}\times\vec{c}=0$

or $\quad\vec{a}\times\vec{b}+\vec{c}\times\vec{b}=0$

or $\quad(\vec{a}+\vec{c})\times\vec{b}=0$

Hence, $\vec{a}+\vec{c}$ is parallel to $\vec{b}$. Thus,

$\vec{a}+\vec{c}=k\vec{b}$

6. (3) $\vec{a}\times\vec{b}=\begin{vmatrix}\hat{i} & \hat{j} & \hat{k}\\ 2 & 3 & -1\\ -1 & 2 & -4\end{vmatrix}=-10\hat{i}+9\hat{j}+7\hat{k}$

$\vec{a}\times\vec{c}=\begin{vmatrix}\hat{i} & \hat{j} & \hat{k}\\ 2 & 3 & -1\\ 1 & 1 & 1\end{vmatrix}=4\hat{i}-3\hat{j}-\hat{k}$

$\Rightarrow\quad (\vec{a}\times\vec{b})\cdot(\vec{a}\times\vec{c})=-40-27-7=-74$

7. We have $\vec{a}\cdot\vec{b}=\vec{a}\cdot\vec{c}$. Therefore,

$\vec{a}\cdot\vec{b}-\vec{a}\cdot\vec{c}=0$ or $\vec{a}\cdot(\vec{b}-\vec{c})=0$

Therefore, there are three possibilities: (i) $\vec{a}=\vec{0}$, (ii) $\vec{b}-\vec{c}=\vec{0}$ and (iii) $\vec{a}$ is perpendicular to $\vec{b}-\vec{c}$.

Again, $\vec{a}\times\vec{b}=\vec{a}\times\vec{c}$. Therefore,

$\vec{a}\times\vec{b}-\vec{a}\times\vec{c}=0$

or $\quad\vec{a}\times(\vec{b}-\vec{c})=\vec{0}$

Therefore, again there are three possibilities:

(i) $\vec{a}=0$, (ii) $\vec{b}-\vec{c}=\vec{0}$ and (iii) $\vec{a}$ is parallel to $\vec{b}-\vec{c}$.

Now $\vec{a}$ is given to be a non-zero vector. Therefore, we have the following possibilities left:

(i) $\vec{b}-\vec{c}=0$.

(ii) $\vec{a}$ is perpendicular to $\vec{b}-\vec{c}$ and $\vec{a}$ is parallel to $\vec{b}-\vec{c}$, which is absurd.

Therefore, the only possibility left is $\vec{b}-\vec{c}=\vec{0}$ or $\vec{b}=\vec{c}$.

8. $(\vec{a}-\vec{b})\times(\vec{a}+\vec{b})$

$=\vec{a}\times\vec{a}+\vec{a}\times\vec{b}-\vec{b}\times\vec{a}-\vec{b}\times\vec{b}$

$=\vec{a}\times\vec{a}+\vec{a}\times\vec{b}+\vec{a}\times\vec{b}-\vec{b}\times\vec{b}$

$=\vec{0}+2\vec{a}\times\vec{b}-\vec{0}=2\vec{a}\times\vec{b}$

Geometrically, the vector area of a parallelogram whose sides are along vectors $\vec{a}$ and $\vec{b}$ is $\vec{a}\times\vec{b}$.

Also diagonals are along vectors $\vec{a}-\vec{b}$ and $\vec{a}+\vec{b}$ and the vector area in terms of diagonal vectors is $\frac{1}{2}[(\vec{a}-\vec{b})\times(\vec{a}+\vec{b})]$.

9. (3) $\vec{z}+\vec{z}\times\vec{x}=\vec{y}$ or $|\vec{z}+\vec{z}\times\vec{x}|^2=|\vec{y}|^2$

$\Rightarrow\quad |\vec{z}|^2+|\vec{z}|^2|\vec{x}|^2\sin^2\theta=1$ (because $\vec{z}\cdot(\vec{z}\times\vec{x})=0$)

$\Rightarrow\quad |\vec{z}|^2(1+\sin^2\theta)=1$

$\Rightarrow\quad |\vec{z}|=\frac{1}{\sqrt{1+\sin^2\theta}}=\frac{2}{\sqrt{7}}$

$\Rightarrow\quad \sin\theta=\sqrt{3}/2$

$\Rightarrow\quad \theta=60°$

10. Let $\vec{a}=a_1\hat{i}+a_2\hat{j}+a_3\hat{k}$. Therefore,

$\vec{a}\cdot\hat{i}=a_1,\ \vec{a}\cdot\hat{j}=a_2$ and $\vec{a}\cdot\hat{k}=a_3$

and $\quad\vec{a}\times\hat{i}=(a_1\hat{i}+a_2\hat{j}+a_3\hat{k})\times\hat{i}=-a_2\hat{k}+a_3\hat{j}$

Similarly, $\vec{a}\times\hat{j}=a_1\hat{k}-a_3\hat{i}$ and $\vec{a}\times\hat{k}=-a_1\hat{j}+a_2\hat{i}$

$\therefore\quad (\vec{a}\cdot\hat{i})(\vec{a}\times\hat{i})+(\vec{a}\cdot\hat{j})(\vec{a}\times\hat{j})+(\vec{a}\cdot\hat{k})(\vec{a}\times\hat{k})$

$=-a_1a_2\hat{k}+a_1a_3\hat{j}+a_1a_2\hat{k}-a_3a_2\hat{i}+a_3a_2\hat{i}-a_3a_1\hat{j}$

$=\vec{0}$

11. $\vec{a}+\vec{b}+\vec{c}=\vec{0}$

$\Rightarrow\quad \vec{a}\times\vec{b}=\vec{b}\times\vec{c}=\vec{c}\times\vec{a}$

$\Rightarrow \quad \vec{a}\times\vec{b}+\vec{b}\times\vec{c}+\vec{c}\times\vec{a}=3\vec{a}\times\vec{b}$

$\Rightarrow \quad 2\vec{b}\times\vec{a}+\vec{b}\times\vec{c}+\vec{c}\times\vec{a}=0 \Rightarrow \lambda=2$

12. We have $\vec{a}\cdot\vec{b}=\vec{a}\cdot\vec{c}=0.$

This implies that $\vec{a}$ is perpendicular to both $\vec{b}$ and $\vec{c}$.

Thus, $\vec{a}$ is a unit vector perpendicular to both $\vec{b}$ and $\vec{c}$. Hence,

$$\vec{a}=\pm\frac{\vec{b}\times\vec{c}}{|\vec{b}\times\vec{c}|}=\pm\frac{\vec{b}\times\vec{c}}{|\vec{b}||\vec{c}|\sin\pi/3}$$

$$=\pm 2(\vec{b}\times\vec{c})$$

13. (1) Since $(\vec{a}\times\vec{b})^2+(\vec{a}\cdot\vec{b})^2=144$, if the angle between $\vec{a}$ and $\vec{b}$ is θ, then

$$|\vec{a}|^2|\vec{b}|^2\sin^2\theta+|\vec{a}|^2|\vec{b}|^2\cos^2\theta=144$$

or $\quad |\vec{a}|^2|\vec{b}|^2=144$

or $\quad |\vec{a}||\vec{b}|=12$

or $\quad 4|\vec{b}|=12$

or $\quad |\vec{b}|=3$

14. (2) We have, $|\vec{a}+\vec{b}|=\sqrt{3}$

or $\quad |\vec{a}+\vec{b}|^2=3$

or $\quad |\vec{a}|^2+|\vec{b}|^2+2(\vec{a}\cdot\vec{b})=3$

or $\quad 1+1+2(\vec{a}\cdot\vec{b})=3$

or $\quad \vec{a}\cdot\vec{b}=1/2 \qquad \ldots(i)$

Now, $\vec{c}-\vec{a}-2\vec{b}=3(\vec{a}\times\vec{b})$

or $\quad (\vec{c}-\vec{a}-2\vec{b})\cdot\vec{b}=3\{(\vec{a}\times\vec{b})\cdot\vec{b}\}$

or $\quad \vec{c}\cdot\vec{b}-\vec{a}\cdot\vec{b}-2(\vec{b}\cdot\vec{b})=0$

or $\quad \vec{c}\cdot\vec{b}-\frac{1}{2}-2\times1=0 \qquad$ [Using (i)]

or $\quad \vec{c}\cdot\vec{b}=5/2$

15. (3) Normal vector to the plane containing vectors $\hat{i}+\hat{j}+\hat{k}$ and $\hat{i}+2\hat{j}+3\hat{k}$ is

$$\vec{n}=\begin{vmatrix}\hat{i} & \hat{j} & \hat{k}\\ 1 & 1 & 1\\ 1 & 2 & 3\end{vmatrix}$$

or $\quad \vec{n}=\hat{i}-2\hat{j}+\hat{k}$

So, projection of $(2\hat{i}+3\hat{j}+\hat{k})$ on $\vec{n}$

$$=\left|\frac{(2\hat{i}+3\hat{j}+\hat{k}).(\hat{i}-2\hat{j}+\hat{k})}{\sqrt{1+4+1}}\right|$$

$$=\frac{3}{\sqrt{6}}=\sqrt{\frac{3}{2}}$$

16. (3) For quadrilateral, the given points must be coplanar.

Therefore, vectors $\overrightarrow{AB}, \overrightarrow{AC}, \overrightarrow{AD}$ are coplanar.

$$\therefore \quad \begin{vmatrix}-1 & 1 & 4\\ -4 & -4 & 4\\ \lambda-2 & -7 & -8\end{vmatrix}=0$$

$\Rightarrow \quad 20(\lambda-2)+7(12)-8(8)=0$

$\Rightarrow \quad 5(\lambda-2)+5=0$

$\Rightarrow \quad \lambda=1$

Hence, the diagonal vectors are $\overrightarrow{AC}=-4\hat{i}-4\hat{j}+4\hat{k}$ and $\overrightarrow{BD}=0\hat{i}-8\hat{j}-12\hat{k}$

$$\text{Now, } \overrightarrow{AC}\times\overrightarrow{BD}=\begin{vmatrix}\hat{i} & \hat{j} & \hat{k}\\ -4 & -4 & 4\\ 0 & -8 & -12\end{vmatrix}$$

$$=80\hat{i}-48\hat{j}+32\hat{k}$$

$$=16(5\hat{i}-3\hat{j}+2\hat{k})$$

Therefore, the area of quadrilateral

$$=\frac{1}{2}|\overrightarrow{AC}\times\overrightarrow{BD}|$$

$$=\frac{1}{2}\times16\sqrt{5^2+3^2+2^2}.$$

$$=8\sqrt{38}$$

17. (1) $\vec{a}\times\vec{b}=\begin{vmatrix}\hat{i} & \hat{j} & \hat{k}\\ -5 & 1 & -3\\ 1 & 2 & -4\end{vmatrix}=2\hat{i}-23\hat{j}-11\hat{k}$

$\Rightarrow \quad (\vec{a}\times\vec{b})\times\hat{i}=23\hat{k}-11\hat{j}$

$\Rightarrow \quad ((\vec{a}\times\vec{b})\times\hat{i})\times\hat{i}=23\hat{j}+11\hat{k}$

$\Rightarrow \quad \vec{c}=(((\vec{a}\times\vec{b})\times\hat{i})\times\hat{i})\times\hat{i}=-23\hat{k}+11\hat{j}$

$\Rightarrow \quad \vec{c}\cdot(-\hat{i}+\hat{j}+\hat{k})=11-23=-12$

18. (2) We have $\frac{1}{2}\left\|\begin{matrix}\hat{i} & \hat{j} & \hat{k}\\ 1 & -1 & 2\\ 2 & 3 & \alpha\end{matrix}\right\|=\frac{\sqrt{93}}{2}$

$\Rightarrow \quad |(-\alpha-6)\hat{i}+(4-\alpha)\hat{j}+5\hat{k}|=\sqrt{93}$

$\Rightarrow \quad (\alpha+6)^2+(4-\alpha)^2+25=93$

$\Rightarrow \quad 2\alpha^2+4\alpha-16=0$

$\Rightarrow \quad \alpha^2+2\alpha-8=0$

$\Rightarrow \quad (\alpha+4)(\alpha-2)=0$

$\Rightarrow \quad \alpha=-2, 4$

19. Area of ΔABC

$$=3\times\text{area of }\Delta OBC$$

$$=3\times\frac{1}{2}|\overrightarrow{OB}\times\overrightarrow{OC}|$$

$$=\frac{3}{2}\left\|\begin{matrix}\hat{i} & \hat{j} & \hat{k}\\ 1 & -2 & 3\\ 2 & 1 & -4\end{matrix}\right\|$$

$$=\frac{3}{2}|5\hat{i}+10\hat{j}+5\hat{k}|$$

$$=\frac{15}{2}\sqrt{6}\text{ square units}$$

20. (3) We have $\frac{1}{2}|\vec{a}\times\vec{b}| = 10$

$\Rightarrow \quad |\vec{a}\times\vec{b}| = 20$

Now, area of parallelogram

$= |\vec{a}\times(\vec{a}\times\vec{b})|$

$= |\vec{a}|\cdot|\vec{a}\times\vec{b}| \qquad (\because \vec{a}\perp\vec{a}\times\vec{b})$

$= 4\times 20$

$= 80$ sq. units.

Exercise 3.3

1. $((\vec{a}\times\hat{j})\cdot\hat{k})\hat{i}+((\vec{a}\times\hat{k})\cdot\hat{i})\hat{j}+((\vec{a}\times\hat{i})\cdot\hat{j})\hat{k}$

$= (\vec{a}\cdot(\hat{j}\times\hat{k}))\hat{i}+(\vec{a}\cdot(\hat{k}\times\hat{i}))\hat{j}+(\vec{a}\cdot(\hat{i}\times\hat{j}))\hat{k}$

$= (\vec{a}\cdot\hat{i})\hat{i}+(\vec{a}\cdot\hat{j})\hat{i}+(\vec{a}\cdot\hat{k})\hat{k}$

$= \vec{a}$

2. $\vec{a}\cdot(\vec{b}\times\hat{i})\,\hat{i} = ((\vec{a}\times\vec{b})\cdot\hat{i})\,\hat{i}$

If $\vec{a}\times\vec{b} = x\hat{i}+y\hat{j}+z\hat{k}$, then $(\vec{a}\times\vec{b})\cdot\hat{i} = x$

Similarly, $\vec{a}\cdot(\vec{b}\times\hat{j}) = y$ and $\vec{a}\cdot(\vec{b}\times\hat{k}) = z$

$\Rightarrow \quad (\vec{a}\cdot(\vec{b}\times\hat{i}))\,\hat{i}+(\vec{a}\cdot(\vec{b}\times\hat{j}))\,\hat{j}+(\vec{a}\cdot(\vec{b}\times\hat{k}))\,\hat{k}$

$= x\hat{i}+y\hat{j}+z\hat{k} = \vec{a}\times\vec{b}$

3. Clearly $\vec{a},\vec{b},\vec{c}$ are non–coplanar.

$\therefore \quad [\vec{a}\;\vec{b}\;\vec{c}]\neq 0.$

Now, given equation is

$$2[\vec{a}\vec{b}\vec{c}]x^2+\alpha[\vec{a}\vec{b}\vec{c}]x+\left(\alpha+\frac{5}{2}\right)[\vec{a}\;\vec{b}\;\vec{c}]=0$$

$\Rightarrow \quad 2x^2+\alpha x+\left(\alpha+\frac{5}{2}\right)=0$, where $x=[\vec{a}\;\vec{b}\;\vec{c}]$

This equation has real roots.

$\therefore \; \alpha^2-4(2)\left(\alpha+\frac{5}{2}\right)\geq 0$

$\Rightarrow \quad \alpha^2-8\alpha-20\geq 0$

$\Rightarrow \quad (\alpha-10)(\alpha+2)\geq 0$

$\Rightarrow \quad \alpha\leq -2$ or $\alpha\geq 10$

4. Since $\vec{x}$ is a non-zero vector, the given conditions will be satisfied if either (i) vectors $\vec{a}$, $\vec{b}$ and $\vec{c}$ are zero or (ii) $\vec{x}$ is perpendicular to vectors $\vec{a}$, $\vec{b}$ and $\vec{c}$.

In case (ii) $\vec{a}$, $\vec{b}$ and $\vec{c}$ are coplanar and so $[\vec{a}\;\vec{b}\;\vec{c}]=0.$

(4) $\begin{vmatrix} 2 & 3 & 4 \\ 1 & \alpha & 2 \\ 1 & 2 & \alpha \end{vmatrix} = \pm 15$

Let $\;2(\alpha^2-4)+3(2-\alpha)+4(2-\alpha)=15$

$\Rightarrow \quad 2\alpha^2-8+6-3\alpha+8-4\alpha=15$

$\Rightarrow \quad 2\alpha^2-7\alpha-9=0$

$\Rightarrow \quad 2\alpha^2-9\alpha+2\alpha-9=0$

$\Rightarrow \quad (\alpha+1)(2\alpha-9)=0$

$\Rightarrow \quad \alpha=-1, 9/2$

For –15 on R.H.S, we get non-real roots.

6. (2) We have $\frac{1}{6}[\vec{a}\;\vec{b}\;\vec{c}]=12$

$\Rightarrow \quad [\vec{a}\;\vec{b}\;\vec{c}]=72$

Projection $\vec{a}$ on $\vec{b}\times\vec{c} = \dfrac{\vec{a}\cdot(\vec{b}\times\vec{c})}{|\vec{b}\times\vec{c}|} = 4$ (given)

$\therefore \quad \dfrac{72}{|\vec{b}\times\vec{c}|}=4 \Rightarrow |\vec{b}\times\vec{c}|=18$

7. (4) We have $\vec{a}\cdot\vec{b}=\vec{b}\cdot\vec{c}=\vec{c}\cdot\vec{a}=2\times 2\times\cos 60°=2$

$$\therefore \quad [\vec{a}\;\vec{b}\;\vec{c}]^2 = \begin{vmatrix} \vec{a}.\vec{a} & \vec{a}.\vec{b} & \vec{a}.\vec{c} \\ \vec{b}.\vec{a} & \vec{b}.\vec{b} & \vec{b}.\vec{c} \\ \vec{c}.\vec{a} & \vec{c}.\vec{b} & \vec{c}.\vec{c} \end{vmatrix} = \begin{vmatrix} 4 & 2 & 2 \\ 2 & 4 & 2 \\ 2 & 2 & 4 \end{vmatrix} = 32$$

$\Rightarrow \quad |[\vec{a}\;\vec{b}\;\vec{c}]| = 4\sqrt{2}$

$\Rightarrow \quad |[\vec{a}+\vec{b}\;\;\vec{b}+\vec{c}\;\;\vec{c}+\vec{a}]| = |2[\vec{a}\;\vec{b}\;\vec{c}]| = 8\sqrt{2}$

8. (1) $\quad [\hat{n}\;\hat{p}\;\hat{m}]=[\hat{p}\;\hat{m}\;\hat{n}]$

$= \hat{p}\cdot(\hat{m}\times\hat{n})$

$= |p|\cdot|m\times n|\cos\alpha$

$= |m\times n|\cos\alpha$

$= \sin\alpha\cos\alpha = 1/4 \qquad$ (given)

$\Rightarrow \quad \sin 2\alpha = 1/2$

$\Rightarrow \quad 2\alpha = 30°$

$\Rightarrow \quad \alpha = 15°$

9. (3) $\vec{a}\times\vec{b} = p\vec{a}+q\vec{b}+r\vec{c}$

Taking dot product with $\vec{a}, \vec{b}$ and $\vec{c}$, we get

$0 = p+\frac{3}{5}q+\frac{3}{5}r \qquad$ ...(i)

$0 = \frac{3}{5}p+q+\frac{3}{5}r \qquad$ ...(ii)

$[\vec{a}\;\vec{b}\;\vec{c}] = \frac{3}{5}p+\frac{3}{5}q+r \qquad$ ...(iii)

$$\text{Also} \quad [\vec{a}\;\vec{b}\;\vec{c}]^2 = \begin{vmatrix} \vec{a}\cdot\vec{a} & \vec{a}\cdot\vec{b} & \vec{a}\cdot\vec{c} \\ \vec{b}\cdot\vec{a} & \vec{b}\cdot\vec{b} & \vec{b}\cdot\vec{c} \\ \vec{c}\cdot\vec{a} & \vec{c}\cdot\vec{b} & \vec{c}\cdot\vec{c} \end{vmatrix}$$

$$= \begin{vmatrix} 1 & 3/5 & 3/5 \\ 3/5 & 1 & 3/5 \\ 3/5 & 3/5 & 1 \end{vmatrix} = \frac{44}{125}$$

Solving (i), (ii) and (iii) for q, we get

$\frac{5}{11}[\vec{a}\vec{b}\vec{c}] = -\frac{2}{3}q$

$\therefore \quad \frac{5}{11}\times\sqrt{\frac{44}{125}} = -\frac{2}{3}q$

$\therefore \quad q = -\frac{3}{\sqrt{55}}$

10. Let $\vec{l} = l_1\hat{i}+l_2\hat{j}+l_3\hat{k}$, $\vec{m} = m_1\hat{i}+m_2\hat{j}+m_3\hat{k}$, $\vec{n} = n_1\hat{i}+n_2\hat{j}+n_3\hat{k}$, $\vec{a} = a_1\hat{i}+a_2\hat{j}+a_3\hat{k}$ and $\vec{b} = b_1\hat{i}+b_2\hat{j}+b_3\hat{k}$. Therefore,

$\vec{l}\cdot\vec{a} = l_1a_1 + l_2a_2 + l_3a_3 = \sum l_1a_1$

Similarly, $\vec{l}\cdot\vec{b} = \sum l_1b_1$, etc.

Now, $[\vec{l}\ \vec{m}\ \vec{n}](\vec{a}\times\vec{b})$

$$= \begin{vmatrix} l_1 & l_2 & l_3 \\ m_1 & m_2 & m_3 \\ n_1 & n_2 & n_3 \end{vmatrix} \times \begin{vmatrix} \hat{i} & \hat{j} & \hat{k} \\ a_1 & a_2 & a_3 \\ b_1 & b_2 & b_3 \end{vmatrix}$$

$$= \begin{vmatrix} \sum l_1\hat{i} & \sum l_1a_1 & \sum l_1b_1 \\ \sum m_1\hat{i} & \sum m_1a_1 & \sum m_1b_1 \\ \sum n_1\hat{i} & \sum n_1a_1 & \sum n_1b_1 \end{vmatrix}$$

$$= \begin{vmatrix} \vec{l} & \vec{l}\cdot\vec{a} & \vec{l}\cdot\vec{b} \\ \vec{m} & \vec{m}\cdot\vec{a} & \vec{m}\cdot\vec{b} \\ \vec{n} & \vec{n}\cdot\vec{a} & \vec{n}\cdot\vec{b} \end{vmatrix} = \begin{vmatrix} \vec{l}\cdot\vec{a} & \vec{l}\cdot\vec{b} & \vec{l} \\ \vec{m}\cdot\vec{a} & \vec{m}\cdot\vec{b} & \vec{m} \\ \vec{n}\cdot\vec{a} & \vec{n}\cdot\vec{b} & \vec{n} \end{vmatrix}$$

11. $\begin{vmatrix} a_1 & a_2 & a_3 \\ b_1 & b_2 & b_3 \\ c_1 & c_2 & c_3 \end{vmatrix}^2 = [\vec{a}\ \vec{b}\ \vec{c}]^2$

$$= ((\vec{a}\times\vec{b})\cdot\vec{c})^2$$

$$= |\vec{a}\times\vec{b}|^2|\vec{c}|^2$$

$$= a^2b^2\sin^2\frac{\pi}{6}\cdot c^2$$

$$= \frac{1}{4}(a_1^2+a_2^2+a_3^2)(b_1^2+b_2^2+b_3^2)$$

12. $\vec{a}\times\vec{b} = \vec{c}$, i.e. $\vec{c}$ is perpendicular to $\vec{a}$ and $\vec{b}$.

$\vec{b}\times\vec{c} = \vec{a}$, i.e. $\vec{a}$ is perpendicular to $\vec{b}$ and $\vec{c}$.

So, $\vec{a}$, $\vec{b}$ and $\vec{c}$ are mutually perpendicular.

$\vec{a}\times\vec{b} = \vec{c} \Rightarrow |\vec{a}||\vec{b}| = |\vec{c}| \Rightarrow |\vec{b}| = |\vec{c}|/2$

$\vec{a}\times\vec{b} = \vec{c} \Rightarrow |\vec{b}\times\vec{c}| = |\vec{a}| \Rightarrow |\vec{b}||\vec{c}| = 2$

$\Rightarrow |\vec{c}| = 2$ and $|\vec{b}| = 1$

Therefore, $[\vec{a}\ \vec{b}\ \vec{c}] + [\vec{c}\ \vec{a}\ \vec{b}]$

$= 2[\vec{a}\ \vec{b}\ \vec{c}] = 2|\vec{a}||\vec{b}||\vec{c}| = 2(2)(1)(2) = 8$

Now, $\vec{a}$ is parallel to $(\vec{b}\times\vec{c})$

So, the scalar projection $\vec{a}$ on $(\vec{b}\times\vec{c})$ is 2.

And, $|3\vec{a}+\vec{b}-2\vec{c}|^2 = 9|\vec{a}|^2 + |\vec{b}|^2 + 4|\vec{c}|^2$

$= (9\times4) + 1 + (4\times4)$

$= 36 + 1 + 16 = 53$

Exercise 3.4

1. $\vec{d}\cdot(\vec{a}\times(\vec{b}\times(\vec{c}\times\vec{d})))$

$$= \vec{d}\cdot(\vec{a}\times((\vec{b}\cdot\vec{d})\vec{c} - (\vec{b}\cdot\vec{c})\vec{d}))$$

$$= \vec{d}\cdot((\vec{b}\cdot\vec{d})(\vec{a}\times\vec{c}) - (\vec{b}\cdot\vec{c})(\vec{a}\times\vec{d}))$$

$$= (\vec{b}\cdot\vec{d})(\vec{d}\cdot(\vec{a}\times\vec{c})) - (\vec{b}\cdot\vec{c})(\vec{d}\cdot(\vec{a}\times\vec{d}))$$

$$= (\vec{b}\cdot\vec{d})[\vec{d}\ \vec{a}\ \vec{c}]$$

$$= (\vec{b}\cdot\vec{d})[\vec{a}\ \vec{c}\ \vec{d}]$$

2. $\vec{a}\times(\vec{a}\times\vec{b}) = \frac{1}{2}\vec{b} \Rightarrow (\vec{a}\cdot\vec{b})\vec{a} - (\vec{a}\cdot\vec{a})\vec{b} = \frac{1}{2}\vec{b}$

$\Rightarrow \vec{a}\cdot\vec{b} = 0$, $\vec{a}\cdot\vec{a} = -\frac{1}{2}$ (which is not possible).

Thus, the given information is inconsistent.

3. We are given that

$(\vec{a}\cdot\vec{c})\vec{b} - (\vec{b}\cdot\vec{c})\vec{a} = (\vec{a}\cdot\vec{c})\vec{b} - (\vec{a}\cdot\vec{b})\vec{c}$.

$\therefore\quad (\vec{a}\cdot\vec{b})\vec{c} - (\vec{b}\cdot\vec{c})\vec{a} = \vec{0}$

$\Rightarrow\quad (\vec{a}\times\vec{c})\times\vec{b} = \vec{0}$ …(i

Thus, either $\vec{b} = \vec{0}$ or $\vec{a}\times\vec{c} = \vec{0}$

But $\vec{b} \neq \vec{0}$, therefore, $\vec{a}\times\vec{c} = \vec{0}$ and hence

$\vec{c} = t\,\vec{a}$, i.e., $\vec{a}$ and $\vec{c}$ are collinear.

4. (3) We have $(\vec{a}\times\vec{b})\times\vec{c} = \frac{1}{3}|\vec{b}||\vec{c}|\vec{a}$

$\Rightarrow\quad (\vec{a}\cdot\vec{c})\vec{b} - (\vec{b}\cdot\vec{c})\vec{a} = \frac{1}{3}|\vec{b}||\vec{c}|\vec{a}$

$\Rightarrow\quad (\vec{a}\cdot\vec{c})\vec{b} - \left\{(\vec{b}\cdot\vec{c}) + \frac{1}{3}|\vec{b}||\vec{c}|\right\}\vec{a} = 0$

$\Rightarrow\quad \vec{a}\cdot\vec{c} = 0$ and $\vec{b}\cdot\vec{c} + \frac{1}{3}|\vec{b}||\vec{c}| = 0$

($\vec{a}$ and $\vec{b}$ are non-collinea

$\Rightarrow\quad |\vec{b}||\vec{c}|\cos\theta + \frac{1}{3}|\vec{b}||\vec{c}| = 0$

$\Rightarrow\quad \cos\theta = -1/3$

$\Rightarrow\quad \sin\theta = \sqrt{\frac{8}{9}} = \frac{2\sqrt{2}}{3}$

5. $\vec{a}\times\vec{b} = \vec{a}\times(\vec{a}\times\vec{c}) = (\vec{a}\cdot\vec{c})\vec{a} - (\vec{a}\cdot\vec{a})\vec{c}$

$= 2\vec{a} - 3\vec{c}$

But $\vec{a}\times\vec{b} = \begin{vmatrix} \hat{i} & \hat{j} & \hat{k} \\ 1 & 1 & 1 \\ 1 & -2 & 1 \end{vmatrix} = 3\hat{i} - 3\hat{k}$

Hence, $3\vec{c} = 2\vec{a} - (3\hat{i} - 3\hat{k})$

$= (2\hat{i} + 2\hat{j} + 2\hat{k}) - (3\hat{i} - 3\hat{k})$

$= -\hat{i} + 2\hat{j} + 5\hat{k}$

or $\vec{c} = \frac{1}{3}(-\hat{i} + 2\hat{j} + 5\hat{k})$

6. Here $\vec{p} = \vec{b}\times\vec{c}$, $\vec{q} = \vec{c}\times\vec{a}$, $\vec{r} = \vec{a}\times\vec{b}$

Now $\vec{a}\times(\vec{q}\times\vec{r}) = (\vec{a}\cdot\vec{r})\vec{q} - (\vec{a}\cdot\vec{q})\vec{r}$

$= \{\vec{a}\cdot(\vec{a}\times\vec{b})\}\vec{q} - \{\vec{a}\cdot(\vec{c}\times\vec{a})\}\vec{r}$

$= (0)\vec{q} - (0)\vec{r} = \vec{0}$

Therefore, $\vec{a}$ is parallel to $\vec{q}\times\vec{p}$

Again $\vec{b} \times (\vec{r} \times \vec{p}) = (\vec{b} \cdot \vec{p})\, \vec{r} - (\vec{b} \cdot \vec{r})\, \vec{p}$

$= \{\vec{b} \cdot (\vec{b} \times \vec{c})\}\, \vec{r} - \{\vec{b} \cdot (\vec{a} \times \vec{b})\}\, \vec{p}$

$= (0)\, \vec{r} - (0)\, \vec{p} = \vec{0}$

Therefore, $\vec{b}$ is parallel to $\vec{r} \times \vec{p}$

Also $\vec{c} \times (\vec{p} \times \vec{q}) = (\vec{c} \cdot \vec{q})\, \vec{p} - (\vec{c} \cdot \vec{p})\, \vec{q}$

$= \{\vec{c} \cdot (\vec{c} \times \vec{a})\}\, \vec{p} - \{\vec{c} \cdot (\vec{b} \times \vec{c})\}\, \vec{q}$

$= (0)\, \vec{p} - (0)\, \vec{q} = \vec{0}$

Therefore, $\vec{c}$ is parallel to $\vec{p} \times \vec{q}$.

7. $\vec{a} = \vec{p} + \vec{q}$

or $\vec{a} \times \vec{b} = \vec{p} \times \vec{b} + \vec{q} \times \vec{b}$

or $\vec{a} \times \vec{b} = \vec{q} \times \vec{b}$ $\quad (\because \vec{p} \times \vec{b} = \vec{0})$

or $\vec{b} \times (\vec{a} \times \vec{b}) = \vec{b} \times (\vec{q} \times \vec{b})$

$= (\vec{b} \cdot \vec{b})\, \vec{q} - (\vec{b} \cdot \vec{q})\, \vec{b}$

$= (\vec{b} \cdot \vec{b})\, \vec{q}$ $\quad (\because \vec{b} \cdot \vec{q} = 0)$

or $\dfrac{\vec{b} \times (\vec{a} \times \vec{b})}{\vec{b} \cdot \vec{b}} = \vec{q}$

. (1) We have $\vec{u} \times (\vec{u} \times \vec{w}) + 3\vec{v} = 0$

$\Rightarrow \quad (\vec{u} \cdot \vec{w})\vec{u} - (\vec{u} \cdot \vec{u})w = -3\vec{v}$

$\Rightarrow \quad (2\sqrt{3}\cos\theta)\vec{u} - 3\vec{w} = -3\vec{v}$

Square both sides, we get

$|\vec{u}|^2 \cdot (2\sqrt{3}\cos\theta)^2 + 9|\vec{w}|^2 - 12\sqrt{3}\cos\theta |\vec{u}|\, |\vec{w}| \cos\theta = 9|\vec{v}|^2$

$\Rightarrow \quad 36\cos^2\theta + 36 - 72\cos^2\theta = 9$

$\Rightarrow \quad \cos^2\theta = \dfrac{3}{4}$

$\Rightarrow \quad \cos\theta = \dfrac{\sqrt{3}}{2}$

$\Rightarrow \quad \theta = 30°$

. (4) $[\vec{a} \times (\vec{b} + \vec{c})\ \ \vec{b} \times (\vec{c} + 2\vec{a})\ \ \vec{c} \times (\vec{a} + 3\vec{b})]$

$= [\vec{a} \times \vec{b} - \vec{c} \times \vec{a}\ \ -2\vec{a} \times \vec{b} + \vec{b} \times \vec{c}\ \ -3\vec{b} \times \vec{c} + \vec{c} \times \vec{a}]$

$= \begin{vmatrix} 1 & 0 & -1 \\ -2 & 1 & 0 \\ 0 & -3 & 1 \end{vmatrix} [\vec{a} \times \vec{b}\ \vec{b} \times \vec{c}\ \vec{c} \times \vec{a}]$

$= -5[\vec{a}\ \vec{b}\ \vec{c}]^2$

$= -5(-3)^2 = -45$

0. (1) $\vec{a} \times (\vec{b} \times \vec{c}) = (\vec{a} \cdot \vec{c})\vec{b} - (\vec{a} \cdot \vec{b})\vec{c} = 3\vec{b} - \vec{c}$

$\vec{b} \times (\vec{c} \times \vec{a}) = (\vec{b} \cdot \vec{a})\vec{c} - (\vec{b} \cdot \vec{c})\vec{a} = \vec{c} - 2\vec{a}$

$\vec{c} \times (\vec{b} \times \vec{a}) = (\vec{c} \cdot \vec{a})\vec{b} - (\vec{c} \cdot \vec{b})\, \vec{a} = 3\vec{b} - 2\vec{a}$

$\therefore \quad [3\vec{b} - \vec{c}, \vec{c} - 2\vec{a}, 3\vec{b} - 2\vec{a}]$

$= [3\vec{b} - \vec{c}, -2\vec{a} + \vec{c}\ , -2\vec{a} + 3\vec{b}\ ]$

$= \begin{vmatrix} 0 & 3 & -1 \\ -2 & 0 & 1 \\ -2 & 3 & 0 \end{vmatrix} [\vec{a}\ \vec{b}\ \vec{c}]$

$= 0$

11. $[\vec{a} \times \vec{b}\ \ \vec{b} \times \vec{c}\ \ \vec{c} \times \vec{a}] + [\vec{a} + \vec{b}\ \ \vec{b} + \vec{c}\ \ \vec{c} + \vec{a}] = 8$

$\Rightarrow \quad [\vec{a}\ \vec{b}\ \vec{c}]^2 + 2[\vec{a}\ \vec{b}\ \vec{c}] - 8 = 0$

$\Rightarrow \quad [\vec{a}\ \vec{b}\ \vec{c}] = 2, -4$

$\Rightarrow \quad \dfrac{1}{6} |[\vec{a}\ \vec{b}\ \vec{c}]| = \dfrac{1}{3}, \dfrac{2}{3}$

12. The volume of the parallelopiped $= [\vec{a}\ \vec{b}\ \vec{c}]$

$[\vec{a}\ \vec{b}\ \vec{c}]^2 = \begin{vmatrix} \vec{a} \cdot \vec{a} & \vec{a} \cdot \vec{b} & \vec{a} \cdot \vec{c} \\ \vec{b} \cdot \vec{a} & \vec{b} \cdot \vec{b} & \vec{b} \cdot \vec{c} \\ \vec{a} \cdot \vec{c} & \vec{b} \cdot \vec{c} & \vec{c} \cdot \vec{c} \end{vmatrix}$

$= |\vec{a}|^6 \begin{vmatrix} 1 & \cos\theta & \cos\theta \\ \cos\theta & 1 & \cos\theta \\ \cos\theta & \cos\theta & 1 \end{vmatrix}$

$= |\vec{a}|^6 \left[2\cos^3\theta - 3\cos^2\theta + 1\right] = |\vec{a}|^6 (1 - \cos\theta)^2 (1 + 2\cos\theta)$

$\Rightarrow \quad [\vec{a}\ \vec{b}\ \vec{c}] = |\vec{a}|^3 \sqrt{1 + 2\cos\theta}\, (1 - \cos\theta)$.

13. (4) Since each side of tetrahedron is 1 unit, it is regular tetrahedron. So, angle between each pair of edges is 60°.

$\therefore \quad \vec{a} \cdot \vec{b} = \vec{b} \cdot \vec{c} = \vec{c} \cdot \vec{a} = \dfrac{1}{2}$

Now, $\vec{c} = \alpha\left(\vec{a} + \vec{b}\right) + \beta\left(\vec{a} \times \vec{b}\right)$ ...(i)

Taking dot product with $\vec{a}$, we get

$\dfrac{1}{2} = \alpha\left(1 + \dfrac{1}{2}\right) \Rightarrow \alpha = \dfrac{1}{3}$

Now, taking dot product of (1), with $a \times b$, we get

$\therefore \quad \left[\vec{a}\vec{b}\vec{c}\right] = \beta\left|\vec{a} \times \vec{b}\right|^2 = \beta \cdot \dfrac{3}{4}$

Now, $\left[\vec{a}\vec{b}\vec{c}\right]^2 = \begin{vmatrix} 1 & \frac{1}{2} & \frac{1}{2} \\ \frac{1}{2} & 1 & \frac{1}{2} \\ \frac{1}{2} & \frac{1}{2} & 1 \end{vmatrix} = \dfrac{3}{4} - \dfrac{1}{8} - \dfrac{1}{8} = \dfrac{1}{2}$

$\therefore \quad \beta = \dfrac{4}{3\sqrt{2}}$

14. $\vec{a} \times \vec{a}' = \dfrac{\vec{a} \times (\vec{b} \times \vec{c})}{[\vec{a}\ \vec{b}\ \vec{c}]} = \dfrac{(\vec{a} \cdot \vec{c})\vec{b} - (\vec{a} \cdot \vec{b})\vec{c}}{[\vec{a}\ \vec{b}\ \vec{c}]}$

Similarly, $\vec{b} \times \vec{b}' = \dfrac{(\vec{b} \cdot \vec{a})\vec{c} - (\vec{b} \cdot \vec{c})\vec{a}}{[\vec{a}\ \vec{b}\ \vec{c}]}$

and $\vec{c} \times \vec{c}' = \dfrac{(\vec{c} \cdot \vec{b})\vec{a} - (\vec{c} \cdot \vec{a})\vec{b}}{[\vec{a}\ \vec{b}\ \vec{c}]}$

So $\vec{a} \times \vec{a}' + \vec{b} \times \vec{b}' + \vec{c} \times \vec{c}' = 0$

15. (3) $\vec{a}\times(\vec{c}\times\vec{d})=(\vec{a}\cdot\vec{d})\vec{c}-(\vec{a}\cdot\vec{c})\vec{d}=\frac{\vec{c}}{2}+\frac{\vec{d}}{3}$

$\Rightarrow \quad \vec{a}\cdot\vec{d}=\frac{1}{2}$ and $\vec{a}\cdot\vec{c}=-\frac{1}{3}$

$(\vec{c}\times\vec{d})\times\vec{b}=(\vec{c}\cdot\vec{b})\vec{d}-(\vec{b}\cdot\vec{d})\vec{c}=\frac{\vec{d}}{4}-\frac{\vec{c}}{5}$

$\Rightarrow \quad \vec{c}\cdot\vec{b}=\frac{1}{4}$ and $\vec{b}\cdot\vec{d}=\frac{1}{5}$

$\therefore \quad (\vec{a}\times\vec{b})\cdot(\vec{c}\times\vec{d}) = (\vec{a}\cdot\vec{c})(\vec{b}\cdot\vec{d})-(\vec{b}\cdot\vec{c})(\vec{a}\cdot\vec{d})$

$=\left(-\frac{1}{3}\right)\left(\frac{1}{5}\right)-\left(\frac{1}{4}\right)\left(\frac{1}{2}\right)$

$=-\frac{1}{15}-\frac{1}{8}=-\frac{23}{120}$

16. $(\vec{a}\times\vec{b})\times(\vec{c}\times\vec{d})=\frac{1}{6}\hat{i}-\frac{1}{3}\hat{j}+\frac{1}{3}\hat{k}$

$\Rightarrow \quad ((\vec{a}\times\vec{b})\cdot\vec{d})\vec{c}-((\vec{a}\times\vec{b})\cdot\vec{c})\vec{d}=\frac{1}{6}\hat{i}-\frac{1}{3}\hat{j}+\frac{1}{3}\hat{k}$

$(\because (\vec{a}\times\vec{b})\cdot\vec{c}=0)$

$\Rightarrow \quad |\vec{a}\times\vec{b}|\cdot|\vec{d}|\vec{c}=\frac{1}{6}\hat{i}-\frac{1}{3}\hat{j}+\frac{1}{3}\hat{k}$

$(\because \vec{a}\times\vec{b}$ and $\vec{d}$ are parallel)

$\Rightarrow \quad |\vec{a}||\vec{b}|\sin 30°\vec{c}=\frac{1}{6}\hat{i}-\frac{1}{3}\hat{j}+\frac{1}{3}\hat{k}$

$\Rightarrow \quad \vec{c}=\frac{1}{3}(\hat{i}-2\hat{j}+2\hat{k})$

Exercises

Single Correct Answer Type

1. (2) We have $|\hat{a}+\hat{c}|^2=|\sqrt{3}\hat{b}|^2$

$\Rightarrow$ $1+1+2\cos\theta=3$, where θ is the angle between vectors $\hat{a}$ and $\hat{c}$

$\Rightarrow \quad \cos\theta=\frac{1}{2}$

$\Rightarrow \quad \theta=60°$

2. (2) If $\vec{a}(x)$ and $\vec{b}(x)$ are $\perp$, then $\vec{a}\cdot\vec{b}=0$

$\Rightarrow \quad \sin x\cos 2x+\cos x\sin 2x=0$

$\sin(3x)=0=\sin 0$

$3x=n\pi \quad$ or $\quad x=\frac{n\pi}{3}$

Therefore, the two vectors are perpendicular for infinite values of x.

3. (4) $\vec{f}\left(\frac{5}{4}\right)=\left[\frac{5}{4}\right]\hat{i}+\left(\frac{5}{4}-\left[\frac{5}{4}\right]\right)\hat{j}+\left[\frac{5}{4}+1\right]\hat{k}$

$=\hat{i}+\left(\frac{5}{4}-1\right)\hat{j}+2\hat{k}$

$=\hat{i}+\frac{1}{4}\hat{j}+2\hat{k}$

When $0<t<1$, $\vec{f}(t)=0\hat{i}+(t-0)\hat{j}+\hat{k}$

$=t\hat{j}+\hat{k}$

$\vec{f}\left(\frac{5}{4}\right)\cdot\vec{f}(t)=2+\frac{t}{4}$

So $\quad \cos\theta=\dfrac{2+\frac{t}{4}}{\left|\hat{i}+\frac{1}{4}\hat{j}+2\hat{k}\right||t\hat{j}+\hat{k}|}$

$=\dfrac{2+\frac{t}{4}}{\sqrt{1+\frac{1}{16}+4}\sqrt{1+t^2}}=\dfrac{8+t}{9\sqrt{1+t^2}}$

4. (4) The angle between $\vec{a}$ and $\vec{b}$ is obtuse. Therefore,

$\vec{a}\cdot\vec{b}<0$

$\Rightarrow \quad 14x^2-8x+x<0$

or $\quad 7x(2x-1)<0$

or $\quad 0<x<1/2 \quad$ ...(i

The angle between $\vec{b}$ and the z-axis is acute and less than $\pi/6$. Therefore,

$\dfrac{\vec{b}\cdot\hat{k}}{|\vec{b}||\hat{k}|}>\cos\pi/6 \quad (\because \theta<\pi/6 \Rightarrow \cos\theta>\cos\pi/6)$

$\Rightarrow \quad \dfrac{x}{\sqrt{x^2+53}}>\dfrac{\sqrt{3}}{2}$

or $\quad 4x^2>3x^2+159$

or $\quad x^2>159$

$\Rightarrow \quad x>\sqrt{159}$ or $x<-\sqrt{159}$ ...(ii

Clearly, (i) and (ii) cannot hold together.

5. (2) Let $\vec{\alpha}=\dfrac{\vec{a}}{|\vec{a}|}+\dfrac{\vec{b}}{|\vec{b}|}+\dfrac{\vec{c}}{|\vec{c}|}$

Since $\vec{a},\vec{b}$ and $\vec{c}$ are mutually perpendicular vectors, if $\vec{\alpha}$ make angles θ, ϕ, ψ with $\vec{a},\vec{b}$ and $\vec{c}$, respectively, then

$\vec{\alpha}\cdot\vec{a}=\dfrac{\vec{a}\cdot\vec{a}}{|\vec{a}|}$

or $\quad |\vec{\alpha}|\cdot|\vec{a}|\cos\theta=|\vec{a}|$

or $\quad \cos\theta=\dfrac{1}{|\vec{\alpha}|}$

Similarly, $\cos\phi=\dfrac{1}{|\vec{\alpha}|}, \cos\psi=\dfrac{1}{|\vec{\alpha}|}$

$\therefore \quad \theta=\phi=\psi$

6. (4) $(3\vec{a}+\vec{b})\cdot(\vec{a}-4\vec{b})$

$=3|\vec{a}|^2-11\vec{a}\cdot\vec{b}-4|\vec{b}|^2$

$=3\times 36-11\times 6\times 8\cos\pi-4\times 64>0$

Therefore, the angle between $\vec{a}$ and $\vec{b}$ is acute.

The longer diagonal is given by

$\vec{\alpha}=(3\vec{a}+\vec{b})+(\vec{a}-4\vec{b})=4\vec{a}-3\vec{b}$

Now, $|\vec{\alpha}|^2 = |4\vec{a} - 3\vec{b}|^2$

$= 16|\vec{a}|^2 + 9|\vec{b}|^2 - 24\vec{a}\cdot\vec{b}$

$= 16 \times 36 + 9 \times 64 - 24 \times 6 \times 8 \cos \pi$

$= 16 \times 144$

or $|4\vec{a} - 3\vec{b}| = 48$

7. (3) Given $\vec{v}\cdot\vec{u} = \vec{w}\cdot\vec{u}$

and $\vec{v} \perp \vec{w} \Rightarrow \vec{v}\cdot\vec{w} = 0$

Now, $|\vec{u} - \vec{v} + \vec{w}|^2$

$= |\vec{u}|^2 + |\vec{v}|^2 + |\vec{w}|^2 - 2\vec{u}\cdot\vec{v}$

$- 2\vec{w}\cdot\vec{v} + 2\vec{u}\cdot\vec{w}$

$= 1 + 4 + 9$

so $|\vec{u} - \vec{v} + \vec{w}| = \sqrt{14}$

8. (3) $|\vec{a} + \vec{b} + \vec{c}|^2 = 6$

or $|\vec{a}|^2 + |\vec{b}|^2 + |\vec{c}|^2$

$+ 2(\vec{a}\cdot\vec{b} + \vec{b}\cdot\vec{c} + \vec{c}\cdot\vec{a}) = 6$

Now, $|\vec{a}| = |\vec{b}| = |\vec{c}|$ and

$\vec{a}\cdot\vec{b} = |\vec{a}||\vec{b}| \cdot \cos\frac{\pi}{3}$

i.e., $\vec{a}\cdot\vec{b} = \frac{1}{2}|\vec{a}|^2$

$\therefore$ $3|\vec{a}|^2 + 3|\vec{a}|^2 = 6$

$\Rightarrow$ $|\vec{a}|^2$ or $|\vec{a}| = 1$

9. (1) Let l, m and n be the direction cosines of the required vector. Then, $l = m$ (given). Therefore

Required vector $\vec{r} = l\hat{i} + m\hat{j} + n\hat{k} = l\hat{i} + l\hat{j} + n\hat{k}$

Now, $l^2 + m^2 + n^2 = 1$ or $2l^2 + n^2 = 1$ ...(i)

Since, $\hat{r}$ is perpendicular to $-\hat{i} + 2\hat{j} + 2\hat{k}$,

$\vec{r}\cdot(-\hat{i} + 2\hat{j} + 2\hat{k}) = 0$

$\Rightarrow$ $-l + 2l + 2n = 0$ or $l + 2n = 0$ ...(ii)

From (i) and (ii), we get: $n = \mp\frac{1}{3}, l = \pm\frac{2}{3}$

Hence, required vector

$\vec{r} = \frac{1}{3}(\pm 2\hat{i} \pm 2\hat{j} \mp \hat{k})$

$= \pm\frac{1}{3}(2\hat{i} + 2\hat{j} - \hat{k})$

10. (2) Let the required vector be $\vec{r}$. Then

$\vec{r} = x_1\vec{b} + x_2\vec{c}$ and $\vec{r}\cdot\vec{a} = \sqrt{\frac{2}{3}}$ $(|\vec{a}|) = 2$

Now, $\vec{r}\cdot\vec{a} = x_1\vec{a}\cdot\vec{b} + x_2\vec{a}\cdot\vec{c}$

$\Rightarrow$ $2 = x_1(2 - 2 - 1) + x_2(2 - 1 - 2)$

$\Rightarrow$ $x_1 + x_2 = -2$

$\Rightarrow$ $\vec{r} = x_1(\hat{i} + 2\hat{j} - \hat{k}) + x_2(\hat{i} + \hat{j} - 2\hat{k})$

$= \hat{i}(x_1 + x_2) + \hat{j}(2x_1 + x_2) - \hat{k}(2x_2 + x_1)$

$= -2\hat{i} + \hat{j}(x_1 - 2) - \hat{k}(-4 - x_1)$,

where $x_1 \in R$

11. (2) Let $\vec{r}$ be the new position. Then $\vec{r} = \lambda\hat{k} + \mu(\hat{i} + \hat{j}), \mu > 0$ for internal bisector.

$\vec{r}\cdot\hat{k} = -\frac{1}{\sqrt{2}} \Rightarrow \lambda = -\frac{1}{\sqrt{2}}$

Also, $\lambda^2 + 2\mu^2 = 1 \Rightarrow 2\mu^2 = \frac{1}{2}$ or $\mu = \frac{1}{2}$

$\therefore$ $\vec{r} = \frac{1}{2}(\hat{i} + \hat{j}) - \frac{\hat{k}}{\sqrt{2}}$

12. (2) Let $\vec{a} = \lambda\vec{b} + \mu\vec{c}$

$\vec{a}$ is equally inclined to $\vec{b}$ and $\vec{d}$, where

$\vec{d} = \hat{j} + 2\hat{k}$.

$\Rightarrow$ $\frac{\vec{a}\cdot\vec{b}}{ab} = \frac{\vec{a}\cdot\vec{d}}{ad}$

$\Rightarrow$ $\frac{(\lambda\vec{b} + \mu\vec{c})\cdot\vec{b}}{b} = \frac{(\lambda\vec{b} + \mu\vec{c})\cdot\vec{d}}{d}$

$\Rightarrow$ $\frac{[\lambda(2\hat{i} + \hat{j}) + \mu(\hat{i} - \hat{j} + \hat{k})]\cdot(2\hat{i} + \hat{j})}{\sqrt{5}}$

$= \frac{[\lambda(2\hat{i} + \hat{j}) + \mu(\hat{i} - \hat{j} + \hat{k})]\cdot(\hat{j} + 2\hat{k})}{\sqrt{5}}$

$\Rightarrow$ $\lambda(4 + 1) + \mu(2 - 1) = \lambda(1) + \mu(-1 + 2)$

$\Rightarrow$ $4\lambda = 0$, i.e., $\lambda = 0$

$\therefore$ $\hat{a} = \frac{\hat{i} - \hat{j} + \hat{k}}{\sqrt{3}}$

13. (1) $(3\vec{a} - 5\vec{b})\cdot(2\vec{a} + \vec{b}) = 0$

or $6|\vec{a}|^2 - 5|\vec{b}|^2 = 7\vec{a}\cdot\vec{b}$

Also, $(\vec{a} + 4\vec{b})\cdot(\vec{b} - \vec{a}) = 0$

$\Rightarrow$ $-|\vec{a}|^2 + 4|\vec{b}|^2 = 3\vec{a}\cdot\vec{b}$

$\Rightarrow$ $\frac{6}{7}|\vec{a}|^2 - \frac{5}{7}|\vec{b}|^2 = -\frac{1}{3}|\vec{a}|^2 + \frac{4}{3}|\vec{b}|^2$

$\Rightarrow$ $25|\vec{a}|^2 = 43|\vec{b}|^2$

$\Rightarrow$ $3\vec{a}\cdot\vec{b} = -|\vec{a}|^2 + 4|\vec{b}|^2 = \frac{57}{25}|\vec{b}|^2$

$\Rightarrow$ $3|\vec{a}||\vec{b}|\cos\theta = \frac{57}{25}|\vec{b}|^2$

$\Rightarrow \quad 3\sqrt{\frac{43}{25}}\,|\vec{b}|^2\cos\theta = \frac{57}{25}|\vec{b}|^2$

$\Rightarrow \quad \cos\theta = \frac{19}{5\sqrt{43}}$

14. (3) $(\hat{a}+\hat{b}+\hat{c})^2 \ge 0$

$\Rightarrow \quad 3 + 2(\vec{a}\cdot\vec{b}+\vec{b}\cdot\vec{c}+\vec{c}\cdot\vec{a}) \ge 0$

$\Rightarrow \quad 3 + 6\cos\theta \ge 0$

$\Rightarrow \quad \cos\theta \ge -\frac{1}{2}$

$\Rightarrow \quad \theta = \frac{2\pi}{3}$

15. (1) $\vec{b} - 2\vec{c} = \lambda\vec{a}$

$\Rightarrow \quad \vec{b} = 2\vec{c} + \lambda\vec{a}$

$\Rightarrow \quad |\vec{b}|^2 = |2\vec{c} + \lambda\vec{a}|^2$

$\Rightarrow \quad 16 = 4|\vec{c}|^2 + \lambda^2|\vec{a}|^2 + 4\lambda\,\vec{a}\cdot\vec{c}$

$= 4 + \lambda^2 + 4\lambda\frac{1}{4}$

$\Rightarrow \quad \lambda^2 + \lambda - 12 = 0$

$\Rightarrow \quad \lambda = 3, -4$

16. (3)

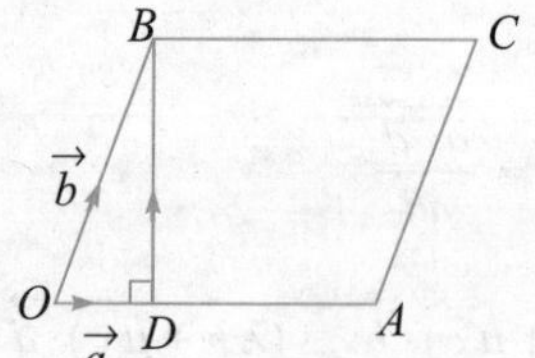

Let $\overrightarrow{OD} = t\vec{a}$

$\therefore \quad \overrightarrow{DB} = \vec{b} - t\vec{a}$

Since $BD \perp OA$

$(\vec{b} - t\vec{a})\cdot\vec{a} = 0 \qquad (\because \overrightarrow{DB} \perp \overrightarrow{OA})$

$\Rightarrow \quad t = \frac{\vec{b}\cdot\vec{a}}{|\vec{a}|^2}$

$\therefore \quad \overrightarrow{DB} = \vec{b} - \frac{(\vec{b}\cdot\vec{a})\,\vec{a}}{|\vec{a}|^2}$

17. (3) We have

$\overrightarrow{AB}\cdot\overrightarrow{AC} + \overrightarrow{BC}\cdot\overrightarrow{BA} + \overrightarrow{CA}\cdot\overrightarrow{CB}$

$= (AB)(AC)\cos\theta + (BC)(BA)\sin\theta + 0$

$= AB(AC\cos\theta + BC\sin\theta)$

$= AB\left(\frac{(AC)^2}{AB} + \frac{(BC)^2}{AB}\right)$

$= AC^2 + BC^2 = AB^2 = p^2$

18. (2) $\vec{a} + \vec{b} = \lambda\vec{c}$ …(i)

and $\vec{b} + \vec{c} = \mu\vec{a}$ …(ii)

$\therefore \quad (\lambda\vec{c} - \vec{a}) + \vec{c} = \mu\vec{a}$ (putting $\vec{b} = \lambda\vec{c} - \vec{a}$)

$\Rightarrow \quad (\lambda + 1)\vec{c} = (\mu + 1)\vec{a}$

$\Rightarrow \quad \lambda = \mu = -1$

$\Rightarrow \quad \vec{a} + \vec{b} + \vec{c} = 0$

or $\quad |\vec{a}|^2 + |\vec{b}|^2 + |\vec{c}|^2 + 2(\vec{a}\cdot\vec{b} + \vec{b}\cdot\vec{c} + \vec{c}\cdot\vec{a}) = 0$

or $\quad \vec{a}\cdot\vec{b} + \vec{b}\cdot\vec{c} + \vec{c}\cdot\vec{a} = -3$

19. (3) $|2\vec{a} + 3\vec{b}|^2 = |3\vec{a} + \vec{b}|^2$

$\Rightarrow \quad 4a^2 + 9b^2 + 12\vec{a}.\vec{b} = 9a^2 + b^2 + 6\vec{a}.\vec{b}$

$\Rightarrow \quad 5a^2 - 8b^2 - 6\vec{a}.\vec{b} = 0$

$\Rightarrow \quad 5(8)^2 - 8b^2 - 6(8)b\cos 60° = 0$

$\Rightarrow \quad b^2 + 3b - 40 = 0$

$\Rightarrow \quad (b-5)(b+8) = 0$

$\Rightarrow \quad b = 5$

20. (1) Let P.V. of P, A, B and C be $\vec{p}$, $\vec{a}$, $\vec{b}$ and $\vec{c}$, respectively, and $O(\vec{0})$ be the circumcentre of equilateral triangle ABC. Then

$|\vec{p}| = |\vec{b}| = |\vec{a}| = |\vec{c}| = \frac{l}{\sqrt{3}}$

Now $|\overrightarrow{PA}|^2 = |\vec{a} - \vec{p}|^2 = |\vec{a}|^2 + |\vec{p}|^2 - 2\vec{p}\cdot\vec{a}$

Similarly, $|\overrightarrow{PB}|^2 = |\vec{b}|^2 + |\vec{p}|^2 - 2\vec{p}\cdot\vec{b}$

and $|\overrightarrow{PC}|^2 = |\vec{c}|^2 + |\vec{p}|^2 - 2\vec{p}\cdot\vec{c}$

$\Rightarrow \quad \Sigma|\overrightarrow{PA}|^2 = 6\cdot\frac{l^2}{3} - 2\vec{p}\cdot(\vec{a} + \vec{b} + \vec{c})$

$= 2l^2 \quad (\text{as } (\vec{a} + \vec{b} + \vec{c})/3 = \vec{0})$

21. (3) Let $\vec{a} = 3\hat{i} + 4\hat{j} - 5\hat{k}$ and $\vec{b} = \ell\hat{i} + m\hat{j} + n\hat{k}$

$|\vec{a}\cdot\vec{b}| = |3\ell + 4m - 5n| \le \sqrt{3^2 + 4^2 + (-5)^2}\sqrt{\ell^2 + m^2 + n^2}$

$\Rightarrow \quad |3\ell + 4m - 5n| \le 5\sqrt{2}$

$\Rightarrow \quad |3\ell + 4m - 5n|_{max} = 5\sqrt{2}$, when vectors $\vec{a}$ and $\vec{b}$ are collinear.

$\Rightarrow \quad \frac{l}{3} = \frac{m}{4} = \frac{n}{-5} = \frac{\sqrt{\ell^2 + m^2 + n^2}}{\sqrt{3^2 + 4^2 + (-5)^2}} = \frac{1}{5\sqrt{2}}$

$\Rightarrow \quad |lm + mn + nl| = \left|\frac{12 - 20 - 15}{50}\right| = \frac{23}{50}$

22. (3) Let the projection be x, then

$\vec{a} = \frac{x(\hat{i} + \hat{j})}{\sqrt{2}} + \frac{x(-\hat{i} + \hat{j})}{\sqrt{2}} + x\,\hat{k}$

$\therefore \quad \vec{a} = \frac{2x\,\hat{j}}{\sqrt{2}} + x\,\hat{k}$ or $\hat{a} = \frac{\sqrt{2}}{\sqrt{3}}\hat{j} + \frac{\hat{k}}{\sqrt{3}}$

23. (2) We have

$\vec{p}\cdot\vec{q} = 0$

$\Rightarrow \quad (5\vec{a} - 3\vec{b})\cdot(-\vec{a} - 2\vec{b}) = 0$

$\Rightarrow \quad 6|\vec{b}|^2 - 5|\vec{a}|^2 - 7\vec{a}\cdot\vec{b} = 0$ ...(i)

Also $\vec{r}\cdot\vec{s} = 0$

$\Rightarrow \quad (-4\vec{a} - \vec{b})\,(-\vec{a} + \vec{b}) = 0$

$\Rightarrow \quad 4|\vec{a}|^2 - |\vec{b}|^2 - 3\vec{a}\cdot\vec{b} = 0$ ...(ii)

Now $\vec{x} = \frac{1}{3}(\vec{p} + \vec{r} + \vec{s})$

$= \frac{1}{3}(5\vec{a} - 3\vec{b} - 4\vec{a} - \vec{b} - \vec{a} + \vec{b}) = -\vec{b}$

and $\vec{y} = \frac{1}{5}(\vec{r} + \vec{s}) = \frac{1}{5}(-5\vec{a}) = -\vec{a}$

Angle between $\vec{x}$ and $\vec{y}$, i.e.,

$\cos\theta = \frac{\vec{x}\cdot\vec{y}}{|\vec{x}||\vec{y}|} = \frac{\vec{a}\cdot\vec{b}}{|\vec{a}||\vec{b}|}$ ...(iii)

From (i) and (ii),

$|\vec{a}| = \sqrt{\frac{25}{19}}\sqrt{\vec{a}\cdot\vec{b}}$

and $|\vec{b}| = \sqrt{\frac{43}{19}}\sqrt{\vec{a}\cdot\vec{b}}$

$\therefore \quad |\vec{a}||\vec{b}| = \frac{\sqrt{25\times 43}}{19}\,\vec{a}\cdot\vec{b}$

$\Rightarrow \quad \theta = \cos^{-1}\left(\frac{19}{5\sqrt{43}}\right)$

24. (3) $\vec{d}\cdot\hat{a} = \vec{d}\cdot\hat{b} = \vec{d}\cdot\hat{c}$

$\Rightarrow \quad \lambda(\hat{a}\cdot\hat{b} + \hat{a}\cdot\hat{c}) = \lambda(1 + \hat{b}\cdot\hat{c}) = \lambda(1 + \hat{b}\cdot\hat{c})$

$\Rightarrow \quad 1 + \hat{b}\cdot\hat{c} = \hat{a}\cdot\hat{b} + \hat{a}\cdot\hat{c}$

$\Rightarrow \quad 1 - \hat{a}\cdot\hat{b} + \hat{b}\cdot\hat{c} - \hat{a}\cdot\hat{c} = 0$

$\Rightarrow \quad 1 - \hat{a}\cdot\hat{b} + (\hat{b} - \hat{a})\cdot\hat{c} = 0$

$\Rightarrow \quad \hat{a}\cdot(\hat{a} - \hat{b}) + (\hat{b} - \hat{a})\cdot\hat{c} = 0$

$\Rightarrow \quad (\hat{a} - \hat{c})\cdot(\hat{a} - \hat{b}) = 0$ Hence $\hat{a} - \hat{c}$ is perpendicular to $(\hat{a} - \hat{b})$ i.e., The triangle is right angled.

25. (3)

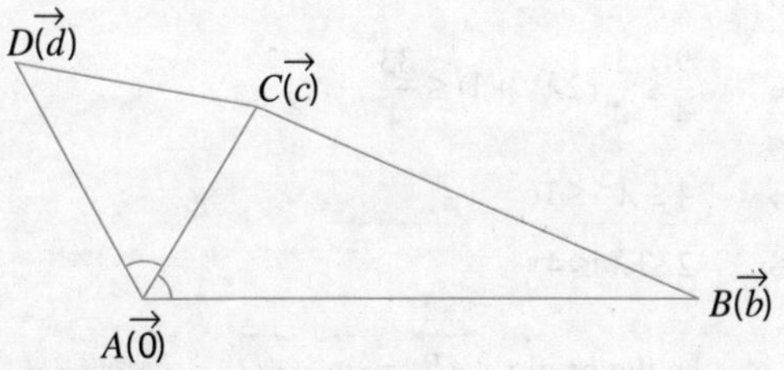

Let $|\overrightarrow{AC}| = \lambda > 0$

Then from $15|\overrightarrow{AC}| = 3|\overrightarrow{AB}| = 5|\overrightarrow{AD}|$

$|\overrightarrow{AB}| = 5\lambda$ and $|\overrightarrow{AD}| = 3\lambda$

Let θ be the angle between $\overrightarrow{BA}$ and $\overrightarrow{CD}$. Then

$\cos\theta = \frac{\overrightarrow{BA}\cdot\overrightarrow{CD}}{|\overrightarrow{BA}||\overrightarrow{CD}|} = \frac{-\vec{b}\cdot(\vec{d} - \vec{c})}{|\vec{b}||\vec{d} - \vec{c}|}$ ...(i)

Now $-\vec{b}\cdot(\vec{d} - \vec{c}) = \vec{b}\cdot\vec{c} - \vec{b}\cdot\vec{d}$

$= |\vec{b}||\vec{c}|\cos\frac{\pi}{3} - |\vec{b}||\vec{d}|\cos\frac{2\pi}{3}$

$= (5\lambda)(\lambda)\frac{1}{2} + (5\lambda)(3\lambda)\frac{1}{2}$

$= \frac{5\lambda^2 + 15\lambda^2}{2}$

$= 10\lambda^2$

Now $|\vec{d} - \vec{c}|^2 = |\vec{d}|^2 + |\vec{c}|^2 - 2\vec{c}\cdot\vec{d}$

$= 9\lambda^2 + \lambda^2 - 2(\lambda)(3\lambda)(1/2)$

$= 10\lambda^2 - 3\lambda^2$

$= 7\lambda^2$

$\therefore$ Denominator of (i), $|\vec{b}||\vec{d} - \vec{c}| = (5\lambda)(\sqrt{7}\lambda) = 5\sqrt{7}\lambda^2$

$\therefore \quad \cos\theta = \frac{10\lambda^2}{5\sqrt{7}\lambda^2} = \frac{2}{\sqrt{7}}$

26. (4) $|\vec{a}\cdot\vec{b}| = |\vec{a}\times\vec{b}|$

$\Rightarrow \quad |\vec{a}||\vec{b}||\cos\theta| = |\vec{a}||\vec{b}||\sin\theta|$

(where θ is the angle between $\vec{a}$ and $\vec{b}$)

$\Rightarrow \quad |\cos\theta| = |\sin\theta|$

$\Rightarrow \quad \theta = \frac{\pi}{4}$ or $\frac{3\pi}{4}$ (as $0 \le \theta \le \pi$)

But $\vec{a}\cdot\vec{b} < 0$, therefore $\theta = \frac{3\pi}{4}$

27. (3) $R(\vec{r})$ moves on PQ.

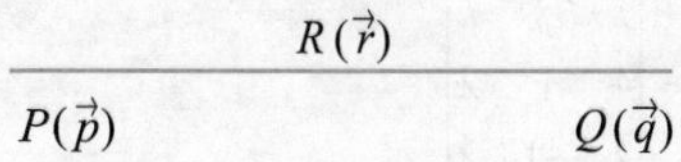

28. (2) $|\overrightarrow{AC}\times\overrightarrow{BD}|$

$= 2|\overrightarrow{AB}\times\overrightarrow{AD}|$

$= 2\begin{vmatrix} \hat{i} & \hat{j} & \hat{k} \\ 2 & 4 & -5 \\ 1 & 2 & 3 \end{vmatrix}$

$= |2[\hat{i}(12+10) - \hat{j}(6+5) + \hat{k}(4-4)]|$

$= |2[22\hat{i} - 11\hat{j}]|$

$= 22|[2\hat{i} - \hat{j}]|$

$= 22\times\sqrt{5}$

29. (3) $|\vec{a}+\vec{b}|^2 = |\vec{a}\times\vec{b}|$

$\Rightarrow \quad |\vec{a}|^2+|\vec{b}|^2+2\vec{a}\cdot\vec{b} = |\vec{a}|\,|\vec{b}|\sin\theta$

$\Rightarrow \quad 2(1+\cos\theta) = \sin\theta$

$\Rightarrow \quad \tan\frac{\theta}{2} = 2$

$\Rightarrow \quad \cos\theta = \dfrac{1-\tan^2\frac{\theta}{2}}{1+\tan^2\frac{\theta}{2}} = -\frac{3}{5}$

$\Rightarrow \vec{a}\cdot\vec{b} = (1)(1)\cos\theta = -\frac{3}{5}$

30. (3) $\vec{a}\times\vec{b}$ is a vector perpendicular to the plane containing $\vec{a}$ and $\vec{b}$. Similarly, $\vec{c}\times\vec{d}$ is a vector perpendicular to the plane containing $\vec{c}$ and $\vec{d}$.

Thus, the two planes will be parallel if their normals, i.e., $\vec{a}\times\vec{b}$ and $\vec{c}\times\vec{d}$, are parallel. Thus,

$(\vec{a}\times\vec{b})\times(\vec{c}\times\vec{d}) = \vec{0}$

31. (3) $4\vec{a}+5\vec{b}+9\vec{c} = 0$

$\Rightarrow$ Vectors $\vec{a}$, $\vec{b}$ and $\vec{c}$ are coplanar.

$\Rightarrow$ $\vec{b}\times\vec{c}$ and $\vec{c}\times\vec{a}$ are collinear

$\Rightarrow$ $(\vec{b}\times\vec{c})\times(\vec{c}\times\vec{a}) = \vec{0}$.

32. (3) $\vec{a}\times(\hat{i}+2\hat{j}+\hat{k}) = \hat{i}-\hat{k} = (\hat{j}\times(\hat{i}+2\hat{j}+\hat{k}))$

$\Rightarrow \quad (\vec{a}-\hat{j})\times(\hat{i}+2\hat{j}+\hat{k}) = \vec{0}$

$\Rightarrow \quad \vec{a}-\hat{j} = \lambda(\hat{i}+2\hat{j}+\hat{k})$

$\Rightarrow \quad \vec{a} = \lambda\hat{i}+(2\lambda+1)\hat{j}+\lambda\hat{k},\ \lambda\in R$

33. (1) A vector perpendicular to the plane of O, P and Q is $\overrightarrow{OP}\times\overrightarrow{OQ}$.

Now, $\overrightarrow{OP}\times\overrightarrow{OQ} = \begin{vmatrix}\hat{i} & \hat{j} & \hat{k}\\ 4 & 1 & \lambda\\ 2 & -1 & \lambda\end{vmatrix}$

$= 2\lambda\hat{i} - 2\lambda\hat{j} - 6\hat{k}$

Therefore, $\hat{i}-\hat{j}+6\hat{k}$ is parallel to

$2\lambda\hat{i} - 2\lambda\hat{j} - 6\hat{k}$

Hence, $\dfrac{1}{2\lambda} = \dfrac{-1}{-2\lambda} = \dfrac{6}{-6}$

$\lambda = -\frac{1}{2}$

34. (3) $k = |2(\vec{a}\times\vec{b})|+|3(\vec{a}\cdot\vec{b})|$

$= 12\sin\theta + 18\cos\theta$

$\Rightarrow$ Maximum value of $k = \sqrt{12^2+18^2} = 6\sqrt{13}$

35. (3) $1+9(\vec{a}\cdot\vec{b})^2 - 6(\vec{a}\cdot\vec{b}) + 4|\vec{a}|^2 + |\vec{b}|^2 + 9|\vec{a}\times\vec{b}|^2 + 4\vec{a}\cdot\vec{b} = 47$

$\Rightarrow \quad 1+4+4+36-4\cos\theta = 47$

$\Rightarrow \quad \cos\theta = -\frac{1}{2}$

Hence, the angle between $\vec{a}$ and $\vec{b}$ is $\frac{2\pi}{3}$.

36. (1) Let $\vec{c} = \lambda(\vec{a}\times\vec{b}) = \lambda\begin{vmatrix}\hat{i} & \hat{j} & \hat{k}\\ 2 & -3 & 1\\ 1 & -2 & 3\end{vmatrix}$

$= \lambda(-7\hat{i}-5\hat{j}-\hat{k})$

Hence, $\lambda(\vec{a}\times\vec{b})\cdot(\hat{i}+2\hat{j}-7\hat{k}) = 10$

$\Rightarrow \quad \lambda(-7-10+7) = 10$

or $\quad \lambda = -1$

$\Rightarrow \quad \vec{c} = -(\vec{a}\times\vec{b})$

37. (4) $\vec{0} = (\vec{a}+\vec{b})\cdot(2\vec{a}+3\vec{b})\times(3\vec{a}-2\vec{b})$

$= (\vec{a}+\vec{b})\cdot(-4\vec{a}\times\vec{b}-9\vec{a}\times\vec{b})$

$= -13(\vec{a}+\vec{b})\cdot(\vec{a}\times\vec{b})$

which is true for all values of $\vec{a}$ and $\vec{b}$.

38. (4) $\vec{a}\times(\vec{a}\times\vec{b}) = \vec{c} \Rightarrow |\vec{a}|\,|\vec{a}\times\vec{b}| = |\vec{c}|$

$(\because \vec{a}\perp(\vec{a}\times\vec{b}))$

$\Rightarrow 1(1\times5)\sin\theta = 3 \Rightarrow \sin\theta = \frac{3}{5} \Rightarrow \tan\theta = \frac{3}{4}$.

39. (2) $|\vec{a}\times\vec{b} - \vec{a}\times\vec{c}|^2$

$= |\vec{a}\times(\vec{b}-\vec{c})|^2$

$= |\vec{a}|^2|\vec{b}-\vec{c}|^2$

$= |\vec{b}|^2+|\vec{c}|^2-2|\vec{b}||\vec{c}|\cos\frac{\pi}{3} = 1$

40. (4) $\Delta = \frac{1}{2}|(\hat{j}+\lambda\hat{k})\times(\hat{i}+\lambda\hat{k})| = \frac{1}{2}|-\hat{k}+\lambda\hat{i}+\lambda\hat{j}|$

$= \frac{1}{2}\sqrt{2\lambda^2+1}$

$\Rightarrow \quad \frac{9}{4} \le \frac{1}{4}(2\lambda^2+1) \le \frac{33}{4}$

$\Rightarrow \quad 4 \le \lambda^2 \le 16$

$\Rightarrow \quad 2 \le |\lambda| \le 4$

41. (1) Let A be the origin. $\overrightarrow{AB} = \vec{a}$, $\overrightarrow{AD} = \vec{b}$

So, $\overrightarrow{AE} = \vec{b}+\frac{3}{2}\vec{a}$, $\overrightarrow{AG} = \vec{a}+3\vec{b}$.

So, required ratio is:

$$= \frac{\frac{1}{2}\left|(\vec{a}+3\vec{b})\times\left(\vec{b}+\frac{3}{2}\vec{a}\right)\right|}{\frac{1}{2}|\vec{a}\times\vec{b}|}$$

$$= \frac{7}{2}$$

42. (3) $|\vec{a}\times\vec{r}| = |\vec{c}|$

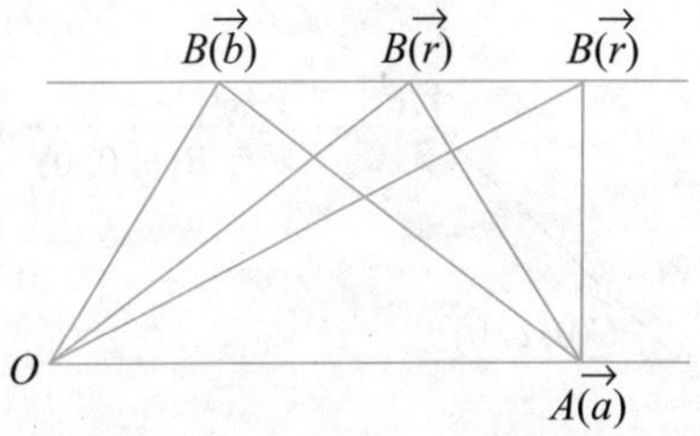

Triangles on the same base and between the same parallel lines have the same area.

43. (1) Let $\vec{r} = x_1\hat{a} + x_2\hat{b} + x_3(\hat{a}\times\hat{b})$

$\Rightarrow \quad \vec{r}\cdot\hat{a} = x_1 + x_2\hat{a}\cdot\hat{b} + x_3\hat{a}\cdot(\hat{a}\times\hat{b}) = x_1$

Also, $\vec{r}\cdot\hat{b} = x_1\hat{a}\cdot\hat{b} + x_2 + x_3\hat{b}\cdot(\hat{a}+\hat{b}) = x_2$

and $\vec{r}\cdot(\hat{a}\times\hat{b}) = x_1\hat{a}\cdot(\hat{a}\times\hat{b}) + x_2\hat{b}\cdot(\hat{a}\times\hat{b})$

$+x_3(\hat{a}\times\hat{b})\cdot(\hat{a}\times\hat{b}) = x_3$

$\Rightarrow \quad \vec{r} = (\vec{r}\cdot\hat{a})\hat{a} + (\vec{r}\cdot\hat{b})\hat{b} + (\vec{r}\cdot(\hat{a}\times\hat{b}))(\hat{a}\times\hat{b})$

44. (4) $|\vec{a}| = 1, |\vec{b}| = 4, \vec{a}.\vec{b} = 2$

$\vec{c} = (2\vec{a}\times\vec{b}) - 3\vec{b}$

$\Rightarrow \vec{c} + 3\vec{b} = 2\vec{a}\times\vec{b}$

Taking dot product with $\vec{b}$, we get

$\vec{b}\cdot\vec{c} + 3|\vec{b}|^2 = 0$

$\Rightarrow \quad \vec{b}\cdot\vec{c} = -48$

Now $|\vec{c}|^2 = |2\vec{a}\times\vec{b} - 3\vec{b}|^2$

$= 4|\vec{a}\times\vec{b}|^2 + 9|\vec{b}|^2$

$= 4(|\vec{a}|^2|\vec{b}|^2 - (\vec{a}\cdot\vec{b})^2) + 9|\vec{b}|^2$

$= 4(16-4) + 9\times 16$

$\therefore \quad |\vec{c}|^2 = 192$

Now $|\vec{c} + 3\vec{b}|^2 = 4|\vec{a}\times\vec{b}|^2$

$\Rightarrow \quad |\vec{c}|^2 + 9|\vec{b}|^2 + 6\vec{b}\cdot\vec{c} = 4\times 12$

$\Rightarrow \quad 192 + 144 + 6\times 4\times\sqrt{192}\ \cos\theta = 48$

$\Rightarrow \quad \cos\theta = -\frac{\sqrt{3}}{2} \Rightarrow \theta = \frac{5\pi}{6}$

45. (3) $\vec{r}\times\vec{a} = \vec{b}\times\vec{a}$ or $(\vec{r}-\vec{b})\times\vec{a} = 0$

$\vec{r}\times\vec{b} = \vec{a}\times\vec{b}$ or $(\vec{r}-\vec{a})\times\vec{b} = 0$

If $\vec{r} = x\hat{i} + y\hat{j} + z\hat{k}$, then

$$\begin{vmatrix} \hat{i} & \hat{j} & \hat{k} \\ x-2 & y & z+1 \\ 1 & 1 & 0 \end{vmatrix} = 0 \text{ and } \begin{vmatrix} \hat{i} & \hat{j} & \hat{k} \\ x-1 & y-1 & z \\ 2 & 0 & -1 \end{vmatrix} = 0$$

$\Rightarrow \quad z+1 = 0, x-y = 2$

and $\quad y-1 = 0, x-1+2z = 0$

$\Rightarrow \quad x = 3, y = 1, z = -1$

46. (2) $\vec{c} = m\vec{a} + n\vec{b} + p(\vec{a}\times\vec{b})$

Taking dot product with $\vec{a}$ and $\vec{b}$, we have

$m = n = \cos\theta$

$\Rightarrow \quad |\vec{c}| = |\cos\theta\,\vec{a} + \cos\theta\,\vec{b} + p(\vec{a}\times\vec{b})| = 1$

Squaring both sides, we get

$\cos^2\theta + \cos^2\theta + p^2 = 1$

or $\quad \cos\theta = \pm\frac{\sqrt{1-p^2}}{\sqrt{2}}$

Now $-\frac{1}{\sqrt{2}} \le \cos\theta \le \frac{1}{\sqrt{2}}$ (for real value of θ)

$\therefore \quad \frac{\pi}{4} \le \theta \le \frac{3\pi}{4}$

47. (3) $\vec{a} = 2\hat{i} + \hat{j} - 2\hat{k}, \vec{b} = \hat{i} + \hat{j}$

$\Rightarrow \quad \vec{a}\times\vec{b} = 2\hat{i} - 2\hat{j} + \hat{k}$

$\Rightarrow \quad |\vec{a}\times\vec{b}| = 3$

Now $|(\vec{a}\times\vec{b})\times\vec{c}| = 3$

$\Rightarrow \quad 3\cdot|\vec{c}|\cdot\sin 30° = 3$

$\Rightarrow \quad |\vec{c}| = 2$

Also $|\vec{c} - \vec{a}| = 3$

$\Rightarrow \quad c^2 + a^2 - 2\vec{a}\cdot\vec{c} = 9$

$\Rightarrow \quad 4 + 9 - 2\vec{a}\cdot\vec{c} = 9$

$\Rightarrow \quad \vec{a}\cdot\vec{c} = 2$

48. (3) We have $(\vec{a} + 2\vec{b})\times\vec{c} = 3(\vec{c}\times\vec{a})$

$\Rightarrow \quad (\vec{a} + 2\vec{b})\times\vec{c} + 3(\vec{a}\times\vec{c}) = 0$

$\Rightarrow \quad (\vec{a} + 2\vec{b} + 3\vec{a})\times\vec{c} = 0$

$\Rightarrow \quad (4\vec{a} + 2\vec{b})\times\vec{c} = 0$

$\Rightarrow \quad \vec{c} = \lambda(4\vec{a} + 2\vec{b}) = \lambda(8\hat{i} - 14\hat{j} + 30\hat{k})$

Now, $\vec{a}.\vec{c} = 130$

$\Rightarrow \quad 8\lambda + 42\lambda + 210\lambda = 130$

$\Rightarrow \quad \lambda = \frac{1}{2}$

$\Rightarrow \quad \vec{c} = 4\hat{i} - 7\hat{j} + 15\hat{k}$

$\Rightarrow \quad \vec{b}.\vec{c} = 8 + 7 + 15 = 30$

49. (2) $\vec{r}\times\vec{a} = \vec{b}\times\vec{r}$

$\Rightarrow \quad \vec{r}\times(\vec{a} + \vec{b}) = 0$

$\Rightarrow \quad \vec{r} = \lambda(\vec{a} + \vec{b})$

$\Rightarrow \quad \vec{r} = \lambda(3\hat{i} - \hat{j} + 2\hat{k})$ ...(i)

Also, $\vec{r}.(\alpha\hat{i} + 2\hat{j} + \hat{k}) = 3$

$\therefore \quad \alpha\lambda = 1$ ... (ii)

Further, $\vec{r}.(2\hat{i} + 5\hat{j} - \alpha\hat{k}) = -1$

$\therefore \quad 2\lambda\alpha - \lambda = 1$ ...(iii)

Solving (ii) and (iii), we get $\alpha = 1, \lambda = 1$

$\therefore \quad \vec{r} = 3\hat{i} - \hat{j} + 2\hat{k}$

$\Rightarrow \quad |\vec{r}|^2 = 14$

$\Rightarrow \quad \alpha + |\vec{r}|^2 = 15$

50. (3) We have, $\vec{a} = 6\hat{i} + 9\hat{j} + 12\hat{k}$

and $\vec{b} = \alpha\hat{i} + 11\hat{j} - 2\hat{k}$

Now, $\vec{a} \times \vec{c} - \vec{a} \times \vec{b} = 0$

$\Rightarrow \quad \vec{a} \times (\vec{c} - \vec{b}) = 0$

$\Rightarrow \quad \vec{c} - \vec{b} = \lambda\vec{a}$

$\Rightarrow \quad \vec{c} = \vec{b} + \lambda\vec{a} \qquad \ldots \text{(i)}$

Now, $\vec{c} \cdot (\hat{i} - 2\hat{j} + \hat{k}) = 5$

$\Rightarrow \quad (\vec{b} + \lambda\vec{a}) \cdot (\hat{i} - 2\hat{j} + \hat{k}) = 5$

$\Rightarrow \quad (\alpha - 22 - 2) + \lambda(6 - 18 + 12) = 5$

$\Rightarrow \quad \alpha = 29$

Also, $\vec{a} \cdot \vec{c} = -12$,

$\Rightarrow \quad \vec{b} \cdot \vec{a} + \lambda(\vec{a} \cdot \vec{a}) = -12$

$\Rightarrow \quad 174 + 99 - 24 + \lambda(261) = -12$

$\Rightarrow \quad \lambda = -1$

Thus, $\vec{c} = 23\hat{i} + 2\hat{j} - 14\hat{k}$

Hence $\vec{c} \cdot (\hat{i} + \hat{j} + \hat{k}) = 23 + 2 - 14 = 11$

51. (3) Let the required vector $\vec{r}$ be such that

$$\vec{r} = x_1\vec{a} + x_2\vec{b} + x_3\vec{a} \times \vec{b}$$

We must have $\vec{r} \cdot \vec{a} = \vec{r} \cdot \vec{b} = \vec{r} \cdot (\vec{a} \times \vec{b})$ (as $\vec{r}, \vec{a}, \vec{b}$ and $\vec{a} \times \vec{b}$ are unit vectors and $\vec{r}$ is equally inclined to $\vec{a}, \vec{b}$ and $\vec{a} \times \vec{b}$)

Now $\vec{r} \cdot \vec{a} = x_1$, $\vec{r} \cdot \vec{b} = x_2$, $\vec{r} \cdot (\vec{a} \times \vec{b}) = x_3$

$\Rightarrow \quad \vec{r} = \lambda(\vec{a} + \vec{b} + (\vec{a} \times \vec{b}))$

Also, $\vec{r} \cdot \vec{r} = 1$

$\Rightarrow \quad \lambda^2(\vec{a} + \vec{b} + \vec{a} \times \vec{b}) \cdot (\vec{a} + \vec{b} + (\vec{a} \times \vec{b})) = 1$

$\Rightarrow \quad \lambda^2(|\vec{a}|^2 + |\vec{b}|^2 + |\vec{a} \times \vec{b}|^2) = 1$

$\Rightarrow \quad \lambda^2 = \dfrac{1}{3}$

$\Rightarrow \quad \lambda = \pm\dfrac{1}{\sqrt{3}}$

$\Rightarrow \quad \vec{r} = \pm\dfrac{1}{\sqrt{3}}(\vec{a} + \vec{b} + \vec{a} \times \vec{b})$

52. (3) $\vec{a}_1 = (\vec{a} \cdot \hat{b})\hat{b} = \dfrac{(\vec{a} \cdot \vec{b})\vec{b}}{|\vec{b}|^2}$

$\Rightarrow \quad \vec{a}_2 = \vec{a} - \vec{a}_1 = \vec{a} - \dfrac{(\vec{a} \cdot \vec{b})\vec{b}}{|\vec{b}|^2}$

Thus, $\vec{a}_1 \times \vec{a}_2 = \dfrac{(\vec{a} \cdot \vec{b})\vec{b}}{|\vec{b}|^2} \times \left(\vec{a} - \dfrac{(\vec{a} \cdot \vec{b})\vec{b}}{|\vec{b}|^2}\right)$

$$= \frac{(\vec{a} \cdot \vec{b})(\vec{b} \times \vec{a})}{|\vec{b}|^2}$$

53. (1) Area of $\Delta BCD = \dfrac{1}{2}|\overrightarrow{BC} \times \overrightarrow{BD}|$

$$= \frac{1}{2}|(b\hat{i} - c\hat{j}) \times (b\hat{i} - d\hat{k})|$$

$$= \frac{1}{2}|bd\,\hat{j} + bc\,\hat{k} + dc\,\hat{i}|$$

$$= \frac{1}{2}\sqrt{b^2c^2 + c^2d^2 + d^2b^2} \qquad \ldots \text{(i)}$$

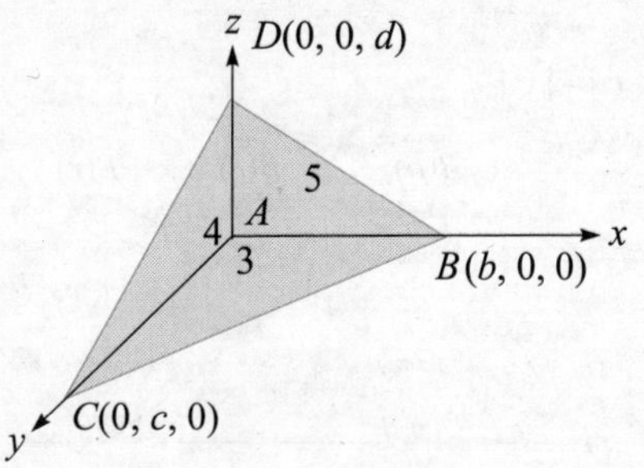

Now, $6 = bc$; $8 = cd$; $10 = bd$

$\therefore \quad b^2c^2 + c^2d^2 + d^2b^2 = 200$

Substituting the value in (i), we get

$$\Delta BCD = \frac{1}{2}\sqrt{200} = 5\sqrt{2}$$

54. (3) $3 = \dfrac{1}{6}[\vec{a}\ \vec{b}\ \vec{c}]$

or $\quad [\vec{a}\ \vec{b}\ \vec{c}] = 18$

Volume of the required parallelepiped

$$= [\vec{a} + \vec{b}\ \ \vec{b} + \vec{c}\ \ \vec{c} + \vec{a}]$$

$$= 2[\vec{a}\ \vec{b}\ \vec{c}] = 36$$

55. (2) Here $[\vec{a}\ \vec{b}\ \vec{c}] = \pm 1$

$\therefore \quad [\vec{a} + \vec{b} + \vec{c}\ \ \vec{a} + \vec{b}\ \ \vec{b} + \vec{c}]$

$$= (\vec{a} + \vec{b} + \vec{c}) \times (\vec{a} + \vec{b}) \cdot (\vec{b} + \vec{c})$$

$$= \vec{c} \times (\vec{a} + \vec{b}) \cdot (\vec{b} + \vec{c})$$

$$= (\vec{c} \times \vec{a} + \vec{c} \times \vec{b}) \cdot (\vec{b} + \vec{c})$$

$$= \vec{c} \times \vec{a} \cdot \vec{b} = [\vec{a}\ \vec{b}\ \vec{c}] = \pm 1$$

56. (4) $\vec{a} \perp \vec{b} \Rightarrow x - y + 2 = 0$

$\vec{a} \cdot \vec{c} = 4 \Rightarrow x + 2y = 4$

Solving we get $x = 0$; $y = 2$

$\Rightarrow \quad \vec{a} = 2\hat{j} + 2\hat{k}$

$\Rightarrow \quad [\vec{a}\ \vec{b}\ \vec{c}] = \begin{vmatrix} 0 & 2 & 2 \\ 1 & -1 & 1 \\ 1 & 2 & 0 \end{vmatrix} = 8 = |\vec{a}|^2$

57. (3) $\vec{r} \cdot \vec{a} = 0$, $|\vec{r} \times \vec{b}| = |\vec{r}||\vec{b}|$ and $|\vec{r} \times \vec{c}|$

$$= |\vec{r}||\vec{c}|$$

$\Rightarrow \quad \vec{r} \perp \vec{a}, \vec{b}, \vec{c}$

$\therefore \quad [\vec{a}\ \vec{b}\ \vec{c}] = 0$

58. (4) $(\hat{i}+\hat{j})\times(\hat{j}+\hat{k}) = \hat{i}-\hat{j}+\hat{k}$ so that unit vector perpendicular to the plane of $\hat{i}+\hat{j}$ and $\hat{j}+\hat{k}$ is

$$\frac{1}{\sqrt{3}}(\hat{i}-\hat{j}+\hat{k}).$$

Similarly, the other two unit vectors are

$$\frac{1}{\sqrt{3}}(\hat{i}+\hat{j}-\hat{k}) \text{ and } \frac{1}{\sqrt{3}}(-\hat{i}+\hat{j}+\hat{k}).$$

$$\text{Required volume} = \frac{1}{3\sqrt{3}}\begin{vmatrix}1 & -1 & 1\\ 1 & 1 & -1\\ -1 & 1 & 1\end{vmatrix} = \frac{4}{3\sqrt{3}}$$

59. (2) Taking dot product of $a(\vec{\alpha}\times\vec{\beta})+b(\vec{\beta}\times\vec{\gamma})+c(\vec{\gamma}\times\vec{\alpha}) = 0$ with $\vec{\gamma}, \vec{\alpha}$ and $\vec{\beta}$, respectively, we have

$$a[\vec{\alpha}\,\vec{\beta}\,\vec{\gamma}] = 0$$

$$b[\vec{\alpha}\,\vec{\beta}\,\vec{\gamma}] = 0$$

$$c[\vec{\alpha}\,\vec{\beta}\,\vec{\gamma}] = 0$$

Since at least one of a, b and $c \neq 0$, we have

$$[\vec{\alpha}\,\vec{\beta}\,\vec{\gamma}] = 0$$

Hence, $\vec{\alpha}, \vec{\beta}$ and $\vec{\gamma}$ are coplanar.

60. (1) We have,

$$\vec{a}\cdot\vec{p} = \vec{a}\cdot\frac{(\vec{b}\times\vec{c})}{[\vec{a}\,\vec{b}\,\vec{c}]} = \frac{\vec{a}\cdot(\vec{b}\times\vec{c})}{[\vec{a}\,\vec{b}\,\vec{c}]} = \frac{[\vec{a}\,\vec{b}\,\vec{c}]}{[\vec{a}\,\vec{b}\,\vec{c}]} = 1$$

$$\vec{a}.\vec{q} = \vec{a}\cdot\frac{\vec{c}\times\vec{a}}{[\vec{a}\,\vec{b}\,\vec{c}]} = \frac{[\vec{a}\,\vec{c}\,\vec{a}]}{[\vec{a}\,\vec{b}\,\vec{c}]} = 0$$

Similarly, $\vec{a}\cdot\vec{r} = 0$, $\vec{b}\cdot\vec{p} = 0, \vec{b}\cdot\vec{q} = 1, \vec{b}\cdot\vec{r} = 0$, $\vec{c}\cdot\vec{p} = 0$, $\vec{c}\cdot\vec{q} = 0$ and $\vec{c}\cdot\vec{r} = 1$

$$\therefore \quad (\vec{a}+\vec{b}+\vec{c})\cdot(\vec{p}+\vec{q}+\vec{r})$$

$$= \vec{a}\cdot\vec{p}+\vec{a}\cdot\vec{q}+\vec{a}\cdot\vec{r}+\vec{b}\cdot\vec{p}+\vec{b}\cdot\vec{q}+\vec{b}\cdot\vec{r}+\vec{c}\cdot\vec{p}+\vec{c}\cdot\vec{q}+\vec{c}\cdot\vec{r}$$

$$= 1+1+1 = 3$$

61. (3) Consider a tetrahedron with vertices $O(0, 0, 0)$, $A(a, 0, 0)$, $B(0, b, 0)$ and $C(0, 0, c)$.

$$\text{Volume } V = \frac{1}{6}[\vec{a}\,\vec{b}\,\vec{c}]$$

Now centroids of the faces OAB, OAC, OBC and ABC are $G_1(a/3, b/3, 0)$, $G_2(a/3, 0, c/3)$, $G_3(0, b/3, c/3)$ and $G_4(a/3, b/3, c/3)$, respectively.

$$\overrightarrow{G_4G_1} = \vec{c}/3,\ \overrightarrow{G_4G_2} = \vec{b}/3,\ \overrightarrow{G_4G_3} = \vec{a}/3.$$

Volume of tetrahedron by centroids

$$V' = \frac{1}{6}\left[\frac{\vec{a}}{3}\ \frac{\vec{b}}{3}\ \frac{\vec{c}}{3}\right] = \frac{1}{27}V$$

$$\Rightarrow \quad K = 27$$

62. (4) $\vec{r} = x_1(\vec{a}\times\vec{b})+x_2(\vec{b}\times\vec{c})+x_3(\vec{c}\times\vec{a})$

$$\Rightarrow \quad \vec{r}\cdot\vec{a} = x_2[\vec{a}\,\vec{b}\,\vec{c}],\ \vec{r}\cdot\vec{b} = x_3[\vec{b}\,\vec{c}\,\vec{a}]$$

$$\text{and} \quad \vec{r}\cdot\vec{c} = x_1[\vec{c}\,\vec{a}\,\vec{b}] = x_1[\vec{a}\,\vec{b}\,\vec{c}]$$

$$\Rightarrow \quad x_1+x_2+x_3 = 4\vec{r}.(\vec{a}+\vec{b}+\vec{c})$$

63. (2) $\vec{c} = \lambda(\vec{a}\times\vec{b})$

$$\Rightarrow \quad \vec{c}\cdot\vec{c} = \lambda(\vec{a}\times\vec{b})\cdot\vec{c}$$

$$\Rightarrow \quad \frac{1}{3} = \lambda$$

Also $|\vec{c}|^2 = \lambda^2|\vec{a}\times\vec{b}|^2$

$$\Rightarrow \quad \frac{1}{3} = \frac{1}{9}(a^2b^2\sin^2\theta) = \frac{1}{9}\times 2\times 3\sin^2\theta$$

$$\Rightarrow \quad \sin^2\theta = \frac{1}{2}$$

$$\Rightarrow \quad \theta = \frac{\pi}{4}$$

64. (3) $\vec{d}\cdot\vec{a} = \vec{d}\cdot\vec{b} = \vec{d}\cdot\vec{c} = [\vec{a}\,\vec{b}\,\vec{c}]$

Then $|(\vec{d}\cdot\vec{c})(\vec{a}\times\vec{b})+(\vec{d}\cdot\vec{a})(\vec{b}\times\vec{c})+(\vec{d}\cdot\vec{b})(\vec{c}\times\vec{a})| = 0$

or $\quad [\vec{a}\,\vec{b}\,\vec{c}]\,|\vec{a}\times\vec{b}+\vec{b}\times\vec{c}+\vec{c}\times\vec{a}| = 0$

or $\quad [\vec{a}\,\vec{b}\,\vec{c}] = 0 \quad (\because \vec{d}$ is non-zero)

Hence, $\vec{a}, \vec{b}, \vec{c}$ are coplanar.

65. (4) Let $\vec{a} = 6\hat{i}+6\hat{k}$, $\vec{b} = 4\hat{j}+2\hat{k}$, $\vec{c} = 4\hat{j}-8\hat{k}$

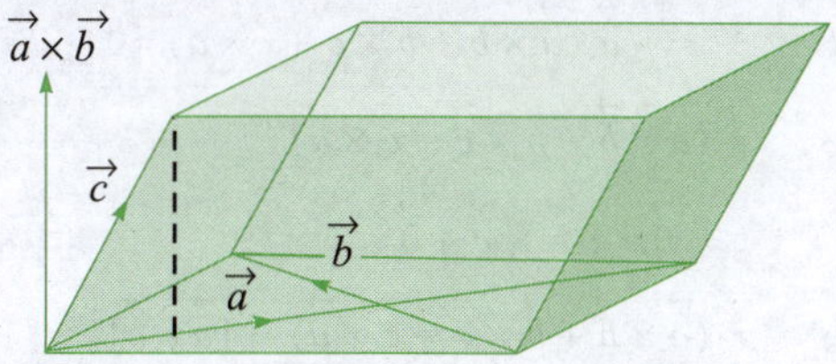

Then $\vec{a}\times\vec{b} = -24\hat{i}-12\hat{j}+24\hat{k}$

$$= 12(-2\hat{i}-\hat{j}+2\hat{k})$$

$\therefore$ Area of the base of the parallelepiped

$$= \frac{1}{2}|\vec{a}\times\vec{b}|$$

$$= \frac{1}{2}(12\times 3)$$

$$= 18$$

Height of the parallelepiped

= Length of projection of $\vec{c}$ on $\vec{a}\times\vec{b}$

$= \dfrac{|\vec{c}\cdot\vec{a}\times\vec{b}|}{|\vec{a}\times\vec{b}|}$

$= \dfrac{|12(-4-16)|}{36}$

$= \dfrac{20}{3}$

$\therefore$ Volume of the parallelepiped $= 18 \times \dfrac{20}{3} = 120$

66. (3) Any vector $\vec{r}$ can be represented in terms of three non-coplanar vectors $\vec{a}, \vec{b}$ and $\vec{c}$ as

$\vec{r} = x(\vec{a}\times\vec{b}) + y(\vec{b}\times\vec{c}) + z(\vec{c}\times\vec{a})$ ...(i)

Taking dot product with $\vec{a}, \vec{b}$ and c, respectively, we have

$x = \dfrac{\vec{r}\cdot\vec{c}}{[\vec{a}\,\vec{b}\,\vec{c}]}$, $y = \dfrac{\vec{r}\cdot\vec{a}}{[\vec{a}\,\vec{b}\,\vec{c}]}$ and $z = \dfrac{\vec{r}\cdot\vec{b}}{[\vec{a}\,\vec{b}\,\vec{c}]}$

From (i), we have

$[\vec{a}\,\vec{b}\,\vec{c}]\vec{r} = \dfrac{1}{2}(\vec{a}\times\vec{b} + \vec{b}\times\vec{c} + \vec{c}\times\vec{a})$

$\therefore$ Area of ΔABC

$= \dfrac{1}{2}|\vec{a}\times\vec{b} + \vec{b}\times\vec{c} + \vec{c}\times\vec{a}|$

$= |[\vec{a}\,\vec{b}\,\vec{c}]\vec{r}|$

67. (2) A vector perpendicular to the plane of $A(\vec{a})$, $B(\vec{b})$ and $C(\vec{c})$ is

$(\vec{b}-\vec{a})\times(\vec{c}-\vec{a}) = \vec{a}\times\vec{b} + \vec{b}\times\vec{c} + \vec{c}\times\vec{a}$.

Now for any point $R(\vec{r})$ in the plane of A, B and C is

$(\vec{r}-\vec{a})\cdot(\vec{a}\times\vec{b} + \vec{b}\times\vec{c} + \vec{c}\times\vec{a}) = 0$

$\vec{r}\cdot(\vec{a}\times\vec{b} + \vec{b}\times\vec{c} + \vec{c}\times\vec{a})$

$- \vec{a}\cdot(\vec{a}\times\vec{b} + \vec{b}\times\vec{c} + \vec{c}\times\vec{a}) = 0$

$\Rightarrow \vec{r}\cdot(\vec{a}\times\vec{b} + \vec{b}\times\vec{c} + \vec{c}\times\vec{a})$

$= 0 + \vec{a}\cdot\vec{b}\times\vec{c} + 0$

$\Rightarrow \vec{r}\cdot(\vec{a}\times\vec{b} + \vec{b}\times\vec{c} + \vec{c}\times\vec{a}) = [\vec{a}\,\vec{b}\,\vec{c}]$

68. (1) Let $\vec{v} = x\vec{a} + y\vec{b} + z\vec{a}\times\vec{b}$

Given: $\vec{a}\cdot\vec{b} = 0$, $\vec{v}\cdot\vec{a} = 0$, $\vec{v}\cdot\vec{b} = 1$, $[\vec{v}\;\vec{a}\;\vec{b}] = 1$

$\Rightarrow \vec{v}\cdot\vec{a} = x\,\vec{a}\cdot\vec{a} = x|\vec{a}|^2$

$(\because \vec{a}\cdot\vec{b} = 0,\ \vec{a}\cdot\vec{a}\times\vec{b} = 0)$

$\Rightarrow x = 0$

Again, $\vec{v}\cdot\vec{b} = y|\vec{b}|^2$ or $1 = yb^2$

$\therefore y = \dfrac{1}{b^2}$ ...(ii)

Again, $\vec{v}\cdot(\vec{a}\times\vec{b}) = z(\vec{a}\times\vec{b})^2$

$\Rightarrow 1 = z(\vec{a}\times\vec{b})^2$ or $z = \dfrac{1}{|\vec{a}\times\vec{b}|^2}$

Hence, $\vec{v} = \dfrac{1}{|\vec{b}|^2}\vec{b} + \dfrac{1}{|\vec{a}\times\vec{b}|^2}\vec{a}\times\vec{b}$

69. (1) $(\vec{a}\times(\vec{a}\times(\vec{a}\times(\vec{a}\times\vec{b}))))$

$= (\vec{a}\times(\vec{a}\times((\vec{a}\cdot\vec{b})\vec{a} - (\vec{a}\cdot\vec{b})\vec{b})))$

$= (\vec{a}\times(\vec{a}\times(-4\vec{b})))$

$= -4(\vec{a}\times(\vec{a}\times\vec{b}))$

$= -4((\vec{a}\cdot\vec{b})\vec{a} - (\vec{a}\cdot\vec{a})\vec{b})$

$= -4(-4\vec{b}) = 16\vec{b} = 48\hat{b}$

70. (1) A vector coplanar with $\vec{a}$ and $\vec{b}$ and perpendicular to $\vec{c}$ is $\lambda((\vec{a}\times\vec{b})\times\vec{c})$.

But $\lambda((\vec{a}\times\vec{b})\times\vec{c}) = \lambda[(\vec{a}\cdot\vec{c})\vec{b} - (\vec{b}\cdot\vec{c})\vec{a}]$

$= \lambda[4\vec{b} - 4\vec{a}]$

$= 4\lambda[\hat{j} - \hat{k}]$

Now $4|\lambda|\sqrt{2} = \sqrt{2}$ (Given) $\Rightarrow \lambda = \pm\dfrac{1}{4}$

Hence, the required vector is $\hat{j} - \hat{k}$ or $-\hat{j} + \hat{k}$

71. (1) $\vec{\alpha} \parallel (\vec{\beta}\times\vec{\gamma}) \Rightarrow \vec{\alpha}\perp\vec{\beta}$ and $\vec{\alpha}\perp\vec{\gamma}$

Now, $(\vec{\alpha}\times\vec{\beta})\cdot(\vec{\alpha}\times\vec{\gamma})$

$= |\vec{\alpha}|^2(\vec{\beta}\cdot\vec{\gamma}) - (\vec{\alpha}\cdot\vec{\beta})(\vec{\alpha}\cdot\vec{\gamma})$

$= |\vec{\alpha}|^2\cdot(\vec{\beta}\cdot\vec{\gamma})$

72. (1) $(\vec{a}\times\vec{b})\cdot(\vec{a}\times\vec{c})$

$= \vec{a}\cdot[\vec{b}\times(\vec{a}\times\vec{c})]$

$= \vec{a}\cdot[(\vec{b}\cdot\vec{c})\vec{a} - (\vec{a}\cdot\vec{b})\vec{c}]$

$= \vec{a}\cdot(\vec{b}\cdot\vec{c})\vec{a}$ $(\because \vec{a}\cdot\vec{b} = 0)$

$= |\vec{a}|^2(\vec{b}\cdot\vec{c})$

73. (1) $[\vec{a}\times\vec{b}\;\;\vec{a}\times\vec{c}\;\;\vec{d}]$

$= (\vec{a}\times\vec{b})\cdot((\vec{a}\times\vec{c})\times\vec{d})$

$= (\vec{a}\times\vec{b})\cdot((\vec{a}\cdot\vec{d})\vec{c} - (\vec{c}\cdot\vec{d})\vec{a})$

$= (\vec{a}\cdot\vec{d})[\vec{a}\,\vec{b}\,\vec{c}]$

74. (1) $[\vec{a} + (\vec{a}\times\vec{b})\;\;\vec{b} + (\vec{a}\times\vec{b})\;\;\vec{a}\times\vec{b}]$

$= (\vec{a} + (\vec{a}\times\vec{b}))\cdot((\vec{b} + (\vec{a}\times\vec{b}))\times(\vec{a}\times\vec{b}))$

$= (\vec{a} + (\vec{a}\times\vec{b}))\cdot(\vec{b}\times(\vec{a}\times\vec{b}))$

$= (\vec{a} + (\vec{a}\times\vec{b}))\cdot(\vec{a} - (\vec{a}\cdot\vec{b})\vec{b})$

$= \vec{a}\cdot\vec{a} = 1$ (as $\vec{a}\cdot\vec{b} = 0$, $\vec{a}\cdot(\vec{a}\times\vec{b}) = 0$)

75. (4) $((\vec{a}\times\vec{b})+(\vec{a}\times\vec{c}))\times(\vec{b}\times\vec{c})$

$= (\vec{a}\times\vec{b})\times(\vec{b}\times\vec{c})+(\vec{a}\times\vec{c})\times(\vec{b}\times\vec{c})$

$= ((\vec{a}\times\vec{b})\cdot\vec{c})\vec{b} - ((\vec{a}\times\vec{b})\cdot\vec{b})\vec{c}$

$+ ((\vec{a}\times\vec{c})\cdot\vec{c})\vec{b} - ((\vec{a}\times\vec{c})\cdot\vec{b})\vec{c}$

$= [\vec{a}\,\vec{b}\,\vec{c}](\vec{b}+\vec{c})$

$\Rightarrow \quad ((\vec{a}\times\vec{b})+(\vec{a}\times\vec{c}))\times(\vec{b}\times\vec{c}))\cdot(\vec{b}-\vec{c})$

$= [\vec{a}\,\vec{b}\,\vec{c}](\vec{b}+\vec{c}).(\vec{b}-\vec{c})$

$= [\vec{a}\,\vec{b}\,\vec{c}](|\vec{b}|^2-|\vec{c}|^2)=0$

76. (1) $\vec{a}\times\vec{b} = \vec{c}$

$\Rightarrow \quad \vec{a}\times(\vec{a}\times\vec{b}) = \vec{a}\times\vec{c}$

$\Rightarrow \quad (\vec{a}\cdot\vec{b})\vec{a} - |\vec{a}|^2\vec{b} = \vec{a}\times\vec{c}$

$\Rightarrow \quad \vec{b} = \dfrac{\beta\vec{a} - \vec{a}\times\vec{c}}{|\vec{a}|^2} \qquad (\because\ \vec{a}\cdot\vec{b} = \beta)$

77. (3) $(\vec{a}\times\vec{b})\times(\vec{b}\times\vec{c}) = \vec{b}$

or $\quad [\vec{a}\,\vec{b}\,\vec{c}]\vec{b} = \vec{b}$

or $\quad [\vec{a}\,\vec{b}\,\vec{c}] = 1$

Therefore, $\vec{a}$, $\vec{b}$ and $\vec{c}$ cannot be coplanar.

78. (1) $\{\vec{a}\times(\vec{b}+\vec{a}\times\vec{b})\}\cdot\vec{b}$

$= \{\vec{a}\times\vec{b}+\vec{a}\times(\vec{a}\times\vec{b})\}\cdot\vec{b}$

$= [\vec{a}\,\vec{b}\,\vec{b}]+\{(\vec{a}\cdot\vec{b})\vec{a}-(\vec{a}\cdot\vec{a})\vec{b}\}\cdot\vec{b}$

$= 0 + (\vec{a}\cdot\vec{b})^2 - (\vec{a}\cdot\vec{a})(\vec{b}\cdot\vec{b})$

$= \cos^2\dfrac{\pi}{3} - 1 = -\dfrac{3}{4}$

79. (3) The given equation reduces to

$[\vec{a}\ \vec{b}\ \vec{c}]^2 x^2 + 2[\vec{a}\ \vec{b}\ \vec{c}]x + 1 = 0 \Rightarrow D = 0$

80. (1) $\vec{a}\times(\vec{b}\times\vec{c}) = (\vec{a}\cdot\vec{c})\vec{b} - (\vec{a}\cdot\vec{b})\vec{c}$

$= 5(\hat{i}+2\hat{j}+2\hat{k}) - 6(\hat{i}+\hat{j}+2\hat{k}) = -\hat{i}+9\hat{j}-2\hat{k}$

$\Rightarrow \quad (1+\alpha)\hat{i} + \beta(1+\alpha)\hat{j} + \gamma(1+\alpha)(1+\beta)\hat{k}$

$= -\hat{i}+4\hat{j}-2\hat{k}$

$\Rightarrow \quad 1+\alpha = -1, \beta = -4$ and $\gamma(-1)(-3) = -2$

$\Rightarrow \quad \gamma = -\dfrac{2}{3}$

81. (3) Given that $\vec{a}$, $\vec{b}$ and $\vec{c}$ are non-coplanar. Thus,

$[\vec{a}\,\vec{b}\,\vec{c}] \neq 0 \qquad \ldots\text{(i)}$

Again $\vec{a}\times(\vec{b}\times\vec{c})\cdot(\vec{a}\times\vec{c}) = 0$

$\Rightarrow \quad [(\vec{a}\cdot\vec{c})\vec{b}-(\vec{a}\cdot\vec{b})\vec{c}]\cdot(\vec{a}\times\vec{c}) = 0$

$\Rightarrow \quad (\vec{a}\cdot\vec{c})[\vec{b}\,\vec{a}\,\vec{c}] = 0$

$\Rightarrow \quad (\vec{a}\cdot\vec{c}) = 0$

Hence, $\vec{a}$ and $\vec{c}$ are perpendicular. $\qquad \ldots\text{(ii)}$

$\therefore \quad \vec{a}\times(\vec{b}\times\vec{c}) = (\vec{a}\cdot\vec{c})\vec{b}-(\vec{a}\cdot\vec{b})\vec{c}$

$\Rightarrow \quad [\vec{a}\times(\vec{b}\times\vec{c})]\times\vec{c} = \vec{0}$

82. (3) $[(\vec{a}\times\vec{b})\times(\vec{b}\times\vec{c})\ \ (\vec{b}\times\vec{c})\times(\vec{c}\times\vec{a})\ \ (\vec{c}\times\vec{a})\times(\vec{a}\times\vec{b})]$

$= [[\vec{a}\ \vec{b}\ \vec{c}]\vec{b}\ \ [\vec{a}\ \vec{b}\ \vec{c}]\ \vec{c}\ \ [\vec{a}\ \vec{b}\ \vec{c}]\cdot\vec{a}]$

$= [\vec{a}\ \vec{b}\ \vec{c}]^3\ [\vec{b}\ \vec{c}\ \vec{a}] = [\vec{a}\ \vec{b}\ \vec{c}]^4$

83. (4) $\vec{a}' = \dfrac{\vec{b}\times\vec{c}}{[\vec{a}\,\vec{b}\,\vec{c}]} = \dfrac{\hat{i}+\hat{j}-\hat{k}}{2}$

84. (4) Volume of the parallelepiped formed by $\vec{a}'$, $\vec{b}'$ and $\vec{c}'$ is 4.

Therefore, the volume of the parallelepiped formed by $\vec{a}$, $\vec{b}$ and $\vec{c}$ is $\dfrac{1}{4}$.

$\vec{b}\times\vec{c} = [\vec{a}\ \vec{b}\ \vec{c}]\vec{a}' = \dfrac{1}{4}\vec{a}'$

$|\vec{b}\times\vec{c}| = \dfrac{\sqrt{2}}{4} = \dfrac{1}{2\sqrt{2}}$

Length of altitude $= \dfrac{1}{4}\times 2\sqrt{2} = \dfrac{1}{\sqrt{2}}$

85. (2) $(\vec{a}\times\vec{b})\times(\vec{r}\times\vec{c})$

$= ((\vec{a}\times\vec{b})\cdot\vec{c})\vec{r} - ((\vec{a}\times\vec{b})\cdot\vec{r})\cdot\vec{c}$

$= [\vec{a}\,\vec{b}\,\vec{c}]\vec{r} - [\vec{a}\,\vec{b}\,\vec{r}]\vec{c}$

Similarly, $(\vec{b}\times\vec{c})\times(\vec{r}\times\vec{a})$

$= [\vec{b}\,\vec{c}\,\vec{a}]\vec{r} - [\vec{b}\,\vec{c}\,\vec{r}]\vec{a}$

and $(\vec{c}\times\vec{a})\times(\vec{r}\times\vec{b}) = [\vec{c}\,\vec{a}\,\vec{b}]\vec{r} - [\vec{c}\,\vec{a}\,\vec{r}]\vec{b}$

$\Rightarrow \quad (\vec{a}\times\vec{b})\times(\vec{r}\times\vec{c})+(\vec{b}\times\vec{c})\times(\vec{r}\times\vec{a})$

$+(\vec{c}\times\vec{a})\times(\vec{r}\times\vec{b})$

$= 3[\vec{a}\,\vec{b}\,\vec{c}]\vec{r} - ([\vec{b}\,\vec{c}\,\vec{r}]\vec{a}$

$+[\vec{c}\,\vec{a}\,\vec{r}]\vec{b} + [\vec{a}\,\vec{b}\,\vec{r}]\vec{c})$

$= 3[\vec{a}\,\vec{b}\,\vec{c}]\vec{r} - [\vec{a}\,\vec{b}\,\vec{c}]\vec{r} = 2[\vec{a}\,\vec{b}\,\vec{c}]\vec{r}$

86. (1) $\vec{r}\times\vec{a} = \lambda\vec{a} + \mu\vec{b} + \gamma\,\vec{a}\times\vec{b}$

$\therefore \quad [\vec{r}\,\vec{a}\,\vec{a}] = \lambda\,\vec{a}\cdot\vec{a} + \mu\,\vec{b}\cdot\vec{a} + \gamma[\vec{a}\,\vec{b}\,\vec{a}]$

$0 = \lambda|\vec{a}|^2 + 0 + 0$

$\lambda = 0$

Also $[\vec{r}\,\vec{a}\,\vec{b}] = \lambda\,\vec{a}\cdot\vec{b} + \mu\,\vec{b}\cdot\vec{b} + \gamma[\vec{a}\,\vec{b}\,\vec{b}] = \mu$

$\therefore \quad (\vec{r}\times\vec{a})\times\vec{b} = \gamma(\vec{a}\times\vec{b})\times\vec{b}$

or $(\vec{r}\cdot\vec{b})\vec{a}-(\vec{a}\cdot\vec{b})\vec{r} = \gamma\{(\vec{a}\cdot\vec{b})\vec{b}-(\vec{b}\cdot\vec{b})\vec{a}\}$

or $(\vec{r}\cdot\vec{b})\vec{a} = -\gamma\vec{a}$ or $\gamma = -(\vec{r}\cdot\vec{b})$

87. (2) $(\vec{a}\times\hat{i})\cdot(\vec{b}\times\hat{i}) = \begin{vmatrix}\vec{a}\cdot\vec{b} & \vec{a}\cdot\hat{i}\\ \vec{b}\cdot\hat{i} & \hat{i}\cdot\hat{i}\end{vmatrix}$

$= (\vec{a}\cdot\vec{b})-(\vec{a}\cdot\hat{i})(\vec{b}\cdot\hat{i})$

Similarly, $(\vec{a}\times\hat{j})\cdot(\vec{b}\times\hat{j}) = (\vec{a}\cdot\vec{b})-(\vec{a}\cdot\hat{j})(\vec{b}\cdot\hat{j})$

and $(\vec{a}\times\hat{k})\cdot(\vec{b}\times\hat{k}) = \vec{a}\cdot\vec{b}-(\vec{a}\cdot\hat{k})(\vec{b}\cdot\hat{k})$

Let $\vec{a} = a_1\hat{i}+a_2\hat{j}+a_3\hat{k},\ \vec{b} = b_1\hat{i}+b_2\hat{j}+b_3\hat{k}$.

$\therefore\ (\vec{a}\cdot\hat{i}) = a_1,\ \vec{a}\cdot\hat{j} = a_2,\ \vec{a}\cdot\hat{k} = a_3,$

$\vec{b}\cdot\hat{i} = b_1,\ \vec{b}\cdot\hat{j} = b_2,\ (\vec{b}\cdot\hat{k}) = b_3$

$\Rightarrow (\vec{a}\times\hat{i})\cdot(\vec{b}\times\hat{i})+(\vec{a}\times\hat{j})\cdot(\vec{b}\times\hat{j})$

$+(\vec{a}\times\hat{k})\cdot(\vec{b}\times\hat{k})$

$= 3\vec{a}\cdot\vec{b} - (a_1b_1+a_2b_2+a_3b_3)$

$= 3\vec{a}\cdot\vec{b} - \vec{a}\cdot\vec{b} = 2\vec{a}\cdot\vec{b}$

88. (4) $\vec{a}+\vec{b} = \mu\vec{p}$ $\quad\vec{b}\cdot\vec{q} = 0,\ |\vec{b}|^2 = 1$

$\because\ \vec{a}+\vec{b} = \mu\vec{p}$

$\Rightarrow (\vec{a}+\vec{b})\times\vec{a} = \mu\vec{p}\times\vec{a},$

$\Rightarrow \vec{b}\times\vec{a} = \mu\vec{p}\times\vec{a}$

$\Rightarrow \vec{q}\times(\vec{b}\times\vec{a}) = \mu\vec{q}\times(\vec{p}\times\vec{a})$

$\Rightarrow (\vec{q}\cdot\vec{a})\vec{b}-(\vec{q}\cdot\vec{b})\vec{a} = \mu\vec{q}\times(\vec{p}\times\vec{a})$

$\Rightarrow (\vec{q}\cdot\vec{a})\vec{b} = \mu\vec{q}\times(\vec{p}\times\vec{a})$

$\because\ \vec{a}+\vec{b} = \mu\vec{p}$

$\Rightarrow \vec{q}\cdot(\vec{a}+\vec{b}) = \mu\vec{q}\cdot\vec{p}$

$\Rightarrow \vec{q}\cdot\vec{a}+\vec{q}\cdot\vec{b} = \mu\vec{p}\cdot\vec{q}$

$\Rightarrow \mu = \dfrac{\vec{q}\cdot\vec{a}}{\vec{p}\cdot\vec{q}}$

$\Rightarrow (\vec{q}\cdot\vec{a})\vec{b} = \dfrac{\vec{q}\cdot\vec{a}}{\vec{p}\cdot\vec{q}}[(\vec{q}\cdot\vec{a})\cdot\vec{p}-(\vec{q}\cdot\vec{p})\vec{a}]$

$\Rightarrow |(\vec{q}\cdot\vec{a})\vec{p}-(\vec{q}\cdot\vec{p})\vec{a}|$

$= |(\vec{p}\cdot\vec{q})\vec{b}| = |(\vec{p}\cdot\vec{q})|\cdot|\vec{b}|$

$\Rightarrow |(\vec{q}\cdot\vec{a})\vec{p}-(\vec{q}\cdot\vec{p})\vec{a}| = |\vec{p}\cdot\vec{q}|$

89. (3) $\vec{r}\times\vec{a} = \vec{b}$

$\Rightarrow \vec{d}\times(\vec{r}\times\vec{a}) = \vec{d}\times\vec{b}$

$\Rightarrow (\vec{a}\cdot\vec{d})\vec{r}-(\vec{d}\cdot\vec{r})\vec{a} = \vec{d}\times\vec{b}$ ...(i)

$\vec{r}\times\vec{c} = \vec{d}$

$\Rightarrow \vec{b}\times(\vec{r}\times\vec{c}) = \vec{b}\times\vec{d}$

$\Rightarrow (\vec{b}\cdot\vec{c})\vec{r}-(\vec{b}\cdot\vec{r})\vec{c} = \vec{b}\times\vec{d}$ ...(ii)

Adding (i) and (ii), we get

$(\vec{a}\cdot\vec{d}+\vec{b}\cdot\vec{c})\vec{r}-(\vec{d}\cdot\vec{r})\vec{a}-(\vec{b}\cdot\vec{r})\vec{c} = \vec{0}$

Now $\vec{r}\cdot\vec{d} = 0$ and $\vec{b}\cdot\vec{r} = 0$ as $\vec{d}$ and $\vec{r}$ as well as $\vec{b}$ and $\vec{r}$ are mutually perpendicular.

Hence, $(\vec{b}\cdot\vec{c}+\vec{a}\cdot\vec{d})\vec{r} = \vec{0}$

Numerical Value Type

1. (19.5) We have, $(2\vec{a}+5\vec{b})\cdot(3\vec{a}+\vec{b}+\vec{a}\times\vec{b})$

$= 6\vec{a}\cdot\vec{a}+17\vec{a}\cdot\vec{b}+5\vec{b}\cdot\vec{b} = 11+17\vec{a}\cdot\vec{b}$

$[\because \vec{a}\cdot(\vec{a}\times\vec{b}) = \vec{b}\cdot(\vec{a}\times\vec{b}) = 0$, since $\vec{a}$ and $\vec{b}$ are perpendicular to $\vec{a}\times\vec{b}]$

Now $|\vec{a}+\vec{b}| = \sqrt{3}$

$\Rightarrow |\vec{a}+\vec{b}|^2 = 3$

$\Rightarrow |\vec{a}|^2+|\vec{b}|^2+2\vec{a}\cdot\vec{b} = 3$

$\Rightarrow \vec{a}\cdot\vec{b} = \dfrac{1}{2}$

$\Rightarrow (2\vec{a}+5\vec{b})\cdot(3\vec{a}+\vec{b}+\vec{a}\times\vec{b}) = 11+\dfrac{17}{2} = \dfrac{39}{2}$

2. (9.2)

$\vec{a}\times\vec{b} = \begin{vmatrix}\hat{i} & \hat{j} & \hat{k}\\ \alpha & 1 & \beta\\ 3 & -5 & 4\end{vmatrix}$

$= (4+5\beta)\hat{i}+(3\beta-4\alpha)\hat{j}+(-5\alpha-3)\hat{k}$

$= -\hat{i}+9\hat{j}+12\hat{k}$ (given)

$\beta = -1,\ \alpha = -3$

$\therefore\ \vec{a} = -3\hat{i}+\hat{j}-\hat{k},\ \vec{b} = 3\hat{i}-5\hat{j}+4\hat{k}$

$\Rightarrow \vec{a}+\vec{b} = -4\hat{j}+3\hat{k}$

Now, projection of $(\vec{b}-2\vec{a})$ on $\vec{a}+\vec{b}$

$= \dfrac{(\vec{b}-2\vec{a}).(\vec{a}+\vec{b})}{|\vec{a}+\vec{b}|}$

$= \dfrac{|\vec{b}|^2-2|\vec{a}|^2-(\vec{a}.\vec{b})}{|\vec{a}+\vec{b}|}$

$= \dfrac{50-22-(-18)}{5} = \dfrac{46}{5}$

3. (237.5)

$\vec{v}$ lies in the plane of $\vec{a}$ and $\vec{b}$

$\vec{v} = \lambda\vec{a}+\mu\vec{b} = (2\lambda+\mu)\hat{i}+(2\mu-\lambda)\hat{j}+(2\lambda-\mu)\hat{k}$

$\vec{v}\cdot(3\hat{i}+2\hat{j}-\hat{k}) = 0$

$\Rightarrow \lambda = -4\mu$ ...(1)

Projection of $\vec{v}$ on $\vec{a}$ is 19

$\vec{v}\cdot\dfrac{\vec{a}}{|\vec{a}|} = 19$

$\Rightarrow 9\lambda-2\mu = 57$ ...(2)

Solving (1) and (2), we get $\lambda = 6,\ \mu = -\dfrac{3}{2}$

$\therefore\ 2\vec{v} = 21\hat{i}-18\hat{j}+27\hat{k}$

$|2\vec{v}|^2 = 441+324+729 = 1494$

or $|\vec{v}|^2 = 441+324+729 = 373.5$

4. (9.5)

$\vec{a} \times \vec{c} = -\vec{b}$

$\Rightarrow \quad (\vec{a} \times \vec{c}) \times \vec{a} = -\vec{b} \times \vec{a}$

$\Rightarrow \quad (\vec{a}.\vec{a})\,\vec{c} - (\vec{c}.\vec{a})\vec{a} = \vec{a} \times \vec{b}$

$\Rightarrow \quad 2\vec{c} - 4\vec{a} = \vec{a} \times \vec{b}$ ...(i)

Now $\vec{a} \times \vec{b} = \begin{vmatrix} \hat{i} & \hat{j} & \hat{k} \\ 1 & -1 & 0 \\ 1 & 1 & 1 \end{vmatrix} = -\hat{i} - \hat{j} + 2\hat{k}$

So, $2\vec{c} = 3\hat{i} - 5\hat{j} + 2\hat{k}$ (from (i))

$\Rightarrow \quad \vec{c} = \frac{3}{2}\hat{i} - \frac{5}{2}\hat{j} + \hat{k}$

$\Rightarrow \quad |\vec{c}| = \sqrt{\frac{9}{4} + \frac{25}{4} + 1} = \sqrt{\frac{38}{4}} = \sqrt{\frac{19}{2}}$

$|\vec{c}|^2 = \frac{19}{2}$

. (1.5)

$|(\vec{a} \times \vec{b}) \times \vec{c}| = |\vec{a} \times \vec{b}||\vec{c}| \sin\frac{\pi}{6}$ ...(i)

Now, $\vec{a} \times \vec{b} = (2\hat{i} + \hat{j} - 2\hat{k}) \times (\hat{i} + \hat{j})$

$= 2\hat{i} - 2\hat{j} + \hat{k}$

From, $|\vec{c} - \vec{a}| = 2\sqrt{2}$, we get

$c^2 + a^2 - 2\vec{c}.\vec{a} = 8$

$\Rightarrow \quad c^2 + 9 - 2(c) = 8$

$\Rightarrow \quad c^2 - 2c + 1 = 0$

$\Rightarrow \quad c = 1 = |\vec{c}|$

$\therefore$ from (i), $|(\vec{a} \times \vec{b}) \times \vec{c}| = (3)(1)(1/2) = 3/2$

(0.5)

$\vec{a} = (\vec{b} \cdot \vec{c})\vec{b} - (\vec{b} \cdot \vec{b})\vec{c}$

$= 1.2\cos\theta\vec{b} - \vec{c}$

$\therefore \quad \vec{a} = 2\cos\theta\vec{b} - \vec{c}$

$\Rightarrow \quad |\vec{a}|^2 = (2\cos\theta)^2 + 2^2 - 2\times 2\cos\theta\vec{b} \cdot \vec{c}$

$\Rightarrow \quad 2 = 4\cos^2\theta + 4 - 4\cos\theta \,.\, 2\cos\theta$

$\Rightarrow \quad 2 = 4\cos^2\theta$

$\Rightarrow \quad \cos^2\theta = \frac{1}{2}$

(71.25)

$\vec{b} \times (\vec{a} \times \vec{c}) = -3\vec{a}$ (as $\vec{a}.\vec{b} = 0$)

Let θ be angle between $\vec{b}, \vec{a} \times \vec{c}$

Then $|\vec{b}|.|\vec{a} \times \vec{c}| \sin\theta = 3\sqrt{14}$

Also, $\vec{b}.(\vec{a} \times \vec{c}) = 27$

$\Rightarrow \quad |\vec{b}|.|\vec{a} \times \vec{c}| \cos\theta = 27$

$\Rightarrow \quad \tan\theta = \frac{\sqrt{14}}{9}$

$\Rightarrow \quad \sin\theta = \frac{\sqrt{14}}{\sqrt{95}}$

$\Rightarrow \quad |\vec{b}| \times |\vec{a} \times \vec{c}| = 3\sqrt{95}$

$\Rightarrow \quad |\vec{a} \times \vec{c}| = \sqrt{3} \times \sqrt{95} = \sqrt{285}$

$\Rightarrow \quad \frac{1}{4}|\vec{a} \times \vec{c}|^2 = \frac{285}{4} = 71.25$

8. (6.6)

$\left[\left[\frac{\vec{a} \times \vec{b}}{5} \quad \vec{b} \times \vec{c} \quad \frac{\vec{a} + \vec{b} + \vec{c}}{\sqrt{3}}\right]\right]$

$= \frac{\vec{a} \times \vec{b}}{5} \cdot \left((\vec{b} \times \vec{c}) \times \frac{(\vec{a} + \vec{b} + \vec{c})}{\sqrt{3}}\right)$

$= \frac{\vec{a} \times \vec{b}}{5\sqrt{3}} \cdot ((\vec{b} \times \vec{c}) \times \vec{a} + (\vec{b} \times \vec{c}) \times \vec{b} + (\vec{b} \times \vec{c}) \times \vec{c})$

$= \frac{\vec{a} \times \vec{b}}{5\sqrt{3}} \cdot (\vec{c} - 3\vec{b} + 4\vec{c} - 6\vec{b} + 6\vec{c} - 9\vec{b})$

$= \frac{\vec{a} \times \vec{b}}{5\sqrt{3}} \cdot (11\vec{c} - 18\vec{b})$

$= \frac{11\vec{a} \times \vec{b} \cdot \vec{c}}{5\sqrt{3}}$

$= \frac{11|\vec{a} \times \vec{b}||\vec{c}|}{5\sqrt{3}}$

$= \frac{11\sqrt{3} \times 3}{5\sqrt{3}} = \frac{33}{5} = 6.6$

9. (2.5)

Unit vectors $\hat{u} = x\hat{i} + y\hat{j} + z\hat{k}$

$\vec{p}_1 = \frac{1}{\sqrt{2}}\hat{i} + \frac{1}{\sqrt{2}}\hat{k}, \vec{p}_2 = \frac{1}{\sqrt{2}}\hat{j} + \frac{1}{\sqrt{2}}\hat{k}$

and $\vec{p}_3 = \frac{1}{\sqrt{2}}\hat{i} + \frac{1}{\sqrt{2}}\hat{j}$

Angle between $\hat{u}$ and $\vec{p}_1$ is $\frac{\pi}{2}$

$\therefore \quad \hat{u} \cdot \vec{p}_1 = 0 \Rightarrow \frac{x}{\sqrt{2}} + \frac{z}{\sqrt{2}} = 0$ or $x + z = 0$ ...(i)

Angle between $\hat{u}$ and $\vec{p}_2$ is $\frac{\pi}{3}$.

$\therefore \quad \hat{u} \cdot \vec{p}_2 = |\hat{u}| \cdot |\vec{p}_2| \cos\frac{\pi}{3}$

$\Rightarrow \quad \frac{x}{\sqrt{2}} + \frac{z}{\sqrt{2}} = \frac{1}{2} \Rightarrow y + z = \frac{1}{\sqrt{2}}$ ...(ii)

Angle between $\hat{u}$ and $\vec{p}_2$ is $\frac{2\pi}{3}$

$\therefore \quad \hat{u} \cdot \vec{p}_3 = |\hat{u}| \cdot |\vec{p}_3| \cos\frac{2\pi}{3}$

$\Rightarrow \quad \frac{x}{\sqrt{2}} + \frac{y}{\sqrt{2}} = \frac{-1}{2} \Rightarrow x + y = \frac{-1}{\sqrt{2}}$ ...(iii)

Solving (i), (ii) and (iii), we get

$x = \frac{-1}{\sqrt{2}}, y = 0, z = \frac{1}{\sqrt{2}}$

Thus $\hat{u} - \hat{v} = \frac{-1}{\sqrt{2}}\hat{i} + \frac{1}{\sqrt{2}}\hat{k} - \frac{1}{\sqrt{2}}\hat{i} - \frac{1}{\sqrt{2}}\hat{j} - \frac{1}{\sqrt{2}}\hat{k}$

$\Rightarrow \quad \hat{u} - \hat{v} = \frac{-2}{\sqrt{2}}\hat{i} - \frac{1}{\sqrt{2}}\hat{j}$

$\Rightarrow \quad |\hat{u} - \hat{v}|^2 = \left(\sqrt{\frac{4}{2} + \frac{1}{2}}\right)^2 = \frac{5}{2}$

Integer Answer Type

1. (5) Let angle between $\vec{a}$ and $\vec{b}$ be θ.

We have $|\vec{a}|=|\vec{b}|=1$

Now $|\vec{a}+\vec{b}| = 2\cos\frac{\theta}{2}$ and $|\vec{a}-\vec{b}| = 2\sin\frac{\theta}{2}$

Consider $F(\theta) = \frac{3}{2}\left(2\cos\frac{\theta}{2}\right)+2\left(2\sin\frac{\theta}{2}\right)$

$\therefore \quad F(\theta) = 3\cos\frac{\theta}{2} + 4\sin\frac{\theta}{2},\ \theta \in [0, \pi]$

2. (1) Since angle between $\vec{u}$ and $\hat{i}$ is 60°, we have

$$\vec{u}\cdot\hat{i} = |\vec{u}||\hat{i}|\cos 60^\circ = \frac{|\vec{u}|}{2}$$

Given that $|\vec{u}-\hat{i}|, |\vec{u}|, |\vec{u}-2\hat{i}|$ are in G.P., so $|\vec{u}-\hat{i}|^2 = |\vec{u}||\vec{u}-2\hat{i}|$

Squaring both sides,

$$[|\vec{u}|^2+|\hat{i}|^2-2\vec{u}\cdot\hat{i}]^2 = |\vec{u}|^2[|\vec{u}|^2+4|\hat{i}|^2-4\vec{u}\cdot\hat{i}]$$

$\Rightarrow \quad [|\vec{u}|^2+1-\frac{2|\vec{u}|}{2}]^2 = |\vec{u}|^2[|\vec{u}|^2+4-4\frac{|\vec{u}|}{2}]$

$\Rightarrow \quad |\vec{u}|^2+2|\vec{u}|-1=0 \Rightarrow |\vec{u}| = -\frac{2\pm 2\sqrt{2}}{2}$

$\Rightarrow \quad |\vec{u}| = \sqrt{2}-1$

3. (2) $\overrightarrow{AB} = 2\hat{i}+\hat{j}+\hat{k}$, $\overrightarrow{AC} = (t+1)\hat{i}+0\hat{j}-\hat{k}$

$$\overrightarrow{AB}\times\overrightarrow{AC} = \begin{vmatrix}\hat{i} & \hat{j} & \hat{k}\\ 2 & 1 & 1\\ t+1 & 0 & -1\end{vmatrix}$$

$$= -\hat{i}+(t+3)\hat{j}-(t+1)\hat{k}$$

$\therefore \quad |\overrightarrow{AB}\times\overrightarrow{AC}| = \sqrt{1+(t+3)^2+(t+1)^2}$

$= \sqrt{2t^2+8t+11}$

$\therefore \quad$ Area of $\Delta ABC = \frac{1}{2}|\overrightarrow{AB}\times\overrightarrow{AC}|$

$= \frac{1}{2}\sqrt{2t^2+8t+1}$

Let $f(t) = \Delta^2 = \frac{1}{4}(2t^2+8t+1)$

$f'(t) = 0 \Rightarrow t = -2$

At $t = -2$, $f''(t) > 0$

So Δ is minimum at $t = -2$

4. (28) $\vec{a} = a_1\hat{i}+a_2\hat{j}+a_3\hat{k}$

$\vec{b} = b_1\hat{i}+b_2\hat{j}+b_3\hat{k}$

$\vec{c} = c_1\hat{i}+c_2\hat{j}+c_3\hat{k}$

L.H.S. $= [3\vec{a}+\vec{b}\quad 3\vec{b}+\vec{c}\quad 3\vec{c}+\vec{a}]$

$$= \begin{vmatrix}3 & 1 & 0\\ 0 & 3 & 1\\ 1 & 0 & 3\end{vmatrix}[\vec{a}\,\vec{b}\,\vec{c}]$$

$= 28[\vec{a}\,\vec{b}\,\vec{c}]$

5. (20) $\vec{a} = \alpha\hat{i}+2\hat{j}-3\hat{k}$, $\vec{b} = \hat{i}+2\alpha\hat{j}-2\hat{k}$,

$\vec{c} = 2\hat{i}-\alpha\hat{j}+\hat{k}$

$\{(\vec{a}\times\vec{b})\times(\vec{b}\times\vec{c})\}\times(\vec{c}\times\vec{a}) = \vec{0}$

$\Rightarrow \quad \{[\vec{a}\,\vec{b}\,\vec{c}]\vec{b}-[\vec{a}\,\vec{b}\,\vec{b}]\vec{c}\}\times(\vec{c}\times\vec{a}) = \vec{0}$

$\Rightarrow \quad [\vec{a}\,\vec{b}\,\vec{c}]\vec{b}\times(\vec{c}\times\vec{a}) = \vec{0}$

$\Rightarrow \quad [\vec{a}\,\vec{b}\,\vec{c}]\ ((\vec{a}\cdot\vec{b})\vec{c}-(\vec{b}\cdot\vec{c})\vec{a}) = \vec{0}$

$\Rightarrow \quad [\vec{a}\,\vec{b}\,\vec{c}] = 0 \qquad (\because \vec{a}$ and $\vec{c}$ are not collinear

$$\Rightarrow \quad \begin{vmatrix}\alpha & 2 & -3\\ 1 & 2\alpha & -2\\ 2 & -\alpha & 1\end{vmatrix}$$

$\Rightarrow \quad \alpha(2\alpha-2\alpha)-2(1+4)-3(-\alpha-4\alpha) = 0$

$\Rightarrow \quad 10-15\alpha = 0$

$\therefore \quad \alpha = 2/3$

6. (7) Let the vertices be A, B, C, D and O be the origin.

$\therefore \quad \overrightarrow{OA} = \hat{i}-6\hat{j}+10\hat{k},\ \overrightarrow{OB} = -\hat{i}-3\hat{j}+7\hat{k},$

$\overrightarrow{OC} = 5\hat{i}-\hat{j}+\lambda\hat{k},\ \overrightarrow{OD} = 7\hat{i}-4\hat{j}+7\hat{k}$

$\therefore \quad \overrightarrow{AB} = \overrightarrow{OB}-\overrightarrow{OA} = -2\hat{i}+3\hat{j}-3\hat{k}$

$\overrightarrow{AC} = \overrightarrow{OC}-\overrightarrow{OA} = 4\hat{i}+5\hat{j}+(\lambda-10)\hat{k}$

$\overrightarrow{AD} = \overrightarrow{OD}-\overrightarrow{OA} = 6\hat{i}+2\hat{j}-3\hat{k}$

Volume of tetrahedron

$$= \frac{1}{6}[\overrightarrow{AB}\ \overrightarrow{AC}\ \overrightarrow{AD}] = \frac{1}{6}\begin{vmatrix}-2 & 3 & -3\\ 4 & 5 & \lambda-10\\ 6 & 2 & -3\end{vmatrix}$$

$= \frac{1}{6}\{-2(-15-2\lambda+20)-3(-12-6\lambda+60)$

$-3(8-30)$

$= \frac{1}{6}\{4\lambda-10-144+18\lambda+66\}$

$= \frac{1}{6}(22\lambda-88) = \pm 11 \quad$ (Given)

$\Rightarrow \quad 2\lambda-8 = \pm 6$

$\Rightarrow \quad \lambda = 7$

7. (45) Let $\vec{R} = x\hat{i}+y\hat{j}+z\hat{k}$

$\vec{u} = \hat{i}-2\hat{j}+3\hat{k};\ \vec{v} = 2\hat{i}+\hat{j}+4\hat{k};\ \vec{w} = \hat{i}+3\hat{j}+3\hat{k}$

$(\vec{u}\cdot\vec{R}-15)\hat{i}+(\vec{v}\cdot\vec{R}-30)\hat{j}+(\vec{w}\cdot\vec{R}-25)\hat{k} = \vec{0}$

(give

So $\quad \vec{u}\cdot\vec{R} = 15 \Rightarrow x-2y+3z = 15$ ...

$\vec{v}\cdot\vec{R} = 30 \Rightarrow 2x+y+4z = 30$ ...(

$\vec{w}\cdot\vec{R} = 25 \Rightarrow x+3y+3z = 25$ ...(i

Solving, we get

$x = 4$

$y = 2$

$z = 5$

8. (1) $\vec{a}\cdot\vec{b} = 0 \Rightarrow \vec{a}\perp\vec{b}$

$\Rightarrow \quad \vec{a}\cdot\vec{c} = 0 \Rightarrow \vec{a}\perp\vec{c}$

$\Rightarrow \quad \vec{a}\perp\left(\vec{b}-\vec{c}\right)$

$$|\vec{a}\times\vec{b}-\vec{a}\times\vec{c}|=|\vec{a}\times(\vec{b}-\vec{c})|$$

$$\Rightarrow \quad =|\vec{a}||\vec{b}-\vec{c}|=|\vec{b}-\vec{c}|$$

Now $|\vec{b}-\vec{c}|^2=|\vec{b}|^2+|\vec{c}|^2-2|\vec{b}||\vec{c}|\cos\frac{\pi}{3}$

$$=2-2\times\frac{1}{2}=1$$

$$|\vec{b}-\vec{c}|=1$$

9. (6) We have $\vec{a}\times\vec{b}=-4\hat{i}+2\hat{j}+5\hat{k}$ and $\vec{a}.\vec{b}=0$

$\vec{a},\vec{b}$ and $\vec{c}$ are mutually perpendicular.

$\therefore \quad \vec{c}=\lambda(\vec{a}\times\vec{b})$

$\Rightarrow \quad \vec{c}=-4\lambda\hat{i}+2\lambda\hat{j}+5\lambda\hat{k}$

$|x|\le 0\Rightarrow|4\lambda|\le 10\Rightarrow -2.5\le\lambda\le 2.5$

Now, $|x+y+z|=|-4\lambda+2\lambda+5\lambda|=|3\lambda|$

So, minimum value of $|x+y+z|$ is 6 when $\lambda=-2$ or 2

10. (4) Let $\vec{r}=x\hat{i}+y\hat{j}+z\hat{k}$

Given that $\frac{|\vec{r}\cdot(2\hat{i}+\hat{j})|}{\sqrt{5}}=\frac{|\vec{r}\cdot(-2\hat{i}+\hat{j})|}{\sqrt{5}}=\frac{|\vec{r}\cdot\hat{k}|}{1}$

$\Rightarrow \quad |2x+y|=|-2x+y|=\sqrt{5}|z|$

$$\Rightarrow \vec{r}=\left(0,\frac{\sqrt{5}}{\sqrt{6}},\frac{1}{\sqrt{6}}\right),\left(0,\frac{\sqrt{5}}{\sqrt{6}},-\frac{1}{\sqrt{6}}\right),\left(\frac{\sqrt{5}}{3},0,\frac{2}{3}\right),\left(\frac{\sqrt{5}}{3},0,-\frac{2}{3}\right)$$

1. (98) $\vec{d}\cdot(\vec{a}\times(\vec{b}\times(\vec{c}\times\vec{d})))$

$=\vec{d}\cdot(\vec{a}\times((\vec{b}\cdot\vec{d})\vec{c}-(\vec{b}\cdot\vec{c})\vec{d}))$

$=\vec{d}\cdot((\vec{b}\cdot\vec{d})\vec{a}\times\vec{c}-(\vec{b}\cdot\vec{c})\vec{a}\times\vec{d})$

$=[\vec{d}\ \vec{a}\ \vec{c}](\vec{b}\cdot\vec{d})$

$$=\begin{vmatrix}0&2&1\\5&-1&1\\-3&1&1\end{vmatrix}(-6-1)$$

$=(-14)(-7)=98$

2. (1) Volume, $V=\begin{vmatrix}1&1&0\\1&2\operatorname{cosec}\alpha&1\\0&1&2\operatorname{cosec}\alpha\end{vmatrix}$

$=4\operatorname{cosec}^2\alpha-1-2\operatorname{cosec}\alpha$

$$=4\left[\operatorname{cosec}^2\alpha-\frac{1}{2}\operatorname{cosec}\alpha\right]-1$$

$$=4\left(\operatorname{cosec}\alpha-\frac{1}{4}\right)^2-\frac{5}{4}$$

$$\therefore \quad V_{\text{least}}\left(\alpha=\frac{\pi}{2}\right)=4\times\frac{9}{16}-\frac{5}{4}=1$$

3. (6) Here $\vec{OA}=\vec{a},\ \vec{OB}=10\vec{a}+2\vec{b},\ \vec{OC}=\vec{b}$

q = Area of parallelogram with OA and OC as adjacent sides.

$\therefore \quad q=|\vec{a}\times\vec{b}| \quad \ldots$(i)

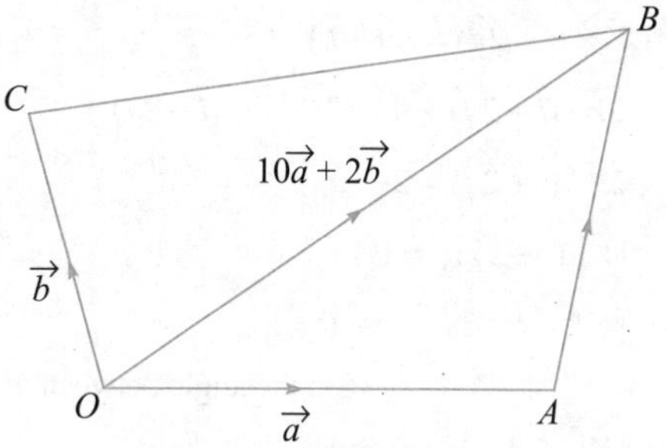

p = Area of quadrilateral $OABC$

= Area of ΔOAB + Area of ΔOBC

$$=\frac{1}{2}|\vec{a}\times(10\vec{a}+2\vec{b})|+\frac{1}{2}|(10\vec{a}+2\vec{b})\times\vec{b}|$$

$=|\vec{a}\times\vec{b}|+5|\vec{a}\times\vec{b}|$

$\therefore \quad p=6|\vec{a}\times\vec{b}|$

or $\quad p=6q$ [From Eq. (i)]

$\therefore \quad k=6$

14. (3) $\left|\vec{a}+\vec{b}+3\vec{c}\right|^2=14$

$\Rightarrow \quad |\vec{a}|^2+|\vec{b}|^2+9|\vec{c}|^2+2\vec{a}.\vec{b}+6\vec{b}.\vec{c}+6\vec{a}.\vec{c}=14$

$\Rightarrow \quad 11+2\cos\alpha+6\cos\beta+6\cos\gamma=14$

$\Rightarrow \quad 2(\cos\alpha+3\cos\beta)=3-6\cos\gamma\le 6\left\{\because \gamma\in\left[\frac{\pi}{6},\frac{2\pi}{3}\right]\right\}$

$\Rightarrow \quad \cos\alpha+3\cos\beta\le 3$

15. (13) Since $\vec{x}$ and $\vec{y}$ are non-collinear vectors, therefore $\vec{x},\vec{y}$ and $\vec{x}\times\vec{y}$ are non-coplanar vectors.

$[(a-2)\alpha^2+(b-3)\alpha+c]\ \vec{x}+[(a-2)\beta^2+(b-3)\ \beta\beta+c]\ \vec{y}$
$+[(a-2)\gamma^2+(b-3)\ \gamma+c]\ (\vec{x}\times\vec{y})=0$

Coefficient of each vector $\vec{x},\vec{y}$ and $\vec{x}\times\vec{y}$ is zero.

$(a-2)\alpha^2+(b-3)\alpha+c=0$

$(a-2)\beta^2+(b-3)\beta+c=0$

$(a-2)\ \gamma^2+(b-3)\ \beta+c=0$

The above three equations will satisfy if the coefficients of α, β and γ are zero because α, β and γ are three distinct real numbers

$\therefore \quad a-2=0$ or $a=2$,

$b-3=0$ or $b=3$ and $c=0$

$\therefore \quad a^2+b^2+c^2=2^2+3^2+0^2=4+9=13$

16. (1) Given, $\vec{u}\times\vec{v}+\vec{u}=\vec{w}$ and $\vec{w}\times\vec{u}=\vec{v}$

$\Rightarrow \quad (\vec{u}\times\vec{v}+\vec{u})\times\vec{u}=\vec{w}\times\vec{u}=\vec{v}$

$\Rightarrow \quad (\vec{u}\times\vec{v})\times\vec{u}=\vec{v}$

$\Rightarrow \quad \vec{v}-(\vec{u}\cdot\vec{v})=\vec{v}$

$\Rightarrow \quad (\vec{u}\cdot\vec{v})\vec{u}=0\Rightarrow(\vec{u}\cdot\vec{v})=0$

Now, $[\vec{u}\ \vec{v}\ \vec{w}]=\vec{u}\cdot(\vec{v}\times\vec{w})$

$=\vec{u}\cdot(\vec{v}\times(\vec{u}\times\vec{v}+\vec{u}))$

$=\vec{u}\cdot(\vec{v}\times(\vec{u}\times\vec{v})+\vec{v}\times\vec{u})$

$=\vec{u}(v^2\vec{u}-(\vec{u}\cdot\vec{v})\vec{v}+\vec{v}\times\vec{u})=v^2u^2=1$

17. (6) $2\vec{V}+\vec{V}\times(\hat{i}+2\hat{j})=(2\hat{i}+\hat{k})$ ...(i)

or $2\vec{V}\cdot(\hat{i}+2\hat{j})+0=(2\hat{i}+\hat{k})\cdot(\hat{i}+2\hat{j})$

or $2\vec{V}\cdot(\hat{i}+2\hat{j})=2$

or $|\vec{V}\cdot(\hat{i}+2\hat{j})|^2=1$

or $|\vec{V}|^2\cdot|\hat{i}+2\hat{j}|^2\cos^2\theta=1$

(θ is the angle between $\vec{V}$ and $\hat{i}+2\hat{j}$)

or $|\vec{V}|^2\ 5(1-\sin^2\theta)=1$

or $|\vec{V}|^2 5\sin^2\theta=5|\vec{V}|^2-1$ ...(ii)

From Eq. (i), we have

$$|2\vec{V}+\vec{V}\times(\hat{i}+2\hat{j})|^2=|2\hat{i}+\hat{k}|^2$$

or $4|\vec{V}|^2+|\vec{V}\times(\hat{i}+2\hat{j})|^2=5$

or $4|\vec{V}|^2+|\vec{V}|^2\cdot|\hat{i}+2\hat{j}|^2\sin^2\theta=5$

or $4|\vec{V}|^2+5|\vec{V}|^2\sin^2\theta=5$

or $4|\vec{V}|^2+5|\vec{V}|^2-1=5$

or $9|\vec{V}|^2=6$

or $3|\vec{V}|=\sqrt{6}$

$=\sqrt{6}=\sqrt{m}$

$\therefore$ $m=6$

Multiple Correct Answers Type

1. (1), (2)

We have $|\vec{a}-\vec{b}|^2=|\vec{a}|^2+|\vec{b}|^2-2(\vec{a}\cdot\vec{b})$

$\Rightarrow$ $|\vec{a}-\vec{b}|^2=|\vec{a}|^2+|\vec{b}|^2-2|\vec{a}||\vec{b}|\cos 2\theta$

$\Rightarrow$ $|\vec{a}-\vec{b}|^2=2-2\cos 2\theta$ $(\because |\vec{a}|=|\vec{b}|=1)$

$=4\sin^2\theta$

$\Rightarrow$ $|\vec{a}-\vec{b}|=2|\sin\theta|$

Now, $|\vec{a}-\vec{b}|<1$

$\Rightarrow$ $2|\sin\theta|<1$

$\Rightarrow$ $|\sin\theta|<\frac{1}{2}$

$\Rightarrow$ $\theta\in[0,\pi/6)$ or $\theta\in(5\pi/6,\pi]$

2. (1), (3)

$\vec{a}\times(\vec{b}\times\vec{c})+(\vec{a}\cdot\vec{b})\vec{b}$

$=(4-2x-\sin y)\vec{b}+(x^2-1)\vec{c}$

or $(\vec{a}\cdot\vec{c})\vec{b}-(\vec{a}\cdot\vec{b})\vec{c}+(\vec{a}\cdot\vec{b})\vec{b}$

$=(4-2x-\sin y)\vec{b}+(x^2-1)\vec{c}$ ...(i)

Now, $(\vec{c}\cdot\vec{c})\vec{a}=\vec{c}$. Therefore,

$(\vec{c}\cdot\vec{c})\ (\vec{a}\cdot\vec{c})=(\vec{c}\cdot\vec{c})$ or $\vec{a}\cdot\vec{c}=1$

So, comparing coefficients of $\vec{b}$ and $\vec{c}$ in (i), we get

$1+\vec{a}\cdot\vec{b}=4-2x-\sin y,\ x^2-1=-(\vec{a}\cdot\vec{b})$

$\Rightarrow$ $1=4-2x-\sin y+x^2-1$

$\Rightarrow$ $\sin y=x^2-2x+2=(x-1)^2+1$

But $\sin y\le 1\Rightarrow x=1,\ \sin y=1$

$\Rightarrow$ $y=(4n+1)\frac{\pi}{2},\quad n\in I$

3. (1), (2), (3), (4)

We have, $\vec{a}\cdot\vec{b}=0$,

$|\vec{a}|=|\vec{b}|=|\vec{c}|=1$ and $\cos\theta=\vec{a}\cdot\vec{c}=\vec{b}\cdot\vec{c}$

Now, $\vec{c}=\alpha\vec{a}+\beta\vec{b}+\gamma(\vec{a}\times\vec{b})$ ...(i)

$\Rightarrow$ $\vec{a}\cdot\vec{c}=\alpha(\vec{a}\cdot\vec{a})+\beta(\vec{a}\cdot\vec{b})+\gamma\{\vec{a}\cdot(\vec{a}\times\vec{b})\}$

$\Rightarrow$ $\cos\theta=\alpha|\vec{a}|^2$ $(\because \vec{a}\cdot\vec{b}=0,\ \vec{a}\cdot(\vec{a}\times\vec{b})=0)$

$\Rightarrow$ $\cos\theta=\alpha$

Similarly, by taking dot product on both sides of (i) by $\vec{b}$, we get $\beta=\cos\theta$

$\therefore$ $\alpha=\beta$

Again, $\vec{c}=\alpha\vec{a}+\beta\vec{b}+\gamma(\vec{a}\times\vec{b})$

$\Rightarrow$ $|\vec{c}|^2=|\alpha\vec{a}+\beta\vec{b}+\gamma(\vec{a}\times\vec{b})|^2$

$\alpha^2|\vec{a}|^2+\beta^2|\vec{b}|^2+\gamma^2|\vec{a}\times\vec{b}|^2$

$=+2\alpha\beta(\vec{a}\cdot\vec{b})+2\alpha\gamma\{\vec{a}\cdot(\vec{a}\times\vec{b})\}$

$+2\beta\gamma(\vec{b}\cdot\{\vec{a}\times\vec{b}\})$

$\Rightarrow$ $1=\alpha^2+\beta^2+\gamma^2|\vec{a}\times\vec{b}|^2$

$=2\alpha^2+\gamma^2\{|\vec{a}|^2|\vec{b}|^2\sin^2\pi/2\}$

$=2\alpha^2+\gamma^2$

or $\alpha^2=\frac{1-\gamma^2}{2}$

But $\alpha=\beta=\cos\theta$.

$\Rightarrow$ $1=2\alpha^2+\gamma^2\Rightarrow\gamma^2=1-2\cos^2\theta=-\cos 2\theta$

and $\beta^2=\frac{1-\gamma^2}{2}=\frac{1+\cos 2\theta}{2}$

4. (1), (2), (3)

We have

$AM=$ projection of $\vec{b}$ on $\vec{a}=\frac{\vec{a}.\vec{b}}{|\vec{a}|}$

$\therefore$ $\overrightarrow{AM}=\left(\frac{\vec{a}.\vec{b}}{|\vec{a}|^2}\right)\vec{a}$

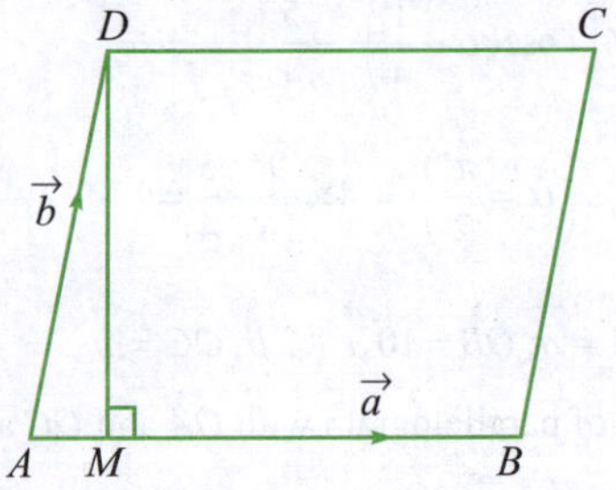

Now, in ΔADM

$$\overrightarrow{AD} = \overrightarrow{AM} + \overrightarrow{MD} \text{ or } \overrightarrow{DM} = \overrightarrow{AM} - \overrightarrow{AD}$$

$$\Rightarrow \quad DM = \frac{(\vec{a}\cdot\vec{b})\vec{a}}{|\vec{a}|^2} - \vec{b}$$

$$= \frac{1}{|\vec{a}|^2}[(\vec{a}\cdot\vec{b})\vec{a} - |\vec{a}|^2\vec{b}]$$

$$\Rightarrow \quad \overrightarrow{MD} = \frac{1}{|\vec{a}|^2}[|\vec{a}|^2\vec{b} - (\vec{a}\cdot\vec{b})\vec{a}]3$$

$$\text{Now, } \frac{\vec{a}\times(\vec{a}\times\vec{b})}{|\vec{a}|^2} = \frac{1}{|\vec{a}|^2}[(\vec{a}\cdot\vec{b})\vec{a} - (\vec{a}\cdot\vec{a})\vec{b}] = \overrightarrow{DM}$$

5. (1), (3)

$$\vec{a}\times(\vec{b}\times\vec{c}) = (\vec{a}\cdot\vec{c})\vec{b} - (\vec{a}\cdot\vec{b})\vec{c}$$

$$\text{and} \quad (\vec{a}\times\vec{b})\times\vec{c} = -(\vec{c}\cdot\vec{b})\vec{a} + (\vec{a}\cdot\vec{c})\vec{b}$$

$$\text{Now,} \quad (\vec{a}\times(\vec{b}\times\vec{c}))\cdot((\vec{a}\times\vec{b})\times\vec{c}) = 0$$

$$\therefore \quad ((\vec{a}\cdot\vec{c})\vec{b} - (\vec{a}\cdot\vec{b})\vec{c})\cdot((\vec{a}\cdot\vec{c})\vec{b} - (\vec{c}\cdot\vec{b})\vec{a}) = 0$$

$$\Rightarrow \quad (\vec{a}\cdot\vec{c})^2|\vec{b}|^2 - (\vec{a}\cdot\vec{c})(\vec{b}\cdot\vec{c})(\vec{a}\cdot\vec{b})$$

$$- (\vec{a}\cdot\vec{b})(\vec{a}\cdot\vec{c})(\vec{b}\cdot\vec{c}) + (\vec{a}\cdot\vec{b})(\vec{b}\cdot\vec{c})(\vec{c}\cdot\vec{a}) = 0$$

$$\Rightarrow \quad (\vec{a}\cdot\vec{c})^2|\vec{b}|^2 = (\vec{a}\cdot\vec{c})(\vec{a}\cdot\vec{b})(\vec{b}\cdot\vec{c})$$

$$\Rightarrow \quad (\vec{a}\cdot\vec{c})((\vec{a}\cdot\vec{c})(\vec{b}\cdot\vec{b}) - (\vec{a}\cdot\vec{b})(\vec{b}\cdot\vec{c})) = 0$$

$$\Rightarrow \quad \vec{a}\cdot\vec{c} = 0 \text{ or } (\vec{a}\cdot\vec{c})\,|\vec{b}|^2 = (\vec{a}\cdot\vec{b})(\vec{b}\cdot\vec{c})$$

6. (1), (3)

We have $[\vec{p}\ \vec{q}\ \vec{r}] = \dfrac{1}{[\vec{a}\ \vec{b}\ \vec{c}]}$. Therefore,

$[\vec{p}\ \vec{q}\ \vec{r}] > 0$

(1) Since $x > 0$, $x[\vec{a}\ \vec{b}\ \vec{c}] + \dfrac{[\vec{p}\ \vec{q}\ \vec{r}]}{x} \geq 2$

(using A.M. ≥ G.M.)

$$\textbf{(2)} \quad \frac{x^4[\vec{a}\,\vec{b}\,\vec{c}]^2 + \dfrac{[\vec{p}\,\vec{q}\,\vec{r}]}{2x^2} + \dfrac{[\vec{p}\,\vec{q}\,\vec{r}]}{2x^2}}{3} \geq \left(\frac{1}{4}\right)^{1/3}$$

(using A.M. ≥ G.M.)

7. (1), (2), (3), (4)

$a_1 + a_2\cos 2x + a_3\sin^2 x = 0 \ \forall\, x \in R$

or $(a_1 + a_2) + \sin^2 x\,(a_3 - 2a_2) = 0$

$\Rightarrow$ $a_1 + a_2 = 0$ and $a_3 - 2a_2 = 0$

or $\dfrac{a_1}{-1} = \dfrac{a_2}{1} = \dfrac{a_3}{2} = \lambda\ (\neq 0)$

$\Rightarrow$ $a_1 = -\lambda,\ a_2 = \lambda,\ a_3 = 2\lambda$

$|\vec{a}|^2 = 6$

$\Rightarrow$ $a_1^2 + a_2^2 + a_3^2 = 6$

$\Rightarrow$ $6\lambda^2 = 6$

$\Rightarrow$ $\lambda = \pm 1$

$\Rightarrow$ $(a_1, a_2, a_3) = (1, -1, -2)$ or $(-1, 1, 2)$

$2a_1 + 3a_2 + 6a_3 = 26$

$\Rightarrow$ $13\lambda = 26 \Rightarrow \lambda = 2$

$\Rightarrow$ $|\vec{a}_1\hat{i} + a_2\hat{j} + a_3\hat{k}|$

$= \sqrt{6}\lambda = 2\sqrt{6}$

8. (1), (2), (3), (4)

$\vec{a}\times\vec{b} = |\vec{a}||\vec{b}|\sin\theta\,\hat{n}$

or $|\vec{a}\times\vec{b}| = |\vec{a}||\vec{b}|\sin\theta$

or $\sin\theta = \dfrac{|\vec{a}\times\vec{b}|}{|\vec{a}||\vec{b}|}$ ...(i)

$\vec{a}\cdot\vec{b} = |\vec{a}||\vec{b}|\cos\theta$

$\Rightarrow$ $\cos\theta = \dfrac{|\vec{a}\cdot\vec{b}|}{|\vec{a}||\vec{b}|}$ ...(ii)

From (i) and (ii),

$\sin^2\theta + \cos^2\theta = 1$

$\Rightarrow$ $|\vec{a}\times\vec{b}|^2 + (\vec{a}\cdot\vec{b})^2 = |\vec{a}|^2|\vec{b}|^2$

If $\theta = \pi/4$, then $\sin\theta = \cos\theta = 1/\sqrt{2}$. Therefore,

$|\vec{a}\times\vec{b}| = \dfrac{|\vec{a}||\vec{b}|}{\sqrt{2}}$ and $\vec{a}\cdot\vec{b} = \dfrac{|\vec{a}||\vec{b}|}{\sqrt{2}}$

$\Rightarrow$ $|\vec{a}\times\vec{b}| = \vec{a}\cdot\vec{b}$

$\Rightarrow$ $\vec{a}\times\vec{b} = |\vec{a}||\vec{b}|\sin\theta\,\hat{n} = \dfrac{|\vec{a}||\vec{b}|}{\sqrt{2}}\hat{n}$

$= (\vec{a}\cdot\vec{b})\hat{n}$

9. (1), (2), (3), (4)

Since $\vec{a}$, $\vec{b}$ and $\vec{a}\times\vec{b}$ are non-coplanar,

$\therefore$ $\vec{r} = x\vec{a} + y\vec{b} + z(\vec{a}\times\vec{b})$

$\Rightarrow$ $\vec{r}\times\vec{b} = x\vec{a}\times\vec{b} + z(\vec{a}\times\vec{b})\times\vec{b}$

$\Rightarrow$ $x\vec{a}\times\vec{b} + z\{(\vec{a}\cdot\vec{b})\vec{b} - (\vec{b}\cdot\vec{b})\vec{a}\} = \vec{a}$

$\Rightarrow$ $-(1 + z|\vec{b}|^2)\vec{a} + x\vec{a}\times\vec{b} = 0$

(since $\vec{a}\cdot\vec{b} = 0$)

$\therefore$ $x = 0$ and $z = -\dfrac{1}{|\vec{b}|^2}$

Thus, $\vec{r} = y\vec{b} - \dfrac{\vec{a}\times\vec{b}}{|\vec{b}|^2}$, where y is the parameter.

10. (1), (3)

$\vec{x} + \vec{c}\times\vec{y} = \vec{a}$ ...(i)

$\vec{y} + \vec{c}\times\vec{x} = \vec{b}$ ...(ii)

Taking cross with $\vec{c}$, we have

$\vec{c}\times\vec{y} + \vec{c}\times(\vec{c}\times\vec{x}) = \vec{c}\times\vec{b}$

$\Rightarrow \quad (\vec{a}-\vec{x})+(\vec{c}\cdot\vec{x})\vec{c}-(\vec{c}\cdot\vec{c})\vec{x}=\vec{c}\times\vec{b}$

Also $\vec{x}+\vec{c}\times\vec{y}=\vec{a}$

$\Rightarrow \quad \vec{c}\cdot\vec{x}+\vec{c}\cdot(\vec{c}\times\vec{y})=\vec{c}\cdot\vec{a}$

or $\quad \vec{c}\cdot\vec{x}+0=\vec{c}\cdot\vec{a}$

or $\quad \vec{c}\cdot\vec{x}=\vec{c}\cdot\vec{a}$

$\Rightarrow \quad \vec{a}-\vec{x}+(\vec{c}\cdot\vec{a})\vec{c}-(\vec{c}\cdot\vec{c})\vec{x}=\vec{c}\times\vec{b}$

or $\quad \vec{x}(1+(\vec{c}\cdot\vec{c}))=\vec{b}\times\vec{c}+\vec{a}+(\vec{c}\cdot\vec{a})\cdot\vec{c}$

or $\quad \vec{x}=\dfrac{\vec{b}\times\vec{c}+\vec{a}+(\vec{c}\cdot\vec{a})\vec{c}}{1+\vec{c}\cdot\vec{c}}$

Putting $\vec{x}$ in (i), we get

$$\vec{y}=\frac{\vec{a}\times\vec{c}+\vec{b}+(\vec{c}\cdot\vec{b})\vec{c}}{1+\vec{c}\cdot\vec{c}}$$

11. (2), (4)

$$\vec{a}\times(\vec{r}\times\vec{a}) = \vec{a}\times\vec{b}$$

$\Rightarrow \quad 3\vec{r}-(\vec{a}\cdot\vec{r})\,\vec{a} = \vec{a}\times\vec{b}$

Also $|\vec{r}\times\vec{a}| = |\vec{b}|$

$\Rightarrow \quad \sin^2\theta=\dfrac{2}{3}$

$\Rightarrow \quad (1-\cos^2\theta)=\dfrac{2}{3}$

$\Rightarrow \quad \dfrac{1}{3}=\cos^2\theta$

$\Rightarrow \quad \vec{a}.\vec{r} = \pm 1$

$\Rightarrow \quad 3\,\vec{r}\pm\vec{a} = \vec{a}\times\vec{b}$

$\Rightarrow \quad \vec{r} = \dfrac{1}{3}(\vec{a}\times\vec{b}\pm\vec{a})$

12. (2), (4)

$$(\vec{a}-\vec{b})\times[(\vec{b}+\vec{a})\times(2\vec{a}+\vec{b})] = \vec{b}+\vec{a}$$

$\Rightarrow \quad \{(\vec{a}-\vec{b})\cdot(2\vec{a}+\vec{b})\}(\vec{b}+\vec{a})$

$\quad -\{(\vec{a}-\vec{b})\cdot(\vec{b}+\vec{a})\}(2\vec{a}+\vec{b}) = \vec{b}+\vec{a}$

$\Rightarrow \quad (2-\vec{a}.\vec{b}-1)(\vec{b}+\vec{a}) = \vec{b}+\vec{a}$

$\Rightarrow \quad$ either $\vec{b}+\vec{a} = \vec{0}$ or $1-\vec{a}\cdot\vec{b} = 1$

$\Rightarrow \quad$ either $\vec{b} = -\vec{a}$ or $\vec{a}\cdot\vec{b} = 0$

$\Rightarrow \quad$ either $\theta=\pi$ or $\theta=\pi/2$

13. (1), (4)

Given $\vec{c} = \lambda_1\vec{a}+\lambda_2\vec{b}+\lambda_3(\vec{a}\times\vec{b})$ ...(i)

and $\vec{a}\cdot\vec{b} = 0$, $|\vec{a}| = 1$, $|\vec{b}| = 1$

From (i), $\vec{a}\cdot\vec{c} = \lambda_1$, $\vec{c}\cdot\vec{b}=\lambda_2$

and $\vec{c}\cdot(\vec{a}\times\vec{b}) = |\vec{a}\times\vec{b}|^2\,\lambda_3$

$\quad = (1\cdot 1\sin 90°)^2\,\lambda_3 = \lambda_3$

Hence, $\lambda_1+\lambda_2+\lambda_3 = (\vec{a}+\vec{b}+\vec{a}\times\vec{b})\cdot\vec{c}$

14. (2), (3), (4)

Obviously, $\dfrac{\vec{a}}{|\vec{a}|}+\dfrac{\vec{b}}{|\vec{b}|}$ is a vector in the plane of $\vec{a}$ and $\vec{b}$ and hence perpendicular to $\vec{a}\times\vec{b}$. It is also equally inclined to $\vec{a}$ and $\vec{b}$ as it is along the angle bisector.

15. (1), (4)

$$|\vec{a}+\vec{b}| = |\vec{a}-2\vec{b}|$$

$\Rightarrow \quad \vec{a}\cdot\vec{b} = \dfrac{|\vec{b}|^2}{2}$

Also $\vec{a}\cdot\vec{b}+\dfrac{1}{|\vec{b}|^2+2}$

$\quad = \dfrac{|\vec{b}|^2+2}{2}+\dfrac{1}{|\vec{b}|^2+2}-1$

$\quad \geq \sqrt{2}-1$ (using A.M. ≥ G.M.)

16. (2), (4)

$$\vec{V_1} = \vec{V_2}$$

$$\vec{a}\times(\vec{b}\times\vec{c}) = (\vec{a}\times\vec{b})\times\vec{c}$$

or $\quad (\vec{a}\cdot\vec{c})\vec{b}-(\vec{a}\cdot\vec{b})\vec{c} = (\vec{a}\cdot\vec{c})\vec{b}-(\vec{b}\cdot\vec{c})\vec{a}$

or $\quad (\vec{a}\cdot\vec{b})\vec{c} = (\vec{b}\cdot\vec{c})\vec{a}$

Thus, either $\vec{c}$ and $\vec{a}$ are collinear or $\vec{b}$ is perpendicular to both $\vec{a}$ and $\vec{c}$ $\Rightarrow \vec{b} = \lambda(\vec{a}\times\vec{c})$

17. (2), (3)

We have $\vec{A}+\vec{B} = \vec{a}$

$\Rightarrow \quad \vec{A}\cdot\vec{a}+\vec{B}\cdot\vec{a} = \vec{a}\cdot\vec{a}$

$\Rightarrow \quad 1+\vec{B}\cdot\vec{a} = a^2$ (given $\vec{A}\cdot\vec{a}=1$)

$\Rightarrow \quad \vec{B}\cdot\vec{a} = a^2-1$ ...(i)

Also $\vec{A}\times\vec{B} = \vec{b}$

$\Rightarrow \quad \vec{a}\times(\vec{A}\times\vec{B}) = \vec{a}\times\vec{b}$

$\Rightarrow \quad (\vec{a}\cdot\vec{B})\vec{A}-(\vec{a}\cdot\vec{A})\vec{B} = \vec{a}\times\vec{b}$

$\Rightarrow \quad (a^2-1)\vec{A}-\vec{B} = \vec{a}\times\vec{b}$ (using (i) and $\vec{a}\cdot\vec{A}=1$) ...(ii)

and $\quad \vec{A}+\vec{B} = \vec{a}$ ...(iii)

From (ii) and (iii), we have

$$\vec{A} = \frac{(\vec{a}\times\vec{b})+\vec{a}}{a^2}$$

and $\quad \vec{B} = \vec{a}-\left\{\dfrac{(\vec{a}\times\vec{b})+\vec{a}}{a^2}\right\}$

$$= \frac{(\vec{b} \times \vec{a}) + \vec{a}(a^2 - 1)}{a^2}$$

Thus, $\vec{A} = \dfrac{(\vec{a} \times \vec{b}) + \vec{a}}{a^2}$

and $\vec{B} = \dfrac{(\vec{b} \times \vec{a}) + \vec{a}(a^2 - 1)}{a^2}$

18. (3), (4)

Since $[\vec{a}\,\vec{b}\,\vec{c}] = 0$; $\vec{a}, \vec{b}$ and $\vec{c}$ are coplanar vectors.

Further, since $\vec{d}$ is equally inclined to $\vec{a}, \vec{b}$ and $\vec{c}$,

$$\vec{d}\cdot\vec{a} = \vec{d}\cdot\vec{b} = \vec{d}\cdot\vec{c} = 0$$

$$\vec{d}\cdot\vec{x} = \vec{d}\cdot\vec{y} = \vec{d}\cdot\vec{z} = 0$$

$$\vec{d}\cdot\vec{r} = 0$$

19. (2), (4)

Let $\vec{\alpha} = \hat{i} - \hat{j} - \hat{k}$, $\vec{\beta} = \hat{i} + \hat{j} + \hat{k}$

and $\vec{\gamma} = -\hat{i} + \hat{j} + \hat{k}$.

Let required vector $\vec{a} = x\hat{i} + y\hat{j} + z\hat{j}$.

Then $\vec{\alpha}, \vec{\beta}, \vec{\gamma}$ are coplanar, i.e.,

$$\begin{vmatrix} x & y & z \\ 1 & 1 & 1 \\ -1 & 1 & 1 \end{vmatrix} = 0 \text{ or } y = z$$

Also, $\vec{a}$ and $\vec{\alpha}$ are perpendicular

$\Rightarrow \quad x - y - z = 0$

$\Rightarrow \quad x = 2y$

Hence, options (2) and (4) are correct.

20. (2), (4)

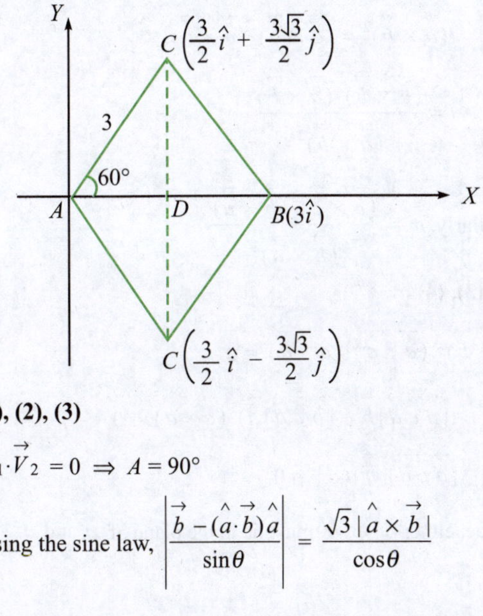

21. (1), (2), (3)

$\vec{V}_1\cdot\vec{V}_2 = 0 \Rightarrow A = 90°$

Using the sine law, $\left|\dfrac{\vec{b} - (\hat{a}\cdot\vec{b})\hat{a}}{\sin\theta}\right| = \dfrac{\sqrt{3}\,|\hat{a} \times \vec{b}|}{\cos\theta}$

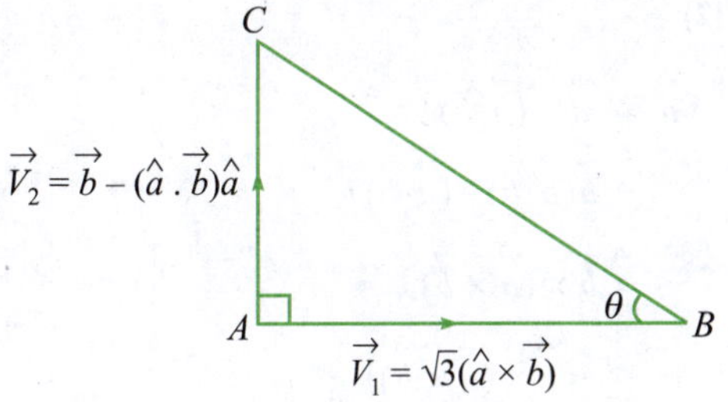

or $\quad \tan\theta = \dfrac{1}{\sqrt{3}}\dfrac{|\vec{b} - (\hat{a}\cdot\vec{b})\hat{a}|}{|\hat{a} \times \vec{b}|}$

$$= \frac{1}{\sqrt{3}}\frac{|(\hat{a} \times \vec{b}) \times \hat{a}|}{|\hat{a} \times \vec{b}|}$$

$$= \frac{1}{\sqrt{3}}\frac{|\hat{a} \times \vec{b}||\hat{a}|\sin 90°}{|\hat{a} \times \vec{b}|} = \frac{1}{\sqrt{3}}$$

or $\quad \theta = \dfrac{\pi}{6}$

22. (1), (2)

Given, $\frac{1}{6}\hat{i} - \frac{1}{3}\hat{j} + \frac{1}{3}\hat{k} = (\vec{a} \times \vec{b}) \times (\vec{c} \times \vec{d})$

$$= [\vec{a}\,\vec{b}\,\vec{d}]\vec{c} - [\vec{a}\,\vec{b}\,\vec{c}]\vec{d}$$

$$= [\vec{a}\,\vec{b}\,\vec{d}]\vec{c} \quad \text{(i)}$$

$[\because \vec{a}, \vec{b}$ and $\vec{c}$ are coplanar$]$

$$[\vec{a}\,\vec{b}\,\vec{d}] = (\vec{a} \times \vec{b})\cdot\vec{d}$$

$$= |\vec{a} \times \vec{b}||\vec{d}|\cos\theta$$

$(\because \vec{d} \perp \vec{a}, \vec{d} \perp \vec{b}, \therefore \vec{d} \parallel \vec{a} \times \vec{b})$

$$= ab\sin 30°\cdot 1\cdot(\pm 1) \quad (\because \theta = 0 \text{ or } \pi)$$

$$= 1\cdot 1\cdot\frac{1}{2}.1(\pm 1) = \pm\frac{1}{2}$$

From (i), we have

$$\vec{c} = \pm\left(\frac{1}{3}\hat{i} - \frac{2}{3}\hat{j} + \frac{2}{3}\hat{k}\right) = \pm\frac{\hat{i} - 2\hat{j} + 2\hat{k}}{3}$$

23. (1), (2), (3)

We know that $\vec{a} + \vec{b} + \vec{c} = \vec{0}$, then $\vec{a} \times \vec{b} = \vec{b} \times \vec{c} = \vec{c} \times \vec{a}$

Given $\vec{a} + 2\vec{b} + 3\vec{c} = \vec{0}$

$\Rightarrow \quad 2\vec{a} \times \vec{b} = 6\vec{b} \times \vec{c} = 3\vec{c} \times \vec{a}$

Hence $\vec{a} \times \vec{b} + \vec{b} \times \vec{c} + \vec{c} \times \vec{a}$

$$= 2(\vec{a} \times \vec{b}) \text{ or } 6(\vec{b} \times \vec{c}) \text{ or } 3(\vec{c} \times \vec{a})$$

24. (1), (2)

$$\vec{u} = \vec{a} - (\vec{a}\cdot\vec{b})\vec{b}$$

$$= \vec{a}(\vec{b}\cdot\vec{b}) - (\vec{a}\cdot\vec{b})\vec{b}$$

$$= \vec{b}\times(\vec{a}\times\vec{b})$$

$$\Rightarrow \quad |\vec{u}| = |\vec{b}\times(\vec{a}\times\vec{b})|$$

$$= |\vec{b}||\vec{a}\times\vec{b}|\sin 90°$$

$$= |\vec{a}\times\vec{b}|$$

$$= |\vec{v}|$$

Also $\vec{u}\cdot\vec{b} = \vec{b}\cdot\vec{b}\times(\vec{a}\times\vec{b})$

$$= [\vec{b}\ \vec{b}\ \vec{a}\times\vec{b}]$$

$$= 0$$

$$\Rightarrow \quad |\vec{v}| = |\vec{u}| + |\vec{u}\cdot\vec{b}|$$

25. (1), (3)

$$\vec{a}\times\vec{b} = \vec{c},\ \vec{b}\times\vec{c} = \vec{a}$$

Taking cross with $\vec{b}$ in the first equation, we get

$$\vec{b}\times(\vec{a}\times\vec{b}) = \vec{b}\times\vec{c} = \vec{a}$$

$$\Rightarrow \quad |\vec{b}|^2\vec{a} - (\vec{a}\cdot\vec{b})\vec{b} = \vec{a}$$

$$\Rightarrow \quad |\vec{b}| = 1 \text{ and } \vec{a}\cdot\vec{b} = 0$$

Also $|\vec{a}\times\vec{b}| = |\vec{c}|$ or $|\vec{a}||\vec{b}|\sin\frac{\pi}{2} = |\vec{c}|$

or $|\vec{a}| = |\vec{c}|$

26. (2), (4)

$$\vec{d}\cdot\vec{a} = [\vec{a}\ \vec{b}\ \vec{c}]\cos y = -\vec{d}\cdot(\vec{b}+\vec{c})$$

or $\cos y = -\dfrac{\vec{d}\cdot(\vec{b}+\vec{c})}{[\vec{a}\cdot\vec{b}\cdot\vec{c}]}$

Similarly, $\sin x = \dfrac{\vec{d}\cdot(\vec{a}+\vec{b})}{[\vec{a}\ \vec{b}\ \vec{c}]}$ and $\dfrac{\vec{d}\cdot(\vec{a}+\vec{c})}{[\vec{a}\ \vec{b}\ \vec{c}]} = -2$

$\therefore \quad \sin x + \cos y + 2 = 0$

$\Rightarrow \quad \sin x + \cos y = -2$

$\Rightarrow \quad \sin x = -1, \cos y = -1$

Since we want the minimum value of $x^2 + y^2$,

$x = -\pi/2, y = \pi$

Therefore, the minimum value of $x^2 + y^2$ is $5\pi^2/4$.

27. (2), (3)

$$\vec{a}\times(\vec{b}\times\vec{c}) = \frac{1}{2}\vec{b}$$

$$\Rightarrow \quad (\vec{a}\cdot\vec{c})\vec{b} - (\vec{a}\cdot\vec{b})\vec{c} = \frac{1}{2}\vec{b}$$

$$\Rightarrow \quad \vec{a}\cdot\vec{c} = \frac{1}{2} \text{ and } \vec{a}\cdot\vec{b} = 0$$

$$\Rightarrow \quad 1\cdot 1\cos\alpha = \frac{1}{2} \text{ and } \vec{a}\perp\vec{b}$$

$$\Rightarrow \quad \alpha = \frac{\pi}{3} \text{ and } \vec{a}\perp\vec{b}$$

28. (1), (2), (3)

$$\overrightarrow{AB} + \overrightarrow{BC} = \overrightarrow{AC}$$

$$\therefore \quad \overrightarrow{BC} = \frac{2\vec{u}}{|\vec{u}|} - \frac{\vec{u}}{|\vec{u}|} + \frac{\vec{v}}{|\vec{v}|} = \frac{\vec{u}}{|\vec{u}|} + \frac{\vec{v}}{|\vec{v}|}$$

$$\overrightarrow{AB}\cdot\overrightarrow{BC} = \left(\frac{\vec{u}}{|\vec{u}|} - \frac{\vec{v}}{|\vec{v}|}\right)\left(\frac{\vec{u}}{|\vec{u}|} + \frac{\vec{v}}{|\vec{v}|}\right)$$

$$= (\hat{u} - \hat{v})\cdot(\hat{u} + \hat{v}) = 1 - 1 = 0$$

$$\Rightarrow \quad \angle B = 90°$$

$$\Rightarrow \quad 1 + \cos 2A + \cos 2B + \cos 2C = 0$$

29. (1), (2), (3)

Let $\vec{A} = \vec{a}\times\vec{b}$, $\vec{B} = \vec{c}\times\vec{d}$ and $\vec{C} = \vec{e}\times\vec{f}$

We know that $\vec{A}\cdot(\vec{B}\times\vec{C}) = \vec{B}\cdot(\vec{C}\times\vec{A}) = \vec{C}\cdot(\vec{A}\times\vec{B})$

$$= (\vec{a}\times\vec{b})\cdot[(\vec{c}\times\vec{d})\times(\vec{e}\times\vec{f})]$$

$$= (\vec{a}\times\vec{b})\cdot[\{(\vec{c}\times\vec{d})\cdot\vec{f}\}\vec{e} - \{(\vec{c}\times\vec{d})\cdot\vec{e}\}\vec{f}\}$$

$$= [\vec{c}\ \vec{d}\ \vec{f}][\vec{a}\ \vec{b}\ \vec{e}] - [\vec{c}\ \vec{d}\ \vec{e}][\vec{a}\ \vec{b}\ \vec{f}]$$

Similarly, other parts can be obtained.

30. (1), (3)

Here $(l\vec{a} + m\vec{b})\times\vec{b} = \vec{c}\times\vec{b}$

$$\Rightarrow \quad l\vec{a} \square \vec{b} = \vec{c} \ \square\vec{b}$$

$$\Rightarrow \quad l(\vec{a}\times\vec{b})^2 = (\vec{c}\times\vec{b})\cdot(\vec{a}\times\vec{b})$$

$$\Rightarrow \quad l = \frac{(\vec{c}\times\vec{b})\cdot(\vec{a}\times\vec{b})}{(\vec{a}\times\vec{b})^2}$$

Similarly, $m = \dfrac{(\vec{c}\times\vec{a})\cdot(\vec{b}\times\vec{a})}{(\vec{b}\times\vec{a})^2}$

31. (2), (3), (4)

$$(\vec{a}\times\vec{b})\times(\vec{c}\times\vec{d})\cdot(\vec{a}\times\vec{d}) = 0$$

or $([\vec{a}\ \vec{c}\ \vec{d}]\vec{b} - [\vec{b}\ \vec{c}\ \vec{d}]\vec{a})\cdot(\vec{a}\times\vec{d}) = 0$

or $[\vec{a}\ \vec{c}\ \vec{d}][\vec{b}\ \vec{a}\ \vec{d}] = 0$

Hence, either $\vec{c}$ or $\vec{b}$ must lie in the plane of $\vec{a}$ and $\vec{d}$.

32. (1), (2), (4)

$$\left.\begin{aligned}\vec{V}_1 &= l\vec{a}+m\vec{b}+n\vec{c}\\ \vec{V}_2 &= n\vec{a}+l\vec{b}+m\vec{c}\\ \vec{V}_3 &= m\vec{a}+n\vec{b}+l\vec{c}\end{aligned}\right\}\text{ where } \vec{a}, \vec{b} \text{ and } \vec{c} \text{ are non-coplanar.}$$

Therefore,

$$[\vec{V}_1\,\vec{V}_2\,\vec{V}_3] = \begin{vmatrix} l & m & n\\ n & l & m\\ m & n & l\end{vmatrix} = 0$$

$\Rightarrow \quad (l+m+n)\,[(l-m)^2+(m-n)^2+(n-l)^2] = 0$

$\Rightarrow \quad l+m+n = 0 \qquad \text{(i)}$

Obviously, $lx^2+mx+n=0$ is satisfied by $x=1$ due to (i).

$l^3+m^3+n^3 = 3lmn$

$\Rightarrow \quad (l+m+n)(l^2+m^2+n^2-lm-mn-ln)=0$, which is true

33. (1), (2), (3)

It is given that $\vec{\alpha}, \vec{b}$ and $\vec{\gamma}$ are coplanar vectors. Therefore,

$[\vec{\alpha}\,\vec{\beta}\,\vec{\gamma}] = 0$

$$\Rightarrow \quad \begin{vmatrix} a & b & c\\ b & c & a\\ c & a & b\end{vmatrix} = 0$$

$\Rightarrow \quad 3abc-a^3-b^3-c^3=0$

$\Rightarrow \quad a^3+b^3+c^3-3abc=0$

$\Rightarrow \quad (a+b+c)(a^2+b^2+c^2-ab-bc-ca)=0$

$\Rightarrow \quad a+b+c=0$

$[\because a^2+b^2+c^2-ab-bc-ca \neq 0]$

$\Rightarrow \quad \vec{v}\cdot\vec{\alpha} = \vec{v}\cdot\vec{\beta} = \vec{v}\cdot\vec{\gamma} = 0$

Hence, $\vec{v}$ is perpendicular to $\vec{\alpha}, \vec{\beta}$ and $\vec{\gamma}$

34. (2), (4)

For $\vec{A}, \vec{B}$ and $\vec{C}$ to form a left-handed system

$[\vec{A}\vec{B}\vec{C}] < 0 \qquad \text{...(i)}$

$$\vec{A}\times\vec{B} = \begin{vmatrix} \hat{i} & \hat{j} & \hat{k}\\ 2 & 3 & 4\\ 1 & 1 & 5\end{vmatrix} = 11\hat{i}-6\hat{j}-\hat{k}$$

(i) is satisfied by options (2) and (4).

35. (1), (2), (3), (4)

$\vec{a}\times(\vec{b}\times\vec{c}) = (\vec{a}\cdot\vec{c})\vec{b}-(\vec{a}\cdot\vec{b})\vec{c}$

$= (yz+yx+zx)\{(y-z)\hat{i}$

$+(z-x)\hat{j}+(x-y)\hat{k}\}$

Clearly this vector is parallel to $(y-z)\hat{i}+(z-x)\hat{j}+(x-y)\hat{k}$.

It is orthogonal to $\hat{i}+\hat{j}+\hat{k}$ as $(y-z)(1)+(z-x)(1)+(x-y)(1) = 0$.

It is orthogonal to $(y+z)\hat{i}+(z+x)\hat{j}+(x+y)\hat{k}$

as $\quad (y-z)(y+z)+(z-x)(z+x)+(x-y)(x+y)$

$= y^2-z^2+z^2-x^2+x^2-y^2 = 0$

Also it is orthogonal to $x\hat{i}+y\hat{j}+z\hat{k}$

36. (2), (3)

$\overrightarrow{AC} = \vec{a}+\vec{b}$

$\therefore \quad |\overrightarrow{AC}| = |\vec{a}+\vec{b}|$

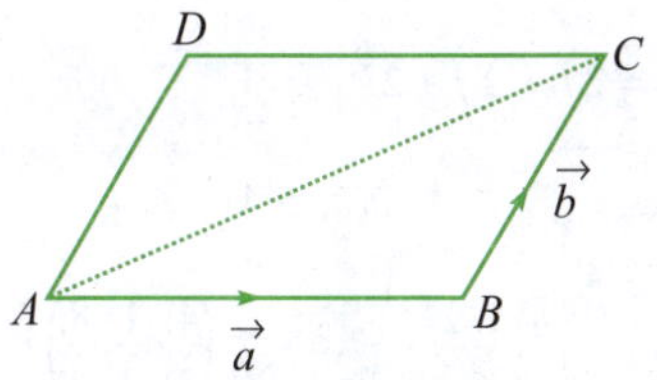

$\therefore \quad |\overrightarrow{AC}|^2 = |\vec{a}|^2+|\vec{b}|^2+2\vec{a}\cdot\vec{b}$

$= |3\vec{\alpha}-\vec{\beta}|^2+|\vec{\alpha}+3\vec{\beta}|^2$

$+2(3\vec{\alpha}-\vec{\beta})\cdot(\vec{\alpha}+3\vec{\beta})$

$= 16|\vec{\alpha}|^2+4|\vec{\beta}|^2+16\vec{\alpha}\cdot\vec{\beta}$

$= 80+16(2)(2)(1/2)$

$= 112$

$\therefore \quad |\overrightarrow{AC}| = 4\sqrt{7}$

$|\overrightarrow{BD}| = |\vec{a}-\vec{b}|$

$|\overrightarrow{BD}|^2 = |\vec{a}|^2+|\vec{b}|^2-2\vec{a}\cdot\vec{b}$

$= |3\vec{\alpha}-\vec{\beta}|^2+|\vec{\alpha}+3\vec{\beta}|^2$

$-2(3\vec{\alpha}-\vec{\beta})\cdot(\vec{\alpha}+3\vec{\beta})$

$= 4|\vec{\alpha}|^2+4|\vec{\beta}|^2-16\vec{\alpha}\cdot\vec{\beta}$

$= 80-16(2)(2)(1/2)$

$= 48$

$\therefore \quad |\overrightarrow{BD}| = 4\sqrt{3}$

Linked Comprehension Type

1. (2) $\quad \vec{a}_1 = \left[(2\hat{i}+3\hat{j}-6\hat{k})\cdot\dfrac{(2\hat{i}-3\hat{j}+6\hat{k})}{7}\right]\dfrac{2\hat{i}-3\hat{j}+6\hat{k}}{7}$

$= \dfrac{-41}{49}(2\hat{i}-3\hat{j}+6\hat{k})$

$\vec{a}_2 = \dfrac{-41}{49}\left((2\hat{i}-3\hat{j}+6\hat{k})\cdot\dfrac{(-2\hat{i}+3\hat{j}+6\hat{k})}{7}\right)$

$\times\dfrac{(-2\hat{i}+3\hat{j}+6\hat{k})}{7}$

$= \dfrac{-41}{(49)^2}(-4-9+36)(-2\hat{i}+3\hat{j}+6\hat{k})$

$= \dfrac{943}{49^2}(2\hat{i}-3\hat{j}-6\hat{k})$

2. (1) $\quad \vec{a}_1\cdot\vec{b} = \dfrac{-41}{49}(2\hat{i}-3\hat{j}+6\hat{k})\cdot(2\hat{i}-3\hat{j}+6\hat{k}) = -41$

3. (3) $\quad \vec{a}, \vec{a}_1$ and $\vec{b}$ are coplanar because $\vec{a}_1$ and $\vec{b}$ are collinear.

4. (51), 5. (96), 6. (14)

Point G is $\left(\dfrac{4}{3}, \dfrac{1}{3}, \dfrac{8}{3}\right)$. Therefore,

$\left|\overrightarrow{AG}\right|^2 = \left(\dfrac{5}{3}\right)^2+\dfrac{1}{9}+\left(\dfrac{5}{3}\right)^2 = \dfrac{51}{9}$

or $\left|\overrightarrow{AG}\right| = \dfrac{\sqrt{51}}{3}$

$\overrightarrow{AB} = -4\hat{i} + 4\hat{j} + 0\hat{k}$

$\overrightarrow{AC} = 2\hat{i} + 2\hat{j} + 2\hat{k}$

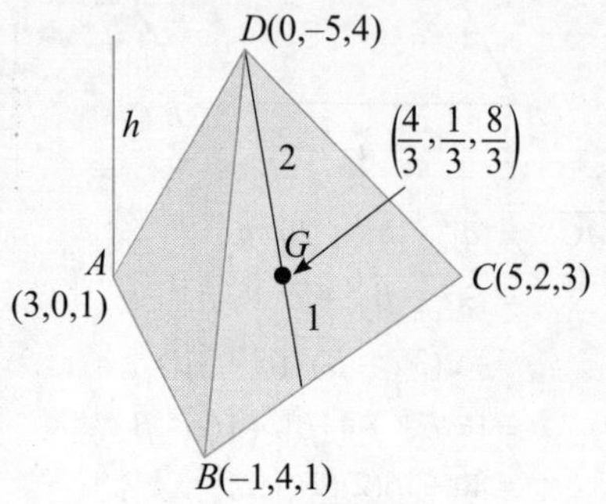

$$\therefore \quad \overrightarrow{AB} \times \overrightarrow{AC} = -8\begin{vmatrix} \hat{i} & \hat{j} & \hat{k} \\ 1 & -1 & 0 \\ 1 & 1 & 1 \end{vmatrix}$$

$$= 8(\hat{i} + \hat{j} - 2\hat{k})$$

Area of $\Delta ABC = \dfrac{1}{2}\left|\overrightarrow{AB} \times \overrightarrow{AC}\right| = 4\sqrt{6}$

$\overrightarrow{AD} = -3\hat{i} - 5\hat{j} + 3\hat{k}$

The length of the perpendicular from the vertex D on the opposite face

$= \left|\text{Projection of } \overrightarrow{AD} \text{ on } \overrightarrow{AB} \times \overrightarrow{AC}\right|$

$$= \left|\frac{(-3\hat{i} - 5\hat{j} + 3\hat{k})(\hat{i} + \hat{j} - 2\hat{k})}{\sqrt{6}}\right|$$

$$= \left|\frac{-3-5-6}{\sqrt{6}}\right| = \frac{14}{\sqrt{6}}$$

7. (8) Let point D be (a_1, a_2, a_3)

$a_1 + 1 = 3$ or $a_1 = 2$

$a_2 + 0 = 1$ or $a_2 = 1$

$a_3 - 1 = 7$ or $a_3 = 8$

$\therefore \quad D(2, 1, 8)$

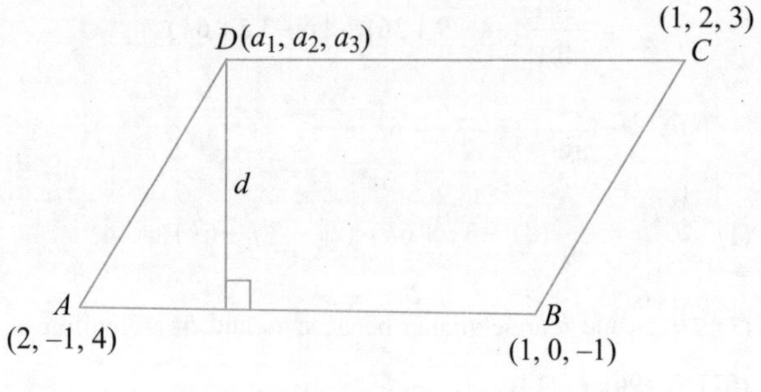

$$\vec{d} = \left|\frac{(\overrightarrow{AB}) \times (\overrightarrow{AD})}{\left|\overrightarrow{AB}\right|}\right|$$

$\overrightarrow{AB} = -\hat{i} + \hat{j} - 5\hat{k}$

$\overrightarrow{AD} = 0\hat{i} + 2\hat{j} + 4\hat{k}$

$$\overrightarrow{AB} \times \overrightarrow{AD} = \begin{vmatrix} \hat{i} & \hat{j} & \hat{k} \\ -1 & 1 & -5 \\ 0 & 2 & 4 \end{vmatrix}$$

$$= 14\hat{i} + 4\hat{j} - 2\hat{k}$$

$$= 2(7\hat{i} + 2\hat{j} - \hat{k})$$

$\Rightarrow \quad d = 2\sqrt{2}$

8. (32) $\vec{n} = 7\hat{i} + 2\hat{j} - \hat{k}$ is normal to the plane $P \equiv (8, 2, -12)$.

$\left|\overrightarrow{AP}\right| = 6\hat{i} + 3\hat{j} - 16\hat{k}$

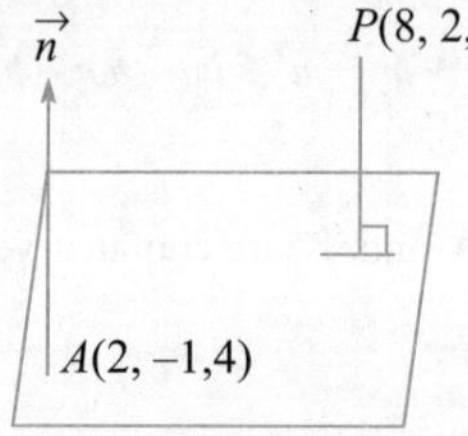

$$\therefore \quad \text{Distance, } d = \left|\frac{\overrightarrow{AP} \cdot \vec{n}}{|\vec{n}|}\right|$$

$$= \left|\frac{42+6+16}{\sqrt{49+4+1}}\right|$$

$$= \frac{64}{\sqrt{54}}$$

$$= \frac{64}{3\sqrt{6}} = \frac{64\sqrt{6}}{18} = \frac{32\sqrt{6}}{9}$$

9. (9), 10. (18), 11. (3)

Let $\vec{r} = x\hat{i} + y\hat{j}$

$x^2 + y^2 + 8x - 10y + 40 = 0$, which is a circle centre $C(-4, 5)$, radius $r = 1$

$p_1 = \max\{(x + 2)^2 + (y - 3)^2\}$

$p_2 = \min\{(x + 2)^2 + (y - 3)^2\}$

Let P be $(-2, 3)$. Then

$CP = \sqrt{2}, r = 1$

$\therefore \quad p_2 = (2\sqrt{2} - 1)^2$

and $\quad p_1 = (2\sqrt{2} + 1)^2$

$p_1 + p_2 = 18$

Slope $= AB = \left(\dfrac{dy}{dx}\right)_{(2,\,2)} = -2$

Equation of AB, $2x + y = 6$

$\therefore \quad \overrightarrow{OA} = 2\hat{i} + 2\hat{j}, \ \overrightarrow{OB} = 3\hat{i}$

$\Rightarrow \quad \overrightarrow{AB} = \hat{i} - 2\hat{j}$

$\Rightarrow \quad \overrightarrow{AB} \cdot \overrightarrow{OB} = (\hat{i} - 2\hat{j})(3\hat{i}) = 3$

12. (1), 13. (2), 14. (3)

$\vec{a} = \overrightarrow{AP} = \overrightarrow{AB} + \overrightarrow{AC} + \overrightarrow{AD}$ ...(i)

$$\overrightarrow{AB} \times \overrightarrow{AC} = \vec{b}$$

$$\overrightarrow{AD} \times \overrightarrow{AB} = \vec{c}$$

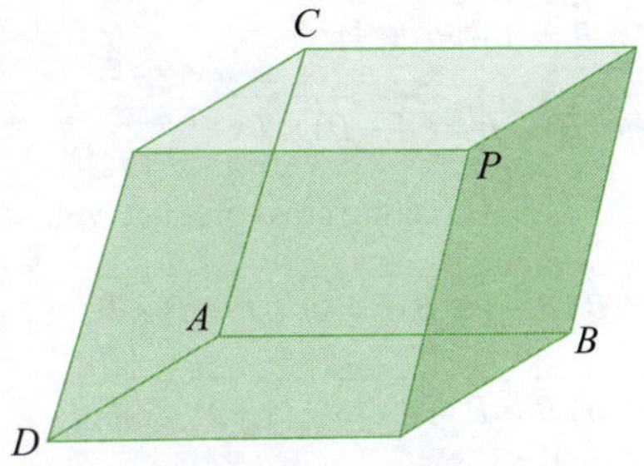

$$\overrightarrow{AB} . \frac{\vec{a}}{|\vec{a}|} = \frac{|\vec{a}|}{3} \Rightarrow \overrightarrow{AB} \cdot \vec{a} = \frac{|\vec{a}|^2}{3}$$

$$\overrightarrow{AC} . \frac{\vec{a}}{|\vec{a}|} = \frac{|\vec{a}|}{3} \Rightarrow \overrightarrow{AC} \cdot \vec{a} = \frac{|\vec{a}|^2}{3}$$

$$\therefore \quad (\overrightarrow{AB} \times \overrightarrow{AC}) \times \vec{a} = \vec{b} \times \vec{a}$$

$$\Rightarrow \quad \frac{|\vec{a}|^2}{3} \overrightarrow{AC} - \frac{|\vec{a}|^2}{3} \overrightarrow{AB} = \vec{b} \times \vec{a}$$

$$\Rightarrow \quad \overrightarrow{AC} - \overrightarrow{AB} = 3\frac{\vec{b} \times \vec{a}}{|\vec{a}|^2} \quad \text{...(ii)}$$

Now, $|\vec{a}|^2 = \overrightarrow{AB} \cdot \vec{a} + \overrightarrow{AC} \cdot \vec{a} + \overrightarrow{AD} \cdot \vec{a}$

$$\Rightarrow \quad \frac{|\vec{a}|^2}{3} = \overrightarrow{AD} \cdot \vec{a}$$

$$(\overrightarrow{AD} \times \overrightarrow{AB}) \times \vec{a} = \vec{c} \times \vec{a}$$

$$\Rightarrow \quad \overrightarrow{AB} - \overrightarrow{AD} = 3\frac{\vec{c} \times \vec{a}}{|\vec{a}|^2} \quad \text{...(iii)}$$

from (i), (ii) and (iii), we get

$$\overrightarrow{AB} = \frac{1}{3}\vec{a} + \frac{\vec{a} \times (\vec{b} - \vec{c})}{|\vec{a}|^2}$$

Now from (ii) and (iii), we get $\overrightarrow{AC}$ and $\overrightarrow{AD}$ as

$$\overrightarrow{AC} = \frac{1}{3}\vec{a} + \frac{\vec{a} \times (\vec{b} - \vec{c})}{|\vec{a}|^2} + \frac{3(\vec{b} \times \vec{a})}{|\vec{a}|^2}$$

$$\overrightarrow{AD} = \frac{1}{3}\vec{a} + \frac{\vec{a} \times (\vec{b} - \vec{c})}{|\vec{a}|^2} - \frac{3(\vec{c} \times \vec{a})}{|\vec{a}|^2}$$

15. (2), 16. (3), 17. (4)

Taking dot product of $\vec{u} + \vec{v} + \vec{w} = \vec{a}$ with $\vec{u}$, we have

$$1 + \vec{u} \cdot \vec{v} + \vec{u} \cdot \vec{w} = \vec{a} \cdot \vec{u} = \frac{3}{2} \text{ or } \vec{u} \cdot \vec{v} + \vec{u} \cdot \vec{w} = \frac{1}{2} \quad \text{...(i)}$$

Similarly, taking dot product with $\vec{v}$, we have

$$\vec{u} \cdot \vec{v} + \vec{w} \cdot \vec{v} = \frac{3}{4} \quad \text{...(ii)}$$

Also, $\vec{a} \cdot \vec{u} + \vec{a} \cdot \vec{v} + \vec{a} \cdot \vec{w} = \vec{a} \cdot \vec{a} = 4$

$$\Rightarrow \quad \vec{a} \cdot \vec{w} = 4 - \left(\frac{3}{2} + \frac{7}{4}\right) = \frac{3}{4}$$

Again, taking dot product with $\vec{w}$, we have

$$\vec{u} \cdot \vec{w} + \vec{v} \cdot \vec{w} = \frac{3}{4} - 1 = -\frac{1}{4} \quad \text{...(iii)}$$

Adding (i), (ii) and (iii), we have

$$2(\vec{u} \cdot \vec{v} + \vec{u} \cdot \vec{w} + \vec{v} \cdot \vec{w}) = 1$$

or
$$\vec{u} \cdot \vec{v} + \vec{u} \cdot \vec{w} + \vec{v} \cdot \vec{w} = \frac{1}{2} \quad \text{...(iv)}$$

Subtracting (i), (ii) and (iii) from (iv) alternatively, we have

$$\vec{v} \cdot \vec{w} = 0, \ \vec{u} \cdot \vec{w} = -\frac{1}{4} \text{ and } \vec{u} \cdot \vec{v} = \frac{3}{4}$$

Now, the equations $\vec{u} \times (\vec{v} \times \vec{w}) = \vec{b}$ and $(\vec{u} \times \vec{v}) \times \vec{w} = \vec{c}$ can be written as

$$(\vec{u} \cdot \vec{w})\vec{v} - (\vec{u} \cdot \vec{v})\vec{w} = \vec{b}$$

and
$$(\vec{u} \cdot \vec{w})\vec{v} - (\vec{v} \cdot \vec{w})\vec{u} = \vec{c}$$

$$\Rightarrow \quad -\frac{1}{4}\vec{v} - \frac{3}{4}\vec{w} = \vec{b}, \ -\frac{1}{4}\vec{v} = \vec{c}, \text{ i.e., } \vec{v} = -4\vec{c}$$

$$\Rightarrow \quad \vec{c} - \frac{3}{4}\vec{w} = \vec{b} \Rightarrow \vec{w} = \frac{4}{3}(\vec{c} - \vec{b})$$

and
$$\vec{u} = \vec{a} - \vec{v} - \vec{w} = \vec{a} + 4\vec{c} - \frac{4}{3}\vec{c} + \frac{4}{3}\vec{b}$$

$$= \vec{a} + \frac{4}{3}\vec{b} + \frac{8}{3}\vec{c}$$

18. (4), 19. (3), 20. (2)

Given that $|\vec{x}| = |\vec{y}| = |\vec{z}| = \sqrt{2}$ and they are inclined at an angle of 60° with each other.

$$\therefore \quad \vec{x} \cdot \vec{y} = \vec{y} \cdot \vec{z} = \vec{z} \cdot \vec{x} = \sqrt{2} \cdot \sqrt{2} \cos 60° = 1$$

$$\vec{x} \times (\vec{y} \times \vec{z}) = \vec{a}$$

or
$$(\vec{x} \cdot \vec{z})\vec{y} - (\vec{x} \cdot \vec{y})\vec{z} = \vec{a} \text{ or } \vec{y} - \vec{z} = \vec{a} \quad \text{...(i)}$$

Similarly, $\vec{y} \times (\vec{z} \times \vec{x}) = \vec{b} \Rightarrow \vec{z} - \vec{x} = \vec{b}$ ...(ii)

$$\vec{y} = \vec{a} + \vec{z}, \ \vec{x} = \vec{z} - \vec{b}$$

[from (i) and (ii)] (iii)

Now, $\vec{x} \times \vec{y} = \vec{c}$

$$\Rightarrow \quad (\vec{z} - \vec{b}) \times (\vec{z} + \vec{a}) = \vec{c}$$

$$\Rightarrow \quad \vec{z} \times \vec{a} - \vec{b} \times \vec{z} - \vec{b} \times \vec{a} = \vec{c}$$

$$\Rightarrow \quad \vec{z} \times (\vec{a} + \vec{b}) = \vec{c} + (\vec{b} \times \vec{a}) \quad \text{...(iv)}$$

$$\Rightarrow \quad (\vec{a} + \vec{b}) \times \{\vec{z} \times (\vec{a} + \vec{b})\}$$

$$= (\vec{a} + \vec{b}) \times \vec{c} + (\vec{a} + \vec{b}) \times (\vec{b} \times \vec{a})$$

$$\Rightarrow \quad (\vec{a} + \vec{b})^2 \vec{z} - \{(\vec{a} + \vec{b}) \cdot \vec{z}\}(\vec{a} + \vec{b})$$

$$= (\vec{a} + \vec{b}) \times \vec{c} + |\vec{a}|^2 \vec{b} - |\vec{b}|^2 \vec{a}$$

$$+ (\vec{a} \cdot \vec{b})(\vec{b} - \vec{a}) \quad \text{...(v)}$$

Now, (i) $\Rightarrow |\vec{a}|^2 = |\vec{y} - \vec{z}|^2 = 2 + 2 - 2 = 2$

Similarly, (ii) $\Rightarrow |\vec{b}|^2 = 2$

Also (i) and (ii) $\Rightarrow \vec{a}+\vec{b} = \vec{y}-\vec{x}$

$$\Rightarrow \quad |\vec{a}+\vec{b}|^2 = 2 \quad \ldots(vi)$$

Also $\quad (\vec{a}+\vec{b})\cdot\vec{z} = (\vec{y}-\vec{x})\cdot\vec{z} = \vec{y}\cdot\vec{z} - \vec{x}\cdot\vec{z}$

$$= 1 - 1 = 0$$

and $\quad \vec{a}\cdot\vec{b} = (\vec{y}-\vec{z})\cdot(\vec{z}-\vec{x})$

$$= \vec{y}\cdot\vec{z} - \vec{x}\cdot\vec{y} - |\vec{z}|^2 + \vec{x}\cdot\vec{z} = -1$$

Thus from (v), we have

$$2\vec{z} = (\vec{a}+\vec{b})\times\vec{c} + 2(\vec{b}-\vec{a}) - (\vec{b}-\vec{a})$$

or $\quad \vec{z} = (1/2)[(\vec{a}+\vec{b})\times\vec{c} + \vec{b} - \vec{a}]$

$\therefore \quad \vec{y} = \vec{a}+\vec{z} = (1/2)[(\vec{a}+\vec{b})\times\vec{c} + \vec{b} + \vec{a}]$ and

$\vec{x} = \vec{z}-\vec{b} = (1/2)[(\vec{a}+\vec{b})\times\vec{c} - (\vec{a}+\vec{b})]$

21. (2), 22. (1), 23. (3)

Given

$$\vec{x}\times\vec{y} = \vec{a} \quad \ldots(i)$$

$$\vec{y}\times\vec{z} = \vec{b} \quad \ldots(ii)$$

$$\vec{x}\cdot\vec{b} = \gamma \quad \ldots(iii)$$

$$\vec{x}\cdot\vec{y} = 1 \quad \ldots(iv)$$

$$\vec{y}\cdot\vec{z} = 1 \quad \ldots(v)$$

From (ii), $\vec{x}\cdot(\vec{y}\times\vec{z}) = \vec{x}\cdot\vec{b} = \gamma \Rightarrow [\vec{x}\ \vec{y}\ \vec{z}] = \gamma$

From (i) and (ii), $(\vec{x}\times\vec{y})\times(\vec{y}\times\vec{z}) = \vec{a}\times\vec{b}$

$\therefore \quad [\vec{x}\ \vec{y}\ \vec{z}]\vec{y} - [\vec{y}\ \vec{y}\ \vec{z}]\vec{x} = \vec{a}\times\vec{b}$

or $\quad \vec{y} = \dfrac{\vec{a}\times\vec{b}}{\gamma} \quad \ldots(vi)$

Also from (i), we get $(\vec{x}\times\vec{y})\times\vec{y} = \vec{a}\times\vec{y}$

$\Rightarrow \quad (\vec{x}\cdot\vec{y})\vec{y} - (\vec{y}\cdot\vec{y})\vec{x} = \vec{a}\times\vec{y}$

$\Rightarrow \quad \vec{x} = (1/|\vec{y}|^2)(\vec{y} - \vec{a}\times\vec{y})$

$$= \frac{\gamma^2}{|\vec{a}\times\vec{b}|^2}\left[\frac{\vec{a}\times\vec{b}}{\gamma} - \frac{\vec{a}\times(\vec{a}\times\vec{b})}{\gamma}\right]$$

$$= \frac{\gamma}{|\vec{a}\times\vec{b}|^2}[\vec{a}\times\vec{b} - \vec{a}\times(\vec{a}\times\vec{b})]$$

Also from (ii), $(\vec{y}\times\vec{z})\times\vec{y} = \vec{b}\times\vec{y}$

$\Rightarrow \quad |\vec{y}|^2\vec{z} - (\vec{z}\cdot\vec{y})\vec{y} = \vec{b}\times\vec{y}$

$\Rightarrow \quad \vec{z} = \dfrac{1}{|\vec{y}|^2}[\vec{y} + \vec{b}\times\vec{y}]$

$$= \frac{\gamma}{|\vec{a}\times\vec{b}|^2}[\vec{a}\times\vec{b} + \vec{b}\times(\vec{a}\times\vec{b})]$$

24. (2), 25. (2), 26. (4)

$\vec{P}\times\vec{B} = \vec{A}-\vec{P}$ and $|\vec{A}| = |\vec{B}| = 1$ and $\vec{A}\cdot\vec{B} = 0$ is given

Now $\quad \vec{P}\times\vec{B} = \vec{A}-\vec{P} \quad \ldots(i)$

$\Rightarrow \quad (\vec{P}\times\vec{B})\times\vec{B} = (\vec{A}-\vec{P})\times\vec{B}$

(taking cross product with $\vec{B}$ on both sides)

$\Rightarrow \quad (\vec{P}\cdot\vec{B})\vec{B} - (\vec{B}\cdot\vec{B})\vec{P} = \vec{A}\times\vec{B} - \vec{P}\times\vec{B}$

$\Rightarrow \quad (\vec{P}\cdot\vec{B})\vec{B} - \vec{P} = \vec{A}\times\vec{B} - \vec{A} + \vec{P}$

$\Rightarrow \quad 2\vec{P} = \vec{A} - \vec{A}\times\vec{B} - (\vec{P}\cdot\vec{B})\vec{B}$

$\Rightarrow \quad \vec{P} = \dfrac{\vec{A} - \vec{A}\times\vec{B} - (\vec{P}\cdot\vec{B})\vec{B}}{2} \quad \ldots(ii)$

Taking dot product with $\vec{B}$ on both sides of (i), we get

$$\vec{P}\times\vec{B}\cdot\vec{B} = \vec{A}\cdot\vec{B} - \vec{P}\cdot\vec{B}$$

$\Rightarrow \quad \vec{P}\cdot\vec{B} = 0 \quad \ldots(iii)$

$\Rightarrow \quad \vec{P} = \dfrac{\vec{A} + \vec{B}\times\vec{A}}{2}$

Now

$$(\vec{P}\times\vec{B})\times\vec{B} = (\vec{P}\cdot\vec{B})\vec{B} - (\vec{B}\cdot\vec{B})\vec{P} = -\vec{P}$$

$\vec{P}, \vec{A}, \vec{P}\times\vec{B}\ (= \vec{A}-\vec{P})$ are dependent

Also $\quad \vec{P}\cdot\vec{B} = 0$

and $\quad |\vec{P}|^2 = \left|\dfrac{\vec{A} - \vec{A}\times\vec{B}}{2}\right|^2$

$$= \frac{|\vec{A}|^2 + |\vec{A}\times\vec{B}|^2}{4}$$

$$= \frac{1+1}{4} = \frac{1}{2} \quad \text{or} \quad |\vec{P}| = \frac{1}{\sqrt{2}}$$

27. (1), 28. (4), 29. (1)

Given that $\vec{a}, \vec{b}$ and $\vec{c}$ are three unit vectors inclined at an angle θ with each other.

Also $\vec{a}, \vec{b}$ and $\vec{c}$ are non-coplanar. Therefore, $[\vec{a}\ \vec{b}\ \vec{c}] \neq 0$.

Also given that $\vec{a}\times\vec{b} + \vec{b}\times\vec{c} = p\vec{a} + q\vec{b} + r\vec{c}$.

Taking dot product on both sides with $\vec{a}$, we get

$$p + q\cos\theta + r\cos\theta = [\vec{a}\ \vec{b}\ \vec{c}] \quad \ldots(i)$$

Similarly, taking dot product on both sides with $\vec{b}$ and $\vec{c}$, we get, respectively,

$$p\cos\theta + q + r\cos\theta = 0 \quad \ldots(ii)$$

and $\quad p\cos\theta + q\cos\theta + r = [\vec{a}\ \vec{b}\ \vec{c}] \quad \ldots(iii)$

Adding (i), (ii) and (iii), we get

$$p + q + r = \frac{2[\vec{a}\ \vec{b}\ \vec{c}]}{2\cos\theta + 1} \quad \ldots(iv)$$

Multiplying (iv) by cos θ and subtracting (i) from it, we get

$$p(\cos\theta - 1) = \frac{2[\vec{a}\ \vec{b}\ \vec{c}]\cos\theta}{2\cos\theta + 1} - [\vec{a}\ \vec{b}\ \vec{c}] = \frac{-[\vec{a}\ \vec{b}\ \vec{c}]}{2\cos\theta + 1}$$

or $$p = \frac{[\vec{a}\ \vec{b}\ \vec{c}]}{(1-\cos\theta)(1+2\cos\theta)}$$

Similarly, (iv) × cos θ – (ii) gives

$$q = \frac{-2[\vec{a}\ \vec{b}\ \vec{c}]\cos\theta}{(1+2\cos\theta)(1-\cos\theta)}$$

and (iv) × cos θ – (iii) gives

$$r(\cos\theta - 1) = \frac{2[\vec{a}\ \vec{b}\ \vec{c}]\cos\theta}{2\cos\theta + 1} - [\vec{a}\ \vec{b}\ \vec{c}]$$

or $$r = \frac{-[\vec{a}\ \vec{b}\ \vec{c}]}{(2\cos\theta + 1)(\cos\theta - 1)}$$

But we have to find p, q and r in terms of θ only.

So let us find the value of $[\vec{a}\ \vec{b}\ \vec{c}]$

We know that

$$[\vec{a}\ \vec{b}\ \vec{c}]^2 = \begin{vmatrix} \vec{a}\cdot\vec{a} & \vec{a}\cdot\vec{b} & \vec{a}\cdot\vec{c} \\ \vec{b}\cdot\vec{a} & \vec{b}\cdot\vec{b} & \vec{b}\cdot\vec{c} \\ \vec{c}\cdot\vec{a} & \vec{c}\cdot\vec{b} & \vec{c}\cdot\vec{c} \end{vmatrix} = \begin{vmatrix} 1 & \cos\theta & \cos\theta \\ \cos\theta & 1 & \cos\theta \\ \cos\theta & \cos\theta & 1 \end{vmatrix}$$

$$= \begin{vmatrix} 1+2\cos\theta & \cos\theta & \cos\theta \\ 1+2\cos\theta & 1 & \cos\theta \\ 1+2\cos\theta & \cos\theta & 1 \end{vmatrix}$$

(On operating $C_1 \to C_1 + C_2 + C_3$)

$$= (1+2\cos\theta)\begin{vmatrix} 1 & \cos\theta & \cos\theta \\ 1 & 1 & \cos\theta \\ 1 & \cos\theta & 1 \end{vmatrix}$$

$$= (1+2\cos\theta)\begin{vmatrix} 0 & \cos\theta - 1 & 0 \\ 0 & 1-\cos\theta & \cos\theta - 1 \\ 1 & \cos\theta & 1 \end{vmatrix}$$

(Operating $R_1 \to R_1 - R_2$ and $R_2 \to R_2 - R_3$)

$$= (1+2\cos\theta)(1-\cos\theta)^2 \quad \text{(Expanding along } C_1)$$

$$\therefore [\vec{a}\ \vec{b}\ \vec{c}] = (1-\cos\theta)\sqrt{1+2\cos\theta}$$

Thus, we get

$$p = \frac{1}{\sqrt{1+2\cos\theta}},\ q = \frac{-2\cos\theta}{\sqrt{1+2\cos\theta}},\ r = \frac{1}{\sqrt{1+2\cos\theta}}$$

Matrix Match Type

1. (3)

(P) $|\vec{a}+\vec{b}+\vec{c}| = \sqrt{6}$

$$\Rightarrow \vec{a}^2 + \vec{b}^2 + \vec{c}^2 + 2\vec{a}\cdot\vec{b} + 2\vec{b}\cdot\vec{c} + 2\vec{c}\cdot\vec{a} = 6$$

$$\therefore \quad |\vec{a}| = 1$$

(Q) $\vec{a}$ is perpendicular to $\vec{b}+\vec{c}$

$$\Rightarrow \vec{a}\cdot\vec{b} + \vec{a}\cdot\vec{c} = 0 \quad \text{...(i)}$$

$\vec{b}$ is perpendicular to $\vec{a}+\vec{c}$

$$\Rightarrow \vec{b}\cdot\vec{c} + \vec{b}\cdot\vec{c} = 0 \quad \text{...(ii)}$$

$\vec{c}$ is perpendicular to $\vec{a}+\vec{b}$

$$\Rightarrow \vec{c}\cdot\vec{a} + \vec{c}\cdot\vec{b} = 0 \quad \text{...(iii)}$$

From (i), (ii) and (iii), we get

$$\vec{a}\cdot\vec{b} = \vec{b}\cdot\vec{c} = \vec{c}\cdot\vec{a} = 0$$

$$\therefore \quad |\vec{a}+\vec{b}+\vec{c}| = 7$$

(R) $(\vec{a}\cdot\vec{c})(\vec{b}\cdot\vec{d}) - (\vec{b}\cdot\vec{c})(\vec{a}\cdot\vec{d}) = 21$

(S) We know that

$$[\vec{a}\times\vec{b}\ \ \vec{b}\times\vec{c}\ \ \vec{c}\times\vec{a}] = [\vec{a}\ \vec{b}\ \vec{c}]^2$$

and $$[\vec{a}\ \vec{b}\ \vec{c}]^2 = \begin{vmatrix} \vec{a}\cdot\vec{a} & \vec{a}\cdot\vec{b} & \vec{a}\cdot\vec{c} \\ \vec{b}\cdot\vec{a} & \vec{b}\cdot\vec{b} & \vec{b}\cdot\vec{c} \\ \vec{c}\cdot\vec{a} & \vec{c}\cdot\vec{b} & \vec{c}\cdot\vec{c} \end{vmatrix}$$

$$= \begin{vmatrix} 4 & 2 & 2 \\ 2 & 4 & 2 \\ 2 & 2 & 4 \end{vmatrix}$$

$$= 32$$

$$\therefore \quad [\vec{a}\ \vec{b}\ \vec{c}] = 4\sqrt{2}$$

2. (2)

(P) $$\vec{a}\times\vec{b} = \begin{vmatrix} \hat{i} & \hat{j} & \hat{k} \\ -1 & 1 & 2 \\ -1 & -2 & -1 \end{vmatrix} = 3\hat{i} - 3\hat{j} + 3\hat{k}$$

Hence, the area of the triangle is $\frac{3\sqrt{3}}{2}$.

(Q) The area of the parallelogram is $3\sqrt{3}$.

(R) The area of a parallelogram whose diagonals are $2\vec{a}$ and $4\vec{b}$ is $\frac{1}{2}|2\vec{a}\times 4\vec{b}| = 12\sqrt{3}$.

(S) Volume of the parallelepiped

$$= |(\vec{a}\times\vec{b})\cdot\vec{c}| = |3 - 3 + 3| = 3$$

3. (4)

(P) $|\vec{a}+\vec{b}| = |\vec{a}+2\vec{b}|$

$$\Rightarrow a^2 + b^2 + 2\vec{a}\cdot\vec{b} = a^2 + 4b^2 + 4\vec{a}\cdot\vec{b}$$

$$\Rightarrow 2\vec{a}\cdot\vec{b} = -3b^2 < 0$$

Hence, angle between $\vec{a}$ and $\vec{b}$ is obtuse.

(Q) $|\vec{a}+\vec{b}| = |\vec{a}-2\vec{b}|$

$$\Rightarrow a^2 + b^2 + 2\vec{a}\cdot\vec{b} = a^2 + 4b^2 - 4\vec{a}\cdot\vec{b}$$

$\Rightarrow \quad 6\vec{a}\cdot\vec{b} = 3b^2$

Hence, angle between $\vec{a}$ and $\vec{b}$ is acute.

(R) $|\vec{a}+\vec{b}| = |\vec{a}-\vec{b}|$

$\Rightarrow \quad \vec{a}\cdot\vec{b}$

Hence, $\vec{a}$ is perpendicular to $\vec{b}$.

(S) $\vec{c}\times(\vec{a}\times\vec{b})$ lies in the plane of vectors $\vec{a}$ and $\vec{b}$.

A vector perpendicular to this plane is parallel to $\vec{a}\times\vec{b}$

Hence, angle is 0°.

4. (1)

$[\vec{a}\times\vec{b}\ \ \vec{b}\times\vec{c}\ \ \vec{c}\times\vec{a}] = 36$

$\Rightarrow \quad |[\vec{a}\ \vec{b}\ \vec{c}]| = 6$

$\Rightarrow$ Volume of tetrahedron formed by vectors

$\vec{a}, \vec{b}$ and $\vec{c}$ is $\frac{1}{6}\left|[\vec{a}\ \vec{b}\ \vec{c}]\right| = 1.$

$\Rightarrow$ Volume of Parallelepiped

$|[\vec{a}+\vec{b}\ \ \vec{b}+\vec{c}\ \ \vec{c}+\vec{a}]| = |2[\vec{a}\ \vec{b}\ \vec{c}]| = 12$

$\vec{a}-\vec{b},\ \vec{b}-\vec{c}$ and $\vec{c}-\vec{a}$ are coplanar

$\Rightarrow \quad [\vec{a}-\vec{b}\ \ \vec{b}-\vec{c}\ \ \vec{c}-\vec{a}] = 0$

Archives

Single Correct Answer Type

1. (3)

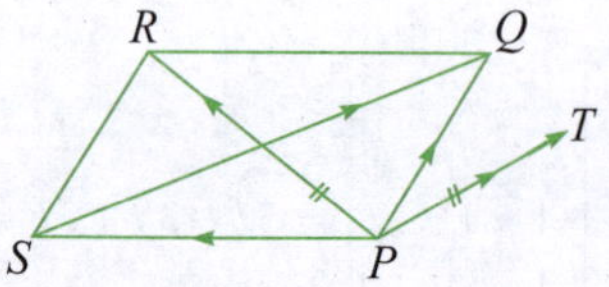

Area of base $(PQRS)$

$$= \frac{1}{2}|\overrightarrow{PR}\times\overrightarrow{SQ}| = \frac{1}{2}\begin{vmatrix}\hat{i} & \hat{j} & \hat{k}\\ 3 & 1 & -2\\ 1 & -3 & -4\end{vmatrix}$$

$$= \frac{1}{2}|-10\hat{i}+10\hat{j}-10\hat{k}|$$

$$= 5|\hat{i}-\hat{j}+\hat{k}| = 5\sqrt{3}$$

Height = Projection of PT on $\hat{i}-\hat{j}+\hat{k}$

$$= \left|\frac{1-2+3}{\sqrt{3}}\right| = \frac{2}{\sqrt{3}}$$

$\therefore$ Volume $= (5\sqrt{3})\left(\frac{2}{\sqrt{3}}\right) = 10$ cu. unit

2. (4) $\overrightarrow{OP}\cdot\overrightarrow{OQ}+\overrightarrow{OR}\cdot\overrightarrow{OS} = \overrightarrow{OR}\cdot\overrightarrow{OP}+\overrightarrow{OQ}\cdot\overrightarrow{OS}$

$\Rightarrow \quad \overrightarrow{OP}\cdot(\overrightarrow{OQ}-\overrightarrow{OR}) = \overrightarrow{OS}\cdot(\overrightarrow{OQ}-\overrightarrow{OR})$

$\Rightarrow \quad (\overrightarrow{OP}-\overrightarrow{OS}).(\overrightarrow{RQ}) = 0 \Rightarrow \overrightarrow{SP}\cdot\overrightarrow{RQ} = 0$

$\Rightarrow \quad \overrightarrow{SP}\perp\overrightarrow{RQ}$

Similarly, $\overrightarrow{SR}\perp\overrightarrow{PQ}$ and $\overrightarrow{SQ}\perp\overrightarrow{PR}$.

Hence, S is orthocenter.

3. (2) We have $\overrightarrow{OP} = \vec{a} = \hat{i}+2\hat{j}-5\hat{k},\ \overrightarrow{OQ} = \vec{b} = 3\hat{i}+6\hat{j}+3\hat{k}$

$$\overrightarrow{OR} = \vec{c} = \frac{17}{5}\hat{i}+\frac{16}{5}\hat{j}+7\hat{k},\ \overrightarrow{OS} = \vec{d} = 2\hat{i}+\hat{j}+\hat{k}$$

$$\text{Now } [\overrightarrow{PQ}\ \overrightarrow{PR}\ \overrightarrow{PS}] = \begin{vmatrix}2 & 4 & 8\\ \frac{12}{5} & \frac{6}{5} & 12\\ 1 & -1 & 6\end{vmatrix} = \frac{12}{5}\begin{vmatrix}1 & 2 & 4\\ 2 & 1 & 10\\ 1 & -1 & 6\end{vmatrix}$$

$$= \frac{12}{5}[16-2\times2+4(-3)] = 0$$

$\therefore \quad P, Q, R, S$ are coplanar.

So, option (1) is not true.

Position vector of point dividing PR internally in the ratio 5 : 4 is

$$\frac{4(\hat{i}+2\hat{j}-5\hat{k})+5\left(\frac{17}{5}\hat{i}+\frac{16}{5}\hat{j}+7\hat{k}\right)}{4+5} = \frac{7\hat{i}+8\hat{j}+5\hat{k}}{3}$$

Also, $\frac{\vec{b}+2\vec{d}}{3} = \frac{7\hat{i}+8\hat{j}+5\hat{k}}{3}$

So, option (2) is true but (3) is not.

Also, $\vec{b}\times\vec{d} = \begin{vmatrix}\hat{i} & \hat{j} & \hat{k}\\ 3 & 6 & 3\\ 2 & 1 & 1\end{vmatrix} = 3\hat{i}+3\hat{j}-9\hat{k}$

$\therefore \quad |\vec{b}\times\vec{d}|^2 = 9+9+81 = 99$

So, option (4) is also not true.

Multiple Correct Answers Type

1. (1), (2), (3)

According to the question

$$\vec{x}\cdot\vec{z} = \vec{x}\cdot\vec{y} = \vec{y}\cdot\vec{z} = \sqrt{2}\cdot\sqrt{2}\cdot\cos\frac{\pi}{3} = 1$$

Given $\vec{a}$ is perpendicular to $\vec{x}$ and $\vec{y}\times\vec{z}$

$\therefore \quad \vec{a} = \lambda_1(\vec{x}\times(\vec{y}\times\vec{z}))$

$\Rightarrow \quad \vec{a} = \lambda_1((\vec{x}\cdot\vec{z})\vec{y}-(\vec{x}\cdot\vec{y})\vec{z})$

$\Rightarrow \quad \vec{a} = \lambda_1(\vec{y}-\vec{z}) \quad (1)$

Now $\vec{a}\cdot\vec{y} = \lambda_1(\vec{y}\cdot\vec{y}-\vec{y}\cdot\vec{z}) = \lambda_1(2-1)$

$\Rightarrow \quad \lambda_1 = \vec{a}\cdot\vec{y} \quad (2)$

From (1) and (2), $\vec{a} = (\vec{a}\cdot\vec{y})(\vec{y}-\vec{z})$

Similarly, $\vec{b} = (\vec{b}\cdot\vec{z})(\vec{z}-\vec{x})$

Now, $\vec{a}\cdot\vec{b} = (\vec{a}\cdot\vec{y})(\vec{b}\cdot\vec{z})[(\vec{y}-\vec{z})\cdot(\vec{z}-\vec{x})]$

$$= (\vec{a}\cdot\vec{y})(\vec{b}\cdot\vec{z})[1-1-2+1]$$

$$= -(\vec{a}\cdot\vec{y})(\vec{b}\cdot\vec{z})$$

2. (1), (3), (4)

$\vec{a}+\vec{b}+\vec{c} = 0$

$\Rightarrow \quad \vec{b}+\vec{c} = -\vec{a}$

$\Rightarrow \quad |\vec{b}|^2 + |\vec{c}|^2 + 2\vec{b}\cdot\vec{c} = |\vec{a}|^2$

$\Rightarrow \quad 48 + |\vec{c}|^2 + 48 = 144$

$\Rightarrow \quad |\vec{c}|^2 = 48$

$\Rightarrow \quad |\vec{c}| = 4\sqrt{3}$

$\therefore \quad \dfrac{|\vec{c}|^2}{2} - |\vec{a}| = 24 - 12 = 12$

$\dfrac{|\vec{c}|^2}{2} + |\vec{a}| = 36$

Further,

$\vec{a} + \vec{b} = -\vec{c}$

$\Rightarrow \quad |\vec{a}|^2 + |\vec{b}|^2 + 2\vec{a}\cdot\vec{b} = |\vec{c}|^2$

$\Rightarrow \quad 144 + 48 + 2\vec{a}\cdot\vec{b} = 48$

$\Rightarrow \quad \vec{a}\cdot\vec{b} = -72$

$\because \quad \vec{a} + \vec{b} + \vec{c} = 0$

$\Rightarrow \quad \vec{a}\times\vec{b} + \vec{a}\times\vec{c} = 0$

$|\vec{a}\times\vec{b} + \vec{c}\times\vec{a}|$

$= 2|\vec{a}\times\vec{b}|$

$\therefore \quad = 2\sqrt{a^2b^2 - (\vec{a}\cdot\vec{b})^2}$

$= 2\sqrt{(144)(48) - (72)^2} = 48\sqrt{3}$

3. (2), (3)

$|\hat{u}\times\vec{v}| = 1$

$\Rightarrow \quad |\vec{v}|\sin\theta = 1$, where θ is angle between $\hat{u}$ and $\vec{v}$

Also $\hat{w}.(\hat{u}\times\vec{v}) = 1$

$\Rightarrow \quad |\hat{w}||\hat{u}||\vec{v}|\ \sin\theta\cos\alpha = 1$, where α is angle between $\hat{w}$ and $(\hat{u}\times\vec{v})$

$\Rightarrow \quad 1\cdot 1(1)\cos\alpha = 1$

$\Rightarrow \quad \alpha = 0$

$\Rightarrow \quad \hat{u}\times\vec{v} = \lambda\hat{w}$ where $\lambda > 0$

$$\Rightarrow \quad \begin{vmatrix} \hat{i} & \hat{j} & \hat{k} \\ u_1 & u_2 & u_3 \\ v_1 & v_2 & v_3 \end{vmatrix} = \frac{\lambda}{\sqrt{6}}(\hat{i} + \hat{j} + 2\hat{k})$$

$$\Rightarrow \quad (u_2v_3 - u_3v_2)\,\hat{i} + (u_3v_1 - u_1v_3)\,\hat{j} + (u_1v_2 - u_2v_1)\,\hat{k} = \frac{\lambda}{\sqrt{6}}(\hat{i} + \hat{j} + 2\hat{k})$$

$\vec{v}$ is a vector such that $(\hat{u}\times\vec{v})$ is parallel to $\hat{w}$

$u_3 = 0 \ \Rightarrow u_2v_3 = \dfrac{\lambda}{\sqrt{6}}$ and $-u_1v_3 = \dfrac{\lambda}{\sqrt{6}}$

$\Rightarrow \quad |u_2| = |u_1|$

$u_2 = 0 \ \Rightarrow -u_3v_2 = \dfrac{\lambda}{\sqrt{6}}$ and $u_1v_2 = \dfrac{2\lambda}{\sqrt{6}}$

$\Rightarrow \quad |u_1| = 2|u_3|$

4. (1), (3)

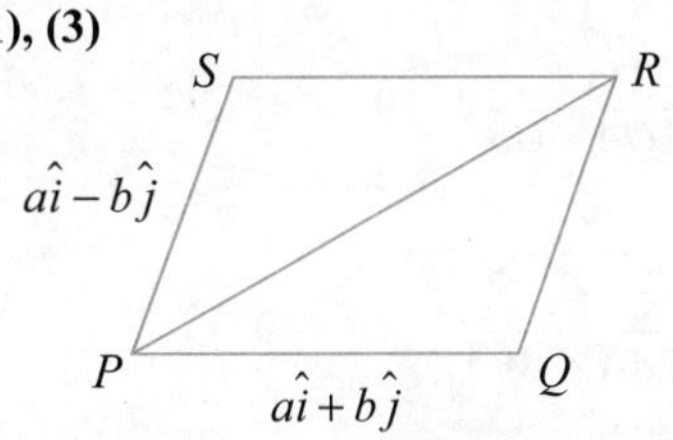

$|\vec{u}| = \dfrac{|(\hat{i}+\hat{j})\cdot\overrightarrow{PQ}|}{|\overrightarrow{PQ}|} = \dfrac{|a+b|}{a^2+b^2}$

$|\vec{v}| = \dfrac{|(\hat{i}+\hat{j})\cdot\overrightarrow{PS}|}{|\overrightarrow{PS}|} = \dfrac{|a-b|}{a^2+b^2}$

Now, $|\vec{u}| + |\vec{v}| = |\vec{w}|$

$\Rightarrow \quad \dfrac{|a+b| + |a-b|}{\sqrt{a^2+b^2}} = \sqrt{2}$

For $a > b$,

$2a = \sqrt{2}.\sqrt{a^2+b^2}$

$\Rightarrow \quad 4a^2 = 2a^2 + 2b^2$

$\Rightarrow \quad a^2 = b^2$

$\therefore \quad a = b$ ...(i)

$(a > 0, b > 0)$

Similarly, for $a < b$ also we will get $a = b$.

Now, area of parallelogram $= |\overrightarrow{PQ}\times\overrightarrow{PS}|$

$= |(a\hat{i}+b\hat{j})\times(a\hat{i}-b\hat{j})| = 2ab$

$\therefore \quad 2ab = 8$ (given)

$\Rightarrow \quad ab = 4$ ...(ii)

From (i) and (ii), $a = 2, b = 2$

$\therefore a + b = 4$

Length of diagonal $= |2a\hat{i}| = |4\hat{i}| = 4$

Clearly, $\vec{w} = \hat{i} + \hat{j}$ is not the angle bisector of vectors $\overrightarrow{PQ}$ and $\overrightarrow{PS}$.

5. (1), (2), (3)

$|\overrightarrow{OB}\times\overrightarrow{OC}| = \dfrac{9}{2}$ and $\overrightarrow{OC} = \dfrac{1}{2}(\overrightarrow{OB} - \lambda\overrightarrow{OA})$

$\therefore \quad |\overrightarrow{OB}\times\dfrac{1}{2}(\overrightarrow{OB} - \lambda\overrightarrow{OA})| = \dfrac{9}{2}$

$\therefore \quad |\lambda|\,|\overrightarrow{OB}\times\overrightarrow{OA}| = 9$ ...(i)

$$\text{Now } \overrightarrow{OB}\times\overrightarrow{OA} = \begin{vmatrix} \hat{i} & \hat{j} & \hat{k} \\ 1 & -2 & 2 \\ 2 & 2 & 1 \end{vmatrix} = -6\hat{i} + 3\hat{j} + 6\hat{k}$$

$\therefore \quad |\overrightarrow{OB}\times\overrightarrow{OA}| = 9$

So, from (i), $\lambda = 1$

$\therefore \quad \overrightarrow{OC} = \dfrac{1}{2}(\overrightarrow{OB} - \overrightarrow{OA}) = \dfrac{1}{2}(-\hat{i} - 4\hat{j} + \hat{k})$

Projection of $\overrightarrow{OC}$ on $\overrightarrow{OA} = \dfrac{\overrightarrow{OC}.\overrightarrow{OA}}{|\overrightarrow{OA}|}$

$$= \frac{\frac{1}{2}(-2-8+1)}{3} = -\frac{3}{2}$$

So, option (1) is true.

$$\text{Area of } \Delta OAB = \frac{1}{2}|\overrightarrow{OA} \times \overrightarrow{OB}| = \frac{9}{2}$$

So, option (2) is true.

$$\text{Area of } \Delta ABC = \frac{1}{2}|\overrightarrow{AB} \times \overrightarrow{AC}|$$

$$= \frac{1}{2}\left|(-\hat{i} - 4\hat{j} + \hat{k}) \times \frac{1}{2}(-5\hat{i} - 8\hat{j} - \hat{k})\right|$$

$$= \frac{1}{4}\left\|\begin{matrix} \hat{i} & \hat{j} & \hat{k} \\ 1 & 4 & -1 \\ 5 & 8 & 1 \end{matrix}\right\|$$

$$= \frac{1}{4}|12\hat{i} - 6\hat{j} - 12\hat{k}| = \frac{3}{2}\sqrt{4+1+4} = \frac{9}{2}$$

So, option (3) is true.

Diagonals of parallelogram whose adjacent sides are $\overrightarrow{OA}$ and $\overrightarrow{OC}$ are

$$\vec{n}_1 = \overrightarrow{OA} + \overrightarrow{OC} = \frac{1}{2}(3\hat{i} + 0\hat{j} + 3\hat{k})$$

$$\vec{n}_2 = \overrightarrow{OA} - \overrightarrow{OC} = \frac{1}{2}(5\hat{i} + 8\hat{j} + \hat{k})$$

$$\cos\theta = \frac{\vec{n}_1 \cdot \vec{n}_2}{|\vec{n}_1||\vec{n}_2|} = \frac{\frac{1}{4}(15+0+3)}{\frac{1}{4}\sqrt{18}\sqrt{90}}$$

$$= \frac{\sqrt{18}}{\sqrt{90}} = \frac{1}{\sqrt{5}}$$

So, option (4) is not true.

6. (2), (3), (4)

$$\vec{a} = 3\hat{i} + \hat{j} - \hat{k}$$

$$\vec{b} = \hat{i} + b_2\hat{j} + b_3\hat{k}$$

$$\vec{c} = c_1\hat{i} + c_2\hat{j} + c_3\hat{k}$$

$$\begin{pmatrix} 0 & -c_3 & c_2 \\ c_3 & 0 & -c_1 \\ -c_2 & c_1 & 0 \end{pmatrix}\begin{pmatrix} 1 \\ b_2 \\ b_3 \end{pmatrix} = \begin{pmatrix} 3-c_1 \\ 1-c_2 \\ -1-c_3 \end{pmatrix}$$

On Multiplying and comparing, we get

$$b_2c_3 - b_2c_2 = c_1 - 3 \quad \text{...(i)}$$

$$c_3 - b_3c_1 = 1 - c_2 \quad \text{...(ii)}$$

$$c_2 - b_2c_1 = 1 + c_3 \quad \text{....(iii)}$$

From (i)$\hat{i}$ – (ii)$\hat{j}$ + (iii)$\hat{k}$, we get

$$\hat{i}(b_2c_3 - b_3c_2) - \hat{j}(c_3 - b_3c_1) + \hat{k}(c_2 - b_2c_1)$$

$$= c_1\hat{i} + c_2\hat{j} + c_2\hat{k} - 3\hat{i} - \hat{j} + \hat{k}$$

$$\therefore \vec{b} \times \vec{c} = \vec{c} - \vec{a}$$

Taking dot product with $\vec{b}$, we get

$$\Rightarrow \quad 0 = \vec{c}.\vec{b} - \vec{a}.\vec{b}$$

$$\Rightarrow \quad \vec{b}.\vec{c} = 0$$

$$\Rightarrow \quad \vec{b} \perp \vec{c}$$

Taking dot product with $\vec{c}$, we get

$$0 = |\vec{c}|^2 - \vec{a}.\vec{c}$$

$$\Rightarrow \vec{a}.\vec{c} = |\vec{c}|^2$$

$$\Rightarrow \vec{a}.\vec{c} \neq 0$$

Now, $\vec{b} \times \vec{c} = \vec{c} - \vec{a}$

So, by squaring, we get

$$|\vec{b}|^2|\vec{c}|^2 = |\vec{c}|^2 + |\vec{a}|^2 - 2\vec{c}.\vec{a}$$

$$\Rightarrow \quad |\vec{b}|^2|\vec{c}|^2 = |\vec{c}|^2 + 11 - 2|\vec{c}|^2$$

$$\Rightarrow \quad |\vec{b}|^2|\vec{c}|^2 = 11 - |\vec{c}|^2$$

$$\Rightarrow \quad |\vec{c}|^2(|\vec{b}|^2 + 1) = 11$$

$$\Rightarrow \quad |\vec{c}|^2 = \frac{11}{|\vec{b}|^2 + 1} \quad \Rightarrow |\vec{c}| \leq \sqrt{11}$$

given $\vec{a}.\vec{b} = 0$

$$\therefore \quad b_2 - b_3 = -3$$

$$\Rightarrow \quad b_2^2 + b_3^2 - 2b_2b_3 = 9$$

$$\Rightarrow \quad b_2^2 + b_3^2 = 9 + 2b_2b_3$$

$$\Rightarrow \quad b_2^2 + b_3^2 + 1 = 10 + 2b_2b_3 > 10 \text{ as } b_2b_3 > 0$$

$$\Rightarrow \quad |\vec{b}| = \sqrt{1 + b_2^2 + b_3^2} > 10$$

Linked Comprehension Type

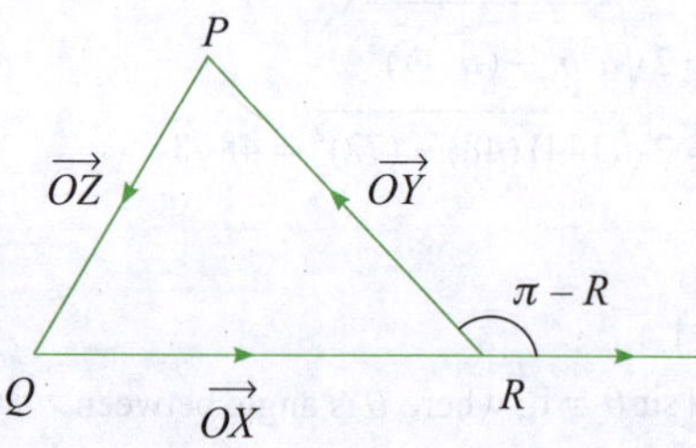

1. (1) $|\overrightarrow{OX} \times \overrightarrow{OY}| = |\overrightarrow{OX}||\overrightarrow{OY}|\sin(\pi - R)$

$= \sin R = \sin(P + Q)$

2. (2) $-(\cos P + \cos Q + \cos R)$

$$= \overrightarrow{OX} \cdot \overrightarrow{OY} + \overrightarrow{OY} \cdot \overrightarrow{OZ} + \overrightarrow{OZ} \cdot \overrightarrow{OX}$$

$$= \frac{(\overrightarrow{OX} + \overrightarrow{OY} + \overrightarrow{OZ})^2 - (|\overrightarrow{OX}|^2 + |\overrightarrow{OY}|^2 + |\overrightarrow{OZ}|^2)}{2}$$

$$\geq -\frac{3}{2}$$

$(\because |\overrightarrow{OX}| = |\overrightarrow{OY}| = |\overrightarrow{OZ}| = 1 \text{ and } (\overrightarrow{OX} + \overrightarrow{OY} + \overrightarrow{OZ})^2 \geq 0)$

Matrix Match Type

1. (P) → (III); (Q) → (IV); (R) → (I); (S) → (II)

(P) $|[\vec{a}\,\vec{b}\,\vec{c}]| = 2$

$$\Rightarrow \quad [2\vec{a} \times \vec{b} \quad 3\vec{b} \times \vec{c} \quad \vec{c} \times \vec{a}] = 6[\vec{a}\,\vec{b}\,\vec{c}]^2$$

$$= 6 \times 4 = 24$$

(Q) $|[\vec{a}\,\vec{b}\,\vec{c}]| = 5$

$\Rightarrow \quad [3(\vec{a}+\vec{b})\,\vec{b}+\vec{c}\;2(\vec{c}+\vec{a})]$

$= 6[(\vec{a}+\vec{b})\,\vec{b}+\vec{c}\,(\vec{c}+\vec{a})] = 12[\vec{a}\;\vec{b}\;\vec{c}] = 60$

(R) Given $\frac{1}{2}|\vec{a}\times\vec{b}| = 20$

Now $\frac{1}{2}|(2\vec{a}+3\vec{b})\times(\vec{a}-\vec{b})|$

$= \frac{1}{2}|-2(\vec{a}\times\vec{b}) - 3(\vec{a}\times\vec{b})|$

$\frac{5}{2}\left|\vec{a}\times\vec{b}\right|$

$= \frac{5}{2}\times 40 = 100$

(S) Given $|\vec{a}\times\vec{b}| = 30 \Rightarrow |(\vec{a}+\vec{b})\times\vec{a}| = |\vec{b}\times\vec{a}| = 30$

2. (1)

(Q) $(\vec{a}_k \times \vec{a}_{k+1}) = r^2 \sin\frac{2\pi}{n}$

and $\vec{a}_k \cdot \vec{a}_{k+1} = r^2 \cos\frac{2\pi}{n}$

Given $\left|\sum_{k=1}^{n-1} \vec{a}_k \times \vec{a}_{k+1}\right| = \left|\sum_{k=1}^{n-1} a_k \cdot a_{k+1}\right|$

$\Rightarrow \quad r^2(n-1)\sin\frac{2\pi}{n} = r^2(n-1)\cos\frac{2\pi}{n}$

$\Rightarrow \quad \tan\frac{2\pi}{n} = 1$

$\Rightarrow \quad \frac{2\pi}{n} = k\pi + \frac{\pi}{4}, k\in Z$

$\Rightarrow \quad n = \frac{8}{4k+1}$

$\Rightarrow \quad n = 8$ (when $k = 0$)

Note: Solutions of the remaining parts are given in their respective chapters.

3. (P) → (I), (II)

Projection of $\alpha\hat{i}+\beta\hat{j}$ on $\sqrt{3}\hat{i}+\hat{j}$ is $\sqrt{3}$.

So, $\left|(\alpha\hat{i}+\beta\hat{j})\cdot\left(\frac{\sqrt{3}\hat{i}+\hat{j}}{2}\right)\right| = \sqrt{3}$

$\Rightarrow \quad \sqrt{3}\alpha+\beta = \pm 2\sqrt{3}$

$\Rightarrow \quad \sqrt{3}\alpha + \left(\frac{\alpha-2}{\sqrt{3}}\right) = \pm 2\sqrt{3}$

$\Rightarrow \quad 3\alpha+\alpha-2 = \pm 6$

$\Rightarrow \quad 4\alpha = 8, -4$

$\Rightarrow \quad \alpha = 2, -1$

Note: Solutions of the remaining parts are given in their respective chapters.

. (1) $\vec{u}\times\vec{v} = \vec{w}$

$\therefore \quad \vec{w}\perp\vec{u}$ *and* $\vec{w}\perp\vec{v}$

Also, $\vec{v}\times\vec{w} = \vec{u}$

$\therefore \quad \vec{u}\perp\vec{v}$

So, $\vec{u}, \vec{v}$ *and* $\vec{w}$ are mutually perpendicular

Now, $|\vec{w}| = \sqrt{6}$

$\vec{u}\times\vec{v} = \vec{w}$

$\Rightarrow \quad |\vec{u}||\vec{v}| = \sqrt{6}$ ...(i)

and $\vec{v}\times\vec{w} = \vec{u}$

$\Rightarrow \quad \sqrt{6}|\vec{v}| = |\vec{u}|$ ...(ii)

So, from (i) and (ii), $|\vec{v}| = 1$ and $|\vec{u}| = \sqrt{6}$

Now, $-t\alpha+\beta+\gamma = 0$

$\alpha - t\beta + \gamma = 0$

and $\alpha+\beta-t\gamma = 0$

Since $(\alpha,\beta,\gamma)\neq(0,0,0)$, system has non–trivial solution.

$\therefore \quad \begin{vmatrix} -t & 1 & 1 \\ 1 & -t & 1 \\ 1 & 1 & -t \end{vmatrix} = 0$

Applying $C_1 \to C_1 + C_2 + C_3$, we get

$\begin{vmatrix} 2-t & 1 & 1 \\ 2-t & -t & 1 \\ 2-t & 1 & -t \end{vmatrix} = 0$

$\Rightarrow \quad \begin{vmatrix} 0 & 1+t & 0 \\ 0 & -t-1 & 1+t \\ 1-t & 1 & -t \end{vmatrix} = 0$

$\Rightarrow \quad (2-t)(1+t)^2 = 0$

$\Rightarrow \quad t = 2, -1$

For $t = -1$, we have $\alpha+\beta+\gamma = 0$

Also, for $t = 2$, we have

$-2\alpha+\beta+\gamma = 0$

$\alpha-2\beta+\gamma = 0$

$\alpha+\beta-2\gamma = 0$

So, we get $\alpha = \beta = \gamma$.

Case I: $t = 2$

$\alpha = \beta = \gamma$

$\therefore \quad \alpha^2+\beta^2+\gamma^2 = 6$ or $3\alpha^2 = 6$ or $\alpha = \pm\sqrt{2}$

So, for $\alpha = \sqrt{2}$, $t+3 = 2+3 = 5$

Case I: $t = -1$

$\alpha+\beta+\gamma = 0$

also, $\alpha+\beta-\gamma = 0$

$\therefore \quad \gamma = 0$ and $a+b = 0$

For, $\alpha = \sqrt{3}$, $\beta = -\sqrt{3}$.

So, $(\beta+\gamma)^2 = 3$

Numerical Value Type

1. (0.5)

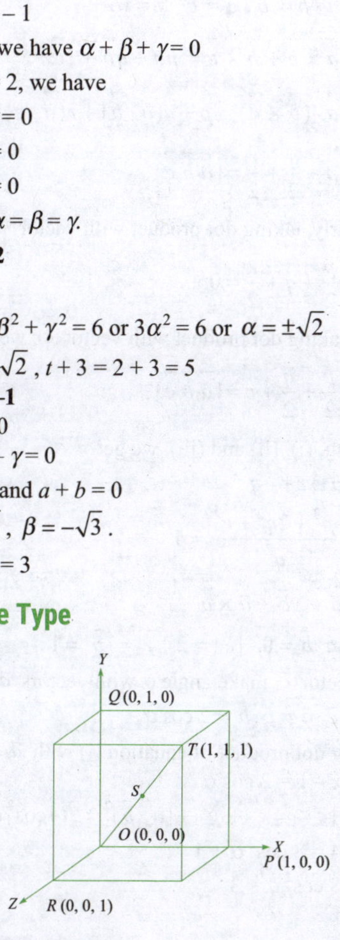

Point $S \equiv (1/2, 1/2, 1/2)$

Point $T \equiv (1, 1, 1)$

$\vec{p} = \overrightarrow{SP} = \frac{\hat{i}-\hat{j}-\hat{k}}{2}$

$$\vec{q} = \overrightarrow{SQ} = \frac{-\hat{i}+\hat{j}-\hat{k}}{2}$$

$$\vec{r} = \overrightarrow{SR} = \frac{-\hat{i}-\hat{j}+\hat{k}}{2}$$

$$\vec{t} = \overrightarrow{ST} = \frac{\hat{i}+\hat{j}+\hat{k}}{2}$$

Now

$$\vec{p}\times\vec{q} = \frac{1}{4}\times\begin{vmatrix}\hat{i} & \hat{j} & \hat{k}\\ 1 & -1 & -1\\ -1 & 1 & -1\end{vmatrix} = \frac{1}{4}(2\hat{i}+2\hat{j}) = \frac{\hat{i}+\hat{j}}{2}$$

$$\vec{r}\times\vec{t} = \frac{1}{4}\times\begin{vmatrix}\hat{i} & \hat{j} & \hat{k}\\ -1 & -1 & 1\\ 1 & 1 & 1\end{vmatrix} = \frac{-2\hat{i}+2\hat{j}}{4} = \frac{-\hat{i}+\hat{j}}{2}$$

Now, $(\vec{p}\times\vec{q})\times(\vec{r}\times\vec{t}) = \frac{1}{4}\times\begin{vmatrix}\hat{i} & \hat{j} & \hat{k}\\ 1 & 1 & 0\\ -1 & 1 & 0\end{vmatrix} = \frac{\hat{k}}{2}$

$$\Rightarrow |(\vec{p}\times\vec{q})\times(\vec{r}\times\vec{t})| = \frac{1}{2} = 0.5$$

Integer Answer Type

1. (4) $|\vec{a}| = |\vec{b}| = |\vec{c}| = 1$

Also, $\vec{a}\cdot\vec{b} = \vec{b}\cdot\vec{c} = \vec{c}\cdot\vec{a} = 1/2$

$$\vec{a}\times\vec{b}+\vec{b}\times\vec{c} = p\vec{a}+q\vec{b}+r\vec{c}$$

$$\Rightarrow \vec{a}\cdot(\vec{b}\times\vec{c}) = p+q(\vec{a}\cdot\vec{b})+r(\vec{a}\cdot\vec{c})$$

$$\therefore \quad p+\frac{q}{2}+\frac{r}{2} = [\vec{a}\,\vec{b}\,\vec{c}] \qquad \ldots\text{(i)}$$

Similarly, taking dot product with vector $\vec{b}$, we get

$$\frac{p}{2}+q+\frac{r}{2} = 0 \qquad \ldots\text{(ii)}$$

And, taking dot product with vector $\vec{c}$, we get

$$\frac{p}{2}+\frac{q}{2}+r = [\vec{a}\,\vec{b}\,\vec{c}] \qquad \ldots\text{(iii)}$$

Solving, (i), (ii) and (iii), we get

$$p = r = -q$$

$$\Rightarrow \frac{p^2+2q^2+r^2}{q^2} = 4$$

2. (3) $\vec{c} = x\vec{a}+y\vec{b}+\vec{a}\times\vec{b} \qquad \ldots\text{(i)}$

and $\vec{a}\cdot\vec{b} = 0,\ |\vec{c}| = 2, |\vec{a}| = |\vec{b}| = 1$

Let vector $\vec{c}$ make angle α with vectors $\vec{a}$ and $\vec{b}$.

$\therefore \quad \vec{c}\cdot\vec{a} = \vec{c}\cdot\vec{b} = 2\cos\alpha$

Taking dot product of equation (i) with $\vec{a}$ and $\vec{b}$, we have

$\Rightarrow \quad x = y = 2\cos\alpha$

Now, $|\vec{c}|^2 = x^2+y^2+|\vec{a}\times\vec{b}|^2 = 2(4\cos^2\alpha)+1$

$\Rightarrow \quad 4 = 8\cos^2\alpha+1$

$\Rightarrow \quad 8\cos^2\alpha = 3$

3. (18.00)

$$\vec{c} = \alpha(2\hat{i}+\hat{j}-\hat{k})+\beta(\hat{i}+2\hat{j}+\hat{k})$$

$$\Rightarrow \vec{c} = (2\alpha+\beta)\hat{i}+(\alpha+2\beta)\hat{j}+(\beta-\alpha)\hat{k}$$

Now, projection of $\vec{c}$ on the vector $(\vec{a}+\vec{b})$ is $3\sqrt{2}$.

$$\Rightarrow \frac{\vec{c}\cdot(\vec{a}+\vec{b})}{|\vec{a}+\vec{b}|} = 3\sqrt{2}$$

$\Rightarrow \quad 9(\alpha+\beta) = 18$

$\Rightarrow \quad \alpha+\beta = 2$

So, $(\vec{c}-(\vec{a}\times\vec{b}))\cdot\vec{c}$

$= (\alpha\vec{a}+\beta\vec{b}-\vec{a}\times\vec{b})\cdot(\alpha\vec{a}+\beta\vec{b})$

$= 6\alpha^2+6\alpha\beta+6\beta^2$

$= 6[\alpha^2+\alpha(2-\alpha)+(2-\alpha)^2]$

$= 6(\alpha-1)^2+3$, which has minimum value 18.

4. (108)

We have $\vec{a}+\vec{b}+\vec{c} = \vec{0}$

$\Rightarrow \quad \vec{c} = -\vec{a}-\vec{b}$

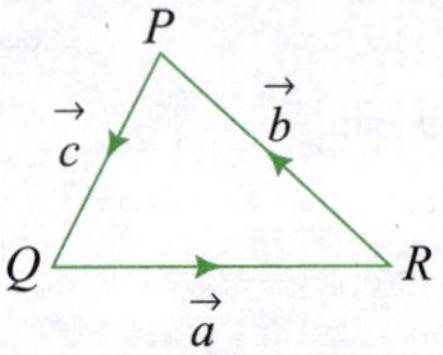

Given that $\dfrac{\vec{a}\cdot(\vec{c}-\vec{b})}{\vec{c}\cdot(\vec{a}-\vec{b})} = \dfrac{|\vec{a}|}{|\vec{a}|+|\vec{b}|}$

$$\Rightarrow \frac{\vec{a}\cdot(-\vec{a}-2\vec{b})}{(-\vec{a}-\vec{b})\cdot(\vec{a}-\vec{b})} = \frac{3}{7}$$

$$\Rightarrow \frac{9+2\vec{a}\cdot\vec{b}}{9-16} = \frac{3}{7}$$

$\Rightarrow \quad \vec{a}\cdot\vec{b} = -6$

$\Rightarrow \quad |\vec{a}\times\vec{b}|^2 = a^2b^2-(\vec{a}\cdot\vec{b})^2$

$= 9\times16-36 = 108$

5. (7) Volume of parallelepiped $= [\vec{u}\,\vec{v}\,\vec{w}] = \sqrt{2}$

Now, $[\vec{u}\,\vec{v}\,\vec{w}]^2 = \begin{vmatrix}\vec{u}.\vec{u} & \vec{u}.\vec{v} & \vec{u}.\vec{w}\\ \vec{v}.\vec{u} & \vec{v}.\vec{v} & \vec{v}.\vec{w}\\ \vec{w}.\vec{u} & \vec{w}.\vec{v} & \vec{w}.\vec{w}\end{vmatrix}$

$$\therefore \begin{vmatrix}1 & \vec{u}.\vec{v} & 1\\ \vec{v}.\vec{u} & 1 & 1\\ 1 & 1 & 4\end{vmatrix} = 2$$

$$\therefore \quad \vec{u}.\vec{v} = \frac{1}{2}$$

$\Rightarrow \quad |3\vec{u}+5\vec{v}|^2 = 9|\vec{u}|^2+25|\vec{v}|^2+30\,\vec{u}.\vec{v}$

$= 34+15 = 49$

6. (5)

$(\overrightarrow{OP}\times\overrightarrow{OQ})\cdot\overrightarrow{OR} = 0$

$$\therefore \begin{vmatrix}\frac{\alpha-1}{\alpha} & 1 & 1\\ 1 & \frac{\beta-1}{\beta} & 1\\ 1 & 1 & \frac{1}{2}\end{vmatrix} = 0$$

Applying $C_1\to C_1-C_2$ and $C_2\to C_2-C_3$, we get

$$\begin{vmatrix} -\frac{1}{\alpha} & 0 & 1 \\ \frac{1}{\beta} & -\frac{1}{\beta} & 1 \\ 0 & \frac{1}{2} & \frac{1}{2} \end{vmatrix} = 0$$

$\Rightarrow \quad \frac{1}{2\alpha\beta} + \frac{1}{2\beta} + \frac{1}{2\alpha} = 0$

$\Rightarrow \quad \alpha + \beta + 1 = 0 \qquad \ldots(i)$

Also, point $(\alpha, \beta, 2)$ lies on the plane $3x + 3y - z + l = 0$

$\therefore \quad 3\alpha + 3\beta - 2 + l = 0$

From (i), $\alpha + \beta = -1$

$\therefore \quad -3 - 2 + l = 0 \quad \text{or} \quad l = 5.$

7. (2)

$2\vec{p} + \vec{q} = 5\hat{i} + \hat{j} + 7\hat{k}$

$\vec{p} - 2\vec{q} = 3\hat{j} + \hat{k}$

$$\vec{p} \times \vec{q} = \begin{vmatrix} \hat{i} & \hat{j} & \hat{k} \\ 2 & 1 & 3 \\ 1 & -1 & 1 \end{vmatrix} = 4\hat{i} + \hat{j} - 3\hat{k}$$

Now,

$15\hat{i} + 10\hat{j} + 6\hat{k} = \alpha(2\vec{p} + \vec{q}) + \beta(\vec{p} - 2\vec{q}) + \gamma(\vec{p} \times \vec{q})$

$\Rightarrow \quad 15\hat{i} + 10\hat{j} + 6\hat{k}$

$= \alpha(5\hat{i} + \hat{j} + 7\hat{k}) + \beta(3\hat{j} + \hat{k}) + \gamma(4\hat{i} + \hat{j} - 3\hat{k})$

$\Rightarrow \quad 5\alpha + 4\gamma = 15 \qquad \ldots(i)$

$\alpha + 3\beta + \gamma = 10 \qquad \ldots(ii)$

and $\quad 7\alpha + \beta - 3\gamma = 6 \qquad \ldots(iii)$

From (ii) – 3(iii), we get

$-20\alpha + 10\gamma = -8$

or $\quad 5\alpha - \frac{5}{2}\gamma = 2 \qquad \ldots(iv)$

From (i) – (iv), we get

$\frac{13}{2}\gamma = 13 \Rightarrow \gamma = 2$

Line and Plane

Concept Application Exercises

Exercise 4.1

1. According to the question, $\frac{a+2}{6}=\frac{b-1}{2}=\frac{c+8}{3}=\lambda$

or $a=6\lambda-2, b=2\lambda+1, c=3\lambda-8, \lambda \in R$

2. Since line is equally inclined to axes it is along the vector $\lambda(\hat{i}+\hat{j}+\hat{k})$.

So, equation of line is $\frac{x-2}{1}=\frac{y-3}{1}=\frac{z-4}{1}$

3. Equation of line is $\frac{x-1}{\cos\frac{\pi}{4}}=\frac{y-2}{\cos\frac{\pi}{4}}=\frac{z-3}{\cos\frac{\pi}{2}}$

or $\frac{x-1}{1}=\frac{y-2}{1}=\frac{z-3}{0}$

4. We have, $\vec{r}=(3\lambda-2)\hat{i}+(\lambda+4)\hat{j}+(5-7\lambda)\hat{k}$

or $\vec{r}=-2\hat{i}+4\hat{j}+5\hat{k}+\lambda(3\hat{i}+\hat{j}-7\hat{k})$ → Vector form

or $(\vec{r}-(-2\hat{i}+4\hat{j}+5\hat{k}))\times(3\hat{i}+\hat{j}-7\hat{k})=0$

→ Cross product form

or $\frac{x+2}{3}=\frac{y-4}{1}=\frac{z-5}{-7}$ → Symmetric form

Direction ratios of the line are 3, 1, –7.

5. Common diameter line passes through the centres of the spheres. Centres of spheres are (0, 0, 0) and (1, –2, 3).

So, equation of common diameter is $\frac{x}{1}=\frac{y}{-2}=\frac{z}{3}$.

6. The given line is $-6x-2=3y+1=2z-2$

or $\frac{x+(1/3)}{-1/6}=\frac{y+(1/3)}{1/3}=\frac{z-1}{1/2}$

The direction ratios are $-\frac{1}{6},\frac{1}{3}$ and $\frac{1}{2}$ or –1, 2 and 3.

So, the required equation is $\frac{x-2}{-1}=\frac{y+1}{2}=\frac{z+1}{3}$

7. (4) The line is along the vector $3\hat{i}+\hat{j}$ which is perpendicular to the z-axis as $(3\hat{i}+\hat{j})\cdot\hat{k}=0$.

8. For point of intersection with z-axis, we put $x=y=0$.

∴ $\frac{0-2}{1}=\frac{0-4}{2}=\frac{z-5}{3}$ or $z-5=-6$ or $z=-1$

So, point is (0, 0, –1)

9. Line intersects x-axis at (4, 0, 0) and y-axis at (0, 3, 0).

So, direction ratios of line are 4 – 0, 0 – 3, 0 – 0 or 4, –3, 0.

Hence equation of line is $\frac{x-4}{4}=\frac{y-0}{-3}=\frac{z-0}{0}$

or $\frac{x-4}{4}=\frac{y}{-3}=\frac{z}{0}$

10. Midpoint of BC is $D(3, 3, -2)$

Direction ratios of AD are 1 – 3, 2 – 3, –1 –(–2) or –2, –1, 1

So, equation of median AD is $\frac{x-1}{-2}=\frac{y-2}{-1}=\frac{z+1}{1}$

Now, centroid of triangle ABC is

$$\left(\frac{1+0+6}{3},\frac{2+1+5}{3},\frac{-1-4+0}{3}\right)\equiv\left(\frac{7}{3},\frac{8}{3},\frac{-5}{3}\right)$$

Putting this point in the equation of median AD, we get

$$\frac{\frac{7}{3}-1}{-2}=\frac{\frac{8}{3}-2}{-1}=\frac{-\frac{5}{3}+1}{1} \text{ or } \frac{-2}{3}=\frac{-2}{3}=\frac{-2}{3}.$$

Hence it is verified that centroid lies on median AD.

11. (4) Clearly, line in option (4) is perpendicular to the given line. Also, it passes through the point (0, 1, 2).

12. We have line $\frac{x-3}{2}=\frac{y-3}{1}=\frac{z+2}{-2}=\lambda, \lambda \in R$

Variable points on the line is $(2\lambda+3, \lambda+3, -2\lambda-2)$ ...(i)

Line meets xy-plane where $z=0$,

So, $-2\lambda-2=0$.

∴ $\lambda=-1$

Putting this value in (i), we get (1, 2, 0).

This is the point where line meets xy-plane.

$2\lambda+3=0 \Rightarrow \lambda=\frac{-3}{2}$. So, point, where line meets yz-plane is $\left(0,\frac{3}{2},1\right)$

$\lambda+3=0 \Rightarrow \lambda=-3$. So, point, where line meets xz-plane is (–3, 0, 4)

13. We have $A(1, -1, 0)$, $B(3, 5, 3)$ and $C(-11, -5, 6)$

$$AB=\sqrt{(1-3)^2+(-1-5)^2+(0-3)^2}$$
$$=\sqrt{4+36+9}=\sqrt{49}=7$$
$$AC=\sqrt{(1-(-11))^2+(-1-(-5))^2+(0-6)^2}$$
$$=\sqrt{144+16+36}=\sqrt{196}=14$$

Let internal angle bisector of angle A meets the side BC at D.

Then, $\frac{BD}{CD}=\frac{AB}{AC}=\frac{7}{14}=\frac{1}{2}$

∴ $D\equiv\left(\frac{1(-11)+2(3)}{1+2},\frac{1(-5)+2(5)}{1+2},\frac{1(6)+2(3)}{1+2}\right)$

$\equiv\left(\frac{-5}{3},\frac{5}{3},\frac{12}{3}\right)$

∴ Equation of angle bisector AD is

$$\frac{x-1}{-\frac{5}{3}-1}=\frac{y+1}{\frac{5}{3}+1}=\frac{z-0}{\frac{12}{3}-0}$$

or $\frac{x-1}{-8}=\frac{y+1}{8}=\frac{z}{9}$

14. (2) $A\equiv(3, 1, 2)$ and $B\equiv(1, 0, 2)$

L: $\frac{x-1}{2}=\frac{y}{1}=\frac{z-2}{0}$

So, plane containing points A and B and the line $L=0$, is $z=2$

Also, $AB=\sqrt{4+1+0}=\sqrt{5}$

15. (3) The line through point (–1, 2, 3) is perpendicular to the lines $\frac{x}{2}=\frac{y-1}{-3}=\frac{2+2}{-2}$ and $\frac{x+3}{-1}=\frac{y+3}{2}=\frac{z-1}{3}$. Therefore, the line is along the vector $\begin{vmatrix} \hat{i} & \hat{j} & \hat{k} \\ 2 & -3 & -2 \\ -1 & 2 & 3 \end{vmatrix}$ or $-5\hat{i}-4\hat{j}+\hat{k}$.

Hence, equation of the line is $\frac{x+1}{5}=\frac{y-2}{4}=\frac{z-3}{-1}$.

16. Eliminating t from the given equations, we get equation of the path.

$$\frac{x}{2}=\frac{y}{-4}=\frac{z}{4}$$

or $\frac{x}{1}=\frac{y}{-2}=\frac{z}{2}$

Thus, the path of the rocket represents a straight line passing through the origin.

For $t = 10$ s, we have

$x = 20, y = -40$ and $z = 40$

and $|\vec{r}|=|\overrightarrow{OM}|=\sqrt{x^2+y^2+z^2}$

$= \sqrt{400+1600+1600} = 60$ km

Exercise 4.2

1. (4) The lines are $\frac{x}{3}=\frac{y}{2}=\frac{z}{-6}$ and $\frac{x}{2}=\frac{y}{-12}=\frac{z}{-3}$.

Since $a_1a_2+b_1b_2+c_1c_2 = 6-24+18=0$, we have

$\theta = 90°$

2. (1) The lines are perpendicular if

$-3(3k)+2k(1)+2(-5)=0$ or $k=-\frac{10}{7}$.

3. We have, $\ell + m - 2n = 0$

or $\ell = 2n - m$

Putting this value in $\ell m + mn + 3n\ell = 0$, we get

$mn + \ell(m+2n) = 0$

$\Rightarrow$ $mn + (2n-m)(m+2n) = 0$

$\Rightarrow$ $6n^2 - m^2 = 0$

$\Rightarrow$ $m = \pm\sqrt{6}n$

For $m = \sqrt{6}n$, we have

$\ell : m : n = (2n-m) : m : n$

$= (2-\sqrt{6}) : \sqrt{6} : 1$

For $m = -\sqrt{6}n$, we have

$\ell : m : n = (2+\sqrt{6}) : -\sqrt{6} : 1$

Therefore, the angle between the lines is given by

$$\cos\theta = \frac{(2-\sqrt{6})(2+\sqrt{6})+(\sqrt{6})(-\sqrt{6})+(1)(1)}{\sqrt{(2-\sqrt{6})^2+6+1}\sqrt{(2+\sqrt{6})^2+6+1}}$$

$$= \frac{4-6-6+1}{\sqrt{17-4\sqrt{6}}\sqrt{17+4\sqrt{6}}}$$

$$= \frac{-7}{\sqrt{289-96}}$$

$$= \frac{-7}{\sqrt{193}}$$

or $\theta = \cos^{-1}\left(\frac{-7}{\sqrt{193}}\right) = \pi - \cos^{-1}\frac{7}{\sqrt{193}}$

Hence, the acute angle between the lines is $\cos^{-1}\frac{7}{\sqrt{193}}$.

4. (3) Let the foot of the perpendicular from point $A(1, -1, 2)$ on the given line L is D.

$\therefore$ $D \equiv (3\lambda-2, 1, 4\lambda)$.

Since, AD is perpendicular to the line L,

$3(3\lambda-2-1)+0(1-(-1))+4(4\lambda-2)=0$

$\Rightarrow 25\lambda = 17$ or $\lambda = 17/25$

Thus, $AD = \sqrt{(3\lambda-3)^2+4+(4\lambda-2)^2}$

$$= \sqrt{\left(\frac{51}{25}-3\right)^2+4+\left(\frac{68}{25}-2\right)^2}$$

$$= \frac{1}{25}\sqrt{576+2500+324}$$

$$= \frac{1}{5}\sqrt{136}$$

Therefore, the area of triangle $ABC = \frac{1}{2}\times 5\times\sqrt{\frac{136}{25}} = \sqrt{34}$ sq. units.

5. The equation of the given line is $2(x+1) = 2 - y = z+3$.

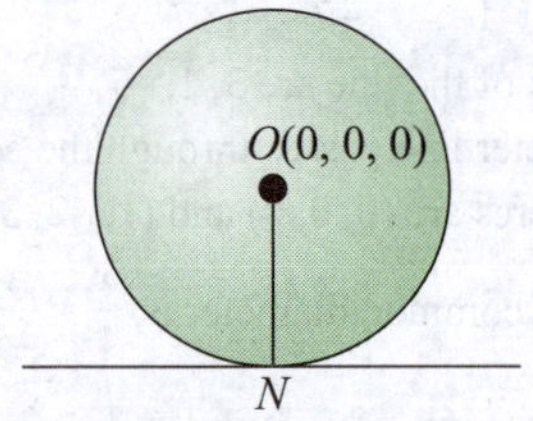

or $\frac{x+1}{1/2}=\frac{y-2}{-1}=\frac{z+3}{1}=\lambda$ (say)

Centre of the sphere is origin 'O'.

Since, the sphere is touching the line and the smallest, foot of perpendicular N from centre upon the line lies on the sphere at the point of contact.

Thus, ON is perpendicular to the given line.

Point N on the given line is $\left(-1+\frac{1}{2}\lambda, 2-\lambda, -3+\lambda\right)$.

$\therefore$ $\frac{1}{2}\left(-1+\frac{1}{2}\lambda\right)+(-1)(2-\lambda)+1(-3+\lambda)=0$

$\Rightarrow$ $\lambda = \frac{22}{9}$

$\Rightarrow$ $N \equiv \left(\frac{2}{9}, -\frac{4}{9}, -\frac{5}{9}\right)$

So, equation of the sphere is

$$x^2+y^2+z^2 = \frac{4}{81}+\frac{16}{81}+\frac{25}{81}$$

or $9(x^2+y^2+z^2) = 5$.

6. (3) Let A' be the image of point A (1, 2, 3) in the line l whose equation is

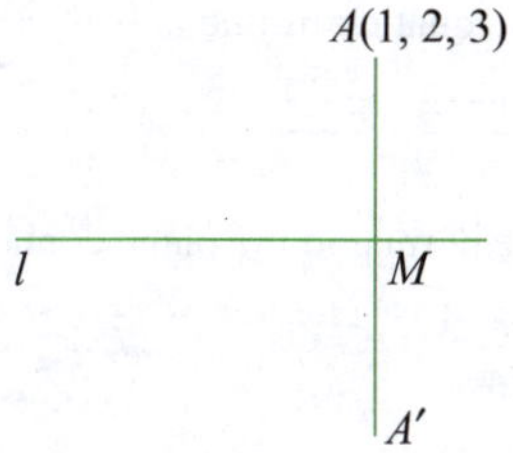

$$\frac{x-6}{3}=\frac{y-7}{2}=\frac{z-7}{-2}=k \text{ (say)} \qquad \ldots\text{(i)}$$

Then AA' is perpendicular to l and the point of intersection of AA' and l is the midpoint of AA'. Note that M is the foot of perpendicular from A on l.

The coordinates of any point on the given line are $(3k + 6, 2k + 7, -2k + 7)$. Therefore, the coordinates of M are $3k + 6$, $2k + 7$ and $-2k + 7$ for some value of k. The direction ratios of AM are $3k + 6 - 1$, $2k + 7 - 2$ and $-2k + 7 - 3$ or $3k + 5$, $2k + 5$, $-2k + 4$.

Also, the direction ratios of l are 3, 2 and -2.

Since $AM \perp l$, we have

$$3(3k + 5) + 2(2k + 5) - 2(-2k + 4) = 0$$

or $\quad 17k + 17 = 0 \quad$ or $\quad k = -1$

$\therefore \quad A' \equiv (5, 8, 15)$

$\quad M \equiv (3, 5, 9)$

Exercise 4.3

1. Intersection point of the lines

$$\frac{x-1}{2}=\frac{y-2}{3}=\frac{z-3}{4} \text{ and } \frac{x-4}{5}=\frac{y-1}{2}=z$$

is $(-1, -1, -1)$ (on solving). Therefore, the equation of the line passing through the points $(-1, -1, -1)$ and $(2, 1, -2)$ is $\frac{x+1}{3}=\frac{y+1}{2}=\frac{z+1}{-1}$.

. (1) The given lines $\frac{x+1}{1}=\frac{y-3}{-\lambda}=\frac{z-1}{\lambda}=s$

and $\frac{x-0}{1/2}=\frac{y-1}{1}=\frac{y-2}{-1}=t$

are coplanar if $\begin{vmatrix} 0+1 & 1-3 & 2-1 \\ 1 & -\lambda & \lambda \\ 1/2 & 1 & -1 \end{vmatrix}=0$

$$\Rightarrow \quad \begin{vmatrix} 1 & -2 & 1 \\ 1 & -\lambda & \lambda \\ 1/2 & 1 & -1 \end{vmatrix}=0$$

$$\Rightarrow \quad 1(\lambda-\lambda)+2\left(-1-\frac{\lambda}{2}\right)+1\left(1+\frac{\lambda}{2}\right)=0$$

$$\Rightarrow \quad \lambda=-2$$

3. Given lines are $\frac{x-1}{2}=\frac{y+2}{3}=\frac{z-3}{4}$ and $\frac{x}{2}=\frac{y-1}{3}=\frac{z+1}{4}$ are parallel.

So, distance between the lines $=\frac{|(\vec{c}-\vec{a})\times\vec{b}|}{|\vec{b}|}$

$$=\frac{\left\|\begin{vmatrix} \hat{i} & \hat{j} & \hat{k} \\ 1-0 & -2-1 & 3-(-1) \\ 2 & 3 & 4 \end{vmatrix}\right\|}{\sqrt{2^2+3^2+4^2}}$$

$$=\frac{\left\|\begin{vmatrix} \hat{i} & \hat{j} & \hat{k} \\ 1 & -3 & 4 \\ 2 & 3 & 4 \end{vmatrix}\right\|}{\sqrt{29}}$$

$$=\frac{|-24\hat{i}+4\hat{j}+9\hat{k}|}{\sqrt{29}}$$

$$=\frac{\sqrt{576+16+81}}{\sqrt{29}}=\frac{\sqrt{673}}{\sqrt{29}} \text{ units.}$$

So, area of square is $\frac{673}{29}$ sq. units.

4. (2) The lines are $\vec{r}=(1-\lambda)\hat{i}+(\lambda-2)\hat{j}+(3-2\lambda)\hat{k}$ and $\vec{r}=(\mu+1)\hat{i}+(2\mu-1)\hat{j}-(2\mu+1)\hat{k}$

or $\quad \vec{r}=(\hat{i}-2\hat{j}+3\hat{k})+\lambda(-\hat{i}-2\hat{j}-2\hat{k})$

and $\quad \vec{r}=(\hat{i}-\hat{j}-\hat{k})+\mu(\hat{i}+2\hat{j}-2\hat{k})$.

Line (i) passes through the point $(x_1, y_1, z_1) \equiv (1, -2, 3)$ and is parallel to the vector

$$a_1\hat{i}+b_1\hat{j}+c_1\hat{k} \equiv -\hat{i}-2\hat{j}-2\hat{k}.$$

Line (ii) passes through the point $(x_2, y_2, z_2) \equiv (1, -1, -1)$ and is parallel to the vector

$$a_2\hat{i}+b_2\hat{j}+c_2\hat{k} \equiv \hat{i}+2\hat{j}-2\hat{k}.$$

Hence, the shortest distance between the lines

$$=\frac{\left|\begin{vmatrix} x_2-x_1 & y_2-y_1 & z_2-z_1 \\ a_1 & b_1 & c_1 \\ a_2 & b_2 & c_2 \end{vmatrix}\right|}{\left|\begin{vmatrix} \hat{i} & \hat{j} & \hat{k} \\ a_1 & b_1 & c_1 \\ a_2 & b_2 & c_2 \end{vmatrix}\right|}$$

$$=\frac{\left\|\begin{vmatrix} 1-1 & -1+2 & -1-3 \\ -1 & -2 & -2 \\ 1 & 2 & -2 \end{vmatrix}\right\|}{\left\|\begin{vmatrix} \hat{i} & \hat{j} & \hat{k} \\ -1 & -2 & -2 \\ 1 & 2 & -2 \end{vmatrix}\right\|}=\frac{4}{\sqrt{80}}=\frac{1}{\sqrt{5}}$$

5. (4) Given lines are

$$L_1 \equiv \frac{x+1}{2}=\frac{y-2}{-1}=\frac{z-1}{1}$$

$$L_2 \equiv \frac{x+2}{\alpha}=\frac{y+1}{5-\alpha}=\frac{z+1}{1}$$

Line L_1 passes through the point $A(-1, 2, 1)$ and line L_2 passes through the point $B(-2, -1, -1)$.

Lines L_1 and L_2 are coplanar

$$\therefore \begin{vmatrix} -1-(-2) & 2-(-1) & 1-(-1) \\ 2 & -1 & 1 \\ \alpha & 5-\alpha & 1 \end{vmatrix} = 0$$

$\Rightarrow \quad \alpha(3+2) - (5-\alpha)(1-4) + 1(-1-6) = 0$

$\Rightarrow \quad 5\alpha + 15 - 3\alpha - 7 = 0$

$\Rightarrow \quad \alpha = -4$

$$\therefore \quad L_2 \equiv \frac{x+2}{-4} = \frac{y+1}{9} = \frac{z+1}{1}$$

Clearly, L_2 passes through the point (2, –10, –2).

6. (1) Given lines are

$$L_1 : \frac{x-1}{2} = \frac{y-2}{1} = \frac{z-1}{3}$$

$$L_2 : \frac{x-2}{1} = \frac{y-\lambda}{2} = \frac{z-3}{4}$$

Line L_1 passes through the point $A(1, 2, 1)$ and parallel to the vector $\vec{b} = 2\hat{i} + \hat{j} + 3\hat{k}$

Line L_2 passes through the point $B(2, \lambda, 3)$ and parallel to the vector $\vec{d} = \hat{i} + 2\hat{j} + 4\hat{k}$

Shortest distance between lines

$= \text{projection of } \overrightarrow{AB} \text{ on } \vec{b}\times\vec{d}$

$$= \frac{|\overrightarrow{AB}\cdot(\vec{b}\times\vec{d})|}{|\vec{b}\times\vec{d}|}$$

$$\text{Now } \overrightarrow{AB}\cdot(\vec{b}\times\vec{d}) = \begin{vmatrix} 1 & \lambda-2 & 2 \\ 2 & 1 & 3 \\ 1 & 2 & 4 \end{vmatrix} = |5\lambda - 14|$$

$$\text{and } \quad |\vec{b}\times\vec{d}| = \left\|\begin{matrix} \hat{i} & \hat{j} & \hat{k} \\ 2 & 1 & 3 \\ 1 & 2 & 4 \end{matrix}\right\| = \sqrt{38}$$

$$\therefore \quad \frac{|14-5\lambda|}{\sqrt{38}} = \frac{1}{\sqrt{38}} \text{ (given)}$$

$\Rightarrow \quad |14 - 5\lambda| = 1$

$\Rightarrow \quad 14 - 5\lambda = 1 \text{ or } 14 - 5\lambda = -1$

$$\Rightarrow \quad \lambda = \frac{13}{5} \text{ or } 3$$

Exercise 4.4

1. (2) Normal vector of plane containing two intersecting lines is parallel to vector.

$$\vec{n_1} = \begin{vmatrix} \hat{i} & \hat{i} & \hat{k} \\ 3 & 0 & 1 \\ -3 & 2 & -1 \end{vmatrix}$$

$= -2\hat{i} + 6\hat{k}$

So, required line is parallel to vector

$$\vec{n_2} = \begin{vmatrix} \hat{i} & \hat{j} & \hat{k} \\ 1 & 2 & -1 \\ -2 & 0 & 6 \end{vmatrix} = 12\hat{i} - 4\hat{j} + 4\hat{k}$$

$\Rightarrow$ Required equation of line is

$$\frac{x+4}{3} = \frac{y-3}{-1} = \frac{z-1}{1}$$

2. (2) Vector along the normal to the plane containing the lines $\frac{x}{3} = \frac{y}{4} = \frac{z}{2}$ and $\frac{x}{4} = \frac{y}{2} = \frac{z}{3}$ is

$$\begin{vmatrix} \hat{i} & \hat{j} & \hat{k} \\ 3 & 4 & 2 \\ 4 & 2 & 3 \end{vmatrix}$$

$= 8\hat{i} - \hat{j} - 10\hat{k}$

Now, vector perpendicular to the vectors $2\hat{i} + 3\hat{j} + 4\hat{k}$ and $8\hat{i} - \hat{j} - 10\hat{k}$ is

$$\begin{vmatrix} \hat{i} & \hat{j} & \hat{k} \\ 8 & -1 & -10 \\ 2 & 3 & 4 \end{vmatrix}$$

$= 26\hat{i} - 52\hat{j} + 26\hat{k}$

So, required plane is $26x - 52y + 26z = 0$ or $x - 2y + z = 0$

3. Vector normal to the plane is $\vec{n} = 2\hat{i} + \hat{j} + 2\hat{k}$

Hence, unit vector perpendicular to the plane is

$$\frac{\vec{n}}{|\vec{n}|} = \frac{2\hat{i} + \hat{j} + 2\hat{k}}{\sqrt{2^2+1^2+2^2}} = \frac{1}{3}(2\hat{i} + \hat{j} + 2\hat{k})$$

4. Equation of plane is $3x - 4y + 5z = 8$

So, vector normal to the plane is $3\hat{i} - 4\hat{j} + 5\hat{k}$.

Since required line is perpendicular to the plane it is parallel to the vector $3\hat{i} - 4\hat{j} + 5\hat{k}$.

Line passes through the point $3\hat{i} - 5\hat{j} + 7\hat{k}$.

Hence required equation of line is

$$\vec{r} = 3\hat{i} - 5\hat{j} + 7\hat{k} + \lambda(3\hat{i} - 4\hat{j} + 5\hat{k}) \text{ or } \frac{x-3}{3} = \frac{y+5}{-4} = \frac{z-7}{5}$$

5. Since $P(1, 2, -3)$ is the foot of the perpendicular from the origin to the plane, OP is normal to the plane. Thus, the direction ratios of normal to the plane are 1, 2 and –3.

Now, since the plane passes through (1, 2, –3), its equation is given by

$$1(x-1) + 2(y-2) - 3(z+3) = 0$$

or $\quad x + 2y - 3z - 14 = 0$

6. Plane is parallel to the vectors $\vec{a} = \hat{i} - 2\hat{j} + 2\hat{k}$ and $\vec{b} = 3\hat{j} + 4\hat{k}$

So, plane will be perpendicular to the vector $\vec{a}\times\vec{b}$, which is

$$\begin{vmatrix} \hat{i} & \hat{j} & \hat{k} \\ 0 & 3 & 4 \\ 1 & -2 & 2 \end{vmatrix} = 14\hat{i} + 4\hat{j} - 3\hat{k}$$

Plane passes through the point (1, 2, –3).

So, equation of plane is $14(x-1) + 4(y-2) - 3(z+3) = 0$

or $\quad 14x + 4y - 3z - 31 = 0$

7. Any plane passing through the origin is $a(x-0) + b(y-0) + c(z-0) = 0$

Plane is perpendicular to the given line. Therefore, the normal to the plane is parallel to the given line. Thus,

$$\frac{a}{2} = \frac{b}{-1} = \frac{c}{2}$$

Therefore, the required plane is

$$2(x-0) - 1(y-0) + 2(z-0) = 0$$

or $2x - y + 2z = 0$

8. (2) Let the equation of plane is $\frac{x}{a} + \frac{y}{b} + \frac{z}{c} = 1$

It meets the axes at $A \equiv (a, 0, 0)$, $B \equiv (0, b, 0)$, $C \equiv (0, 0, c)$

Centroid of triangle ABC, $\left(\frac{a}{3}, \frac{b}{3}, \frac{c}{3}\right) = (1, 1, 2)$ (given)

$\therefore \quad a = 3, b = 3$ and $c = 6$

So, equation of plane is $\frac{x}{3} + \frac{y}{3} + \frac{z}{6} = 1$ or $2x + 2y + z = 6$

$\therefore$ Line perpendicular to the plane is along the vector $2\hat{i} + 2\hat{j} + \hat{k}$

So, equation of line through (1, 1, 2) and perpendicular to the plane is

$$\frac{x-1}{2} = \frac{y-1}{2} = \frac{z-2}{1}$$

9. The equation of any plane through the point (1, 0, –1) is

$$A(x-1) + B(y-0) + C(z+1) = 0 \quad \text{...(i)}$$

Since it passes through the point (3, 2, 2), we get

$$2A + 2B + 3C = 0 \quad \text{...(ii)}$$

Since plane (i) is parallel to the line

$$\frac{x-1}{1} = \frac{y-1}{-2} = \frac{z-2}{3}$$

we have $1A + (-2)B + 3C = 0$ ...(iii)

From (i) and (iii),

$$A : B : C = 4 : -1 : -2$$

Substituting these values in (i), we get

$$4(x-1) - 1(y-0) - 2(z+1) = 0$$

or $4x - y - 2z - 6 = 0$

10. (3) Plane passes through the points (2, 3, 1) and (4, – 5, 3)

Therefore, plane is parallel to vector

$$(4-2)\hat{i} + (-5-3)\hat{j} + (3-1)\hat{k} = 2\hat{i} - 8\hat{j} + 2\hat{k}$$

Also plane is parallel to the x-axis, so it parallel to vector $\hat{i}$

$\therefore$ Vector Normal to the plane is $\begin{vmatrix} \hat{i} & \hat{j} & \hat{k} \\ 2 & -8 & 2 \\ 1 & 0 & 0 \end{vmatrix} = 2\hat{j} + 8\hat{k}$

Hence, equation of plane through point (2, 3, 1) and normal to vector $2\hat{j} + 8\hat{k}$ is

$$0(x-2) + 2(y-3) + 8(z-1) = 0$$

or $y + 4z - 7 = 0$

11. The required equation of plane is

$$\begin{vmatrix} x-5 & y-7 & z+3 \\ 4 & 4 & -5 \\ 7 & 1 & 3 \end{vmatrix} = 0$$

or $17(x-5) - (47)(y-7) + (-24)(z+3) = 0$

or $17x - 47y - 24z + 172 = 0$

12. Any plane through $\frac{x-1}{5} = \frac{y+2}{6} = \frac{z-3}{4}$ is

$$A(x-1) + B(y+2) + C(z-3) = 0 \quad \text{...(i)}$$

where $5A + 6B + 4C = 0$ ...(ii)

Also, the plane passes through (4, 3, 7). Therefore,

$$3A + 5B + 4C = 0 \quad \text{...(iii)}$$

From (ii) and (iii), $\frac{A}{4} = \frac{B}{-8} = \frac{C}{7}$

Therefore, the plane is

$$4(x-1) - 8(y+2) + 7(z-3) = 0$$

or $4x - 8y + 7z = 41$.

13. The plane passes through the point A (1, 2, 3) and is at the maximum distance from point B (–1, 0, 2); then the plane is perpendicular to line AB. Therefore, the direction ratios of the normal to the plane are 2, 2 and 1.

Hence, the equation of the plane is

$$2(x-1) + 2(y-2) + 1(z-3) = 0$$

or $2x + 2y + z = 9$

14. In two-dimensional plane we can draw only one line through the given point and perpendicular to the given line.

But in three-dimensional space we can draw infinite lines through given point and perpendicular to the given line of which one line will intersect the given line. All these lines lie in the plane perpendicular to the given line and passing through the given point.

Now given line is $\frac{x-1}{2} = \frac{y-2}{-1} = \frac{z}{3}$ and point is $A(2, 3, -1)$.

So, required lines will lie in the plane

$$2(x-2) - 1(y-3) + 3(z+1) = 0 \text{ or } 2x - y + 3z + 2 = 0$$

15. Equation of plane is $\begin{vmatrix} x-2 & y-1 & z-0 \\ 3-2 & -2-1 & -2-0 \\ 3-2 & 1-1 & 7-0 \end{vmatrix} = 0$

$$\Rightarrow \begin{vmatrix} x-2 & y-1 & z \\ 1 & -3 & -2 \\ 1 & 0 & 7 \end{vmatrix} = 0$$

$\Rightarrow \quad (x-2)(-21) - (y-1)(9) + z(3) = 0$

$\therefore \quad 7x + 3y - z = 17$

16. We have $\vec{r} \cdot (2\hat{i} + \hat{j} - 2\hat{k}) = -4$

or $\vec{r} \cdot (-2\hat{i} - \hat{j} + 2\hat{k}) = 4$

or $\vec{r} \cdot \vec{n} = 4$, where $\vec{n} = -2\hat{i} - \hat{j} + 2\hat{k}$

$\therefore \quad |\vec{n}| = \sqrt{(-2)^2 + (-1)^2 + 2^2} = 3$

Dividing given equation both sides by '3', we get

$$\vec{r} \cdot \left(\frac{-2\hat{i} - \hat{j} + 2\hat{k}}{3}\right) = \frac{4}{3}$$

or $\vec{r} \cdot \hat{n} = \frac{4}{3}$.

Distance of plane from the origin is $\frac{4}{3}$.

17. Equation of the plane ABC is

$$\begin{vmatrix} x-1 & y-2 & z-3 \\ 2-1 & 3-2 & 1-3 \\ 3-1 & 1-2 & 2-3 \end{vmatrix} = 0$$

or $\begin{vmatrix} x-1 & y-2 & z-3 \\ 1 & 1 & -2 \\ 2 & -1 & -1 \end{vmatrix} = 0$

or $-3(x-1) - 3(y-2) - 3(z-3) = 0$

or $x + y + z = 6$

Clearly, the points which are equidistant from the vertices of triangle lie on the straight line passing through the circumcentre of ABC and perpendicular to plane ABC.

Since, $AB = BC = AC$, triangle is equilateral.

So, circumcentre of triangle is

$$\left(\frac{1+2+3}{3}, \frac{2+3+1}{3}, \frac{3+1+2}{3}\right) \equiv (2, 2, 2)$$

So, equation of locus is $\frac{x-2}{1} = \frac{y-2}{1} = \frac{z-2}{1}$ or $x = y = z$

Exercise 4.5

1. Line is parallel to the planes $\vec{r}\cdot(\hat{i}-\hat{j}+2\hat{k}) = 5$ and $\vec{r}\cdot(3\hat{i}+\hat{j}+\hat{k}) = 6$, i.e., line is perpendicular to the vectors $\hat{i}-\hat{j}+2\hat{k}$ and $3\hat{i}+\hat{j}+\hat{k}$

   $\therefore$ Line is parallel to the vector $(\hat{i}-\hat{j}+2\hat{k}) \times (3\hat{i}+\hat{j}+\hat{k}) = \begin{vmatrix} \hat{i} & \hat{j} & \hat{k} \\ 1 & -1 & 2 \\ 3 & 1 & 1 \end{vmatrix} = -3\hat{i}+5\hat{j}+4\hat{k}$

   Line passes through (1, 2, 3).

   $\therefore$ Equation of line is

   $\vec{r} = (\hat{i}+2\hat{j}+3\hat{k}) + \lambda(-3\hat{i}+5\hat{j}+4\hat{k})$

2. Required plane contains the line $x - y = 2$, $z = 2$.

   So, its equation is taken as $(x - y - 2) + \lambda(z - 2) = 0$.

   Now vector along line intersection of planes $3x - 2z = 6$ and $y - 2 = 0$ is

   $$\begin{vmatrix} \hat{i} & \hat{j} & \hat{k} \\ 3 & 0 & -2 \\ 0 & 1 & 0 \end{vmatrix} = 2\hat{i}+3\hat{k}$$

   Given that required plane is parallel to vector $2\hat{i}+3\hat{k}$

   So, $2 + 3\lambda = 0$ or $\lambda = -2/3$.

   Therefore, equation of required plane is $3(x - y - 2) - 2(z - 2) = 0$ or $3x - 3y - 2z - 2 = 0$

3. x-axis is the line of intersection of the plans $y = 0$ and $z = 0$.

   So, equation of family of planes passing through x-axis is $y + \lambda z = 0$, $\lambda \in R$ ...(i)

   Plane passes through the point $(-2, 3, 5)\cdot\lambda$

   $\therefore$ $3 + 5\lambda = 0$ or $\lambda = -3/5$.

   Putting this value of λ in (i), we get equation of plane as $5y - 3z = 0$

4. Equation of plane through the line of intersection of planes P_1 and P_2 is

   $$x + \alpha y - 3z + 4 + \lambda(2x - y + 4z - 5) = 0$$

   or $(1 + 2\lambda)x + (\alpha - \lambda)y + (-3 + 4\lambda)z + 4 - 5\lambda = 0$ ...(i)

   One of the members of this family passes through the points (0, –11, 0) and (1, 5, 1)

   $\therefore$ $-11(\alpha - \lambda) + 4 - 5\lambda = 0$

   or $11\alpha - 6\lambda - 4 = 0$ ...(ii)

   and $(1 + 2\lambda) + 5(\alpha - \lambda) + (-3 + 4\lambda) + 4 - 5\lambda = 0$

   or $5\alpha - 4\lambda + 2 = 0$ ...(iii)

   Solving (ii) and (iii), we get $\alpha = 2$ and $\lambda = 3$.

   Putting these values in (i)

   Putting $\lambda = 3$ in (1), we get equation of plane P_3 as

   $$7x - y + 9z - 11 = 0$$

5. **(2)** Equation of plane containing line of intersection of plane $x + 2y + 3z - 4 = 0$ and $2x + y - z + 5 = 0$ is

   $x + 2y + 3z - 4 + \lambda(2x + y - z + 5) = 0$

   $(1 + 2\lambda)x + (2 + \lambda)y + (3 - \lambda)z + (5\lambda - 4) = 0$ ...(i)

   This plane is perpendicular to the plane

   $\vec{r} = (\hat{i} - \hat{j}) + \lambda(\hat{i} + \hat{j} + \hat{k}) + \mu(\hat{i} - 2\hat{j} + 3\hat{k})$ ...(ii)

   Normal to this plane is $\begin{vmatrix} \hat{i} & \hat{j} & \hat{k} \\ 1 & 1 & 1 \\ 1 & -2 & 3 \end{vmatrix} = 5\hat{i} - 2\hat{j} - 3\hat{k}$

   Planes (i) and (ii) are perpendicular

   $\therefore$ $5(1 + 2\lambda) - 2(2 + \lambda) - 3(3 - \lambda) = 0$

   $\Rightarrow$ $\lambda = \frac{8}{11}$

   Putting this value in (i), we get equation of plane as $27x + 30y + 25z = 4$.

   Comparing with $ax + by + cz = 4$, we get $a = 27$, $b = 30$ and $c = 25$

   Hence, $a - b + c = 27 - 30 + 25 = 22$.

6. **(2)** Equation of plane through the line $4ax - y + 5z - 7a = 0 = 2x - 5y - z - 3$ is

   $4ax - y + 5z - 7a + \lambda(2x - 5y - z - 3) = 0$ ...(i)

   Point (4, –1, 0) on the line $\frac{x-4}{1} = \frac{y+1}{-2} = \frac{z}{1}$ lies on above plane.

   $\therefore$ $16a + 1 - 7a + \lambda(8 + 5 - 3) = 0$

   or $9a + 1 + 10\lambda = 0$ ...(ii)

   Direction ratios of normal of plane (i) are $(4a + 2\lambda, -1 - 5\lambda, 5 - \lambda)$.

   Vector $\vec{b} = \hat{i} - 2\hat{j} + \hat{k}$ is perpendicular to the normal vector.

   $\therefore$ $4a + 2\lambda + 2 + 10\lambda + 5 - \lambda = 0$

   $\Rightarrow$ $11\lambda + 4a + 7 = 0$ ...(iii)

   Solving (i) and (ii), we get $a = 1$, $\lambda = -1$

   So, equation of plane is

   $x + 2y + 3z - 2 = 0$ ...(iv)

   Let the point on the line $\frac{x-3}{7} = \frac{y-2}{-1} = \frac{z-3}{-4}$ is $(7\mu + 3, -\mu + 2, -4\mu + 3)$.

   If this point lies on the plane (iv), then

   $\therefore$ $7\mu + 3 - 2\mu + 4 + 9 - 12\mu - 2 = 0$ or $\mu = 2$

   So $\alpha + \beta + \gamma = 2\mu + 8 = 12$

Exercise 4.6

1. The equation of the plane parallel to

   $x - 2y + 2z = 5$ is $x - 2y + 2z + k = 0$. ...(i)

   Now, according to the equation,

   $$\frac{1 - 4 + 6 + k}{\sqrt{9}} = \pm 1$$

$k+3=\pm 3 \Rightarrow k=0$ or -6

Then the planes are $x-2y+2z-6=0$ or $x-2y+2z=0$.

2. (2) The radius of the sphere is 5

The given plane is $x+y-z=4\sqrt{3}$

The length of the perpendicular from the centre (0, 0, 0) of the sphere on the plane $=\dfrac{4\sqrt{3}}{\sqrt{1+1+1}}=4$

Hence, radius of the circular section is $\sqrt{25-16}=\sqrt{9}=3$

3. (3) Equations of two closed ends are $x+y-z=4$ and $x+y-z=10$.

So, height of the cylinder, h

= distance between two parallel planes

$=\dfrac{|10-4|}{\sqrt{3}}=2\sqrt{3}$ units.

Also, radius of the cylinder, $r=3$ units.

So, volume of the cylinder, $V=\pi r^2 h=\pi(3)^2(2\sqrt{3})=18\pi\sqrt{3}$ cu. units.

4. Given planes $x+y+z=3$ and $x+y+z=9$ are parallel.

Clearly required locus is plane which is equidistant from $x+y+z=3$ and $x+y+z=9$

So, let the equation of required locus is $x+y+z=\lambda$

$\therefore \quad \dfrac{|k-3|}{\sqrt{3}}=\dfrac{|9-k|}{\sqrt{3}}$

$\Rightarrow \quad k=6$

Hence, equation of required locus is $x+y+z=6$.

5. Point P on the line $\dfrac{x+1}{-1}=\dfrac{y-3}{2}=\dfrac{z-2}{1}$ can be taken as $(-\lambda-1, 2\lambda+3, \lambda+2)$.

Foot of perpendicular from point P on the plane is Q.

Let $Q\equiv(x_1, y_1, z_1)$

$$\frac{x_1-(-\lambda-1)}{1}=\frac{y_1-(2\lambda+3)}{2}=\frac{z_1-(\lambda+2)}{1}$$

$$=-\frac{(-\lambda-1)+2(2\lambda+3)+\lambda+2-5}{1+4+1}=\frac{-1-2\lambda}{3}$$

$$\Rightarrow \quad x_1=\frac{-4-5\lambda}{3},\ y_1=\frac{7+2\lambda}{3},\ z_1=\frac{\lambda+5}{3}$$

Point Q lies on the plane $x-y+z-2=0$

$$\therefore \quad \frac{-4-5\lambda}{3}-\frac{7+2\lambda}{3}+\frac{5+\lambda}{3}-2=0$$

$\Rightarrow \quad \lambda=-2$

$\therefore \quad Q\equiv(2, 1, 1)$ and $P\equiv(1, -1, 0)$

6. (4) Vector normal to the plane $23x-10y-2z+48=0$ is $23\hat{i}-10\hat{j}-2\hat{k}$.

Vector normal the plane containing lines $\dfrac{x+1}{2}=\dfrac{y-3}{4}=\dfrac{z+1}{3}$ and $\dfrac{x+3}{2}=\dfrac{y+2}{6}=\dfrac{z-1}{\lambda}$ $(\lambda\in R)$ is

$$\begin{vmatrix}\hat{i} & \hat{j} & \hat{k}\\ 2 & 4 & 3\\ 2 & 6 & \lambda\end{vmatrix}=(4\lambda-18)\hat{i}+(6-2\lambda)\hat{j}+4\hat{k}.$$

The planes will be parallel if $\dfrac{4\lambda-18}{23}=\dfrac{6-2\lambda}{-10}=\dfrac{4}{-2}$ or $\lambda=-7$.

Distance between the planes is distance of the point (–1, 3, –1) from the given plane, which is $\dfrac{3}{\sqrt{633}}$

Thus, $k=3$

Exercise 4.7

1. (4) Acute angle between given planes is

$$\cos^{-1}\frac{|\alpha-1-2|}{\sqrt{1+1+1}\sqrt{\alpha^2+1+4}}=\cos^{-1}\frac{1}{3\sqrt{3}}\text{ (given)}$$

$\therefore \quad 3|\alpha-3|=\sqrt{\alpha^2+5}$

$\Rightarrow \quad 9\alpha^2-54\alpha+81=\alpha^2+5$

$\Rightarrow \quad 8\alpha^2-54\alpha+76=0$

$\Rightarrow \quad 4\alpha^2-27\alpha+38=0$

$\Rightarrow \quad (4\alpha-19)(\alpha-2)=0$

$\Rightarrow \quad \alpha=2, \dfrac{19}{4}$

2. (3) Given planes are

$x+2y+3z=5$ ...(i)

$3x+3y+z=0$ ...(ii)

Normal to planes (i) and (ii) are $\hat{i}+2\hat{j}+3\hat{k}$ and $3\hat{i}+3\hat{j}+\hat{k}$, respectively.

Required plane is perpendicular to planes (i) and (ii), i.e., perpendicular to the line of intersection of planes.

Vector along the line of intersection of planes (i) and (ii) is

$$\begin{vmatrix}\hat{i} & \hat{j} & \hat{k}\\ 1 & 2 & 3\\ 3 & 3 & 1\end{vmatrix}=-7\hat{i}+8\hat{j}-3\hat{k}$$

Therefore, normal to given plane is $-7\hat{i}+8\hat{j}-3\hat{k}$

Plane is passing through the point (–1, 3, 2)

Then equation of required plane is

$-7(x+1)+8(y-3)-3(z-2)=0$

or $\quad 7x-8y+3z+25=0$

3. (3) Let the equation of a plane containing the line be $l(x-1)+m(y+2)+nz=0$. Then

$2l-3m+5n=0$ and $l-m+n=0$

$\therefore \quad \dfrac{l}{2}=\dfrac{m}{3}=\dfrac{n}{1}$

Hence, the plane is $2(x-1)+3(y+2)+z=0$

or $\quad 2x+3y+z+4=0$

4. Plane which is equally inclined to the given planes is parallel to the angle bisector of the given planes.

Now the angle bisectors of the given planes are

$$\frac{x-2y+2z-3}{3}=\pm\frac{8x-4y+z-7}{9}.$$

or $\quad 5x+2y-5z+2=0$

and $\quad 11x-10y+7z-16=0.$

The equation of the required planes are $5x+2y-5z+p=0$ and $11x-10y+7z+q=0$.

Since both the planes are passing through point (1, 2, 3), $p=6$ and $q=-12$.

Hence, The planes are $5x+2y-5z+6=0$ and $11x-10y+7z-12=0$

5. We have $f_1(x, y, z) = x - 2y + 2z - 3$ and $f_2(x, y, z) = 2x + y - 2z + 4$.
Now $f_1(1, 2, -3) = 1 - 4 - 6 - 3 < 0$ and $f_2(1, 2, -3) = 2 + 2 + 6 + 4 > 0$.
Thus, $f_1 f_2 < 0$
Also, $a_1a_2 + b_1b_2 + c_1c_2 = 2 - 2 - 4 < 0$
So, point (1, 2, 3) lies in the obtuse angle region.
Therefore, equation of obtuse angle bisector is

$$\frac{x-2y+2z-3}{\sqrt{1+4+4}} = -\frac{2x+y-2z+4}{\sqrt{4+1+4}}$$

or $\quad 3x - y + 1 = 0$

Now $f_1(-10, 3, -5) = -10 - 6 - 10 - 3 < 0$ and $f_2(-10, 3, -5) = -20 + 3 + 10 + 4 < 0$
Also, $a_1a_2 + b_1b_2 + c_1c_2 < 0$
So, point (–10, 3, –5) lies in the acute angle region.
Therefore, equation of acute angle bisector is

$$\frac{x-2y+2z-3}{\sqrt{1+4+4}} = +\frac{2x+y-2z+4}{\sqrt{4+1+4}}$$

or $\quad x + 3y - 4z + 7 = 0$

6. Equation of family of planes through the line of intersection of planes $x - y + z - 5 = 0$ and $x + y - z + 4 = 0$ is

$$(x - y + z - 5) + \lambda(x + y - z + 4) = 0$$

or $\quad (1 + \lambda)x + (-1 + \lambda)y + (1 - \lambda)z - 5 + 4\lambda = 0$

This plane is perpendicular to the plane $x - y + z - 5 = 0$

$\therefore \quad 1 + \lambda + 1 + \lambda + 1 - \lambda = 0$

$\Rightarrow \quad \lambda = -3$

So, equation of required plane is $-2x - 4y + 4z - 17 = 0$

7. (2) Equation of family of planes through line L is
$(\alpha x + 2y - 5z + 6) + \lambda(3x - (\alpha + 3)y + z - 7) = 0$,
One of the members of this family say plane P is perpendicular to the plane $x + y - z - 3 = 0$

$\therefore \quad \alpha + 3\lambda + 2 - (\alpha + 3)\lambda + 5 - \lambda = 0$

$\Rightarrow \quad 2 - (\alpha + 3)\lambda = -(\alpha + 2\lambda + 5)$

Also, given equation of P is $13x - 14y - z - 22 = 0$

$$\therefore \quad \frac{\alpha+3\lambda}{13} = \frac{-(\alpha+2\lambda+5)}{-14} = \frac{\lambda-5}{-1} = \frac{6-7\lambda}{-22}$$

Solving, we get $\alpha = 1$ and $\lambda = 4$

Exercise 4.8

1. (4) Any point on the line $\frac{x-2}{3} = \frac{y+1}{4} = \frac{z-2}{12} = \lambda$ is
$(3\lambda + 2, 4\lambda - 1, 12\lambda + 2)$.
This lies on plane $x - y + z = 5$.
So, $3\lambda + 2 - 4\lambda + 1 + 12\lambda + 2 = 5$ or $\lambda = 0$, then the point is (2, –1, 2).
Its distance from (–1, –5, – 10) is

$$\sqrt{(2+1)^2 + (-1+5)^2 + (2+10)^2}$$

$$= \sqrt{9+16+144} = 13$$

2. (3) The line is along the vector $\vec{a} = -3\hat{i} + 2\hat{j} + \hat{k}$ and vector normal to the plane is $\vec{b} = \hat{i} + \hat{j} + \hat{k}$.

Since $\vec{a} \cdot \vec{b} = 0$, the line is parallel to the plane.
Hence, the distance between the line and the plane is the distance of point (–1, 3, 2) from the plane,

$$\frac{|-1+3+2+3|}{\sqrt{1+1+1}} = \frac{7}{\sqrt{3}}$$

3. A line parallel to $\vec{c}$ and passing through $A(\vec{a})$ is $\vec{r} = \vec{a} + \lambda\vec{c}$
Solving this line and given plane, we get

$$(\vec{a} + \lambda\vec{c}) \cdot \hat{n} = d$$

or $\quad \lambda = \dfrac{d - \vec{a}\cdot\hat{n}}{\vec{c}\cdot\hat{n}}$

$\therefore \quad$ Required Distance $= \dfrac{|d - \vec{a}\cdot\hat{n}|}{|\vec{c}\cdot\hat{n}|}|\vec{c}|$

4. (2) Given line is the line of intersection of planes $3x - 2y + z + 3 = 0$ and $4x - 3y + 4z + 1 = 0$.
Now vector along line of intersection is

$$\begin{vmatrix} \hat{i} & \hat{j} & \hat{k} \\ 3 & -2 & 1 \\ 4 & -3 & 4 \end{vmatrix} = -5\hat{i} - 8\hat{j} - \hat{k}$$

Now, line is parallel to the plane $2x - y + mz - 2 = 0$

$\therefore \quad (2)(-5) + (-1)(-8) + (m)(-1) = 0$

or $\quad m = -2$

5. Line is parallel to the vector $3\hat{i} + 2\hat{j} + 4\hat{k}$
Vector normal to the plane is $2\hat{i} + \hat{j} - 3\hat{k}$
So, required angle,

$$\theta = \sin^{-1}\frac{|(3)(2) + (2)(1) + (4)(-3)|}{\sqrt{29}\sqrt{14}} = \sin^{-1}\frac{4}{\sqrt{406}}$$

6. The given line is $\vec{r} = (\hat{i} + 2\hat{j} - \hat{k}) + \lambda(\hat{i} - \hat{j} + \hat{k})$

Here, $\vec{b} = \hat{i} - \hat{j} + \hat{k} \qquad (\vec{r} = \vec{a} + \lambda\vec{b})$

The given plane is $\vec{r} \cdot (2\hat{i} - \hat{j} + \hat{k}) = 4 \qquad (\vec{r}\cdot\vec{n} = p)$

Here $\vec{n} = 2\hat{i} - \hat{j} + \hat{k}$

Now $\cos\theta = \dfrac{\vec{n}\cdot\vec{b}}{|\vec{n}||\vec{b}|}$

(θ is the angle between the line and the normal to the plane)

$$= \frac{2+1+1}{\sqrt{6}\sqrt{3}} = \frac{4}{\sqrt{2}\cdot 3} = \frac{2\sqrt{2}}{3}$$

$$\therefore \quad \theta = \cos^{-1}\left(\frac{2\sqrt{2}}{3}\right)$$

7. Any point on the line $\frac{x-1}{1} = \frac{y+1}{-2} = \frac{z-2}{3} = t$ is $(t + 1, -2t - 1, 3t + 2)$, which lies on the given plane if $t + 1 + 2t + 1 + 6t + 4 - 3 = 0$ or $t = -1/3$.
Hence, the point of intersection of the line and the plane is $P(2/3, -1/3, 1)$.
Also, if the foot of the perpendicular from A (1, – 1, 2) on the plane is Q (x, y, z), then

$$\frac{x-1}{1} = \frac{y+1}{-1} = \frac{z-2}{2}$$

$$= -\frac{(1+1+4-3)}{1+1+4} = -\frac{1}{2}$$

Therefore, $Q \equiv (1/2, -1/2, 1)$.

Hence, the direction ratios of PQ are $\frac{2}{3} - \frac{1}{2}, -\frac{1}{3} + \frac{1}{2}$ and $1 - 1$ or $\frac{1}{6}, \frac{1}{6}$ and 0 or 1, 1, 0.

If the image of point A (1, – 1, 2) in the plane is R, then Q is the midpoint of AR. Therefore, point R is (0, 0, 0).
Hence, the direction ratios of PR (image) of the line in the plane are 2/3, – 1/3 and 1 or 2, –1, 3.

8. (2) Given points are $A(2, 3, 4)$ and $B(5, 6, 7)$

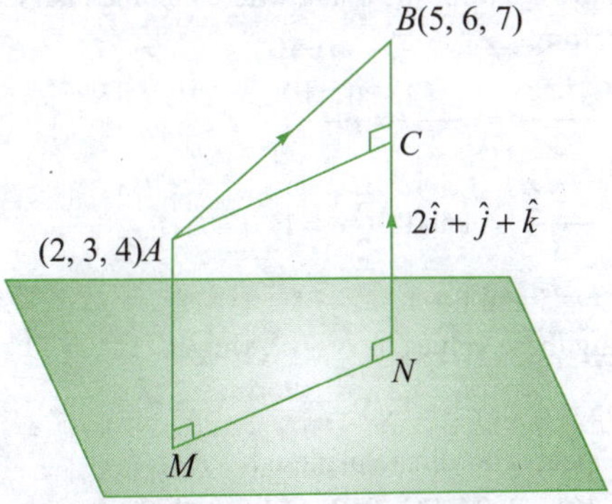

In the figure,

Now BC = projection of $\overrightarrow{AB}$ on $2\hat{i}+\hat{j}+\hat{k}$

$$=\frac{3\cdot 2+3\cdot 1+3\cdot 1}{\sqrt{6}}=2\sqrt{6}$$

Now in triangle ABC, $AC^2 = AB^2 - BC^2 = 27 - 24 = 3$

$\therefore \quad AC = \sqrt{3}$

9. Line of shortest distance between given lines is parallel to the vector

$$\begin{vmatrix} \hat{i} & \hat{j} & \hat{k} \\ 2 & 3 & -1 \\ 1 & 4 & 2 \end{vmatrix} = 10\hat{i}-5\hat{j}+5\hat{k}$$

So, angle between the line of shortest distance and plane is

$$\sin^{-1}\frac{1}{\sqrt{6}} = \sin^{-1}\frac{|2\alpha-3+6|}{\sqrt{4+1+1}\sqrt{\alpha^2+9+36}}$$

$\Rightarrow \quad \dfrac{|2\alpha+3|}{\sqrt{\alpha^2+45}}=1$

$\Rightarrow \quad (\alpha^2+45) = 4\alpha^2+12\alpha+9$

$\Rightarrow \quad 3\alpha^2+12\alpha-36=0$

$\Rightarrow \quad \alpha^2+4\alpha-12=0$

$\Rightarrow \quad (\alpha-2)(\alpha+6)=0$

$\Rightarrow \quad \alpha = 2, -6$

Exercises

Single Correct Answer Type

1. (4) Given line is $\dfrac{x+1}{3}=\dfrac{y-3}{-2}=\dfrac{z-4}{1}$

Clearly line L is parallel to the above line.

$\therefore$ equation of line L is $\dfrac{x-1}{3}=\dfrac{y-2}{-2}=\dfrac{z-3}{1}$

Clearly line L does not pass through the point (–5, 6, 2).

2. (2) The required line passes through the point $\hat{i}+3\hat{j}+2\hat{k}$ and is perpendicular to the lines $\vec{r}=(\hat{i}+2\hat{j}-\hat{k})+\lambda(2\hat{i}+\hat{j}+\hat{k})$ and $\vec{r}=(2\hat{i}+6\hat{j}+\hat{k})+\mu(\hat{i}+2\hat{j}+3\hat{k})$; therefore, it is parallel to the vector $\vec{b}=(2\hat{i}+\hat{j}+\hat{k})\times(\hat{i}+2\hat{j}+3\hat{k})=(\hat{i}-5\hat{j}+3\hat{k})$

Hence, the equation of the required line is

$$\vec{r}=(\hat{i}+3\hat{j}+2\hat{k})+\lambda(\hat{i}-5\hat{j}+3\hat{k})$$

3. (4) Given lines are

$$L_1: \frac{x-3}{1}=\frac{y+1}{\alpha}=\frac{z-2}{-\alpha^2} \text{ and}$$

$$L_2: \frac{x-1}{1}=\frac{y+1}{2}=\frac{z-1}{3}$$

Line L is parallel to L_1 and perpendicular to L_2

$\therefore \quad 1(1)+\alpha(2)+(-\alpha^2)(3)=0$

$\Rightarrow \quad 3\alpha^2-2\alpha-1=0$

$\Rightarrow \quad (3\alpha+1)(\alpha-1)=0$

$\Rightarrow \quad \alpha=1, (\because \alpha>0)$

So, equation of line L is $\dfrac{x+1}{1}=\dfrac{y}{1}=\dfrac{z-2}{-1}$

Clearly, line does not pass through the point (–3, –2, 3).

4. (3)

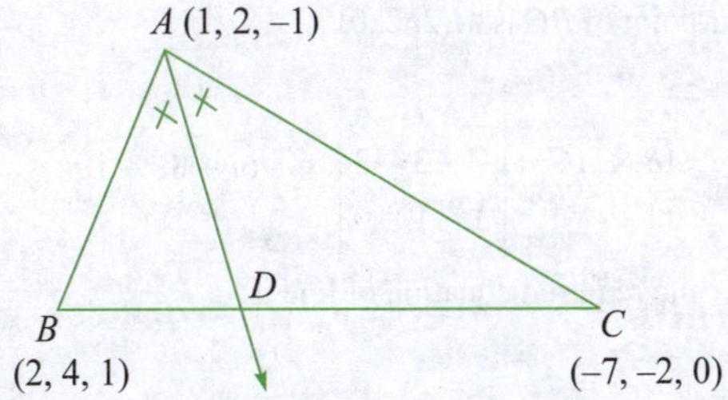

We have $AB=\sqrt{1+4+4}=\sqrt{9}=3$

and $AC=\sqrt{64+16+1}=\sqrt{81}=9$

So, angle bisector AD meets BC at D, such that

$$\frac{BD}{DC}=\frac{AB}{AC}=\frac{3}{9}=\frac{1}{3}.$$

$$\therefore \quad D\equiv\left(\frac{1(-7)+3(2)}{1+3},\frac{1(-2)+3(4)}{1+3},\frac{1(0)+3(1)}{1+3}\right)$$

$$\equiv\left(\frac{-1}{4},\frac{5}{2},\frac{3}{4}\right)$$

So, equation of AD is

$$\frac{x-1}{-\frac{1}{4}-1}=\frac{y-2}{\frac{5}{2}-2}=\frac{z+1}{\frac{3}{4}+1} \text{ or } \frac{x-1}{-5}=\frac{y-2}{2}=\frac{z+1}{7}$$

5. (3) We have $l+3m+5n=0$ and $5lm-2mn+6nl=0$, is :
Eliminating l, we have $-(3m+5n)(5m+6n)-2mn=0$

$\therefore \quad m^2+3mn+2n^2=0$

$\Rightarrow \quad (m+n)(m+2n)=0$

If $m+n=0$, we have $\dfrac{l}{-2}=\dfrac{m}{-1}=\dfrac{n}{1}$

If $m+2n=0$, we have $\dfrac{l}{1}=\dfrac{m}{-2}=\dfrac{n}{1}$

$\therefore \quad$ Direction ratios of lines are –2, – 1, 1 and 1, –2, 1

Angle between lines is given by

$$\theta = \cos^{-1}\frac{(-2)(1)+(-1)(-2)+(1)(1)}{\sqrt{6}.\sqrt{6}} = \cos^{-1}\frac{1}{6}$$

6. (3) Diagonal vector through point A is

$$(\hat{i} - 2\hat{j} + \hat{k}) + (2\hat{i} - \hat{j} + \hat{k}) + (\hat{i} + \hat{j} + \hat{k}) = 4\hat{i} - 2\hat{j} + 3\hat{k}$$

So, equation of diagonal is $\dfrac{x-1}{4} = \dfrac{y+1}{-2} = \dfrac{z}{3}$

7. (2) Mid–point of diagonal BD is $M(0, -1, 4)$

$\therefore$ Equation of diagonal AC is

$$\frac{x-2}{-2} = \frac{y}{-1} = \frac{z+3}{7}$$

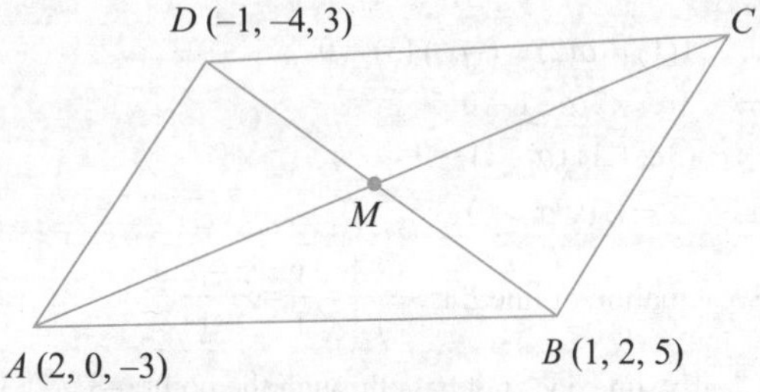

AC meets yz–plane at (0, –1 , 4).

8. (4) Mid-point of BC is $M(2, 2, 0)$

$$\overrightarrow{AB} \times \overrightarrow{AC} = \begin{vmatrix} \hat{i} & \hat{j} & \hat{k} \\ -1 & 3 & 3 \\ 3 & 1 & -1 \end{vmatrix} = -6\hat{i} + 8\hat{j} - 10\hat{k}$$

$\therefore$ the required equation of line is

$$\frac{x-2}{3} = \frac{y-2}{-4} = \frac{z}{5}$$

Line meets xz–plane at (7/2, 0, 5/2).

9. (2) Direction cosines of the given line are $\dfrac{1}{3}, -\dfrac{2}{3}, \dfrac{2}{3}$

Hence, the equation of line can be a point in the form

$$\frac{x-2}{1/3} = \frac{y+3}{-2/3} = \frac{z+5}{-2/3} = r$$

Therefore, any point on the line is

$$\left(2 + \frac{r}{3}, -3 - \frac{2r}{3}, -5 - \frac{2r}{3}\right), \text{ where } r = \pm 6.$$

Points are (4, – 7, – 9) and (0, 1, –1).

10. (2) Let any point on the second line be $(\lambda, 2\lambda, 3\lambda)$

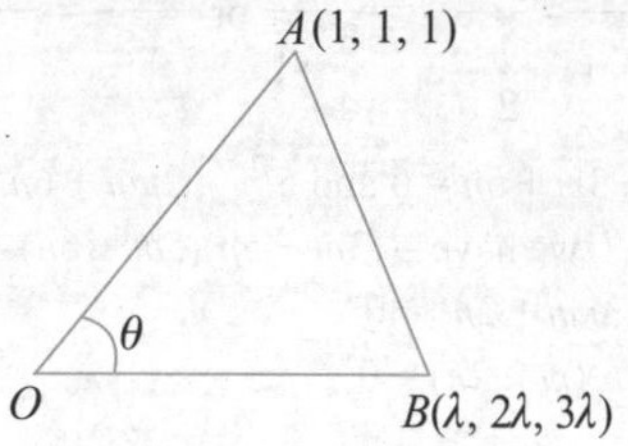

$$\cos\theta = \frac{6}{\sqrt{42}}, \ \sin\theta = \frac{\sqrt{6}}{\sqrt{42}}$$

$$\Delta_{OAB} = \frac{1}{2}(OA)\ OB \sin\theta$$

$\Rightarrow \ \dfrac{1}{2}\sqrt{3}\ \lambda\sqrt{14} \times \dfrac{\sqrt{6}}{\sqrt{42}} = \sqrt{6}$

or $\lambda = 2$

So B is (2, 4, 6)

11. (3) We have $z = 0$ for the point, where the line intersects the curve. Therefore,

$$\frac{x-2}{3} = \frac{y+1}{2} = \frac{0-1}{-1}$$

$\Rightarrow \ \dfrac{x-2}{3} = 1$ and $\dfrac{y+1}{2} = 1$

$\Rightarrow \ x = 5$ and $y = 1$

Putting these values in $xy = c^2$, we get

$$5 = c^2 \text{ or } c = \pm\sqrt{5}$$

12. (3) Line meets coordinate planes at

$A(0, -2, 3)$, $B(1, 0, 2)$ and $C(3, 4, 0)$

$$AB = \sqrt{1+4+1} = \sqrt{6}$$

$$AC = \sqrt{9+36+9} = 3\sqrt{6}$$

and $BC = \sqrt{4+16+4} = 2\sqrt{6}$

Now, $AB + BC = AC$.

$$A \underset{\lambda}{-\!-} B \underset{2\lambda}{-\!-\!-\!-\!-\!-\!-} C$$

So, B divides AC internally in the ratio $AB : BC = 1 : 2$.

Point C divides AB externally in the ratio $AC : BC = 3 : 2$.

Point A divides BC externally in the ratio $AB : AC = 1 : 3$.

13. (3) Point C on L_1 is $(2\lambda + 1, \lambda + 3, 2\lambda + 2)$ and point D on line L_2 is $(\mu + 2, 2\mu + 2, 3\mu + 3)$.

$\therefore \ \overrightarrow{CD} = (\mu - 2\lambda + 1)\hat{i} + (2\mu - \lambda - 1)\hat{j} + (3\mu - 2\lambda + 1)\hat{k}$

$\overrightarrow{CD}$ is parallel to x-axis

$\therefore \ 2\mu - \lambda - 1 = 0$ and $3\mu - 2\lambda + 1 = 0$

Solving we get $\lambda = 5$, $\mu = 3$

$\therefore \ \overrightarrow{CD} = -6\hat{i}$

14. (2) Line L_2 is perpendicular to vectors $\vec{a} = \hat{i} + \hat{j} - \hat{k}$ and $\vec{b} = -2\hat{i} + \hat{j} + 5\hat{k}$

So, line L_2 is parallel to $\vec{a} \times \vec{b}$.

$$\vec{a} \times \vec{b} = \begin{vmatrix} \hat{i} & \hat{j} & \hat{k} \\ 1 & 1 & -1 \\ -2 & 1 & 5 \end{vmatrix} = 6\hat{i} - 3\hat{j} + 3\hat{k}$$

Line L_1 is $\dfrac{x - \frac{1}{2}}{\frac{3}{2}} = \dfrac{y-3}{-1} = \dfrac{z - \frac{4}{3}}{-\frac{1}{3}}$

or $\dfrac{x - \frac{1}{2}}{9} = \dfrac{y-3}{-6} = \dfrac{z - \frac{4}{3}}{-2}$

Thus line L_1 is parallel to vector

$$\vec{c} = 9\hat{i} - 6\hat{j} - 2\hat{k}$$

Now, line L is perpendicular to lines L_1 and L_2

So, L is perpendicular to vectors $\vec{a} \times \vec{b}$ and $\vec{c}$.

Therefore, L is parallel to vector $(\vec{a} \times \vec{b}) \times \vec{c}$.

Now, $(\vec{a}\times\vec{b})\times\vec{c}=\begin{vmatrix}\hat{i} & \hat{j} & \hat{k}\\ 2 & -1 & 1\\ 9 & -6 & -2\end{vmatrix}$

$= 8\hat{i}+13\hat{j}-3\hat{k}$

So, equation of line L is

$$\frac{x-1}{8}=\frac{y-1}{13}=\frac{z-1}{-3}$$

15. (3) Line does not meet the xy–plane.

So, line is parallel to xy–plane, i.e. perpendicular to z–axis.

$\therefore\quad (\alpha-3)\cdot 1=0$ or $\alpha=3$

So, equation of line is $\dfrac{x-3}{2}=\dfrac{y-9}{4}=\dfrac{z+2}{0}=\lambda$

So, line meets yz–plane and zx–plane at

$A(0, 3, -2)$ and $B\left(\dfrac{-3}{2}, 0, -2\right)$ respectively

$$\therefore\quad AB=\sqrt{\frac{9}{4}+9+0}=\frac{3}{2}\sqrt{5}$$

16. (3)

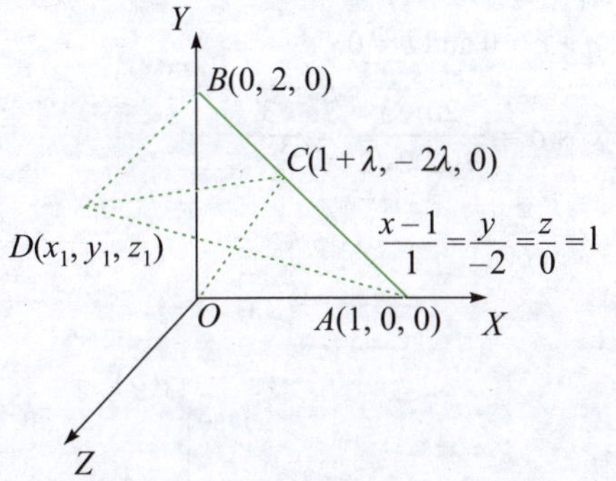

Equation of line AB is $\dfrac{x-1}{1}=\dfrac{y}{-2}=\dfrac{z}{0}=\lambda$

Now $AB\perp OC \Rightarrow 1\,(\lambda+1)+(-2\lambda)\,(-2)=0$

or $\quad 5\lambda=-1$ or $\lambda=-\dfrac{1}{5}$

$\Rightarrow\quad C$ is $\left(\dfrac{4}{5},\dfrac{2}{5},0\right)$

Now, $x_1^2+(y_1-2)^2+z_1^2=4$ ...(i)

and $(x_1-1)^2+y_1^2+z_1^2=1$ ...(ii)

Also $OC\perp CD$

$$\Rightarrow\quad\left(x_1-\frac{4}{5}\right)\frac{4}{5}+\left(y_1-\frac{2}{5}\right)\frac{2}{5}+(z_1-0)\,0=0 \qquad \text{...(iii)}$$

From (i) and (ii), we get

$-4y_1+2x_1=0$ or $x_1=2y_1$

In (iii), putting $x_1=2y_1$ we get $y_1=\dfrac{2}{5}$ and $x_1=\dfrac{4}{5}$.

Putting this value of x_1 and y_1 in (i), we get

$$z_1=\pm\frac{2}{\sqrt{5}}$$

17. (4) Let AD be the perpendicular and D be the foot of the perpendicular which divides BC in the ratio $\lambda : 1$, then

$$D\left(\frac{10\lambda-9}{\lambda+1},\frac{4}{\lambda+1},\frac{-\lambda+5}{\lambda+1}\right). \qquad \text{...(i)}$$

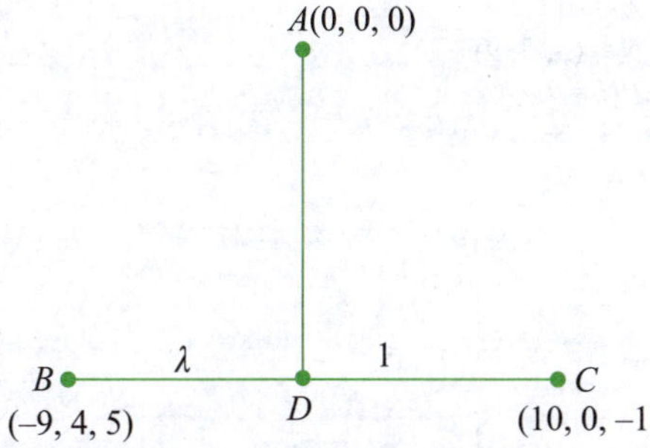

The direction ratios of AD are $\dfrac{10\lambda-9}{\lambda+1},\dfrac{4}{\lambda+1}$ and $\dfrac{-\lambda+5}{\lambda+1}$ and direction ratios of BC are 19, – 4 and – 6.

Since $AD\perp BC$, we get

$$19\left(\frac{10\lambda-9}{\lambda+1}\right)-4\left(\frac{4}{\lambda+1}\right)-6\left(\frac{-\lambda+5}{\lambda+1}\right)=0$$

or $\quad\lambda=\dfrac{31}{28}$

Hence, on putting the value of λ in (i), we get required foot of the perpendicular as $\left(\dfrac{58}{59},\dfrac{112}{59},\dfrac{109}{59}\right)$.

18. (2) Here, $\alpha=\beta=\gamma$

$\because\quad\cos^2\alpha+\cos^2\beta+\cos^2\gamma=1$

$\therefore\quad\cos\alpha=\dfrac{1}{\sqrt{3}}$

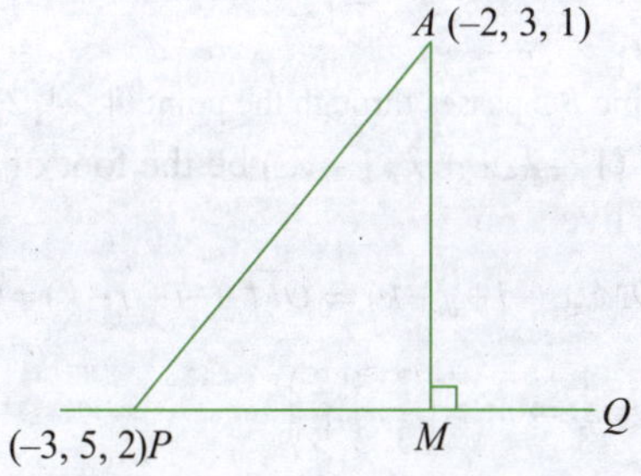

Direction cosines of PQ are $\left(\dfrac{1}{\sqrt{3}},\dfrac{1}{\sqrt{3}},\dfrac{1}{\sqrt{3}}\right)$

$PM=$ Projection of AP on PQ

$$=\left|(-2+3)\frac{1}{\sqrt{3}}+(3-5)\frac{1}{\sqrt{3}}+(1-2)\frac{1}{\sqrt{3}}\right|$$

$$=\frac{2}{\sqrt{3}}$$

and $\quad AP=\sqrt{(-2+3)^2+(3-5)^2+(1-2)^2}=\sqrt{6}$

$$\therefore\quad AM=\sqrt{(AP)^2-(PM)^2}=\sqrt{6-\frac{4}{3}}=\sqrt{\frac{14}{3}}$$

19. (4) Let P be the point (1, 2, 3) and PN be the length of the perpendicular from P on the given line.

Coordinates of point N are $(3\lambda+6,\ 2\lambda+7,\ -2\lambda+7)$.

Now PN is perpendicular to the given line or vector $3\hat{i}+2\hat{j}-2\hat{k}$. Thus,

$$3(3\lambda+6-1)+2(2\lambda+7-2)-2(-2\lambda+7-3)=0$$

or $\lambda=-1$

$\therefore$ $N\equiv(3, 5, 9)$

$\Rightarrow$ $PN=7$

20. (4)

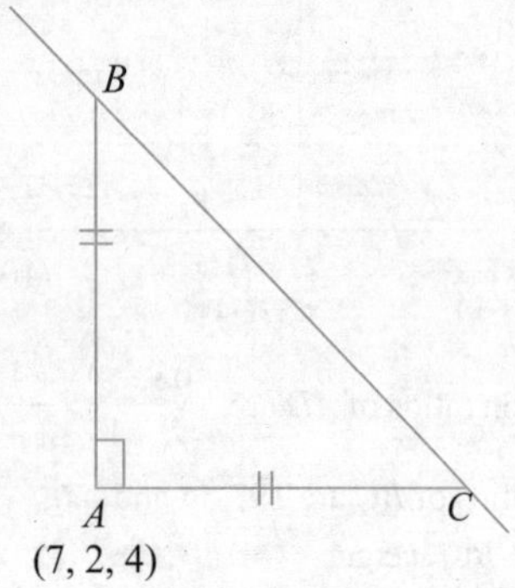

Given one vertex $A(7, 2, 4)$ and line

$$BC \equiv \frac{x+6}{5}=\frac{y+10}{3}=\frac{z+14}{8}$$

Variable point on the above line $B\equiv(5\lambda-6, 3\lambda-10, 8\lambda-14)$

Direction ratios of line AB are $<5\lambda-13, 3\lambda-12, 8\lambda-18>$

Direction ratios of line BC are $<5, 3, 8>$

Since angle between AB and BC is $\pi/4$, we have

$$\cos\frac{\pi}{4}=\frac{|(5\lambda-13)5+3(3\lambda-12)+8(8\lambda-18)|}{\sqrt{5^2+3^2+8^2}\cdot\sqrt{(5\lambda-13)^2+(3\lambda-12)^2+(8\lambda-18)^2}}$$

Squaring and solving, we get $\lambda=3, 2$

Hence, equation of lines are $\frac{x-7}{2}=\frac{y-2}{-3}=\frac{z-4}{6}$ and $\frac{x-7}{3}=\frac{y-2}{6}=\frac{z-4}{2}$

Also, line BC passes through the point $(4, -4, 2)$.

21. (1) Let $M(1, -\lambda, 1+\lambda, 1-\lambda)$ be the foot of perpendicular on the live

$\therefore$ $\overrightarrow{OM}\perp(-\hat{i}+\hat{j}-\hat{k})\Rightarrow\overrightarrow{OM}\cdot(-\hat{i}+\hat{j}-\hat{k})=0$

$\Rightarrow$ $\lambda=\frac{1}{3}\Rightarrow M\equiv\left(\frac{2}{3},\frac{4}{3},\frac{2}{3}\right)$

22. (2) Let foot of perpendicular from $P(\beta, 0, \beta)$ on the line is M

Let point M be $(\lambda, 1, \lambda-1)$

Now PM is perpendicular to the line.

$\therefore$ $1(\lambda-\beta)+0(1-0)-1(-\lambda-1-\beta)=0$

$\Rightarrow$ $\lambda=-\frac{1}{2}$

$\therefore$ $M\left(-\frac{1}{2}, 1, -\frac{3}{2}\right)$

Now, $PM=\sqrt{\frac{3}{2}}$

$\Rightarrow$ $\left(\beta+\frac{1}{2}\right)^2+1+\left(\beta+\frac{3}{2}\right)^2=\frac{3}{2}$

$\Rightarrow$ $\beta^2+\beta+\frac{1}{4}+1+\beta^2+3\beta+\frac{9}{4}=\frac{3}{2}$

$\Rightarrow$ $2\beta^2+4\beta+2=0$

$\Rightarrow$ $2(\beta+1)^2=0$

$\Rightarrow$ $\beta=-1$

23. (1) Both the lines pass through the origin. Line L_1 is parallel to the vector

$$\vec{V_1}=(\cos\theta+\sqrt{3})\hat{i}+(\sqrt{2}\sin\theta)\hat{j}+(\cos\theta-\sqrt{3})\hat{k}$$

and L_2 is parallel to the vector

$$\vec{V_2}=a\hat{i}+b\hat{j}+c\hat{k}$$

$$\therefore\quad \cos\alpha=\frac{|\vec{V_1}\cdot\vec{V_2}|}{|\vec{V_1}||\vec{V_2}|}$$

$$=\frac{|a(\cos\theta+\sqrt{3})+(b\sqrt{2})\sin\theta+c(\cos\theta-\sqrt{3})|}{\sqrt{a^2+b^2+c^2}\sqrt{(\cos\theta+\sqrt{3})^2+2\sin^2\theta+(\cos\theta-\sqrt{3})^2}}$$

$$=\frac{|(a+c)\cos\theta+b\sqrt{2}\sin\theta+(a-c)\sqrt{3}|}{\sqrt{a^2+b^2+c^2}\sqrt{2+6}}$$

For $\cos\alpha$ to be independent of θ, we get

$a+c=0$ and $b=0$

$$\therefore\quad \cos\alpha=\frac{2a\sqrt{3}}{a\sqrt{2}\,2\sqrt{2}}=\frac{\sqrt{3}}{2}$$

or $\alpha=\frac{\pi}{6}$

24. (3)

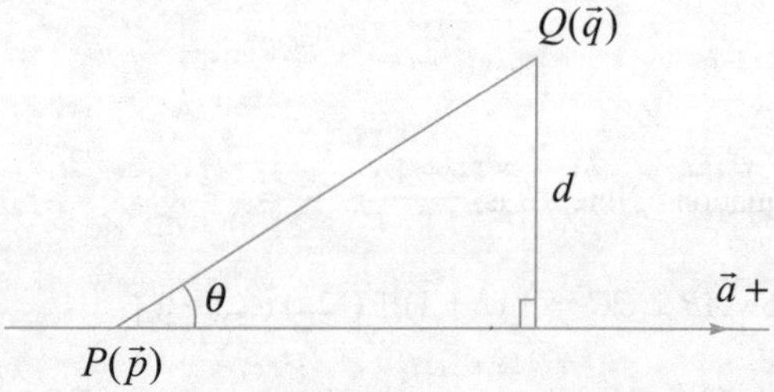

Since line is equally inclined to unit vectors $\vec{a}$ and $\vec{b}$, in the plane of $\vec{a}$ and $\vec{b}$, it is parallel to vector $\vec{a}+\vec{b}$.

So, equation of line is $\vec{r}=\vec{p}+\lambda(\vec{a}+\vec{b})$

From the figure, $d=|\vec{q}-\vec{p}|\sin\theta$

Now, $(\vec{q}-\vec{p})\times(\vec{a}+\vec{b})=|\vec{q}-\vec{p}||\vec{a}+\vec{b}|\sin\theta$

$$\therefore\quad d=\frac{|(\vec{q}-\vec{p})\times(\vec{a}+\vec{b})|}{|\vec{a}+\vec{b}|}$$

25. (2) The lines $\frac{x-2}{1}=\frac{y-3}{1}=\frac{z-4}{-k}$ ...(i

and $\frac{x-1}{k}=\frac{y-4}{2}=\frac{z-5}{1}$ ...(ii

are coplanar if

$$\text{or}\quad \begin{vmatrix} 2-1 & 3-4 & 4-5 \\ 1 & 1 & -k \\ k & 2 & 1 \end{vmatrix}=0$$

or $k^2 + 3k = 0$

$\Rightarrow \quad k = 0 \text{ or } -3$

26. (1) Given lines are

$$\frac{x-5}{3} = \frac{y-7}{-1} = \frac{z+2}{1} = r_1 \text{ (say)}$$

and $$\frac{x+3}{-36} = \frac{y-3}{2} = \frac{z-6}{4} = r_2 \text{ (say)}$$

$\therefore \quad x = 3r_1 + 5 = -36r_2 - 3,$

$y = -r_1 + 7 = 3 + 2r_2$

and $z = r_1 - 2 = 4r_2 + 6$

On solving, we get

$$x = 21, y = \frac{5}{3}, z = \frac{10}{3}$$

27. (3) The given lines are parallel.

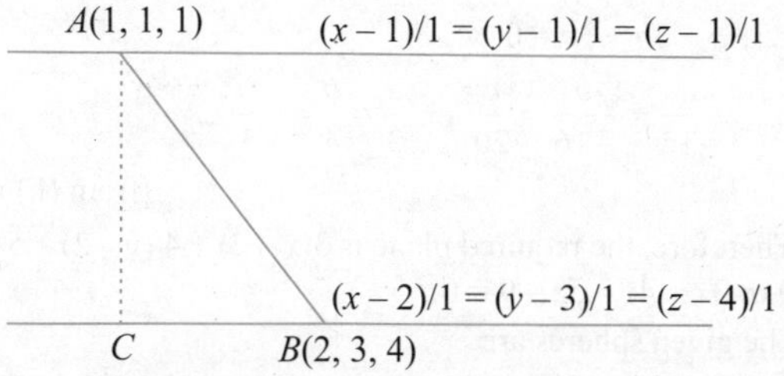

From the figure, we get

$$BC = \frac{|\overrightarrow{AB}\cdot(\hat{i}+\hat{j}+\hat{k})|}{|\hat{i}+\hat{j}+\hat{k}|}$$

$$= \frac{|1+2+3|}{\sqrt{3}} = 2\sqrt{3}$$

Also, $AB = \sqrt{1+4+9} = \sqrt{14}$

$\therefore$ The shortest distance $= AC = \sqrt{14-12} = \sqrt{2}$

28. (4) Line of intersection of planes

$x + y + z + 1 = 0$ ...(i)

$2x - y + z + 3 = 0$ ...(ii)

Eliminating y, we get

$3x + 2z + 4 = 0$

or $x = \dfrac{-2z-4}{3}$ ...(iii)

Eliminating z from (i) and (ii), we get

$x = 2y - 2$ ...(iv)

From (iii) and (iv), we get line of intersection of planes as

$$x = 2y - 2 = \frac{-2z-4}{3}$$

$$\frac{x}{2} = \frac{y-1}{1} = \frac{z+2}{-3} \quad \text{...(v)}$$

Given line is

$$\frac{x-1}{0} = \frac{y+1}{-1} = \frac{z}{1} \quad \text{...(vi)}$$

Shortest distance between lines (5) and (6) is

$$= \left|\frac{(\vec{b}-\vec{a}).(\vec{c}\times\vec{d})}{|\vec{c}\times\vec{d}|}\right|$$

where $\vec{a} = (0, 1, -2)$

$\vec{b} = (1, -1, 0)$

$\vec{c} = (2, 1, -3)$

$\vec{d} = (0, -1, 1)$

$\Rightarrow \quad S.D = \dfrac{1}{\sqrt{3}}$

29. (2) The lines $\vec{r} = \vec{a} + \lambda(\vec{b}\times\vec{c})$ and $\vec{r} = \vec{b} + \mu(\vec{c}\times\vec{a})$ pass through points $\vec{a}$ and $\vec{b}$, respectively, and are parallel to the vectors $\vec{b}\times\vec{c}$ and $\vec{c}\times\vec{a}$, respectively. Therefore, they intersect if $\vec{a}-\vec{b}$, $\vec{b}\times\vec{c}$ and $\vec{c}\times\vec{a}$ are coplanar and so

$$(\vec{a}-\vec{b})\cdot\{(\vec{b}\times\vec{c})\times(\vec{c}\times\vec{a})\} = 0$$

$$\Rightarrow (\vec{a}-\vec{b})\cdot([\vec{b}\ \vec{c}\ \vec{a}]\vec{c} - [\vec{b}\ \vec{c}\ \vec{c}]\vec{a}) = 0$$

$$\Rightarrow ((\vec{a}-\vec{b})\cdot\vec{c})[\vec{b}\ \vec{c}\ \vec{a}] = 0$$

$$\Rightarrow \vec{a}\cdot\vec{c} - \vec{b}\cdot\vec{c} = 0 \text{ or } \vec{a}\cdot\vec{c} = \vec{b}\cdot\vec{c}$$

30. (2) $l = \left|(\vec{b}-\vec{a})\cdot\dfrac{(\vec{p}\times\vec{q})}{|\vec{p}\times\vec{q}|}\right|$

$\Rightarrow \quad |\vec{b}-\vec{a}|\cos 60° = 1 \Rightarrow AB = 2$

31. (2) Let $\vec{r}\times\vec{a} = \vec{b}\times\vec{a}$

$\Rightarrow \quad (\vec{r}-\vec{b})\times\vec{a} = 0 \Rightarrow \vec{r} = \vec{b} + t\vec{a}$

Similarly, the other line is $\vec{r} = \vec{a} + k\vec{b}$, where t and k are scalars.

Now for the point of intersection $\vec{a} + k\vec{b} = \vec{b} + t\vec{a}$

$\Rightarrow \quad t = 1, k = 1$

(equating the coefficients of $\vec{a}$ and $\vec{b}$)

$\therefore \quad \vec{r} = \vec{a} + \vec{b} = \hat{i} + \hat{j} + 2\hat{i} - \hat{k} = 3\hat{i} + \hat{j} - \hat{k}$

i.e., $(3, 1, -1)$

32. (3) Direction ratios of OP are (a, b, c)

Therefore, equation of the plane is

$a(x-a) + b(y-b) + c(z-c) = 0$

i.e., $xa + yb + zc = a^2 + b^2 + c^2$

33. (3) Given plane is

$$\vec{r} = (1+\lambda-\mu)\hat{i} + (2-\lambda)\hat{j} + (3-2\lambda+2\mu)\hat{k}$$

$$= (\hat{i}+2\hat{j}+3\hat{k}) + \lambda(\hat{i}-\hat{j}-2\hat{k}) + \mu(-\hat{i}+2\hat{k})$$

which is a plane passing through $\vec{a} = \hat{i}+2\hat{j}+3\hat{k}$ and parallel to the vectors $\vec{b} = \hat{i}-\hat{j}-2\hat{k}$ and $\vec{c} = -\hat{i}+2\hat{k}$.

Therefore, it is perpendicular to the vector $\vec{n} = \vec{b}\times\vec{c} = -2\hat{i} - \hat{k}$

Hence, equation of plane is $-2(x-1) + (0)(y-2) - (z-3) = 0$

or $2x + z = 5$

34. (1) Plane contains y-axis.

So, normal to the plane must be perpendicular to y-axis.

Let equation of plane is, $ax + cz = 0$.

Plane passes through the point $(3, 2, 1)$

$\Rightarrow \quad 3a + c = 0$

$\Rightarrow \quad c = -3a$

So, equation of plane is $ax - 3az = 0$ or $x = 3z$.

35. (2) Let the equation of plane is $\frac{x}{a}+\frac{y}{b}+\frac{z}{c}=1$

It meets the axes at $A\equiv(a, 0, 0)$, $B\equiv(0, b, 0)$, $C\equiv(0, 0, c)$

Centroid of triangle ABC, $\left(\frac{a}{3},\frac{b}{3},\frac{c}{3}\right)=(1, 1, 2)$ (given)

$\therefore\quad a=3, b=3$ and $c=6$

So, equation of plane is $\frac{x}{3}+\frac{y}{3}+\frac{z}{6}=1$ or $2x+2y+z=6$

$\therefore$ Line perpendicular to the plane is along the vector $2\hat{i}+2\hat{j}+\hat{k}$

So, equation of line through (1, 1, 2) and perpendicular to the plane is $\frac{x-1}{2}=\frac{y-1}{2}=\frac{z-2}{1}$.

36. (3) For x-intercept, $\vec{r}=a_i\hat{i}\Rightarrow \vec{n}.\hat{i}=\frac{q}{a_1}$

Similarly, $\vec{n}.\hat{j}=\frac{q}{a_2}$ and $\vec{n}.\hat{k}=\frac{q}{a_3}$

$\therefore\quad \vec{n}=(\vec{n}.\hat{i})\hat{i}+(\vec{n}.\hat{j})\hat{j}+(\vec{n}.\hat{k})\hat{k}$

or $\quad \vec{n}=\frac{q}{a_1}\hat{i}+\frac{q}{a_2}\hat{j}+\frac{q}{a_3}\hat{k}$

37. (3) Plane meets axes at $A(2, 0, 0)$, $B(0, 3, 0)$ and $C(0, 0, 6)$.

Then area of ΔABC is

$\frac{1}{2}|\overrightarrow{AB}\times\overrightarrow{AC}|$

$=\frac{1}{2}|(-2\hat{i}+3\hat{j})\times(-2\hat{i}+6\hat{k})|$

$=3\sqrt{14}$ sq units

38. (1) The plane is perpendicular to the line

$$\frac{x-a}{\cos\theta}=\frac{y+2}{\sin\theta}=\frac{z-3}{0}$$

Hence, the direction ratios of the normal of the plane are $\cos\theta$, $\sin\theta$ and 0. ...(i)

Now, the required plane passes through the z-axis. Hence, the point (0, 0, 0) lies on the plane.

From Eqs. (i) and (ii), we get equation of the plane as

$\cos\theta(x-0)+\sin\theta(y-0)+0(z-0)=0$

$\cos\theta\, x+\sin\theta\, y=0$

$x+y\tan\theta=0$

39. (4) Equation of plane is $\begin{vmatrix}x-1 & y-1 & z+1\\ 3-1 & 0-1 & 2+1\\ -2-1 & 1-1 & 0+1\end{vmatrix}=0$

or $\quad x+11y+3z-9=0$

Since plane passes through the point (9, 0, 0), its equation can be written as

$((x-9)\hat{i}+y\hat{j}+z\hat{k})\cdot(\hat{i}+11\hat{j}+3\hat{k})=0$

Also, $(2\hat{i}-\hat{j}+3\hat{k})\times(3\hat{i}-\hat{k})=\hat{i}+11\hat{j}+3\hat{k}$

Its equation can be written as

$((2\hat{i}-\hat{j}+3\hat{k})\times(3\hat{i}-\hat{k}))\cdot(x\hat{i}+y\hat{j}+z\hat{k})=9$

40. (4) Let the plane $\vec{r}\cdot(\hat{i}-2\hat{j}+3\hat{k})=17$ divide the line joining the points $-2\hat{i}+4\hat{j}+7\hat{k}$ and $3\hat{i}-5\hat{j}+8\hat{k}$ in the ratio $t:1$ at point P.

Therefore, point P is

$$\frac{3t-2}{t+1}\hat{i}+\frac{-5t+4}{t+1}\hat{j}+\frac{8t+7}{t+1}\hat{k}$$

This lies on the given plane

$\therefore\quad \frac{3t-2}{t+1}\cdot(1)+\frac{-5t+4}{t+1}(-2)+\frac{8t+7}{t+1}(3)=17$

Solving, we get

$t=\frac{3}{10}$

41. (2) Any plane through (2, 2, 1) is

$a(x-2)+b(y-2)+c(z-1)=0$ ...(i)

It passes through (9, 3, 6) if $7a+b+5c=0$. ...(ii)

Also (i) is perpendicular to $2x+6y+6z-1=0$, we have

$2a+6b+6c=0$

$\therefore\quad a+3b+3c=0$ ...(iii)

$\therefore\quad \frac{a}{-12}=\frac{b}{-16}=\frac{c}{20}$ or $\frac{a}{3}=\frac{b}{4}=\frac{c}{-5}$

[from (ii) and (iii)]

Therefore, the required plane is $3(x-2)+4(y-2)-5(z-1)=0$ or $3x+4y-5z-9=0$.

42. (4) The given spheres are

$x^2+y^2+z^2+7x-2y-z-13=0$ ...(i)

and $x^2+y^2+z^2-3x+3y+4z-8=0$ ...(ii)

Subtracting (ii) from (i), we get

$10x-5y-5z-5=0$

or $\quad 2x-y-z=1$

43. (2) Coordinates of L and M are $(0, b, c)$ and $(a, 0, c)$, respectively. Therefore, the equation of the plane passing through (0, 0, 0), $(0, b, c)$ and $(a, 0, c)$ is

$$\begin{vmatrix}x-0 & y-0 & z-0\\ 0 & b & c\\ a & 0 & c\end{vmatrix}=0 \text{ or } \frac{x}{a}+\frac{y}{b}-\frac{z}{c}=0$$

44. (4) Required equation of plane is

$a(x-4)+b(y-3)+c(z-2)=0$,

where $a+b+2c=0$ and $a-4b+5c=0$

Solving, we have

$\frac{a}{5+8}=\frac{b}{2-5}=\frac{c}{-4-1}$

or $\quad \frac{a}{13}=\frac{b}{-3}=\frac{c}{-5}$

Therefore, the required equation of plane is

$-13x+3y+5z+33=0$

45. (1) Direction ratios of the line joining points $P(1, 2, 3)$ and $Q(-3, 4, 5)$ are $-4, 2, 2$ which are direction ratios of the normal to the plane.

Then, equation of plane is $-4x+2y+2z=k$.

Also this plane passes through the midpoint of $PQ(-1, 3, 4)$. Thus,

$-4(-1)+2(3)+2(4)=k$

or $\quad k=18$

Hence, equation of plane is $2x-y-z=-9$

Therefore, intercepts are $(-9/2)$, 9 and 9

46. (1) The required plane is $\begin{vmatrix} x-3 & y-6 & z-4 \\ 3-3 & 2-6 & 0-4 \\ 1 & 5 & 4 \end{vmatrix} = 0$

or $\begin{vmatrix} x-3 & y-z-2 & z-4 \\ 0 & 0 & -4 \\ 1 & 1 & 4 \end{vmatrix} = 0$

(Operating $C_2 \to C_2 - C_3$)

or $4(x-3-y+z+2) = 0$

or $x - y + z = 1$

47. (2) The equation of the line through the centre $\hat{j} + 2\hat{k}$ and normal to the given plane is

$\vec{r} = \hat{j} + 2\hat{k} + \lambda(\hat{i} + 2\hat{j} + 2\hat{k})$ ...(i)

This meets the plane for which

$[\hat{j} + 2\hat{k} + \lambda(\hat{i} + 2\hat{j} + 2\hat{k})] \cdot (\hat{i} + 2\hat{j} + 2\hat{k}) = 15$

or $6 + 9\lambda = 15$ or $\lambda = 1$

Putting in (i), we get

$\vec{r} = \hat{j} + 2\hat{k} + (\hat{i} + 2\hat{j} + 2\hat{k}) = \hat{i} + 3\hat{j} + 4\hat{k}$

Hence, centre is (1, 3, 4).

48. (2) Let a point $(3\lambda+1, \lambda+2, 2\lambda+3)$ of the first line also lies on the second line. Then

$\frac{3\lambda+1-3}{1} = \frac{\lambda+2-1}{2} = \frac{2\lambda+3-2}{3}$ or $\lambda = 1$

Hence, the point of intersection P of the two lines is (4, 3, 5). Equation of plane perpendicular to OP, where O is (0, 0, 0), and passing through P is $4x + 3y + 5z = 50$.

49. (2) Let the equation of the plane be $\frac{x}{a} + \frac{y}{b} + \frac{z}{c} = 1$

Plane passes through point (1, 1, 1)

$\Rightarrow \frac{1}{a} + \frac{1}{b} + \frac{1}{c} = 1$

Therefore, Volume of tetrahedron $OABC = V = \frac{1}{6}(abc)$

Now $(abc)^{1/3} \geq \frac{3}{\frac{1}{a}+\frac{1}{b}+\frac{1}{c}} \geq 3$ (G.M. ≥ H.M.)

$\Rightarrow abc \geq 27 \Rightarrow V \geq \frac{9}{2}$

50. (4) Given line is common to original and new plane. New plane contains the given line and point (4, 3, 7).

So, equation of new plane is $\begin{vmatrix} x-3 & y-1 & z-2 \\ 2 & 4 & 5 \\ 4-3 & 3-1 & 7-2 \end{vmatrix} = 0$

or $2x - y - 5 = 0$

Now original plane is perpendicular to the above plane and passes through the given line.

So, its equation is $\begin{vmatrix} x-3 & y-1 & z-2 \\ 2 & 4 & 5 \\ 2 & -1 & 0 \end{vmatrix} = 0$ or $x + 2y - 2z = 1$

51. (1) Let the foot of the perpendicular from the origin on the given plane be $P(\alpha, \beta, \gamma)$. Since the plane passes through $A(a, b, c)$,

$AP \perp OP \Rightarrow \overrightarrow{AP} \cdot \overrightarrow{OP} = 0$

$\Rightarrow [(\alpha-a)\hat{i} + (\beta-b)\hat{j} + (\gamma-c)\hat{k}] \cdot (\alpha\hat{i} + \beta\hat{j} + \gamma\hat{k}) = 0$

or $\alpha(\alpha-a) + \beta(\beta-b) + \gamma(\gamma-c) = 0$

Hence, the locus of (α, β, γ) is

$x(x-a) + y(y-b) + z(z-c) = 0$

$x^2 + y^2 + z^2 - ax - by - cz = 0$

which is a sphere of radius $\frac{1}{2}\sqrt{a^2+b^2+c^2}$

52. (3) Given equation of planes are

$x + ay + (b+c)z + d = 0$ ...(i)

$x + by + (c+a)z + d = 0$ ...(ii)

$x + y + (a+b)z + d = 0$ ...(iii)

From (i) – (ii) and (iii) – (iv) we get the same equation at plane which is $y - z = 0$. So, planes intersect in a line.

53. (1) Any plane through the line of intersection of the given planes is

$x + 2y + 3z - 4 + \lambda(4x + 3y + 2z + 1) = 0$

It passes through (0, 0, 0). Therefore,

$-4 + \lambda = 0$

or $\lambda = 4$

Therefore, the required plane is

$x + 2y + 3z - 4 + 4(4x + 3y + 2z + 1) = 0$ or $17x + 14y + 11z = 0$.

54. (2) Plane passing through the line of intersection of planes $4y + 6z = 5$ and $2x + 3y + 5z = 5$ is

$(4y + 6z - 5) + \lambda(2x + 3y + 5z - 5) = 0$

or $2\lambda x + (3\lambda+4)y + (5\lambda+6)z - 5\lambda - 5 = 0$

Clearly, for $\lambda = -3$, we get the plane $6x + 5y + 9z = 10$.

Hence, the given three planes have common line of intersection.

55. (3) Given variable plane is

$x + 3y + z - 4 + \lambda(2x - y) = 0$

This is equation of family of planes containing the line intersection of planes

$2x - y = 0$ and $x + 3y + z - 4 = 0$

Let $x = a$

$\therefore\ y = 2a$ and $z = 4 - 7a$

$\therefore\ \frac{x}{1} = \frac{y}{2} = \frac{z-4}{-7}$

56. (2) The equation of family of planes through the line of intersection of the planes $\vec{r} \cdot \vec{a} = \lambda$ and $\vec{r} \cdot \vec{b} = \mu$ is

$(\vec{r} \cdot \vec{a} - \lambda) + k(\vec{r} \cdot \vec{b} - \mu) = 0$

or $\vec{r} \cdot (\vec{a} + k\vec{b}) = \lambda + k\mu$ ...(i)

This passes through the origin, therefore

$\vec{0}(\vec{a} + k\vec{b}) = \lambda + \mu k$ or $k = \frac{-\lambda}{\mu}$

Putting the value of k in (i), we get the equation of the required plane as

$\vec{r} \cdot (\mu\vec{a} - \lambda\vec{b}) = 0$ or $\vec{r} \cdot (\lambda\vec{b} - \mu\vec{a}) = 0$

57. (3) $\hat{a} = \pm \frac{\vec{n}_1 \times \vec{n}_2}{|\vec{n}_1 \times \vec{n}_2|} = \pm \frac{2\hat{i} + 5\hat{j} + 3\hat{k}}{\sqrt{38}}$

(where $\vec{n}_1$ and $\vec{n}_2$ are vectors normal to the planes)

58. (4) Since line of intersection is perpendicular to both the planes, direction ratios of the line of intersection are given by

$$\begin{vmatrix} \hat{i} & \hat{j} & \hat{k} \\ 2 & 3 & 1 \\ 1 & 3 & 2 \end{vmatrix} = 3\hat{i} - 3\hat{j} + 3\hat{k}$$

Hence, $\cos\alpha = \dfrac{3}{\sqrt{9+9+9}} = \dfrac{1}{\sqrt{3}}$

59. (4) Given

$$P_1 : \vec{r}\cdot\vec{n}_1 - d_1 = 0,\ P_2 : \vec{r}\cdot\vec{n}_2 - d_2 = 0,\ P_3 : \vec{r}\cdot\vec{n}_3 - d_3 = 0$$

and $\vec{n}_1, \vec{n}_2, \vec{n}_3$ are non-coplanar $\Rightarrow$ they are linearly independent

$\Rightarrow$ the three planes have a unique common point A.

Now the lines $P_1 = 0$, $P_2 = 0$, $P_3 = 0$, and $P_3 = 0$, $P_1 = 0$ are the pairwise intersections of the planes, and each pairwise intersection contains the common point A.

$\therefore$ all three lines pass through A, i.e. they are **concurrent.**

60. (3) Let variable point on line L is $P(x, y, z)$

We have,

$$AP^2 = BP^2 = CP^2$$

$$\Rightarrow (x-1)^2 + (y-2)^2 + (z+1)^2 = x^2 + (y-1)^2 + (z+3)^2$$
$$= (x+1)^2 + (y+2)^2 + (z-3)^2$$

$$\Rightarrow x + y + 2z + 2 = 0 \text{ and } x + 3y - 6z - 2 = 0$$

So, line L is the line of intersection of above two planes.

$\therefore$ direction ratio of line L are given by

$$\begin{vmatrix} \hat{i} & \hat{j} & \hat{k} \\ 1 & 1 & 2 \\ 1 & 3 & -6 \end{vmatrix} = -12\hat{i} + 8\hat{j} + 2\hat{k}$$

61. (3) The equation of a plane through the line of intersection of the planes $ax + by + cz + d = 0$ and $a'x + b'y + c'z + d' = 0$ is

$$(ax + by + cz + d) + \lambda\,(a'x + b'y + c'z + d') = 0$$

or $x(a + \lambda a') + y(b + \lambda b') + z(c + \lambda c') + d + \lambda d' = 0$ ...(i)

This is parallel to the x-axis, i.e., $y = 0$, $z = 0$. Therefore,

$$1\,(a + \lambda a') + 0(b + \lambda b') + 0(c + \lambda c') = 0$$

or $\lambda = -\dfrac{a}{a'}$

Putting the value of λ in (i), we get

$$y\,(a'b - ab') + z\,(a'c - ac') + a'd - ad' = 0$$

or $(ab' - a'b)\,y + (ac' - a'c)z + ad' - a'd = 0$

62. (1) Equation of family of planes through the intersection of planes $\vec{r}\cdot\vec{n}_1 = q_1$ and $\vec{r}\cdot\vec{n}_2 = q_2$ is

$$(\vec{r}\cdot\vec{n}_1 - q_1) + \lambda(\vec{r}\cdot\vec{n}_2 - q_2) \quad \text{...(1)}$$

or $\vec{r}\cdot(\vec{n}_1 + \lambda\vec{n}_2) = q_1 + \lambda q_2$

Vector along line of intersection of planes

$\vec{r}\cdot\vec{n}_3 = q_3$ and $\vec{r}\cdot\vec{n}_4 = q_4$ is $\vec{n}_3 \times \vec{n}_4$.

If plane (1) is parallel to this vector then,

$$(\vec{n_1} + \lambda\vec{n_2})\cdot(\vec{n_3} \times \vec{n_4}) = 0$$

$$\Rightarrow \lambda = -\frac{[\vec{n_1}\ \vec{n_3}\ \vec{n_4}]}{[\vec{n_2}\ \vec{n_3}\ \vec{n_4}]}$$

Putting this value in (1), we get equation of plane as

$$(\vec{r}\cdot\vec{n_1} - q_1) - \frac{[\vec{n_1}\ \vec{n_3}\ \vec{n_4}]}{[\vec{n_2}\ \vec{n_3}\ \vec{n_4}]}(\vec{r}\cdot\vec{n_2} - q_2) = 0$$

or $[\vec{n_2}\ \vec{n_3}\ \vec{n_4}](\vec{r}\cdot\vec{n_1} - q_1) = [\vec{n_1}\ \vec{n_3}\ \vec{n_4}](\vec{r}\cdot\vec{n_2} - q_2)$

63. (1) Points A, B, C are lying on xy-plane.

So, distance of $P(\vec{p})$ from xy plane is $|\vec{p}.\hat{k}|$

64. (4) The given sphere is

$$x^2 + y^2 + z^2 + 4x - 2y - 6z - 155 = 0$$

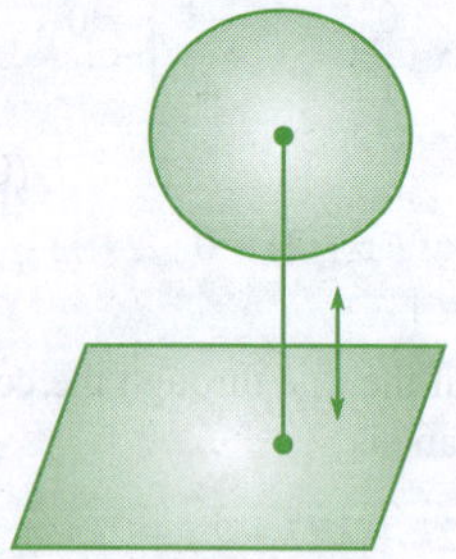

Its centre is $(-2, 1, 3)$ and

radius $= \sqrt{4+1+9+155} = \sqrt{169} = 13$

The distance of centre $(-2, 1, 3)$ from the plane $12x + 4y + 3z$ 327 is

$$\frac{|12(-2) + 4(1) + 3(3) - 327|}{\sqrt{144+16+9}} = 26$$

Hence, the shortest distance is 13.

65. (1) $\overrightarrow{AB} = \vec{\beta} - \vec{\alpha} = -2\hat{i} - 3\hat{j} - 6\hat{k}$

Equation of the plane passing through B and perpendicular to A is

$$(\vec{r} - \overrightarrow{OB})\cdot\overrightarrow{AB} = 0$$

or $\vec{r}\,.\,(2\hat{i} + 3\hat{j} + 6\hat{k}) + 28 = 0$

Hence, the distance of point $-\hat{i} + \hat{j} + \hat{k}$ from the plane

$$= \left|\frac{(-\hat{i} + \hat{j} + \hat{k})\cdot(2\hat{i} + 3\hat{j} + 6\hat{k}) + 28}{|2\hat{i} + 3\hat{j} + 6\hat{k}|}\right|$$

$$= \left|\frac{-2+3+6+28}{7}\right|$$

$= 5$ units

66. (2)

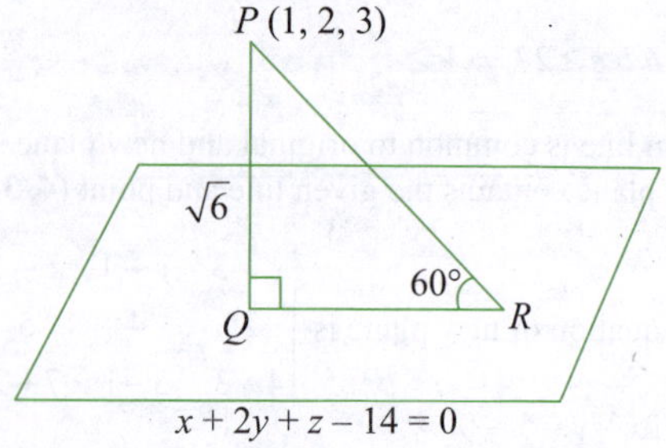

From the figure,

$$PQ = \left|\frac{1+4+3-14}{\sqrt{6}}\right| = \sqrt{6}$$

$\therefore$ $QR = (PQ)\cot 60° = \sqrt{2}$

So, area of $\Delta PQR = \dfrac{1}{2}(PQ)\,(QR) = \sqrt{3}$

67. (1) $A\,(1, 1, 1)$, $B\,(2, 3, 5)$, $C\,(-1, 0, 2)$ direction ratios of AB a $< 1, 2, 4 >$.

Direction ratios of AC are <–2, –1, 1>.
Therefore, direction ratios of normal to plane ABC are < 2, –3, 1 >
As a result, equation of the plane ABC is $2x - 3y + z = 0$.
Let the equation of the required plane be $2x - 3y + z = k$. Then

$$\left|\frac{k}{\sqrt{4+9+1}}\right| = 2 \quad \text{or} \quad k = \pm 2\sqrt{14}$$

Hence, equation of the required plane is $2x - 3y + z + 2\sqrt{14} = 0$

68. (3) The plane containing the given line is

$$(2x + 3y + 5z + 1) + \lambda(3x + 4y + 6z + 2) = 0$$

or $(2 + 3\lambda)x + (3 + 4\lambda)y + (5 + 6\lambda)z + (1 + 2\lambda) = 0$

This plane must be parallel to y-axis

$\therefore \quad 3 + 4\lambda = 0$ or $\lambda = -\dfrac{3}{4}$

A point on y-axis is the origin.
So, perpendicular distance of the origin from the plane $x - 2z + 2 = 0$ is $\dfrac{2}{\sqrt{5}}$.

$\therefore \quad k = 5$

69. (4) Let $P(\alpha, \beta, \gamma)$ be the image of the point $Q(-1, 3, 4)$.
Midpoint of PQ lies on $x - 2y = 0$. Then,

$$\frac{\alpha - 1}{2} - 2\left(\frac{\beta + 3}{2}\right) = 0$$

or $\alpha - 1 - 2\beta - 6 = 0$ or $\alpha - 2\beta = 7$ ...(i)

Also PQ is perpendicular to the plane. Then,

$$\frac{\alpha + 1}{1} = \frac{\beta - 3}{-2} = \frac{\gamma - 4}{0} \quad \text{...(ii)}$$

Solving (i) and (ii), we get

$$\alpha = \frac{9}{5}, \beta = -\frac{13}{5}, \gamma = 4$$

Therefore, image is

$$\left(\frac{9}{5}, -\frac{13}{5}, 4\right)$$

Alternative method:
For image,

$$\frac{\alpha - (-1)}{1} = \frac{\beta - 3}{-2} = \frac{\gamma - 4}{0} = \frac{-2(-1 - 2(3))}{(1)^2 + (-2)^2}$$

or $\alpha = \dfrac{9}{5}, \beta = -\dfrac{13}{5}, \gamma = 4$

70. (2) Centre of the sphere is $(-1, 1, 2)$ and its radius $= \sqrt{1 + 1 + 4 + 19} = 5$

perpendicular distance of centre from the plane, is

$$\left|\frac{-1 + 2 + 4 + 7}{\sqrt{1 + 4 + 4}}\right| = 4$$

Hence, radius of the circle $= \sqrt{5^2 - 4^2} = 3$

71. (3) The given plane passes through $\vec{a}$ and is parallel to the vectors $\vec{b} - \vec{a}$ and $\vec{c}$. So it is normal to the vector $(\vec{b} - \vec{a}) \times \vec{c}$. Hence, equation of plane is

$$(\vec{r} - \vec{a}) \cdot ((\vec{b} - \vec{a}) \times \vec{c}) = 0$$

or $\vec{r} \cdot (\vec{b} \times \vec{c} + \vec{c} \times \vec{a}) = [\vec{a}\ \vec{b}\ \vec{c}]$

The length of the perpendicular from the origin to this plane is

$$\frac{\left|[\vec{a}\ \vec{b}\ \vec{c}]\right|}{|\vec{b} \times \vec{c} + \vec{c} \times \vec{a}|}$$

72. (3) Let $Q(\vec{q})$ be the foot of altitude drawn from P on the plane $\vec{r} \cdot \vec{n} = 0$. Then

$$\vec{q} - \vec{p} = \lambda \vec{n} \quad \text{or} \quad \vec{q} = \vec{p} + \lambda \vec{n}$$

Also $\vec{q} \cdot \vec{n} = 0 \Rightarrow (\vec{p} + \lambda \vec{n}) \cdot \vec{n} = 0$

or $\lambda = -\dfrac{\vec{p} \cdot \vec{n}}{|\vec{n}|^2}$. Therefore $\vec{q} - \vec{p} = -\dfrac{(\vec{p} \cdot \vec{n})}{|\vec{n}|^2}\vec{n}$

Thus, required distance

$$= |\vec{q} - \vec{p}| = \frac{|\vec{p} \cdot \vec{n}|}{|\vec{n}|} = |\vec{p} \cdot \hat{n}|$$

73. (2) Given plane is $\vec{r} \cdot \vec{n} = q$ ...(i)

Let the image of $A(\vec{a})$ in the plane be $B(\vec{b})$.

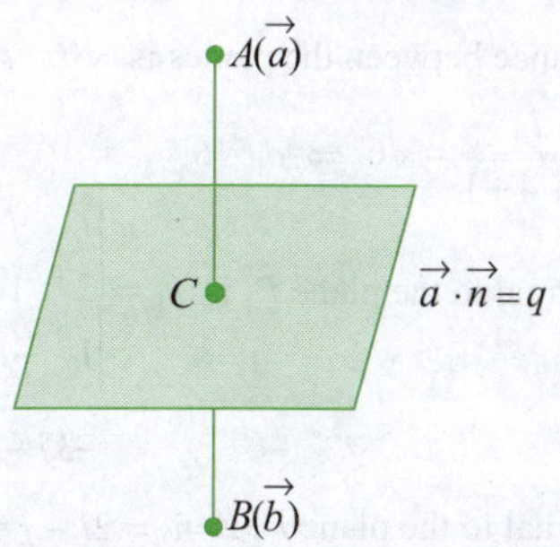

Equation of AC is $\vec{r} = \vec{a} + \lambda \vec{n}$

($\because AC$ is normal to the plane) ...(ii)

Solving (i) and (ii), we get

$$(\vec{a} + \lambda \vec{n}) \cdot \vec{n} = q$$

or $\lambda = \dfrac{q - \vec{a} \cdot \vec{n}}{|\vec{n}|^2}$

$\therefore \quad \overrightarrow{OC} = \vec{a} + \dfrac{(q - \vec{a} \cdot \vec{n})}{|\vec{n}|^2} \cdot \vec{n}$

But $\overrightarrow{OC} = \dfrac{\vec{a} + \vec{b}}{2}$

$\therefore \quad \vec{a} + \dfrac{(q - \vec{a} \cdot \vec{n})\vec{n}}{|\vec{n}|^2} = \dfrac{\vec{a} + \vec{b}}{2}$

or $\vec{b} = \vec{a} + 2\left(\dfrac{q - \vec{a} \cdot \vec{n}}{|\vec{n}|^2}\right)\vec{n}$

74. (2) We have $\vec{s} - \vec{p} = \lambda \vec{n}$ and $\vec{s} \cdot \vec{n} = q$. Thus,

$$(\lambda \vec{n} + \vec{p}) \cdot \vec{n} = q$$

or $\lambda = \dfrac{q - \vec{p}\cdot\vec{n}}{|\vec{n}|^2}$

$\Rightarrow \vec{s} = \vec{p} + \dfrac{(q - \vec{p}\cdot\vec{n})\vec{n}}{|\vec{n}|^2}$

75. (1) Foot of the perpendicular from point $A(\vec{a})$ on the plane $\vec{r}\cdot\vec{n} = d$ is $\vec{a} + \dfrac{\left(d - \vec{a}\cdot\vec{n}\right)}{|\vec{n}|^2}\vec{n}$

Therefore, equation of the line parallel to $\vec{r} = \vec{a} + \lambda\vec{b}$ in the plane $\vec{r}\,.\,\vec{n} = d$ is given by

$$\vec{r} = \vec{a} + \frac{\left(d - \vec{a}\,.\,\vec{n}\right)}{|\vec{n}|^2}\vec{n} + \lambda\vec{b}$$

76. (4) Equation of the plane containing the given lines is

$$\begin{vmatrix} x-1 & y-2 & z-3 \\ 2 & 3 & 4 \\ 3 & 4 & 5 \end{vmatrix} = 0 \text{ or } x - 2y + z = 0$$

Since this plane must be parallel to the plane $\alpha x - 2y + z = d$, we have $\alpha = 1$

Now distance between the planes is $\sqrt{6}$.

$\therefore \dfrac{|d|}{\sqrt{1+4+1}} = \sqrt{6} \Rightarrow |d| = 6$

77. (2) Vector normal to the plane P_1 is $\vec{n}_1 = \begin{vmatrix} \hat{i} & \hat{j} & \hat{k} \\ 2 & 1 & -1 \\ 1 & -1 & 1 \end{vmatrix}$

$= -3\hat{j} - 3\hat{k}$

Vector normal to the plane P_2 is $\vec{n}_2 = 2\hat{i} - \hat{j} + 2\hat{k}$

So, angle between planes, $\theta = \cos^{-1}\dfrac{|3-6|}{\sqrt{18}.\sqrt{9}} = \cos^{-1}\dfrac{1}{3\sqrt{2}}$

78. (4) Vector perpendicular to the face OAB is

$\overrightarrow{OA} \times \overrightarrow{OB} = (\hat{i} + 2\hat{j} + \hat{k}) \times (2\hat{i} + \hat{j} + 3\hat{k})$

$= 5\hat{i} - \hat{j} - 3\hat{k}$

Vector perpendicular to face ABC is

$\overrightarrow{AB} \times \overrightarrow{AC} = (\hat{i} - \hat{j} + 2\hat{k}) \times (-2\hat{i} - \hat{j} + \hat{k})$

$\therefore = \hat{i} - 5\hat{j} - 3\hat{k}$

The angle between the face OAB and ABC is given by

$\cos\theta = \dfrac{5+5+9}{\sqrt{35}\sqrt{35}} = \dfrac{19}{35}$ or $\theta = \cos^{-1}\left(\dfrac{19}{35}\right)$

79. (2) We have $\cos^{-1}\dfrac{|1 - 2\alpha + 6|}{\sqrt{14}\,.\sqrt{\alpha^2 + 5}} = \cos^{-1}\dfrac{1}{14}$

$\Rightarrow \dfrac{|2\alpha - 7|}{\sqrt{\alpha^2 + 5}} = \dfrac{1}{\sqrt{14}}$

$\Rightarrow 14(4\alpha^2 - 28\alpha + 49) = \alpha^2 + 5$

$\Rightarrow 55\alpha^2 - 392\alpha + 681 = 0$

$\Rightarrow 55\alpha^2 - 165\alpha - 227\alpha + 681 = 0$

$\Rightarrow (55\alpha - 227)(\alpha - 3) = 0$

$\Rightarrow \alpha = \dfrac{227}{55}, 3$

80. (1) The equation of the plane through the line of intersection of the planes $4x + 7y + 4z + 81 = 0$ and $5x + 3y + 10z = 25$ is

$(4x + 7y + 4z + 81) + \lambda(5x + 3y + 10z - 25) = 0$

or $(4 + 5\lambda)x + (7 + 3\lambda)y + (4 + 10\lambda)z + 81 - 25\lambda = 0$ ...(i)

This plane is perpendicular to the plane $4x + 7y + 4z + 81 = 0$

$\Rightarrow 4(4 + 5\lambda) + 7(7 + 3\lambda) + 4(4 + 10\lambda) = 0$

or $81\lambda + 81 = 0$

or $\lambda = -1$

Hence, the plane is $x - 4y + 6z = 106$

81. (2) Any plane through (1, 0, 0) is

$a(x - 1) + by + cz = 0$ ...(i)

It passes through (0, 1, 0). Therefore,

$a(0 - 1) + b(1) + c(0) = 0$ or $-a + b = 0$ ...(ii)

(i) makes an angle of $\dfrac{\pi}{4}$ with $x + y = 3$, therefore

$\cos\dfrac{\pi}{4} = \dfrac{a(1) + b(1) + c(0)}{\sqrt{a^2 + b^2 + c^2}\sqrt{1+1+0}}$

or $\dfrac{1}{\sqrt{2}} = \dfrac{a+b}{\sqrt{2}\sqrt{a^2 + b^2 + c^2}}$

or $a + b = \sqrt{a^2 + b^2 + c^2}$

Squaring, we get

$a^2 + b^2 + 2ab = a^2 + b^2 + c^2$

or $2ab = c^2$ or $2a^2 = c^2$ [using (ii)]

or $c = \sqrt{2}\,a$

Hence, $a : b : c = a : a : \sqrt{2}\,a$

$= 1 : 1 : \sqrt{2}$

82. (3) We have $1(1) + 2(-1) + 3(1) = 2 > 0$

So, point $(2, 3, \alpha)$ will in the acute angle region if

$(2 + 6 + 3\alpha - 5)(2 - 3 + \alpha - 7) < 0$

$\therefore (3\alpha + 3)(\alpha - 8) < 0$

$\Rightarrow -1 < \alpha < 8$

83. (4) We have planes,

$P_1 : x + 2y + z + 18 = 0$

$P_2 : x - 2y + 2z - 5 = 0$

and $P_3 : ax + by + cz + d = 0$

Equation of family planes through line of intersection of planes P_1 and P_2 is

$x + 2y + z + 18 + \lambda(x - 2y + 2z - 5) = 0, \lambda \in R$

or $P : (1 + \lambda)x + (2 - 2\lambda)y + (1 + 2\lambda)z + 18 - 5\lambda = 0$

Plane P_1 makes equal angle with planes P_2 and P.

$\therefore \dfrac{|1(1) + 2(-2) + 1(2)|}{\sqrt{6}\cdot\sqrt{9}} = \dfrac{|1(1 + \lambda) + 2(2 - 2\lambda) + 1(1 + 2\lambda)|}{\sqrt{6}\sqrt{9\lambda^2 - 2\lambda + 6}}$

$\Rightarrow \dfrac{1}{3} = \dfrac{|-\lambda + 6|}{\sqrt{9\lambda^2 - 2\lambda + 6}}$

$\Rightarrow 9\lambda^2 - 2\lambda + 6 = 9\lambda^2 - 108\lambda + 324$

$\Rightarrow 106\lambda = 318$

$\Rightarrow \lambda = 3$

So, the required equation of plane is

$4x - 4y + 7z + 3 = 0$

84. (1) We have $4(2) - 2(3) - 1(2) = 0$
Also, point $(-3, 4, -5)$ does not lie on the plane.
Therefore, the line is parallel to the plane.

85. (1) Vector $\left((3\hat{i} - 2\hat{j} + \hat{k}) \times (4\hat{i} - 3\hat{j} + 4\hat{k})\right)$ is perpendicular to $2\hat{i} - \hat{j} + m\hat{k}$. Thus,

$$\begin{vmatrix} 3 & -2 & 1 \\ 4 & -3 & 4 \\ 2 & -1 & m \end{vmatrix} = 0 \quad \text{or} \quad m = -2$$

86. (1) Since line is parallel to the plane, vector $2\hat{i} + 3\hat{j} + \lambda\hat{k}$ is perpendicular to the normal to the plane $2\hat{i} + 3\hat{j} + 4\hat{k}$

$\Rightarrow \quad 2 \times 2 + 3 \times 3 + 4\lambda = 0$

or $\quad \lambda = -\dfrac{13}{4}$

87. (1) Clearly, the given line and plane are parallel.
Given point on the line is $A(2, -2, 3)$.
Equation of plane is $x + 5y + z - 5 = 0$.
So, the required distance
= distance of point A from the plane

$$= \frac{|2 - 10 + 3 - 5|}{\sqrt{1 + 25 + 1}} = \frac{10}{3\sqrt{3}}$$

88. (2) The line is $\dfrac{x+1}{1} = \dfrac{y-1}{2} = \dfrac{z-2}{2}$ and the plane is $2x - y + \sqrt{\lambda}\, z + 4 = 0$.
If θ is the angle between the line and the plane, then $90° - \theta$ is the angle between the line and normal to the plane. Thus,

$$\cos(90° - \theta) = \frac{(1)(2) + (2)(-1) + (2)(\sqrt{\lambda})}{\sqrt{1+4+4}\,\sqrt{4+1+\lambda}}$$

$\Rightarrow \quad \sin\theta = \dfrac{2 - 2 + 2\sqrt{\lambda}}{3\sqrt{5+\lambda}}$ or $\dfrac{1}{3} = \dfrac{2\sqrt{\lambda}}{3\sqrt{5+\lambda}}$

$\Rightarrow \quad \sqrt{5+\lambda} = 2\sqrt{\lambda}$

$\Rightarrow \quad 5 + \lambda = 4\lambda$

$\Rightarrow \quad 3\lambda = 5$

$\Rightarrow \quad \lambda = \dfrac{5}{3}$

89. (3) Angle between planes P_1 and P_2 is θ.

So, $\quad \cos\theta = \dfrac{(\hat{i} + \hat{j} + 2\hat{k}) \cdot (2\hat{i} - \hat{j} + \hat{k})}{\sqrt{6}\sqrt{6}} = \dfrac{2 - 1 + 2}{6} = \dfrac{1}{2}$

$\therefore \quad \theta = 30°.$

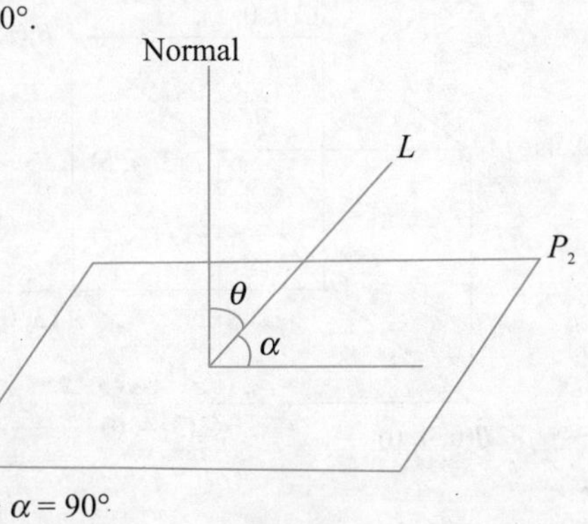

Clearly, $\theta + \alpha = 90°$

$\therefore \quad (\tan^2\theta)(\cot^2\alpha) = (\tan^2 60°)(\cot^2 3°) = 3 \times 3 = 9$

90. (2) Let direction ratios of the line be (a, b, c); then
$2a - b + c = 0$
and $a - b - 2c = 0$, i.e., $\dfrac{a}{3} = \dfrac{b}{5} = \dfrac{c}{-1}$

Therefore, direction ratios of the line are $(3, 5, -1)$.
Any point on the given line is $(2 + \lambda, 2 - \lambda, 3 - 2\lambda)$; it lies on the given plane π if

$2(2 + \lambda) - (2 - \lambda) + (3 - 2\lambda) = 4$

or $\quad 4 + 2\lambda - 2 + 1 + 3 - 2\lambda = 4$ or $\lambda = -1$

Therefore, the point of intersection of the line and the plane is $(1, 3, 5)$.
Therefore, equation of the required line is

$$\frac{x-1}{3} = \frac{y-3}{5} = \frac{z-5}{-1}$$

91. (1) Equation of the plane through $(-1, 0, 1)$ is

$a(x + 1) + b(y - 0) + c(z - 1) = 0$ ...(i)

which is parallel to the given line and perpendicular to the given plane

$\therefore \quad -a + 2b + 3c = 0$ ...(ii)

and $a - 2b + c = 0$ ...(iii)

From Eqs. (ii) and (iii), we get

$c = 0, a = 2b$

From Eq. (i), $2b(x + 1) + by = 0$

$\Rightarrow \quad 2x + y + 2 = 0$

92. (4)

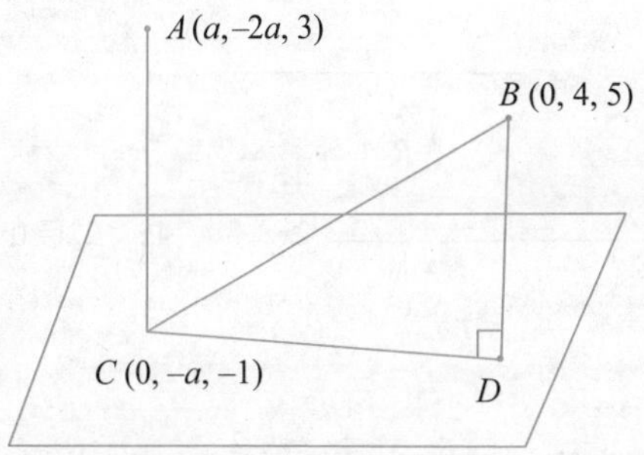

Clearly, $\overline{CA}$ is normal to the plane.
So, equation of plane is $ax - ay + 4z = 0$.
Point C lies on the plane.
So, $0 + a^2 - 4 = 0$

$\therefore \quad a = 2$ (as $a > 0$)

So, equation of plane is $2x - 2y + 4z = 0$ or $x - y + 2z = 0$

If $D \equiv (p, q, r)$ then $\dfrac{p-0}{1} = \dfrac{q-4}{-1} = \dfrac{r-5}{2} = \dfrac{-(0-4+10)}{1+1+4}$

So, $D \equiv (-1, 5, 3)$
Also, $C \equiv (0, -2, -1)$

$\therefore \quad CD = \sqrt{66}$

93. (2)

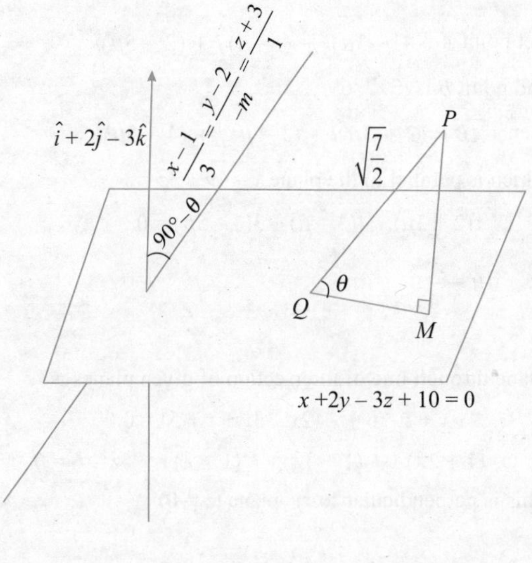

Distance of point P from the plane is $PM = \dfrac{|1-4-9+10|}{\sqrt{1+4+9}} = \dfrac{2}{\sqrt{14}}$

Also, $PQ = \sqrt{\dfrac{7}{2}} = \dfrac{7}{\sqrt{14}}$

So, $\sin\theta = \dfrac{2}{7} = \dfrac{|3-2m-3|}{\sqrt{14}\sqrt{10+m^2}}$

$\Rightarrow \quad |m| = 2$

Numerical Value Type

1. (2.25)

The point where the lin meets the z-axis, $x = y = 0$.

$\therefore \quad \dfrac{2(0)-3}{4} = \dfrac{0-\alpha}{3}$

$\Rightarrow \quad \alpha = \dfrac{9}{4} = 2.25$

2. (1.2)

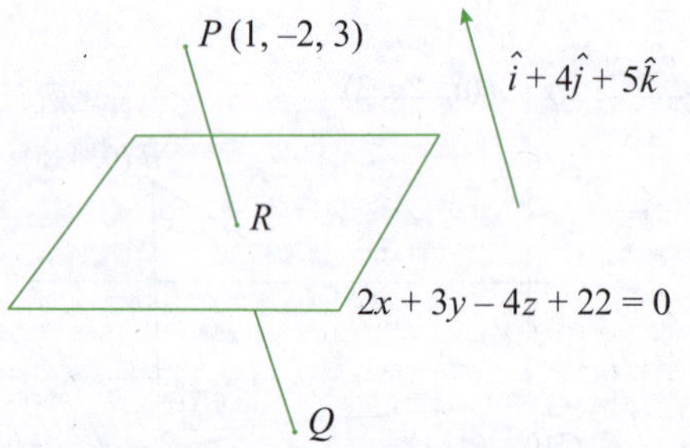

Equation of line PQ is $\dfrac{x-1}{1} = \dfrac{y+2}{4} = \dfrac{z-3}{5}$

Let $R(\lambda+1, 4\lambda-2, 5\lambda+3)$

R lies on the plane,

$$2(\lambda+1)+3(4\lambda-2)-4(5\lambda+3)+22=0$$

$\Rightarrow \quad -6\lambda+6=0$

$\Rightarrow \quad \lambda = 1$

$\therefore \quad R \equiv (2, 2, 8)$

Hence, $Q \equiv (3, 6, 13)$

3. (0.25)

Let point A is $(1-3\mu)\hat{i} + (\mu-1)\hat{j} + (2+5\mu)\hat{k}$

and point B is (3, 2, 6)

then $\quad \overrightarrow{AB} = (2+3\mu)\hat{i} + (3-\mu)\hat{j} + (4-5\mu)\hat{k}$

which is parallel to the plane $x - 4y + 3z = 1$

$\therefore \quad 1(2+3\mu) - 4(3-\mu) + 3(4-5\mu) = 0$

$\Rightarrow \quad \mu = \dfrac{1}{4}.$

4. (24.2)

Plane through line of intersection of given planes is

$$x + y + z - 6 + \lambda(2x + 3y + z + 5) = 0$$

or $\quad (1+2\lambda)x + (1+3\lambda)y + (1+\lambda)z + 5\lambda - 6 = 0$

This is perpendicular to xy–plane ($z = 0$)

$\therefore \quad 1 + \lambda = 0 \Rightarrow \lambda = -1$

$\therefore \quad$ So, equation of plane P is $x + 2y + 11 = 0$

Distance of plane from point (0, 0, 256) is $= \dfrac{0+0+11}{\sqrt{1+4}} = \dfrac{11}{\sqrt{5}}$

5. (1.75)

Midpint of PQ must lie on the given line.

$\therefore \quad \dfrac{\frac{20+a}{2}-3}{7} = \dfrac{\frac{b+6}{2}-2}{5} = \dfrac{-\frac{a}{2}-1}{-9}$

$\Rightarrow \quad \dfrac{14+a}{7} = \dfrac{b+2}{5} = \dfrac{a+2}{9}$

$\Rightarrow \quad a = -56$ and $b = -32$

$\Rightarrow \quad \dfrac{a}{b} = \dfrac{56}{32} = \dfrac{7}{4} = 1.75$

6. (5.5)

Line of shortest distance between lines $\vec{r} = (\hat{i} - \hat{j}) + \lambda(2\hat{i} + \hat{k})$

and $\vec{r} = (2\hat{i} - \hat{j}) + \mu(\hat{i} - \hat{j} + \hat{k})$ is along the vector $\begin{vmatrix} \hat{i} & \hat{j} & \hat{k} \\ 2 & 0 & 1 \\ 1 & -1 & 1 \end{vmatrix}$

$= \hat{i} - \hat{j} - 2\hat{k}$

Line through the point $P(-1, 2, 3)$ and parallel to the above vector is

$$\frac{x+1}{1} = \frac{y-2}{-1} = \frac{z-3}{-2} = \lambda$$

Any point on this line is $Q(\lambda-1, 2-\lambda, 3-2\lambda)$

If this point lies on the plane $\vec{r}.(\hat{i} - 2\hat{j} + 3\hat{k}) = 10$, then

$\Rightarrow \quad (\lambda-1) - 2(2-\lambda) + 3(3-2\lambda) = 10$

$\Rightarrow \quad \lambda = -2$

$\therefore \quad Q \equiv (-3, 4, 7)$

$\Rightarrow \quad PQ = \alpha = \sqrt{4+4+16} = 2\sqrt{6}$

$\Rightarrow \quad \dfrac{132}{\alpha^2} = \dfrac{132}{24} = 5.5$

7. (2.4)

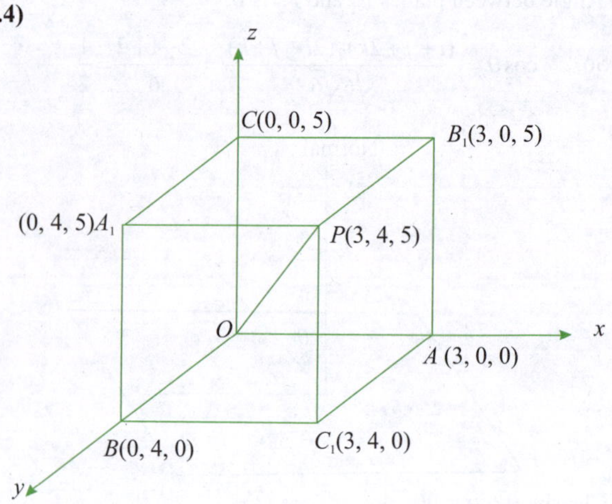

Equation of OP is $\dfrac{x}{3} = \dfrac{y}{4} = \dfrac{z}{5} : L_1$

Equation of AB_1 is $\dfrac{x-3}{0} = \dfrac{y}{0} = \dfrac{z}{1} : L_2$

Shortest distance between L_1 and L_2 is

$$\frac{\begin{vmatrix} 3-0 & 0 & 0 \\ 3 & 4 & 5 \\ 0 & 0 & 1 \end{vmatrix}}{\left\|\begin{vmatrix} \hat{i} & \hat{j} & \hat{k} \\ 3 & 4 & 5 \\ 0 & 0 & 1 \end{vmatrix}\right\|} = \frac{12}{|4\hat{i}-3\hat{j}|} = \frac{12}{5}$$

8. (0.25)

Plane through line of intersection of planes $x + 2y + az = 2$ and $x - y + z = 3$ is

$$(x + 2y + az - 2) + \lambda(x - y + z - 3) = 0 \quad \ldots(1)$$

$$5x - 11y + bz = 6a - 1 \quad \ldots(2)$$

Comparing (1) and (2), we get

$$\frac{1+\lambda}{5} = \frac{2-\lambda}{-11} = \frac{a+\lambda}{b} = \frac{2+3\lambda}{6a-1}$$

Solving we get $\lambda = -\frac{7}{2}$, $a = 3$, $b = 1$

Distance of point $(a, -c, c)$ from the plane (2) is

$$\left|\frac{5a + 11c + bc - 6a + 1}{\sqrt{25+121+1}}\right| = \frac{2}{\sqrt{a}} \text{ (given)}$$

$\therefore \quad c = -1$

So, $\left|\frac{c}{a+b}\right| = \left|\frac{-1}{3+1}\right| = \frac{1}{4}$

9. (2.1)

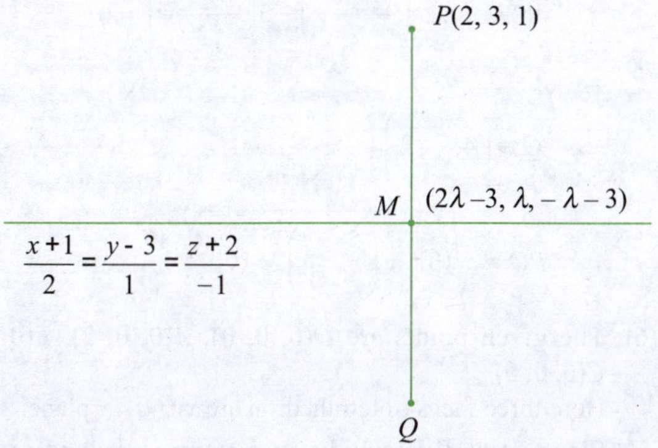

$$\overline{PM} = (2\lambda - 3, \lambda, -\lambda - 3)$$

$$\overline{PM} \perp (2\hat{i} + \hat{j} - \hat{k})$$

$\therefore \quad 4\lambda - 6 + \lambda + \lambda + 3 = 0 \Rightarrow \lambda = \frac{1}{2}$

$\therefore \quad M \equiv \left(0, \frac{7}{2}, \frac{-5}{2}\right)$

So, reflection, $Q \equiv (-2, 4, -6)$

Therefore, equation of plane passing through Q and containing line $\frac{x-2}{3} = \frac{1-y}{2} = \frac{z+1}{1}$ is

$$\begin{vmatrix} x-2 & y-1 & z+1 \\ 3 & -2 & 1 \\ -2-2 & 4-1 & -6-(-1) \end{vmatrix} = 0$$

or $\quad 7x + 11y + z = 24$

$\therefore \quad \alpha = 7, \beta = 11, \gamma = 1$

Hence, $\frac{3\alpha}{\beta - \gamma} = \frac{21}{11-1} = 2.1$

Integer Answer Type

1. (1) If image of point (2, – 3, 3) in the plane x – 2y – z + 1 = 0 is (a, b, c), then

$$\frac{a-2}{1} = \frac{b+3}{-2} = \frac{c-3}{-1} = \frac{-2(2-2(-3)-3+1)}{(1)^2+(-2)^2+(-1)^2} = -2$$

Hence, the image is (0, 1, 5)

Obviously distance of image of the point from the z-axis is 1.

2. (2) Vector normal to the plane is $\vec{n} = \hat{i} - 3\hat{j} + 2\hat{k}$ and vector along the line is $\vec{b} = 2\hat{i} + \hat{j} - 3\hat{k}$

Now, $\sin\theta = \frac{|\vec{b}\cdot\vec{n}|}{|\vec{b}|\,|\vec{n}|} = \left|\frac{2-3-6}{\sqrt{14}\sqrt{14}}\right| = \left|\frac{7}{14}\right|$

Hence, $\operatorname{cosec}\theta = 2$

3. (7) Clearly minimum value of $a^2 + b^2 + c^2$

= distance of origin from the plane

$$= \left(\frac{|(3(0) + 2(0) + (0) - 7|}{\sqrt{(3)^2+(2)^2+(1)^2}}\right)^2 = \frac{49}{14} = \frac{7}{2} \text{ units}$$

4. (53) $4x + 7y + 4z + 81 = 0 \quad \ldots$(i)

$5x + 3y + 10z = 25 \quad \ldots$(ii)

Equation of plane passing through their line of intersection is

$$(4x + 7y + 4z + 81) + \lambda(5x + 3y + 10z - 25) = 0$$

or $\quad (4 + 5\lambda)x + (7 + 3\lambda)y + (4 + 10\lambda)z + 81 - 25\lambda = 0 \quad \ldots$(iii)

Plane (iii) $\perp$ to (i), so

$$4(4 + 5\lambda) + 7(7 + 3\lambda) + 4(4 + 10\lambda) = 0$$

$\therefore \quad \lambda = -1$

From (iii), equation of plane is

$$-x + 4y - 6z + 106 = 0 \quad \ldots\text{(iv)}$$

Distance of (iv) from (0,0,0) $= \frac{106}{\sqrt{1+16+36}} = \frac{106}{\sqrt{53}}$

5. (7) As the line $\frac{x-4}{1} = \frac{y-2}{1} = \frac{z-k}{2}$ lies in the plane $2x - 4y + z = 7$, the point $(4, 2, k)$ through which it passes must also lie on the given plane, and hence $2 \times 4 - 4 \times 2 + k = 7$ or $k = 7$.

6. (8) The equation of the plane passing through the point (1, – 2, 1) and perpendicular to the planes $2x - 2y + z = 0$ and $x - y + 2z = 4$ is given by $\begin{vmatrix} x-1 & y+2 & z-1 \\ 2 & -2 & 1 \\ 1 & -1 & 2 \end{vmatrix} = 0$

or $\quad x + y + 1 = 0$

Its distance from the point (1, 2, 2) is

$$\left|\frac{1+2+1}{\sqrt{2}}\right| = 2\sqrt{2}$$

7. (5) Vector normal to the plane $2x + y + z - 3 = 0$ is $\vec{n}_1 = 2\hat{i} + \hat{j} + \hat{k}$

Vector normal to the plane $x - y + 2z + 5 = 0$ is $\vec{n}_2 = \hat{i} - \hat{j} + 2\hat{k}$

So, vector along line L, contained by the planes is $(\vec{n}_1 \times \vec{n}_2)$

Now, line L is parallel to the plane $\alpha x + 2y + 3z + 1 = 0$.

So, $(\vec{n}_1 \times \vec{n}_2)\cdot\vec{n}_3 = 0$, where $\vec{n}_3 = \alpha\hat{i} + 2\hat{j} + 3\hat{k}$

$$\begin{vmatrix} 2 & 1 & 1 \\ 1 & -1 & 2 \\ \alpha & 2 & 3 \end{vmatrix} = 0$$

$\Rightarrow \quad \alpha = 5$

8. (5) Plane P passes through (1, 5, 3) and perpendicular to $\vec{n} = 2\hat{i} + 3\hat{j} + 6\hat{k}$

So, equation of plane is $2(x-1) + 3(y-5) + 6(z-3) = 0$ or $2x + 3y + 6z = 35$.

Distance of plane from (0, 0, 0) $= \left|\dfrac{-35}{7}\right| = 5$

9. (3) We have

$A \equiv (3a + 7, 2a + 5, a + 3)$

$B \equiv (2b + 1, 4b - 1, 3b - 1)$

Direction ratios of AB are $\langle 3a - 2b + 6, 2a - 4b + 6, a - 3b + 4\rangle$

Now AB is parallel to line L whose direction ratios are 2, 2, and 1.

So, $\dfrac{3a-2b+6}{2} = \dfrac{2a-4b+6}{2} = \dfrac{a-3b+4}{1}$

$\Rightarrow \quad 3a - 2b + 6 = 2a - 4b + 6$ and $2a - 4b + 6 = 2a - 6b + 8$

$\Rightarrow \quad a + 2b = 0$ and $2b = 2$

$\Rightarrow \quad a = -2, b = 1$

So, $A \equiv (1, 1, 1)$ and $B \equiv (3, 3, 2)$

Therefore, $AB = \sqrt{2^2 + 2^2 + 1^2} = 3$

10. (6) Lines will lie in the plane parallel to the plane $x - 2y + 2z - 3 = 0$.

Equation of plane parallel to this plane and passing through the point (2, –2, 6) is $x - 2y + 2z - 18 = 0$.

Distance of this plane form origin is 6 units.

11. (474) Let $A(1, 0, -1)$, $B(-1, 2, 2)$

Direction ratios of AB are (2, – 2, – 3)

Let θ be the angle between the line and normal to plane; then

$$\cos\theta = \frac{|1(2) + 3(-2) - 5(-3)|}{\sqrt{1+9+25}\sqrt{4+4+9}}$$

$$= \frac{11}{\sqrt{17}\sqrt{35}} = \frac{11}{\sqrt{595}}$$

$\therefore$ Length of projection

$= (AB)\sin\theta$

$= \sqrt{(2)^2 + (2)^2 + (3)^2} \times \sqrt{1 - \dfrac{121}{595}}$

$= \sqrt{\dfrac{474}{35}}$ units

12. (6) A plane containing the line of intersection of the given planes is

$x - y - z - 4 + \lambda(x + y + 2z - 4) = 0$

i.e., $(\lambda + 1)x + (\lambda - 1)y + (2\lambda - 1)z - 4(\lambda + 1) = 0$

Vector normal to this plane,

$$\vec{V} = (\lambda + 1)\hat{i} + (\lambda - 1)\hat{j} + (2\lambda - 1)\hat{k} \quad \ldots\text{(i)}$$

Now the vector along the line of intersection of the planes $2x + 3y + z - 1 = 0$ and $x + 3y + 2z - 2 = 0$ is given by

$$\vec{n} = \begin{vmatrix} \hat{i} & \hat{j} & \hat{k} \\ 2 & 3 & 1 \\ 1 & 3 & 2 \end{vmatrix} = 3(\hat{i} - \hat{j} + \hat{k})$$

As $\vec{n}$ is parallel to the plane (i), we have

$\vec{n}\cdot\vec{V} = 0$

$(\lambda + 1) - (\lambda - 1) + (2\lambda - 1) = 0$

$2 + 2\lambda - 1 = 0$ or $\lambda = \dfrac{-1}{2}$

Hence, the required plane is

$\dfrac{x}{2} - \dfrac{3y}{2} - 2z - 2 = 0$

$x - 3y - 4z - 4 = 0$

Hence, $|A + B + C| = 6$

13. (17) Given line is $\dfrac{x+2}{3} = \dfrac{2y+3}{4} = \dfrac{3z+4}{5}$

or $\dfrac{x+2}{3} = \dfrac{y+\frac{3}{2}}{2} = \dfrac{z+\frac{4}{3}}{\frac{5}{3}} = \lambda$

or $\dfrac{x+2}{9} = \dfrac{y+\frac{3}{2}}{6} = \dfrac{z+\frac{4}{3}}{5} = \lambda$ (say)

Variable point on this line is

$$Q\left(3\lambda - 2, 2\lambda - \frac{3}{2}, \frac{5}{3}\lambda - \frac{4}{3}\right)$$

Direction ratios of PQ are $\left(3\lambda, \dfrac{4\lambda - 9}{2}, \dfrac{5\lambda + 8}{3}\right)$

Now PQ is parallel to the given plane

$4x + 12y - 3z + 1 = 0$

Hence, line is perpendicular to the vector normal to the plane. Thus,

$$4(3\lambda) + 12\left(\frac{4\lambda - 9}{2}\right) - 3\left(\frac{5\lambda + 8}{3}\right) = 0$$

$\Rightarrow \quad \lambda = 2$

$\Rightarrow \quad Q \equiv \left(4, \dfrac{5}{2}, 2\right)$

$\Rightarrow \quad PQ = \sqrt{(6)^2 + \left(\dfrac{5}{2} - 3\right)^2 + (6)^2} = \dfrac{17}{2}$

14. (6) The given points are $O(0, 0, 0)$, $A(0, 0, 2)$, $B(0, 4, 0)$ and $C(6, 0, 0)$

Here three faces of tetrahedron are xy, yz, zx plane.

Since point P is equidistance from zx, xy and yz planes, its coordinates are (r, r, r)

Equation of plane ABC is

$2x + 3y + 6z = 12$ (using intercept form)

P is also at distance r from plane ABC. Thus,

$$\frac{|2r + 3r + 6r - 12|}{\sqrt{4 + 9 + 36}} = r$$

$\Rightarrow \quad |11r - 12| = 7r$

$\Rightarrow \quad 11r - 12 = \pm 7r$

$\Rightarrow \quad r = \dfrac{12}{18}, 3$

$\Rightarrow \quad r = 2/3$ (as $r < 2$)

15. (9) Let the equation of the variable plane be $\dfrac{x}{a} + \dfrac{y}{b} + \dfrac{z}{c} = 1$, which meets the axes at $A(a, 0, 0)$, $B(0, b, 0)$ and $C(0, 0, c)$.

The centroid of ΔABC is $\left(\dfrac{a}{3}, \dfrac{b}{3}, \dfrac{c}{3}\right)$ and it satisfies the relation $\dfrac{1}{x^2} + \dfrac{1}{y^2} + \dfrac{1}{z^2} = k$. Thus,

$$\frac{9}{a^2}+\frac{9}{b^2}+\frac{9}{c^2}=k$$

or $\frac{1}{a^2}+\frac{1}{b^2}+\frac{1}{c^2}=\frac{k}{9}$ ...(i)

Also it is given that the distance of the plane $\frac{x}{a}+\frac{y}{b}+\frac{z}{c}=1$ from (0, 0, 0) is 1 unit. Therefore,

$$\frac{1}{\sqrt{\frac{1}{a^2}+\frac{1}{b^2}+\frac{1}{c^2}}}=1 \text{ or } \frac{1}{a^2}+\frac{1}{b^2}+\frac{1}{c^2}=1 \quad \text{...(ii)}$$

From (i) and (ii), we get $k/9 = 1$, i.e., $k = 9$

ultiple Correct Answers Type

1. (2), (3), (4)

If P be (x, y, z), then from the figure,

$x = r\sin\theta\cos\phi$, $y = r\sin\theta\sin\phi$ and $z = r\cos\theta$

$1 = r\sin\theta\cos\phi$, $2 = r\sin\theta\sin\phi$ and $3 = r\cos\theta$

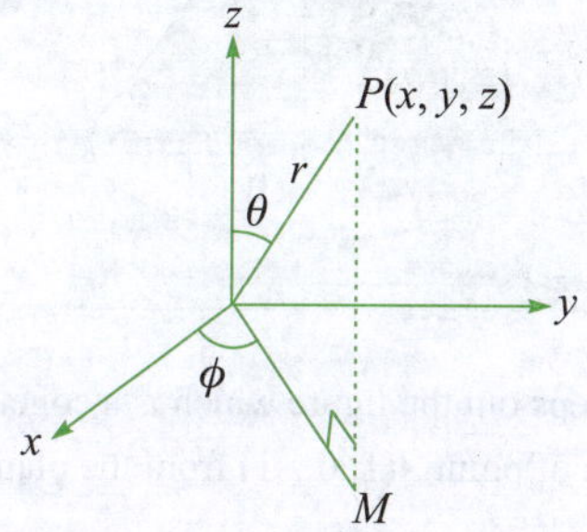

$\Rightarrow \quad 1^2+2^2+3^2=r^2$ or $r=\pm\sqrt{14}$

$\therefore \quad \sin\theta\cos\phi=\frac{1}{\sqrt{14}},$

$\sin\theta\sin\phi=\frac{2}{\sqrt{14}}$ and $\cos\theta=\frac{3}{\sqrt{14}}$

(neglecting negative sign as θ and ϕ are acute)

$\frac{\sin\theta\sin\phi}{\sin\theta\cos\phi}=\frac{2}{1}$ or $\tan\phi=2$

Also, $\tan\theta=\sqrt{5}/3$

2. (1), (2)

Given lines are L_1: $\frac{x}{1}=\frac{y+5}{1}=\frac{z+1}{1}$ and L_2: $\frac{x+5}{6}=\frac{y}{2}=\frac{z}{3}$

Points C and D on lines L_1 and L_2 respectively can be taken as $(\lambda, \lambda-5, \lambda-1)$ and $(6\mu-5, 2\mu, 3\mu)$.

CD is parallel to the line having direction ratios 1, –5, –2.

$$\therefore \quad \frac{6\mu-6-\lambda}{1}=\frac{2\mu-\lambda+5}{-5}=\frac{3\mu-\lambda+1}{-2}$$

$$\Rightarrow \quad \frac{6\mu-5-\lambda}{1}=\frac{2\mu-\lambda+5}{-5}=\frac{3\mu-\lambda+1}{-2}$$

Solving we get $\lambda = 2$ and $\mu = 1$.

3. (2), (3), (4)

$x+y+z-1=0$

$4x+y-2z+2=0$

Let $z = k$. Then $x = k-1$ and $y = 2-2k$

Therefore, $(k-1, 2-2k, k)$ is a variable point on the line $\frac{x+1}{1}=\frac{y-2}{-2}=\frac{z}{1}$

Hence, (–1, 2, 0), (0, 0, 1) and (–1/2, 1, 1/2) are the points on the line.

4. (1), (4)

The given lines intersect if

$$\begin{vmatrix} 2-1 & 3-4 & 4-5 \\ 1 & 1 & \lambda \\ \lambda & 2 & 1 \end{vmatrix}=0 \Rightarrow \lambda=0,-1.$$

5. (1), (3)

Clearly, given lines form a cube.

6. (1), (2), (4)

Any point on the line is $(\lambda+1, 2\lambda+2, 3\lambda+3)$

If it lies in the plane $x-2y+z=0$ then $\lambda+1-2(2\lambda+2)+3\lambda+3=0$ or $0=0$, which is true.

So, line lies in the plane.

Since origin satisfies the given line, it is same as the line $\frac{x}{1}=\frac{y}{2}=\frac{z}{3}$.

Putting point (2, 3, 5) in the equation of given line, we get $\frac{1}{1}=\frac{1}{2}=\frac{2}{3}$, which is not true.

If line is parallel to the plane $x-2y+z-6=0$, then $1(1)+2(-2)+3(1)=0$, which is true.

So, line is parallel to the plane.

7. (1), (3)

Plane P_1 contains the line $\vec{r}=\hat{i}+\hat{j}+\hat{k}+\lambda(\hat{i}-\hat{j}-\hat{k})$, hence contains the point $\hat{i}+\hat{j}+\hat{k}$ and is normal to vector $(\hat{i}+\hat{j})$.

Hence, equation of plane is

$$(\vec{r}-(\hat{i}+\hat{j}+\hat{k}))\cdot(\hat{i}+\hat{j})=0$$

or $\quad x+y=2$

Plane P_2 contains the line

$$\vec{r}=\hat{i}+\hat{j}+\hat{k}+\lambda(\hat{i}-\hat{j}-\hat{k}) \text{ and point } \hat{j}$$

Hence, equation of plane is

$$\begin{vmatrix} x-0 & y-1 & z-0 \\ 1-0 & 1-1 & 1-0 \\ 1 & -1 & -1 \end{vmatrix}=0$$

or $\quad x+2y-z=2$

If θ is the acute angle between P_1 and P_2, then

$$\cos\theta=\frac{|\vec{n}_1\cdot\vec{n}_2|}{|\vec{n}_1||\vec{n}_2|}=\left|\frac{(\hat{i}+\hat{j}).(\hat{i}+2\hat{j}-\hat{k})}{\sqrt{2}.\sqrt{6}}\right|$$

$$=\frac{3}{\sqrt{2}\cdot\sqrt{6}}=\frac{\sqrt{3}}{2}$$

$$\therefore \quad \theta=\cos^{-1}\frac{\sqrt{3}}{2}=\frac{\pi}{6}$$

As L is contained in P_2, angle between P_2 and L is 0.

8. (1), (4)

The equation of the plane passing through the line of intersection of the planes $2x - y = 0$ and $3z - y = 0$ is

$$2x - y + \lambda(3z - y) = 0 \quad \text{...(i)}$$

or $\quad 2x - y(\lambda + 1) + 3\lambda z = 0$

Plane (i) is perpendicular to $4x + 5y - 3z = 8$. Therefore,

$$4 \times 2 - 5(\lambda + 1) - 9\lambda = 0$$

$\Rightarrow \quad 8 - 5\lambda - 5 - 9\lambda = 0$

$\Rightarrow \quad 3 - 14\lambda = 0$

$\Rightarrow \quad \lambda = 3/14$

$\therefore \quad 2x - y + \frac{3}{14}(3z - y) = 0$

$$28x - 17y + 9z = 0$$

9. (1), (2), (4)

$$3x - 6y + 2z + 5 = 0 \quad \text{...(i)}$$

$$-4x + 12y - 3z + 3 = 0 \quad \text{...(ii)}$$

Bisectors are $\frac{3x - 6y + 2z + 5}{\sqrt{9+36+4}} = \pm \frac{-4x + 12y - 3z + 3}{\sqrt{16+144+9}}$

The plane which bisects the angle between the planes that contains the origin is

$$13(3x - 6y + 2z + 5) = 7(-4x + 12y - 3z + 3)$$

$$67x - 162y + 47z + 44 = 0 \quad \text{...(iii)}$$

Further, $3 \times (-4) + (-6)(12) + 2 \times (-3) < 0$

Hence, the origin lies in the acute angle.

Also, $3(1) - 6(1) + 2(1) + 5 > 0$ and $-4(1) + 12(1) - 3(1) + 3 > 0$.

So, (1, 1, 1) lies in the acute angle.

10. (1), (3)

The required plane is parallel to the bisector of the given planes.

Bisectors are $\frac{x - y + z - 3}{\sqrt{3}} = \pm \frac{x + y + z + 4}{\sqrt{3}}$

or $2y + 7 = 0$ and $2x + 2y + 1 = 0$. Hence, the planes are $y = 0$ and $x + y = 0$.

11. (1), (4)

The equation of a plane passing through the line of intersection of the x–y and y–z planes is

$$z + \lambda x = 0, \quad \lambda \in R$$

This plane makes an angle 45° with the x–y plane ($z = 0$). Thus,

$$\cos 45^\circ = \frac{1}{\sqrt{1}\sqrt{\lambda^2 + 1}}$$

or $\quad \lambda = \pm 1$

12. (1), (2)

The plane is equally inclined to the lines. Hence, it is perpendicular to the angle bisector of the vectors $2\hat{i} - 2\hat{j} - \hat{k}$ and $8\hat{i} + \hat{j} - 4\hat{k}$.

Vector along the angle bisectors of the vectors are

$$\frac{2\hat{i} - 2\hat{j} - \hat{k}}{3} \pm \frac{8\hat{i} + \hat{j} - 4\hat{k}}{9}$$

or $\quad \frac{14\hat{i} - 5\hat{j} - 7\hat{k}}{9}$ and $\frac{-2\hat{i} - 7\hat{j} + \hat{k}}{9}$.

Hence, the equations of the planes are

$$14x - 5y - 7z = 0 \text{ or } 2x + 7y - z = 0$$

13. (1), (3)

For line $\frac{x-1}{1} = \frac{y}{-1} = \frac{z-5}{-1}$, point (1, 0, 5) lies on the plane. Also, the vector along the line $\hat{i} - \hat{j} - \hat{k}$ is perpendicular to the normal $\hat{i} + 2\hat{j} - \hat{k}$ to the plane. For line $\vec{r} = 2\hat{i} - \hat{j} + 4\hat{k} + \lambda(3\hat{i} + \hat{j} + 5\hat{k})$, point (2, –1, 4) lies on the plane and vector $3\hat{i} + \hat{j} + 5\hat{k}$ is perpendicular to the normal $\hat{i} + 2\hat{j} - \hat{k}$.

Line $x - y + z = 2x + y - z = 0$ passes through the origin, which is not on the given plane.

14. (1), (3), (4)

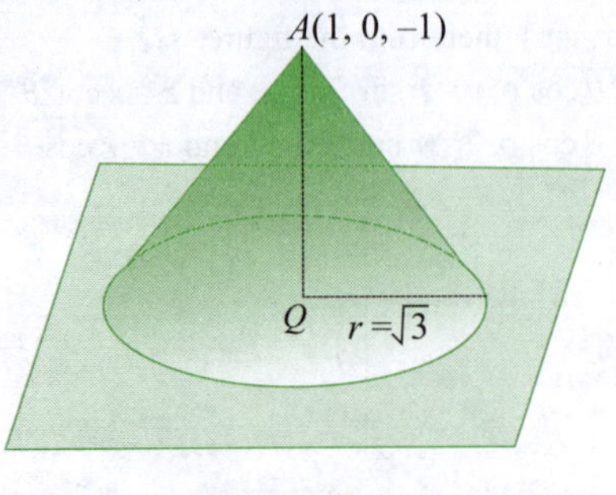

The rod sweeps out the figure which is a cone.

The distance of point $A(1, 0, -1)$ from the plane is

$$\frac{|1 - 2 + 4|}{\sqrt{9}} = 1 \text{ unit}$$

The slant height l of the cone is 2 units.

Then the radius of the base of the cone is

$$\sqrt{l^2 - 1} = \sqrt{4 - 1} = \sqrt{3}$$

Hence, the volume of the cone is

$$\frac{\pi}{3}(\sqrt{3})^2 \cdot 1 = \pi \text{ cubic units}$$

Area of the circle on the plane which the rod traces is 3π.

Also, the centre of the circle is $Q(x, y, z)$. Then

$$\frac{x-1}{1} = \frac{y-0}{-2} = \frac{z+1}{2} = \frac{-(1 - 0 - 2 + 4)}{1^2 + (-2)^2 + 2^2}$$

or $\quad Q(x, y, z) \equiv \left(\frac{2}{3}, \frac{2}{3}, \frac{-5}{3}\right)$

15. (2), (3)

Distance between the planes is $h = 5/\sqrt{6}$

Also the figure formed is a cylinder, whose radius $r = 2$ units.

Hence, the volume of the cylinder is

$$\pi r^2 h = \pi(2)^2 \cdot \frac{5}{\sqrt{6}} = \frac{20\pi}{\sqrt{6}} \text{ cubic units.}$$

Also the curved surface area is

$$2\pi r h = 2\pi(2) \cdot \frac{5}{\sqrt{6}} = \frac{20\pi}{\sqrt{6}}$$

16. (1), (2)

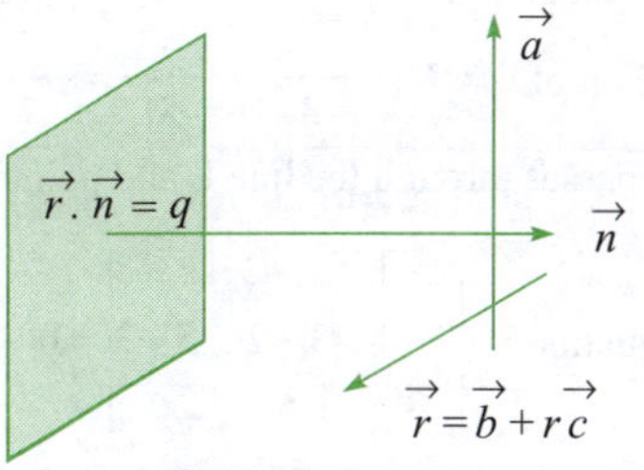

Required line is parallel to $\vec{n}\times\vec{c}$

The equation of line is $\vec{r}=\vec{a}+\lambda(\vec{n}\times\vec{c})$

or $(\vec{r}-\vec{a})=\lambda(\vec{n}\times\vec{c})$

$\therefore$ $(\vec{r}-\vec{a})\times(\vec{n}\times\vec{c})=0$

17. (2), (3)

Volume of tetrahedron $ABCD=\frac{1}{6}|[\overrightarrow{AB}\ \overrightarrow{AC}\ \overrightarrow{AD}]|$

$= 1$ cubic units.

$$\Rightarrow \begin{vmatrix} -1 & 1 & -1 \\ 1 & 1 & -1 \\ x-0 & y-1 & z-2 \end{vmatrix} = \pm 6$$

or $-2(y-1)-2(z-2)=\pm 6$

or $y-1+z-2=\pm 3$

$\Rightarrow$ $y+z=6$ or $y+z=0$

Linked Comprehension Type

1. (2), 2. (3), 3. (4)

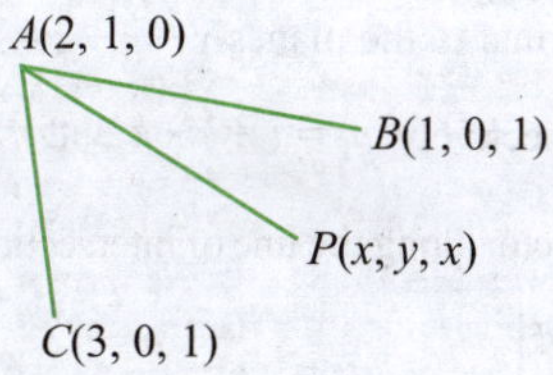

$$\begin{vmatrix} x-2 & y-1 & z \\ 1-2 & 0-1 & 1-0 \\ 3-2 & 0-1 & 1-0 \end{vmatrix} = 0$$

or $2(y-1)+2z=0$

or $y+z-1=0$

The vector normal to the plane is $\vec{n}=0\hat{i}+\hat{j}+\hat{k}$.

The equation of the line through (0, 0, 2) and parallel to $\vec{n}$ is $\vec{r}=2\hat{k}+\lambda(\hat{j}+\hat{k})$.

The perpendicular distance of $D(0, 0, 2)$ from plane ABC is $\left|\frac{2-1}{\sqrt{1^2+1^2}}\right|$

$=\frac{1}{\sqrt{2}}$.

4. (2) Let $A'\equiv(x_2, y_2, z_2)$ is the image of $A(2, 1, 6)$ about mirror $x+y-2z=3$. Then,

$$\frac{x_2-2}{1}=\frac{y_2-1}{1}=\frac{z_2-6}{-2}$$

$$=\frac{-2(2+1-12-3)}{1^2+1^2+2^2}=4$$

$\Rightarrow$ $(x_2, y_2, z_2)\equiv(6, 5, -2)$.

5. (3) $\frac{x-2}{3}=\frac{y-1}{4}=\frac{z-6}{5}=\lambda$

Let $x=2+3\lambda, y=1+4\lambda, z=6+5\lambda$ lies on plane

$x+y-2z=3$

$\therefore$ $2+3\lambda+1+4\lambda-2(6+5\lambda)=3$

or $3+7\lambda-12-10\lambda=3$ or $-3\lambda=12$ or $\lambda=-4$

So, Point $B\equiv(-10, -15, -14)$

6. (3) The equation of the reflected ray L_1 is the line joining A' and $B(-10, -15, -14)$.

$\therefore$ L_1: $\frac{x+10}{16}=\frac{y+15}{20}=\frac{z+14}{12}$

or $\frac{x+10}{4}=\frac{y+15}{5}=\frac{z+14}{3}$

7. (4) The line $\frac{x-1}{3}=\frac{y-2}{-1}=\frac{z-3}{4}=r$

Let point B is $(3r+1, 2-r, 3+4r)$

$\therefore$ $\overrightarrow{AB}=(3r)\hat{i}+(-r)\hat{j}+(4r+6)\hat{k}$

Now, $\overrightarrow{AB}$ is parallel to $x+y-z=1$

$\Rightarrow$ $3r-r-4r-6=0$ or $r=-3$

$\therefore$ B is $(-8, 5, -9)$

8. (2) The equation of plane containing the line L is

$A(x-1)+B(y-2)+C(z-3)=0$,

where $3A-B+4C=0$ ...(i)

(i) also contains point $A(1, 2, -3)$.

Hence, $C=0$ and $3A=B$.

The equation of plane is $x-1+3(y-2)=0$ or $x+3y-7=0$

9. (4) The distance of point $(1+3r, 2-r, 3+4r)$ from the plane is

$$\frac{|1+3r+2-r-3-4r-1|}{\sqrt{1+1+1}}=\frac{|2r+1|}{\sqrt{3}}=\frac{4}{\sqrt{3}}$$

$\Rightarrow$ $r=\frac{3}{2}, -\frac{5}{2}$

Hence, the points are $A\left(\frac{11}{2}, \frac{1}{2}, \frac{18}{2}\right)$ and $B\left(\frac{-13}{2}, \frac{9}{2}, \frac{-14}{2}\right)$.

$\Rightarrow$ $AB=2\sqrt{104}$

10. (2), 11. (1)

Equation of family of planes passing through the line of intersection of planes $3x-2y+z-6=0$ and $x+y-4z+5=0$ is

$3x-2y+z-6+\lambda(x+y-4z+5)=0$ ...(1)

Vector along the lines $x-y+z-3=0$, $2x+y-z-7=0$ is

$$\begin{vmatrix} \hat{i} & \hat{j} & \hat{k} \\ 1 & -1 & 1 \\ 2 & 1 & -1 \end{vmatrix} = 3\hat{j}+3\hat{k}$$

Plane (1) is parallel to the vector $3\hat{j}+3\hat{k}$.

$\therefore \quad 3(-2+\lambda)+3(1-4\lambda)=0$

$\Rightarrow \quad -9\lambda-3=0$

$\Rightarrow \quad \lambda=-1/3$

Putting the value of λ in (1), we get equation of plane as

$8x-7y+7z-23=0$

Distance of this plane from the origin is

$$\frac{23}{\sqrt{64+49+49}}=\frac{23}{\sqrt{162}}=\frac{23}{9\sqrt{2}}$$

Vector normal to the plane $\vec{r}=\hat{i}+\hat{j}-\hat{k}+\lambda(2\hat{i}-\hat{j}+3\hat{k})+\mu(-\hat{i}+2\hat{j}+\alpha\hat{k})$ is

$$\begin{vmatrix} \hat{i} & \hat{j} & \hat{k} \\ 2 & -1 & 3 \\ -1 & 2 & \alpha \end{vmatrix} = (-\alpha-6)\hat{i}-(2\alpha+3)\hat{j}+3\hat{k}$$

Plane P is perpendicular this plane.

$\therefore \quad 8(-\alpha-6)+7(2\alpha+3)+7\times 3=0$

$\Rightarrow \quad 6\alpha-6=0$

$\Rightarrow \quad \alpha=1$

12. (2), 13. (2)

Given lines are $\frac{x+1}{-3}=\frac{y-3}{2}=\frac{z+2}{1}$ and $\frac{x}{1}=\frac{y-7}{-3}=\frac{z+7}{\alpha}$

Since plane passes through the given lines, they must be coplanar.

$$\therefore \quad \begin{vmatrix} -1 & -4 & 5 \\ -3 & 2 & 1 \\ 1 & -3 & \alpha \end{vmatrix}=0$$

$\Rightarrow \quad \alpha=2$

Equation of plane through the lines is

$$\begin{vmatrix} x+1 & y-3 & z+2 \\ -3 & 2 & 1 \\ 1 & -3 & 2 \end{vmatrix}=0$$

or $\quad x+y+z=0$

Given that points $(2, -9, \lambda)$ and $(\lambda, -1, -3)$ lie on the opposite sides of the of the plane.

$\therefore \quad (2-9+\lambda)(\lambda-1-3)<0$

$\Rightarrow \quad (\lambda-4)(\lambda-7)<0$

$\Rightarrow \quad 4<\lambda<7$

Therefore, integral values of λ are 5 and 6.

14. (126), 15. (12)

Coordinates of any point on the line L are $(3\lambda+\alpha, 5\lambda+2, \lambda\beta+3)$

Line L lies in the plane P_1.

$\therefore \quad (3\lambda+\alpha)-2(5\lambda+2)+\lambda\beta+3=0, \forall \lambda \in R$

$\Rightarrow \quad (-7+\beta)\lambda+\alpha-1=0, \forall \lambda \in R$

$\Rightarrow \quad \beta=7$ and $\alpha=1$

So, equation of line L is $\frac{x-1}{3}=\frac{y-2}{5}=\frac{z-3}{7}$

Plane P_2 passes through the line L and point (2, 3, 5).

$$\text{So, its equation is } \begin{vmatrix} x-1 & y-2 & z-3 \\ 2-1 & 3-2 & 5-3 \\ 3 & 5 & 7 \end{vmatrix}=0$$

or $\quad -3(x-1)-(y-2)+2(z-3)=0$

or $\quad 3x+y-2z+1=0$

Distance of point (1, 4, –3) from the plane P_2 is

$$\frac{|3+4+6+1|}{\sqrt{9+1+4}}=\sqrt{14}$$

Let foot of perpendicular from origin O upon the line L is $(3\lambda+1, 5\lambda+2, 7\lambda+3)$.

$\overrightarrow{OL}\perp(3\hat{i}+5\hat{j}+7\hat{k})$

$\Rightarrow \quad 3(3\lambda+1)+5(5\lambda+2)+7(7\lambda+3)=0$

$\Rightarrow \quad \lambda=-\frac{34}{83}$

$\therefore \quad p+q+r=3\lambda+1+5\lambda+2+7\lambda+3=15\lambda+6$

$$=15\times\left(-\frac{34}{83}\right)+6=\frac{-510+498}{83}=\frac{-12}{83}$$

Matrix Match Type

1. (1)

(P) Line $x=2t+1, y=t+2, z=-t-3$ or $\frac{x-1}{2}=\frac{y-2}{1}=\frac{z+3}{-1}$

which is along the vector $2\hat{i}+\hat{j}-\hat{k}$. Vector $\hat{i}+3\hat{j}+5\hat{k}$ is perpendicular to the line.

(Q) Vectors normal to the planes $x+y+z-3=0$ and $2x-y+3z=0$ are respectively $\vec{n}_1=\hat{i}+\hat{j}+\hat{k}$ and $\vec{n}_2=2\hat{i}-\hat{j}+3\hat{k}$

Then the vector along the line of intersection of planes is $\vec{n}_1\times\vec{n}_2$

$$=\begin{vmatrix} \hat{i} & \hat{j} & \hat{k} \\ 1 & 1 & 1 \\ 2 & -1 & 3 \end{vmatrix}=4\hat{i}-\hat{j}-3\hat{k}$$

(R) The shortest distance between the lines $\frac{x}{2}=\frac{y}{-3}=\frac{z}{-1}$ and $\vec{r}=(3\hat{i}-\hat{j}+\hat{k})+t(\hat{i}+\hat{j}-2\hat{k})$ occurs along the vector

$$(2\hat{i}-3\hat{j}-\hat{k})\times(\hat{i}+\hat{j}-2\hat{k})=\begin{vmatrix} \hat{i} & \hat{j} & \hat{k} \\ 2 & -3 & -1 \\ 1 & 1 & -2 \end{vmatrix}$$

$$=7\hat{i}+3\hat{j}+5\hat{k}$$

(S) Vector normal to the plane $\vec{r}=-\hat{i}+4\hat{j}-6\hat{k}+\lambda(\hat{i}+3\hat{j}-2\hat{k})+\mu(-\hat{i}+2\hat{j}-5\hat{k})$ is

$$\begin{vmatrix} \hat{i} & \hat{j} & \hat{k} \\ 1 & 3 & -2 \\ -1 & 2 & -5 \end{vmatrix}=-11\hat{i}+7\hat{j}+5\hat{k}$$

2. (3)

(P) The given line is $x = 4y + 5, z = 3y - 6$

or $\frac{x-5}{4} = y, \frac{z+6}{3} = y$

or $\frac{x-5}{4} = \frac{y}{1} = \frac{z+6}{3} = \lambda$ (say)

Any point on the line is $(4\lambda + 5, \lambda, 3\lambda - 6)$.

The distance between points $(4\lambda + 5, \lambda, 3\lambda - 6)$ and $(5, 3, -6)$ is 3 units (given). Therefore

$(4\lambda + 5 - 5)^2 + (\lambda - 3)^2 + (3\lambda - 6 + 6)^2 = 9$

$\Rightarrow 16\lambda^2 + \lambda^2 + 9 - 6\lambda + 9\lambda^2 = 9$

$\Rightarrow 26\lambda^2 - 6\lambda = 0$

$\Rightarrow \lambda = 0, 3/13$

The point is $(5, 0, -6)$.

(Q) The equation of the plane containing the line $\frac{x-2}{3} = \frac{y+3}{5} = \frac{z+5}{7}$ and parallel to $\hat{i} + 4\hat{j} + 7\hat{k}$ is

$$\begin{vmatrix} x-2 & y+3 & z+5 \\ 1 & 4 & 7 \\ 3 & 5 & 7 \end{vmatrix} = 0$$

or $x - 2y + z - 3 = 0$

Point $(-1, -2, 0)$ lies on this plane.

(R) The line passing through points $A(2, -3, -1)$ and $B(8, -1, 2)$ is $\frac{x-2}{8-2} = \frac{y+3}{-1+3} = \frac{z+1}{2+1}$

or $\frac{x-2}{6} = \frac{y+3}{2} = \frac{z+1}{3} = \lambda$ (say).

Variable point on this line is $P(6\lambda + 2, 2\lambda - 3, 3\lambda - 1)$, whose distance from point $A(2, -3, -1)$ is 14 units.

Therefore,

$PA = 14$

or $PA^2 = (14)^2$

$\Rightarrow (6\lambda)^2 + (2\lambda)^2 + (3\lambda)^2 = 196$

$\Rightarrow 49\lambda^2 = 196$

$\Rightarrow \lambda^2 = 4$

$\Rightarrow \lambda = \pm 2$

Therefore, the required points are $(14, 1, 5)$ and $(-10, -7 -7)$.

The point nearer to the origin is $(-10, -7, -7)$.

(S) Any point on line AB, $\frac{x}{2} = \frac{y-2}{3} = \frac{z-3}{4} = \lambda$ is $M(2\lambda, 3\lambda + 2, 4\lambda + 3)$. Therefore, the direction ratios of PM are $2\lambda - 3$, $3\lambda + 3$ and $4\lambda - 8$.

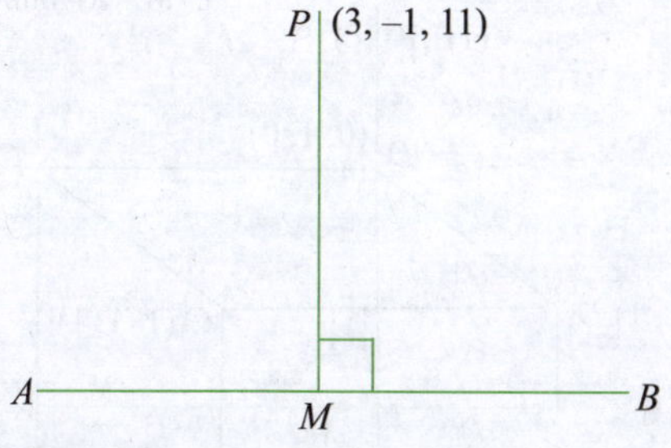

But $PM \perp AB$

$\therefore 2(2\lambda - 3) + 3(3\lambda + 3) + 4(4\lambda - 8) = 0$

$\Rightarrow 4\lambda - 6 + 9\lambda + 9 + 16\lambda - 32 = 0$

$\Rightarrow 29\lambda - 29 = 0$

$\Rightarrow \lambda = 1$

Therefore, foot of the perpendicular is $M(2, 5, 7)$.

3. (4)

(P) The given line and plane are $\vec{r} = (2\hat{i} - 2\hat{j} + 3\hat{k}) + \lambda(\hat{i} - \hat{j} + 4\hat{k})$ and $\vec{r} \cdot (\hat{i} + 5\hat{j} + \hat{k}) = 5$, respectively.

Since $(\hat{i} - \hat{j} + 4\hat{k}) \cdot (\hat{i} + 5\hat{j} + \hat{k}) = 0$, line and plane are parallel.

Hence, the required distance is distance of point $(2, -2, 3)$ from the plane $x + 5y + z - 5 = 0$, which is $\frac{|2 - 10 + 3 - 5|}{\sqrt{1 + 25 + 1}} = \frac{10}{3\sqrt{3}}$

(Q) The distance between two parallel planes $\vec{r} \cdot (2\hat{i} - \hat{j} + 3\hat{k}) = 4$ and $\vec{r} \cdot (6\hat{i} - 3\hat{j} + 9\hat{k}) + 13 = 0$ is

$$d = \frac{|4 - (-13/3)|}{\sqrt{(2)^2 + (-1)^2 + (3)^2}} = \frac{(25/3)}{\sqrt{14}}$$

$$= \frac{25}{3\sqrt{14}}$$

(R) The perpendicular distance of the point $(2, 5, -3)$ from the plane $\vec{r} \cdot (6\hat{i} - 3\hat{j} + 2\hat{k}) = 4$ or $6x - 3y + 2z - 4 = 0$ is

$$d = \frac{|12 - 15 - 6 - 4|}{\sqrt{36 + 9 + 4}}$$

$$= 13/\sqrt{49} = 13/7$$

(S) The equation of the line AB is

$$\frac{x-2}{2} = \frac{y+2}{3} = \frac{z-6}{-6}$$

The equation of line passing through $Q(1, 0, -3)$ and parallel to AB is

$$\frac{x-1}{2} = \frac{y}{3} = \frac{z+3}{-6} = r \text{ (say)}$$

This line intersects plane at P.

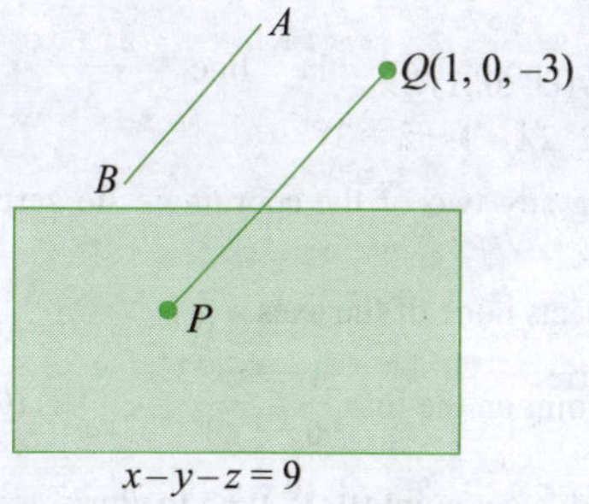

$\therefore P \equiv (2r + 1, 3r, -6r - 3)$ which lies on the plane

$\therefore (2r + 1) - (3r) - (-6r - 3) = 9$

$\Rightarrow r = 1$

$\Rightarrow P \equiv (3, 3, -9)$

So, Required distance

$$PQ = \sqrt{(3-1)^2 + (3-0)^2 + (-9+3)^2}$$

$$= \sqrt{4 + 9 + 36} = 7$$

4. (2)

(P) If the required image is (x, y, z), then

$$\frac{x-3}{2} = \frac{y-5}{1} = \frac{z-7}{1} = -\frac{2(6+5+7+18)}{2^2+1^2+1^2} = -12$$

or $(x, y, z) \equiv (-21, -7, -5)$.

(Q) Any point on the line $\frac{x-2}{-3}=\frac{y-1}{2}=\frac{z-3}{2}=\lambda$ is $(-3\lambda+2,$ $2\lambda+1, 2\lambda+3)$, which lies on the plane $2x+y-z=3$. Therefore

$$-6\lambda+4+2\lambda+1-2\lambda-3=3$$
$$-6\lambda=1$$
$$\lambda=-1/6$$

Therefore, the point is $\left(\frac{5}{2},\frac{2}{3},\frac{8}{3}\right)$.

(R) If (x, y, z) is the required foot of the perpendicular, then

$$\frac{x-1}{2}=\frac{y-1}{-2}=\frac{z-2}{4}=-\frac{(2-2+8+5)}{2^2+(-2)^2+4^2}=\frac{-13}{24}$$

or $(x, y, z)\equiv\left(\frac{-1}{12},\frac{25}{12},\frac{-2}{12}\right)$

(S) Any point on the line $\frac{x-1}{2}=\frac{y-2}{3}=\frac{z-3}{4}=\lambda$ is $P(2\lambda+1,$ $3\lambda+2, 4\lambda+3)$, which satisfies the line $\frac{x-4}{5}=\frac{y-1}{2}=\frac{z}{1}$

$$\therefore\ \frac{2\lambda+1-4}{5}=\frac{3\lambda+2-1}{2}=\frac{4\lambda+3}{1}$$

$\Rightarrow\ \lambda=-1$

So, The required point is $(-1,-1,-1)$

5. (4)

(P) Any point on the line $\frac{x-3}{2}=\frac{y-6}{4}=\frac{z+3}{1}$ is $(2\lambda+3,$ $4\lambda+6, \lambda-3)$

From $2\lambda+3=0$ and $4\lambda+6=0$ we get the same value of λ.

So, line meets z-axis.

From $4\lambda+6=0$ and $\lambda-3=0$, we get different values of λ.

So, line does not meet y-axis.

Similarly, line does not meet x-axis also.

(Q) Any point on the line $\frac{2x+5}{3}=\frac{y+1}{2}=\frac{z-3}{6}$ is $((3\lambda-5)/2, 2\lambda-1, 6\lambda+3)$

Comparing any two of the coordinates to zero we get different values of λ.

So, line meets none of the axes.

(R) Any point on the line $\frac{x}{0}=\frac{y-3}{3}=\frac{z}{-4}$ is $(0, 3\lambda+3, -4\lambda)$

For $\lambda=0$, we get point $(0, 3, 0)$ on the line.

For $3\lambda+3=0$, we have point $(0, 0, 4)$ on the line.

Thus line intersect y-axis and z-axis.

(S) Line $\frac{x}{-3}=\frac{y}{5}=\frac{z}{6}$ passes through origin.

So, line meets all axis.

Archives

Single Correct Answer Type

1. (4) Variable point B on line is $(2\lambda-2, -\lambda-1, 3\lambda)$

Point B lies on the plane for some λ

$\Rightarrow\ (2\lambda-2)+(-\lambda-1)+3\lambda=3$ or $\lambda=3/2$

$\Rightarrow\ B\equiv(1, -5/2, 9/2)$

The foot of the perpendicular from point $(-2, -1, 0)$ on the plane is the point $A\ (0, 1, 2)$

$\Rightarrow$ Direction ratio of AB are $\left(1, \frac{-7}{2}, \frac{5}{2}\right)\equiv(2, -7, 5)$.

Hence, feet of perpendicular lie on the line $\frac{x}{2}=\frac{y-1}{-7}=\frac{z-2}{5}$

2. (3) Let the image is (x, y, z).

$$\therefore\ \frac{x-3}{1}=\frac{y-1}{-1}=\frac{z-7}{1}=-2\left(\frac{3-1+7-3}{1^2+1^2+1^2}\right)=-4$$

$\therefore\ P\equiv(-1, 5, 3)$

Plane passing through $P(-1, 5, 3)$ is

$a(x+1)+b(y-5)+c(z-3)=0$ ...(i)

Plane contains line $\frac{x}{1}=\frac{y}{2}=\frac{z}{1}$

So, $(0, 0, 0)$ satisfies the plane (1).

$\Rightarrow\ a-5b-3c=0$ ...(ii)

Also, $a+2b+c=0$ ...(iii)

From (ii) and (iii), $a=\frac{1}{7}c, b=-\frac{4}{7}c.$

Putting in (1) we get

$$(x+1)-4(y-5)+7(z-3)=0$$

or $\quad x-4y+7z=0$

3. (1) Given that required plane is perpendicular to the given two planes. Therefore, the normal vector of required plane is perpendicular to the normal vectors to the given planes.

Therefore, the normal vector of the required plane is parallel to the vector:

$$\begin{vmatrix}\hat{i} & \hat{j} & \hat{k}\\ 2 & 1 & -2\\ 3 & -6 & -2\end{vmatrix}=-14\hat{i}-2\hat{j}-15\hat{k}$$

Thus, the equation of required plane passing through point $(1, 1, 1)$ is

$$-14(x-1)-2(y-1)-15(z-1)=0$$

or $\quad 14x+2y+15z=31$

4. (1) For maximum value of $d(l_1, l_2)$, we should consider non-intersecting lines l_1 and l_2.

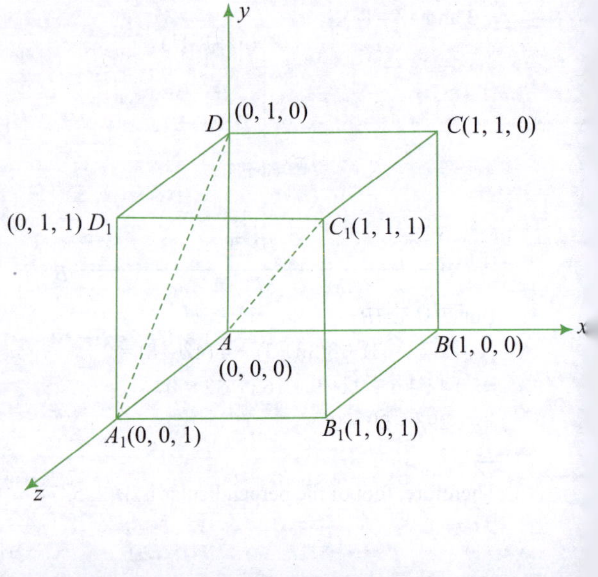

One pair of such lines is diagonal A_1D in set F and diagonal AC_1 in set S.

Now, equation of diagonal A_1D is $\frac{x}{0}=\frac{y}{-1}=\frac{z-1}{1}\equiv l_1$

Equation of diagonal AC_1 is $\frac{x}{1}=\frac{y}{1}=\frac{z}{1}\equiv l_2$

$\therefore$ Shortest distance between A_1D and AC_1 is

$$\frac{\left|\begin{vmatrix}0 & 0 & -1\\ 0 & -1 & 1\\ 1 & 1 & 1\end{vmatrix}\right|}{\left|\begin{vmatrix}\hat{i} & \hat{j} & \hat{k}\\ 0 & -1 & 1\\ 1 & 1 & 1\end{vmatrix}\right|}=\frac{1}{|-2\hat{i}+\hat{j}+\hat{k}|}=\frac{1}{\sqrt{6}}$$

Multiple Correct Answers Type

1. (1), (4)

Lines $\frac{x-5}{0}=\frac{y-0}{3-\alpha}=\frac{z-0}{-2}$ and $\frac{x-\alpha}{0}=\frac{y}{-1}=\frac{z}{2-\alpha}$ are coplanar

$$\therefore \quad \begin{vmatrix}5-\alpha & 0 & 0\\ 0 & 3-\alpha & -2\\ 0 & -1 & 2-\alpha\end{vmatrix}=0$$

$\Rightarrow \quad (5-\alpha)(\alpha^2-5\alpha+4)=0,\ \alpha=1,4,5$

2. (2), (4)

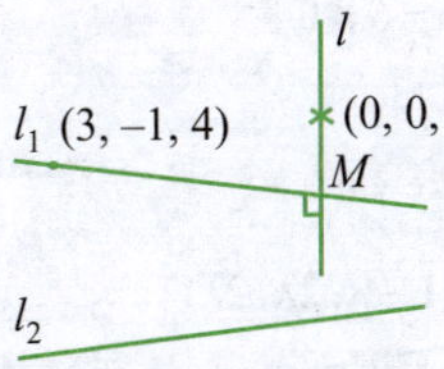

Line l is perpendicular to lines l_1 and l_2.

So, line l is parallel to the vector

$$\begin{vmatrix}\hat{i} & \hat{j} & \hat{k}\\ 1 & 2 & 2\\ 2 & 2 & 1\end{vmatrix}=-2\hat{i}+3\hat{j}-2\hat{k}$$

So, equation of line l is $\frac{x}{2}=\frac{y}{-3}=\frac{z}{2}$

Let $\quad M\equiv(2\lambda,-3\lambda,2\lambda)$

So, $\quad \frac{2\lambda-3}{1}=\frac{-3\lambda+1}{2}=\frac{2\lambda-4}{2}$

$\Rightarrow \quad \lambda=1$

So, $\quad M\equiv(2,-3,2)$

Let the required point on line l_2 be $P(3+2s, 3+2s, 2+s)$.

Given that $PM=\sqrt{17}$

$\Rightarrow \quad (3+2s-2)^2+(3+2s+3)^2+(2+s-2)^2=17$

or $\quad 9s^2+28s+20=0$ or $s=-2,-\frac{10}{9}$.

So, $\quad P\equiv(-1,-1,0)$ or $\left(\frac{7}{9},\frac{7}{9},\frac{8}{9}\right)$

3. (1), (2)

Let $\vec{v}$ be the vector along L.

Clearly, line L is parallel to given planes

$P_1: x+2y-z+1=0$ and $P_2: 2x-y+z-1=0$

$$\therefore \quad \vec{v}=\begin{vmatrix}\hat{i} & \hat{j} & \hat{k}\\ 1 & 2 & -1\\ 2 & -1 & 1\end{vmatrix}=\hat{i}-3\hat{j}-5\hat{k}$$

Since L is passing through origin, any point on line L is $A(\lambda,-3\lambda,-5\lambda)$.

Foot of perpendicular (h,k,l) from A on plane P_1, is

$$\frac{h-\lambda}{1}=\frac{k+3\lambda}{2}=\frac{l+5\lambda}{-1}=-\frac{(\lambda-6\lambda+5\lambda+1)}{1+4+1}=-\frac{1}{6}$$

$$\therefore \quad h=\lambda-\frac{1}{6},\ k=-3\lambda-\frac{1}{3},\ l=-5\lambda+\frac{1}{6}$$

So, the required locus is $\frac{x+\frac{1}{6}}{1}=\frac{y+\frac{1}{3}}{-3}=\frac{z-\frac{1}{6}}{-5}$

Clearly, points (1) and (2) lies on this locus.

4. (2), (4)

Clearly, plane P_3 is $P_2+\lambda P_1=0$.

or $\quad x+\lambda y+z-1=0$

Distance of this plane from point (0, 1, 0) is 1.

$$\Rightarrow \quad \frac{0+\lambda+0-1}{\sqrt{1+\lambda^2+1}}=\pm 1$$

$$\therefore \quad \lambda=-\frac{1}{2}$$

Thus, equation of P_3 is $2x-y+2z-2=0$.

Distance of this plane from point (α,β,γ) is 2.

$$\Rightarrow \quad \left|\frac{2\alpha-\beta+2\gamma-2}{3}\right|=2$$

$\Rightarrow \quad 2\alpha-\beta+2\gamma=2\pm 6$

Thus, options (2), (4) are correct.

5. (2), (3), (4)

Direction ratios of $OQ\equiv(3,3,0)$

Direction ratios of $OS\equiv\left(\frac{3}{2},\frac{3}{2},3\right)$

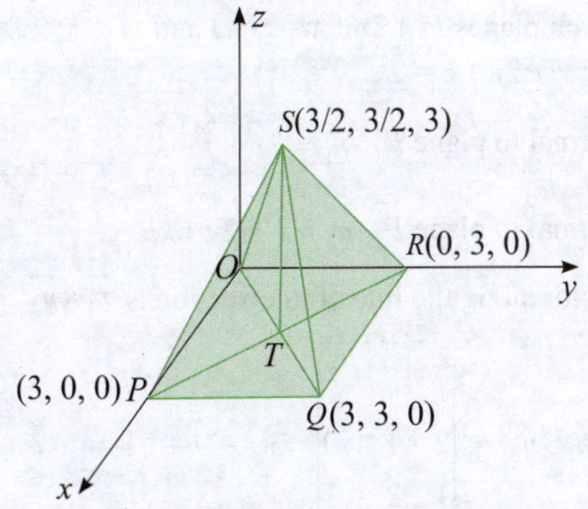

If θ is the angle between OS and OQ, then

$$\cos\theta = \frac{3\times\frac{3}{2}+3\times\frac{3}{2}}{\sqrt{3^2+3^2}\sqrt{\left(\frac{3}{2}\right)^2+\left(\frac{3}{2}\right)^2+3^2}} = \frac{1}{\sqrt{3}}$$

Hence (1) is wrong.

Normal to plane OQS,

$$\overrightarrow{OQ}\times\overrightarrow{OS} = \begin{vmatrix} \hat{i} & \hat{j} & \hat{k} \\ 3 & 3 & 0 \\ \frac{3}{2} & \frac{3}{2} & 3 \end{vmatrix} = (9\hat{i}-9\hat{j})$$

∴ Equation of plane OQS is

$$(x\hat{i}+y\hat{j}+z\hat{k}).(9\hat{i}-9\hat{j}) = 0$$

$\Rightarrow \quad x - y = 0$

Distance of P from plane OQS = distance PT

= distance $OT = \dfrac{3}{\sqrt{2}}$

Equation of RS is $\dfrac{x-0}{\frac{3}{2}-0} = \dfrac{y-3}{\frac{3}{2}-3} = \dfrac{z-0}{3-0}$

or $\quad \dfrac{x}{\frac{3}{2}} = \dfrac{y-3}{-\frac{3}{2}} = \dfrac{z}{3}$

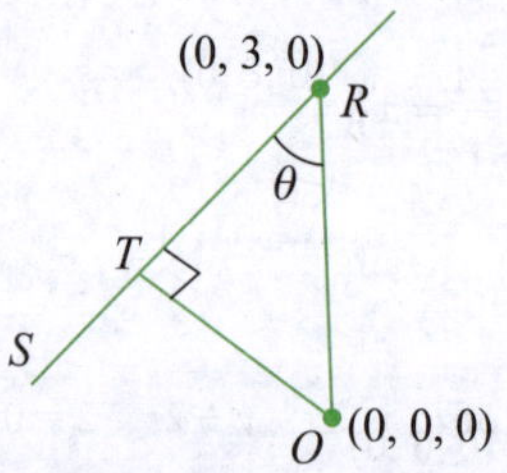

Angle between line RS and OR

$$\cos\theta = \frac{\left|0+3\left(-\frac{3}{2}\right)+0\right|}{3\sqrt{\left(\frac{3}{2}\right)^2+\left(\frac{3}{2}\right)^2+3^2}} = \frac{1}{\sqrt{6}}$$

Distance $= OT = OR\sin\theta = 3\sqrt{1-\frac{1}{6}} = 3\sqrt{\frac{5}{6}} = \sqrt{\frac{15}{2}}$

6. (3), (4)

(1) Given planes $P_1 : 2x + y - z = 3$ and $P_2 : x + 2y + z = 2$

Normal to plane P_1, $\overrightarrow{n_1} = 2\hat{i}+\hat{j}-\hat{k}$

Normal to plane P_2, $\overrightarrow{n_2} = \hat{i}+2\hat{j}+\hat{k}$

Vector along the line of intersection is $\overrightarrow{n_1}\times\overrightarrow{n_2}$.

$$\overrightarrow{n_1}\times\overrightarrow{n_2} = \begin{vmatrix} \hat{i} & \hat{j} & \hat{k} \\ 2 & 1 & -1 \\ 1 & 2 & 1 \end{vmatrix} = 3(\hat{i}-\hat{j}+\hat{k})$$

(2) $\dfrac{3x-4}{9} = \dfrac{1-3y}{9} = \dfrac{z}{3}$

or $\quad \dfrac{x-4/3}{3} = \dfrac{y-1/3}{-3} = \dfrac{z}{3}$

This line is along the vector $3(\hat{i}-\hat{j}+\hat{k})$.

Thus, the line is parallel to the line of intersection of given planes P_1 and P_2.

(3) Angle between planes P_1 and P_2 is given by

$$\cos\theta = \frac{\overrightarrow{n_1}\cdot\overrightarrow{n_2}}{|\overrightarrow{n_1}||\overrightarrow{n_2}|} = \frac{2+2-1}{\sqrt{6}\sqrt{6}} = \frac{3}{6} = \frac{1}{2}$$

$\Rightarrow \theta = 60°$

(4) Plane P_3 is perpendicular to the line of intersection of planes P_1 and P_2.

Also, P_3 passes through the point (4, 2, –2).

So, equation of P_3 is

$$1(x-4) - (y-2) + 1(z+2) = 0$$

or $\quad P_3 : x - y + z = 0$

Distance of (2, 1, 1) from $P_3 = \left|\dfrac{2-1+1}{\sqrt{3}}\right| = \dfrac{2}{\sqrt{3}}$

7. (1), (2), (3)

$L_1 \equiv \vec{r} = \hat{i} + \lambda(-\hat{i}+2\hat{j}+2\hat{k})$

$L_2 \equiv \vec{r} = \mu(2\hat{i}-\hat{j}+2\hat{k})$

$$\therefore \quad \begin{vmatrix} 1 & 0 & 0 \\ -1 & 2 & 2 \\ 2 & -1 & 2 \end{vmatrix} = 6 \neq 0$$

So, L_1 and L_2 are skew lines.

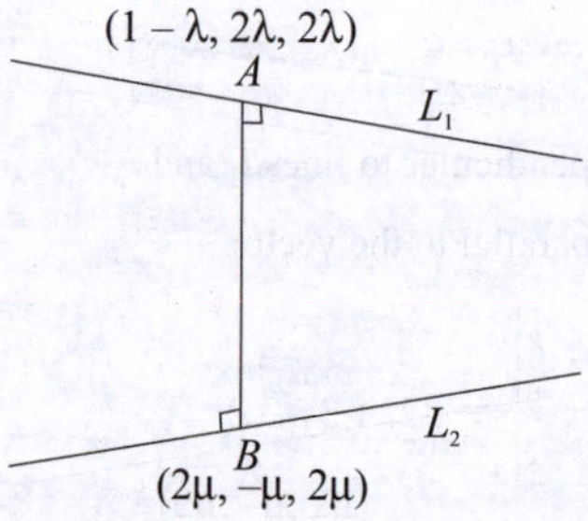

The direction ratios of line AB which is perpendicular to both L_1 and L_2 are given by

$$\begin{vmatrix} \hat{i} & \hat{j} & \hat{k} \\ -1 & 2 & 2 \\ 2 & -1 & 2 \end{vmatrix} = 6\hat{i}+6\hat{j}-3\hat{k}$$

Hence, direction ratios of AB are (2, 2, –1).

Therefore,

$1 - \lambda - 2\mu = 2\alpha$ ...(i)

$2\lambda + \mu = 2\alpha$ ...(ii)

$2\lambda - 2\mu = -\alpha$ ...(iii)

Solving (i) (ii) and (iii), we get $\lambda = 1/9$ and $\mu = 2/9$.

So, $A \equiv \left(\dfrac{8}{9}, \dfrac{2}{9}, \dfrac{2}{9}\right)$ and $B \equiv \left(\dfrac{4}{9}, \dfrac{2}{9}, \dfrac{4}{9}\right)$

Equation of line L_3 passing through A is

$$\vec{r} = \left(\frac{8}{9}\hat{i} + \frac{2}{9}\hat{j} + \frac{2}{9}\hat{k}\right) + t(2\hat{i} + 2\hat{j} - \hat{k}), t \in R$$

Also, equation of line L_3 passing through B is

$$\vec{r} = \left(\frac{4}{9}\hat{i} - \frac{2}{9}\hat{j} + \frac{4}{9}\hat{k}\right) + t(2\hat{i} + 2\hat{j} - \hat{k})$$

Points $\frac{1}{9}(6\hat{i} + 3\hat{k})$, $\frac{8}{9}\hat{i} + \frac{2}{9}\hat{j} + \frac{2}{9}\hat{k}$

and $\frac{4}{9}\hat{i} - \frac{2}{9}\hat{j} + \frac{4}{9}\hat{k}$ are collinear.

So, equation of line is also

$$\vec{r} = \frac{1}{3}(2\hat{i} + \hat{k}) + t(2\hat{i} + 2\hat{j} - \hat{k}), t \in R.$$

Origin is not lying on the line. So, option (4) is not correct.

8. (1), (4)

P.V. of point P on L_1 is $\vec{p} = \lambda\hat{i}$.

P.V. of point Q on L_2 is $\vec{q} = \mu\hat{j} + \hat{k}$.

P.V. of point R on L_3 is $\vec{r} = \hat{i} + \hat{j} + v\hat{k}$.

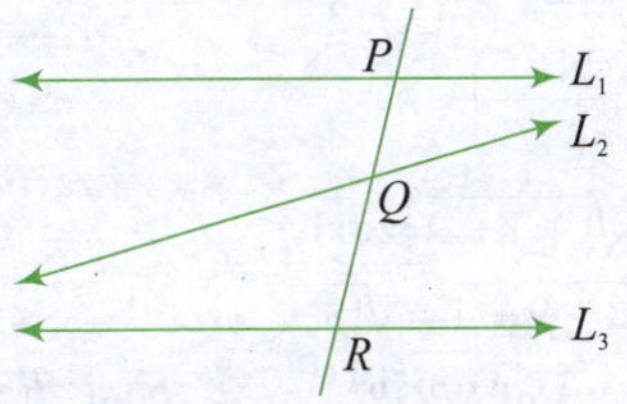

Points P, Q and R are collinear.

$\therefore \quad \overrightarrow{PQ} = \alpha\overrightarrow{PR}$

$\Rightarrow \quad (\hat{k} + \mu\hat{j} - \lambda\hat{i}) = \alpha((1-\lambda)\hat{i} + \hat{j} + v\hat{k})$

$\Rightarrow \quad \frac{\lambda}{\lambda - 1} = \mu = \frac{1}{v}$

Clearly, μ cannot take value 0 and 1.

9. (1), (2) Given lines L_1 and L_2 intersect at (1, 0, 1).

So, line L passes through (1, 0, 1).

$\therefore \quad \frac{1-\alpha}{l} = -\frac{1}{m} = \frac{1-\gamma}{-2}$ ... (i)

$\therefore$ angle bisector of L_1 and L_2 is

$$\vec{r} = \hat{i} + \hat{k} + \lambda\left(\frac{\hat{i} - \hat{j} + 3\hat{k} - 3\hat{i} - \hat{j} + \hat{k}}{\sqrt{11}}\right)$$

or $\quad \vec{r} = \hat{i} + \hat{k} + \mu(\hat{i} + \hat{j} - 2\hat{k})$ ...(ii)

Comparing (i) and (ii), we get

$$\frac{l}{1} = \frac{m}{1} = \frac{-2}{-2}$$

$\Rightarrow \quad l = m = 1$

From (1),

$$\frac{1-\alpha}{1} = -1 \quad \Rightarrow \alpha = 2$$

and $\quad \frac{1-\gamma}{-2} = -1 \quad \Rightarrow \gamma = -1$

10. (1), (2), (3)

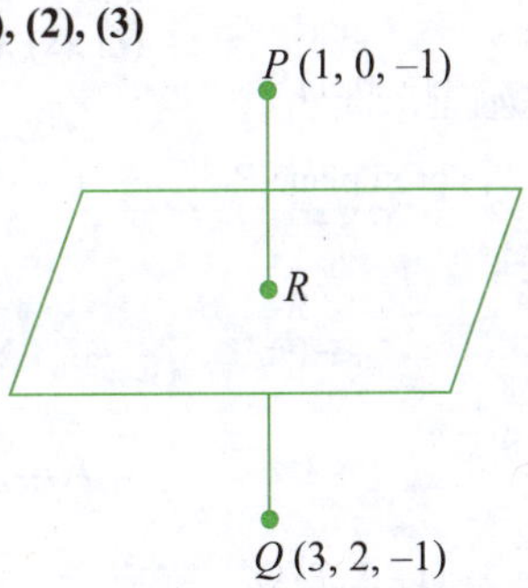

Point $Q(3, 2, -1)$ is the image of point $P(1, 0, -1)$ in the plane.

PQ intersects plane at R.

So, R is mid-point of PQ.

$\therefore \quad R \equiv (2, 1, -1)$.

Point R lies on the plane $\alpha x + \beta y + \gamma z = \delta$

$\therefore \quad 2\alpha + \beta - \gamma = \delta$ ... (i)

Vector normal to the plane is

$$\vec{n} = 2\hat{i} + 2\hat{j}$$

$\therefore \quad \frac{\alpha}{2} = \frac{\beta}{2} = \frac{\gamma}{0} = \lambda$

$\therefore \quad \alpha = 2\lambda, \beta = 2\lambda, \gamma = 0$ ...(ii)

Given that

$\alpha + \gamma = 1$

$\therefore \quad 2\lambda = 1$

$\therefore \quad \alpha = 1, \beta = 1, \gamma = 0$

From (1),

$2(1) + 1 - 0 = \delta$

$\delta = 3$

11. (1), (2), (4)

Given planes are

$P_1: 10x + 15y + 12z - 60 = 0$ ...(i)

$P_2: -2x + 5y + 4z - 20 = 0$ ...(ii)

Putting $x = 0$ in (1) and (2), we get

$5y + 4z - 20 = 0$ or $\frac{y-4}{-4} = \frac{z}{5}$

So, equation of line of intersection is

$L: \frac{x}{0} = \frac{y-4}{-4} = \frac{z}{5}$ ...(iii)

Now, any skew line with above line can be the edge of tetrahedron. Also, any line which is non-parallel to L and lies in either plane P_1 or plane P_2 can be the edge of tetrahedron.

For line $L_1: \frac{x-1}{0} = \frac{y-1}{0} = \frac{z-1}{5}$,

$$\begin{vmatrix} 0-1 & 4-1 & 0-1 \\ 0 & 0 & 5 \\ 0 & -4 & 5 \end{vmatrix} \neq 0$$

So, L and L_1 are skew lines.

For line $L_2: \frac{x-6}{-5} = \frac{y}{2} = \frac{z}{3}$,

$$\begin{vmatrix} 0-6 & 4-0 & 0-0 \\ -5 & 2 & 3 \\ 0 & -4 & 5 \end{vmatrix} \neq 0$$

So, L and L_2 are skew lines.

Clearly, line L_3 and L intersect at (0, 4, 0).

But L_3 lies neither in plane P_1 nor in plane P_2.

For line $L_4 : \frac{x}{1} = \frac{y-4}{-2} = \frac{z}{3}$,

$$\begin{vmatrix} 1-0 & 1-4 & 1-0 \\ 1 & -2 & 3 \\ 0 & -4 & 5 \end{vmatrix} \neq 0$$

So, L and L_4 are skew lines.

12. (1), (2), (3)

Equation of plane is $\vec{r} - \hat{k} = t(-\hat{i} + \hat{j}) + p(-\hat{i} + \hat{k})$

or $\begin{vmatrix} x & y & z-1 \\ -1 & 1 & 0 \\ -1 & 0 & 1 \end{vmatrix} = 0$

or $x + y + z - 1 = 0$

Point $S(\alpha, \beta, \gamma)$ is the reflection of $Q(10, 15, 20)$ in the above plane.

$$\frac{\alpha-10}{1} = \frac{\beta-15}{1} = \frac{\gamma-20}{1} = -\frac{2(10+15+20-1)}{3}$$

$\Rightarrow \quad 3\alpha = -58, 3\beta = -43, 3\gamma = -28$

So, options (1), (2) and (3) are correct.

13. (1), (2), (3)

In set S, let $X = (x, y, z)$

We have $XP^2 - XQ^2 = 50$

$\therefore \quad (x-1)^2 + (y-2)^2 + (z-3)^2 - (x-4)^2 - (y-2)^2 - (z-7)^2 = 50$

$\Rightarrow \quad 6x + 8z - 105 = 0 \quad$ ...(i)

In set T, let $Y = (x, y, z)$

We have, $YQ^2 - YP^2 = 50$

$\Rightarrow \quad 6x + 8z - 5 = 0 \quad$ ...(ii)

Clearly, equation (i) and (ii) represent two parallel planes.

Distance between planes $= \frac{|105-5|}{\sqrt{6^2+8^2}} = 10$

Option (1) is true, as there will be infinity many triangles on plane (1) having area 1 unit.

Option (2) is true as any point T on line LM on the plane (2), lies on the same plane.

Consider rectangle $ABCD$, such that A and B lie on S and C and D lie on T.

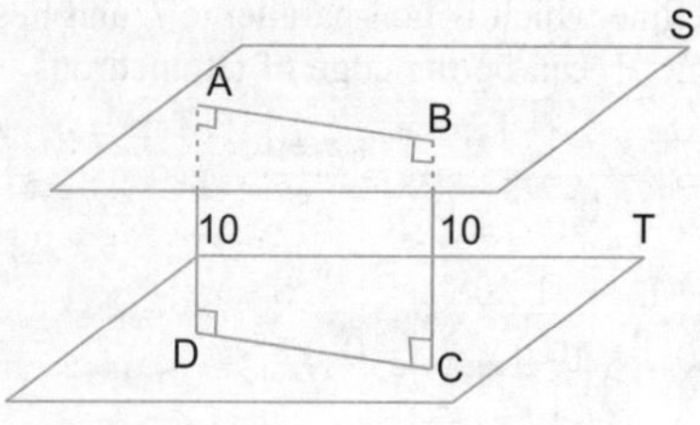

Now $AD = BC = 10$

Perimeter of rectangle is 48 units.

$\therefore \quad AB = CD = 14$ units, which is possible.

Thus, infinitely many rectangles are possible.

So, option (3) is true.

Now, if $ABCD$ is a square of perimeter 48 then $AD = BC = 12$, which is not possible.

So, option (4) is not true.

14. (1), (3)

Given line $P_1 : \frac{x-2}{1} = \frac{y-4}{2} = \frac{z-6}{1}$

Equation of line through point $P(1, 3, 2)$ and parallel to the line P_1 is

$$P_2 : \frac{x-1}{1} = \frac{y-3}{2} = \frac{z-2}{1}$$

Line P_2 intersects plane L_1 at Q.

Let $Q \equiv (\lambda + 1, 2\lambda + 3, \lambda + 2)$

Putting this point in plane L_1, we get

$\lambda + 1 - 2\lambda - 3 + 3\lambda + 6 = 6$ or $\lambda = 1$

$\therefore \quad Q \equiv (2, 5, 3)$

Now, line through point Q and perpendicular to the plane L_1 is

$$P_3 : \frac{x-2}{1} = \frac{y-5}{-1} = \frac{z-3}{3}$$

Line P_3 intersects plane L_2 at R

Let R $\equiv (\mu + 2, -\mu + 5, 3\mu + 3)$

Putting this point in plane L_2, we get

$2\mu + 4 + \mu - 5 + 3\mu + 3 = -4$ or $\mu = -1$

$\therefore \quad R \equiv (1, 6, 0)$

Now, $PQ = \sqrt{1+4+1} = \sqrt{6}$

Centroid of ΔPQR is $\left(\frac{1+2+1}{3}, \frac{3+5+6}{3}, \frac{2+3+0}{3}\right)$

$\equiv \left(\frac{4}{3}, \frac{14}{3}, \frac{5}{3}\right)$

Also, $QR = \sqrt{1+1+9} = \sqrt{11}$

and $PR = \sqrt{0+9+4} = \sqrt{13}$

$\therefore \quad$ Perimeter of $\Delta PQR = \sqrt{6} + \sqrt{11} + \sqrt{13}$

15. (1), (3)

$L_1 : 2x + 3y + z - 4 = 0 = x + 2y + z - 5$

So, line L_1 is parallel to the vector

$$\begin{vmatrix} \hat{i} & \hat{j} & \hat{k} \\ 2 & 3 & 1 \\ 1 & 2 & 1 \end{vmatrix}$$

$= \hat{i} - \hat{j} + \hat{k}$

Now, $P \equiv (2, -1, 3)$

So, $\quad L_2 : \frac{x-2}{1} = \frac{y+1}{-1} = \frac{z-3}{1}$

$\therefore \quad$ any point on L_2 is $(\lambda + 2, -\lambda - 1, \lambda + 3)$

Putting this point in the equation of plane M, we get

$2(\lambda + 2) + (-\lambda - 1) - 2(\lambda + 3) = 6$

$\lambda = -9$

$\therefore \quad Q \equiv (-7, 8, -6)$

Let $R \equiv (\alpha, \beta, \gamma)$

$\therefore \quad \frac{\alpha-2}{2} = \frac{\beta+1}{1} = \frac{\gamma-3}{-2} = \frac{-(2(2)+(-1)-2(3)-6)}{4+1+4} = 1$

So, $R \equiv (4, 0, 1)$

$\therefore \quad \overrightarrow{PQ} = -9\hat{i} + 9\hat{j} - 9\hat{k}$

$\overrightarrow{PR} = 2\hat{i} + \hat{j} - 2\hat{k}$

and $\quad \overrightarrow{QR} = 11\hat{i} - 8\hat{j} + 7\hat{k}$

$\therefore \quad |\overrightarrow{PQ}| = \sqrt{81+81+81} = 9\sqrt{3}$

$|\overrightarrow{QR}| = \sqrt{121+64+49} = \sqrt{234}$

Angle between $\overrightarrow{PQ}$ and $\overrightarrow{PR}$ is given by

$$\cos\alpha = \frac{-18+9+18}{9\sqrt{3}\,.\,3} = \frac{1}{3\sqrt{3}}$$

or $$\alpha = \cos^{-1}\frac{1}{3\sqrt{3}} = \sin^{-1}\frac{\sqrt{26}}{3\sqrt{3}}$$

So, arear of ΔPQR

$$= \frac{1}{2}|\overrightarrow{PQ}|.|\overrightarrow{PR}|.\sin\alpha$$

$$= \frac{1}{2}.9\sqrt{3}.3.\frac{\sqrt{26}}{3\sqrt{3}}$$

$$= \frac{3}{2}\sqrt{234}$$

Matrix Match Type

1. (3)

Plane perpendicular to P_1 and P_2 has direction ratios of normal given by

$$\begin{vmatrix} \hat{i} & \hat{j} & \hat{k} \\ 7 & 1 & 2 \\ 3 & 5 & -6 \end{vmatrix} = -16\hat{i}+48\hat{j}+32\hat{k} \quad \ldots\text{(i)}$$

For point of intersection of given lines

$$(2\lambda_1+1, -\lambda_1, \lambda_1-3) \equiv (\lambda_2+4, \lambda_2-3, 2\lambda_2-3)$$

$\Rightarrow$ $2\lambda_1+1 = \lambda_2+4$ or $2\lambda_1-\lambda_2=3$

and $-\lambda_1 = \lambda_2-3$ or $\lambda_1+\lambda_2=3$

$\Rightarrow$ $\lambda_1 = 2, \lambda_2 = 1$

$\therefore$ Point is $(5, -2, -1)$ ...(ii)

From (i) and (ii), required planes is

$$-1(x-5)+3(y+2)+2(z+1)=0$$

or $x-3y-2z = 13$

$\Rightarrow$ $a=1, b=-3, c=-2, d=13$

2. (2)

Line l_2 may intersect or remain parallel to the plane H.

If l_2 intersects H, then $d(H) = 0$.

If l_2 is parallel to H, then

$d(H)$ = distance between l_2 and H

$= d(H_0)$, which is the maximum value of $d(H)$

So, for $d(H_0)$, l_2 is parallel to H, which is plane H_0.

Hence, equation of plane H_0 is

$$\begin{vmatrix} x & y & z \\ 1 & 1 & 1 \\ 1 & 0 & 1 \end{vmatrix} = 0$$

or $H_0 : x - z = 0$ ...(1)

(P) $d(H_0)$ = distance of line l_2 from plane H_0

= distance of $(0, 1, -1)$ from H_0

$$= \frac{1}{\sqrt{2}}$$

(Q) Distance of $(0, 1, 2)$ from H_0

$$= \frac{2}{\sqrt{2}} = \sqrt{2}$$

(R) Distance of $(0, 0, 0)$ from H_0

$= 0$

(S) Line $y = z, x = 1$ meets plane H_0 at $P \equiv (1, \lambda, \lambda)$.

$\therefore$ $1-\lambda = 0$ or $\lambda = 1$

So, $P \equiv (1, 1, 1)$

$\therefore$ $OP = \sqrt{3}$, where O is origin.

3. (3)

Point on line L_1 can be taken as $(\lambda-11, 2\lambda-21, 3\lambda-29)$.

This point lies on line L_2 as lines intersect.

$$\therefore \quad \frac{\lambda-11+16}{3} = \frac{2\lambda-21+11}{2} = \frac{3\lambda-29+4}{\gamma}$$

or $$\frac{\lambda+5}{3} = \lambda-5 = \frac{3\lambda-25}{\gamma}$$

From first two ratios, we get $\lambda = 10$

From last two ratios, we get

$$10-5 = \frac{3(10)-25}{\gamma} \text{ or } \gamma = 1$$

Also point of intersection is

$$R_1 \equiv (10-11, 20-21, 30-29)$$
$$\equiv (-1, -1, 1)$$

$$\therefore \quad \overrightarrow{OR_1} = -\hat{i}-\hat{j}+\hat{k}$$

Line L_1 is parallel to the vector, $\vec{n}_1 = \hat{i}+2\hat{j}+3\hat{k}$

Line L_2 is parallel to the vector, $\vec{n}_2 = 3\hat{i}+2\hat{j}+\hat{k}$

$\therefore$ Vector normal to the plane of lines L_1 and L_2 is

$$\begin{vmatrix} \hat{i} & \hat{j} & \hat{k} \\ 1 & 2 & 3 \\ 3 & 2 & 1 \end{vmatrix} = -4\hat{i}+8\hat{j}-4\hat{k}$$

So, the possible choice for $\hat{n}$ is $\frac{1}{\sqrt{6}}\hat{i} - \frac{2}{\sqrt{6}}\hat{j} + \frac{1}{\sqrt{6}}\hat{k}$

Hence, the possible value of

$$\overrightarrow{OR_1}.\hat{n} = -\frac{1}{\sqrt{6}} + \frac{2}{\sqrt{6}} + \frac{1}{\sqrt{6}} = \frac{2}{\sqrt{6}} = \sqrt{\frac{2}{3}}$$

Numerical Value Type

1. (0.75)

Putting $(\lambda, 0, 0)$ in $x+y+z=1$, we get $\lambda = 1$.

So, $P \equiv (1, 0, 0)$.

Putting $(\mu, \mu, 0)$, we get $2\mu = 1$. So, $Q \equiv \left(\frac{1}{2}, \frac{1}{2}, 0\right)$.

Putting (γ, γ, γ), we get $\gamma = \frac{1}{3}$. So, $R \equiv \left(\frac{1}{3}, \frac{1}{3}, \frac{1}{3}\right)$.

Area of triangle PQR, $\Delta = \frac{1}{2}|\overrightarrow{PQ} \times \overrightarrow{PR}|$

$$= \frac{1}{2}\left|\left(\frac{\hat{i}-\hat{j}}{2}\right) \times \left(\frac{2\hat{i}-\hat{j}-\hat{j}}{3}\right)\right|$$

$$= \frac{1}{12}|\hat{i}+\hat{j}+\hat{k}| = \frac{\sqrt{3}}{12} \Rightarrow (6\Delta)^2 = 0.75$$

Integer Answer Type

1. (8)

Let $P \equiv (\alpha, \beta, \gamma)$.

R is image of P in xy-plane. So, $R \equiv (\alpha, \beta, -\gamma)$.

$Q(x, y, z)$ is the image of P in plane $x + y - 3 = 0$.

$$\therefore \quad \frac{x-\alpha}{1} = \frac{y-\beta}{1} = \frac{z-\gamma}{0} = \frac{-2(\alpha+\beta-3)}{2}$$

$\therefore \quad x = 3-\beta, y = 3-\alpha, z = \gamma$

Point $Q(3-\beta, 3-\alpha, \gamma)$ lies on z-axis.

$\therefore \quad \beta = 3, \alpha = 3$

$\therefore \quad P \equiv (3, 3, \gamma)$ whose distance from x-axis is 5.

$\therefore \quad 9 + \gamma^2 = 25$

So, $\gamma^2 = 16 \Rightarrow \gamma = 4$

Hence $P \equiv (3, 3, 4)$ and $R \equiv (3, 3, -4)$.

$\therefore \quad PR = 8$

2. (45) From the given information we draw the following figure.

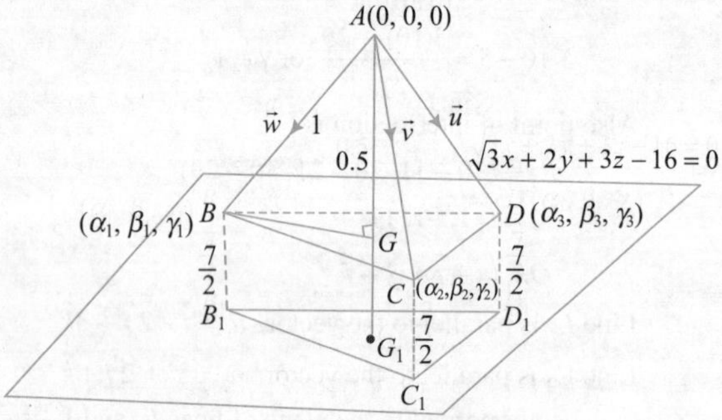

$ABCD$ is tetrahedron having vertex A at origin.

$\overline{AB} = \vec{u}, \overline{AC} = \vec{v}, \overline{AD} = \vec{w}$

Distance of points B, C and D from the given plane $\sqrt{3}x + 2y + 3z = 16$ is $\frac{7}{2}$.

So, given plane is parallel to the base BCD of tetrahedron.

Also, $|\vec{u}| = |\vec{v}| = |\vec{w}| = 1$ as $\alpha^2 + \beta^2 + \gamma^2 = 1$

Now, $|\vec{u} - \vec{v}| = |\vec{v} - \vec{w}| = |\vec{w} - \vec{u}|$

$\therefore \quad |\vec{u}|^2 + |\vec{v}|^2 - 2\vec{u}\cdot\vec{v} = |\vec{v}|^2 + |\vec{w}|^2 - 2\vec{v}\cdot\vec{w} = |\vec{u}|^2 + |\vec{w}|^2 - 2\vec{u}\cdot\vec{w}$

$\Rightarrow \vec{u}\cdot\vec{v} = \vec{v}\cdot\vec{w} = \vec{u}\cdot\vec{w}$

So, $\vec{u}, \vec{v}$ and $\vec{w}$ are equally inclined to each other.

Therefore, BCD is an equilateral triangle.

Distance of A from the given plane, $AG_1 = 4$

So, the height of the tetrahedron, $AG = 4 - \frac{7}{2} = \frac{1}{2}$

Now, in right angled triangle AGB, $BG = \sqrt{1 - \left(\frac{1}{2}\right)^2} = \frac{\sqrt{3}}{2}$

BG is circumradius of equilateral triangle BCD.

So, altitude of triangle BCD, $BM = \frac{3\sqrt{3}}{4}$

$$\therefore \ BC = \frac{\frac{3\sqrt{3}}{4}}{\sin 60°} = \frac{3}{2}$$

So, volume of tetrahedron $= \frac{1}{3} \times$ Area of $BCD \times AG$

$$= \frac{1}{3} \times \frac{\sqrt{3}}{4}\left(\frac{3}{2}\right)^2 \times \frac{1}{2}$$

$$= \frac{3\sqrt{3}}{32}$$

Therefore, volume of parallelopiped formed by $\vec{u}, \vec{v}$ and $\vec{w}$

$$V = 6 \times \frac{3\sqrt{3}}{32} = \frac{9\sqrt{3}}{16}$$

Hence, $\frac{80}{\sqrt{3}} V = \frac{80}{\sqrt{3}} \times \frac{9\sqrt{3}}{16} = 45$

3. (5)

$(\overline{OP} \times \overline{OQ}) \cdot \overline{OR} = 0$

$$\therefore \quad \begin{vmatrix} \frac{\alpha-1}{\alpha} & 1 & 1 \\ 1 & \frac{\beta-1}{\beta} & 1 \\ 1 & 1 & \frac{1}{2} \end{vmatrix} = 0$$

Applying $C_1 \to C_1 - C_2$ and $C_2 \to C_2 - C_3$, we get

$$\begin{vmatrix} -\frac{1}{\alpha} & 0 & 1 \\ \frac{1}{\beta} & -\frac{1}{\beta} & 1 \\ 0 & \frac{1}{2} & \frac{1}{2} \end{vmatrix} = 0$$

$$\Rightarrow \quad \frac{1}{2\alpha\beta} + \frac{1}{2\beta} + \frac{1}{2\alpha} = 0$$

$\Rightarrow \quad \alpha + \beta + 1 = 0$ ...(i)

Also, point $(\alpha, \beta, 2)$ lies on the plane $3x + 3y - z + l = 0$

$\therefore \quad 3\alpha + 3\beta - 2 + l = 0$

From (i), $\alpha + \beta = -1$

$\therefore \quad -3 - 2 + l = 0$ or $l = 5$.

4. (2)

$2\vec{p} + \vec{q} = 5\hat{i} + \hat{j} + 7\hat{k}$

$\vec{p} - 2\vec{q} = 3\hat{j} + \hat{k}$

$$\vec{p} \times \vec{q} = \begin{vmatrix} \hat{i} & \hat{j} & \hat{k} \\ 2 & 1 & 3 \\ 1 & -1 & 1 \end{vmatrix} = 4\hat{i} + \hat{j} - 3\hat{k}$$

Now,

$15\hat{i} + 10\hat{j} + 6\hat{k} = \alpha(2\vec{p} + \vec{q}) + \beta(\vec{p} - 2\vec{q}) + \gamma(\vec{p} \times \vec{q})$

$\Rightarrow \quad 15\hat{i} + 10\hat{j} + 6\hat{k}$

$= \alpha(5\hat{i} + \hat{j} + 7\hat{k}) + \beta(3\hat{j} + \hat{k}) + \gamma(4\hat{i} + \hat{j} - 3\hat{k})$

$\Rightarrow \quad 5\alpha + 4\gamma = 15$ ...(i)

$\alpha + 3\beta + \gamma = 10$ ...(ii)

and $\quad 7\alpha + \beta - 3\gamma = 6$ ...(iii)

From (ii) – 3(iii), we get

$-20\alpha + 10\gamma = -8$

or $\quad 5\alpha - \frac{5}{2}\gamma = 2$ ...(iv)

From (i) – (iv), we get

$\frac{13}{2}\gamma = 13 \Rightarrow \gamma = 2$

CHAPTER 1

Introduction to 3D Geometry

DPP 1.1

Introduction to 3D Geometry

Single Correct Answer Type

1. Let points (1, 2, 3), (0, –4, 3), (2, 3, 5) and (1, –5, –3) be vertices of a tetrahedron. Now, consider the following two statements.

S_1: The point where line joining the mid points of opposite edges are concurrent is (1, –1, 2).

S_2: The point where line segments joining vertex to the centroid of the opposite face are concurrent is (1, –1, 2).

Then

(a) Only S_1 is true

(b) Only S_2 is true

(c) Both S_1 and S_2 are false

(d) Both S_1 and S_2 are true

2. In a three dimensional coordinate system *P*, *Q* and *R* are the images of a point *A*(*a*, *b*, *c*) in the *xy*- the *yz*- and the *zx*-planes, respectively. If *G* is the centroid of triangle *PQR*, then

(a) area of triangle *AOG* is $a^2 + b^2 + c^2$

(b) area of triangle *AOG* is $\frac{2}{3}(a^2 + b^2 + c^2)$

(c) points *A*, *O* and *G* are collinear

(d) none of these

3. A parallelepiped is formed by the planes drawn parallel to the coordinate axes through the points *A* (1, 2, 3) and *B* (9, 8, 5). The volume of that parallelepiped is equal to (in cubic units)

(a) 192 (b) 48

(c) 32 (d) 96

4. The ratio in which the plane $2x - 1 = 0$ divides the line joining (–2, 4, 7) and (3, –5, 8) is

(a) 2 : 3 (b) 4 : 5

(c) 7 : 8 (d) 1 : 1

5. Given two points *A* and *B*. If the area of triangle *ABC* is constant, then the locus of point *C* in space is

(a) sphere (b) cone

(c) cylinder (d) none of these

6. A line segment joining (1, 0, 1) and the origin (0, 0, 0) is revolved about the *x*-axis to form a right circular cone. If (*x*, *y*, *z*) is any point on the cone, other than the origin, then it satisfies the equation

(a) $x^2 - 2y^2 - z^2 = 0$

(b) $x^2 - y^2 - z^2 = 0$

(c) $2x^2 - y^2 - 2z^2 = 0$

(d) $x^2 - 2y^2 - 2z^2 = 0$

7. If the sum of square of distances of point *P* from the three axes is constant, then the locus of point *P* is

(a) plane (b) cylinder

(c) cone (d) sphere

8. If the directions cosines of a line are *k*, *k* and *k*, then

(a) $k > 0$ (b) $0 < k < 1$

(c) $k = 1$ (d) $k = \frac{1}{\sqrt{3}}$ or $-\frac{1}{\sqrt{3}}$

9. Consider four points *A* (0, –1, –1), *B* (4, 5, 1), *C* (3, 9, 4) and *D* (– 4, 4, 4). Now, consider the following two statements:

S_1: *AB* divides *CD* externally in the ratio 5 : 3.

S_2 : *CD* divides *AB* externally in the ratio 1 : 2.

Then

(a) Only S_1 is true

(b) Only S_2 is true

(c) Both S_1 and S_2 are false

(d) Both S_1 and S_2 are true

Multiple Correct Answers Type

10. Consider the line segment joining the points $A \equiv (2, 3, 4)$ and $B \equiv (-1, 2, -5)$. Then

(a) xy-plane divides AB internally in the ratio 4 : 5

(b) xz-plane divides AB externally in the ratio 2 : 3

(c) yz-plane divides AB internally in the ratio 2 : 1

(d) xy-plane divides AB internally in the ratio 2 : 3

11. Consider the sphere having equation $x^2 + y^2 + z^2 - 2x + 4y - 6z - 4 = 0$. Then

(a) Radius of the circle in which sphere cuts the xy-plane is 3.

(b) Radius of the circle in which sphere cuts the yz-plane is $\sqrt{17}$.

(c) Radius of the circle in which sphere cuts the yz-plane is $\sqrt{10}$.

(d) None of these

12. If the equation of one of the faces of the cube whose volume is 125 cu. units is $z = 3$, then the equation of the other faces can be

(a) $z = 4$ (b) $z = -2$

(c) $z = 8$ (d) $z = 10$

13. If the line joining the points A (2, −1, 4) and B (−3, 1, a) is intersected by the x-axis at point P, then

(a) the value of 'a' is −4

(b) the value of 'a' is −2

(c) the abscissa of point P is −1

(d) the abscissa of point P is −1/2

Numerical Value Type

14. The distance of point P from the point (2, 3, −1) is twice the distance of P from the point (−1, 0, 2), then the locus of P is a sphere square of whose radius is ___________.

15. If the combined equation of two faces of a cube is $y^2 - 4y - 5 = 0$, then the volume of the cube (in cu. units) is ___________.

16. If the two ends of diagonal of a cube are (1, 2, 3) and (−2, 0, −3), then the volume of cube is $\frac{\lambda}{3\sqrt{3}}$ cu. units, where λ is ___________.

Answer Key

Single Correct Answer Type

1. (d) **2.** (c) **3.** (d) **4.** (d) **5.** (c) **6.** (b) **7.** (d) **8.** (d) **9.** (d)

Multiple Correct Answers Type

10. (a, b, c) **11.** (a, b, c) **12.** (b, c) **13.** (a, d)

Numerical Value Type

14. (13) **15.** (216) **16.** (343)

CHAPTER 2

Introduction to Vectors

DPP 2.1

Vector Geometry

Single Correct Answer Type

1. A line makes an angle θ with both, the x-axis and the y-axis. Then possible range of values of θ is

(a) $\left[0, \frac{\pi}{4}\right]$ (b) $\left[0, \frac{\pi}{2}\right]$

(c) $\left[\frac{\pi}{4}, \frac{\pi}{2}\right]$ (d) $\left[\frac{\pi}{6}, \frac{\pi}{3}\right]$

2. A line segment has length 63 and has direction ratios 3, –2 and 6. Then the components of the line vector on the axes are

(a) –27, 18, 54 (b) 27, –18, 54

(c) 27, –18, –54 (d) –27, –18, –54

3. Let $\vec{a}$, $\vec{b}$ and $\vec{c}$, be position vectors of A, B and C, respectively, of ΔABC. If $|\vec{a}-\vec{b}|=4$, $|\vec{b}-\vec{c}|=2$ and $|\vec{c}-\vec{a}|=3$, then the distance between the centroid and the incentre of ΔABC is

(a) 1 (b) 1/2

(c) 1/3 (d) 2/3

4. Let position vectors of points A, B and C of triangle ABC, respectively, be $\hat{i}+\hat{j}+2\hat{k}$, $\hat{i}+2\hat{j}+\hat{k}$ and $2\hat{i}+\hat{j}+\hat{k}$. If l_1, l_2 and l_3 are the lengths of perpendiculars drawn from the orthocentre 'O' on the sides AB, BC and CA, respectively, then $(l_1+l_2+l_3)$ equals to

(a) $\frac{2}{\sqrt{6}}$ (b) $\frac{3}{\sqrt{6}}$

(c) $\frac{\sqrt{6}}{2}$ (d) $\frac{\sqrt{6}}{3}$

5. The unit vector parallel to the resultant of the vectors $2\hat{i}+3\hat{j}-\hat{k}$ and $4\hat{i}-3\hat{j}+2\hat{k}$ is

(a) $\frac{1}{\sqrt{37}}(6\hat{i}+\hat{k})$ (b) $\frac{1}{\sqrt{37}}(6\hat{i}+\hat{j})$

(c) $\frac{1}{\sqrt{37}}(6\hat{j}+\hat{k})$ (d) none of these

6. In a regular hexagon $ABCDEF$, $\overrightarrow{AB}+\overrightarrow{AC}+\overrightarrow{AD}+\overrightarrow{AE}+\overrightarrow{AF}=k\overrightarrow{AD}$, where k is equal to

(a) 1 (b) 2

(c) 3 (d) none of these

7. If $\vec{a}+\vec{b}+\vec{c}=0$ and $|\vec{a}|=3$, $|\vec{b}|=5$, $|\vec{c}|=7$, then the angle between $\vec{a}$ and $\vec{b}$ is

(a) $\frac{\pi}{2}$ (b) $\frac{\pi}{3}$

(c) $\frac{\pi}{4}$ (d) $\frac{\pi}{6}$

8. If the sum of two unit vectors is a unit vector, then the magnitude of their difference is

(a) $\sqrt{2}$ (b) $\sqrt{3}$

(c) 1 (d) none of these

9. If $\vec{p}, \vec{q}, \vec{r}$ are unit vectors such that $\vec{p}$ is perpendicular to $\vec{q}$ and $\vec{r}$, and $|\vec{p}+\vec{q}+\vec{r}|=1$, then angle between $\vec{q}$ and $\vec{r}$ is

(a) $\frac{\pi}{2}$ (b) π

(c) 0 (d) $\frac{2\pi}{3}$

10. The position vectors of the points A, B and C are $\hat{i}+2\hat{j}-\hat{k}$, $\hat{i}+\hat{j}+\hat{k}$ and $2\hat{i}+3\hat{j}+2\hat{k}$, respectively. If A is chosen as the origin, then the position vectors of B and C are
(a) $\hat{i}+2\hat{k}, \hat{i}+\hat{j}+3\hat{k}$
(b) $\hat{j}+2\hat{k}, \hat{i}+\hat{j}+3\hat{k}$
(c) $-\hat{j}+2\hat{k}, \hat{i}-\hat{j}+3\hat{k}$
(d) $-\hat{j}+2\hat{k}, \hat{i}+\hat{j}+3\hat{k}$

11. If the position vectors of P and Q are $\hat{i}+2\hat{j}-7\hat{k}$ and $5\hat{i}-3\hat{j}+4\hat{k}$, respectively, then the cosine of the angle between $\overline{OQ}$ and z-axis is
(a) $\frac{4}{\sqrt{162}}$ (b) $\frac{11}{\sqrt{162}}$
(c) $\frac{5}{\sqrt{162}}$ (d) $\frac{-5}{\sqrt{162}}$

12. The non-zero vectors $\vec{a}$, $\vec{b}$ and $\vec{c}$ are related by $\vec{a} = 8\vec{b}$ and $\vec{c} = -7\vec{b}$. Then the angle between $\vec{a}$ and $\vec{c}$ is
(a) $\frac{\pi}{4}$ (b) $\frac{\pi}{2}$
(c) π (d) 0

13. The unit vector bisecting the angle between vectors, positive y-axis and positive z-axis is
(a) $\frac{\hat{i}+\hat{j}+\hat{k}}{\sqrt{3}}$ (b) $\frac{\hat{j}-\hat{k}}{\sqrt{2}}$
(c) $\frac{\hat{j}+\hat{k}}{\sqrt{2}}$ (d) $\frac{-\hat{j}+\hat{k}}{\sqrt{2}}$

14. If $\vec{a}$ and $\vec{b}$ are position vectors of points A and B, respectively, then the position vector of a point C when $\overrightarrow{AB}$ is produced such that $\overrightarrow{AC} = 2015\ \overrightarrow{AB}$ is
(a) $2014\vec{a} - 2015\vec{b}$ (b) $2014\vec{b} + 2015\vec{a}$
(c) $2015\vec{b} + 2014\vec{a}$ (d) $2015\vec{b} - 2014\vec{a}$

15. Given below are two statements. One is labelled as Assertion (A) and the other is labelled as Reason (R).
Assertion (A): If $|\vec{a} + \vec{b}| = |\vec{a} - \vec{b}|$, then $\vec{a}$ and $\vec{b}$ are perpendicular to each other.
Reason (R): If the diagonals of a parallelogram are equal in magnitude, then the parallelogram is rectangle.
(a) (A) is true but (R) is false.
(b) Both (A) and (R) are true and (R) is the true explanation of (A).
(c) (A) is false and (R) is true.
(d) Both (A) and (R) are true but (R) is not the true explanation of (A).

16. Let $\vec{a}$, $\vec{b}$, $\vec{c}$ and $\vec{d}$ be position vectors of four points A, B, C and D such that $3\vec{a} - 2\vec{b} + 5\vec{c} - 6\vec{d} = 0$.
Given below are two statements. One is labelled as Assertion (A) and the other is labelled as Reason (R).
Assertion (A): Points A, B, C and D are coplanar.
Reason (R): Line segments AC and BD intersect.
(a) (A) is true but (R) is false.
(b) Both (A) and (R) are true and (R) is the true explanation of (A).
(c) (A) is false and (R) is true.
(d) Both (A) and (R) are true but (R) is not the true explanation of (A).

Multiple Correct Answers Type

17. If $A(\vec{a})$, $B(\vec{b})$ and $C(\vec{c})$ are three collinear points such that $(\cos^2 \alpha)\vec{a} + (\sin^2 \beta)\vec{b} - (2\sec^2 \gamma)\vec{c} = \vec{0}$, then
(a) C is midpoint of AB
(b) B is midpoint of AC
(c) A divides CB externally in the ratio 1:2
(d) B divides AC externally in the ratio 3:1

18. Let $\vec{a} = (1, 1, -1)$, $\vec{b} = (5, -3, -3)$ and $\vec{c} = (3, -1, 2)$. If $\vec{r}$ is collinear with $\vec{c}$ and has length $\frac{|\vec{a}+\vec{b}|}{2}$, then $\vec{r}$ equals
(a) $-3\vec{c}$ (b) $-\vec{c}$
(c) $\vec{c}$ (d) $\frac{2}{3}\vec{c}$

19. A line passes through the points whose position vectors are $\hat{i}+\hat{j}-2\hat{k}$ and $\hat{i}-3\hat{j}+\hat{k}$. The position vector of a point on it at unit distance from the first point is
(a) $\frac{1}{5}(5\hat{i}+\hat{j}-7\hat{k})$ (b) $\frac{1}{5}(5\hat{i}+9\hat{j}-13\hat{k})$
(c) $(\hat{i}-4\hat{j}+3\hat{k})$ (d) $\frac{1}{5}(\hat{i}-4\hat{j}+3\hat{k})$

Linked Comprehension Type

For Questions 20–22

Three points A, B and C have position vectors $-2\vec{a} + 3\vec{b} + 5\vec{c}$, $\vec{a} + 2\vec{b} + 3\vec{c}$ and $7\vec{a} - \vec{c}$, respectively, with reference to an origin O.

20. Which of the following is true?
(a) $\overrightarrow{AC} = 2\overrightarrow{AB}$ (b) $\overrightarrow{AC} = -3\overrightarrow{AB}$
(c) $\overrightarrow{AC} = 3\overrightarrow{AB}$ (d) None of these

21. Which of following is true?

(a) $2\overrightarrow{OA}-3\overrightarrow{OB}+\overrightarrow{OC}=0$

(b) $2\overrightarrow{OA}+7\overrightarrow{OB}-9\overrightarrow{OC}=0$

(c) $\overrightarrow{OA}+\overrightarrow{OB}+\overrightarrow{OC}=0$

(d) None of these

22. Point B divides AC in the ratio

(a) 1 : 1 (b) 2 : 3

(c) 2 : (–3) (d) 1 : 2

Matrix Match Type

23. $ABCDEF$ is a regular hexagon in the xy-plane with vertices in the anticlockwise direction and $\overrightarrow{AB}=2\hat{i}$. Now, match the List I with the List II.

List I (Function)	List II (Nature of function)
(1) $\overrightarrow{CD}$ =	(i) $3\hat{i}+\sqrt{3}\hat{j}$
(2) $\overrightarrow{AC}$ =	(ii) $-3\hat{i}+\sqrt{3}\hat{j}$
(3) $\overrightarrow{AD}$ =	(iii) $-\hat{i}+\sqrt{3}\hat{j}$
(4) $\overrightarrow{CE}$ =	(iv) $2\hat{i}+2\sqrt{3}\hat{j}$

(a) (1)-(iii), (2)-(iv), (3)-(i), (4)-(ii)

(b) (1)-(ii), (2)-(iii), (3)-(iv), (4)-(i)

(c) (1)-(iii), (2)-(i), (3)-(iv), (4)-(ii)

(d) (1)-(iv), (2)-(i), (3)-(ii), (4)-(iii)

Numerical Value Type

24. Orthocentre of an equilateral triangle ABC is the origin O. If $\overrightarrow{OA}=\vec{a}$, $\overrightarrow{OB}=\vec{b}$ and $\overrightarrow{OC}=\vec{c}$, then $\overrightarrow{AB}+2\overrightarrow{BC}+3\overrightarrow{CA}=\lambda\vec{a}$, where $\lambda=$ _____.

25. Given $\vec{a}$, $\vec{b}$ and $\vec{c}$ are three vectors such that $\vec{b}$ and $\vec{c}$ are unit like parallel vectors and $|\vec{a}|=4$. If $\vec{a}+\lambda\vec{c}=2\vec{b}$, then the sum of all possible values of λ is _______.

26. Let D, E and F be the mid points of the sides BC, CA and AB, respectively, of a triangle ABC. If λ is a scalar, such that $\overrightarrow{AD}+\frac{2}{3}\overrightarrow{BE}+\frac{1}{3}\overrightarrow{CF}=\lambda\overrightarrow{AC}$, then 10λ is equal to _____.

27. Let O be an interior point of ΔABC such that $\overrightarrow{OA}+2\overrightarrow{OB}+3\overrightarrow{OC}=\vec{0}$. Then the ratio of the area of ΔABC to the area of ΔAOC is ______.

Answer Key

Single Correct Answer Type

1. (c) **2.** (b) **3.** (c) **4.** (c) **5.** (a) **6.** (c) **7.** (b) **8.** (b) **9.** (b) **10.** (d)
11. (b) **12.** (c) **13.** (c) **14.** (d) **15.** (b) **16.** (b)

Multiple Correct Answers Type

17. (a, c) **18.** (b, c) **19.** (a, b)

Linked Comprehension Type

20. (c) **21.** (a) **22.** (d)

Matrix Match Type

23. (c)

Numerical Value Type

24. (3) **25.** (4) **26.** (5) **27.** (3)

DPP 2.2

Linear Combination of Vectors

Single Correct Answer Type

1. If $\vec{a}$ and $\vec{b}$ are non - zero vectors which are linearly dependent such that $\dfrac{|\vec{a}+\vec{b}|}{|\vec{a}-\vec{b}|} = 2$, $|\vec{b}| > |\vec{a}|$, then
   (a) $\vec{b} = 3\vec{a}$ (b) $\vec{b} = -3\vec{a}$
   (c) $\vec{b} = 2\vec{a}$ (d) $\vec{b} = -2\vec{a}$

2. The number of integral values of p for which $(p+1)\hat{i}-3\hat{j}+p\hat{k}$, $p\hat{i}+(p+1)\hat{j}-3\hat{k}$ and $-3\hat{i}+p\hat{j}+(p+1)\hat{k}$ are linearly dependent vectors is
   (a) 0 (b) 1
   (c) 2 (d) 3

3. The base vectors $\vec{a}_1, \vec{a}_2, \vec{a}_3$ are given in the form of base vectors $\vec{b}_1, \vec{b}_2, \vec{b}_3$ as $\vec{a}_1 = 2\vec{b}_1 + 3\vec{b}_2 - \vec{b}_3$, $\vec{a}_2 = \vec{b}_1 - 2\vec{b}_2 + 2\vec{b}_3$ and $\vec{a}_3 = -2\vec{b}_1 + \vec{b}_2 - 2\vec{b}_3$. If $\vec{F} = 3\vec{b}_1 - \vec{b}_2 + 2\vec{b}_3$, then vector $\vec{F}$ in the form of $\vec{a}_1, \vec{a}_2$ and $\vec{a}_3$ is
   (a) $\vec{F} = 3\vec{a}_1 + 2\vec{a}_2 + 5\vec{a}_3$
   (b) $\vec{F} = 3\vec{a}_1 - 5\vec{a}_2 - 2\vec{a}_3$
   (c) $\vec{F} = 2\vec{a}_1 + 5\vec{a}_2 + 3\vec{a}_3$
   (d) none of these

4. The number of distinct real values of λ for which the vectors $\lambda^3\hat{i}+\hat{k}$, $\hat{i}-\lambda^3\hat{j}$ and $\hat{i}+(2\lambda-\sin\lambda)\hat{j}-\lambda\hat{k}$ are coplanar is
   (a) 0 (b) 1
   (c) 2 (d) 3

5. If the points $A(2-x, 2, 2)$, $B(2, 2-y, 2)$, $C(2, 2, 2-z)$ and $D(1, 1, 1)$ are coplanar, then locus of $P(x, y, z)$ is
   (a) $\dfrac{1}{x}+\dfrac{1}{y}+\dfrac{1}{z}=1$
   (b) $x+y+z=1$
   (c) $\dfrac{1}{1-x}+\dfrac{1}{1-y}+\dfrac{1}{1-z}=1$
   (d) None of these

6. Let a, b, c be distinct non-negative numbers and the vectors $a\hat{i}+b\hat{j}+c\hat{k}$, $\hat{i}+\hat{k}$, $c\hat{i}+c\hat{j}+b\hat{k}$ lie in a plane. Then the quadratic equation $ax^2 + 2cx + b = 0$ has
   (a) real and equal roots
   (b) real unequal roots
   (c) unreal roots
   (d) both roots real and positive

7. In ΔOAB, M is the mid point of AB and C is a point on OM, such that $2\overrightarrow{OC} = \overrightarrow{CM}$. X is a point on the side OB such that $\overrightarrow{OX} = 2\overrightarrow{XB}$. The line XC is produced to meet OA in Y. Then $\dfrac{OY}{YA} =$
   (a) 1/3 (b) 2/7
   (c) 3/2 (d) 2/5

8. Let $A(\vec{a})$, $B(\vec{b})$ and $C(\vec{c})$ be three points such that $\vec{a}=2\hat{i}+\hat{k}$, $\vec{b}=3\hat{i}-\hat{j}+3\hat{k}$ and $\vec{c}=\hat{i}+\hat{j}-\hat{k}$.
   Assertion (A): Points O, A, B and C form a tetrahedron, where 'O' is origin.
   Reason (R): Four non-coplanar points form a tetrahedron.
   (a) (A) is true but (R) is false.
   (b) Both (A) and (R) are true and (R) is the true explanation of (A).
   (c) (A) is false and (R) is true.
   (d) Both (A) and (R) are true but (R) is not the true explanation of (A).

Multiple Correct Answers Type

9. Given three non-zero non-coplanar vectors $\vec{a}, \vec{b}$ and $\vec{c}$. Then which of the following sets of vectors is(are) coplanar?
   (a) $\vec{a}+\vec{b}, \vec{b}+\vec{c}, \vec{c}+\vec{a}$
   (b) $\vec{a}-\vec{b}, \vec{b}+\vec{c}, \vec{c}+\vec{a}$
   (c) $\vec{a}+\vec{b}, \vec{b}-\vec{c}, \vec{c}+\vec{a}$
   (d) $\vec{a}+\vec{b}, \vec{b}+\vec{c}, \vec{c}-\vec{a}$

10. The value of $\lambda \in R$ such that $(x, y, z) \neq (0, 0, 0)$ and $(\hat{i}+\hat{j}+3\hat{k})x+(3\hat{i}-3\hat{j}+\hat{k})y+z(-4\hat{i}+5\hat{j})$ $= \lambda(x\hat{i}+y3\hat{j}+z\hat{k})$ can be
    (a) 0 (b) –1
    (c) 1 (d) 2

Linked Comprehension Type

For Questions 11 and 12

Points X and Y are taken on the sides QR and RS, respectively, of a parallelogram $PQRS$, so that $\overrightarrow{QX} = 4\overrightarrow{XR}$ and $\overrightarrow{RY} = 4\overrightarrow{YS}$. The line XY cuts the line PR at Z.

11. Then $\overrightarrow{PZ}$ =

(a) $\frac{21}{25}\overrightarrow{PR}$ (b) $\frac{19}{25}\overrightarrow{PR}$

(c) $\frac{3}{5}\overrightarrow{PR}$ (d) $\frac{4}{5}\overrightarrow{PR}$

12. Z divides YX in the ratio

(a) 2 : 1 (b) 3 : 1

(c) 4 : 1 (d) 3 : 2

Numerical Value Type

13. On the xy-plane where O is the origin, given points, $A(1, 0)$, $B(0, 1)$ and $C(1, 1)$. Let P, Q and R be moving points on the line OA, OB, OC, respectively, such that $\overrightarrow{OP} = 45t(\overrightarrow{OA})$, $\overrightarrow{OQ} = 60t(\overrightarrow{OB})$ and $\overrightarrow{OR} = (1-t)(\overrightarrow{OC})$ with $t > 0$. If three points P, Q and R are collinear, then the value of $7/t$ is ______.

14. If $\vec{a} = \lambda x\hat{i} + y\hat{j} + 4z\hat{k}$, $\vec{b} = y\hat{i} + x\hat{j} + 3y\hat{k}$ and $\vec{c} = -z\hat{i} - 2z\hat{j} - (\lambda+1)x\hat{k}$ are the sides of triangle as shown in following figure then the value of λ is (where x, y, z are not all zero) ____.

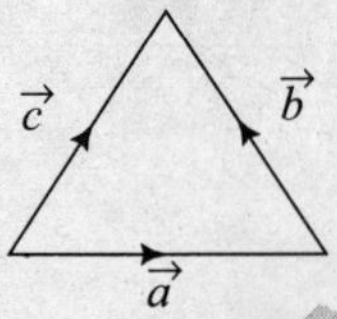

15. If a_1 and a_2 are two values of 'a' for which the unit vector $a\hat{i} + b\hat{j} + \frac{1}{2}\hat{k}$ is linearly dependent with $\hat{i} + 2\hat{j}$ and $\hat{j} - 2\hat{k}$, then $11\left|\frac{1}{a_1} + \frac{1}{a_2}\right| =$ ______.

Answer Key

Single Correct Answer Type

1. (a) **2.** (b) **3.** (c) **4.** (a) **5.** (a) **6.** (a) **7.** (b) **8.** (c)

Multiple Correct Answers Type

9. (b, c, d) **10.** (a, b)

Linked Comprehension Type

11. (a) **12.** (c)

Numerical Value Type

13. (187) **14.** (0) **15.** (16)

CHAPTER 3

Different Products of Vectors

DPP 3.1
Dot Product

Single Correct Answer Type

1. If $a, b > 0$ and $\vec{\alpha} = \frac{\hat{i}}{a} + \frac{4\hat{j}}{b} + b\hat{k}$ and $\vec{\beta} = b\hat{i} + a\hat{j} + \frac{1}{b}\hat{k}$, then the maximum value of $\frac{10}{5 + \vec{\alpha}\cdot\vec{\beta}}$ is

(a) 1 (b) 2
(c) 4 (d) 8

2. If a vector $\vec{r}$ is equally inclined with the vectors $\vec{a} = \cos\theta\hat{i} + \sin\theta\hat{j}$, $\vec{b} = -\sin\theta\hat{i} + \cos\theta\hat{j}$ and $\vec{c} = \hat{k}$, then the angle between $\vec{r}$ and $\vec{a}$ is

(a) $\cos^{-1}\frac{1}{\sqrt{2}}$ (b) $\cos^{-1}\frac{1}{\sqrt{3}}$
(c) $\cos^{-1}\frac{1}{3}$ (d) $\cos^{-1}\frac{1}{2}$

3. Let $\vec{a}, \vec{b}$ and $\vec{c}$ be unit vectors such that $\vec{a}$ is perpendicular to the plane of $\vec{b}$ and $\vec{c}$. If the angle between $\vec{b}$ and $\vec{c}$ is $\frac{\pi}{3}$, then $|\vec{a} + \vec{b} + \vec{c}| =$

(a) 1 (b) 2
(c) 3 (d) 4

4. A unit vector $\vec{a}$ in the plane of $\vec{b} = 2\hat{i} + \hat{j}$ and $\vec{c} = \hat{i} - \hat{j} + \hat{k}$ such that the angle between $\vec{a}$ and $\vec{b}$ is same as the angle between $\vec{a}$ and $\vec{d}$ where $\vec{d} = \hat{j} + 2\hat{k}$ is

(a) $\frac{\hat{i}+\hat{j}+\hat{k}}{\sqrt{3}}$ (b) $\frac{\hat{i}-\hat{j}+\hat{k}}{\sqrt{3}}$
(c) $\frac{2\hat{i}+\hat{j}}{\sqrt{5}}$ (d) $\frac{2\hat{i}-\hat{j}}{\sqrt{5}}$

5. The vectors $\vec{x}$ and $\vec{y}$ satisfy the equation $p\vec{x} + q\vec{y} = \vec{a}$ (where p, q are scalar constants and $\vec{a}$ is a known vector). If it is given that $\vec{x}\cdot\vec{y} \ge \frac{|\vec{a}|^2}{4pq}$, then $\frac{|\vec{x}|}{|\vec{y}|}$ is equal to $(pq > 0)$

(a) 1 (b) $\frac{p^2}{q^2}$
(c) $\frac{p}{q}$ (d) $\frac{q}{p}$

6. If $\vec{a}, \vec{b}, \vec{c}$ are non-zero vectors such that $\vec{a}$ is perpendicular to $\vec{b}$ and $\vec{c}$, $|\vec{a}| = 1$, $|\vec{b}| = 2$, $|\vec{c}| = 1$ and $\vec{b}\cdot\vec{c} = 1$. There is a non-zero vector $\vec{d}$ coplanar with $\vec{a} + \vec{b}$ and $2\vec{b} - \vec{c}$. If $\vec{d}\cdot\vec{a} = 1$, then the minimum value of $|\vec{d}|$ is

(a) $\frac{2}{\sqrt{13}}$ (b) $\frac{3}{\sqrt{13}}$
(c) $\frac{4}{\sqrt{5}}$ (d) $\frac{4}{\sqrt{13}}$

7. Four vectors $\vec{a}, \vec{b}, \vec{c}$ and $\vec{x}$ satisfy the relation $(\vec{a}\cdot\vec{x})\vec{b} = \vec{c} + \vec{x}$, where $\vec{b}\cdot\vec{a} \ne 1$. Then $\vec{x} =$

(a) $\frac{(\vec{a}\cdot\vec{c})\vec{b} - \vec{c}(\vec{a}\cdot\vec{b} - 1)}{(\vec{a}\cdot\vec{b} - 1)}$ (b) $\frac{\vec{c}}{\vec{a}\cdot\vec{b} - 1}$
(c) $\frac{2(\vec{a}\cdot\vec{c})\vec{b} + \vec{c}}{\vec{a}\cdot\vec{b} - 1}$ (d) $\frac{2(\vec{a}\cdot\vec{c})\vec{c} + \vec{c}}{\vec{a}\cdot\vec{b} - 1}$

8. If $\vec{a}$, $\vec{b}$ and $\vec{c}$ are three unit vectors, equally inclined to each other at an angle α, then the angle between $\vec{a}$, and plane of $\vec{b}$ and $\vec{c}$ is

(a) $\theta = \cos^{-1}\left(\dfrac{\cos\alpha}{\cos\dfrac{\alpha}{2}}\right)$ (b) $\theta = \sin^{-1}\left(\dfrac{\cos\alpha}{\cos\dfrac{\alpha}{2}}\right)$

(c) $\theta = \cos^{-1}\left(\dfrac{\sin\dfrac{\alpha}{2}}{\sin\alpha}\right)$ (d) $\theta = \sin^{-1}\left(\dfrac{\sin\dfrac{\alpha}{2}}{\sin\alpha}\right)$

Multiple Correct Answers Type

9. If $OABC$ is a tetrahedron such that $OA^2 + BC^2 = OB^2 + CA^2 = OC^2 + AB^2$, then which of the following is/are true?

(a) $OA \perp BC$ (b) $OB \perp AC$

(c) $OC \perp AB$ (d) $AB \perp AC$

10. If $\vec{a} = 3\hat{i} - \hat{j} + 5\hat{k}$ and $\vec{b} = \hat{i} + 2\hat{j} - 3\hat{k}$ are given vectors. A vector $\vec{c}$,perpendicular to z-axis, satisfies $\vec{c} \cdot \vec{a} = 9$ and $\vec{c} \cdot \vec{b} = -4$. If inclinations of $\vec{c}$ with x-axis and y-axis are α and β, respectively, then

(a) $\alpha < \dfrac{\pi}{4}$ (b) $\alpha > \dfrac{\pi}{4}$

(c) $\beta > \dfrac{\pi}{2}$ (d) $\beta < \dfrac{\pi}{2}$

11. The volume of a right triangular prism $ABC\,A_1B_1C_1$ is equal to 3 cubic unit. Then the coordinates of the vertex A_1, if the coordinates of the base vertices of the prism are $A(1, 0, 1)$, $B(2, 0, 0)$ and $C(0, 1, 0)$ are

(a) (–2, 0, 2) (b) (0, –2, 0)

(c) (0, 2, 0) (d) (2, 2, 2)

Matrix Match Type

12. In triangle ABC, $\overrightarrow{AB} = \vec{p} = 2\hat{i} - 4\hat{j} + 4\hat{k}$ and $\overrightarrow{AC} = \vec{q} = 2\hat{i} + \hat{j} + 2\hat{k}$. Then match the List I with the List II.

List I	List II
(1) If the component of vector $\vec{p}$ in the direction of vector $\vec{q}$ is $\dfrac{16}{9}\hat{i} + \lambda\hat{j} + \mu\hat{k}$, then $(7\lambda + \mu) =$	(i) 7
(2) If the component of vector $\vec{p}$ in the direction which is perpen-dicular to the vector $\vec{q}$ is $\lambda\hat{i} - \dfrac{44}{9}\hat{j} + \mu\hat{k}$, then $(8\lambda + \mu) =$	(ii) 2
(3) If the vector along the internal angle bisector of $\angle BAC$ is $\lambda\hat{i} - \hat{j} + \mu\hat{k}$, then $3(\lambda + \mu) =$	(iii) 8
(4) If the vector along the median through vertex A is $\lambda\hat{i} + \mu\hat{j} + 6\hat{k}$, then $(\lambda + \mu) =$	(iv) 4

(a) (1)-(iii), (2)-(iv), (3)-(i), (4)-(ii)

(b) (1)-(ii), (2)-(iii), (3)-(iv), (4)-(i)

(c) (1)-(iii), (2)-(i), (3)-(iv), (4)-(ii)

(d) (1)-(iv), (2)-(i), (3)-(ii), (4)-(iii)

Numerical Value Type

13. Let G be the centroid of ΔABC whose sides are a, b and c. If P is a point in the plane of ΔABC such that $PA = 1$, $PB = 3$, $PC = 4$, and $PG = 2$, then the value of $a^2 + b^2 + c^2$ is ______.

14. Let two non-collinear vectors $\vec{a}$ and $\vec{b}$ inclined at an angle $\dfrac{2\pi}{3}$ be such that $|\vec{a}| = 3$ and $|\vec{b}| = 2$. If point P moves in such a way that at any time t, its position vector $\overline{OP}$ (where O is the origin) is given as $\overline{OP} = \left(t + \dfrac{1}{t}\right)\vec{a} + \left(t - \dfrac{1}{t}\right)\vec{b}$, then the least distance of P from the origin is $\sqrt{\lambda + 2\sqrt{\mu}}$, where $(\lambda + \mu) =$ ______.

15. In a tetrahedron $OABC$, $|OA| = |BC| = a$, $|OB| = |AC| = b$ and $|OC| = |AB| = c$. Let G_1 and G_2 be the centroids of the triangles ABC and AOC such that $OG_1 \perp BG_2$. The value of $\dfrac{a^2 + c^2}{b^2}$ is ______.

16. If $\vec{p}$ and $\vec{q}$ are two diagonals of a quadrilateral such that $|\vec{p} - \vec{q}| = \vec{p} \cdot \vec{q}$, $|\vec{p}| = 1$ and $|\vec{q}| = \sqrt{2}$, then the area of quadrilateral is equal to ______.

17. Let $\vec{a}$, $\vec{b}$ and $\vec{c}$ be three non-zero vectors such that projection of $(\vec{b} + \vec{c})$ on $\vec{a}$ is equal to $2|\vec{a}|$, projection of $(\vec{a} + \vec{c})$ on $\vec{b}$ is $3|\vec{b}|$ and projection of $(\vec{a} + \vec{b})$ on $\vec{c}$ is $4|\vec{c}|$, (where $|\vec{a}|, |\vec{b}|, |\vec{c}| \in N$), then the minimum value of $|\vec{a} + \vec{b} + \vec{c}|$ is $\sqrt{\lambda}$, where $\lambda =$ ______.

Answer Key

Single Correct Answer Type

1. (a) **2.** (b) **3.** (b) **4.** (b) **5.** (d) **6.** (d) **7.** (a) **8.** (a)

Multiple Correct Answers Type

9. (a, b, c) **10.** (b, c) **11.** (b, d)

Matrix Match Type

12. (a)

Numerical Value Type

13. (42) **14.** (143) **15.** (3) **16.** (0.5) **17.** (12)

DPP 3.2
Cross Product

Single Correct Answer Type

1. Let $\vec{a}$ and $\vec{b}$ be two vectors of equal magnitudes i.e., 5 units. Also, let $\vec{p}$, $\vec{q}$ be vectors such that $\vec{p} = \vec{a} + \vec{b}$ and $\vec{q} = \vec{a} - \vec{b}$. If $|\vec{p} \times \vec{q}| = 2\left\{\lambda - (\vec{a} \cdot \vec{b})^2\right\}^{\frac{1}{2}}$, then the value of λ is
   (a) 25 (b) 125
   (c) 625 (d) none of these

2. Let $\vec{u} = 2\hat{i} - \hat{j} + \hat{k}$ and $\vec{v} = -3\hat{j} + 2\hat{k}$ be two vectors. If $\vec{w}$ is a unit vector in the *xy*-plane, then the square of maximum possible value of $|(\vec{u} \times \vec{v}) \cdot \vec{w}|$ is
   (a) 5 (b) 12
   (c) 13 (d) 17

3. Let $\vec{a}$, $\vec{b}$ and $\vec{c}$ be three unit vectors in a plane such that they are equally inclined to each other. Then the value of $(\vec{a} \times \vec{b}) \cdot (\vec{b} \times \vec{c}) + (\vec{b} \times \vec{c}) \cdot (\vec{c} \times \vec{a}) + (\vec{a} \times \vec{b}) \cdot (\vec{c} \times \vec{a})$ can be
   (a) 9/4 (b) –9/4
   (c) 3/4 (d) –3/4

4. The coordinates of the midpoints of the sides of ΔPQR are (3*a*, 0, 0), (0, 3*b*, 0) and (0, 0, 3*c*). The area of ΔPQR is
   (a) $18\sqrt{b^2c^2 + c^2a^2 + a^2b^2}$
   (b) $9\sqrt{b^2c^2 + c^2a^2 + a^2b^2}$
   (c) $\frac{9}{2}\sqrt{b^2c^2 + c^2a^2 + a^2b^2}$
   (d) $18\sqrt{ab + bc + ca}$

5. If $\vec{a} = \hat{i} + \hat{j} + \hat{k}$, $\vec{a} \cdot \vec{b} = 1$ and $a \times b = \hat{j} - \hat{k}$, then $\vec{b} \cdot (3\hat{i} + 4\hat{j} + 5\hat{k}) =$
   (a) 3 (b) 9
   (c) 10 (d) 12

6. If $\vec{a}, \vec{b}, \vec{c}$ are unit vectors such that $\vec{a} \cdot \vec{b} = 0$, $(\vec{a} - \vec{c}) \cdot (\vec{b} + \vec{c}) = 0$ and $\vec{c} = \lambda\vec{a} + \mu\vec{b} + \omega(\vec{a} \times \vec{b})$ where λ, μ, ω are scalars, then
   (a) $\mu^2 + \omega^2 = 1$
   (b) $\lambda + \mu = 1$
   (c) $(\mu + 1)^2 + \mu^2 + \omega^2 = 1$
   (d) $\lambda^2 + \mu^2 = 1$

7. Let ΔABC be a given triangle. If $|\overrightarrow{BA} - t\overrightarrow{BC}| \geq |\overrightarrow{AC}|$ for all $t \in R$, then ΔABC is
   (a) equilateral (b) right angled
   (c) isosceles (d) none of these

8. If $\vec{a}$ and $\vec{b}$ are non -zero, non -parallel vectors, then the value of $|\vec{a} + \vec{b} + \vec{a} \times \vec{b}|^2 + |\vec{a} + \vec{b} - \vec{a} \times \vec{b}|^2$ equals to
   (a) $(1 + \vec{a} \cdot \vec{a})(1 + \vec{b} \cdot \vec{b})$
   (b) $2(1 + \vec{a} \cdot \vec{a})(1 + \vec{b} \cdot \vec{b})$
   (c) $2((1 + \vec{a} \cdot \vec{a})(1 + \vec{b} \cdot \vec{b}) - (1 - \vec{a} \cdot \vec{b})^2)$
   (d) $2((1 - \vec{a} \cdot \vec{a})(1 - \vec{b} \cdot \vec{b}) + (1 - \vec{a} \cdot \vec{b})^2)$

9. If $a^2 + b^2 + c^2 = 1$; $a, b, c \in R$, then the maximum value of $(4a - 3b)^2 + (5b - 4c)^2 + (3c - 5a)^2$ is
   (a) 25 (b) 50
   (c) 144 (d) none of these

Multiple Correct Answers Type

10. Three vectors $\vec{a}$ ($|\vec{a}| \neq 0$), $\vec{b}$ and $\vec{c}$ are such that $\vec{a} \times \vec{b} = 3\vec{a} \times \vec{c}$. Also, $|\vec{a}| = |\vec{b}| = 1$ and $|\vec{c}| = \frac{1}{3}$. If the angle between $\vec{b}$ and $\vec{c}$ is 60°, then
    (a) $\vec{b} = 3\vec{c} + \vec{a}$ (b) $\vec{b} = 3\vec{c} - \vec{a}$
    (c) $\vec{a} = 6\vec{c} + 2\vec{b}$ (d) $\vec{a} = 6\vec{c} - 2\vec{b}$

11. Let $2\vec{a}$, $3\vec{b}$ and $2(\vec{a} \times \vec{b})$ be position vectors of the vertices *A*, *B* and *C*, respectively, of ΔABC. If $|\vec{a}| = |\vec{b}| = 1$ and $\overrightarrow{OA} \cdot \overrightarrow{OB} = -3$ (where *O* is the origin), then
    (a) triangle *ABC* is right angled triangle
    (b) angle *B* is 90°
    (c) $A = \cos^{-1}\left(\sqrt{\frac{7}{19}}\right)$
    (d) the position vector of orthocentre of triangle is $2\vec{a} \times \vec{b}$

Linked Comprehension Type

For Questions 12 and 13

Let $\vec{a}$ and $\vec{b}$ be two vectors such that $|\vec{a}| = 1$, $|\vec{b}| = 4$ and $\vec{a} \cdot \vec{b} = 2$. If $\vec{c} = (2\vec{a} \times \vec{b}) - 3\vec{b}$.

12. The angle between $\vec{b}$ and $\vec{c}$ is
    (a) $\frac{\pi}{3}$ (b) $\frac{\pi}{6}$
    (c) $\frac{3\pi}{4}$ (d) $\frac{5\pi}{6}$

13. The value of $|\vec{c}|^2$ is
(a) 48 (b) 96
(c) 144 (d) 192

Numerical Value Type

14. If $\vec{a}$ and $\vec{b}$ are two vectors perpendicular to each other such that $|\vec{a}| = 2$, $|\vec{b}| = 3$ and $\vec{c} \times \vec{a} = \vec{b}$, then the least value of $2|\vec{c} - \vec{a}|$ is _____.

15. If a, b, c and $A, B, C \in R - \{0\}$ are such that $aA + bB + cC + \sqrt{(a^2+b^2+c^2)(A^2+B^2+C^2)} = 0$, then value of $\frac{aB}{bA} + \frac{bC}{cB} + \frac{cA}{aC}$ is _____.

Answer Key

Single Correct Answer Type

1. (c) 2. (d) 3. (a) 4. (a) 5. (a) 6. (c) 7. (b) 8. (c) 9. (b)

Multiple Correct Answers Type

10. (a, b) 11. (a, c, d)

Linked Comprehension Type

12. (d) 13. (d)

Numerical Value Type

14. (3) 15. (3)

DPP 3.3

Scalar Triple Product

Single Correct Answer Type

1. Number of integral value(s) of λ for which the vectors $x^2\hat{i} - \hat{j} + x\hat{k}$, $(\lambda-1)\hat{i} - 2\lambda\hat{j} + \hat{k}$ and $\hat{i} - \hat{j} + \hat{k}$, in the order form right-handed system $\forall x \in R$, is
   (a) 0 (b) 2
   (c) 4 (d) 6

2. Let $\vec{a}$, $\vec{b}$ and $\vec{c}$ be three linearly independent vectors. Then
   $$\frac{[\vec{a}+2\vec{b}-\vec{c}\ \ 2\vec{a}+\vec{b}+\vec{c}\ \ 4\vec{a}-\vec{b}+5\vec{c}]}{[\vec{a}\ \vec{b}\ \vec{c}]} =$$
   (a) 0 (b) 1
   (c) 2 (d) -1

3. If $\vec{a}$ and $\vec{b}$ are two unit vectors such that $\vec{a} + (\vec{a} \times \vec{b}) = \vec{c}$, where $|\vec{c}| = 2$, then the value of $[\vec{a}\ \vec{b}\ \vec{c}]$ is
   (a) 0 (b) ± 1
   (c) -3 (d) 3

4 Let $\vec{a}$ and $\vec{b}$ be two unit vectors that are per-pendicular to each other. Then $[\vec{a} + (\vec{a} \times \vec{b})\ \vec{b} + (\vec{a} \times \vec{b})\ \vec{a} \times \vec{b}]$ will be equal to
   (a) 1 (b) zero
   (c) -1 (d) 3

5. Let $\vec{a}$, $\vec{b}$ and $\vec{c}$ be any three vectors forming a linearly independent system such that $\forall\ \theta \in R, \vec{p} = \vec{a}\cos\theta + \vec{b}\sin\theta + \vec{c}\cos 2\theta$
   $$\vec{q} = \vec{a}\cos\left(\frac{2\pi}{3}+\theta\right)+\vec{b}\sin\left(\frac{2\pi}{3}+\theta\right)+\vec{c}\cos 2\left(\frac{2\pi}{3}+\theta\right)$$
   and $\vec{r} = \vec{a}\cos\left(\theta-\frac{2\pi}{3}\right)+\vec{b}\sin\left(\theta-\frac{2\pi}{3}\right)+\vec{c}\cos 2\left(\theta-\frac{2\pi}{3}\right)$.
   Then $[\vec{p}\ \vec{q}\ \vec{r}] =$
   (a) $[\vec{a}\ \vec{b}\ \vec{c}]\cos\theta$ (b) $[\vec{a}\ \vec{b}\ \vec{c}]\cos 2\theta$
   (c) $[\vec{a}\ \vec{b}\ \vec{c}]\cos 3\theta$ (d) none of these

6. Let $\vec{r} = (\vec{a} \times \vec{b})\sin x + (\vec{b} \times \vec{c})\cos y + (\vec{c} \times \vec{a})$, where $\vec{a}$, $\vec{b}$ and $\vec{c}$ are non-zero non-coplanar vectors. If $\vec{r}$ is orthogonal to $3\vec{a} + 5\vec{b} + 2\vec{c}$, then the value of $\sec^2 y + \operatorname{cosec}^2 x + \sec y \operatorname{cosec} x$ is
   (a) 3 (b) 4
   (c) 5 (d) 6

7. Let $DABC$ be a tetrahedron such that AD is perpendicular to the base ABC and $\angle ABC = 30°$ and the volume is 18. If the value of $AB + BC + AD$ is minimum, then the length AC is
   (a) $6\sqrt{2-\sqrt{3}}$ (b) $3(\sqrt{6}-\sqrt{2})$
   (c) $6\sqrt{2+\sqrt{3}}$ (d) $3(\sqrt{6}+\sqrt{2})$

8. Let $\bar{r}\cdot\bar{a} = \bar{r}\cdot\bar{b} = \bar{r}\cdot\bar{c} = \frac{1}{2}$ where $\bar{a}, \bar{b}$ and $\bar{c}$ are the position vectors of points A, B and C, respectively. If $\bar{a}$, $\bar{b}$ and $\bar{c}$ are non-coplanar, then the area of the triangle ABC is
   (a) $\frac{1}{2}\left|[\bar{a}\ \bar{b}\ \bar{c}]\bar{r}\right|$ (b) $\left|[\bar{a}\ \bar{b}\ \bar{c}]\bar{r}\right|$
   (c) $\frac{1}{4}\left|[\bar{a}\ \bar{b}\ \bar{c}]\bar{r}\right|$ (d) $\frac{1}{8}\left|[\bar{a}\ \bar{b}\ \bar{c}]\bar{r}\right|$

9. If $|\hat{a}+\hat{b}+\hat{a}\times\hat{b}| = |\hat{a}-\hat{b}|$, then the value of $|\hat{a}-\hat{b}|^2$ is
   (a) 4 sin 18° (b) 8 cos 36°
   (c) 8 cos 18° (d) 4 cos 36°

10. Consider the tetrahedron $ABCD$ having vertices $A(1, -2, 0)$, $B(3, 1, 2)$, $C(-1, 1, -1)$ and $D(1, 0, -1)$. Now, given below are two statements.One is labelled as Assertion (A) and the other is labelled as Reason (R).

    Assertion(A): Distance of point D from the plane ABC is $\frac{8}{\sqrt{229}}$.

    Reason(R): Volume of tetrahedron $ABCD$ is $\frac{8}{3}$ cubic unit.

    (a) (A) is true but (R) is false.
    (b) Both (A) and (R) are true and (R) is the true explanation of (A).
    (c) (A) is false and (R) is true.
    (d) Both (A) and (R) are true but (R) is not the true explanation of (A).

11. Given below are two statements. One is labelled as Assertion (A) and the other is labelled as Reason (R).

    Assertion (A): $\vec{a}\times\vec{b} = ((\hat{i}\times\vec{a})\cdot\vec{b})\hat{i} + ((\hat{j}\times\vec{a})\cdot\vec{b})\hat{j} + ((\hat{k}\times\vec{a})\cdot\vec{b})\hat{k}$

    Reason (R): $\vec{c} = (\hat{i}\cdot\vec{c})\hat{i} + (\hat{j}.\vec{c})\hat{j} + (\hat{k}\cdot\vec{c})\hat{k}$

    (a) (A) is true but (R) is false.
    (b) Both (A) and (R) are true and (R) is the true explanation of (A).

(c) (A) is false and (R) is true.

(d) Both (A) and (R) are true but (R) is not the true explanation of (A).

Multiple Correct Answers Type

12. Let $\vec{a} = \hat{i} - \hat{j}, \vec{b} = \hat{j} - \hat{k}$ and $\vec{c} = \hat{k} - \hat{i}$. If $\vec{d}$ is a unit vector such that $\vec{a} \cdot \vec{d} = 0$ and $[\vec{b}\ \vec{c}\ \vec{d}] = 0$, then $\vec{d} =$

(a) $\pm\dfrac{\hat{i}+\hat{j}}{\sqrt{2}}$ (b) $-\dfrac{\hat{i}+\hat{j}-2\hat{k}}{\sqrt{6}}$

(c) $\dfrac{\hat{i}+\hat{j}+2\hat{k}}{\sqrt{6}}$ (d) $\dfrac{\hat{i}+\hat{j}-2\hat{k}}{\sqrt{6}}$

13. Let $\alpha(\vec{a} \times \vec{b}) + \beta(\vec{b} \times \vec{c}) + \gamma(\vec{c} \times \vec{a}) = 0$. Then

(a) $\vec{a}, \vec{b}$ and $\vec{c}$ will be coplanar if all of α, β and γ are non-zero.

(b) $\vec{a}, \vec{b}$ and $\vec{c}$ will be coplanar if any one of α, β and γ is non-zero.

(c) $\vec{a}, \vec{b}$ and $\vec{c}$ will be non-coplanar if all α, β and γ are non-zero.

(d) none of these

Linked Comprehension Type

For Questions 14 and 15

Let *OABC* be a tetrahedron. Also, let position vectors of *A*, *B* and *C* be $\hat{i}, \hat{i}+\hat{j}$ and $\hat{j}+\hat{k}$, respectively, and *O* be the origin.

14. The distance of the face *ABC* from the origin *O* is

(a) $\dfrac{1}{2}$ (b) $\dfrac{1}{\sqrt{2}}$

(c) $\dfrac{1}{2\sqrt{2}}$ (d) 1

15. Volume of parallelepiped whose coterminous edges are $\overline{OA}$, $\overline{OB}$ and $\overline{OC}$ is

(a) 1/3 (b) 1/2

(c) 1 (d) 2

For Questions 16 and 17

Let $\vec{a}, \vec{b}$ and $\vec{c}$ be three vectors having magnitudes 1, 2 and 3 units, respectively, such that $|[\vec{a}\ \vec{b}\ \vec{c}]| = 6$. If $\vec{r}$ is a unit vector coplanar with $\vec{b}$ and $\vec{c}$, and $\vec{r} \cdot \vec{b} = 1$.

16. The value of $|(\vec{a} \times \vec{c}) \times \vec{r}|$ is

(a) 3/2 (b) 2

(c) $2\sqrt{3}$ (d) $3\sqrt{3}/2$

17. The value of $|[\vec{a}\ \vec{b}\ \vec{r}]|$ is

(a) $1/\sqrt{3}$ (b) $\sqrt{3}$

(c) $2\sqrt{3}$ (d) $\sqrt{3}/2$

Numerical Value Type

18. Let $\vec{a} = a_1\hat{i} + a_2\hat{j} + a_3\hat{k}, \vec{b} = b_1\hat{i} + b_2\hat{j} + b_3\hat{k}$ and $\vec{c} = c_1\hat{i} + c_2\hat{j} + c_3\hat{k}$ be three non-zero vectors such that $\vec{c}$ is a unit vector perpendicular to vectors $\vec{a}$ and $\vec{b}$. If $|\vec{a}| = 2|\vec{b}| = 4$ and the angle between $\vec{a}$ and $\vec{b}$ is $\dfrac{\pi}{3}$, then

$$\begin{vmatrix} a_1 & a_2 & a_3 \\ b_1 & b_2 & b_3 \\ c_1 & c_2 & c_3 \end{vmatrix}^2$$ is equal to ______.

19. If the volume of a parallelopiped having three coterminous edges as $\vec{a}, \vec{b}$ and $\vec{c}$ is 4, then the volume of the parallelepiped having three coterminous edges as $\vec{\alpha} = (\vec{a} \cdot \vec{a})\vec{a} + (\vec{a} \cdot \vec{b})\vec{b} + (\vec{a} \cdot \vec{c})\vec{c}$, $\vec{\beta} = (\vec{b} \cdot \vec{a})\vec{a} + (\vec{b} \cdot \vec{b})\vec{b} + (\vec{b} \cdot \vec{c})\vec{c}$, and $\vec{\gamma} = (\vec{c} \cdot \vec{a})\vec{a} + (\vec{c} \cdot \vec{b})\vec{b} + (\vec{c} \cdot \vec{c})\vec{c}$, is ______.

20. Let $\vec{q}$ and $\vec{r}$ be unit vectors such that $\vec{p} = \vec{q} \times \vec{p} + \vec{r}$. If λ and μ are, respectively, the minimum and the maximum values of $[\vec{p}\ \vec{q}\ \vec{r}]$, then the value of $10\lambda + 20\mu$ is ______.

21. Let $\vec{p}, \vec{q}$ and $\vec{r}$ be three unit vectors such that $\vec{p} \times \vec{q} = \vec{r}$. If $\vec{a}$ is any vector such that $[\vec{a}\ \vec{q}\ \vec{r}] = 1$, $[\vec{a}\ \vec{r}\ \vec{p}] = 2$ and $[\vec{a}\ \vec{p}\ \vec{q}] = 3$, then $\vec{a} = \vec{p} + \lambda\vec{q} + \mu\vec{r}$, where $\lambda + \mu =$ ______.

Answer Key

Single Correct Answer Type

1. (a) **2.** (a) **3.** (d) **4.** (a) **5.** (d) **6.** (a) **7.** (a) **8.** (b) **9.** (c) **10.** (c) **11.** (b)

Multiple Correct Answers Type

12. (b, d) **13.** (a, b)

Linked Comprehension Type

14. (b) **15.** (c) **16.** (d) **17.** (b)

Numerical Value Type

18. (48) **19.** (64) **20.** (10) **21.** (5)

DPP 3.4

Vector Triple Product and Reciprocal System of Vectors

Single Correct Answer Type

1. Let $\vec{a} = 2\hat{i} + \hat{j} + \hat{k}$ and $\vec{b} = b_1\hat{i} + b_2\hat{j} + b_3\hat{k}$ such that $\vec{a} \times \vec{b} = 5\hat{i} + 2\hat{j} - 12\hat{k}$ and $\vec{a} \cdot \vec{b} = 11$. Then $b_1 + b_2 + b_3 =$

(a) 3 (b) 5
(c) 7 (d) 9

2. If $\vec{a}$, $\vec{b}$, $\vec{c}$ and $\vec{d}$ are unit vectors such that $\vec{a} \cdot \vec{b} = \frac{1}{2}$, $\vec{c} \cdot \vec{d} = \frac{1}{2}$ and angle between $\vec{a} \times \vec{b}$ and $\vec{c} \times \vec{d}$ is $\frac{\pi}{6}$, then the value of $|[\vec{a}\ \vec{b}\ \vec{d}]\ \vec{c} - [\vec{a}\ \vec{b}\ \vec{c}]\ \vec{d}\ | =$

(a) 3/2 (b) 3/4
(c) 3/8 (d) 2

3. If $\vec{a}, \vec{b}, \vec{c}$ and $\vec{d}$ are four vectors such that $[\vec{a}, \vec{b}, \vec{c}] = 2$ and $(\vec{a} \times \vec{b}) \times (\vec{c} \times \vec{d}) + (\vec{b} \times \vec{c}) \times (\vec{a} \times \vec{d}) + (\vec{c} \times \vec{a}) \times (\vec{b} \times \vec{d}) = -\mu\vec{d}$, then the value of μ is

(a) 0 (b) 1
(c) 3 (d) 4

4. Let $(\hat{p} \times \vec{q}) \times \hat{p} + (\hat{p} \cdot \vec{q})\vec{q} = (x^2 + y^2)\vec{q} + (14 - 4x - 6y)\hat{p}$, where $\hat{p}$ and $\vec{q}$ are two non-collinear vectors ($\hat{p}$ is a unit vector) and x, y are scalars. Then the value of $(x + y)$ is

(a) 4 (b) 5
(c) 6 (d) 7

5. If $\vec{a}, \vec{b}, \vec{c}$ are three non-coplanar vectors such that $\vec{a} \times \vec{b} = \vec{c}$, $\vec{b} \times \vec{c} = \vec{a}$ and $\vec{c} \times \vec{a} = \vec{b}$, then the value of $|\vec{a}| + |\vec{b}| + |\vec{c}|$ is

(a) 1/3 (b) 1
(c) 3 (d) 6

6. If the angles between the vectors $\vec{a}$ and $\vec{b}$, $\vec{b}$ and $\vec{c}$, and $\vec{c}$ and $\vec{a}$ are, respectively, $\frac{\pi}{6}, \frac{\pi}{4}$ and $\frac{\pi}{3}$, then the angle that vector $\vec{a}$ makes with the plane containing $\vec{b}$ and $\vec{c}$, is

(a) $\cos^{-1}\sqrt{1-\sqrt{2/3}}$ (b) $\cos^{-1}\sqrt{2-\sqrt{3/2}}$
(c) $\cos^{-1}\sqrt{\sqrt{3/2}-1}$ (d) $\cos^{-1}\sqrt{\sqrt{2/3}}$

7. Let $\vec{a}, \vec{b}$ and $\vec{c}$ be three vectors having mag-nitudes 1, 1 and 2, respectively. If $\vec{a} \times (\vec{a} \times \vec{c}) - \vec{b} = 0$, then the acute angle between $\vec{a}$ and $\vec{c}$ is

(a) $\pi/4$ (b) $\pi/6$
(c) $\pi/3$ (d) $\pi/2$

8. If $\vec{a}$, $\vec{b}$ and $\vec{c}$ are three non-coplanar vectors and $\vec{p}$, $\vec{q}$ and $\vec{r}$, respectively, are their reciprocal vectors, then $(\ell\vec{a} + m\vec{b} + n\vec{c}) \cdot (\ell\vec{p} + m\vec{q} + n\vec{r})$ is equal to

(a) $\ell^2 + m^2 + n^2$ (b) $\ell m + mn + n\ell$
(c) 0 (d) None of these

9. If $\vec{a}$, $\vec{b}$ and $\vec{c}$ are non-zero vectors, then $((\vec{a} \times \vec{b}) \times \vec{a}) \cdot ((\vec{b} \times \vec{a}) \times \vec{b}) =$

(a) $-(\vec{a} \cdot \vec{b})\,|\vec{a} \times \vec{b}|^2$ (b) $(\vec{a} \cdot \vec{b})\,|\vec{a} \times \vec{b}|^2$
(c) $(\vec{a} \cdot \vec{b})^2\,|\vec{a} \times \vec{b}|^2$ (d) $|\vec{a}|\,|\vec{b}|\,|\vec{a} \times \vec{b}|^2$

10. Let $\vec{v}$ be a unit vector which follows the equation, $\vec{v} \times \vec{b} = \vec{c}$. If $|\vec{b}| = 2$ and $|\vec{c}| = \sqrt{3}$, then

(a) $\vec{v} = \frac{1}{4}(-\vec{b} + \vec{b} \times \vec{c})$ (b) $\vec{v} = \frac{3}{4}(\vec{b} + \vec{b} \times \vec{c})$
(c) $\vec{v} = \frac{1}{4}(\vec{b} + \vec{b} \times \vec{c})$ (d) $\vec{v} = \frac{3}{4}(\vec{b} - \vec{b} \times \vec{c})$

Multiple Correct Answers Type

11. The volume of the parallelepiped whose coterminous edges are represented by the vectors $2\vec{b} \times \vec{c}$, $3\vec{c} \times \vec{a}$ and $4\vec{a} \times \vec{b}$ where $\vec{a} = (1 + \sin\theta)\,\hat{i} + \cos\theta\,\hat{j} + \sin 2\theta\,\hat{k}$,

$$\vec{b} = \sin\left(\theta + \frac{2\pi}{3}\right)\hat{i} + \cos\left(\theta + \frac{2\pi}{3}\right)\hat{j} + \sin\left(2\theta + \frac{4\pi}{3}\right)\hat{k},$$

$$\vec{c} = \sin\left(\theta - \frac{2\pi}{3}\right)\hat{i} + \cos\left(\theta - \frac{2\pi}{3}\right)\hat{j} + \sin\left(2\theta - \frac{4\pi}{3}\right)\hat{k}$$

is 18 cubic units, then the values of θ, in the interval $\left(0, \frac{\pi}{2}\right)$, is/are

(a) $\frac{\pi}{9}$ (b) $\frac{2\pi}{9}$
(c) $\frac{\pi}{3}$ (d) $\frac{4\pi}{9}$

12. Let $\vec{a}$ and $\vec{b}$ be two given perpendicular vectors, which are non-zero. Then vector $\vec{r}$ satisfying the equation $\vec{r} \times \vec{b} = \vec{a}$, can be

(a) $\vec{b} - \frac{\vec{a} \times \vec{b}}{|\vec{b}|^2}$ (b) $2\vec{b} - \frac{\vec{a} \times \vec{b}}{|\vec{b}|^2}$
(c) $|\vec{a}|\vec{b} - \frac{\vec{a} \times \vec{b}}{|\vec{b}|^2}$ (d) $|\vec{b}|\vec{b} - \frac{\vec{a} \times \vec{b}}{|\vec{b}|^2}$

13. If $\vec{a}, \vec{b}$ and $\vec{c}$ are three non-zero vectors, then which of the following statement(s) is(are) true?
   (a) Vectors $\vec{a} \times (\vec{b} \times \vec{c}), \vec{b} \times (\vec{c} \times \vec{a}), \vec{c} \times (\vec{a} \times \vec{b})$ form a right-handed system
   (b) Vectors $\vec{c}, (\vec{a} \times \vec{b}) \times \vec{c}, \vec{a} \times \vec{b}$ form a right-handed system
   (c) $\vec{a} \cdot \vec{b} + \vec{b} \cdot \vec{c} + \vec{c} \cdot \vec{a} < 0$ if $\vec{a} + \vec{b} + \vec{c} = \vec{0}$
   (d) $\dfrac{(\vec{a}\times\vec{b}).(\vec{b}\times\vec{c})}{(\vec{b}\times\vec{c}).(\vec{a}\times\vec{c})} = -1$ if $\vec{a} + \vec{b} + \vec{c} = \vec{0}$

14. Vectors $\vec{a}, \vec{b}, \vec{c}$ are three unit vectors and $\vec{c}$ is equally inclined to both $\vec{a}$ and $\vec{b}$. If $\vec{a} \times (\vec{b} \times \vec{c}) + \vec{b} \times (\vec{c} \times \vec{a}) = (4+x^2)\vec{b} - (4x\cos^2\theta)\vec{a}$, then ($\vec{a}$ and $\vec{b}$ are non-collinear vectors, $x > 0$)
   (a) $x = 2$ (b) $\theta = 0°$
   (c) $\theta = \pi$ (d) $x = 4$

15. If $\vec{a}$ and $\vec{b}$ are unequal unit vectors such that $(\vec{a} - \vec{b}) \times [(\vec{b} + \vec{a}) \times (2\vec{a} + \vec{b})] = \vec{a} + \vec{b}$, then the angle between $\vec{a}$ and $\vec{b}$ can be
   (a) $\pi/2$ (b) 0
   (c) π (d) $\pi/4$

Linked Comprehension Type

For Questions 16–18

Vectors $\vec{a} = 2\hat{i}+\hat{j}+2\hat{k}$, $\vec{b} = \hat{i}-\hat{j}+\hat{k}$ and non-zero vector $\vec{c}$ are such that $(\vec{a} \times \vec{b}) \times \vec{c} = \vec{a} \times (\vec{b} \times \vec{c})$. Also, $|\vec{c}| = 6$ and $\vec{c} \cdot \hat{i} > 0$.

16. Vector $\vec{c}$ may be
   (a) $4\hat{i} + 2\hat{j} + 4\hat{k}$ (b) $2\hat{i} - 4\hat{j} + 4\hat{k}$
   (c) $4\hat{i} - 2\hat{j} - 4\hat{k}$ (d) $2\hat{i} + 4\hat{j} - 4\hat{k}$

17. Volume of parallelepiped whose adjacent sides are given by $\vec{a}, \vec{b}$ and $\vec{b} \times \vec{c}$ is
   (a) 18 (b) 54
   (c) 12 (d) 36

18. A vector along the bisector of angle between the vectors $\vec{b}$ and $\vec{c}$ is
   (a) $(2+\sqrt{3})\hat{i}+(1-\sqrt{3})\hat{j}+(2+\sqrt{3})\hat{k}$
   (b) $(2+\sqrt{3})\hat{i}+(1-\sqrt{3})\hat{j}-(2+\sqrt{3})\hat{k}$
   (c) $(2-\sqrt{3})\hat{i}+(\sqrt{3}+1)\hat{j}+(2-\sqrt{3})\hat{k}$
   (d) $(2+\sqrt{3})\hat{i}-(1-\sqrt{3})\hat{j}+(2+\sqrt{3})\hat{k}$

For Questions 19 and 20

Let $\vec{a}, \vec{b}$ and $\vec{c}$ be three non-collinear unit vectors such that $\vec{a}\times(\vec{b}\times\vec{c}) = \dfrac{\vec{b}}{3}+\dfrac{\vec{c}}{2}$ and $\vec{b}\times(\vec{c}\times\vec{a}) = -\dfrac{\vec{c}}{2}$.

19. The value of $|\vec{a} + \vec{b} + \vec{c}|$ is
   (a) $\dfrac{2\sqrt{2}}{\sqrt{3}}$ (b) $\dfrac{\sqrt{3}}{2}$
   (c) $\dfrac{2\sqrt{2}}{3}$ (d) $2\sqrt{3}$

20. Volume of parallelepiped whose coterminous edges are $\vec{a}$, $\vec{b}$ and $\vec{c}$ is
   (a) $\dfrac{11}{36}$ (b) $\dfrac{\sqrt{17}}{2}$
   (c) $\dfrac{\sqrt{23}}{4}$ (d) $\dfrac{\sqrt{23}}{6}$

Matrix Match Type

21. Consider the vectors $\vec{a} = \hat{i} + \hat{j} + \hat{k}$, $\vec{b} = -\hat{i} + 2\hat{j}$, $\vec{c} = 2\hat{i} + \hat{k}$ and $\vec{d} = \hat{i} - \hat{j} + 2\hat{k}$. Then match the List I with the List II.

List I	List II
(1) If the vector perpendicular to the plane of $\vec{a}$ and $\vec{b}$ is $\lambda\hat{i} + 2\hat{j} + \mu\hat{k}$, then $(\lambda + \mu)$ is	(i) 1
(2) If the vector coplanar with $\vec{c}$ and $\vec{d}$ and perpendicular to $\vec{a}$ is $-\hat{i} + \lambda\hat{j} + \mu\hat{k}$, then $(\lambda + \mu)$ is	(ii) 3
(3) If the vector parallel to the line of intersection of plane of $\vec{a}$ and $\vec{c}$ and the plane of $\vec{b}$ and $\vec{d}$ is $\lambda\hat{i} - 21\hat{j} + \mu\hat{k}$, then $(\lambda + \mu)$ is	(iii) 4
(4) If the vector $\vec{b} \times (\vec{a} \times \vec{d})$ is along the vector $\lambda\hat{i} - \mu\hat{j} + 10\hat{k}$, then $(\lambda + \mu)$ is	(iv) –2

   (a) (1)-(iii), (2)-(iv), (3)-(i), (4)-(ii)
   (b) (1)-(ii), (2)-(iii), (3)-(iv), (4)-(i)
   (c) (1)-(iii), (2)-(i), (3)-(iv), (4)-(ii)
   (d) (1)-(iv), (2)-(i), (3)-(ii), (4)-(iii)

Numerical Value Type

22. Let $\vec{a}, \vec{b}$ and $\vec{c}$ be non-zero non-collinear vectors such that $\vec{c} = \vec{a} \times (\vec{c} + \vec{b})$. Then the value of $|(1+\vec{a}^2)\vec{c}+\vec{b}\times\vec{a}+(\vec{a}\times\vec{b})\times\vec{a}|$ is ___.

23. If θ is the angle between $\vec{a}$ and $\vec{b}$ such that $\sqrt{3}(\vec{a}\cdot\vec{b})+|\vec{b}-(\vec{a}\cdot\vec{b})\vec{a}|=1+\vec{b}\cdot\vec{b}$ where $|\vec{a}| = 1$, then $\cot^2\theta =$ ______.

24. Let $\vec{a}$ and $\vec{c}$ be unit vectors inclined at angle $\pi/3$ with each other. If $(\vec{a} \times (\vec{b} \times \vec{c})) \cdot (\vec{a} \times \vec{c}) = 5$, then the absolute value of $[\vec{a}\ \vec{b}\ \vec{c}]$ is _____.

25. Let $\vec{A} = \hat{i} - 3\hat{j} + 4\hat{k}$, $\vec{B} = 6\hat{i} + 4\hat{j} - 8\hat{k}$, $\vec{C} = 5\hat{i} + 2\hat{j} + 5\hat{k}$ and a vector $\vec{R}$ satisfy the relations $\vec{R} \times \vec{B} = \vec{C} \times \vec{B}$, $\vec{R} \cdot \vec{A} = 0$, then the value of $\dfrac{|\vec{B}|}{|\vec{R} - \vec{C}|}$ is ______.

26. Let $\hat{a}$ and $\hat{b}$ be two unit vectors such that $\hat{a} \cdot \hat{b} = \dfrac{1}{3}$ and $\hat{a} \times \hat{b} = \hat{c}$. Also, $\vec{F} = \alpha\hat{a} + \beta\hat{b} + \gamma\hat{c}$, where α, β and γ are scalars. If $\alpha = k_1(\vec{F} \cdot \hat{a}) + k_2(\vec{F} \cdot \hat{b})$, then the value of $8(k_1 - k_2)$ is ______.

27. If $[\vec{a} \times \vec{b} \quad \vec{b} \times \vec{c} \quad \vec{c} \times \vec{a}]$ has the maximum value then the value, of $|(\vec{a} \times \vec{b}) \times \vec{c}|^2$ is____.

Answer Key

Single Correct Answer Type

1. (b) **2.** (c) **3.** (d) **4.** (b) **5.** (c) **6.** (b) **7.** (b) **8.** (a) **9.** (a) **10.** (c)

Multiple Correct Answers Type

11. (a, b, d) **12.** (a, b, c, d) **13.** (b, c, d) **14.** (a, b, c) **15.** (a, c)

Linked Comprehension Type

16. (a) **17.** (d) **18.** (a) **19.** (a) **20.** (d)

Matrix Match Type

21. (d)

Numerical Value Type

22. (0) **23.** (3) **24.** (10) **25.** (2) **26.** (12) **27.** (0)

CHAPTER 4

Line and Plane

DPP 4.1

Line in Space

Single Correct Answer Type

1. The direction cosines of a line equally inclined to three mutually perpendicular lines having direction cosines as l_1, m_1, n_1; l_2, m_2, n_2 and l_3, m_3, n_3 are
   (a) $l_1+l_2+l_3, m_1+m_2+m_3, n_1+n_2+n_3$
   (b) $\frac{l_1+l_2+l_3}{\sqrt{3}}, \frac{m_1+m_2+m_3}{\sqrt{3}}, \frac{n_1+n_2+n_3}{\sqrt{3}}$
   (c) $\frac{l_1+l_2+l_3}{3}, \frac{m_1+m_2+m_3}{3}, \frac{n_1+n_2+n_3}{3}$
   (d) None of these

2. Three straight lines mutually perpendicular to each other meet in a point P and one of them intersects the x-axis and another intersects the y-axis, while the third line passes through a fixed point $(0, 0, c)$ on the z-axis. Then the locus of P is
   (a) $x^2+y^2+z^2-2cx=0$
   (b) $x^2+y^2+z^2-2cy=0$
   (c) $x^2+y^2+z^2-2cz=0$
   (d) $x^2+y^2+z^2-2c(x+y+z)=0$

3. The angle between the line whose d.c.'s are connected by the relations $l^2+m^2-n^2=0$ and $l+m+n=0$ is
   (a) $\pi/6$ (b) $\pi/4$
   (c) $\pi/3$ (d) $\pi/2$

4. A ray of light is sent through the point $P(1, 2, 3)$ and is reflected on the XY-plane. If the reflected ray passes through the point $Q(3, 2, 5)$, then the equation of the reflected ray is
   (a) $\frac{x-3}{1}=\frac{y-2}{0}=\frac{z-5}{1}$
   (b) $\frac{x-3}{1}=\frac{y-2}{0}=\frac{z-5}{-4}$
   (c) $\frac{x-3}{1}=\frac{y-2}{0}=\frac{z-5}{4}$
   (d) $\frac{x-1}{1}=\frac{y-2}{0}=\frac{z-3}{4}$

5. The centroid of the triangle formed by the point $(0, 0, 0)$ and the point of intersection of line $\frac{x-1}{1}=\frac{y-1}{2}=\frac{z-1}{1}$ with $x=0$ and $y=0$ is
   (a) $(1, 1, 1)$ (b) $(1/6, -1/3, 1/6)$
   (c) $(-1/6, 1/3, -1/6)$ (d) $(1/3, 1/3, 1/3)$

6. If the lines $x=y=z$ and $x=\frac{y}{2}=\frac{z}{3}$ form a triangle of area $\sqrt{6}$ sq. unit with the line passing through $(1, 1, 1)$, then the point of intersection of third line with the second line is
   (a) $\left(\frac{4}{3}, \frac{8}{3}, \frac{12}{3}\right)$ (b) $(1, 2, 3)$
   (c) $(2, 4, 6)$ (d) $(3, 6, 9)$

Multiple Correct Answers Type

7. A and B are two points with the coordinates (x_1, y_1, z_1) and (x_2, y_2, z_2), respectively, in the space. Let P and Q be feet of the perpendiculars drawn from A and B, respectively, to a line L whose direction ratios are l, m, n. Let θ be the angle between AB and L. Then
   (a) $PQ = AB\cos\theta$
   (b) $PQ = |(x_2-x_1)l+(y_2-y_1)m+(z_2-z_1)n|$
   (c) $PQ = \frac{|(x_2-x_1)l+(y_2-y_1)m+(z_2-z_1)n|}{\sqrt{l^2+m^2+n^2}}$
   (d) AB and PQ are always coplanar

8. The directional cosines of two lines are connected by relation $l + m + n = 0$ and $4l$ is the harmonic mean between m and n. Then

(a) $\frac{l_1}{l_2} + \frac{m_1}{m_2} + \frac{n_1}{n_2} = -3/2$

(b) $l_1 l_2 + m_1 m_2 + n_1 n_2 = -1/2$

(c) $l_1 m_1 n_1 + l_2 m_2 n_2 = -\sqrt{6}/9$

(d) $(l_1 + l_2)(m_1 + m_2)(n_1 + n_2) = \frac{\sqrt{6}}{18}$

9. Line L is passing through the point $8\hat{i} + 10\hat{j} - 8\hat{k}$ and parallel to the vector $2\hat{i} + \hat{j} + 2\hat{k}$. Point(s) on the L which are at distance 12 units from the point $8\hat{i} + 10\hat{j} - 8\hat{k}$ is/are

(a) $16\hat{i} + 14\hat{j}$ (b) $6\hat{i} - 16\hat{k}$

(c) $(16\hat{i} + 18\hat{j} - 4\hat{k})$ (d) none of these

Matrix Match Type

10. Three mutually perpendicular lines L_1, L_2 and L_3 are drawn from the point $(1, 2, -1)$. If L_1 is perpendicular to the x-axis and the direction ratios of L_2 are $(1, 2, -1)$, then match the List I with the List II

List I	List II
(1) Line L_1 passes through the point	(i) (–1/3, –2/3, 1/3)
(2) Line L_2 passes through the point	(ii) (7/3, 5/6, 0)
(3) Line L_3 passes through the point	(iii) (1, 8/3, 1/3)
(4) Centroid of triangle formed by the points of intersection with the xy-plane is	(iv) (28/3, –4/3, 2/3)

(a) (1)-(iii), (2)-(iv), (3)-(i), (4)-(ii)

(b) (1)-(ii), (2)-(iii), (3)-(iv), (4)-(i)

(c) (1)-(iii), (2)-(i), (3)-(iv), (4)-(ii)

(d) (1)-(iv), (2)-(i), (3)-(ii), (4)-(iii)

Numerical Value Type

11. If the line $\frac{x-2}{-1} = \frac{y+2}{1} = \frac{z+k}{4}$ is one of the angle bisectors of the lines $\frac{x}{1} = \frac{y}{-2} = \frac{z}{3}$ and $\frac{x}{-2} = \frac{y}{3} = \frac{z}{1}$, then the value of k is _____ .

12. A line with the direction ratios (2, 1, 2) intersects the lines $\vec{r} = -\hat{j} + \lambda(\hat{i} + \hat{j} + \hat{k})$ and $\vec{r} = -\hat{i} + \mu(2\hat{i} + \hat{j} + \hat{k})$ at A and B, respectively, then the length AB is equal to _____ .

13. Line passing through the origin and perpendicular to the lines $x - 2y + 3z - 4 = 0 = 2x + y - z + 3$ and $x - y + 2z - 5 = 0 = x + y + z + 6$ also passes through the point $(\alpha, -13, \beta)$. Then $\alpha + \beta =$ _____ .

Answer Key

Single Correct Answer Type

1. (b) **2.** (c) **3.** (c) **4.** (c) **5.** (b) **6.** (c)

Multiple Correct Answers Type

7. (a, c) **8.** (a, b, c, d) **9.** (a, b)

Matrix Match Type

10. (c)

Numerical Value Type

11. (8) **12.** (3) **13.** (29)

DPP 4.2

Properties of Line in Space

Single Correct Answer Type

1. If the lines $(\vec{r} - (\hat{i} + \hat{j} + \hat{k})) \times ((1-p)\hat{i} + 3\hat{j} - 2\hat{k}) = 0$ and $(\vec{r} - (3\hat{i} + \hat{j} - 5\hat{k})) \times ((3-p)\hat{i} + 4\hat{j} - 8\hat{k}) = 0$ are coplanar, then the value of p is
   (a) 4/3 (b) 1
   (c) 2/3 (d) 1/3

2. The distance of the point $(-\hat{i} + 2\hat{j} + 6\hat{k})$ from the straight line through the point $(2, 3, -4)$ and parallel to $6\hat{i} + 3\hat{j} - 4\hat{k}$ is
   (a) 6 (b) 7
   (c) 8 (d) 9

3. A variable point P lies on the circle $x^2 + y^2 = 1$ in the xy-plane. From point P, perpendicular PN is drawn to the line $x = y = z$, where N is the foot of perpendicular on the line. Then the minimum length PN is
   (a) $\sqrt{2}$ (b) $\frac{1}{\sqrt{2}}$
   (c) $\sqrt{3}$ (d) $\frac{1}{\sqrt{3}}$

4. Consider two lines $2x + y + z - 1 = 0 = 3x + y + 2z - 2$ and $x = y = z$.
   Now, consider following two statements:
   S_1: The shortest distance between the lines is $\sqrt{2}$ units.
   S_2: Line of the shortest length is perpendicular to the x-axis.
   Then
   (a) Only S_1 is true
   (b) Only S_2 is true
   (c) Both S_1 and S_2 are false
   (d) Both S_1 and S_2 are true

5. Given below are two statements. One is labelled as Assertion (A) and the other is labelled as Reason (R).
   Assertion (A): Lines $L_1 : \vec{r} = \hat{i} - \hat{j} + \lambda(\hat{i} + \hat{j} - \hat{k})$ and $L_2 : \vec{r} = 2\hat{i} - \hat{j} + \mu(\hat{i} + \hat{j} - \hat{k})$ do not intersect.
   Reason (R): Skew lines never intersect
   (a) (A) is true but (R) is false.
   (b) Both (A) and (R) are true and (R) is the true explanation of (A).
   (c) (A) is false and (R) is true.
   (d) Both (A) and (R) are true but (R) is not the true explanation of (A).

6. Given below are two statements. One is labelled as Assertion (A) and the other is labelled as Reason (R).
   Assertion (A): Lines $\vec{r} = \hat{i} + \hat{j} - \hat{k} + \lambda(3\hat{i} - \hat{j})$: $\vec{r} = 4\hat{i} - \hat{k} + \mu(2\hat{i} + 3\hat{k})$ intersect.
   Reason (R): If $\vec{b} \times \vec{d} = 0$, then lines $\vec{r} = \vec{a} + \lambda\vec{b}$ and $\vec{r} = \vec{c} + \lambda\vec{d}$ do not intersect.
   (a) (A) is true but (R) is false.
   (b) Both (A) and (R) are true and (R) is the true explanation of (A).
   (c) (A) is false and (R) is true.
   (d) Both (A) and (R) are true but (R) is not the true explanation of (A).

7. Given below are two statements. One is labelled as Assertion (A) and the other is labelled as Reason (R).
   Assertion (A): There exist two points on the line $\frac{x-1}{1} = \frac{y}{-1} = \frac{z+2}{2}$ which are at distance 2 units from point $(1, 2, -4)$.
   Reason (R): Perpendicular distance of point $(1, 2, -4)$ from the line $\frac{x-1}{1} = \frac{y}{-1} = \frac{z+2}{2}$ is 1 unit.
   (a) (A) is true but (R) is false.
   (b) Both (A) and (R) are true and (R) is the true explanation of (A).
   (c) (A) is false and (R) is true.
   (d) Both (A) and (R) are true but (R) is not the true explanation of (A).

8. Consider the lines $\frac{x}{-3} = \frac{y-1}{1} = \frac{z+1}{-1}$ and $\frac{x-2}{1} = \frac{y-3}{2} = \frac{z-13/7}{-1}$. Given below are two statements. One is labelled as Assertion (A) and the other is labelled as Reason (R).
   Assertion (A): The shortest distance between the lines is zero.
   Reason (R): Given lines are perpendicular.
   (a) (A) is true but (R) is false.
   (b) Both (A) and (R) are true and (R) is the true explanation of (A).
   (c) (A) is false and (R) is true.
   (d) Both (A) and (R) are true but (R) is not the true explanation of (A).

Multiple Correct Answers Type

9. Two lines $\frac{x-1}{1}=\frac{y+1}{-1}=\frac{z}{2}$ and $\frac{x}{2}=\frac{y-1}{-2}=\frac{z-1}{\lambda}$

(a) are parallel if $\lambda = 4$

(b) are perpendicular if $\lambda = -2$

(c) are coplanar if $\lambda = 4$

(d) are skew lines if $\lambda = 5$

Matrix Match Type

10. Match the List I with the List II

List I	List II
(1) $\frac{x-1}{2}=\frac{y-2}{3}=\frac{z-3}{4}$ and $\frac{x-1}{3}=\frac{y-3}{4}=\frac{z-5}{5}$ are	(i) coincident lines
(2) $\frac{x-1}{2}=\frac{y-2}{3}=\frac{z-3}{4}$ and $\frac{x-3}{2}=\frac{y-5}{3}=\frac{z-7}{4}$ are	(ii) parallel lines
(3) $\frac{x-2}{5}=\frac{y+3}{4}=\frac{5-z}{2}$ and $\frac{x-7}{5}=\frac{y-1}{4}=\frac{z-2}{-2}$ are	(iii) non-coplanar lines
(4) $\frac{x-3}{2}=\frac{y+2}{3}=\frac{z-4}{5}$ and $\frac{x-3}{3}=\frac{y-2}{2}=\frac{z-7}{5}$ are	(iv) intersecting lines

(a) (1)-(iii), (2)-(iv), (3)-(i), (4)-(ii)

(b) (1)-(ii), (2)-(iii), (3)-(iv), (4)-(i)

(c) (1)-(iii), (2)-(i), (3)-(iv), (4)-(ii)

(d) (1)-(iv), (2)-(i), (3)-(ii), (4)-(iii)

Numerical Value Type

11. If point (α, β, γ) on the line $\frac{x-2}{1}=\frac{y+1}{-1}=\frac{z}{2}$ is nearest to the origin, then $16\alpha\beta\gamma =$ ______.

12. If the shortest distance between the lines $(\vec{r}-(-\hat{i}+\hat{j}+\hat{k}))\times(2\hat{i}-\hat{j}+2\hat{k})=0$ and $\vec{r}=(3+2\mu)\hat{i}-(1+\mu)\hat{j}+(1+2\mu)\hat{k}$ is $\frac{4\sqrt{\alpha}}{3}$, then $\alpha =$ ______.

Answer Key

Single Correct Answer Type

1. (d) **2.** (b) **3.** (d) **4.** (b) **5.** (d) **6.** (d) **7.** (a) **8.** (d)

Multiple Correct Answers Type

9. (a, b, c, d)

Matrix Match Type

10. (d)

Numerical Value Type

11. (12) **12.** (5)

DPP 4.3
Equation of Plane

Single Correct Answer Type

1. Equation of the plane passing through the origin and perpendicular to the planes $x + 2y + z = 1$, $3x - 4y + z = 5$ is
(a) $x + 2y - 5z = 0$ (b) $x - 2y - 3z = 0$
(c) $x - 2y + 5z = 0$ (d) $3x + y - 5z = 0$

2. A vector $\vec{n}$ is inclined at an angle 45° with the x-axis and at an angle 60° with the y-axis. Also, the angle of $\vec{n}$ with the z-axis is acute. If $\vec{n}$ is normal to the plane passing through the point $(\sqrt{2}, -1, 1)$, then the equation of plane is
(a) $3\sqrt{2}x - 4y - 3z = 7$
(b) $4\sqrt{2}x + 7y + z = 2$
(c) $\sqrt{2}x + y + z = 2$
(d) $\sqrt{2}x - y - z = 2$

3. Let the equation of plane P be $\vec{r} = (1 + \lambda - \mu)\hat{i} + (2 - \lambda)\hat{j} + (3 - 2\lambda + 2\mu)\hat{k}$.
Now, consider the following two statements:
S_1: Equation of plane P is $2x - z = 5$.
S_2: Plane P is parallel to the y-axis.
(a) Only S_1 is true
(b) Only S_2 is true
(c) Both S_1 and S_2 are false
(d) Both S_1 and S_2 are true

4. Let the line L pass through the point $A(1, 0, 3)$, intersect the line $\frac{x}{2} = \frac{y-1}{3} = \frac{z-2}{1}$ and be parallel to the plane $x + y + z = 2$.
Now, consider the following two statements:
S_1: Line L passes through the point $(3, -9/2, 11/2)$.
S_2: Line L is perpendicular to the line $\frac{x+1}{1} = \frac{y-4}{4} = \frac{z}{2}$.
(a) Only S_1 is true
(b) Only S_2 is true
(c) Both S_1 and S_2 are false
(d) Both S_1 and S_2 are true

5. The variable plane $(2\lambda + 1)x + (3 - \lambda)y + z = 4$ always passes through the line
(a) $\frac{x}{0} = \frac{y}{0} = \frac{z+4}{1}$ (b) $\frac{x}{1} = \frac{y}{2} = \frac{z}{-3}$
(c) $\frac{x}{1} = \frac{y}{1} = \frac{z-4}{-7}$ (d) $\frac{x}{1} = \frac{y}{2} = \frac{z-4}{-7}$

6. Equation of the line of projection of the line $3x - y + 2z - 1 = 0 = x + 2y - z - 2$ on the plane $3x + 2y + z = 0$ is
(a) $\frac{x+1}{11} = \frac{y-1}{-9} = \frac{z-1}{-15}$
(b) $3x - 8y + 7z + 4 = 0 = 3x + 2y + z$
(c) $\frac{x+12}{11} = \frac{y+8}{-9} = \frac{z+14}{15}$
(d) $\frac{x+12}{11} = \frac{y+8}{-9} = \frac{z+14}{-15}$

7. A plane cutting the axes at P, Q and R passes through the point $(\alpha - \beta, \beta - \gamma, \gamma - \alpha)$. If O is the origin, then the locus of the centre of the sphere $OPQR$ is
(a) $\alpha x + \beta y + \gamma z = 4$
(b) $(\alpha - \beta)x + (\beta - \gamma)y + (\gamma - \alpha)z = 0$
(c) $(\alpha - \beta)yz + (\beta - \gamma)zx + (\gamma - \alpha)xy = 2xyz$
(d) $\left(\frac{1}{\alpha^2} + \frac{1}{\beta^2} + \frac{1}{\gamma^2}\right)(x^2 + y^2 + z^2) = xyz$

8. A line drawn from the point $P(1, 1, 1)$ is perpendicular to the line having direction ratios (1, 1, 1) and it intersects the plane $x + 2y + 3z = 4$ at Q. The locus of point Q is
(a) $\frac{x}{1} = \frac{y-5}{-2} = \frac{z+2}{1}$
(b) $\frac{x}{-2} = \frac{y-5}{1} = \frac{z+2}{1}$
(c) $x = y = z$
(d) none of these

9. The orthocentre of triangle whose vertices are $A(a, 0, 0)$, $B(0, b, 0)$ and $C(0, 0, c)$ is $\left(\frac{k}{a}, \frac{k}{b}, \frac{k}{c}\right)$. Then k is equal to
(a) $\left(\frac{1}{a^2} + \frac{1}{b^2} + \frac{1}{c^2}\right)^{-1}$
(b) $\left(\frac{1}{a} + \frac{1}{b} + \frac{1}{c}\right)^{-1}$
(c) $\frac{1}{a^2} + \frac{1}{b^2} + \frac{1}{c^2}$
(d) $\frac{1}{a} + \frac{1}{b} + \frac{1}{c}$

Multiple Correct Answers Type

10. Consider equations $E_1 : \vec{r}\times(2\hat{i}-\hat{j}+3\hat{k}) = 3\hat{i}+\hat{k}$ and E_2 : $\vec{r}\times(\hat{i}+2\hat{j}-3\hat{k}) = 2\hat{i}-\hat{j}$.

(a) E_1 represents a line
(b) E_1 represents two parallel planes
(c) E_2 represents a line
(d) E_2 represents two parallel planes

11. A plane is passing through the origin and is parallel to the line $\frac{x-1}{2} = \frac{y+3}{-1} = \frac{z+1}{-2}$ such that the distance between the plane and the line is $\frac{5}{3}$. Then the equation of the plane(s) is(are)

(a) $x - 2y + 2z = 0$ (b) $x - 2y - 2z = 0$
(c) $2x + 2y + z = 0$ (d) $x + y + z = 0$

12. If the line $\frac{x}{1} = \frac{y}{2} = \frac{z}{3}$ intersects the line $3\beta^2 x + 3(1-2\alpha)$ $y + z = 3 = -\frac{1}{2}\{6\alpha^2 x + 3(1-2\beta)\,y + 2z\}$, then the point $(\alpha, \beta, 1)$ lies on the plane

(a) $2x - y + z = 4$ (b) $x + y - z = 2$
(c) $x - 2y = 0$ (d) $2x - y = 0$

Linked Comprehension Type

For Questions 13–15

A line L_1 with the direction ratios $-3, 2, 4$ passes through the point $A(7, 6, 2)$ and a line L_2 with the direction ratios $2, 1, 3$ passes through the point $B(5, 3, 4)$. A line L_3 with the direction ratios $2, -2, -1$ intersects L_1 and L_3 at C and D, respectively.

13. The length CD is equal to

(a) 4 units (b) 6 units
(c) 9 units (d) 11 units

14. The equation of the plane parallel to line L_1 and containing line L_2 is equal to

(a) $x + 3y + 4z = 30$ (b) $x + 2y + z = 15$
(c) $2x - y + z = 11$ (d) $2x + 17y - 7z = 33$

15. The volume (cubic unit) of parallelepiped formed by $\overrightarrow{AB}$, $\overrightarrow{AC}$ and $\overrightarrow{AD}$ is equal to

(a) 140 (b) 138
(c) 134 (d) 130

For Questions 16 and 17

Let $P_1: x + y + 2z - 3 = 0$ and $P_2 : x - 2y + z = 4$ be two planes. Also, let $A(1, 3, 4)$ and $B(3, 2, 7)$ be two points in the space.

16. Equation of plane through the line of intersection of P_1 and P_2 such that the projection of AB on the plane is the greatest is

(a) $2x + 3y + z + 4 = 0$
(b) $3x - 3y + 4z - 11 = 0$
(c) $x + 3y + z + 2 = 0$
(d) $3y + z + 1 = 0$

17. The equation of plane through line of intersection of P_1 and P_2 such that the projection of AB on the plane is the least is

(a) $x + 3y + z + 2 = 0$
(b) $3y + z + 1 = 0$
(c) $2x - y + 3z - 7 = 0$
(d) $3x - 3y + 4z - 11 = 0$

Numerical Value Type

18. The distance of the point $(-1, -5, -10)$ from the point of intersection of the line $\frac{x-2}{3} = \frac{y+1}{4} = \frac{z-2}{12}$ and the plane $x - y + z = 5$ is______.

19. If the projection of the line $\frac{x}{2} = \frac{y-1}{2} = \frac{z-1}{1}$ on a plane P is $\frac{x}{1} = \frac{y-1}{1} = \frac{z-1}{-1}$. Then the distance of plane P from the origin is $\sqrt{\frac{\lambda}{20}}$, where λ = ______.

Answer Key

Single Correct Answer Type

1. (d) **2.** (c) **3.** (b) **4.** (d) **5.** (d) **6.** (b) **7.** (c) **8.** (a) **9.** (a)

Multiple Correct Answers Type

10. (b, c) **11.** (a, c) **12.** (a, b, c)

Linked Comprehension Type

13. (c) **14.** (d) **15.** (b) **16.** (d) **17.** (c)

Numerical Value Type

18. (13) **19.** (30)

DPP 4.4

Properties of Plane

Single Correct Answer Type

1. If the perpendicular distance of a point P other than origin from the plane $x + y + z = p$ is equal to the distance of the plane from the origin, then the coordinates of P are
(a) $(p, 2p, 0)$ (b) $(0, 2p, -p)$
(c) $(2p, p, -p)$ (d) $(2p, -p, 2p)$

2. Sum of the values of k for which the planes $kx + 4y + z = 0$, $4x + ky + 2z = 0$ and $2x + 2y + z = 0$ intersect in a straight line is
(a) 2 (b) 4
(c) 6 (d) 8

3. Angle between the two planes of which one plane is $4x + y + 2z = 0$ and another plane containing the lines $\frac{x-3}{2} = \frac{y-2}{3} = \frac{z-1}{\lambda}$, $\frac{x-2}{3} = \frac{y-3}{2} = \frac{z-2}{3}$ is
(a) $\frac{\pi}{3}$ (b) $\frac{\pi}{2}$
(c) $\frac{\pi}{6}$ (d) $\frac{2\pi}{3}$

4. If $P(\alpha, \beta, \gamma)$ is the vertex of an equilateral triangle PQR, where vertices Q and R are $(-1, 0, 1)$ and $(1, 0, -1)$, respectively, then P can lie on the plane
(a) $x + y + z + 6 = 0$
(b) $2x + 4y + 3z + 20 = 0$
(c) $x - y + z + 12 = 0$
(d) $x + y + z + 3\sqrt{2} = 0$

5. If plane $2x + 3y + 6z + k = 0$ is tangent to the sphere $x^2 + y^2 + z^2 + 2x - 2y + 2z - 6 = 0$, then a value of k is
(a) 26 (b) 16
(c) -26 (d) none of these

6. The shortest distance from $(1, 1, 1)$ to the line of intersection of the pair of planes $xy + yz + zx + y^2 = 0$ is
(a) $\sqrt{\frac{8}{3}}$ (b) $\frac{2}{\sqrt{3}}$
(c) $\frac{1}{\sqrt{3}}$ (d) $\frac{2}{3}$

7. Equation of the image of the sphere $x^2 + y^2 + z^2 = 9$ in the plane $2x + 3y + 4z - 29 = 0$ is
(a) $x^2 + y^2 + z^2 - 8x - 12y - 16z + 107 = 0$
(b) $x^2 + y^2 + z^2 + 8x - 12y - 16z + 107 = 0$
(c) $x^2 + y^2 + z^2 - 8x + 12y - 16z + 107 = 0$
(d) $x^2 + y^2 + z^2 - 8x - 12y + 16z + 107 = 0$

8. The locus of point which moves in such a way that its distance from the line $\frac{x}{1} = \frac{y}{1} = \frac{z}{-1}$ is twice the distance from the plane $x + y + z = 0$ is
(a) $x^2 + y^2 + z^2 - 5x - 3y - 3z = 0$
(b) $x^2 + y^2 + z^2 + 5x + 3y + 3z = 0$
(c) $x^2 + y^2 + z^2 - 5xy - 3yz - 3zx = 0$
(d) $x^2 + y^2 + z^2 + 5xy + 3yz + 3zx = 0$

9. A variable plane intersects the axes at A, B and C. If the volume of tetrahedron $OABC$ is 64 cu. units, then the locus of the foot of perpendicular from origin on this plane is (where O is origin)
(a) $(x^2 + y^2 + z^2)^3 = 384\, xyz$
(b) $xyz = 681$
(c) $(x+y+z)\left(\frac{1}{x}+\frac{1}{y}+\frac{1}{z}\right)^2 = 16$
(d) $xyz\,(x + y + z) = 81$

Multiple Correct Answers Type

10. Let $\vec{a} = \hat{i} + \hat{j} + \hat{k}$, $\vec{b} = -\hat{i} + \hat{j} + \hat{k}$, $\vec{c} = \hat{i} - \hat{j} + \hat{k}$ and $\vec{d} = \hat{i} + \hat{j} - \hat{k}$. Then, the line of intersection of planes one determined by $\vec{a}, \vec{b}$ and the other determined by $\vec{c}, \vec{d}$ is perpendicular to
(a) x-axis (b) y-axis
(c) z-axis (d) none of these

11. The equation of a plane is $2x - y - 3z = 5$. Also, $A(1, 1, 1)$, $B(2, 1, -3)$, $C(1, -2, -2)$ and $D(-3, 1, 2)$ are four points. Then which of the following line segments are intersected by the plane?
(a) AD (b) AB
(c) AC (d) BC

Linked Comprehension Type

For Questions 12 and 13

Let P be the plane consisting of all points that are equidistant from the points $A(-4, 2, 1)$ and $B(2, -4, 3)$. Also, let Q be the plane $x - y + cz = 1$ where $c \in R$.

12. The plane P is parallel to plane Q
(a) for no value of c (b) if $c = 3$
(c) if $c = 1/3$ (d) if $c = 1$

13. If the angle between the planes P and Q is 45°, then the product of all possible values of c is

(a) –17 (b) –2

(c) 17 (d) 24/17

For Questions 14 and 15

Let $A \equiv (1, 1, -1)$ and $B \equiv (0, 2, 1)$ be two points. Also, let $\pi : x + y + z = 0$, respectively, be a plane.

14. If A' and B' be the feet of perpendiculars from points A and B, respectively, on the plane π then $A'B'$ equals to

(a) $\sqrt{\dfrac{14}{3}}$ (b) $\sqrt{\dfrac{5}{3}}$

(c) $\sqrt{3}$ (d) $\sqrt{\dfrac{2}{3}}$

15. The equation of the line perpendicular to $\overrightarrow{AB}$ and lying completely in the plane π is

(a) $\dfrac{x-\frac{3}{2}}{1} = \dfrac{y-\frac{1}{2}}{-3} = \dfrac{z+2}{2}$

(b) $\dfrac{x-3}{2} = \dfrac{y-1}{-6} = \dfrac{z+4}{2}$

(c) $x = y = z$

(d) $x - 3 = y - 1 = z + 4$

Matrix Match Type

16. Match the List I with the List II

List I	List II
(1) If plane P_1: $\vec{r} = (\hat{i} - \hat{j} - 2\hat{k}) + \hat{j} + \lambda(2\hat{i} + \hat{j} + \hat{k}) + \mu(3\hat{i} - \hat{j} + \hat{k})$ and plane P_2: $\vec{r} \cdot (2\hat{i} + \hat{j} - 5\hat{k}) + 4 = 0$, then	(i) P_1 and P_2 are perpendicular
(2) If planes P_1 and P_2 are given by the equation $xy + yz = 0$, then	(ii) P_1 and P_2 are coincident
(3) If planes P_1 and P_2 are given by the equation $2x^2 - 2y^2 + 4z^2 + 6xz + 2yz + 3xy = 0$, then	(iii) P_1 and P_2 are parallel
(4) If P_1 is the plane through the points (1, 2, –1), (0, 1, 1) and (–1, 0, 2) while P_2 is the plane parallel to the z-axis and passing through the points (3, 4, 2) and (–2, –1, 1), then	(iv) P_1 and P_2 intersect but are not perpendicular

(a) (1)-(iii), (2)-(iv), (3)-(i), (4)-(ii)
(b) (1)-(ii), (2)-(iii), (3)-(iv), (4)-(i)
(c) (1)-(iii), (2)-(i), (3)-(iv), (4)-(ii)
(d) (1)-(iv), (2)-(i), (3)-(ii), (4)-(iii)

17. Consider a line L having the equation $(\vec{r} - \vec{p}) \times \vec{a} = 0$ and plane having the equation $\vec{r} = \vec{q} + \lambda\vec{b} + \mu\vec{c}$, where $\vec{p}, \vec{q}, \vec{a}, \vec{b}$ and $\vec{c}$ are given vectors.
Now, match the List I with the List II.

List I	List II
(1) If $[\vec{a}\ \vec{b}\ \vec{c}] = 0$ and $[\vec{p} - \vec{q}\ \vec{b}\ \vec{c}] \neq 0$, then	(i) the line will intersect the plane but not orthogonally.
(2) If $[\vec{a}\ \vec{b}\ \vec{c}] = 0$ and $[\vec{p} - \vec{q}\ \vec{b}\ \vec{c}] = 0$, then	(ii) the line will be parallel to the plane and won't lie in the plane
(3) If $\vec{a} \times (\vec{b} \times \vec{c}) = 0$, then	(iii) the line will lie in the plane
(4) If $\vec{a} \times (\vec{b} \times \vec{c}) \neq 0$ and $[\vec{a}\ \vec{b}\ \vec{c}] = 0$, then	(iv) the line will intersect the plane orthogonally.

(a) (1)-(iii), (2)-(iv), (3)-(i), (4)-(ii)
(b) (1)-(ii), (2)-(iii), (3)-(iv), (4)-(i)
(c) (1)-(iii), (2)-(i), (3)-(iv), (4)-(ii)
(d) (1)-(iv), (2)-(i), (3)-(ii), (4)-(iii)

Numerical Value Type

18. Let $P \equiv (1, 7, \sqrt{2})$ be a point and line L is $2\sqrt{2}(x - 1) = y - 2, z = 0$. If λ is the distance of plane $\sqrt{2}x + y - z = 1$ from point P measured along a line inclined at an angle 45° with the line L and is minimum, then the value of λ is ____.

19. The distance of the line $\dfrac{x-2}{1} = \dfrac{y-3}{2} = \dfrac{z-4}{2}$ from the plane $2x - 2y + z + 34 = 0$ is ______.

20. The distance of the point (1, –2, 3) from the plane $x - y + z - 5 = 0$, measured parallel to the line $\dfrac{x}{2} = \dfrac{y}{3} = \dfrac{z-1}{-6}$ is equal to ______.

21. P is a plane containing the line $x - y + z = 0 = x + y + z - 2$ and it is farthest from the origin. The distance of P from the point (3, 4, 5) is $\dfrac{\alpha}{\sqrt{6}}$, where $\alpha =$ _______.

22. If each point on the line $\dfrac{x+1}{\lambda} = \dfrac{y}{3} = \dfrac{z-2}{4}$ lies at the same distance from the plane $4x - \lambda y - 2z + 6 = 0$, then the value of λ is ____.

23. The non-zero value of k for which the lines $kx - 4y + 7z + 16 = 0 = 4x + 3y - 2z + 3$ and $x - 3y + 4z + 6 = 0 = x - y + z + 1$ intersect is _______.

Answer Key

Single Correct Answer Type

1. (c) **2.** (c) **3.** (b) **4.** (d) **5.** (a) **6.** (a) **7.** (a) **8.** (d) **9.** (a)

Multiple Correct Answers Type

10. (b, c) **11.** (b, c)

Linked Comprehension Type

12. (c) **13.** (b) **14.** (a) **15.** (a)

Matrix Match Type

16. (c) **17.** (b)

Numerical Value Type

18. (3) **19.** (12) **20.** (1) **21.** (13) **22.** (8) **23.** (7)

Solutions

CHAPTER 1

DPP 1.1

Single Correct Answer Type

1. (d) We know that line joining the midpoints of opposite edges of a tetrahedron are concurrent at point which is the mean of the vertices of tetrahedron.

Also, line segments joining vertex to the centroid of the opposite face of tetrahedron are concurrent at the point which is the mean of the vertices of tetrahedron.

2. (c) $A \equiv (a, b, c)$

So, $P \equiv (a, b, -c)$, $Q \equiv (-a, b, c)$ and $R \equiv (a, -b, c)$.

$\therefore$ Centroid of triangle PQR is $(a/3, b/2, c/3) \equiv G$

Now, points A, O, G are collinear

So, area of triangle AOG is zero.

3. (d) Lengths of the edges of the parallelepiped are 8, 6 and 2.

So, the volume of parallelepiped = $8 \times 6 \times 2 = 96$ cu. units

4. (d) The coordinates of the point which divides the segment joining the points $(-2, 4, 7)$ and $(3, -5, 8)$ in the ratio $k : 1$ are $\left(\frac{3k-2}{k+1}, \frac{-5k+4}{k+1}, \frac{8k+7}{k+1}\right)$.

As this point lies on the plane $2x - 1 = 0$, we have

$$\frac{3k-2}{k+1} = \frac{1}{2} \Rightarrow k = 1.$$

So, the required ratio is 1 : 1.

5. (c) Points A and B are given and area of triangle ABC is constant.

$\therefore$ Distance of vertex C from the base AB is constant.

$\therefore$ All such points lie on infinite lines parallel to AB at constant distance.

These infinite lines form a cylinder in the space.

6. (b) ΔOPM and ΔOAC are similar

$$\therefore \frac{OP}{OA} = \frac{OM}{OC} \Rightarrow \frac{\sqrt{x^2+y^2+z^2}}{\sqrt{2}} = \frac{x}{1}$$

$$\Rightarrow x^2 + y^2 + z^2 = 2x^2$$

$$\Rightarrow x^2 - y^2 - z^2 = 0$$

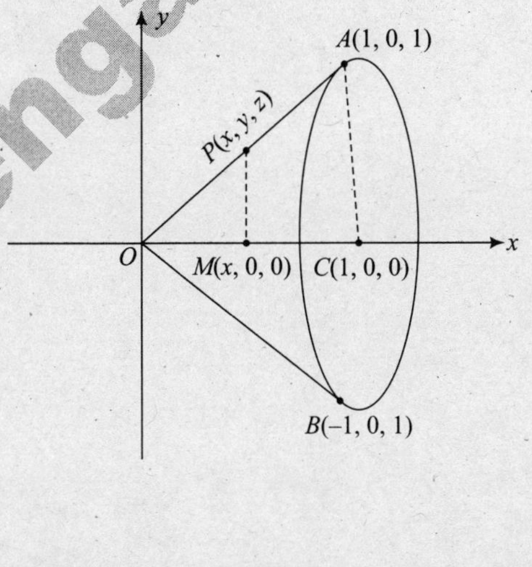

7. (d) We have $x^2 + y^2 + y^2 + z^2 + x^2 + z^2 = k$

$\therefore \quad x^2 + y^2 + z^2 = k/2$

So, the locus is sphere.

8. (d) Since, direction cosines of the given line are k, k and k.

$l = k$, $m = k$ and $n = k$

We know that, $l^2 + m^2 + n^2 = 1$

$$\Rightarrow k^2 + k^2 + k^2 = 1$$

$$\Rightarrow k^2 = \frac{1}{3}$$

$$\therefore k = \pm \frac{1}{\sqrt{3}}$$

9. (d) Point dividing AB in the ratio $\lambda : 1$ is

$$P\left(\frac{4\lambda+0}{\lambda+1}, \frac{5\lambda-1}{\lambda+1}, \frac{\lambda-1}{\lambda+1}\right)$$

Point dividing CD in the ratio $\mu : 1$ is

$$Q\left(\frac{-4\mu+3}{\mu+1}, \frac{4\mu+9}{\mu+1}, \frac{4\mu+4}{\mu+1}\right)$$

Comparing points P and Q, we get

$$\frac{4\lambda}{\lambda+1} = \frac{-4\mu+3}{\mu+1}, \frac{5\lambda-1}{\lambda+1} = \frac{4\mu+9}{\mu+1},$$

$$\frac{\lambda-1}{\lambda+1} = \frac{4\mu+4}{\mu+1}$$

$$\Rightarrow 4 - \frac{4}{\lambda+1} = -4 + \frac{7}{\mu+1},$$

$$5 - \frac{6}{\lambda+1} = 4 + \frac{5}{\mu+1}, \frac{\lambda-1}{\lambda+1} = 4$$

$$\Rightarrow \lambda = -\frac{5}{3}, \mu = -\frac{1}{2}$$

So, AB divides CD externally in the ratio 5 : 3 and CD intersect AB externally in the ratio 1 : 2

Multiple Correct Answers Type

10. (a, b, c)

Let xy-plane divide AB in the ratio $\lambda : 1$.

$$\therefore \frac{-5\lambda+4}{\lambda+1} = 0 \Rightarrow \lambda = 4/5$$

So, xy-plane divides AB internally in the ratio 4 : 5.

Let xz-plane divide AB in the ratio $\lambda : 1$.

$$\therefore \frac{2\lambda+3}{\lambda+1} = 0 \Rightarrow \lambda = -2/3$$

So, xz-plane divides AB externally in the ratio 2 : 3.

Let yz-plane divide AB in the ratio $\lambda : 1$.

$$\therefore \frac{-\lambda+2}{\lambda+1} = 0 \Rightarrow \lambda = 2$$

So, yz-plane divides AB internally in the ratio 2 : 1.

11. (a, b, c)

Sphere cuts the xy-plane in the circle whose equation is $x^2 + y^2 - 2x + 4y - 4 = 0$.

Radius of this circle is $\sqrt{1+4+4} = 3$.

Sphere cuts the yz-plane in the circle whose equation is $y^2 + z^2 + 4y - 6z - 4 = 0$

Radius of this circle is $\sqrt{4+9+4} = \sqrt{17}$

Sphere cuts the zx-plane in the circle whose equation is $x^2 + z^2 - 2x - 6z - 4 = 0$

Radius of this circle is $\sqrt{1+9+4} = \sqrt{10}$

12. (b, c)

Volume of the cube is 125 cu. units.

So, the side of the cube is 5 units.

Therefore, faces parallel to the given face $z = 3$ can be $z = -2$ and $z = 8$.

13. (a, d)

Let point $P(\alpha, 0, 0)$ on x-axis divide AB in the ratio $\lambda : 1$. Then

$$(\alpha, 0, 0) \equiv \left(\frac{-3\lambda+2}{\lambda+1}, \frac{\lambda-1}{\lambda+1}, \frac{a\lambda+4}{\lambda+1}\right)$$

$\therefore \quad \lambda = 1 \quad \Rightarrow \quad \alpha = -1/2$ and $a = -4$.

Numerical Value Type

14. (13)

Let $P \equiv (x, y, z)$

Given points are $A(2, 3; -1)$ and $B(-1, 0, 2)$.

Given that $AP = 2BP$

$\therefore \quad AP^2 = 4BP^2$

$\Rightarrow \quad (x-2)^2 + (y-3)^2 + (z+1)^2$

$= 4[(x+1)^2 + y^2 + (z-2)^2]$

$\Rightarrow \quad x^2 + y^2 + z^2 + 4x + 2y - 6z + 1 = 0$

This is sphere, whose center is $(-2, -1, 3)$ and radius is $\sqrt{4+1+9-1} = \sqrt{13}$.

15. (216)

$y^2 - 4y - 5 = 0$

$\therefore \quad (y-5)(y+1) = 0$

$\Rightarrow \quad y = -1$ and $y = 5$.

Distance between these two parallel planes is 6 units.

So, volume of cube is $6^3 = 216$ cu. units.

16. (343)

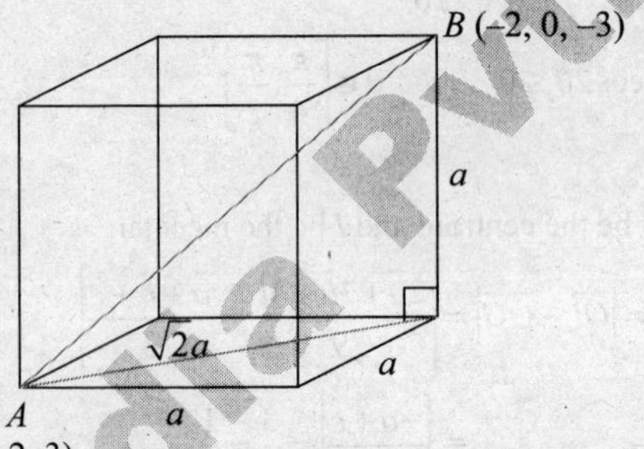

Two ends of diagonal of cube are $A(1, 2, 3)$ and $B(-2, 0, -3)$.

$\therefore \quad AB = 7$.

So, $\quad (\sqrt{2}a)^2 + a^2 = 7^2$

$\Rightarrow \quad a^2 = \frac{49}{3} \quad \Rightarrow \quad a = \frac{7}{\sqrt{3}}$

So, volume of cube $= a^3 = \left(\frac{7}{\sqrt{3}}\right)^3 = \frac{343}{3\sqrt{3}}$

CHAPTER 2

DPP 2.1

Single Correct Answer Type

1. (c) Let the angle with z-axis be γ. Then

$$\cos^2\theta + \cos^2\theta + \cos^2\gamma = 1$$

$$\Rightarrow \cos^2\gamma = -\cos 2\theta$$

$$\Rightarrow \cos 2\theta \le 0 \Rightarrow \theta \in \left[\frac{\pi}{4}, \frac{\pi}{2}\right]$$

2. (b)

3. (c) Let G be the centroid and I be the incenter.

$$|\overrightarrow{GI}| = |\overrightarrow{OI} - \overrightarrow{OG}| = \left|\frac{2\vec{a}+3\vec{b}+4\vec{c}}{9} - \frac{\vec{a}+\vec{b}+\vec{c}}{3}\right|$$

$$= \left|\frac{-\vec{a}+\vec{c}}{9}\right| = \frac{3}{9} = \frac{1}{3}$$

4. (c)

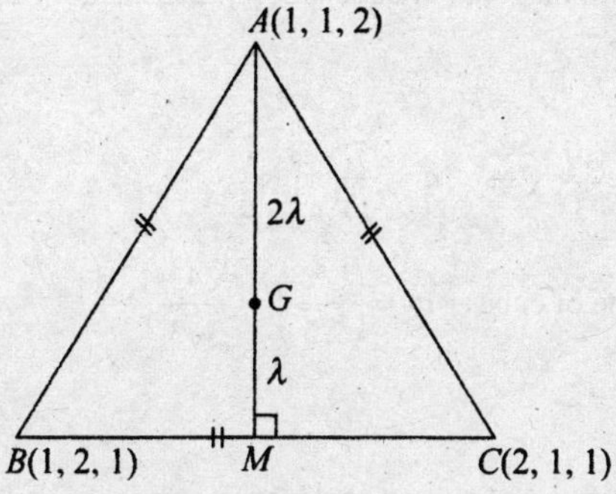

$$AB = BC = AC = \sqrt{2}$$

So, triangle ABC is equilateral.

$\therefore$ Distance of orthocentre 'O' from the sides is equal to distance of centroid from the side.

From the figure, $AD = \sqrt{2}\sin 60° = \sqrt{\frac{3}{2}}$

$$\therefore GD = l_1 = l_2 = l_3 = \frac{1}{3}\sqrt{\frac{3}{2}} = \frac{1}{\sqrt{6}}$$

$$\therefore l_1 + l_2 + l_3 = \frac{3}{\sqrt{6}} = \frac{\sqrt{6}}{2}$$

5. (a) Let $\vec{R}$ be the resultant of the given vectors. Then,

$$\vec{R} = (2\hat{i}+3\hat{j}-\hat{k}) + (4\hat{i}-3\hat{j}+2\hat{k}) = 6\hat{i}+0\hat{j}+\hat{k}$$

The required unit vector parallel to $\vec{R}$

$$= \frac{\vec{R}}{|\vec{R}|} = \frac{6\hat{i}+0\hat{j}+\hat{k}}{\sqrt{6^2+0^2+1^2}} = \frac{6\hat{i}+0\hat{j}+\hat{k}}{\sqrt{37}}$$

$$= \frac{1}{\sqrt{37}}(6\hat{i}+\hat{k}).$$

6. (c)

7. (b)

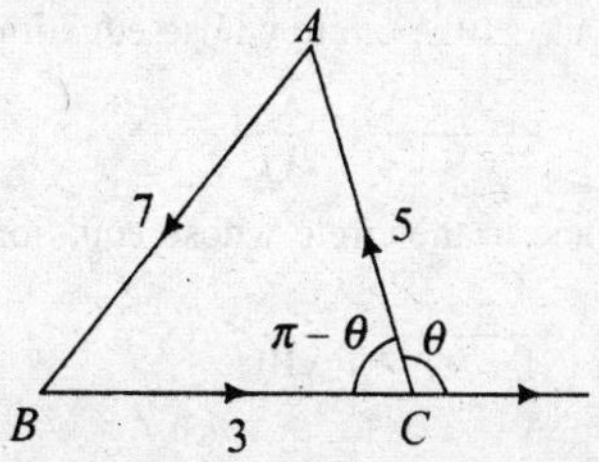

Let θ be the angle between $\vec{a}$ and $\vec{b}$

Then in triangle ABC, $\angle C = \pi - \theta$

$$\therefore \cos(\pi - \theta) = \frac{3^2+5^2-7^2}{2(3)(5)}$$

$$\Rightarrow -\cos\theta = \frac{-1}{2}$$

$$\therefore \theta = 60°$$

8. (b)

9. (b) This is possible only if $\vec{q}$ and $\vec{r}$ are in opposite directions.

$\therefore \theta = \pi$

10. (d) $\overrightarrow{OA} = \hat{i}+2\hat{j}-\hat{k}$, $\overrightarrow{OB} = \hat{i}+\hat{j}+\hat{k}$,

$\overrightarrow{OC} = 2\hat{i}+3\hat{j}+2\hat{k}$

Position vector of B w.r.t. origin at A is

$$\overrightarrow{AB} = \overrightarrow{OB} - \overrightarrow{OA} = -\hat{j} + 2\hat{k}$$

Position vector of C w.r.t. origin at A is

$$\overrightarrow{AC} = \overrightarrow{OC} - \overrightarrow{OA} = \hat{i} + \hat{j} + 3\hat{k}$$

11. (b)

12. (c) Vectors $\vec{a}, \vec{b}$ are in the same direction.

Vectors $\vec{b}$ and $\vec{c}$ are in the opposite directions.

So, vectors $\vec{a}$ and $\vec{c}$ are in the opposite directions.

13. (c) $\hat{j}$ and $\hat{k}$ are unit vectors along y-axis and z-axis, respectively.

Then the required unit vector $= \dfrac{\hat{j}+\hat{k}}{\sqrt{2}}$.

14. (d) $\overrightarrow{AC} = 2015\,\overrightarrow{AB}$

$$\therefore \vec{c} - \vec{a} = 2015(\vec{b} - \vec{a})$$

$$\Rightarrow \vec{c} = 2015\vec{b} - 2014\vec{a}$$

15. (b) $\vec{a}+\vec{b}$ and $\vec{a}-\vec{b}$ are the diagonals of the parallelogram whose sides are $\vec{a}$ and $\vec{b}$.

Given that $|\vec{a}+\vec{b}| = |\vec{a}-\vec{b}|$

So, the diagonals of the parallelogram have same length.

Therefore, parallelogram is rectangle.

Hence, $\vec{a} \perp \vec{b}$

16. (b) $3\vec{a} - 2\vec{b} + 5\vec{c} - 6\vec{d} = \vec{0}$

$$\therefore \frac{3\vec{a}+5\vec{c}}{8} = \frac{2\vec{b}+6\vec{d}}{7}$$

So, the position of vector of point dividing AC in the ratio 5 : 3

is same as that of point dividing BD in the ratio 6 : 2.

Therefore, line segments AC and BD intersect.

Hence, points A, B, C and D are coplanar.

Multiple Correct Answers Type

17. (a, c)

We must have $\cos^2\alpha + \sin^2\beta - 2\sec^2\gamma = 0$

$\Rightarrow \underbrace{\cos^2\alpha + \sin^2\beta}_{\le 2} = \underbrace{2\sec^2\gamma}_{\ge 2}$

$\Rightarrow \cos^2\alpha + \sin^2\beta = 2$ and $\sec^2\gamma = 1$

$\Rightarrow \cos^2\alpha = 1$, $\sin^2\beta = 1$ and $\sec^2\gamma = 1$

$\Rightarrow \vec{a} + \vec{b} - 2\vec{c} = 0$

$\Rightarrow \vec{c} = \dfrac{\vec{a}+\vec{b}}{2}$ or $\vec{a} = \dfrac{2\vec{c}-\vec{b}}{2-1}$

18. (b, c)

We have, $\vec{r} = \lambda\vec{c}$

$\therefore |\vec{r}| = |\lambda||\vec{c}|$

$\Rightarrow \dfrac{\left|\vec{a}+\vec{b}\right|}{2} = |\lambda|\left|\vec{c}\right|$

$\therefore |6\hat{i} - 2\hat{j} - 4\hat{k}| = 2|\lambda||3\hat{i} - \hat{j} + 2\hat{k}|$

$\Rightarrow \sqrt{56} = 2|\lambda|\sqrt{14}$

$\therefore \lambda = \pm 1$

$\therefore \vec{r} = \pm\vec{c}$

19. (a, b)

Linked Comprehension Type

20. (c) $\overrightarrow{AB} = \overrightarrow{OB} - \overrightarrow{OA} = 3\vec{a} - \vec{b} - 2\vec{c}$

$\overrightarrow{AC} = \overrightarrow{OC} - \overrightarrow{OA} = 9\vec{a} - 3\vec{b} - 6\vec{c} = 3\overrightarrow{AB}$

21. (a) $2\overrightarrow{OA} - 3\overrightarrow{OB} + \overrightarrow{OC} = 2(-2\vec{a} + 3\vec{b} + 5\vec{c}) - 3(\vec{a} + 2\vec{b} + 3\vec{c}) + (7\vec{a} - \vec{c}) = 0$

22. (d) As $2\overrightarrow{OA} - 3\overrightarrow{OB} + \overrightarrow{OC} = 0$, we have

$$\frac{2\overrightarrow{OA} + \overrightarrow{OC}}{3} = \overrightarrow{OB}$$

So, B Divides AC in 1 : 2.

Matrix Match Type

23. (c)

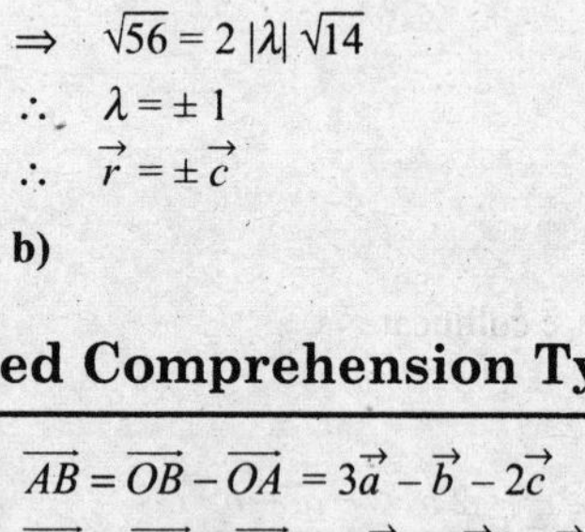

In the figure $\overrightarrow{AB} = 2\hat{i}$

$BM = BC\sin 60° = 2\sin 60° = \sqrt{3}$

$\therefore \overrightarrow{BD} = 2\sqrt{3}\hat{j}$

$\overrightarrow{BC} = BC\cos 60°\,\hat{i} + BC\sin 60°\,\hat{j} = \hat{i} + \sqrt{3}\hat{j}$

$\therefore \overrightarrow{CD} = \overrightarrow{BD} - \overrightarrow{BC} = 2\sqrt{3}\hat{j} - (\hat{i} + \sqrt{3}\hat{j})$

$= -\hat{i} + \sqrt{3}\hat{j}$

$\overrightarrow{AC} = \overrightarrow{AB} + \overrightarrow{BC} = 2\hat{i} + (\hat{i} + \sqrt{3}\hat{j}) = 3\hat{i} + \sqrt{3}\hat{j}$

$\overrightarrow{AD} = \overrightarrow{AB} + \overrightarrow{BD} = 2\hat{i} + 2\sqrt{3}\hat{j}$

$\overrightarrow{CE} = \overrightarrow{CD} + \overrightarrow{DE} = (-\hat{i} + \sqrt{3}\hat{j}) - 2\hat{i} = -3\hat{i} + \sqrt{3}\hat{j}$

Numerical Value Type

24. (3) In an equilateral triangle, centroid is same as orthocentre

$\therefore \dfrac{\overrightarrow{OA} + \overrightarrow{OB} + \overrightarrow{OC}}{3} = \vec{0}$

$\Rightarrow \overrightarrow{OA} + \overrightarrow{OB} + \overrightarrow{OC} = \vec{0}$

$\therefore \overrightarrow{AB} + 2\overrightarrow{BC} + 3\overrightarrow{CA}$

$= \overrightarrow{OB} - \overrightarrow{OA} + 2\overrightarrow{OC} - 2\overrightarrow{OB} + 3\overrightarrow{OA} - 3\overrightarrow{OC}$

$= -\overrightarrow{OB} + 2\overrightarrow{OA} - \overrightarrow{OC}$

$= -(\overrightarrow{OB} + \overrightarrow{OA} + \overrightarrow{OC}) + 3\overrightarrow{OA} = 3\overrightarrow{OA} = 3\vec{a}$

25. (4) Given, $|\vec{a}| = 4$, $|\vec{b}| = 1 = |\vec{c}|$

Also, $\vec{b}$ and $\vec{c}$ are parallel.

$\therefore \vec{b} = \vec{c}$

$\therefore \vec{a} + \lambda\vec{c} = 2\vec{b} \Rightarrow \vec{a} + \lambda\vec{c} = 2\vec{c}$

$\Rightarrow \vec{a} = (2 - \lambda)\vec{c} \Rightarrow |\vec{a}| = |2 - \lambda||\vec{c}|$

$\Rightarrow |2 - \lambda| = 4 \quad \therefore \lambda = -2, 6$

26. (5) Consider the point O as origin.

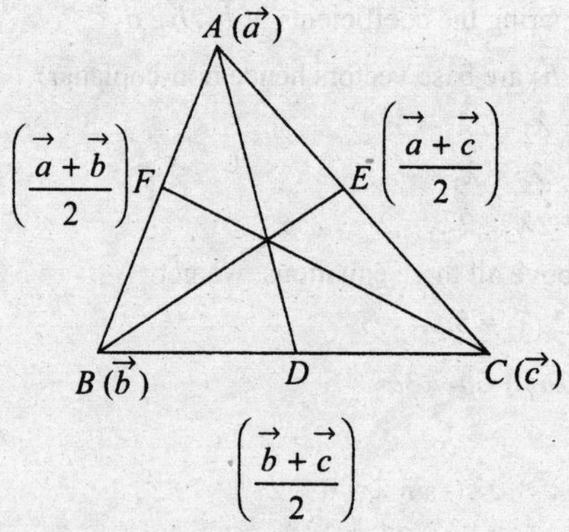

Here $\overrightarrow{AD} + \dfrac{2}{3}\overrightarrow{BE} + \dfrac{1}{3}\overrightarrow{CF}$

$= \dfrac{\vec{b}+\vec{c}}{2} - \vec{a} + \dfrac{2}{3}\left(\dfrac{\vec{a}+\vec{c}}{2} - \vec{b}\right) + \dfrac{1}{3}\left(\dfrac{\vec{a}+\vec{b}}{2} - \vec{c}\right)$

$= \dfrac{\vec{b}+\vec{c}}{2} - \vec{a} + \dfrac{2}{3}\left(\dfrac{\vec{a}+\vec{c}}{2} - \vec{b}\right) + \dfrac{1}{3}\left(\dfrac{\vec{a}+\vec{b}}{2} - \vec{c}\right)$

$= \dfrac{1}{2}(\vec{c} - \vec{a}) = \dfrac{1}{2}(\overrightarrow{AC})$

27. (3)

DPP 2.2

Single Correct Answer Type

1. (a) Since vectors are linearly dependent, then let $\vec{b} = \lambda\vec{a}$.

$$\Rightarrow \quad \frac{|\vec{a} + \lambda\vec{a}|}{|\vec{a} - \lambda\vec{a}|} = 2 \quad \Rightarrow \quad \frac{1+\lambda}{1-\lambda} = \pm 2$$

$$\Rightarrow \quad \lambda = 3, \frac{1}{3} \quad \Rightarrow \quad \vec{b} = 3\vec{a} \qquad (\because \ |\vec{b}| > |\vec{a}|)$$

2. (b) The vectors are linearly dependent

$$\therefore \quad \begin{vmatrix} p+1 & -3 & p \\ p & p+1 & -3 \\ -3 & p & p+1 \end{vmatrix} = 0$$

$\Rightarrow \quad (p-1)(p^2 + 7p + 13) = 0$

$\Rightarrow \quad p = 1$

So, only one integral value of p exists.

3. (c) $\vec{a}_1 = 2\vec{b}_1 + 3\vec{b}_2 - \vec{b}_3$

$\vec{a}_2 = \vec{b}_1 - 2\vec{b}_2 + 2\vec{b}_3$

$\vec{a}_3 = -2\vec{b}_1 + \vec{b}_2 - 2\vec{b}_3$

Let $\vec{F} = \lambda_1\vec{a}_1 + \lambda_2\vec{a}_2 + \lambda_3\vec{a}_3$

$= \lambda_1(2\vec{b}_1 + 3\vec{b}_2 - \vec{b}_3) + \lambda_2(\vec{b}_1 - 2\vec{b}_2 + 2\vec{b}_3) + \lambda_3(-2\vec{b}_1 + \vec{b}_2 - 2\vec{b}_3)$

$\therefore \quad \vec{F} = (2\lambda_1 + \lambda_2 - 2\lambda_3)\vec{b}_1 + (3\lambda_1 - 2\lambda_2 + \lambda_3)\vec{b}_2 + (-\lambda_1 + 2\lambda_2 - 2\lambda_3)\vec{b}_3$

Also given that $\vec{F} = 3\vec{b}_1 - \vec{b}_2 + 2\vec{b}_3$

$\therefore$ Comparing the coefficients of $\vec{b}_1, \vec{b}_2, \vec{b}_3$

($\because$ $\vec{b}_1, \vec{b}_2, \vec{b}_3$ are base vectors hence non-coplanar)

$2\lambda_1 + \lambda_2 - 2\lambda_3 = 3;$

$3\lambda_1 - 2\lambda_2 + \lambda_3 = -1;$

$-\lambda_1 + 2\lambda_2 - 2\lambda_3 = 2$

Solving above all three equations, we get

$\lambda_1 = 2, \lambda_2 = 5, \lambda_3 = 3$

$\therefore \quad \vec{F} = 2\vec{a}_1 + 5\vec{a}_2 + 3\vec{a}_3$

4. (a) Put $\Delta = 0$

$\Rightarrow \quad \lambda^7 + \lambda^3 + 2\lambda - \sin\lambda = 0$

Let $f(\lambda) = \lambda^7 + \lambda^3 + 2\lambda - \sin\lambda$

$\Rightarrow \quad f'(\lambda) = (7\lambda^6 + 3\lambda^2 + 2 - \cos\lambda) > 0$ ® $\lambda \in R$

$\therefore \quad f(\lambda) = 0$ has only one real solution $\lambda = 0$.

5. (a) Here $\overrightarrow{AB} = x\hat{i} - y\hat{j}$

$\overrightarrow{AC} = x\hat{i} - z\hat{k}$

$\overrightarrow{AD} = (x-1)\hat{i} - \hat{j} - \hat{k}$

As these vectors are coplanar, we have

$$\begin{vmatrix} x & -y & 0 \\ x & 0 & -z \\ x-1 & -1 & -1 \end{vmatrix} = 0 \quad \Rightarrow \quad \frac{1}{x} + \frac{1}{y} + \frac{1}{z} = 1$$

6. (a) $a\hat{i} + b\hat{j} + c\hat{k}, \hat{i} + \hat{k}, c\hat{i} + c\hat{j} + b\hat{k}$ are coplanar

$$\therefore \quad \begin{vmatrix} a & a & c \\ 1 & 0 & 1 \\ c & c & b \end{vmatrix} = 0$$

$\Rightarrow \quad c^2 - ab = 0$ ...(1)

For equation $ax^2 + 2cx + b = 0$

$D = 4c^2 - 4ab = 0$ (using (1))

So, roots are real and equal.

7. (b)

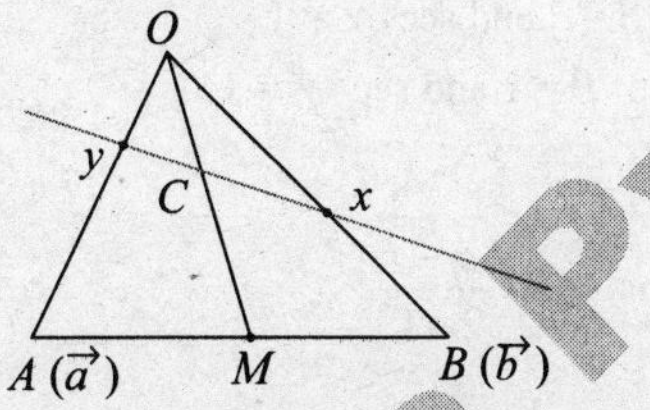

From the figure, $\overrightarrow{OA} = \vec{a}, \overrightarrow{OB} = \vec{b}$

$\therefore \quad \overrightarrow{OM} = \frac{\vec{a} + \vec{b}}{2}$

$\Rightarrow \quad \overrightarrow{OC} = \frac{\vec{a} + \vec{b}}{6}$ and $\overrightarrow{OX} = \frac{2}{3}\vec{b}$

Let $\frac{OY}{YA} = \lambda$

$\therefore \quad \overrightarrow{OY} = \frac{\lambda}{\lambda + 1}\vec{a}$

Now points Y, C, X are collinear

$\therefore \quad \overrightarrow{YC} = m\overrightarrow{CX}$

$$\therefore \quad \frac{\vec{a} + \vec{b}}{6} - \frac{\lambda}{\lambda+1}\vec{a} = m\frac{2\vec{b}}{3} - m\frac{\vec{a} + \vec{b}}{6}$$

Comparing the coefficients of $\vec{a}$ and $\vec{b}$, we get

$$\therefore \quad \frac{1}{6} - \frac{\lambda}{\lambda+1} = -\frac{m}{6} \text{ and } \frac{1}{6} = \frac{2m}{3} - \frac{m}{6}$$

$$\therefore \quad m = \frac{1}{3} \text{ and } \lambda = \frac{2}{7}$$

8. (c) Clearly, Reason is true.

Vectors $\overrightarrow{OA}, \overrightarrow{OB}$ and $\overrightarrow{OC}$ are coplanar.

So, Assertion is false.

Multiple Correct Answers Type

9. (b, c, d)

$\vec{c} + \vec{a} = (\vec{b} + \vec{c}) + (\vec{a} - \vec{b})$

$\vec{a} + \vec{b} = (\vec{b} - \vec{c}) + (\vec{c} + \vec{a})$

$\vec{b} + \vec{c} = (\vec{a} + \vec{b}) + (\vec{c} - \vec{a})$

So, vectors in options (b), (c), (d) are coplanar

10. (a, b)

We have $x + 3y - 4z = \lambda x$

$x - 3y + 5z = \lambda y$ and $3x + y = \lambda$

This system of homogeneous equations must have non-trivial solution.

$$\therefore \begin{vmatrix} 1-\lambda & 3 & -4 \\ 1 & -3-\lambda & 5 \\ 3 & 1 & -\lambda \end{vmatrix} = 0$$

$\Rightarrow \quad \lambda(\lambda+1)^2 = 0$

$\Rightarrow \quad \lambda = 0, -1$

Linked Comprehension Type

11. (a), 12. (c)

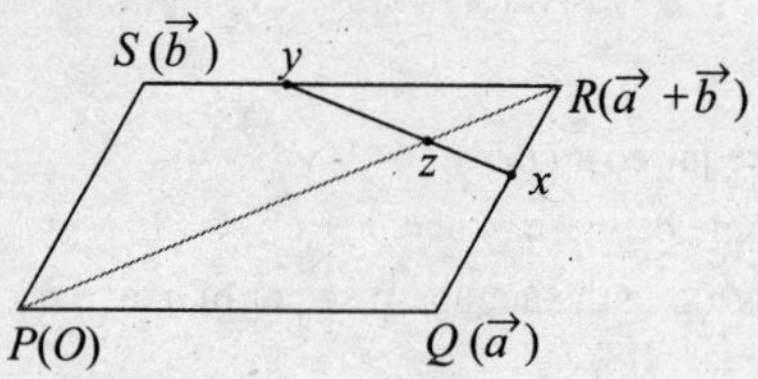

From the figure,

$$\text{P.V. of } X = \frac{4(\vec{a}+\vec{b})+\vec{a}}{5} = \frac{5\vec{a}+4\vec{b}}{5}$$

$$\text{P.V. of } Y = \frac{4\vec{b}+\vec{a}+\vec{b}}{5} = \frac{\vec{a}+5\vec{b}}{5}$$

$\overrightarrow{PZ} = m\overrightarrow{PR}$

$\therefore \quad \overrightarrow{PZ} = m(\vec{a}+\vec{b})$

Let Z divide YX in the ratio $\lambda : 1$.

$$\therefore \quad \overrightarrow{PZ} = \frac{\lambda\left(\frac{5\vec{a}+4\vec{b}}{5}\right)+\frac{\vec{a}+5\vec{b}}{5}}{\lambda+1} = m(\vec{a}+\vec{b})$$

Comparing coefficients of $\vec{a}$ and $\vec{b}$, we get

$$m = \frac{5\lambda+1}{5(\lambda+1)} \text{ and } m = \frac{4\lambda+5}{5(\lambda+1)}$$

$\therefore \quad \lambda = 4$

$$\therefore \quad \overrightarrow{PZ} = \frac{21}{25}(\vec{a}+\vec{b}) = \frac{21}{25}\overrightarrow{PR}$$

Numerical Value Type

13. (187)

It is given that the points P, Q and R are collinear.

$\Rightarrow \quad \overrightarrow{PQ} = \lambda\overrightarrow{QR}$

$\Rightarrow \quad 15t(4\hat{j}-3\hat{i}) = \lambda[(1-t)(\hat{i}+\hat{j})-60t\hat{j}]$

$\qquad = \lambda[(1-t)\hat{i}+(1-61t)\hat{j}]$

$$\Rightarrow \quad \frac{45t}{t-1} = \frac{60t}{1-61t}$$

$$\Rightarrow \quad \frac{3t}{t-1} = \frac{4t}{1-61t}$$

$\Rightarrow \quad 187t = 7 \quad \Rightarrow \quad t = 7/187$

14. (0) We have $\vec{a}+\vec{b} = \vec{c}$

$$(\lambda x\hat{i}+y\hat{j}+4z\hat{k})+(y\hat{i}+x\hat{j}+3y\hat{k})$$
$$= -z\hat{i}-2z\hat{j}-(\lambda+1)x\hat{k}$$

$\Rightarrow \quad \lambda x+y+z = 0, x+y+2z = 0$

and $(\lambda+1)x+3y+4z = 0$

Since, x, y and z are not all zero, system of equations must have non-trivial solution.

$$\therefore \begin{vmatrix} \lambda & 1 & 1 \\ 1 & 1 & 2 \\ \lambda+1 & 3 & 4 \end{vmatrix} = 0$$

$\Rightarrow \quad \lambda(4-6)-1(4-3)+(\lambda+1)(2-1) = 0$

$\Rightarrow \quad \lambda = 0$

15. (16) We have,

$$a\hat{i}+b\hat{j}+\frac{1}{2}\hat{k} = l(\hat{i}+2\hat{j})+m(\hat{j}-2\hat{k})$$

$$\Rightarrow \quad a = l, b = 2l+m \text{ and } m = \frac{-1}{4}$$

Also, $a\hat{i}+b\hat{j}+\frac{1}{2}\hat{k}$ is unit vector

$$\therefore \quad a^2+b^2 = \frac{3}{4}$$

$$\Rightarrow \quad 5a^2-a-\frac{11}{16} = 0$$

Now, a_1 and a_2 are the roots of above equation

$$\Rightarrow \quad \frac{1}{a_1}+\frac{1}{a_2} = \frac{a_1+a_2}{a_1a_2} = -\frac{16}{11}$$

CHAPTER 3

DPP 3.1

Single Correct Answer Type

1. (a) $\vec{a}\cdot\vec{\beta}=\dfrac{b}{a}+\dfrac{4a}{b}+1\geq 5$

So, $\left(\dfrac{10}{5+\vec{a}\cdot\vec{\beta}}\right)_{\max.}=1.$

2. (b) $\vec{a}=\cos\theta\hat{i}+\sin\theta\hat{j},\ \vec{b}=-\sin\theta\hat{i}+\cos\theta\hat{j}$ and $\vec{c}=\hat{k}$

$\therefore\quad |\vec{a}|=|\vec{b}|=|\vec{c}|$

$\vec{a}\cdot\vec{b}=\vec{b}\cdot\vec{c}=\vec{c}\cdot\vec{a}=0$

So, $\vec{a}, \vec{b}$ and $\vec{c}$ are mutually perpendicular.

$\therefore$ Required angle $=\cos^{-1}\dfrac{1}{\sqrt{3}}$

3. (b) Here $|\vec{a}|=1, |\vec{b}|=1, |\vec{c}|=1$

$\vec{a}\cdot\vec{b}=0$ and $\vec{a}\cdot\vec{c}=0$

Also, $\vec{b}\cdot\vec{c}=|\vec{b}|\cdot|\vec{c}|\cos\dfrac{\pi}{3}=1\times1\times\dfrac{1}{2}=\dfrac{1}{2}$

Now, $|\vec{a}+\vec{b}+\vec{c}|^2$

$=\vec{a}^2+\vec{b}^2+\vec{c}^2+2\vec{a}\cdot\vec{b}+2\vec{a}\cdot\vec{c}+2\vec{b}\cdot\vec{c}$

$=1^2+1^2+1^2+2\cdot\dfrac{1}{2}=4$

$\Rightarrow\quad |\vec{a}+\vec{b}+\vec{c}|=2$

4. (b)

5. (d) $p\vec{x}+q\vec{y}=\vec{a}$

Let $p\vec{x}-q\vec{y}=R\hat{n}$

$\therefore\quad R=|p\vec{x}-q\vec{y}|$

$\Rightarrow\quad \left|p\vec{x}-q\vec{y}\right|^2=\left|p\vec{x}+q\vec{y}\right|^2-4pq\,\vec{x}\cdot\vec{y}\geq 0$

$\Rightarrow\quad |\vec{a}|^2\geq 4pq\vec{x}\cdot\vec{y}$

But given that $\vec{x}\cdot\vec{y}\geq\dfrac{|\vec{a}|^2}{4pq}$

$\Rightarrow\quad |\vec{a}|^2=4pq\vec{x}\cdot\vec{y}$

$\Rightarrow\quad |p\vec{x}-q\vec{y}|=0\quad\Rightarrow\quad p\vec{x}=q\vec{y}$

$\Rightarrow\quad \vec{x}=\dfrac{\vec{a}}{2p},\ \vec{y}=\dfrac{\vec{a}}{2q}\quad\Rightarrow\quad \dfrac{|\vec{x}|}{|\vec{y}|}=\dfrac{q}{p}$

6. (d) We have $\vec{a}\cdot\vec{b}=\vec{a}\cdot\vec{c}=0,$

$|\vec{a}|=|\vec{c}|=1, |\vec{b}|=2$ and $\vec{b}\cdot\vec{c}=1$

Let $\vec{d}=x(\vec{a}+\vec{b})+y(2\vec{b}-\vec{c}),$

But $\vec{d}\cdot\vec{a}=1$

$\Rightarrow\quad 1=x(1+0)+0\quad\Rightarrow\quad x=1$

$\Rightarrow\quad \vec{d}=\vec{a}+\vec{b}+y(2\vec{b}-\vec{c})$

$\Rightarrow\quad |\vec{d}|^2=|\vec{a}|^2+|\vec{b}|^2+2\vec{a}\cdot\vec{b}+y^2(2\vec{b}-\vec{c})^2+2y(\vec{a}+\vec{b})\cdot(2\vec{b}-\vec{c})$

$=1+4+y^2(16+1-4)+2y(8-1)$

$=13y^2+14y+5$

$\therefore\quad |\vec{d}|_{\min}=\sqrt{\dfrac{4\times13\times5-14\times14}{4\times13}}=\dfrac{4}{\sqrt{13}}$

7. (a)

8. (a) Let θ be the required angle.

Then, $\theta=$ angle between $\vec{a}$ and $\vec{b}+\vec{c}$

(as $\vec{b}+\vec{c}$ lies along the angular bisector of $\vec{a}$ and $\vec{b}$)

$\therefore\quad \cos\theta=\dfrac{\vec{a}\cdot(\vec{b}+\vec{c})}{|\vec{a}||\vec{b}+\vec{c}|}$

$=\dfrac{2\cos\alpha}{\sqrt{2+2\cos\alpha}}=\dfrac{\cos\alpha}{\cos\dfrac{\alpha}{2}}$

$\therefore\quad \theta=\cos^{-1}\left(\dfrac{\cos\alpha}{\cos\alpha/2}\right)$

Multiple Correct Answers Type

9. (a, b, c)

Let $\overline{OA}=\vec{a}, \overline{OB}=\vec{b}, \overline{OC}=\vec{c}$

Then from the given conditions,

$\vec{a}\cdot\vec{a}+(\vec{b}-\vec{c})\cdot(\vec{b}-\vec{c})$

$=\vec{b}\cdot\vec{b}+(\vec{c}-\vec{a})\cdot(\vec{c}-\vec{a})$

$\Rightarrow\quad -2\vec{b}\cdot\vec{c}=-2\vec{c}\cdot\vec{a}$

$\Rightarrow\quad \overline{BA}\cdot\overline{OC}=0$

Hence, $AB\perp OC$.

Similarly, we have $BC\perp OA$ and $CA\perp OB$.

10. (b, c)

11. (b, d)

Volume = Area of base × height

$3=\dfrac{1}{2}\times\sqrt{2}\times\sqrt{3}\times h$

$h=\sqrt{6}$

$(A_1A)^2=h^2=6$

$\overline{A_1A}\cdot\overline{AB}=0$

$\overline{A_1A}\cdot\overline{AC}=0$

$\overline{AA_1}\cdot\overline{BC}=0$

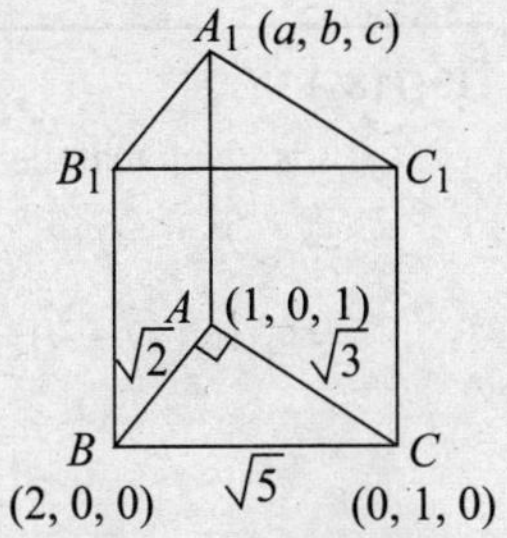

Solving we get position vector of A_1 as (0, –2, 0) or (2, 2, 2).

Matrix Match Type

12. (a) (1) Component of vector $\vec{p}$ in the direction of vector $\vec{q}$ is

$\dfrac{\vec{p}\cdot\vec{q}}{|\vec{q}|^2}\vec{q}=\dfrac{8}{9}(2\hat{i}+\hat{j}+2\hat{k})$

(2) Component of vector $\vec{p}$ in the direction which is perpendicular to the vector $\vec{q}$ is

$$\vec{p}-\frac{\vec{p}\cdot\vec{q}}{|\vec{q}|^2}\vec{q}=2\hat{i}-4\hat{j}+4\hat{k}-\frac{8}{9}(2\hat{i}+\hat{j}+2\hat{k})$$

$$=\frac{2}{9}(\hat{i}-22\hat{j}+10\hat{k})$$

(3) Vector along the internal angle bisector of $\angle BAC$ is

$$\frac{\vec{p}}{|\vec{p}|}+\frac{\vec{q}}{|\vec{q}|}=\frac{2\hat{i}-4\hat{j}+4\hat{k}}{6}+\frac{2\hat{i}+\hat{j}+2\hat{k}}{3}$$

$$=\frac{3\hat{i}-\hat{j}+4\hat{k}}{3}$$

(4) Vector along the median through vertex A is

$\vec{p}+\vec{q}=5\hat{i}-3\hat{j}+6\hat{k}$

Numerical value type

13. (42)

Let the vertices of triangle be $A(\vec{p})$, $B(\vec{q})$ and $C(\vec{r})$ and P be the origin.

$\therefore$ Centroid is $G\left(\frac{\vec{p}+\vec{q}+\vec{r}}{3}\right)$

Now $|\vec{p}+\vec{q}+\vec{r}|^2+|\vec{p}-\vec{q}|^2+|\vec{q}-\vec{r}|^2+|\vec{r}-\vec{p}|^2$
$=3(|\vec{p}|^2+|\vec{q}|^2+|\vec{r}|^2)$

$\therefore$ $9(GP)^2+(AB)^2+(BC)^2+(CA)^2$
$=3((PA)^2+(PB)^2+(PC)^2)$

$\Rightarrow$ $9\times 2^2+(a^2+b^2+c^2)=3(1^2+3^2+4^2)$

$\Rightarrow$ $a^2+b^2+c^2=42.$

14. (143)

Minimum value of $|\overrightarrow{OP}|=\sqrt{10+2\sqrt{133}}$

15. (3)

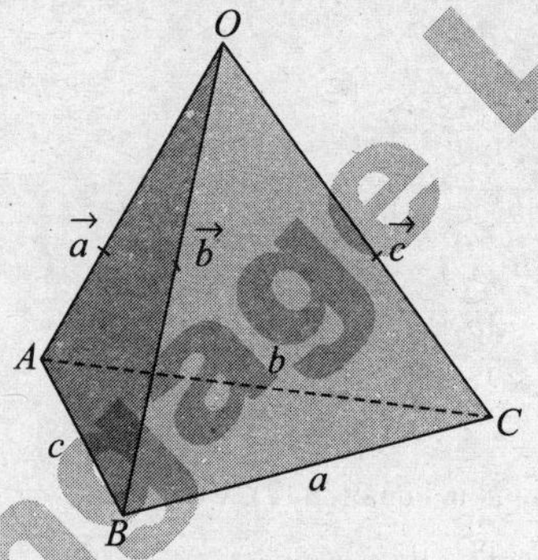

Let vertices of tetrahedron be $O(\vec{0})$, $A(\vec{a})$, $B(\vec{b})$, $C(\vec{c})$.

$\overrightarrow{OG_1}\cdot\overrightarrow{BG_2}=0$

$\Rightarrow$ $\frac{\vec{a}+\vec{b}+\vec{c}}{3}\cdot\frac{\vec{a}+\vec{c}-3\vec{b}}{3}=0$

$\Rightarrow$ $a^2+c^2-3b^2+2\vec{a}\cdot\vec{c}-2\vec{b}\cdot\vec{c}-2\vec{a}\cdot\vec{b}=0$ ...(1)

Now, $|\vec{c}-\vec{a}|^2=b^2$, $|\vec{c}-\vec{b}|=a^2$ and $|\vec{a}-\vec{b}|=c^2$

$\therefore$ $2\vec{a}\cdot\vec{c}=a^2+c^2-b^2$, $2\vec{b}\cdot\vec{c}=b^2+c^2-a^2$

and $2\vec{a}\cdot\vec{b}=a^2+b^2-c^2$

Putting in (1), we get $2a^2+2c^2-6b^2=0$

$\Rightarrow$ $\frac{a^2+c^2}{b^2}=3.$

16. (0.5)

We have $|\vec{p}-\vec{q}|=\vec{p}\cdot\vec{q}$

$\therefore$ $|\vec{p}|^2+|\vec{q}|^2-2\vec{p}\cdot\vec{q}=2\cos^2\theta$

$\Rightarrow$ $1+2-2\sqrt{2}\cos\theta=2\cos^2\theta$

$\Rightarrow$ $2\cos^2\theta+2\sqrt{2}\cos\theta-3=0$

$\Rightarrow$ $(\sqrt{2}\cos\theta-1)(\sqrt{2}\cos\theta+3)=0$

$\Rightarrow$ $\cos\theta=\frac{1}{\sqrt{2}}$

$\therefore$ Area of quadrilateral $=\frac{1}{2}|\vec{p}|\cdot|\vec{q}|\sin\theta=\frac{1}{2}$

17. (12)

We have

$$\frac{(\vec{b}+\vec{c})\cdot\vec{a}}{|\vec{a}|}=2|\vec{a}|,\ \frac{(\vec{a}+\vec{c})\cdot\vec{b}}{|\vec{b}|}=3|\vec{b}|$$

and $\frac{(\vec{a}+\vec{b})\cdot\vec{c}}{|\vec{c}|}=4|\vec{c}|$

$\therefore$ $(\vec{b}+\vec{c})\cdot\vec{a}=2a^2$ ...(1)

$(\vec{c}+\vec{a})\cdot\vec{b}=3b^2$ ...(2)

$(\vec{a}+\vec{b})\cdot\vec{c}=4c^2$ ....(3)

Adding, we get

$2(\vec{a}\cdot\vec{b}+\vec{b}\cdot\vec{c}+\vec{c}\cdot\vec{a})=2a^2+3b^2+4c^2$

Now, $|\vec{a}+\vec{b}+\vec{c}|^2=(a^2+b^2+c^2)+2(\vec{a}\cdot\vec{b}+\vec{a}\cdot\vec{c}+\vec{b}\cdot\vec{c})$

$\Rightarrow$ $|\vec{a}+\vec{b}+\vec{c}|=\sqrt{3a^2+4b^2+5c^2}$

$\Rightarrow$ $|\vec{a}+\vec{b}+\vec{c}|_{min.}=\sqrt{3+4+5}=\sqrt{12}$

DPP 3.2

Single Correct Answer Type

1. (c) $\vec{p}\times\vec{q}=(\vec{a}-\vec{b})\times(\vec{a}+\vec{b})=2(\vec{a}\times\vec{b})$

$\Rightarrow$ $|\vec{p}\times\vec{q}|^2=4(|\vec{b}|^2|\vec{a}|^2-(\vec{b}\cdot\vec{a})^2)$

$\Rightarrow$ $|\vec{p}\times\vec{q}|^2=4(625-\vec{b}\cdot\vec{a})^2$

$\Rightarrow$ $|\vec{p}\times\vec{q}|=2(\{625-(\vec{a}\cdot\vec{b})^2\}^{1/2}$

$\Rightarrow$ $\lambda=625.$

2. (d) $\vec{u}\times\vec{v}=(2\hat{i}-\hat{j}+\hat{k})\times(-3\hat{j}+2\hat{k})=\hat{i}-4\hat{j}-6\hat{k}$

$\vec{w}$ is unit vector in the xy-plane.

$\therefore$ $\vec{w}=(\cos\theta)\hat{i}+(\sin\theta)\hat{j}$

Now $(\vec{u}\times\vec{v})\cdot\vec{w}=\cos\theta-4\sin\theta$

$\therefore$ $|(\vec{u}\times\vec{v})\cdot\vec{w}|_{max.}=\sqrt{1^2+(-4)^2}=\sqrt{17}$

3. (a)

4. (a) Let $L \equiv (3a, 0, 0)$, $M \equiv (0, 3b, 0)$ and $N \equiv (0, 0, 3c)$

$\therefore$ Area of triangle PQR

$= 4 \times$ Area of triangle LMN

$= 4 \times \frac{1}{2} |\overline{MN} \times \overline{ML}|$

$= 2\left|(-3b\hat{j}+3c\hat{k})\times(-3a\hat{i}+3b\hat{j})\right|$

$= 18\,|bc\hat{i}+ca\hat{j}+ab\hat{k}|$

$= 18\sqrt{b^2c^2+c^2a^2+a^2b^2}$

5. (a) Let $\vec{b} = \alpha\hat{i}+\beta\hat{j}+\gamma\hat{k}$, $\vec{a}\times\vec{b} = \hat{j}-\hat{k}$

$$\Rightarrow \begin{vmatrix} \hat{i} & \hat{j} & \hat{k} \\ 1 & 1 & 1 \\ \alpha & \beta & \gamma \end{vmatrix} = \hat{j}-\hat{k}$$

$\Rightarrow \beta-\gamma = 0, \alpha-\gamma = 1, \alpha-\beta = 1$

$\Rightarrow \beta = \gamma, \alpha = 1+\gamma, \alpha = 1+\beta$

Now, $\vec{a}\cdot\vec{b} = 1 \Rightarrow \alpha+\beta+\gamma = 1$

$\Rightarrow \beta+1+\beta+\beta = 1 \Rightarrow \beta = 0$

$\therefore \alpha = 1, \gamma = 0$

$\therefore \vec{b} = \hat{i}$

$\Rightarrow \vec{b}\cdot(3\hat{i}+4\hat{j}+5\hat{k}) = 3$

6. (c) $(\vec{a}-\vec{c})\cdot(\vec{b}+\vec{c}) = 0$

$\Rightarrow \vec{a}\cdot\vec{b}+\vec{c}\cdot(\vec{a}-\vec{b})-|\vec{c}|^2 = 0$

$\Rightarrow (\vec{a}-\vec{b})\cdot\vec{c} = 1$

$\Rightarrow (\vec{a}-\vec{b})\cdot(\lambda\vec{a}+\mu\vec{b}+\omega(\vec{a}\times\vec{b})) = 1$

$\Rightarrow \lambda-\mu = 1$

$\Rightarrow \lambda = \mu+1,$

$\vec{a}\cdot\vec{b} = 0$

So, $\vec{a}, \vec{b}$ and $\vec{a}\times\vec{b}$ are mutually perpendicular.

$|\vec{c}| = 1$

$\Rightarrow \lambda^2+\mu^2+\omega^2 = 1$

$\Rightarrow (\mu+1)^2+\mu^2+\omega^2 = 1$

7. (b)

8. (c) $|\vec{a}+\vec{b}+\vec{a}\times\vec{b}|^2+|\vec{a}+\vec{b}-\vec{a}\times\vec{b}|^2$

$= 2(|\vec{a}+\vec{b}|^2+|\vec{a}\times\vec{b}|^2)$

$= 2(|\vec{a}|^2+|\vec{b}|^2+2|\vec{a}\cdot\vec{b}|+|\vec{a}|^2|\vec{b}|^2-|\vec{a}\cdot\vec{b}|^2)$

$= 2((1+\vec{a}\cdot\vec{a})(1+\vec{b}\cdot\vec{b})-(1-\vec{a}\cdot\vec{b})^2)$

9. (b) Let $\vec{r}_1 = a\hat{i}+b\hat{j}+c\hat{k}$, $\vec{r}_2 = 3\hat{i}+4\hat{j}+5\hat{k}$

Now, $|\vec{r}_1\times\vec{r}_2| \le |\vec{r}_1|^2\,|\vec{r}_2|^2$ ...(1)

$$\vec{r}_1\times\vec{r}_2 = \begin{vmatrix} \hat{i} & \hat{j} & \hat{k} \\ a & b & c \\ 3 & 4 & 5 \end{vmatrix}$$

$= \hat{i}(5b-4c)+\hat{j}(3c-5a)+\hat{k}(4a-3b)$

So, from (1)

$(5b-4c)^2+(3c-5a)^2+(4a-3b)^2 \le 50$

Multiple Correct Answers Type

10. (a, b)

$\vec{a}\times(\vec{b}-3\vec{c}) = \vec{0}$

$\Rightarrow \vec{b}-3\vec{c} = \lambda\vec{a}$

$\Rightarrow |\vec{b}-3\vec{c}| = |\lambda\vec{a}|$

$\Rightarrow 1+1-6\times1\times\frac{1}{3}\times\frac{1}{2} = |\lambda| \Rightarrow \lambda = \pm 1.$

$\therefore \vec{b}-3\vec{c} = \pm\vec{a}$

11. (a, c, d)

$\overline{OA}\cdot\overline{OB} = -3$

$\therefore 2\times 3\cos\theta = -3$

$\Rightarrow \cos\theta = -\frac{1}{2} \Rightarrow \theta = \frac{2\pi}{3}$

$\overline{AC}\cdot\overline{BC} = (2(\vec{a}\times\vec{b})-2\vec{a})\cdot(2(\vec{a}\times\vec{b})-3\vec{b})$

$= 4|\vec{a}|^2|\vec{b}|^2\sin^2\theta+6\vec{a}\cdot\vec{b}$

$= 4\times\frac{3}{4}-6\frac{1}{2} = 0$

Therefore, angle C is 90°.

Linked Comprehension Type

12. (d), 13. (d)

$|\vec{a}| = 1, |\vec{b}| = 4, \vec{a}\cdot\vec{b} = 2$

$\vec{c} = (2\vec{a}\times\vec{b})-3\vec{b}$

$\Rightarrow \vec{c}+3\vec{b} = 2\vec{a}\times\vec{b}$ ...(1)

Now, $\vec{a}\cdot\vec{b} = 2$

$\Rightarrow |\vec{a}|\cdot|\vec{b}|\cos\theta = 2$

$\Rightarrow \cos\theta = \frac{2}{|\vec{a}|.|\vec{b}|} = \frac{2}{4}$

$\Rightarrow \cos\theta = \frac{1}{2}.$

$\therefore \theta = \frac{\pi}{3}$

From (1), $|\vec{c}+3\vec{b}|^2 = |2\vec{a}\times\vec{b}|^2$

$\Rightarrow |\vec{c}|^2+6\vec{c}\cdot\vec{b}+9|\vec{b}|^2 = 4|\vec{a}|^2|\vec{b}|^2\sin^2\theta$

$\Rightarrow |\vec{c}|^2+144+6\vec{b}\cdot\vec{c} = 48$

$\Rightarrow |\vec{c}|^2+96+6(\vec{b}\cdot\vec{c}) = 0$ ...(2)

Now, $\vec{c} = 2\vec{a}\times\vec{b}-3\vec{b}$

$\Rightarrow \vec{b}\cdot\vec{c} = 0-3\times 16$

$\therefore \vec{b}\cdot\vec{c} = -48$

Putting value of $\vec{b}\cdot\vec{c}$ in equation (2), we get

$|\vec{c}|^2+96-6\times 48 = 0$

$\Rightarrow |\vec{c}|^2 = 192$

Again, putting value of $|\vec{c}|$ in equation (2),

$192+96+6|\vec{b}|\cdot|\vec{c}|\cos\alpha = 0$

$\Rightarrow 6\times4\times8\sqrt{3}\cos\alpha = -288$

$\Rightarrow \cos\alpha = -\frac{\sqrt{3}}{2}$

$\therefore \alpha = \frac{5\pi}{6}$

Numerical Value Type

14. (3)

15. (3) Let $\vec{r}_1 = a\hat{i} + b\hat{j} + c\hat{k}$ and $\vec{r}_2 = A\hat{i} + B\hat{j} + C\hat{k}$

$\Rightarrow \quad \vec{r}_1 . \vec{r}_2 = aA + bB + cC$

and $|\vec{r}_1||\vec{r}_2| = \sqrt{(a^2+b^2+c^2)(A^2+B^2+C^2)}$

Given that $aA + bB + cC$

$$+\sqrt{(a^2+b^2+c^2)(A^2+B^2+C^2)} = 0$$

$\therefore \quad \vec{r}_1 \vec{r}_2 = -|\vec{r}_1||\vec{r}_2|$

So, $\vec{r}_1$ and $\vec{r}_2$ are anti-parallel.

$\Rightarrow \quad \dfrac{a}{A} = \dfrac{b}{B} = \dfrac{c}{C} = k$ where k is any constant

$\Rightarrow \quad \dfrac{aB}{bA} + \dfrac{bC}{cB} + \dfrac{cA}{aC} = 3$

DPP 3.3

Single Correct Answer Type

1. (a)

2. (a) Let $\vec{x} = \vec{a} + 2\vec{b} - \vec{c}$

$\vec{y} = 2\vec{a} + \vec{b} + \vec{c}$

$\vec{z} = 4\vec{a} - \vec{b} + 5\vec{c}$

Now $\begin{vmatrix} 1 & 2 & -1 \\ 2 & 1 & 1 \\ 4 & -1 & 5 \end{vmatrix} = 0$

Thus, vectors are coplanar

$\Rightarrow \quad [\vec{a} + 2\vec{b} - \vec{c} \;\; 2\vec{a} + \vec{b} + \vec{c} \;\; 4\vec{a} - \vec{b} + 5\vec{c}] = 0$

3. (d) $\vec{a} + (\vec{a} \times \vec{b}) = \vec{c}$

Taking dot product with $\vec{a}$, we get

$\vec{a}^2 = \vec{a} \cdot \vec{c}$

$\Rightarrow \quad \vec{a} \cdot \vec{c} = 1$

Taking dot product with $\vec{c}$ we get

$\vec{a} \cdot \vec{c} + (\vec{a} \times \vec{b}) \cdot \vec{c} = \vec{c} \cdot \vec{c}$

$\Rightarrow \quad 1 + [\vec{a}\;\vec{b}\;\vec{c}] = 4 \quad \Rightarrow \quad [\vec{a}\;\vec{b}\;\vec{c}] = 3$

4. (a) $\left[\vec{a} + (\vec{a}\times\vec{b}) \;\; \vec{b} + (\vec{a}\times\vec{b}) \;\; \vec{a}\times\vec{b}\right]$

$= (\vec{a} + (\vec{a}\times\vec{b})).((\vec{b} + (\vec{a}\times\vec{b})) \times (\vec{a}\times\vec{b}))$

$= (\vec{a} + (\vec{a}\times\vec{b})).(\vec{b}\times(\vec{a}\times\vec{b})) = \vec{a}\cdot\vec{b}\times(\vec{a}\times\vec{b})$

$= (\vec{a}\times\vec{b})\cdot(\vec{a}\times\vec{b}) = 1$

5. (d) $[\vec{p}\;\vec{q}\;\vec{r}]$

$$= \begin{bmatrix} \cos\theta & \sin\theta & \cos 2\theta \\ \cos\left(\frac{2\pi}{3}+\theta\right) & \sin\left(\frac{2\pi}{3}+\theta\right) & \cos 2\left(\frac{2\pi}{3}+\theta\right) \\ \cos\left(\theta-\frac{2\pi}{3}\right) & \sin\left(\theta-\frac{2\pi}{3}\right) & \cos 2\left(\theta-\frac{2\pi}{3}\right) \end{bmatrix}$$

$$= \begin{bmatrix} \cos\theta + 2\cos\theta\cos\frac{2\pi}{3} & \sin\theta + 2\sin\theta\cos\frac{2\pi}{3} & \cos 2\theta + 2\cos\frac{4\pi}{3}\cos 2\theta \\ \cos\left(\frac{2\pi}{3}+\theta\right) & \sin\left(\frac{2\pi}{3}+\theta\right) & \cos 2\left(\frac{2\pi}{3}+\theta\right) \\ \cos\left(\theta-\frac{2\pi}{3}\right) & \sin\left(\theta-\frac{2\pi}{3}\right) & \cos 2\left(\theta-\frac{2\pi}{3}\right) \end{bmatrix}$$

(Applying $R_1 \to R_1 + R_2 + R_3$)

$$= \begin{bmatrix} 0 & 0 & 0 \\ \cos\left(\frac{2\pi}{3}+\theta\right) & \sin\left(\frac{2\pi}{3}+\theta\right) & \cos 2\left(\frac{2\pi}{3}+\theta\right) \\ \cos\left(\theta-\frac{2\pi}{3}\right) & \sin\left(\theta-\frac{2\pi}{3}\right) & \cos 2\left(\theta-\frac{2\pi}{3}\right) \end{bmatrix} = 0$$

6. (a) $\vec{r}.(3\vec{a} + 5\vec{b} + 2\vec{c}) = 0$

$\Rightarrow \quad \vec{a}.(\vec{b}\times\vec{c})[2\sin x + 3\cos y + 5] = 0$

$\Rightarrow \quad 2\sin x + 3\cos y + 5 = 0$

(As $\vec{a}$, $\vec{b}$ and $\vec{c}$ are non-zero and non-coplanar)

$\Rightarrow \quad 2\sin x + 3\cos y = -5$

$\Rightarrow \quad \sin x = -1, \cos y = 1$

$\Rightarrow \quad \operatorname{cosec} x = -1, \sec y = -1.$

7. (a)

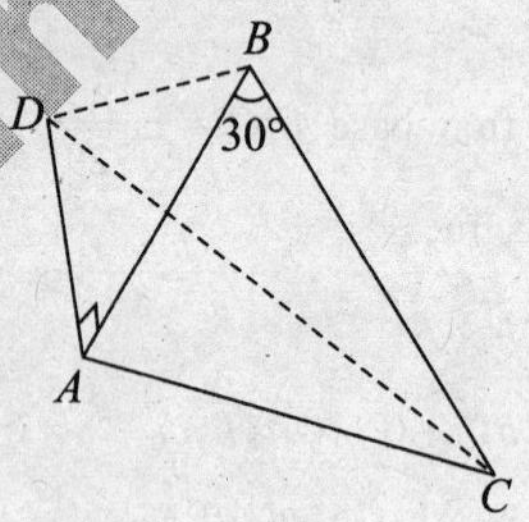

Volume of tetrahedron

$= \dfrac{1}{3} AD\left(\dfrac{1}{2} AB \cdot BC \sin 30°\right)$

$= \dfrac{1}{12}(AD \cdot AB \cdot BC) = 18$ (given)

$\Rightarrow \quad AD \cdot AB \cdot BC = 216$

Now, $AB + BC + AD \geq 3(AD \cdot AB \cdot BC)^{1/3}$

$\Rightarrow \quad AB + BC + AD \geq 18$

So, the minimum value occurs when $AB = BC = AD = 6$.

Hence, $AC = \sqrt{AB^2 + BC^2 - 2AB \,.\, BC.\cos 30°}$

$= 6\sqrt{2-\sqrt{3}}$

8. (b)

9. (c) $|\hat{a} + \hat{b} + \hat{a}\times\hat{b}| = |\hat{a} - \hat{b}|$

Squaring, we get

$(\hat{a}+\hat{b})^2 + (\hat{a}\times\hat{b})^2 + 2(\hat{a}+\hat{b})\cdot(\hat{a}\times\hat{b})$

$= 1 + 1 - 2\hat{a}\cdot\hat{b}$

$\Rightarrow \quad 1 + 1 + 2\cos\theta + \sin^2\theta + 0 = 2 - 2\cos\theta$

$= \cos^2\theta - 4\cos\theta - 1 = 0$

$\therefore \quad \cos\theta = 2 - \sqrt{5}$

$\Rightarrow \quad |\hat{a} - \hat{b}|^2 = 1 + 1 - 2\hat{a}\cdot\hat{b} = 2 - 2(2 - \sqrt{5})$

$= 2(\sqrt{5} - 1) = 8\sin 18°$

10. (c) $\overrightarrow{AD} = 2\hat{j} - \hat{k}$, $\overrightarrow{BD} = -2\hat{i} - \hat{j} - 3\hat{k}$

and $\overrightarrow{CD} = 2\hat{i} - \hat{j}$

$\therefore$ Volume of tetrahedron $= \frac{1}{6}[\overrightarrow{AD}\ \overrightarrow{BD}\ \overrightarrow{CD}]$

$$= \frac{1}{6}\left\|\begin{matrix} 0 & 2 & -1 \\ -2 & -1 & -3 \\ 2 & -1 & 0 \end{matrix}\right\| = \frac{8}{3}$$

Also, area of triangle ABC

$$= \frac{1}{2}|\overrightarrow{AB} \times \overrightarrow{AC}| = \frac{1}{2}\left\|\begin{matrix} \hat{i} & \hat{j} & \hat{k} \\ 2 & 3 & 2 \\ -2 & 3 & -1 \end{matrix}\right\|$$

$$= \frac{1}{2}|-9\hat{i} - 2\hat{j} + 12\hat{k}|$$

$$= \frac{\sqrt{229}}{2}$$

Now, Volume $= \frac{1}{3}$ height $\times$ (area of triangle ABC)

$\therefore \quad \frac{8}{3} = \frac{1}{3}\text{height} \times \frac{\sqrt{229}}{2}$

So, the distance of D from base $ABC = \frac{16}{\sqrt{229}}$

11. (b) Clearly, Reason (R) is true.

Also $(\hat{i} \times \vec{a}) \cdot \vec{b} = \hat{i} \cdot (\vec{a} \times \vec{b})$

$$\therefore \quad \vec{a} \times \vec{b} = ((\hat{i} \cdot (\vec{a} \times \vec{b}))\hat{i} + ((\hat{j} \cdot (\vec{a} \times \vec{b}))\hat{j} + ((\hat{k} \cdot (\vec{a} \times \vec{b}))\hat{k}$$

Multiple Correct Answers Type

12. (b, d)

13. (a, b)

We have,

$$\alpha(\vec{a} \times \vec{b}) + \beta(\vec{b} \times \vec{c}) + \gamma(\vec{c} \times \vec{a}) = 0$$

Taking dot product with $\vec{c}$, we get

$$\alpha[\vec{a}\ \vec{b}\ \vec{c}] = 0$$

Similarly, taking dot product with b and c, we get

$$\gamma[\vec{a}\ \vec{b}\ \vec{c}] = 0, \beta[\vec{a}\ \vec{b}\ \vec{c}] = 0$$

Now, even if one of $\alpha, \beta, \gamma \neq 0$ then we have

$$[\vec{a}\ \vec{b}\ \vec{c}] = 0$$

Therefore, a, b, c are coplanar.

Linked Comprehension Type

14. (b), 15. (c)

16. (d), 17. (b)

$|\vec{a}| = 1, |\vec{b}| = 2, |\vec{c}| = 3$ and $|[\vec{a}\ \vec{b}\ \vec{c}]| = 6$

$\Rightarrow \quad |[\vec{a}\ \vec{b}\ \vec{c}]| = |\vec{a}|\,|\vec{b}|\,|\vec{c}| = 6$

$\Rightarrow \quad \vec{a}, \vec{b}$ and $\vec{c}$ are mutually perpendicular.

So, let $\vec{a} = \hat{i}, \vec{b} = 2\hat{j}$ and $\vec{c} = 3\hat{k}$

$\vec{r}$ coplanar with $\vec{b}$ and $\vec{c}$

$\therefore \quad \vec{r} = \lambda(2\hat{j}) + \mu(3\hat{k}) \qquad \ldots(1)$

Given that $\vec{r} \cdot \vec{b} = 1$ or $\vec{r} \cdot (2\hat{j}) = 1$

So, from (1) taking dot product with $2\hat{j}$, we get

$1 = 4\lambda$

Also, $|\vec{r}| = 1$

$\Rightarrow \quad 4\lambda^2 + 9\mu^2 = 1$

$\Rightarrow \quad \frac{1}{4} + 9\mu^2 = 1 \quad \Rightarrow \quad \mu^2 = \frac{1}{12}$

$\Rightarrow \quad \vec{r} = \frac{1}{2}\hat{j} + \frac{\sqrt{3}}{2}\hat{k}$

$$\Rightarrow \quad |(\vec{a} \times \vec{c}) \times \vec{r}|^2 = \left|(\hat{i} \times 3\hat{k}) \times \left(\frac{1}{2}\hat{j} + \frac{\sqrt{3}}{2}\hat{k}\right)\right|^2$$

$$= \left|3\hat{j} \times \left(\frac{1}{2}\hat{j} + \frac{\sqrt{3}}{2}\hat{k}\right)\right|^2$$

$$= \left|\frac{3\sqrt{3}}{2}\hat{i}\right|^2 = \frac{27}{4}$$

$$|[\vec{a}\ \vec{b}\ \vec{r}]| = \left|(\hat{i} \times 2\hat{j}) \cdot \left(\frac{1}{2}\hat{j} + \frac{\sqrt{3}}{2}\hat{k}\right)\right|$$

$$= \left|2\hat{k} \cdot \left(\frac{1}{2}\hat{j} + \frac{\sqrt{3}}{2}\hat{k}\right)\right| = \sqrt{3}$$

Numerical Value Type

18. (48)

$$\begin{vmatrix} a_1 & a_2 & a_3 \\ b_1 & b_2 & b_3 \\ c_1 & c_2 & c_3 \end{vmatrix}^2 = \begin{vmatrix} a_1 & a_2 & a_3 \\ b_1 & b_2 & b_3 \\ c_1 & c_2 & c_3 \end{vmatrix}\begin{vmatrix} a_1 & a_2 & a_3 \\ b_1 & b_2 & b_3 \\ c_1 & c_2 & c_3 \end{vmatrix}$$

$$= \begin{vmatrix} \vec{a}\cdot\vec{a} & \vec{a}\cdot\vec{b} & \vec{a}\cdot\vec{c} \\ \vec{b}\cdot\vec{a} & \vec{b}\cdot\vec{b} & \vec{b}\cdot\vec{c} \\ \vec{c}\cdot\vec{a} & \vec{c}\cdot\vec{b} & \vec{c}\cdot\vec{c} \end{vmatrix}$$

$$= \begin{vmatrix} 16 & 4 & 0 \\ 4 & 4 & 0 \\ 0 & 0 & 1 \end{vmatrix} = 48$$

19. (64) Let $V = [\vec{a}\ \vec{b}\ \vec{c}]$

$$\therefore \quad [\vec{\alpha}\ \vec{\beta}\ \vec{\gamma}] = \begin{vmatrix} \vec{a}\cdot\vec{a} & \vec{a}\cdot\vec{b} & \vec{a}\cdot\vec{c} \\ \vec{b}\cdot\vec{a} & \vec{b}\cdot\vec{b} & \vec{b}\cdot\vec{c} \\ \vec{c}\cdot\vec{a} & \vec{c}\cdot\vec{b} & \vec{c}\cdot\vec{c} \end{vmatrix}[\vec{a}\ \vec{b}\ \vec{c}]$$

$$= [\vec{a}\ \vec{b}\ \vec{c}]\,[\vec{a}\ \vec{b}\ \vec{c}]\,[\vec{a}\ \vec{b}\ \vec{c}] = V^3$$

20. (10)

21. (5) We have $\vec{p} \times \vec{q} = \vec{r}$

$\therefore \quad [\vec{p}\ \vec{q}\ \vec{r}] = \vec{r} \cdot \vec{r} = 1$

Let $\vec{a} = x\vec{p} + y\vec{q} + z\vec{r}$

$\therefore \quad \vec{a} \cdot (\vec{q} \times \vec{r}) = x[\vec{p}\ \vec{q}\ \vec{r}]$

$\Rightarrow \quad x = 1$

Similarly, we get $y = 2$ and $z = 3$

$\Rightarrow \quad \vec{a} = \vec{p} + 2\vec{q} + 3\vec{r}$

DPP 3.4

Single Correct Answer Type

1. (b) $\vec{b} = \dfrac{(\vec{a}\times\vec{b})\times\vec{a}+(\vec{a}\cdot\vec{b})\vec{a}}{|\vec{a}|^2}$

$= \dfrac{(14\hat{i}-29\hat{j}+\hat{k})+(22\hat{i}+11\hat{j}+11\hat{k})}{6}$

$= 6\hat{i}-3\hat{j}+2\hat{k}$

$\Rightarrow \quad b_1+b_2+b_3 = 6-3+2 = 5$

2. (c) $|[\vec{a}\,\vec{b}\,\vec{d}]\vec{c}-[\vec{a}\,\vec{b}\,\vec{c}]\vec{d}| = (\vec{a}\times\vec{b})\times(\vec{c}\times\vec{d})$

$= |\vec{a}\times\vec{b}||\vec{c}\times\vec{d}|\sin 30°$

$= \dfrac{\sqrt{3}}{2}\times\dfrac{\sqrt{3}}{2}\times\dfrac{1}{2} = \dfrac{3}{8}$

3. (d) $(\vec{a}\times\vec{b})\times(\vec{c}\times\vec{d})+(\vec{b}\times\vec{c})\times(\vec{a}\times\vec{d})$
$+(\vec{c}\times\vec{a})\times(\vec{b}\times\vec{d})$

$= [\vec{a}\vec{b}\vec{d}]\vec{c}+[\vec{b}\,\vec{c}\,\vec{d}]\vec{a}+[\vec{c}\,\vec{a}\,\vec{d}]\vec{b}-3[\vec{a}\vec{b}\,\vec{c}]\vec{d}$

$= [\vec{a}\vec{b}\vec{d}]\vec{c}+[\vec{b}\,\vec{c}\,\vec{d}]\vec{a}+[\vec{c}\,\vec{a}\,\vec{d}]\vec{b}-6\vec{d}$ ...(1)

Now $\vec{d} = x\vec{a}+y\vec{b}+z\vec{c}$; where x, y, z are scalars.

$\therefore \quad (\vec{a}\times\vec{b})\cdot\vec{d} = z[\vec{a}\,\vec{b}\,\vec{c}] = 2z$

$\therefore \quad z = \dfrac{1}{2}[\vec{a}\,\vec{b}\,\vec{d}]$

Similarly, $x = \dfrac{1}{2}[\vec{b}\,\vec{c}\,\vec{d}]$ and $y = \dfrac{1}{2}[\vec{c}\,\vec{a}\,\vec{d}]$

So, (1) reduces to

$\Sigma(\vec{a}\times\vec{b})\times(\vec{c}\times\vec{d})$

$= 2\{z\vec{c}+x\vec{a}+y\vec{b}\}-6\vec{d}$

$= 2\vec{d}-6\vec{d} = -4\vec{d}$

$\Rightarrow \quad \mu = 4$

4. (b)

5. (c) We have $\vec{a}\times\vec{b}\cdot\vec{c} = |\vec{c}|^2 \quad \Rightarrow \quad [\vec{a}\vec{b}\vec{c}] = |\vec{c}|^2$

Similarly, $[\vec{b}\;\vec{c}\;\vec{a}] = |\vec{a}|^2$ and $[\vec{c}\;\vec{a}\;\vec{b}] = |\vec{b}|^2$

$\therefore \quad |\vec{a}| = |\vec{b}| = |\vec{c}|$

Also, $\left[\vec{b}\times\vec{c}\;\;\vec{c}\times\vec{a}\;\;\vec{a}\times\vec{b}\right] = \left[\vec{a}\;\vec{b}\;\vec{c}\right]^2$

$\Rightarrow \quad \left[\vec{a}\;\vec{b}\;\vec{c}\right]^2 = \left[\vec{a}\;\vec{b}\;\vec{c}\right]$

$\Rightarrow \quad \left[\vec{a}\;\vec{b}\;\vec{c}\right] = 1$ (as $\vec{a}, \vec{b}, \vec{c}$ are non-coplanar)

$\Rightarrow \quad |\vec{a}| = |\vec{b}| = |\vec{c}| = 1$

$\Rightarrow \quad |\vec{a}|+|\vec{b}|+|\vec{c}| = 3.$

6. (b)

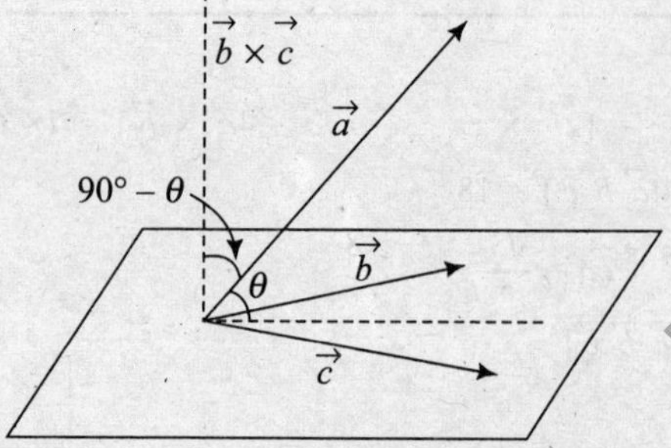

Let angle between vector $\vec{a}$ and plane containing $\vec{b}$ and $\vec{c}$ be θ.

Now, $|\hat{a}\times(\hat{b}\times\hat{c})|^2 = |(\hat{a}\cdot\hat{c})\hat{b}-(\hat{a}\cdot\hat{b})\hat{c}|^2$

$\Rightarrow \quad \sin^2\left(\dfrac{\pi}{2}-\theta\right)\sin^2\dfrac{\pi}{4} = \left|\left(\cos\dfrac{\pi}{3}\right)\hat{b}-\left(\cos\dfrac{\pi}{6}\right)\hat{c}\right|^2$

$\Rightarrow \quad (\cos^2\theta)\times\dfrac{1}{2} = \dfrac{1}{4}+\dfrac{3}{4}-2\times\dfrac{1}{2}\times\dfrac{\sqrt{3}}{2}\times\dfrac{1}{\sqrt{2}} = 1-\dfrac{\sqrt{3}}{2\sqrt{2}}$

$\Rightarrow \quad \cos^2\theta = 2-\sqrt{3/2}.$

7. (b) $\vec{a}\times(\vec{a}\times\vec{c})-\vec{b} = 0$

$\Rightarrow \quad (\vec{a}\cdot\vec{c})\vec{a}-(\vec{a}\cdot\vec{a})\vec{c}-\vec{b} = 0$

$\Rightarrow \quad (\vec{a}\cdot\vec{c})\vec{a}-\vec{c} = \vec{b}$

$\Rightarrow \quad |\vec{c}-(\vec{a}\cdot\vec{c})\vec{a}|^2 = |\vec{b}|^2$

$\Rightarrow \quad |\vec{c}|^2+|\vec{a}\cdot\vec{c}|^2|\vec{a}|^2-2(\vec{a}\cdot\vec{c})(\vec{a}\cdot\vec{c}) = |\vec{b}|^2$

If θ is the angle between $\vec{a}$ and $\vec{c}$, then

$4+(2\cos\theta)^2-2(2\cos\theta)^2 = 1$

$\Rightarrow \quad 4\cos^2\theta = 3 \quad \Rightarrow \quad \cos^2\theta = 3/4$

$\Rightarrow \quad \cos\theta = \sqrt{3}/2 \quad \Rightarrow \quad \theta = \pi/6$

8. (a) The vectors reciprocal to $\vec{a}, \vec{b}$, and $\vec{c}$ are given by

$\vec{p} = \dfrac{\vec{b}\times\vec{c}}{[\vec{a}\;\vec{b}\;\vec{c}]}, \; \dfrac{\vec{c}\times\vec{a}}{[\vec{a}\;\vec{b}\;\vec{c}]}$ and $\dfrac{\vec{a}\times\vec{b}}{[\vec{a}\;\vec{b}\;\vec{c}]}$ such that

$\vec{a}\cdot\vec{p} = 1, \vec{a}\cdot\vec{q} = \vec{a}\cdot\vec{r} = 0$

$\vec{b}\cdot\vec{q} = 1, \vec{c}\cdot\vec{q} = \vec{a}\cdot\vec{q} = 0$

$\vec{a}\cdot\vec{r} = 1, \vec{b}\cdot\vec{r} = \vec{c}\cdot\vec{r} = 0$

This gives $(\ell\vec{a}+m\vec{b}+n\vec{c})\cdot(\ell\vec{p}+m\vec{q}+n\vec{r})$

$= \ell^2+m^2+n^2.$

9. (a) $((\vec{a}\times\vec{b})\times\vec{a})\cdot((\vec{b}\times\vec{a})\times\vec{b}))$

$= (|\vec{a}|^2\vec{b}-(\vec{a}\cdot\vec{b})\vec{a})\cdot(|\vec{b}|^2\vec{a}-(\vec{a}\cdot\vec{b})\vec{b})$

$= |\vec{a}|^2|\vec{b}|^2(\vec{a}\cdot\vec{b})-(\vec{a}\cdot\vec{b})|\vec{a}|^2|\vec{b}|^2$

$-(\vec{a}\cdot\vec{b})|\vec{a}|^2|\vec{b}|^2+(\vec{a}\cdot\vec{b})^3$

$= -(\vec{a}\cdot\vec{b})[|\vec{a}|^2|\vec{b}|^2-(\vec{a}\cdot\vec{b})^2]$

$= -(\vec{a}\cdot\vec{b})|\vec{a}\times\vec{b}|^2$

10. (c)

Multiple Correct Answers Type

11. (a, b, d)

Volume = $[2\vec{b}\times\vec{c} \quad 3\vec{c}\times\vec{a} \quad 4\vec{a}\times\vec{b}] = 18$ (Given)

$\Rightarrow \; 24[\vec{a}\;\vec{b}\;\vec{c}]^2 = 18$

$\Rightarrow \; [\vec{a}\;\vec{b}\;\vec{c}] = \dfrac{\sqrt{3}}{2}$

Now, $[\vec{a}\;\vec{b}\;\vec{c}]$

$$= \begin{vmatrix} 1+\sin\theta & \cos\theta & \sin 2\theta \\ \sin\left(\theta+\frac{2\pi}{3}\right) & \cos\left(\theta+\frac{2\pi}{3}\right) & \sin\left(2\theta+\frac{4\pi}{3}\right) \\ \sin\left(\theta-\frac{2\pi}{3}\right) & \cos\left(\theta-\frac{2\pi}{3}\right) & \sin\left(2\theta-\frac{4\pi}{3}\right) \end{vmatrix}$$

Applying $R_1 \to R_1 + R_2 + R_3$ and expanding, we get

$[\vec{a}\;\vec{b}\;\vec{c}] = 3\cos 3\theta = \dfrac{\sqrt{3}}{2}$

$\Rightarrow \; \cos 3\theta = \dfrac{1}{2}$

$\Rightarrow \; 3\theta = \dfrac{\pi}{3}, \dfrac{2\pi}{3}, \dfrac{4\pi}{3}$

$\Rightarrow \; \theta = \dfrac{\pi}{9}, \dfrac{2\pi}{9}, \dfrac{4\pi}{9}$.

12. (a, b, c, d)

Since $\vec{a}$, $\vec{b}$ and $\vec{a}\times\vec{b}$ are non-coplanar, we have

$\vec{r} = x\vec{a} + y\vec{b} + z(\vec{a}\times\vec{b})$

Given $\vec{r}\times\vec{b} = \vec{a}$

$\Rightarrow \; x\vec{a}\times\vec{b} + z((\vec{a}\cdot\vec{b})\,\vec{b} - (\vec{b}\cdot\vec{b})\vec{a}) = \vec{a}$

$\Rightarrow \; -(1 + z|\vec{b}|^2)\,\vec{a} + x\vec{a}\times\vec{b} = 0 \quad [\because \; \vec{a}\cdot\vec{b} = 0]$

$\therefore \; x = 0, z = -\dfrac{1}{|\vec{b}|^2}$

Hence, $\vec{r} = y\vec{b} - \dfrac{\vec{a}\times\vec{b}}{|\vec{b}|^2}$, where y is the parameter.

13.(b, c, d)

14. (a, b, c)

$|\vec{a}| = |\vec{b}| = |\vec{c}| = 1$

and $\vec{a}\cdot\vec{c} = \vec{b}\cdot\vec{c}$ ...(1)

$\vec{a}\times(\vec{b}\times\vec{c}) + \vec{b}\times(\vec{c}\times\vec{a})$

$= (\vec{a}\cdot\vec{c})\,\vec{b} - (\vec{a}\cdot\vec{b})\vec{c} + (\vec{a}\cdot\vec{b})\,\vec{c} - (\vec{b}\cdot\vec{c})\,\vec{a}$

$= (\vec{a}\cdot\vec{c})\,\vec{b} - (\vec{b}\cdot\vec{c})\,\vec{a}$

$= (4 + x^2)\,\vec{b} - (4x\cos^2\theta)\,\vec{a}$ (given)

$\Rightarrow \; \vec{a}\cdot\vec{c} = 4 + x^2$ and $\vec{b}\cdot\vec{c} = 4x\cos^2\theta$. ...(2)

From (1) and (2), $4 + x^2 = 4x\cos^2\theta$

$\Rightarrow \; 4 + x^2 = 4x - 4x\sin^2\theta$

$\Rightarrow \; (x-2)^2 = -4x\sin^2\theta$

$\Rightarrow \; x = 2$ or $\sin\theta = 0$.

15. (a, c)

Linked Comprehension Type

16. (a) $(\vec{a}\times\vec{b})\times\vec{c} = \vec{a}\times(\vec{b}\times\vec{c})$

$\Rightarrow \; (\vec{a}\cdot\vec{c})\vec{b} - (\vec{b}\cdot\vec{c})\vec{a} = (\vec{a}\cdot\vec{c})\vec{b} - (\vec{a}\cdot\vec{b})\vec{c}$

$\Rightarrow \; (\vec{b}\cdot\vec{c})\vec{a} = (\vec{a}\cdot\vec{b})\vec{c}$

$\Rightarrow \; \vec{a}$ and $\vec{c}$ are parallel.

$\therefore \; \vec{c} = 4\hat{i} + 2\hat{j} + 4\hat{k}$ (as $|\vec{c}| = 6$ and $\vec{c}\cdot\hat{i} > 0$)

17. (d) $\vec{b}\times\vec{c} = \begin{vmatrix} \hat{i} & \hat{j} & \hat{k} \\ 1 & -1 & 1 \\ 4 & 2 & 4 \end{vmatrix} = -6\hat{i} + 6\hat{k}$

$\therefore \; [\vec{a}\;\vec{b}\;\vec{b}\times\vec{c}] = \begin{vmatrix} 2 & 1 & 2 \\ 1 & -1 & 1 \\ -6 & 0 & 6 \end{vmatrix} = -36$

$\therefore$ Volume = 36

18. (a) Vector along the bisectors is

$\dfrac{\hat{i} - \hat{j} + \hat{k}}{\sqrt{3}} + \dfrac{4\hat{i} + 2\hat{j} + 4\hat{k}}{6}$

$= \dfrac{1}{3}((2+\sqrt{3})\hat{i} + (1-\sqrt{3})\hat{j} + (2+\sqrt{3})\hat{k})$

Therefore, vector is

$(2+\sqrt{3})\hat{i} + (1-\sqrt{3})\hat{j} + (2+\sqrt{3})\hat{k}$

19 . (a), 20. (d)

We have $\vec{a}\times(\vec{b}\times\vec{c}) = \dfrac{\vec{b}}{3} + \dfrac{\vec{c}}{2}$

$\Rightarrow \; (\vec{a}\cdot\vec{c})\,\vec{b} - (\vec{a}\cdot\vec{b})\,\vec{c} = \dfrac{\vec{b}}{3} + \dfrac{\vec{c}}{2}$

$\Rightarrow \; \vec{a}\cdot\vec{c} = \dfrac{1}{3}$ and $\vec{a}\cdot\vec{b} = -\dfrac{1}{2}$

Also, $\vec{b}\times(\vec{c}\times\vec{a}) = -\dfrac{\vec{c}}{2}$

$\Rightarrow \; (\vec{b}\cdot\vec{a})\,\vec{c} - (\vec{b}\cdot\vec{c})\,\vec{a} = -\dfrac{\vec{c}}{2}$

$\Rightarrow \; \vec{b}\cdot\vec{c} = 0$

$\therefore \; |\vec{a} + \vec{b} + \vec{c}|^2 = |\vec{a}|^2 + |\vec{b}|^2 + |\vec{c}|^2 + 2\vec{a}\cdot\vec{b} + 2\vec{b}\cdot\vec{c} + 2\vec{a}\cdot\vec{c}$

$= 1 + 1 + 1 + 2\left(-\dfrac{1}{2}\right) + 0 + 2\left(\dfrac{1}{3}\right) = \dfrac{8}{3}$

Now, $[\vec{a}\;\vec{b}\;\vec{c}]^2 = \begin{vmatrix} \vec{a}\cdot\vec{a} & \vec{a}\cdot\vec{b} & \vec{a}\cdot\vec{c} \\ \vec{b}\cdot\vec{a} & \vec{b}\cdot\vec{b} & \vec{b}\cdot\vec{c} \\ \vec{c}\cdot\vec{a} & \vec{c}\cdot\vec{b} & \vec{c}\cdot\vec{c} \end{vmatrix}$

$$= \begin{vmatrix} 1 & -\frac{1}{2} & \frac{1}{3} \\ -\frac{1}{2} & 1 & 0 \\ \frac{1}{3} & 0 & 1 \end{vmatrix} = \frac{23}{36}$$

So, the volume of parallelepiped

$$= [\vec{a}\ \vec{b}\ \vec{c}] = \frac{\sqrt{23}}{6}$$

Matrix Match Type

21. (d) (1) $\vec{a} \times \vec{b} = \begin{vmatrix} \hat{i} & \hat{j} & \hat{k} \\ 1 & 1 & 1 \\ -1 & 2 & 0 \end{vmatrix} = -2\hat{i} - \hat{j} + 3\hat{k} \parallel (4\hat{i} + 2\hat{j} - 6\hat{k})$

$\therefore\ \lambda + \mu = -2$

(2) Vector coplanar with $\vec{c}$ and $\vec{d}$ is $p\vec{c} + q\vec{d}$ where $\vec{d}$ is perpendicular to $\vec{a}$.

$\therefore\ (p\vec{c} + q\vec{d}) \cdot \vec{a} = 0$

$\Rightarrow\ 3p + 2q = 0$

If $p = -2$ then $q = 3$.

So, the required vector is

$$-2\vec{c} + 3\vec{d} = -2(2\hat{i} + \hat{k}) + 3(\hat{i} - \hat{j} + 2\hat{k})$$
$$= -\hat{i} - 3\hat{j} + 4\hat{k}$$

(3) $(\vec{a} \times \vec{c}) \times (\vec{b} \times \vec{d})$

$= [\vec{a}\ \vec{c}\ \vec{d}]\vec{b} - [\vec{a}\ \vec{c}\ \vec{b}]\vec{d} = -4\vec{b} - \vec{d}$

$= -4(-\hat{i} + 2\hat{j}) - (\hat{i} - \hat{j} + 2\hat{k})$

$= 3\hat{i} - 7\hat{j} - 2\hat{k} \parallel (9\hat{i} - 21\hat{j} - 6\hat{k})$

(4) $\vec{b} \times (\vec{a} \times \vec{d}) = (\vec{b} \cdot \vec{d})\vec{a} - (\vec{b} \cdot \vec{a})\vec{d}$

$= -3\vec{a} - \vec{d}$

$= -3(\hat{i} + \hat{j} + \hat{k}) - (\hat{i} - \hat{j} + 2\hat{k})$

$= -4\hat{i} - 2\hat{j} - 5\hat{k} \parallel (8\hat{i} + 4\hat{j} + 10\hat{k})$

Numerical Value Type

22. (0) We have $\vec{c} = \vec{a} \times \vec{c} + \vec{a} \times \vec{b}$ ...(1)

$\Rightarrow\ \vec{a} \cdot \vec{c} = 0$ ...(2)

Also $\vec{a} \times \vec{c} = \vec{a} \times (\vec{a} \times \vec{c}) + \vec{a} \times (\vec{a} \times \vec{b})$

$\therefore\ \vec{c} - \vec{a} \times \vec{b} = -\vec{a}^2\,\vec{c} + \vec{a} \times (\vec{a} \times \vec{b})$

$\Rightarrow\ (1 + \vec{a}^2)\vec{c} + \vec{b} \times \vec{a} + (\vec{a} \times \vec{b}) \times \vec{a} = 0$

23. (3)

24. (10) $(\vec{a} \times (\vec{b} \times \vec{c}))\,(\vec{a} \times \vec{c}) = 5$

$\Rightarrow\ [(\vec{a} \cdot \vec{c})\vec{b} - (\vec{a} \cdot \vec{b})\vec{c}) \cdot (\vec{a} \times \vec{c}) = 5$

$\Rightarrow\ (\vec{a} \cdot \vec{c})\,[\vec{b}\ \vec{a}\ \vec{c}] = 5$

$\Rightarrow\ [\vec{a}\ \vec{b}\ \vec{c}] = -10$

25. (2) $\vec{A} \times (\vec{R} \times \vec{B}) = \vec{A} \times (\vec{C} \times \vec{B})$

$\Rightarrow\ (\vec{A} \cdot \vec{B})\vec{R} - (\vec{A} \cdot \vec{R})\vec{B} = (\vec{A} \cdot \vec{B})\vec{C} - (\vec{A} \cdot \vec{C})\vec{B}$

$\Rightarrow\ (-38)\,\vec{R} - \vec{0} = (-38)\,\vec{C} - 19\vec{B}$

$\Rightarrow\ \dfrac{|\vec{B}|}{|\vec{R} - \vec{C}|} = 2$

26. (12)

27. (0) $[\vec{a} \times \vec{b}\ \ \vec{b} \times \vec{c}\ \ \vec{c} \times \vec{a}] = [\vec{a}\ \vec{b}\ \vec{c}]^2$

$= |\vec{c} \cdot (\vec{a} \times \vec{b})|^2$

$= |\vec{c}|^2\, |\vec{a} \times \vec{b}|^2 \cos^2\theta$

The maximum value of $[\vec{a}\ \vec{b}\ \vec{c}]^2$ is possible only when $\vec{c}$ is parallel to $\vec{a} \times \vec{b}$.

$\Rightarrow\ \theta = 0$ or π

Therefore, $|(\vec{a} \times \vec{b}) \times \vec{c}|^2 = |\vec{a} \times \vec{b}|^2\, |\vec{c}|^2 \sin^2\theta = 0$

CHAPTER 4

DPP 4.1

Single Correct Answer Type

1. (b)

2. (c) Let point P be (p, q, r) from which three mutually perpendicular lines L_1, L_2 and L_3 are drawn.

L_1 cuts the x-axis at $A(a, 0, 0)$, L_2 meets the y-axis at $B(0, b, 0)$ and given that L_3 passes through the fixed point $C(0, 0, c)$.

So, the direction ratios of L_1, L_2 and L_3 are $(p - a, q, r)$, $(p, q - b, r)$ and $(p, q, r - c)$, respectively.

Since lines are mutually perpendicular,

$$p(p-a) + q(q-b) + r^2 = 0 \qquad \ldots(1)$$

$$p^2 + q(q-b) + r(r-c) = 0 \qquad \ldots(2)$$

and $p(p-a) + q^2 + r(r-c) = 0 \qquad \ldots(3)$

From (2) + (3) – (1), we get

$$p^2 + q^2 + 2r(r-c) - r^2 = 0$$

or $p^2 + q^2 + r^2 - 2cr = 0$

Therefore, the locus is $x^2 + y^2 + z^2 - 2cz = 0$.

3. (c) From the given equations eliminating n, we have $(l^2 + m^2) - (l + m)^2 = 0$

$\therefore \quad 2lm = 0$

When $l = 0 \Rightarrow m + n = 0$

$$\therefore \quad \frac{l}{0} = \frac{m}{1} = \frac{n}{-1} \qquad \ldots(1)$$

When $m = 0 \Rightarrow l + n = 0$

$$\therefore \quad \frac{l}{1} = \frac{m}{0} = \frac{n}{-1} \qquad \ldots(2)$$

So, D.R.'s are 0, 1, –1 and 1, 0, –1.

$$\therefore \quad \cos\theta = \frac{0.1 + 1.0 + 1}{\sqrt{1+1}\sqrt{1+1}} = \frac{1}{2} \quad \therefore \quad \theta = \frac{\pi}{3}$$

4. (c) Here, P and Q lie on the same side of XY-plane.

Image of (1, 2, 3) in the XY-plane is $P'(1, 2, -3)$

So, the equation of reflected ray is

$$P'Q: \frac{x-3}{3-1} = \frac{y-2}{2-2} = \frac{z-5}{5-(-3)}$$

$$\Rightarrow \frac{x-3}{1} = \frac{y-2}{0} = \frac{z-5}{4}$$

5. (b) For point of intersection of given line with $x = 0$, we have

$$\frac{0-1}{1} = \frac{y-1}{2} = \frac{z-1}{1}.$$

So, the point of intersection is (0, –1, 0).

For point of intersection of given line with $y = 0$, we have

$$\frac{x-1}{1} = \frac{0-1}{2} = \frac{z-1}{1}$$

So, the point of intersection is (1/2, 0, 1/2).

So, the centroid is (–1/6, 1/3, –1/6).

6. (c) Any point on second line is $(\lambda, 2\lambda, 3\lambda)$

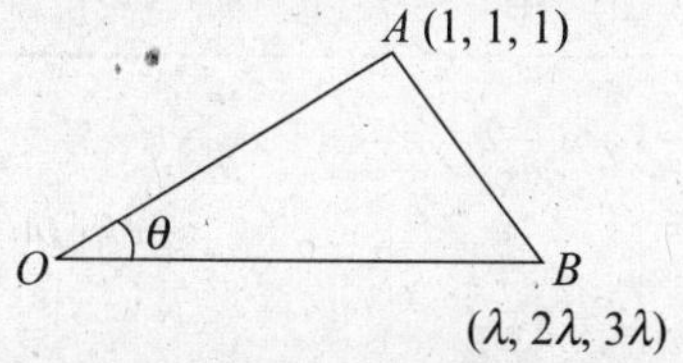

$$\cos\theta = \frac{6}{\sqrt{42}}, \sin\theta = \frac{\sqrt{6}}{\sqrt{42}}$$

$$\text{Area of } \Delta OAB = \frac{1}{2}(OA)(OB)\sin\theta = \sqrt{6}$$

$$\Rightarrow \frac{1}{2}(\sqrt{3})(\sqrt{14}\lambda)\frac{\sqrt{6}}{\sqrt{42}} = \sqrt{6} \Rightarrow \lambda = 2$$

$\therefore \quad B = (2, 4, 6)$

Multiple Correct Answers Type

7. (a, c)

DC's of AB are $\frac{x_2 - x_1}{AB}, \frac{y_2 - y_1}{AB}, \frac{z_2 - z_1}{AB}$

DC's of L are

$$\frac{l}{\sqrt{l^2 + m^2 + n^2}}, \frac{m}{\sqrt{l^2 + m^2 + n^2}}, \frac{n}{\sqrt{l^2 + m^2 + n^2}}$$

$$\therefore \quad \cos\theta = \frac{(x_2 - x_1)l + (y_2 - y_1)m + (z_2 - z_1)n}{AB\sqrt{l^2 + m^2 + n^2}}$$

Also, $PQ = AB \cos\theta$

$$= \frac{(x_2 - x_1)l + (y_2 - y_1)m + (z_2 - z_1)n}{\sqrt{l^2 + m^2 + n^2}}$$

8. (a, b, c, d)

9. (a, b)

Equation of line is

$$\vec{r} = (8\hat{i} + 10\hat{j} - 8\hat{k}) + t(2\hat{i} + \hat{j} + 2\hat{k})$$

$$\Rightarrow \vec{r} - (8\hat{i} + 10\hat{j} - 8\hat{k}) = t(2\hat{i} + \hat{j} + 2\hat{k})$$

$$\Rightarrow \left|\vec{r} - (8\hat{i} + 10\hat{j} - 8\hat{k})\right| = |t|\left|2\hat{i} + \hat{j} + 2\hat{k}\right|$$

$$\Rightarrow |\vec{r} - (8\hat{i} + 10\hat{j} - 8\hat{k})| = 3|t|$$

$$\Rightarrow 12 = \pm t \times 3 \Rightarrow t = \pm 4$$

$$\Rightarrow \vec{r} = (8\hat{i} + 10\hat{j} - 8\hat{k}) \pm 4(2\hat{i} + \hat{j} + 2\hat{k})\vec{r}$$

$$\Rightarrow \vec{r} = 16\hat{i} + 14\hat{j} \text{ or } \vec{r} = 6\hat{j} - 16\hat{k}$$

Matrix Match Type

10. (c) Let the direction ratios of the first and third lines be $(0, m_1, n_1)$ and (l_2, m_2, n_2).

Direction ratios of second line are 1, 2, –1.

Since lines are perpendicular

$2m_1 - n_1 = 0,\ l_2 + 2m_2 - n_2 = 0$

and $m_1 m_2 + n_1 n_2 = 0$

Solving we get

$$\Rightarrow \quad \frac{l_1}{0} = \frac{m_1}{1} = \frac{n_1}{2} \text{ and } \frac{l_2}{5} = \frac{m_2}{-2} = \frac{n_2}{1}.$$

$$\text{So, } L_1 \equiv \frac{x-1}{0} = \frac{y-2}{1} = \frac{z+1}{2}$$

$$L_2 \equiv \frac{x-1}{1} = \frac{y-2}{2} = \frac{z+1}{-1}$$

$$L_3 \equiv \frac{x-1}{5} = \frac{y-2}{-2} = \frac{z+1}{1}$$

L_1, L_2, and L_3 intersect xy-plane at $A(1, 5/2, 0)$, $B(0, 0, 0)$ and $C(6, 0, 0)$ respectively.

So, centroid of triangle ABC is $(7/3, 5/6, 0)$.

Numerical Value Type

11. (8) Since point of intersection of the given lines is $(0, 0, 0)$. It must lie on the angle bisectors.

$$\text{So, } \frac{0-2}{-1} = \frac{0+2}{1} = \frac{0+k}{4} \Rightarrow k = 8$$

12. (3)

13.(29) Line $L_1: x - 2y + 3z - 4 = 0 = 2x + y - z + 3$ is parallel to the vector

$$\begin{vmatrix} \hat{i} & \hat{j} & \hat{k} \\ 1 & -2 & 3 \\ 2 & 1 & -1 \end{vmatrix} = -\hat{i} + 7\hat{j} + 5\hat{k}$$

Line $L_2: x - y + 2z - 5 = 0 = x + y + z + 6$ is parallel to the vector

$$\begin{vmatrix} \hat{i} & \hat{j} & \hat{k} \\ 1 & -1 & 2 \\ 1 & 1 & 1 \end{vmatrix} = -3\hat{i} + \hat{j} + 2\hat{k}$$

The required line is perpendicular to the lines L_1 and L_2.

So, it will be parallel to the vector

$$\begin{vmatrix} \hat{i} & \hat{j} & \hat{k} \\ -1 & 7 & 5 \\ -3 & 1 & 2 \end{vmatrix} = 9\hat{i} - 13\hat{j} + 20\hat{k}$$

Therefore, equation of the required line is

$$\frac{x}{9} = \frac{y}{-13} = \frac{z}{20},$$

which passes through the point $(9, -13, 20)$.

DPP 4.2

Single Correct Answer Type

1. (d) Line $(\vec{r} - (\hat{i} + \hat{j} + \hat{k})) \times ((1-p)\hat{i} + 3\hat{j} - 2\hat{k}) = 0$ is passing through the point $A(1, 1, 1)$ and parallel to the vector $\overrightarrow{n_1} = (1-p)\hat{i} + 3\hat{j} - 2\hat{k}$.

Line $(\vec{r} - (3\hat{i} + \hat{j} - 5\hat{k})) \times ((3-p)\hat{i} + 4\hat{j} - 8\hat{k}) = 0$ is passing through the point $B(3, 1, -5)$ and parallel to the vector $\overrightarrow{n_2} = (3-p)\hat{i} + 4\hat{j} - 8\hat{k}$.

Since, lines are coplanar, vectors $\overrightarrow{n_1}$, $\overrightarrow{n_2}$ and $\overline{AB}$ are coplanar.

$$\therefore \quad \begin{vmatrix} 2 & 0 & -6 \\ 1-p & 3 & -2 \\ 3-p & 4 & -8 \end{vmatrix} = 0 \Rightarrow p = 1/3$$

2. (b)

3. (d)

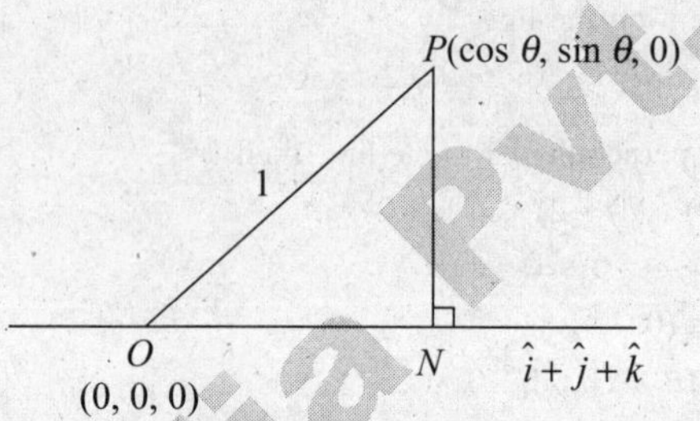

Any point on line circle $x^2 + y^2 = 1$ is $P(\cos\theta, \sin\theta, 0)$.

Line $x = y = z$ passes through the origin 'O'.

Now $OP = 1$.

Projection of OP on line having direction ratios 1, 1, 1 is

$$d = \frac{\cos\theta + \sin\theta}{\sqrt{3}}$$

$$\therefore \quad PN = \sqrt{1 - \left(\frac{\cos\theta + \sin\theta}{\sqrt{3}}\right)^2} = \frac{\sqrt{2 - \sin 2\theta}}{\sqrt{3}}$$

$$\therefore \quad PN_{\min} = \sqrt{\frac{2-1}{3}} = \sqrt{\frac{1}{3}}$$

4. (b) Line L_1 is $2x + y + z - 1 = 0 = 3x + y + 2z - 2$

or $\quad x = -y = -(z-1)$

Line L_2 is $x = y = z$

Let point P on the line L_1 be $(p, -p, -p+1)$ and point Q on the line L_2 be (q, q, q).

Let PQ be perpendicular to L_1 and L_2.

$\therefore \quad (q-p)(1) + (q+p)(-1) + (q+p-1)(-1) = 0$

and $\quad (q-p)(1) + (q+p)(1) + (q+p-1)(1) = 0$

$\Rightarrow \quad -3p - q = -1$ and $p + 3q = 1$

$\Rightarrow \quad p = q = 1/4$

$\Rightarrow \quad P \equiv (1/4, -1/4, 3/4)$ and $Q \equiv (1/4, 1/4, 1/4)$

$$\Rightarrow \quad PQ = \sqrt{\left(\frac{1}{4} - \frac{1}{4}\right)^2 + \left(\frac{1}{4} + \frac{1}{4}\right)^2 + \left(\frac{1}{4} - \frac{3}{4}\right)^2}$$

$$= \frac{1}{\sqrt{2}}$$

Also, equation of line PQ is

$$\frac{x - \frac{1}{4}}{0} = \frac{y - \frac{1}{4}}{1/2} = \frac{z - \frac{1}{4}}{-1/2},$$

which is perpendicular to the x-axis.

5. (d) Given lines are parallel as both are directed along the vector $(\hat{i}+\hat{j}-\hat{k})$.
So, lines do not intersect.
Hence Assertion is true.
Also, Reason is true by the definition of skew lines.
Skew lines are those which are neither parallel nor intersecting.
So, the reason is not the correct explanation of the assertion.

6. (d)

7. (a) Given point $A \equiv (1, 2, -4)$.

Any point on the line $\frac{x-1}{1}=\frac{y}{-1}=\frac{z+2}{2}$ is $B(t+1, -t, 2t-2), t \in R$

If AB is perpendicular to the line, then

$1(t)-(-t-2)+2(2t+2)=0$

$\Rightarrow \quad 6t+6=0 \quad \Rightarrow \quad t=-1$

So, $B \equiv (0, 1, -4)$.

Hence $AB=\sqrt{1+1+0}=\sqrt{2}$

So, there will be two points on the line at distance 2 units from the point A.

8. (d)

Multiple Correct Answers Type

9. (a, b, c, d)

$L_1: \frac{x-1}{1}=\frac{y+1}{-1}=\frac{z}{2}$ passes through the point $A(1,-1,0)$ and parallel to vector $\overrightarrow{n_1}=\hat{i}-\hat{j}+2\hat{k}$

$L_2: \frac{x}{2}=\frac{y-1}{-2}=\frac{z-1}{\lambda}$ passes through the point $B(0, 1, 1)$ and parallel to vector $\overrightarrow{n_2}=2\hat{i}-2\hat{j}+\lambda\hat{k}$

(1) If lines are parallel then $\overrightarrow{n_1}$ is parallel to $\overrightarrow{n_2}$

$\therefore \quad \frac{2}{1}=\frac{-2}{-1}=\frac{\lambda}{2} \Rightarrow \lambda=4$

(2) If lines are perpendicular then $\overrightarrow{n_1}\cdot\overrightarrow{n_2}=0$

$\therefore \quad 2+2+2\lambda=0 \quad \Rightarrow \quad \lambda=-2$

(3) Clearly for $\lambda=4$, lines are parallel and hence coplanar for $\lambda=4$

(4) For $\lambda=5$, $\overrightarrow{AB}\cdot\left(\overrightarrow{n_1}\times\overrightarrow{n_2}\right)\neq 0$
Hence lines are skew.

Matrix Match Type

10. (d) (1) Lines are intersecting as

$$\begin{vmatrix} 1-1 & 2-3 & 3-5 \\ 2 & 3 & 4 \\ 3 & 4 & 5 \end{vmatrix}=0$$

(2) Clearly, lines are in the same direction.
Points on first and second lines are $(2a+1, 3a+2, 4a+3)$ and $(2b+3, 3b+5, 4b+7)$
For $a=0$ and $b=-1$, we get the common point $(1, 2, 3)$
So, lines are coincident

(3) Lines are parallel, but not coincident as points $(5a+2, 4a-3, 5-2a)$ and $(5b+7, 4b+1, 2-2b)$ do not match for any values of a and b.

(4) Lines are non-coplanar as

$$\begin{vmatrix} 3-3 & -2-2 & 4-7 \\ 2 & 3 & 5 \\ 3 & 2 & 5 \end{vmatrix}\neq 0$$

Numerical Value Type

11. (12)

12. (5)

Line L_1 is $(\vec{r}-(-\hat{i}+\hat{j}+\hat{k}))\times(2\hat{i}-\hat{j}+2\hat{k})=0$

or $\quad \vec{r}=(-\hat{i}+\hat{j}+\hat{k})+\lambda(2\hat{i}-\hat{j}+2\hat{k})$

Line L_2 is $\vec{r}=(3+2\mu)\hat{i}-(1+\mu)\hat{j}+(1+2\mu)\hat{k}$

or $\quad \vec{r}=(3\hat{i}-\hat{j}+\hat{k})+\mu(2\hat{i}-\hat{j}+2\hat{k})$

So, lines are parallel.
Hence, distance between lines

$$d=\frac{|[(3\hat{i}-\hat{j}+\hat{k})-(-\hat{i}+\hat{j}+\hat{k})]\times(2\hat{i}-\hat{j}+2\hat{k})|}{|2\hat{i}-\hat{j}+2\hat{k}|}$$

$$=\frac{1}{3}\left\|\begin{matrix} \hat{i} & \hat{j} & \hat{k} \\ 4 & -2 & 0 \\ 2 & -1 & 2 \end{matrix}\right\|$$

$$=\frac{1}{3}|-4\hat{i}-8\hat{j}|$$

$$=\frac{4\sqrt{5}}{3}$$

DPP 4.3

Single Correct Answer Type

1. (d) Required plane is perpendicular to the line of intersection of given two planes.

Vector along line of intersection of planes is

$$\begin{vmatrix} \hat{i} & \hat{j} & \hat{k} \\ 1 & 2 & 1 \\ 3 & -4 & 1 \end{vmatrix}=6\hat{i}+2\hat{j}-10\hat{k}$$

So, the required plane is $3x+y-5z=0$.

2. (c) Let the plane make an angle θ with the z-axis.

$\therefore \quad \cos^2 45°+\cos^2 60°+\cos^2\theta=1$

$\therefore \quad \frac{1}{2}+\frac{1}{4}+\cos^2\theta=1$

$\Rightarrow \cos\theta = 1/2 \Rightarrow \theta = 60°$

So, the equation of plane is

$$\frac{x-\sqrt{2}}{\sqrt{2}}+\frac{y+1}{2}+\frac{z-1}{2}=0$$

or $\sqrt{2}x+y+z=2$

3. (b)

4. (d) Any point on the line $\frac{x}{2}=\frac{y-1}{3}=\frac{z-2}{1}$ is $B(2\lambda, 1+3\lambda, 2+\lambda)$.

Now, AB is parallel to the plane $x+y+z=2$, where $A\equiv(1, 0, 3)$.

$\therefore (2\lambda-1)+(1+3\lambda)+(\lambda-1)=0$

$\Rightarrow \lambda = 1/6$

So, point B is (1/3, 3/2, 13/6).

Direction ratios of AB are $-\frac{2}{3}, \frac{3}{2}, -\frac{5}{6}$.

So, equation of AB is

$$\frac{x-1/3}{-2/3}=\frac{y-3/2}{3/2}=\frac{z-13/6}{-5/6}$$

or $\frac{3x-1}{2}=\frac{2y-3}{-3}=\frac{6z-13}{5}$.

Clearly, line passes through the point (3, –9/2, 11/2).

Also, it is perpendicular to the line

$$\frac{x+1}{1}=\frac{y-4}{4}=\frac{z}{2}.$$

5. (d) Equation of family of planes is $x+3y+z-4+\lambda(2x-y)=0$.

All the planes pass through the line of intersection of the planes $x+3y+z-4=0$ and $2x-y=0$.

Now, $2x-y=0 \Rightarrow \frac{x}{1}=\frac{y}{2}$

$\therefore x+3y+z-4=0$

$\Rightarrow x+3\times 2x+z-4=0 \Rightarrow 7x+z-4=0$

$\therefore 7x=-(z-4) \Rightarrow \frac{x}{1}=\frac{z-4}{-7}$

$\therefore$ line is $\frac{x}{1}=\frac{y}{2}=\frac{z-4}{-7}$

6. (b) Equation of a plane passing through the line $3x-y+2z-1=0=x+2y-z-2$ is

$3x-y+2z-1+\lambda(x+2y-z-2)=0$

$\Rightarrow (3+\lambda)x+(-1+2\lambda)y+(2-\lambda)z+(-1-2\lambda)=0$

Since it is perpendicular to the given plane, we have

$\Rightarrow (3+\lambda)3+(-1+2\lambda)2+(2-\lambda)1=0$

$\Rightarrow 6\lambda=-9 \Rightarrow \lambda=-3/2$

So, the plane is $3x-8y+7z+4=0$.

Therefore, line of projection is the line of intersection of planes $3x-8y+7z+4=0$ and $3x+2y+z=0$.

7. (c)

8. (a) Given plane is $x+2y+3z-4=0$ ...(1)

Plane through the point $P(1, 1, 1)$ having normal (1, 1, 1) is $1(x-1)+1(y-1)+1(z-1)=0$

or $x+y+z-3=0$ ...(2)

Locus of Q is the line of intersection of the planes (1) and (2), which is

$$\frac{x}{1}=\frac{y-5}{-2}=\frac{z+2}{1}$$

9. (a) Let $O(\alpha, \beta, \gamma)$ be orthocentre. Then $AO \perp BC$

$\Rightarrow b\beta = c\gamma$

Similarly, we get $a\alpha = b\beta = c\gamma = k$ (say)

$\therefore \alpha=\frac{k}{a}, \beta=\frac{k}{b}, \gamma=\frac{k}{c}$

Now, equation of plane is $\frac{x}{a}+\frac{y}{b}+\frac{z}{c}=1$, which contains point (α, β, γ).

$\therefore k\left(\frac{1}{a^2}+\frac{1}{b^2}+\frac{1}{c^2}\right)=1$

Multiple Correct Answers Type

10. (b, c)

$E_1: \vec{r}\times(2\hat{i}-\hat{j}+3\hat{k})=3\hat{i}+\hat{k}$

$\Rightarrow \hat{i}(3y+z)-\hat{j}(3x-2z)+\hat{k}(-x-2y)=3\hat{i}+\hat{k}$

$\therefore 3y+z=3, 3x-2z=0, x-2y=1$

From the first two equations

$3x-2(3-3y)=0$

$\Rightarrow 3x+6y=6 \Rightarrow x+2y=2$

Now, $x+2y=-1$, $x+2y=2$ are parallel planes.

$E_2: \vec{r}\times(\hat{i}+2\hat{j}-3\hat{k})=2\hat{i}-\hat{j}$

$\Rightarrow \hat{i}(-3y-2z)-\hat{j}(-3x-z)+\hat{k}(2x-y)=2\hat{i}-\hat{j}$

$\therefore -3y-2z=2, 3x+z=-1, 2x-y=0$

$\Rightarrow -6x-2z=2, 3x+z=-1$

$\therefore$ Straight line, $2x-y=0=3x+z+1$

11. (a, c)

Let the equation of plane be $lx+my+nz=0$, where l, m, n be d.c's.

$\Rightarrow l^2+m^2+n^2=1$ ...(1)

Plane is parallel to the line $\frac{x-1}{2}=\frac{y+3}{-1}=\frac{z+1}{-2}$

So, $2l-m-2n=0$ ...(2)

Also, the distance of the required plane from the given line is 5/3.

$\therefore \frac{l-3m-n}{\sqrt{l^2+m^2+n^2}}=\frac{5}{3}$

$\Rightarrow l-3m-n=\frac{5}{3}$ ...(3)

Solving (1), (2) and (3), we get equation of plane as $x-2y+2z=0$ or $2x+2y+z=0$.

12. (a, b, c)

Any point on the line $\frac{x}{1}=\frac{y}{2}=\frac{z}{3}$ is $(\lambda, 2\lambda, 3\lambda)$.

Putting this point in the equation of the second line, we get

$$\frac{3}{3\beta^2+6(1-2\alpha)+3}=\frac{-6}{6\alpha^2+6(1-2\beta)+6}$$

$\Rightarrow (\beta-1)^2+(\alpha-2)^2=0$

$\Rightarrow \alpha=2, \beta=1$

Linked Comprehension Type

13. (c) $L_1 : \dfrac{x-7}{-3} = \dfrac{y-6}{2} = \dfrac{z-2}{4}$

and $L_2 : \dfrac{x-5}{2} = \dfrac{y-3}{1} = \dfrac{z-4}{3}$

$\therefore \quad C(-3\lambda + 7, 2\lambda + 6, 4\lambda + 2)$

and $D(2\mu + 5, \mu + 3, 3\mu + 4)$

So, $\dfrac{2-3\lambda-2\mu}{2} = \dfrac{3+2\lambda-\mu}{-2} = \dfrac{-2+4\lambda-3\mu}{-1}$

$\therefore \quad \lambda = 2, \mu = 1$

So, $C \equiv (1, 10, 10)$ and $D \equiv (7, 4, 7)$

$\Rightarrow \quad CD = 9.$

14. (d) Let equation of plane parallel to L_1 and containing L_2 is

$a(x-5) + b(y-3) + c(z-4) = 0$

$\therefore \quad 2a + b + 3c = 0$ and $-3a + 2b + 4c = 0$

$\Rightarrow \quad \dfrac{a}{-2} = \dfrac{b}{-17} = \dfrac{c}{7}$

So, the required plane is $2x + 17y - 7z = 33$.

15. (b) Volume $= [\overrightarrow{AB}\ \overrightarrow{AC}\ \overrightarrow{AD}] = \begin{vmatrix} 2 & 3 & -2 \\ 6 & -4 & -8 \\ 0 & 2 & -5 \end{vmatrix} = 138.$

16. (d) Equation of family of planes containing line of intersection of planes P_1 and P_2 is given by

$(x + y + 2z - 3) + \lambda(x - 2y + z - 4) = 0$

or $\quad (1 + \lambda)x + (1 - 2\lambda)y + (2 + \lambda)z - 3 - 4\lambda = 0 \quad$ ...(1)

Now projection AB on the plane will be greatest if AB is parallel to the plane

Now d.r.'s of AB are 2, – 1, 3

$\therefore \quad 2(1 + \lambda) - 1(1 - 2\lambda) + 3(2 + \lambda) = 0$

$\Rightarrow \quad 7\lambda + 7 = 0$

$\Rightarrow \quad \lambda = -1$

$\Rightarrow$ Equation of plane is $3y + z + 1 = 0$

17. (c) Projection of AB on the plane will be least if AB is perpendicular to plane

$\therefore \quad \dfrac{1+\lambda}{2} = \dfrac{1-2\lambda}{-1} = \dfrac{2+\lambda}{3} \quad \Rightarrow \quad \lambda = 1$

So, the equation of plane is $2x - y + 3z - 7 = 0$

Numerical Value Type

18. (13)

19. (30) Projection of line $\dfrac{x}{2} = \dfrac{y-1}{2} = \dfrac{z-1}{1}$ on a plane P is

$\dfrac{x}{1} = \dfrac{y-1}{1} = \dfrac{z-1}{-1}$

So, the plane through this line is perpendicular to the plane P.

Now, vector normal to the plane determined by the given lines is

$$\begin{vmatrix} \hat{i} & \hat{j} & \hat{k} \\ 2 & 2 & 1 \\ 1 & 1 & -1 \end{vmatrix} = -3\hat{i} + 3\hat{j}$$

So, the vector normal to the required plane is

$$\begin{vmatrix} \hat{i} & \hat{j} & \hat{k} \\ -3 & 3 & 0 \\ 1 & 1 & -1 \end{vmatrix} = -3\hat{i} - 3\hat{j} - 6\hat{k}$$

So, Equation of the plane P is $x + y + 2z - 3 = 0$ as it passes through (0, 1, 1)

Its distance from origin is $\sqrt{\dfrac{3}{2}}$.

DPP 4.4

Single Correct Answer Type

1. (c)

2. (c) Planes intersect in a line. Then

$$D = \begin{vmatrix} k & 4 & 1 \\ 4 & k & 2 \\ 2 & 2 & 1 \end{vmatrix} = 0$$

$\Rightarrow \quad k(k-4) - 4(0) + 1(8 - 2k) = 0$

$\Rightarrow \quad k^2 - 6k + 8 = 0$

$\Rightarrow \quad k = 2, 4$

3. (b) Lines $\dfrac{x-3}{2} = \dfrac{y-2}{3} = \dfrac{z-1}{\lambda}, \dfrac{x-2}{3} = \dfrac{y-3}{2} = \dfrac{z-2}{3}$

must be coplanar.

$$\Rightarrow \begin{vmatrix} 2 & 3 & \lambda \\ 3 & 2 & 3 \\ 1 & -1 & -1 \end{vmatrix} = 0$$

$\Rightarrow \quad \lambda = 4$

$\therefore$ Equation of plane is

$$\begin{vmatrix} x-3 & y-2 & z-1 \\ 2 & 3 & 4 \\ 3 & 2 & 3 \end{vmatrix} = 0$$

or $\quad x + 6y - 5z = 10 \quad$ ...(1)

Equation of given plane is $4x + y + 2z = 0 \quad$ ...(2)

$\therefore$ Angle between the planes, 'θ' is given by

$$\cos\theta = \frac{4(1) + 6(1) - 5(2)}{\sqrt{1+36+15}\,\sqrt{16+1+4}} = 0$$

$\Rightarrow \quad \theta = 90°$

4. (d) $Q \equiv (-1, 0, 1)$ and $R \equiv (1, 0, -1)$

$\therefore \quad QR = 2\sqrt{2}$

Now midpoint of QR is $M(0, 0, 0)$

Triangle PQR is equilateral

$\therefore \quad PM = MQ \times \tan 60° = \sqrt{2}, \sqrt{3} = \sqrt{6}$

$\therefore$ P can lie on the plane whose distance from point M is less than or equal to $\sqrt{6}$

Hence P can lie on the plane $x + y + z = 3\sqrt{2} = 0$

5. (a) Centre and radius of the sphere are (–1, 1, –1) and 3, respectively.

Distance of (–1, 1, –1) from the plane

$$=\left|\frac{-2+3-6+k}{\sqrt{4+9+36}}\right|$$

Since the plane is tangent to the sphere,

$$\left|\frac{k-5}{7}\right|=3 \quad \therefore \quad k=-16, 26$$

6. (a)

7. (a) For the sphere $x^2 + y^2 + z^2 = 9$, center is (0, 0, 0) and radius is 3.

Image of (0, 0, 0) in the plane $2x + 3y + 4z - 29 = 0$ is

$$\frac{x-0}{2}=\frac{y-0}{3}=\frac{z-0}{4}=\frac{-2(0+0+0-29)}{2^2+3^2+4^2}$$

$\therefore$ (4, 6, 8)

$\therefore$ Equation of the image of the sphere is

$(x-4)^2 + (y-6)^2 + (z-8)^2 = 9$

or $x^2 + y^2 + z^2 - 8x - 12y - 16z + 107 = 0$

8. (d)

9. (a) Let the equation of the variable plane be

$$\frac{x}{a}+\frac{y}{b}+\frac{z}{c}=1 \quad \text{...(1)}$$

$\therefore$ Volume of tetrahedron $OABC = \frac{abc}{6} = 64$ (given)

$\Rightarrow$ $abc = 384$ ...(2)

Let the foot of perpendicular from origin '*O*' on the plane be $P(\alpha, \beta, \gamma)$.

Then $\overline{OP} = \alpha\hat{i} + \beta\hat{j} + \gamma\hat{k}$

$\therefore$ Equation of plane *ABC* is

$\alpha(x-\alpha) + \beta(y-\beta) + \gamma(z-\gamma) = 0$

or $$\frac{x}{\frac{\alpha^2+\beta^2+\gamma^2}{\alpha}}+\frac{y}{\frac{\alpha^2+\beta^2+\gamma^2}{\beta}}+\frac{z}{\frac{\alpha^2+\beta^2+\gamma^2}{\gamma}}=1 \quad \text{...(3)}$$

Comparing equations (1) and (3), we get

$$a=\frac{\alpha^2+\beta^2+\gamma^2}{\alpha}, b=\frac{\alpha^2+\beta^2+\gamma^2}{\beta},$$

$$c=\frac{\alpha^2+\beta^2+\gamma^2}{\gamma}$$

$$\Rightarrow \quad abc=\frac{(\alpha^2+\beta^2+\gamma^2)^3}{\alpha\beta\gamma}$$

$\Rightarrow$ $(\alpha^2 + \beta^2 + \gamma^2)^3 = 384\alpha\beta\gamma$

So, the required locus is $(x^2 + y^2 + z^2)^3 = 384xyz$

Multiple Correct Answers Type

10. (b, c)

Required line is parallel to the vector

$$(\vec{a}\times\vec{b})\times(\vec{c}\times\vec{d})=[\vec{a}\ \vec{b}\ \vec{d}]\vec{c}-[\vec{a}\ \vec{b}\ \vec{c}]\vec{d}$$

$$=-4\vec{c}-4\vec{d}=-8\hat{i},$$

which is perpendicular to *y*-axis, *z*-axis.

11. (b, c)

For *A* (1, 2, 3), $2x - y - 3z - 5 = 2 - 1 - 3 - 5 < 0$

For *B* (2, 1, –3), $2x - y - 3z - 5 = 4 - 1 + 9 - 5 > 0$

For *C* (1, –2, –2), $2x - y - 3z - 5 = 2 + 2 + 6 - 5 > 0$

For *D* (–3, 1, 2), $2x - y - 3z - 5 = -6 - 1 - 6 - 5 < 0$

Thus, *A*, *D* lie on one side of the plane and *B*, *C* lie on the other side of the plane.

So, the line segments *AB*, *AC*, *BD*, *CD* intersect the plane.

Linked Comprehension Type

12. (c)

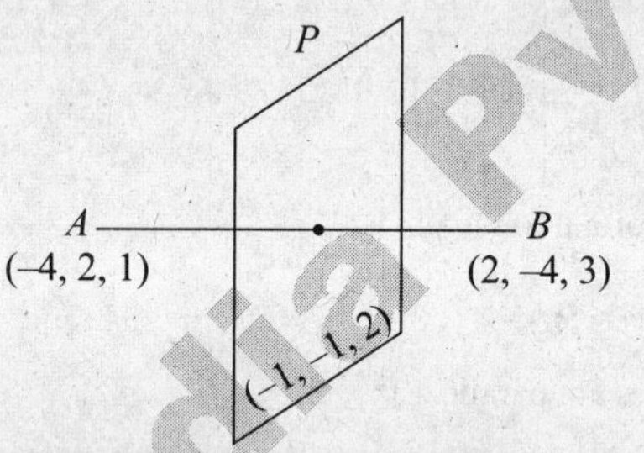

Plane *P* will contain the midpoint of *AB* which is *M*(–1, –1, 2) and will be perpendicular to the *AB*.

So, the equation of the plane *P* is

$6(x + 1) - 6(y + 1) + 2(z - 2) = 0$

or $3x - 3y + z - 2 = 0$ ...(1)

Plane *Q* is $x - y + cz - 1 = 0$ ...(2)

Plane *P* will be parallel to the plane *Q*, if $\frac{3}{1}=\frac{1}{c}$ or $c=\frac{1}{3}$

13. (b) Angle between planes *P* and *Q* is 45°.

$$\therefore \quad \cos 45° = \frac{|3+3+c|}{\sqrt{19}\sqrt{2+c^2}}$$

$\Rightarrow$ $2(6 + c)^2 = 19(2 + c^2)$

$\Rightarrow$ $17c^2 - 24c - 34 = 0$

14. (a) $AB = \sqrt{1+1+4} = \sqrt{6}$

d_1 = distance of *A* from the plane $\pi = \frac{1}{\sqrt{3}}$

d_2 = distance of B from the plane $\pi = \frac{3}{\sqrt{3}}$

$$\therefore \quad A'B' = \sqrt{(AB)^2-(d_1-d_2)^2}$$

$$=\sqrt{6-\frac{4}{3}}=\sqrt{\frac{14}{3}}$$

15. (a) Direction ratios of *AB* are 1, –1, –2.

Direction ratios of normal to the plane are 1, 1, 1.

$\therefore$ Direction ratios of the required line are given by

$$\begin{vmatrix}\hat{i} & \hat{j} & \hat{k}\\ 1 & -1 & -2\\ 1 & 1 & 1\end{vmatrix}=(1,-3,2)$$

So, the equation of the required line is

$$\frac{x-\frac{3}{2}}{1}=\frac{y-\frac{1}{2}}{-3}=\frac{z+2}{2}$$

($\because$ (3/2, 1/2, –2) lies on the plane)

Matrix Match Type

16. (c) (1) Vector normal to the plane

$$\vec{r}=(\hat{i}-\hat{j}-2\hat{k})\,\hat{j}+\lambda(2\hat{i}+\hat{j}+\hat{k})+\mu(3\hat{i}-\hat{j}+\hat{k}) \text{ is}$$

$$\begin{vmatrix}\hat{i} & \hat{j} & \hat{k}\\ 2 & 1 & 1\\ 3 & -1 & 1\end{vmatrix}=2\hat{i}+\hat{j}-5\hat{k}=\vec{n_1}$$

Vector normal to the plane

$$\vec{r}\cdot(2\hat{i}+\hat{j}-5\hat{k})+4=0 \text{ is } \vec{n_2}=2\hat{i}+\hat{j}-5\hat{k}$$

So, planes are parallel

(2) $xy+yz=0$

$\therefore$ $x(y+z)=0$

$\Rightarrow$ $x=0$ and $y+z=0$

So, planes are perpendicular.

(3) $2x^2-2y^2+4z^2+6xz+2yz+3xy=0$

or $2x^2+x(6z+3y)-2y^2+4z^2+2yz=0$

$$x=\frac{-(6z+3y)\pm\sqrt{36z^2+9y^2+36yz-8(-2y^2+4z^2+2yz)}}{4}$$

$$x=\frac{-(6z+3y)\pm\sqrt{(2z+5y)^2}}{4}$$

$$x=\frac{-(6z+3y)\pm(2z+5y)}{4}$$

or $2x-y+2z=0,\ x+2y+2z=0$

$\therefore$ Angle between planes is

$$\cos^{-1}\frac{(2)(1)+(-1)(2)+(2)(2)}{\sqrt{(2)^2+(-1)^2+(2)^2}\sqrt{(1)^2+(2)^2+(2)^2}}$$

$$=\cos^{-1}\left(\frac{4}{9}\right)$$

So, planes intersect but they are not perpendicular.

(4) Equation of plane through the points (1, 2, –1), (0, 1, 1) and (–1, 0, 2) is

$$\begin{vmatrix}x-1 & y-2 & z+1\\ 0-1 & 1-2 & 1-(-1)\\ -1-1 & 0-2 & 2-(-1)\end{vmatrix}=0$$

or $x-y+1=0$

Equation of plane parallel to the z-axis is

$ax+by+d=0$

It passes through the points (3, 4, 2) and (–2, –1, 1).

$\therefore$ $3a+4b+d=0$ and $-2a-b+d=0$

$\therefore$ $5a+5b=0$

$\Rightarrow$ $-a=b=-d$

So, equation of plane is $x-y+1=0$

Therefore, planes are coincident.

17. (b) Line L is $(\vec{r}\times\vec{p})\times\vec{a}=0$ or $(\vec{r}=\vec{p})+\alpha\vec{a}$

So, L is parallel to the vector $\vec{a}$

Plane is $\vec{r}+\vec{q}+\lambda\vec{b}+\mu\vec{c}$ or $(\vec{r}-\vec{q})\cdot(\vec{b}\times\vec{c})=0$

So, vector normal to the plane is $\vec{b}\times\vec{c}$.

If $\vec{a}\cdot(\vec{b}\times\vec{c})=0$ and $[\vec{p}-\vec{q}\ \vec{b}\ \vec{c}]\neq 0$, then line will be parallel to the plane and won't lie in the plane.

If $\vec{a}\cdot(\vec{b}\times\vec{c})=0$ and $[\vec{p}-\vec{q}\ \vec{b}\ \vec{c}]=0$, then line will lie in the plane.

If $\vec{a}\times(\vec{b}\times\vec{c})=0$, then line will intersect plane orthogonally.

If $\vec{a}\times(\vec{b}\times\vec{c})\neq 0$ and $\vec{a}\cdot(\vec{b}\times\vec{c})=0$, then line will intersect the plane but not orthogonally.

Numerical Value Type

18. (3)

19. (12) Clearly line is parallel to the plane.

So, the required distance is distance of point (2, 3, 4) from the plane, which is 12 units.

20. (1)

21. (13) Equation of family of planes containing given line is

$x+y+z-2+\lambda(x-y+z)=0$

or $(1+\lambda)x+(1-\lambda)y+(1+\lambda)z-2=0$

Distance of plane from the origin,

$$d=\frac{2}{\sqrt{(1+\lambda)^2+(1-\lambda)^2+(1+\lambda)^2}}$$

$$=\frac{2}{\sqrt{3\lambda^2+2\lambda+3}}=\frac{2}{\sqrt{3}\sqrt{\lambda^2+\frac{2}{3}\lambda+1}}$$

$$=\frac{2}{\sqrt{3}\sqrt{\left(\lambda+\frac{1}{3}\right)^2+\frac{8}{9}}}$$

'd' is maximum when $\lambda=-\frac{1}{3}$.

So, the required equation of plane is

$x+2y+z-3=0$

Distance of above plane from the point (3, 4, 5)

$$=\frac{3+8+5-3}{\sqrt{6}}=\frac{13}{\sqrt{6}}$$

22. (8) Line must be parallel to the plane.

$\therefore$ $4\lambda-3\lambda-8=0$

$\Rightarrow$ $\lambda=8$

23. (7) If the lines intersect, then all the four planes will have a common point.

Solving $4x+3y-2z+3=0$

and $x-3y+4z+6=0$, we get

$$x=-\frac{1}{3},\ y=-3 \text{ and } z=-\frac{11}{3}$$

Putting it in $kx-4y+7z+16=0$, we get $k=7$.

Contents

Contents

Chapterwise/Topicwise Solved JEE (Main) 2026 Questions (Session 1)

Products of Vectors

Single Correct Answer Type

1. Let $\overrightarrow{AB} = 2\hat{i} + 4\hat{j} - 5\hat{k}$ and $\overrightarrow{AD} = \hat{i} + 2\hat{j} + \lambda\hat{k}$, $\lambda \in R$. Let the projection of the vector $\vec{v} = \hat{i} + \hat{j} + \hat{k}$ on the diagonal $\overrightarrow{AC}$ of the parallelogram $ABCD$ be of length one unit. If α, β, where $\alpha > \beta$, be the roots of the equation $\lambda^2x^2 - 6\lambda x + 5 = 0$, then $2\alpha - \beta$ is equal to

(1) 4 (2) 6

(3) 3 (4) 1

[Dot Product of Vectors]

***2.** Let $\vec{a} = 2\hat{i} - \hat{j} - \hat{k}$, $\vec{b} = \hat{i} + 3\hat{j} - \hat{k}$ and $\vec{c} = 2\hat{i} + \hat{j} + 3\hat{k}$. Let $\vec{v}$ be the vector in the plane of the vectors $\vec{a}$ and $\vec{b}$, such that the length of its projection on the vector $\vec{c}$ is $\frac{1}{\sqrt{14}}$. Then, which of the following is not the possible value of $|\vec{v}|$?

(1) $\frac{\sqrt{35}}{2}$ (2) $\frac{\sqrt{21}}{2}$

(3) $\frac{\sqrt{6}}{2}$ (4) 4

[Dot Product of Vectors]

3. For three unit vectors $\vec{a}, \vec{b}, \vec{c}$ satisfying

$|\vec{a} - \vec{b}|^2 + |\vec{b} - \vec{c}|^2 + |\vec{c} - \vec{a}|^2 = 9$

and $|2\vec{a} + k\vec{b} + k\vec{c}| = 3$, the positive value of k is

(1) 5 (2) 4

(3) 3 (4) 6

[Dot Product of Vectors]

4. Let P be a point in the plane of the vectors $\overrightarrow{AB} = 3\hat{i} + \hat{j} - \hat{k}$ and $\overrightarrow{AC} = \hat{i} - \hat{j} + 3\hat{k}$ such that P is equidistant from the line AB and AC. If $|\overrightarrow{AP}| = \frac{\sqrt{5}}{2}$, then the area of the triangle ABP is

(1) $\frac{\sqrt{30}}{4}$ (2) $\frac{3}{2}$

(3) $\frac{\sqrt{26}}{4}$ (4) 2

[Dot Product of Vectors]

5. Let $\vec{a} = -\hat{i} + 2\hat{j} + 2\hat{k}$, $\vec{b} = 8\hat{i} + 7\hat{j} - 3\hat{k}$ and $\vec{c}$ be a vector such that $\vec{a} \times \vec{c} = \vec{b}$. If $\vec{c} \cdot (\hat{i} + \hat{j} + \hat{k}) = 4$, then $|\vec{a} + \vec{c}|^2$ is equal to:

(1) 33 (2) 30

(3) 27 (4) 35

[Cross Product of Vectors]

6. Let $\vec{c}$ and $\vec{d}$ be vectors such that $|\vec{c} + \vec{d}| = \sqrt{29}$ and $\vec{c} \times (2\hat{i} + 3\hat{j} + 4\hat{k}) = (2\hat{i} + 3\hat{j} + 4\hat{k}) \times \vec{d}$. If λ_1, λ_2 $(\lambda_1 > \lambda_2)$ are the possible values of $(\vec{c} + \vec{d}) \cdot (-7\hat{i} + 2\hat{j} + 3\hat{k})$, then the equation

$$K^2x^2 + (K^2 - 5K + \lambda_1)xy + \left(3K + \frac{\lambda_2}{2}\right)y^2 - 8x + 12y + \lambda_2 = 0$$

represents a circle, for K equal to:

(1) 2 (2) 4

(3) –1 (4) 1

[Cross Product of Vectors]

7. For a triangle ABC, let $\vec{p} = \overrightarrow{BC}$, $\vec{q} = \overrightarrow{CA}$ and $\vec{r} = \overrightarrow{BA}$. If $|\vec{p}| = 2\sqrt{3}$, $|\vec{q}| = 2$ and $\cos\theta = \frac{1}{\sqrt{3}}$, where θ is the angle between $\vec{p}$ and $\vec{q}$, then $|\vec{p} \times (\vec{q} - 3\vec{r})|^2 + 3|\vec{r}|^2$ is equal to:

(1) 200 (2) 410

(3) 220 (4) 340

[Cross Product of Vectors]

8. Let $\vec{a} = 2\hat{i} - \hat{j} + \hat{k}$ and $\vec{b} = \lambda\hat{j} + 2\hat{k}$, $\lambda \in Z$ be two vectors, Let $\vec{c} = \vec{a} \times \vec{b}$ and $\vec{d}$ be a vector of magnitude 2 in yz-plane. If $|\vec{c}| = \sqrt{53}$, then the maximum possible value of $(\vec{c} \cdot \vec{d})^2$ is equal to

(1) 104 (2) 26

(3) 208 (4) 52

[Cross Product of Vectors]

9. Let $\vec{a} = -\hat{i} + \hat{j} + 2\hat{k}$, $\vec{b} = \hat{i} - \hat{j} - 3\hat{k}$, $\vec{c} = \vec{a} \times \vec{b}$ and $\vec{d} = \vec{c} \times \vec{a}$. Then $(\vec{a} - \vec{b}) \cdot \vec{d}$ is equal to

(1) 2 (2) 4

(3) –4 (4) –2

[Cross Product of Vectors]

*The original question has been updated to remove ambiguity while preserving its conceptual intent.

10. Let $\vec{a}, \vec{b}, \vec{c}$ be three vectors such that $\vec{a} \times \vec{b} = 2(\vec{a} \times \vec{c})$. If $|\vec{a}| = 1, |\vec{b}| = 4, |\vec{c}| = 2,$ and the angle between $\vec{b}$ and $\vec{c}$ is 60°, then $|\vec{a} \cdot \vec{c}|$ is equal to:

(1) 1 (2) 0 (3) 4 (4) 2

[Cross Product of Vectors]

11. Let $\vec{a} = \hat{i} - 2\hat{j} + 3\hat{k}, \vec{b} = 2\hat{i} + \hat{j} - \hat{k}, \vec{c} = \lambda\hat{i} + \hat{j} + \hat{k}$ and $\vec{v} = \vec{a} \times \vec{b}$. If $\vec{v} \cdot \vec{c} = 11$ and the length of the projection of $\vec{b}$ and $\vec{c}$ is p, then $9p^2$ is equal to

(1) 6 (2) 12 (3) 9 (4) 4

[Cross Product of Vectors]

12. Let $\vec{a} = 2\hat{i} + \hat{j} - 2\hat{k}, \vec{b} = \hat{i} + \hat{j}$ and $\vec{c} = \vec{a} \times \vec{b}$. Let $\vec{d}$ be a vector such that $|\vec{d} - \vec{a}| = \sqrt{11}, |\vec{c} \times \vec{d}| = 3$ and the angle between $\vec{c}$ and $\vec{d}$ is $\frac{\neq}{4}$. Then $\vec{a} \cdot \vec{d}$ is equal to

(1) 3 (2) 11 (3) 1 (4) 0

[Cross Product of Vectors]

13. Let $\vec{a} = 2\hat{i} - 5\hat{j} + 5\hat{k}$ and $\vec{b} = \hat{i} - \hat{j} + 3\hat{k}$. If $\vec{c}$ is a vector such that $2(\vec{a} \times \vec{c}) + 3(\vec{b} \times \vec{c}) = \vec{0}$ and $(\vec{a} - \vec{b}) \cdot \vec{c} = -97$, then $|\vec{c} \times k|^2$ is equal to

(1) 233 (2) 205 (3) 218 (4) 193

[Cross Product of Vectors]

Numerical Value Type

14. Let a vector $\vec{a} = \sqrt{2}\hat{i} - \hat{j} + \lambda\hat{k}, \lambda > 0$, make an obtuse angle with the vector $\vec{b} = -\lambda^2\hat{i} + 4\sqrt{2}\hat{j} + 4\sqrt{2}\hat{k}$ and an angle θ, $\frac{\pi}{6} < \theta < \frac{\pi}{2}$, with the positive z-axis. If the set of all possible values of λ is $(\alpha, \beta) - \{\gamma\}$, then $\alpha + \beta + \gamma$ is equal to ______.

[Dot Product of Vectors]

***15.** Let PQR be a triangle such that $\overrightarrow{PQ} = -2\hat{i} - \hat{j} + 2\hat{k}$ and $\overrightarrow{PR} = a\hat{i} + b\hat{j} - 4\hat{k}$, $a, b \in Z$. Let S be the point on QR, which is equidistant from the lines PQ and PR. If $|\overrightarrow{PR}| = 9$ and $\overrightarrow{PS} = \frac{1}{4}\hat{i} - \frac{7}{4}\hat{j} + \frac{1}{2}\hat{k}$, then the value of $3a - 4b$ is ______.

[Dot Product of Vectors]

Line and Plane

Single Correct Answer Type

16. Let the line L_1 be parallel to the vector $-3\hat{i} + 2\hat{j} + 4\hat{k}$ and pass through the point (2, 6, 7) and the line L_2 be parallel to the vector $2\hat{i} + \hat{j} + 3\hat{k}$ and pass through the point (4, 3, 5). If the line L_3 is parallel to the vector $-3\hat{i} + 5\hat{j} + 16\hat{k}$ and intersects the lines L_1 and L_2 at the points C and D, respectively, then $|\overrightarrow{CD}|^2$ is equal to:

(1) 171 (2) 89

(3) 312 (4) 290

[Equation of Line]

17. Let L be the line $\frac{x+1}{2} = \frac{y+1}{3} = \frac{z+3}{6}$ and let S be the set of all points (a, b, c) on L, whose distance from the line $\frac{x+1}{2} = \frac{y+1}{3} = \frac{z-9}{0}$ along the line L is 7. Then $\sum_{(a,b,c) \in S} (a + b + c)$ is equal to:

(1) 34 (2) 28

(3) 6 (4) 40

[Equation of Line]

18. Let the direction cosines of two lines satisfy the equations : $4\ell + m - n = 0$ and $2mn + 10n\ell + 3\ell m = 0$. Then the cosine of the acute angle between these lines is :

(1) $\frac{10}{\sqrt{38}}$ (2) $\frac{10}{7\sqrt{38}}$

(3) $\frac{20}{3\sqrt{38}}$ (4) $\frac{10}{3\sqrt{38}}$

[Equation of Line]

19. Let the lines $L_1 : \vec{r} = \hat{i} + 2\hat{j} + 3\hat{k} + \lambda(2\hat{i} + 3\hat{j} + 4\hat{k})$, $\lambda \in R$ and $L_2 : \vec{r} = (4\hat{i} + \hat{j}) + \mu(5\hat{i} + 2\hat{j} + \hat{k})$, $\mu \in R$, intersect at the point R. Let P and Q be the points lying on lines L_1 and L_2, respectively, such that $|\overrightarrow{PR}| = \sqrt{29}$ and $|\overrightarrow{PQ}| = \sqrt{\frac{47}{3}}$.

If the point P lies in the first octant, then $27(QR)^2$ is equal to

(1) 360 (2) 340

(3) 348 (4) 320

[Equation of Line]

20. If the distances of the point (1, 2, a) from the line $\frac{x-1}{1} = \frac{y}{2} = \frac{z-1}{1}$ along the lines

$L_1 : \frac{x-1}{3} = \frac{y-2}{4} = \frac{z-a}{b}$ and $L_2 : \frac{x-1}{1} = \frac{y-2}{4} = \frac{z-a}{c}$

are equal then $a + b + c$ is equal to

(1) 4 (2) 7

(3) 5 (4) 6

[Equation of Line]

21. Let (α, β, γ) be the co-ordinates of the foot of the perpendicular drawn from the point (5, 4, 2) on the line $\vec{r} = (-\hat{i} + 3\hat{j} + \hat{k}) + \lambda(2\hat{i} + 3\hat{j} - \hat{k})$.

Then the length of the projection of the vector $\alpha\hat{i} + \beta\hat{j} + \gamma\hat{k}$ on the vector $6\hat{i} + 2\hat{j} + 3\hat{k}$ is:

(1) $\frac{15}{7}$ (2) 4

(3) 3 (4) $\frac{18}{7}$

[Foot of Perpendicular from a Point on a Line, Distance of a Point from a Line]

---

*The original question has been updated to remove ambiguity while preserving its conceptual intent.

22. Let the line L pass through the point (–3, 5, 2) and make equal angles with the positive coordinate axes. If the distance of L from the point (–2, r, 1) is $\sqrt{\frac{14}{3}}$, then the sum of all possible values of r is :

(1) 10 (2) 16
(3) 6 (4) 12

[Foot of Perpendicular from a Point on a Line, Distance of a Point from a Line]

23. If the image of the point $P(1, 2, a)$ in the line $\frac{x-6}{3}=\frac{y-7}{2}=\frac{7-z}{2}$ is $Q(5, b, c)$, then $a^2+b^2+c^2$ is equal to

(1) 293 (2) 264
(3) 298 (4) 283

[Foot of Perpendicular from a Point on a Line, Distance of a Point from a Line]

24. The vertices B and C of a triangle ABC lie on the line $\frac{x}{1}=\frac{1-y}{-2}=\frac{z-2}{3}$. The coordinates of A and B are (1, 6, 3) and (4, 9, α) respectively and C is at a distance of 10 units from B. The area (in sq. units) of ΔABC is:

(1) $15\sqrt{13}$ (2) $5\sqrt{13}$
(3) $20\sqrt{13}$ (4) $10\sqrt{13}$

[Foot of Perpendicular from a Point on a Line, Distance of a Point from a Line]

25. Let $Q(a, b, c)$ be the image of the point $P(3, 2, 1)$ in the line $\frac{x-1}{1}=\frac{y}{2}=\frac{z-1}{1}$. Then the distance of Q from the line $\frac{x-9}{2}=\frac{y-9}{2}=\frac{z-5}{-2}$ is

(1) 5 (2) 8
(3) 7 (4) 6

[Foot of Perpendicular from a Point on a Line, Distance of a Point from a Line]

26. Let $P(\alpha, \beta, \gamma)$ be the point on the line $\frac{x-1}{2}=\frac{y+1}{-3}=z$ at a distance $4\sqrt{14}$ from the point (1, –1, 0) and nearer to the origin. Then the shortest distance, between the lines $\frac{x-\alpha}{1}=\frac{y-\beta}{2}=\frac{z-\gamma}{3}$ and $\frac{x+5}{2}=\frac{y-10}{1}=\frac{z-3}{1}$, is equal to

(1) $7\sqrt{\frac{5}{4}}$ (2) $2\sqrt{\frac{7}{4}}$
(3) $4\sqrt{\frac{7}{5}}$ (4) $4\sqrt{\frac{5}{7}}$

[Shortest Distance between Lines]

27. The sum of all values of α, for which the shortest distance between the lines $\frac{x+1}{\alpha}=\frac{y-2}{-1}=\frac{z-4}{-\alpha}$ and $\frac{x}{\alpha}=\frac{y-1}{2}=\frac{z-1}{2\alpha}$ is $\sqrt{2}$, is

(1) 8 (2) 6
(3) –6 (4) –8

[Shortest Distance between Lines]

Numerical Value Type

28. Let a line L passing through the point $P(1, 1, 1)$ be perpendicular to the lines $\frac{x-4}{4}=\frac{y-1}{1}=\frac{z-1}{1}$ and $\frac{x-17}{1}=\frac{y-71}{1}=\frac{z}{0}$. Let the line L intersect the yz-plane at the point Q. Another line parallel to L and passing through the point $S(1, 0, -1)$ intersects the yz-plane at the point R. Then the square of the area of the parallelogram $PQRS$ is equal to _____. **[Equation of Line]**

29. If the distance of the point $P(43, \alpha, \beta)$, $\beta < 0$, from the line $\vec{r}=4\hat{i}-\hat{k}+\mu(2\hat{i}+3\hat{k})$, $\mu \in R$ along a line with direction ratios 3, –1, 0 is $13\sqrt{10}$, then $\alpha^2+\beta^2$ is equal to _______.

[Equation of Line]

30. If the image of the point $P(a, 2, a)$ in the line $\frac{x}{2}=\frac{y+a}{1}=\frac{z}{1}$ is Q and the image of Q in the line $\frac{x-2b}{2}=\frac{y-a}{1}=\frac{z+2b}{-5}$ is P, then $a+b$ is equal to _______.

[Foot of Perpendicular from a Point on a Line, Distance of a Point from a Line]

Answer Key

PRODUCTS OF VECTORS

Single Correct Answer Type

1. (3) **2.** (3) **3.** (1) **4.** (1) **5.** (3)
6. (4) **7.** (1) **8.** (3) **9.** (4) **10.** (1)
11. (2) **12.** (4) **13.** (3)

Numerical Value Type

14. (5) **15.** (37)

LINE AND PLANE

Single Correct Answer Type

16. (4) **17.** (1) **18.** (4) **19.** (1) **20.** (2)
21. (4) **22.** (1) **23.** (3) **24.** (2) **25.** (3)
26. (3) **27.** (3)

Numerical Value Type

28. (6) **29.** (170) **30.** (3)

JEE (Main) Mock Test

SECTION 1 (Maximum Marks: 80)

- This section contains **TWENTY (20)** questions.
- Each question has **FOUR** options (1), (2), (3) and (4). **ONLY ONE** of these four options is the correct answer.
- For each question, choose the option corresponding to the correct answer.
- Answer to each question will be evaluated according to the following marking scheme:
  1. Correct answer or the most appropriate answer: Four marks (+4)
  2. Any incorrect option marked will be given minus one mark (–1)
  3. Unanswered/Marked for Review will be given no mark (0)
  4. If more than one option is found to be correct then Four marks (+4) will be awarded to only those who have marked any of the correct options
  5. If all options are found to be correct then Four marks (+4) will be awarded to all those who have attempted the question

1. If A, B, C, D be any four points and E and F be the middle points of AC and BD respectively, then $\vec{AB} + \vec{CB} + \vec{CD} + \vec{AD}$ is equal to

(1) $3\vec{EF}$ (2) $4\vec{EF}$

(3) $4\vec{FE}$ (4) $3\vec{FE}$

2. The centroid of the triangle formed by (0, 0, 0) and the point of intersection of $\frac{x-1}{1} = \frac{y-1}{2} = \frac{z-1}{1}$ with $x = 0$ and $y = 0$ is

(1) (1, 1, 1)

(2) (1/6, –1/3, 1/6)

(3) (–1/6, 1/3, –1/6)

(4) (1/3, 1/3, 1/3)

3. Let $\vec{a}$ and $\vec{b}$ be two unit vectors inclined at an angle θ, then $\sin(\theta/2)$ is equal to

(1) $\frac{1}{2}|\vec{a} - \vec{b}|$ (2) $\frac{1}{2}|\vec{a} + \vec{b}|$

(3) $|\vec{a} - \vec{b}|$ (4) $|\vec{a} + \vec{b}|$

4. The angle between any two diagonals of a cube is:

(1) $\cos^{-1}\frac{1}{2}$ (2) $\cos^{-1}\frac{1}{3}$

(3) $\cos^{-1}\frac{1}{4}$ (4) $\frac{\pi}{2}$

5. Let $\vec{F} = 2\vec{i} + 2\vec{j} + 5\vec{k}$ and $A = (1, 2, 5)$, $B = (-1, -2, -3)$ and $\vec{BA} \times \vec{F} = 4\vec{i} + 6\vec{j} + 2\lambda\,\vec{k}$, then the value of λ is

(1) 0 (2) 1

(3) 2 (4) –2

6. The pair of lines whose direction cosines are given by the equations $3l + m + 5n = 0$, $6mn - 2nl + 5lm = 0$, are

(1) parallel

(2) perpendicular

(3) inclined at $\cos^{-1}\left(\frac{1}{6}\right)$

(4) inclined at $\sin^{-1}\left(\frac{1}{2}\right)$

7. If $\vec{a}, \vec{b}, \vec{c}$ are three vectors mutually perpendicular to each other such that $|\vec{a}| = 1$, $|\vec{b}| = 3$ and $|\vec{c}| = 5$ then $[\vec{a} - 2\vec{b}\;\; \vec{b} - 3\vec{c}\;\; \vec{c} - 4\vec{a}]$ is equal to

(1) 0

(2) –24

(3) 3600

(4) –345

8. The equation of the line which passes through the point (1, 1, 1) and intersecting the lines $\frac{x \aleph 1}{2} = \frac{y \;\; 2}{3} = \frac{z \;\; 3}{4}$ and $\frac{x+2}{1} = \frac{y-3}{2} = \frac{z+1}{4}$ is

(1) $\frac{x-1}{4} = \frac{y-1}{11} = \frac{z-1}{13}$

(2) $\frac{x-1}{17} = \frac{y-1}{-3} = \frac{z-1}{17}$

(3) $\frac{x-1}{13} = \frac{y-1}{5} = \frac{z-1}{-2}$

(4) $\frac{x-1}{3} = \frac{y-1}{10} = \frac{z-1}{17}$

9. If $\vec{a} = \hat{i} + \hat{j} - 2\hat{k}$ then

$[(\vec{a} \times \hat{i}) \times \hat{j}]^2 + [(\vec{a} \times \hat{j}) \times \hat{k}]^2 + [(\vec{a} \times \hat{k}) \times \hat{i}]^2 =$

(1) 6 (2) 9

(3) 16 (4) 25

10. The line $\vec{r} = \vec{a} + \lambda(\vec{b} \times \vec{c})$ and $\vec{r} = \vec{b} + \mu(\vec{c} \times \vec{a})$ will intersect if

(1) $\vec{a} \times \vec{c} = \vec{b} \times \vec{c}$ (2) $\vec{a} \cdot \vec{c} = \vec{b} \cdot \vec{c}$

(3) $\vec{b} \times \vec{a} = \vec{c} \times \vec{a}$ (4) $\vec{a} \cdot \vec{c} - \vec{b} \cdot \vec{c} = 1$

11. The value of a for which $2\vec{i} - \vec{j} + \vec{k}$, $\vec{i} + 2\vec{j} - 3\vec{k}$ and $3\vec{i} + a\vec{j} + 5\vec{k}$ are coplanar is

(1) 2 (2) 4

(3) –4 (4) 3

12. A straight line $\vec{r} = \vec{a} + \lambda\vec{b}$ meets the plane $\vec{r} \cdot \vec{n} = 0$ in P. The position vector of P is

(1) $\vec{a} + \frac{\vec{a} \cdot \vec{n}}{\vec{b} \cdot \vec{n}}\vec{b}$ (2) $\vec{a} - \frac{\vec{a} \cdot \vec{n}}{\vec{n} \cdot \vec{n}}\vec{b}$

(3) $\vec{a} - \frac{\vec{a} \cdot \vec{n}}{\vec{b} \cdot \vec{n}}\vec{b}$ (4) $\vec{a} - \frac{\vec{a} \cdot \vec{n}}{\vec{b} \cdot \vec{n}}$

13. If the vectors $\hat{i} - 2x\hat{j} - 3y\hat{k}$ and $\hat{i} + 3x\hat{j} + 2y\hat{k}$ are orthogonal to each other, then the locus of the point (x, y) is

(1) a circle

(2) an ellipse

(3) a parabola

(4) a straight line

14. The co-ordinates of the foot of perpendicular drawn from the origin to the line joining the points (–9, 4, 5) and (10, 0, –1) will be

(1) (–3, 2, 1) (2) $\left(1, 2, \frac{5}{3}\right)$

(3) (4, 5, 3) (4) $\left(\frac{58}{59}, \frac{112}{59}, \frac{109}{59}\right)$

15. The area (in sq units) of the parallelogram whose diagonals are $\vec{a} = 3\hat{i} + \hat{j} - 2\hat{k}$ and $\vec{b} = \hat{i} - 3\hat{j} + 4\hat{k}$ is

(1) $10\sqrt{3}$ (2) $5\sqrt{3}$

(3) 8 (4) 4

16. A line segment joining (1, 0, 1) and the origin (0, 0, 0) is revolved about the x–axis to form a right circular cone. If (x, y, z) is any point on the cone, other than the origin, then it satisfies the equation

(1) $x^2 - 2y^2 - z^2 = 0$ (2) $x^2 - y^2 - z^2 = 0$

(3) $2x^2 - y^2 - 2z^2 = 0$ (4) $x^2 - 2y^2 - 2z^2 = 0$

17. Let $\vec{a}$ and $\vec{b}$ be two unit vectors that are mutually perpendicular. Then a unit vector that is equally inclined to $\vec{a}$, $\vec{b}$ and $\vec{a} \times \vec{b}$ is

(1) $\frac{1}{\sqrt{2}}(\vec{a} + \vec{b} + \vec{a} \times \vec{b})$

(2) $\frac{1}{2}(\vec{a} \times \vec{b} + \vec{a} + \vec{b})$

(3) $\frac{1}{\sqrt{3}}(\vec{a} + \vec{b} + \vec{a} \times \vec{b})$

(4) $\frac{1}{3}(\vec{a} + \vec{b} + \vec{a} \times \vec{b})$

18. The direction cosines of the normal to the plane $2x + 3y - 6z = 5$ are

(1) 2, 3, 6 (2) $\frac{2}{7}, \frac{3}{7}, -\frac{6}{7}$

(3) $\frac{2}{5}, \frac{3}{5}, -\frac{6}{5}$ (4) $\frac{1}{7}, \frac{2}{7}, -\frac{3}{7}$

19. For any four vectors $\vec{a}, \vec{b}, \vec{c}$ and $\vec{d}$, $\vec{d} \cdot (\vec{a} \times (\vec{b} \times (\vec{c} \times \vec{d})))$ is always equal to

(1) $(\vec{a} \cdot \vec{d})[\vec{a}\ \vec{c}\ \vec{d}]$ (2) $(\vec{b} \cdot \vec{d})[\vec{a}\ \vec{c}\ \vec{d}]$

(3) $(\vec{c} \cdot \vec{d})[\vec{a}\ \vec{c}\ \vec{d}]$ (4) $(\vec{a} \cdot \vec{d})[\vec{a}\ \vec{b}\ \vec{d}]$

20. The point of intersection of lines $\frac{x-4}{5} = \frac{y-1}{2} = \frac{z}{1}$ and $\frac{x-1}{2} = \frac{y-2}{3} = \frac{z-3}{4}$ is

(1) (–1, –1, –1) (2) (–1, –1, 1)

(3) (1, –1, –1) (4) (–1, 1, –1)

SECTION 2 (Maximum Marks: 20)

- This section contains **FIVE (05)** questions.
- The answer to each question is a **NON-NEGATIVE INTEGER**. The answer should be rounded off to the nearest integer.
- For each question, enter the correct integer corresponding to the answer using the mouse and the onscreen virtual numeric keypad in the place designated to enter the answer.
- Answer to each question will be evaluated according to the following marking scheme:
  1. Correct Answer: Four marks (+4)
  2. Incorrect Answer: Minus one mark (–1)
  3. Unanswered/Marked for Review: No mark (0).
  4. If a question is found to be incorrect or the Question is dropped then Four marks (+4) will be awarded to all Candidates. The reason could be due to human error or technical error.
  5. Candidates are advised to do the calculations with the constants given (if any) in the questions

21. The number of spheres of radius r touching the coordinate axes is _______.

22. If $\vec{a} = a_1\hat{i} + a_2\hat{j} + a_3\hat{k}$; $\vec{b} = b_1\hat{i} + b_2\hat{j} + b_3\hat{k}$ and $\vec{c} = c_1\hat{i} + c_2\hat{j} + c_3\hat{k}$ such that $[3\vec{a} + \vec{b}\ 3\vec{b} + \vec{c}\ 3\vec{c} + \vec{a}] = \lambda \begin{vmatrix} \vec{a}\cdot\hat{i} & \vec{a}\cdot\hat{j} & \vec{a}\cdot\hat{k} \\ \vec{b}\cdot\hat{i} & \vec{b}\cdot\hat{j} & \vec{b}\cdot\hat{k} \\ \vec{c}\cdot\hat{i} & \vec{c}\cdot\hat{j} & \vec{c}\cdot\hat{k} \end{vmatrix}$, then the value of λ is______.

23. Let $P(a, b, c)$ be any point on the plane $3x + 2y + z = 7$, then the least value of $2(a^2 + b^2 + c^2)$, is _______.

24. If $\vec{a}, \vec{b}$ are vectors perpendicular to each other and $|\vec{a}| = 2, |\vec{b}| = 3, \vec{c} \times \vec{a} = \vec{b}$, then the least value of $2|\vec{c} - \vec{a}|$ is_______.

25. If the line $\frac{x-2}{3} = \frac{y+1}{2} + \frac{z-1}{-1}$ intersects the curve $xy = c^2$, $z = 0$ then c^2 is equal to _______.

Answer Key

1. (2)	**2.** (2)	**3.** (1)	**4.** (2)	**5.** (4)	**6.** (3)
7. (4)	**8.** (4)	**9.** (1)	**10.** (2)	**11.** (3)	**12.** (3)
13. (1)	**14.** (4)	**15.** (2)	**16.** (2)	**17.** (3)	**18.** (2)
19. (2)	**20.** (1)	**21.** (8)	**22.** (28)	**23.** (7)	**24.** (3)
25. (5)					

JEE (Advanced) Mock Test

(Paper 1)

SECTION 1 (Maximum Marks: 12)

- This section contains **FOUR (04)** questions.
- Each question has **FOUR** options (1), (2), (3) and (4). **ONLY ONE** of these four options is the correct answer.
- For each question, choose the option corresponding to the correct answer.
- Answer to each question will be evaluated according to the following marking scheme:

Full Marks : +3 If **ONLY** the correct option is chosen;

Zero Marks : 0 If none of the options is chosen (i.e., the question is unanswered);

Negative Marks : −1 In all other cases.

1. The minimum area of the triangle whose vertices are $A(-1, 1, 2)$, $B(1, 2, 3)$ and $C(t, 1, 1)$ where t is a real number is

(1) $\tan 60^\circ$ (2) $\sin 60^\circ$

(3) $\cos 60^\circ$ (4) $\sin 15^\circ$

2. If the perpendicular distance of a point P other than the origin from the plane $x + y + z = p$ is equal to the distance of the plane from the origin, then the coordinates of P are

(1) $(p, 2p, 0)$ (2) $(0, 2p, -p)$

(3) $(2p, p, -p)$ (4) $(2p, -p, 2p)$

3. If the vectors $2\hat{i} - 3\hat{j}, \hat{i} + \hat{j} - \hat{k}$ and $3\hat{i} - \hat{k}$ form three concurrent edges of a parallelepiped, then the volume (in cubic units) of the parallelepiped is

(1) 8 (2) 10

(3) 4 (4) 14

4. Let $P \equiv (1, 7, \sqrt{2})$ be a point and line L be $2\sqrt{2}(x - 1) = y - 2, z = 0$. If PQ is the distance of plane $\sqrt{2}x + y - z = 1$ from point P measured along a line inclined at an angle of 45° with the line L and it is minimum, then the value of PQ is

(1) 3 (2) 4

(3) 6 (4) 8

SECTION 2 (Maximum Marks: 12)

- This section contains **THREE (03)** questions.
- Each question has **FOUR** options (1), (2), (3) and (4). **ONE OR MORE THAN ONE** of these four options is (are) correct answer(s).
- For each question, choose the option(s) corresponding to (all) the correct answer(s).
- Answer to each question will be evaluated according to the following marking scheme:

Full Marks : +4 ONLY if (all) the correct option(s) is(are) chosen;
Partial Marks : +3 If all the four options are correct but ONLY three options are chosen;
Partial Marks : +2 If three or more options are correct but ONLY two options are chosen, both of which are correct;
Partial Marks : +1 If two or more options are correct but ONLY one option is chosen and it is a correct option;
Zero Marks : 0 If none of the options is chosen (i.e., the question is unanswered);
Negative Marks : −2 In all other cases.

- For example, in a question, if (1), (2) and (4) are the ONLY three options corresponding to correct answers, then
choosing ONLY (1), (2) and (4) will get +4 marks;
choosing ONLY (1) and (2) will get +2 marks;
choosing ONLY (1) and (4) will get + 2 marks;
choosing ONLY (2) and (4) will get +2 marks;
choosing ONLY (1) will get +1 mark;
choosing ONLY (2) will get +1 mark;
choosing ONLY (4) will get +1 mark;
choosing no option(s) (i.e., the question is unanswered) will get 0 marks and
choosing any other option(s) will get −2 marks.

5. If a line has a vector equation $\vec{r} = 2\hat{i} + 6\hat{j} + \lambda(\hat{i} - 3\hat{j})$, then which of the following statements hold good?

(1) The line is parallel to $\vec{r} = 2\hat{i} + 6\hat{j} + \lambda(\hat{i} - 3\hat{j})$.

(2) The line passes through the point $3\hat{i} + 3\hat{j}$.

(3) The line passes through the point $\hat{i} + 9\hat{j}$.

(4) The line is parallel to xy plane.

6. Let three vectors $\vec{a}, \vec{b}, \vec{c}$ be such that $\vec{a} \times \vec{b} = 4(\vec{a} \times \vec{c})$ and $|\vec{a}| = |\vec{b}| = 1$ and $|\vec{c}| = \frac{1}{4}$. If the angle between $\vec{b}$ and $\vec{c}$ is $\frac{\pi}{3}$, then $\vec{b}$ is

(1) $\vec{a} + 4\vec{c}$ (2) $\vec{a} - 4\vec{c}$

(3) $4\vec{c} - \vec{a}$ (4) $2\vec{c} - \vec{a}$

7. The planes $ax + 4y + z = 0$, $2y + 3z - 1 = 0$, $3x - bz + 2 = 0$

(1) will meet at a point if $ab \neq 15$

(2) will meet on a line if $ab = 15$, $a = 3$

(3) will have no common point if $ab = 15$, $a \neq 3$

(4) will have no common point $ab = 15$, $a \neq 5$

SECTION 3 (Maximum Marks: 24)

- This section contains **SIX (06)** questions.
- The answer to each question is a **NON-NEGATIVE INTEGER**.
- For each question, enter the correct integer corresponding to the answer using the mouse and the onscreen virtual numeric keypad in the place designated to enter the answer.
- Answer to each question will be evaluated according to the following marking scheme:

Full Marks : +4 If **ONLY** the correct integer is entered;

Zero Marks : 0 In all other cases.

8. If the projections of a line segment on the x, y and z axes are 2, 3 and 6, respectively, then the length of the line segment is ______.

9. The distance of the z-axis from the image of the point $M(2, -3, 3)$ in the plane $x - 2y - z + 1 = 0$, is ______.

10. Let $\vec{a}, \vec{b}, \vec{c}$ be three vectors having magnitudes 1, 2, 3, respectively, satisfying the relation $|[\vec{a}\,\vec{b}\,\vec{c}]| = 6$. Then the value of $|(\vec{a} \times \vec{c}) \cdot \vec{b}|^2 + |(\vec{a} \times \vec{c}) \times \vec{b}|^2$ is______.

11. If the angle between the plane $x - 3y + 2z = 1$ and the line $\frac{x-1}{2} = \frac{y-1}{1} = \frac{z-1}{-3}$ is θ, then the value of cosec θ is______.

12. The resultant of vectors $\vec{a}$ and $\vec{b}$ is $\vec{c}$. If $\vec{c}$ trisects the angle between $\vec{a}$ and $\vec{b}$, $|\vec{a}| = 6, |\vec{b}| = 4$, then $|\vec{c}| =$______.

13. The distance of the point $-\hat{i} + 2\hat{j} + 6\hat{k}$ from the straight line through the point (2, 3, –4) and parallel to $6\hat{i} + 3\hat{j} - 4\hat{k}$ is______.

SECTION 4 (Maximum Marks: 12)

- This section contains **FOUR (04)** Matching List Sets..
- Each set has **ONE** Multiple Choice Question.
- Each set has **TWO** lists: **List I** and **List II**.
- **List I** has **Four** entries (P), (Q), (R) and (S) and **List II** has **Four** entries (I), (II), (III) and (IV).
- **FOUR** options are given in each Multiple Choice Question based on List I and List II and ONLY ONE of these four options satisfies the condition asked in the Multiple Choice Question.
- Answer to each question will be evaluated according to the following marking scheme:

Full Marks : +3 ONLY if the option corresponding to the correct combination is chosen;
Zero Marks : 0 If none of the options is chosen (i.e., the question is unanswered);
Negative Marks : −1 In all other cases.

14. Match each entry in List I to the correct entry in List II.

	List I		List II
(P)	The distance between the parallel planes $x + 2y - 2z + 1 = 0$ and $2x + 4y - 4z + 5 = 0$ is	(I)	$\frac{4}{\sqrt{3}}$
(Q)	The radius of the circular section of the sphere $\lvert \vec{r} \rvert = 5$ by the plane $\vec{r}.(\hat{i} + 2\hat{j} - \hat{k}) = 4\sqrt{3}$ is	(II)	13
(R)	The distance between the line $\frac{x+1}{-3} = \frac{y-3}{2} = \frac{z-2}{1}$ and the plane $x + y + z + 3 = 0$	(III)	3
(S)	The distance of the point $(-1, -5, -10)$ from the point of intersection of the line $\frac{x-2}{3} = \frac{y+1}{4} = \frac{z-2}{2}$ and plane $x - y + z = 5$ is	(IV)	$\frac{1}{2}$

The correct option is

(1) (P) → (III); (Q) → (II); (R) → (IV); (S) → (I)
(2) (P) → (II); (Q) → (III); (R) → (I); (S) → (IV)
(3) (P) → (IV); (Q) → (III); (R) → (I); (S) → (II)
(4) (P) → (IV); (Q) → (II); (R) → (I); (S) → (III)

15. Match each entry in List I to the correct entry in List II.

	List I		List II
(P)	If $\vec{a}, \vec{b}, \vec{c}$ are three mutually perpendicular vectors where $\lvert \vec{a} \rvert = \lvert \vec{b} \rvert = 2, \lvert \vec{c} \rvert = 1$ then $[\vec{a} \times \vec{b} \ \ \vec{b} \times \vec{c} \ \ \vec{c} \times \vec{a}]$ is	(I)	−12
(Q)	If $\vec{a}, \vec{b}$ are two unit vectors inclined at $\frac{\pi}{3}$, then $16[\vec{a} \ \ \vec{b} + \vec{a} \times \vec{b} \ \ \vec{b}]$ is	(II)	0
(R)	If $\vec{b}, \vec{c}$ are orthogonal unit vectors and $\vec{b} \times \vec{c} = \vec{a}$, then $[\vec{a} + \vec{b} + \vec{c} \ \ \vec{a} + \vec{b} \ \ \vec{b} + \vec{c}]$ is	(III)	16
(S)	If $[\vec{x}\ \vec{y}\ \vec{a}] = [\vec{x}\ \vec{y}\ \vec{b}] = [\vec{a}\ \vec{b}\ \vec{c}] = 0$ each vector being a non-zero vectors, then $[\vec{x}\ \vec{y}\ \vec{c}]$ is	(IV)	1

The correct option is

(1) (P) → (III); (Q) → (II); (R) → (I); (S) → (IV)
(2) (P) → (III); (Q) → (I); (R) → (IV); (S) → (II)
(3) (P) → (III); (Q) → (IV); (R) → (I); (S) → (II)
(4) (P) → (I); (Q) → (II); (R) → (IV); (S) → (III)

16. Consider three planes

$$P_1 \equiv 2x + y + z = 1$$
$$P_2 \equiv x - y + z = 2$$
$$P_3 \equiv ax - y + 3z = 5$$

The three planes intersects each other at point P on XOY plane and at point Q on YOZ plane. O is the origin.

Match each entry in List I to the correct entry in List II.

	List I	List II	
(P)	The value of α is	(I)	1
(Q)	The length of projection of PQ on x-axis is	(II)	2
(R)	If the coordinates of point R situated at minimum distance from point 'O' on the line PQ are (a, b, c), then the value of $7a + 14b + 14c$ is	(III)	3
(S)	If the area of ΔPOQ is $\sqrt{\frac{a}{b}}$. Then the value of $a - b$ can be	(IV)	4

The correct option is

(1) (P) → (III); (Q) → (II); (R) → (IV); (S) → (I)
(2) (P) → (II); (Q) → (III); (R) → (I); (S) → (IV)
(3) (P) → (IV); (Q) → (III); (R) → (I); (S) → (II)
(4) (P) → (IV); (Q) → (I); (R) → (II); (S) → (III)

17. Consider two vectors

$\vec{a} = -\hat{i} + \hat{j} + 2\hat{k}, \vec{b} = -\hat{i} - 2\hat{j} - \hat{k}$

Match each entry in List I to the correct entry in List II.

	List I	List II	
(P)	Area of triangle formed by $\vec{a}, \vec{b}$ is	(I)	$3\sqrt{6}$
(Q)	Area of parallelogram having sides $\vec{a}$ and $\vec{b}$ is	(II)	$12\sqrt{3}$
(R)	Area of parallelogram having diagonals $2\vec{a}$ and $4\vec{b}$ is	(III)	$3\sqrt{3}$
(S)	Volume of parallelepiped formed by $\vec{a}$ and $\vec{b}$ and $\vec{c} = \hat{i} + \hat{j} + \hat{k}$ is	(IV)	$\frac{3\sqrt{3}}{2}$

The correct option is

(1) (P) → (IV); (Q) → (II); (R) → (III); (S) → (I)
(2) (P) → (I); (Q) → (II); (R) → (III); (S) → (IV)
(3) (P) → (IV); (Q) → (III); (R) → (II); (S) → (I)
(4) (P) → (IV); (Q) → (I); (R) → (II); (S) → (III)

Answer Key

1. (2) **2.** (3) **3.** (3) **4.** (1) **5.** (2), (3), (4) **6.** (1), (3)
7. (1), (2), (3) **8.** (7) **9.** (1) **10.** (36) **11.** (2) **12.** (5)
13. (7) **14.** (3) **15.** (2) **16.** (4) **17.** (3)

JEE (Advanced) Mock Test

(Paper 2)

SECTION 1 (Maximum Marks: 12)

- This section contains **FOUR (04)** questions.
- Each question has **FOUR** options (1), (2), (3) and (4). **ONLY ONE** of these four options is the correct answer.
- For each question, choose the option corresponding to the correct answer.
- Answer to each question will be evaluated according to the following marking scheme:

Full Marks : +3 If **ONLY** the correct option is chosen;

Zero Marks : 0 If none of the options is chosen (i.e., the question is unanswered);

Negative Marks : −1 In all other cases.

1. Given $\vec{\alpha} = 3\hat{i} + \hat{j} + 2\hat{k}$, $\vec{\beta} = \hat{i} - 2\hat{j} - 4\hat{k}$ are the position vectors of point A and B. Then the distance of point $-\hat{i} + \hat{j} + \hat{k}$ from the plane passing through B and perpendicular to AB is

(1) 5 (2) 10

(3) 15 (4) 20

2. If $\vec{a}, \vec{b}, \vec{c}$ are three non-coplanar vectors such that $\vec{a} \times \vec{b} = \vec{c}$, $\vec{b} \times \vec{c} = \vec{a}$, $\vec{c} \times \vec{a} = \vec{b}$, then the value of $|\vec{a}| + |\vec{b}| + |\vec{c}|$ is

(1) 1/3 (2) 1

(3) 3 (4) 6

3. If $P(\alpha, \beta, \gamma)$ be a vertex of an equilateral triangle PQR where vertex Q and R are (–1, 0, 1) and (1, 0, –1) respectively, then P can lie on the plane

(1) $x + y + z + 6 = 0$ (2) $2x + 4y + 3z + 20 = 0$

(3) $x - y + z + 12 = 0$ (4) $x + y + z + 3\sqrt{2} = 0$

4. Let $\vec{a}, \vec{b}$, and $\vec{c}$ be unit vectors inclined to each other at an angle of $\frac{2\pi}{3}$. If $\hat{p}, \hat{q}$ and $\hat{r}$ are unit vectors along the angle bisectors of $\vec{a}$ and $\vec{b}$, $\vec{b}$ and $\vec{c}$, $\vec{c}$ and $\vec{a}$ respectively, then $[\hat{p}\ \hat{q}\ \hat{r}]$ is equal to

(1) 1 (2) 0

(3) $\frac{1}{\sqrt{3}}$ (4) $\sqrt{3}$

SECTION 2 (Maximum Marks: 12)

- This section contains **THREE (03)** questions.
- Each question has **FOUR** options (1), (2), (3) and (4). **ONE OR MORE THAN ONE** of these four options is (are) correct answer(s).
- For each question, choose the option(s) corresponding to (all) the correct answer(s).
- Answer to each question will be evaluated according to the following marking scheme:

Full Marks : +4 ONLY if (all) the correct option(s) is(are) chosen;
Partial Marks : +3 If all the four options are correct but ONLY three options are chosen;
Partial Marks : +2 If three or more options are correct but ONLY two options are chosen, both of which are correct;
Partial Marks : +1 If two or more options are correct but ONLY one option is chosen and it is a correct option;
Zero Marks : 0 If none of the options is chosen (i.e., the question is unanswered);
Negative Marks : −2 In all other cases.

- For example, in a question, if (1), (2) and (4) are the ONLY three options corresponding to correct answers, then
choosing ONLY (1), (2) and (4) will get +4 marks;
choosing ONLY (1) and (2) will get +2 marks;
choosing ONLY (1) and (4) will get + 2 marks;
choosing ONLY (2) and (4) will get +2 marks;
choosing ONLY (1) will get +1 mark;
choosing ONLY (2) will get +1 mark;
choosing ONLY (4) will get +1 mark;
choosing no option(s) (i.e., the question is unanswered) will get 0 marks and
choosing any other option(s) will get −2 marks.

5. If $\vec{A}, \vec{B}, \vec{C}$ and $\vec{D}$ are four non zero vectors in the same plane no two of which are collinear then which of the following hold(s) good?

(1) $\left(\vec{A}\times\vec{B}\right)\cdot\left(\vec{C}\times\vec{D}\right)=0$

(2) $\left(\vec{A}\times\vec{C}\right)\cdot\left(\vec{B}\times\vec{D}\right)\neq 0$

(3) $\left(\vec{A}\times\vec{B}\right)\times\left(\vec{C}\times\vec{D}\right)=\vec{0}$

(4) $\left(\vec{A}\times\vec{C}\right)\times\left(\vec{B}\times\vec{D}\right)\neq\vec{0}$

6. Let $P = 0$ be the equation of the plane passing through the line of intersection of the planes $2x - y = 0$ and $3z - y = 0$ and perpendicular to the plane $4x + 5y - 3z = 8$. Then the points which lie on the planes $P = 0$ is/are

(1) (0, 9, 17) (2) $\left(\frac{1}{7}, 2, \frac{1}{9}\right)$

(3) (1, 3, −4) (4) $\left(\frac{1}{2}, 1, \frac{1}{3}\right)$

7. Let $\vec{r}$ be a unit vector satisfying $\vec{r}\times\vec{a}=\vec{b}$, where $|\vec{a}|=\sqrt{3}$ and $|\vec{b}|=\sqrt{2}$, then

(1) $\vec{r}=\frac{2}{3}(\vec{a}+\vec{a}\times\vec{b})$ (2) $\vec{r}=\frac{1}{3}(\vec{a}+\vec{a}\times\vec{b})$

(3) $\vec{r}=\frac{2}{3}(\vec{a}-\vec{a}\times\vec{b})$ (4) $\vec{r}=\frac{1}{3}(-\vec{a}+\vec{a}\times\vec{b})$

SECTION 3 (Maximum Marks: 24)

- This section contains **SIX (06)** questions.
- The answer to each question is a **NON-NEGATIVE INTEGER**.
- For each question, enter the correct integer corresponding to the answer using the mouse and the onscreen virtual numeric keypad in the place designated to enter the answer.
- Answer to each question will be evaluated according to the following marking scheme:

Full Marks : +4 If **ONLY** the correct integer is entered;

Zero Marks : 0 In all other cases.

8. Let $\vec{u}$ be a vector on rectangular coordinate system with sloping angle 60°. Suppose that $|\vec{u}-\hat{i}|$ is geometric mean of $|\vec{u}|$ and $|\vec{u}-2\hat{i}|$ where $\hat{i}$ is the unit vector along x-axis then the value of $(\sqrt{2}+1)\ |\vec{u}|$ is______.

9. If the length of the shortest distance between two lines $\frac{1}{2}(x-1)=\frac{1}{4}(y-3)=z+2$ and $3x-y-2z+4=0=2x+y+z+1$ is d, then the value of $\sqrt{14}d$ is_______.

10. If $\vec{a},\vec{b},\vec{c}$ are unit vectors such that $\vec{a}\cdot\vec{b}=0=\vec{a}\cdot\vec{c}$ and the angle between $\vec{b}$ and $\vec{c}$ is $\frac{\pi}{3}$. Then the value of $|\vec{a}\times\vec{b}-\vec{a}\times\vec{c}|$ is______.

11. A line with direction ratios (2, 1, 2) intersects the lines $\vec{r}=-\hat{j}+\lambda(\hat{i}+\hat{j}+\hat{k})$ and $r=-\hat{i}+\mu(2\hat{i}+\hat{j}+\hat{k})$ at A and B, then length AB is equal to _______.

12. Let $\vec{r}=(\vec{a}\times\vec{b})\sin x+(\vec{b}\times\vec{c})\cos y+(\vec{c}\times\vec{a})$, where $\vec{a},\vec{b}$ and $\vec{c}$ are non-zero non-coplanar vectors. If $\vec{r}$ is orthogonal to $3\vec{a}+5\vec{b}+2\vec{c}$, then the value of $\sec^2 y+\operatorname{cosec}^2 x+\sec y\operatorname{cosec} x$ is______.

13. If the straight line $3x-2y+z+3=0=4x-3y+4z+1$ is parallel to the plane $2x-y+mz-2=0$, then the value of $|m|=$ _______.

SECTION 4 (Maximum Marks: 12)

- This section contains **TWO (02)** paragraphs.
- Based on each paragraph, there are **TWO (02)** questions.
- The answer to each question is a **NUMERICAL VALUE**.
- For each question, enter the correct numerical value of the answer using the mouse and the on-screen virtual numeric keypad in the place designated to enter the answer.
- If the numerical value has more than two decimal places, **truncate/round-off** the value to **TWO** decimal places.
- Answer to each question will be evaluated according to the following marking scheme:

Full Marks : +3 If ONLY the correct numerical value is entered in the designated place;

Zero Marks : 0 In all other cases.

PARAGRAPH I

Let a vector $\vec{v}$ be defined as $\vec{v} = \lambda\vec{a} + \mu\vec{b} + \eta(\vec{a}\times\vec{b})$, where $\vec{a}\perp\vec{b}$ and given that $\vec{v}\cdot\vec{a} = 0, \vec{v}\cdot\vec{b} = 1[\vec{v}\ \vec{a}\ \vec{b}] = 1$. Then

14. $\lambda =$ _______.

15. $|b|^2\mu =$ _______.

PARAGRAPH II

Consider the line $L: \dfrac{x-1}{2} = \dfrac{y}{1} = \dfrac{z+1}{-2}$ and a point $A(1, 1, 1)$. Let P be the foot of the perpendicular from A on L and Q be the image of the point A in the line L, 'O' being the origin.

16. The distance of the origin from the plane passing through the point A and containing the line L, is _______.

17. The distance of the point A from the line L, is _______.

Answer Key

1. (1)	**2.** (3)	**3.** (4)	**4.** (2)	**5.** (2, 3)	**6.** (1, 4)
7. (2, 4)	**8.** (1)	**9.** (8)	**10.** (1)	**11.** (3)	**12.** (3)
13. (2)	**14.** (0)	**15.** (1)	**16.** (0.33)	**17.** (2)	

Solutions

JEE (MAIN) 2026 (SESSION 1)

Products of Vectors

Single Correct Answer Type

1. (3) $\overrightarrow{AC}=3\hat{i}+6\hat{j}+(\lambda-5)\hat{k}$

Projection of $\vec{v}$ on $\overrightarrow{AC}$ is 1.

$\therefore\ \vec{v}\cdot AC=1 \Rightarrow \dfrac{3+6+\lambda-5}{\sqrt{9+36+(\lambda-5)^2}}=1$

$\Rightarrow \lambda=3$

Hence, the given quadratic equation becomes,

$9x^2-18x+5=0$

$\Rightarrow x=\dfrac{1}{3},\dfrac{5}{3}$

$\therefore\ 2\alpha-\beta=\dfrac{10-1}{3}=3$

2. (3) $\vec{v}=x\vec{a}+y\vec{b}=x(2\hat{i}-\hat{j}-\hat{k})+y(\hat{i}+3\hat{j}-\hat{k})$

or $\vec{v}=(2x+y)\hat{i}+(3y-x)\hat{j}+(-x-y)\hat{k}$

Now, $\dfrac{|\vec{v}.\vec{c}|}{|\vec{c}|}=\dfrac{1}{\sqrt{14}}$

$|\vec{v}\cdot\vec{c}|=|2y|=1$

$\therefore\ |\vec{v}|=\sqrt{(2x+y)^2+(3y-x)^2+(x+y)}$

$=\sqrt{6x^2+11y^2+4xy-6xy+2xy}$

$=\sqrt{6x^2+\dfrac{11}{4}}=\dfrac{\sqrt{24x^2+11}}{2}$

3. (1) $|\vec{a}-\vec{b}|^2+|\vec{b}-\vec{c}|^2+|\vec{c}-\vec{a}|^2=9$

$\Rightarrow \vec{a}\cdot\vec{b}+\vec{b}\cdot\vec{c}+\vec{c}\cdot\vec{a}=-\dfrac{3}{2}$

$\Rightarrow \vec{a}^2+\vec{b}^2+\vec{c}^2+2\vec{a}\cdot\vec{b}+2\vec{b}\cdot\vec{c}+2\vec{c}\cdot\vec{a}=0$

$\Rightarrow \vec{a}+\vec{b}+\vec{c}=0 \Rightarrow \vec{b}+\vec{c}=-\vec{a}$

So, $|2\vec{a}+k\vec{b}+k\vec{c}|=3$

$\Rightarrow |2\vec{a}+k(\vec{b}+\vec{c})|=3$

$\Rightarrow |\vec{a}(2-k)|=3$

$\Rightarrow k=5$ (as k is positive)

4. (1) Point P must lie on the bisector of the angle $\angle BAC$.

$\therefore\ \overrightarrow{AP}=\lambda(3\hat{i}+\hat{j}-\hat{k}+\hat{i}-\hat{j}+3\hat{k})=\lambda(4\hat{i}+2\hat{k})$

$|\overrightarrow{AP}|=|\lambda|\sqrt{20}=\dfrac{\sqrt{5}}{2}$ (given)

$\therefore\ |\lambda|=\dfrac{1}{4}$

For $\lambda=\dfrac{1}{4}$, $\overrightarrow{AP}=\dfrac{1}{2}(2\hat{i}+\hat{k})$

Therefore, the area of triangle ABP is

$\dfrac{1}{2}|\overrightarrow{AB}\times\overrightarrow{AP}|=\dfrac{1}{4}\left\|\begin{matrix}\hat{i} & \hat{j} & \hat{k}\\ 3 & 1 & -1\\ 2 & 0 & 1\end{matrix}\right\|$

$=\dfrac{1}{4}|\hat{i}-5\hat{j}-2\hat{k}|=\dfrac{1}{4}\sqrt{30}$

5. (3) Let $\vec{c}=c_1\hat{i}+c_2\hat{j}+c_3\hat{k}$

$\vec{a}\times\vec{c}=\vec{b}$

$\Rightarrow (2c_3-2c_2)\hat{i}+(c_3+2c_1)\hat{j}-(c_2+2c_1)\hat{k}=8\hat{i}+7\hat{j}-3\hat{k}$

$\Rightarrow 2c_3-2c_2=8,\ c_3+2c_1=7,\ c_2+2c_1=3$

Also, $(c_1\hat{i}+c_2\hat{j}+c_3\hat{k}).(\hat{i}+\hat{j}+\hat{k})=4$

$\Rightarrow c_1+c_2+c_3=4$

Solving we get $c_1=2,\ c_2=-1,\ c_3=3$

$|\vec{a}+\vec{c}|^2=|\hat{i}+\hat{j}+\hat{k}|^2=27$

6. (4) $\vec{c}\times(2\hat{i}+3\hat{j}+4\hat{k})=(2\hat{i}+3\hat{j}+4\hat{k})\times\vec{d}$.

$\Rightarrow \vec{c}+\vec{d}=\lambda(2\hat{i}+3\hat{j}+4\hat{k})$

$\therefore\ |\vec{c}+\vec{d}|=\sqrt{29}$ give, $\lambda=\pm 1$

$\Rightarrow (\vec{c}+\vec{d})\cdot(-7\hat{i}+2\hat{j}+3\hat{k})=\lambda(-14+6+12)=4\lambda$

$\Rightarrow \lambda_1=4,\ \lambda_2=-4$

Therefore, the given equation represents circle, if

$k^2-5k+4=0 \Rightarrow k=1, 4$ and,

$k^2=3k-2 \Rightarrow k=1, 2$

Hence $k=1$ (common value)

7. (1)

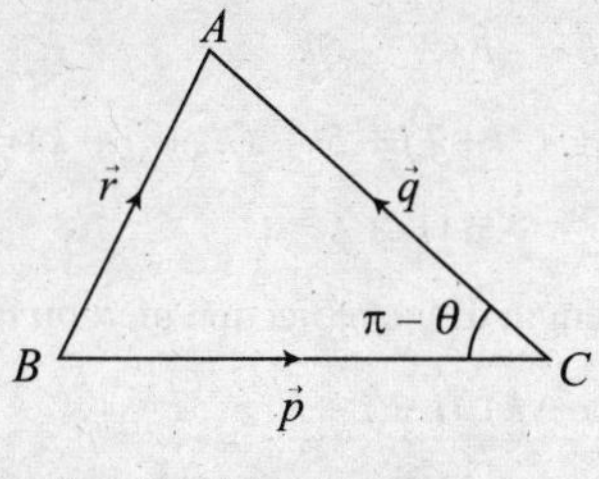

We have, $\vec{p}+\vec{q}=\vec{r}$, and

$\cos(\pi-\theta)=\dfrac{|\vec{p}|^2+|\vec{q}|^2-|\vec{r}|^2}{2|\vec{p}||\vec{q}|}$

$\Rightarrow \dfrac{-1}{\sqrt{3}}=\dfrac{12+4-|\vec{r}|^2}{2\cdot 2\sqrt{3}\cdot 2}$ or $|\vec{r}|^2=24$

$\therefore\ |\vec{p}\times(\vec{q}-3\vec{r})|^2+3|\vec{r}|^2$

$=|\vec{p}\times(\vec{q}-3\vec{p}-3\vec{q})|^2+72$

$=|-2\vec{p}\times\vec{q}|^2+72=4|\vec{p}|^2|\vec{q}|^2\times\sin^2\theta+72$

$=4\cdot 12\cdot 4\cdot\dfrac{2}{3}+72=200$

8. (3) $\vec{c} = \vec{a} \times \vec{b} = (-2 - \lambda)\hat{i} - 4\hat{j} + 2\lambda\hat{k}$

$|\vec{c}| = \sqrt{53}$

$\Rightarrow 5\lambda^2 + 4\lambda - 33 = 0$

$\Rightarrow \lambda = -3$ (as $\lambda \in Z$)

$\therefore\ \vec{c} = \hat{i} - 4\hat{j} - 6\hat{k}$

Let $\vec{d} = y\hat{j} + z\hat{k}$

$|\vec{d}| = 2 \Rightarrow y^2 + z^2 = 4$

Hence,

$(\vec{c}\cdot\vec{d})^2 = (4y + 6z)^2 \le (\sqrt{4^2 + 6^2} \times \sqrt{y^2 + z^2})^2 \le 208$

9. (4) $\vec{c} = \vec{a} \times \vec{b}$

$= (-\hat{i} + \hat{j} + 2\hat{k}) \times (\hat{i} - \hat{j} - 3\hat{k}) = -\hat{i} - \hat{j}$

$= (\vec{a} - \vec{b})\cdot\vec{d}$

$= (-2\hat{i} + 2\hat{j} + 5\hat{k})\cdot(-2\hat{i} + 2\hat{j} - 2\hat{k})$

$= 4 + 4 - 10 = -2$

10. (1) $\vec{a} \times \vec{b} = 2(\vec{a} \times \vec{c})$

$\Rightarrow \vec{a} \times (\vec{b} - 2\vec{c}) = 0$

$\Rightarrow \vec{b} - 2\vec{c} = \lambda\vec{a}$ ...(i)

$\Rightarrow |\lambda a|^2 = |\vec{b} - 2\vec{c}|^2$

$\Rightarrow \lambda^2 |\vec{a}|^2 = |\vec{b}|^2 + 4|\vec{c}|^2 - 4\vec{b}.\vec{c}$

$\Rightarrow \lambda^2 = 16 + 16 - 4 \times 4 \times 2 \times \frac{1}{2}$

$\Rightarrow \lambda^2 = 16$ or $\lambda = \pm 4$

$\Rightarrow \vec{b} - 2\vec{c} = \pm 4\vec{a}$

Taking dot product with $\vec{c}$, we get

$\vec{b}.\vec{c} - 2|\vec{c}|^2 = \pm 4(\vec{a}.\vec{c})$

or $4\cdot 2\cdot\frac{1}{2} - 2\cdot 4 = \pm 4(\vec{a}\cdot\vec{c})$

$\Rightarrow |\vec{a}\cdot\vec{c}| = 1$

11. (2) $\vec{v} = \vec{a} \times \vec{b} = -\hat{i} + 7\hat{j} + 5\hat{k}$

$\vec{v}\cdot\vec{c} = 1 = (-\hat{i} + 7\hat{j} + 5\hat{k})\cdot(\lambda\hat{i} + \hat{j} + \hat{k}) = 1$ (given)

$\Rightarrow -\lambda + 7 + 5 = 11$ or $\lambda = 1$.

So, the length of the projection of $\vec{b}$ on $\vec{c}$

$= \frac{(2\hat{i} + \hat{j} - \hat{k})\cdot(\hat{i} + \hat{j} + \hat{k})}{\sqrt{3}} = \frac{2}{\sqrt{3}} = p$

$\Rightarrow 9p^2 = 9\left(\frac{4}{3}\right) = 12$

12. (4) $\vec{c} = \begin{vmatrix} \hat{i} & \hat{j} & \hat{k} \\ 2 & 1 & -2 \\ 1 & 1 & 0 \end{vmatrix} = 2\hat{i} - 2\hat{j} + \hat{k}$

$\therefore |c| = 3$

Now, $|\vec{c} \times \vec{d}| = 3$

$\Rightarrow |\vec{c}||\vec{d}| \sin\frac{\pi}{4} = 3 \Rightarrow |\vec{d}| = \sqrt{2}$

Also, $|\vec{d} - \vec{a}| = \sqrt{11}$

$\Rightarrow |\vec{a}|^2 + |\vec{d}|^2 - 2\vec{a}\cdot\vec{d} = 11$

$\Rightarrow 9 + 2 - 2\vec{a}\cdot\vec{d} = 11$

$\Rightarrow \vec{a}\cdot\vec{d} = 0$

13. (3) We have, $2(\vec{a} \times \vec{c}) + 3(\vec{b} \times \vec{c}) = 0$

$\Rightarrow (2\vec{a} + 3\vec{b}) \times \vec{c} = 0$

$\Rightarrow \vec{c} = \lambda(2\vec{a} + 3\vec{b})$

$\Rightarrow \vec{c} = \lambda(7\hat{i} - 13\hat{j} + 19\hat{k})$

Now $(\vec{a} - \vec{b})\cdot\vec{c} = \lambda(7 + 52 + 38) = -97$

$\Rightarrow \lambda = -1$

$\vec{c} = -7\hat{i} + 13\hat{j} - 19\hat{k}$

$\Rightarrow \vec{c} \times \hat{k} = 7\hat{j} + 13\hat{i} \Rightarrow |\vec{c} \times \hat{k}|^2 = 7^2 + 13^2 = 218$

Numerical Value Type

14. (5) $\frac{\vec{a}\cdot\hat{k}}{|\vec{a}|} = \cos\theta \Rightarrow \frac{\lambda}{\sqrt{3 + \lambda^2}} = \cos\theta,\ \frac{\pi}{6} < \theta < \frac{\pi}{2}$

$\Rightarrow 0 < \frac{\lambda}{\sqrt{3 + \lambda^2}} < \frac{\sqrt{3}}{2}$

$\Rightarrow \lambda > 0$ and $4\lambda^2 < 9 + 3\lambda^2$ or $\lambda^2 < 9$

$\Rightarrow \lambda \in (0, 3)$ ...(i)

Also, $\vec{a}\cdot\vec{b} < 0 \Rightarrow -\sqrt{2}\lambda^2 - 4\sqrt{2} + 4\sqrt{2}\lambda < 0$

$\Rightarrow \lambda^2 - 4\lambda + 4 > 0$ or $(\lambda - 2)^2 > 0$

$\Rightarrow \lambda \ne 2$ ...(ii)

From (i) and (ii)

$\lambda \in (0, 3) - \{2\}$

$\therefore \alpha = 0, \beta = 3, \gamma = 2$ and hence $\alpha + \beta + \gamma = 5$

15. (37)

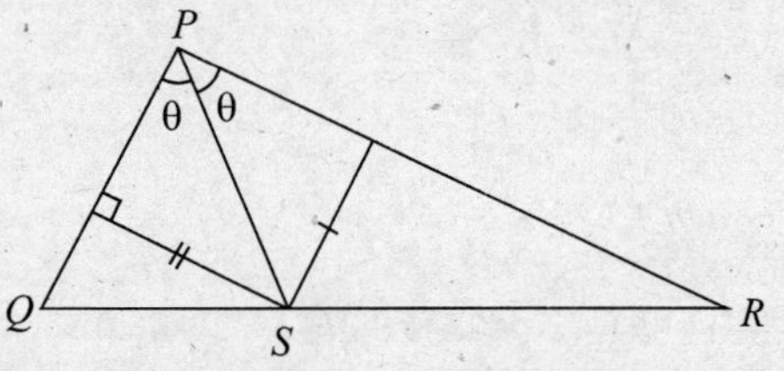

Since, $|\overrightarrow{PR}| = 9$, $a^2 + b^2 + 16 = 81$

or $a^2 + b^2 = 65$ ...(i)

Clearly, point S lies on the bisector of an angle QPR.

$\cos\theta = \frac{\overrightarrow{PQ}\cdot\overrightarrow{PS}}{|\overrightarrow{PQ}||\overrightarrow{PS}|} = \frac{\overrightarrow{PS}\cdot\overrightarrow{PR}}{|\overrightarrow{PS}||\overrightarrow{PR}|}$

$\frac{1}{\sqrt{6}} = \frac{a - 7b - 8}{3(9\sqrt{6})}$

$\Rightarrow a - 7b = 35$ ...(ii)

Solving (i) and (ii), we get

$\Rightarrow a = 7, b = -4$

$\therefore\ 3a - 4b = 21 + 16 = 37$

Line and Plane

Single Correct Answer Type

16. (4) $L_1 : \dfrac{x-2}{-3} = \dfrac{y-6}{2} = \dfrac{z-7}{4}$

$\therefore$ Point C on L_1 is $(-3\lambda_1 + 2, 2\lambda_1 + 6, 4\lambda_1 + 7)$

$L_2 : \dfrac{x-4}{2} = \dfrac{y-3}{1} = \dfrac{z-5}{3}$

$\therefore$ Point D on L_2 is $(2\lambda_2 + 4, \lambda_2 + 3, 3\lambda_2 + 5)$

Therefore, comparing the direction ratios of CD with the given direction ratios, we get

$\dfrac{2\lambda_2 + 3\lambda_1 + 2}{-3} = \dfrac{\lambda_2 - 2\lambda_1 - 3}{5} = \dfrac{3\lambda_2 - 4\lambda_1 - 2}{16}$

Solving, we get $\lambda_1 = -3, \lambda_2 = 2$.

$\therefore C \equiv (11, 0, -5)$ and $D \equiv (8, 5, 11)$

Hence, $|\overline{CD}|^2 = 3^2 + 5^2 + 16^2 = 290$

17. (1) Any point on the line L is $(2\lambda - 1, 3\lambda - 1, 6\lambda - 3)$

Putting this point in the second line, we get

$\dfrac{2\lambda - 1 + 1}{2} = \dfrac{3\lambda - 1 + 1}{3} = \dfrac{6\lambda - 3 - 9}{0}$

Solving we get $\lambda = 2$

So, the point of intersection of lines is $P(3, 5, 9)$.

Again, consider any point on L as $(2\mu - 1, 3\mu - 1, 6\mu - 3)$.

It's distance from point P is 7.

$\therefore (2\mu - 4)^2 + (3\mu - 6)^2 + (6\mu - 12)^2 = 49$

$\Rightarrow \mu^2 - 4\mu + 3 = 0$

$\Rightarrow (\mu - 1)(\mu - 3) = 0$

$\Rightarrow \mu = 1, 3$

So, points on L are $A(1, 2, 3)$ and $B(5, 8, 15)$

The sum of all coordinates of A and B is 34.

18. (4) We have, $4\ell + m - n = 0$ ...(i)

$2mn + 10n\ell + 3\ell m = 0$ ...(ii)

Also, $\ell^2 + m^2 + n^2 = 1$ ...(iii)

From (i), putting $n = 4\ell + m$ in (ii), we get

$2m(4\ell + m) + 10\ell(4\ell + m) + 3\ell m = 0$

$\Rightarrow 40\ell^2 + 21\ell m + 2m^2 = 0$

$\Rightarrow (8\ell + m)(5\ell + 2m) = 0$

If $m = -8\ell$, then $n = -8\ell$ (from (i))

If $m = \dfrac{-5}{2}\ell$, then $n = \dfrac{3\ell}{2}$ (from (i))

So, the direction ratio of lines are $\ell, -8\ell, -4\ell$ and $\ell, \dfrac{-5\ell}{2}, \dfrac{3\ell}{2}$

Hence, the cosine of angle between the lines is given by,

$$\cos\theta = \frac{|\ell^2 + 20\ell^2 - 6\ell^2|}{\sqrt{\ell^2 + 64\ell^2 + 16\ell^2}\sqrt{\ell^2 + \frac{25\ell^2}{4} + \frac{9\ell^2}{4}}} = \frac{10}{3\sqrt{38}}$$

19. (1) Solving lines L_1 and L_2, we get

$2\lambda + 1 = 5\mu + 4, 3\lambda + 2 = 2\mu + 1, 4\lambda + 3 = \mu$

$\Rightarrow \lambda = \mu = -1$

$\therefore R(-1, -1, -1)$

Now, $P \equiv (2p + 1, 3p + 2, 4p + 3)$

$\therefore PR^2 = 29$

$\Rightarrow (2p + 2)^2 + (3p + 3)^2 + (4p + 4)^2 = 29$

$\Rightarrow p = 0$ or $\lambda = -2$ (not possible as P lies in first octant)

$\Rightarrow P \equiv (1, 2, 3)$

Let $Q \equiv (5q + 4, 2q + 1, q)$

$\therefore PQ^2 = \dfrac{47}{3}$

$\Rightarrow (5q + 3)^2 + (2q - 1)^2 + (q - 3)^2 = \dfrac{47}{3}$

$\Rightarrow \mu = -\dfrac{1}{3}$ (as Q lies in the first octant)

$\Rightarrow Q = \left(\dfrac{7}{3}, \dfrac{1}{3}, -\dfrac{1}{3}\right)$

$(QR)^2 = \left(\dfrac{7}{3} + 1\right)^2 + \left(\dfrac{1}{3} + 1\right)^2 + \left(-\dfrac{1}{3} + 1\right)^2 = \dfrac{120}{9}$

$\Rightarrow 27 \times (QR)^2 = 360$

20. (2)

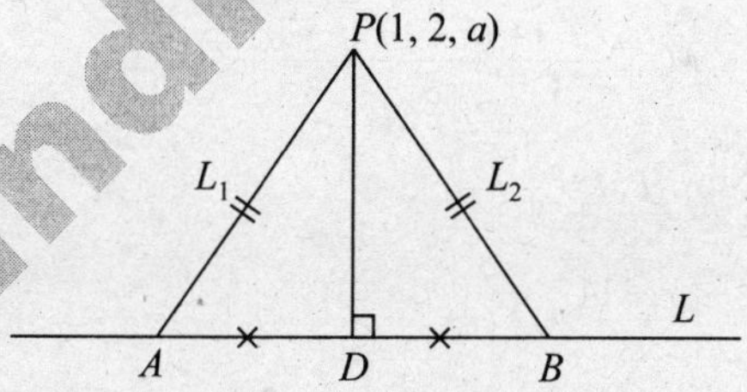

Let point A on line L_1 is $(3\lambda + 1, 4\lambda + 2, b\lambda + a)$.

It lies on L

$\therefore \dfrac{3\lambda}{1} = \dfrac{4\lambda + 2}{2} = \dfrac{b\lambda + a - 1}{1}$

$\Rightarrow \lambda = 1$ and $a + b - 1 = 3$

$\Rightarrow A(4, 6, 4)$ and $a + b = 4$ ...(i)

Let point B on line L_2 is $(\mu + 1, 4\mu + 2, c\mu + a)$

It lies on L

$\therefore \dfrac{\mu}{1} = \dfrac{4\mu + 2}{2} = \dfrac{c\mu + a - 1}{1}$

$\Rightarrow \mu = -1$ and $a - c - 1 = -1$

$\Rightarrow B(0, -2, 0)$ and $a = c$ ...(ii)

Also $PA = PB$

$\Rightarrow 9 + 16 + (a - 4)^2 = 1 + 16 + a^2$

$\Rightarrow a = 3$

$\Rightarrow b = 1$ and $c = 3$

Hence, $a + b + c = 7$

21. (4)

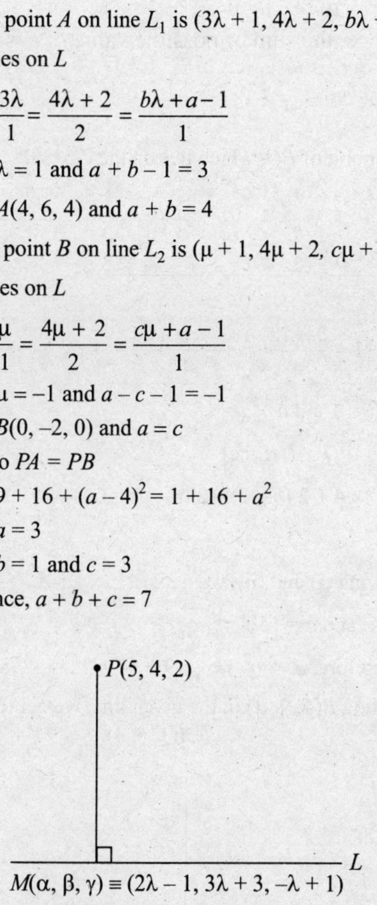

From the figure, $PM \perp L$ (line)

$\therefore 2(2\lambda - 6) + 3(3\lambda - 1) - 1(-\lambda - 1) = 0$

$\Rightarrow \lambda = 1$

Therefore, $(\alpha, \beta, \gamma) \equiv (1, 6, 0)$

Hence the required projection $= \dfrac{6\alpha + 2\beta + 3\gamma}{\sqrt{6^2 + 2^2 + 3^2}} = \dfrac{18}{7}$

22. (1) Equation line L is : $\dfrac{x+3}{1} = \dfrac{y-5}{1} = \dfrac{z-2}{1} = \lambda$

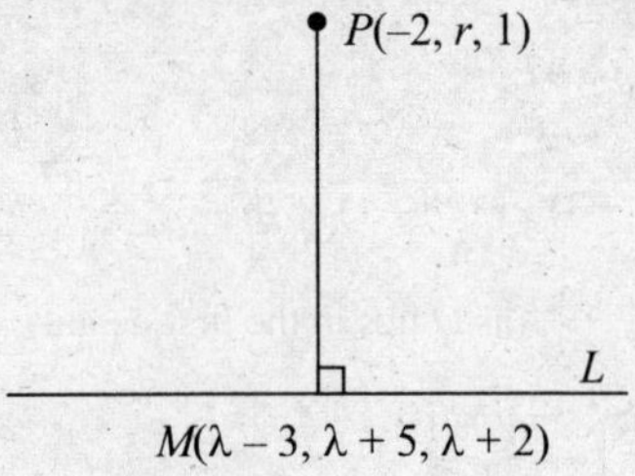

Now, $\overrightarrow{PM} \perp L$ (Line)

$\therefore (\lambda - 1)1 + (\lambda + 5 - r)1 + (\lambda + 1)1 = 0$

$\Rightarrow \lambda = \dfrac{r-5}{3}$

$\therefore M \equiv \left(\dfrac{r-14}{3}, \dfrac{r+10}{3}, \dfrac{r+1}{3}\right)$

Now, $PM = \sqrt{\dfrac{14}{3}}$

$\Rightarrow \left(\dfrac{r-14}{3} + 2\right)^2 + \left(\dfrac{r+10}{3} - r\right)^2 + \left(\dfrac{r+1}{3} - 1\right)^2 = \dfrac{14}{3}$

$\Rightarrow r^2 - 10r + 21 = 0$

Hence, the sum of possible value of r is 10.

23. (3) Point $M\left(3, \dfrac{b}{2} + 1, \dfrac{c+a}{2}\right)$ is the midpoint of PQ, which lies on the line.

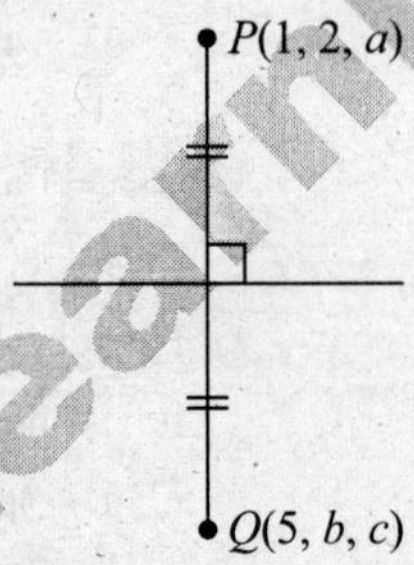

$\therefore \dfrac{3-6}{3} = \dfrac{\frac{b}{2} + 1 - 7}{2} = \dfrac{\frac{c+a}{2} - 7}{-2}$

$\Rightarrow -1 = \dfrac{b-12}{4} = \dfrac{c+a-14}{-4}$

$\Rightarrow b = 8$ and $c + a = 18$ ...(i)

Also, $PQ \perp L$ (Line)

$\Rightarrow 3 \times 4 + 2(b-2) - 2(c-a) = 0$

$\Rightarrow c - a = 12$ ...(ii)

Solving (i) and (ii), we get

$a = 3$ and $c = 15$

Therefore, $a^2 + b^2 + c^2 = 9 + 64 + 225 = 298$

24 (2) Putting $B(4, 9, \alpha)$ in the given line, we get $\alpha = 14$

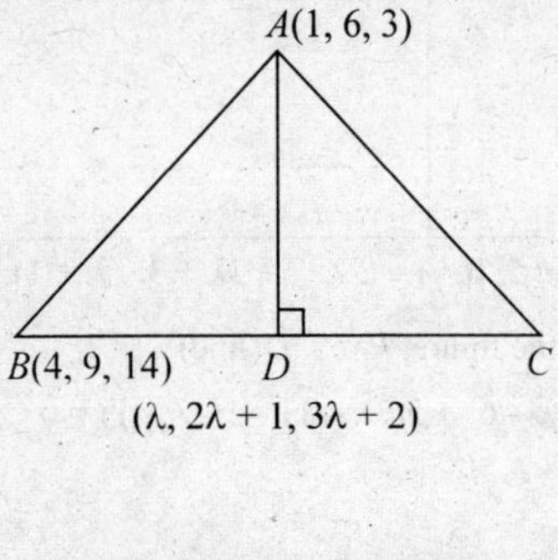

$\overrightarrow{AD} \cdot (\hat{i} + 2\hat{j} + 3\hat{k}) = 0$

$\Rightarrow \lambda - 1 + 2(2\lambda - 5) + 3(3\lambda - 1) = 0$

$\Rightarrow \lambda = 1$

$\Rightarrow D = (1, 3, 5)$

$\Rightarrow AD = \sqrt{3^2 + 2^2} = \sqrt{13}$

So, $Ar(\Delta ABC) = \dfrac{1}{2} \times \sqrt{13} \times 10 = 5\sqrt{13}$

25. (3)

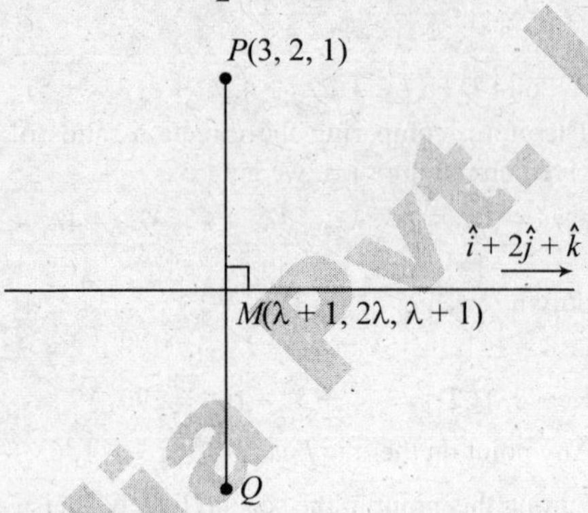

$PM \perp$ Line.

$\therefore 1(\lambda - 2) + 2(2\lambda - 2) + 1(\lambda) = 0$

$\Rightarrow \lambda = 1$

$\therefore M \equiv (2, 2, 2) \Rightarrow Q \equiv (1, 2, 3)$

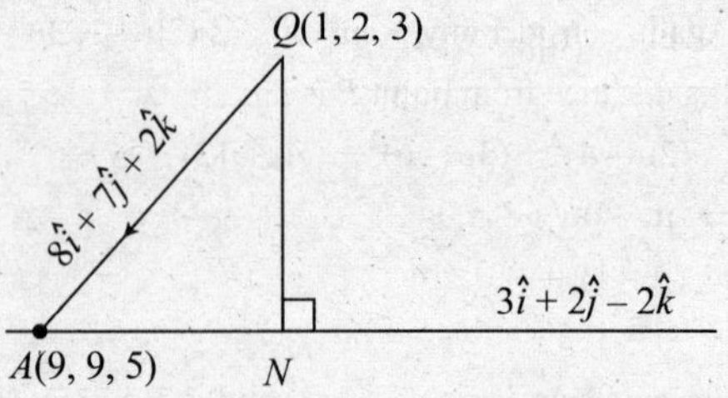

$AQ = \sqrt{64 + 49 + 4} = \sqrt{117}$

AN = projection of $\overrightarrow{AQ}$ on $3\hat{i} + 2\hat{j} - 2\hat{k}$

$\therefore AN = \dfrac{|24 + 14 - 4|}{\sqrt{9 + 4 + 4}} = \dfrac{34}{\sqrt{17}} = 2\sqrt{17}$

Hence, $QN = \sqrt{117 - 68} = \sqrt{49} = 7$

26. (3) Let $P(2\lambda + 1, -3\lambda - 1, \lambda)$

Distance of this point from the point $(1, -1, 0)$ is $4\sqrt{14}$.

$\therefore 4\lambda^2 + 9\lambda^2 + \lambda^2 = 16 \times 14$

$\Rightarrow \lambda = -4$ (as P is nearer to origin).

$\therefore P \equiv (-7, 11, -4)$

$\therefore$ Therefore, the shortest distance

$$= \dfrac{\begin{vmatrix} -5-(-7) & 10-11 & 3-(-4) \\ 1 & 2 & 3 \\ 2 & 1 & 1 \end{vmatrix}}{\left|\begin{vmatrix} \hat{i} & \hat{j} & \hat{k} \\ 1 & 2 & 3 \\ 2 & 1 & 1 \end{vmatrix}\right|}$$

$$= \dfrac{28}{\sqrt{1 + 25 + 9}} = \dfrac{4\sqrt{7}}{\sqrt{5}}$$

27. (3) The shortest distance between the lines

$$\frac{\left|\begin{vmatrix} -1-0 & 2-1 & 4-1 \\ \alpha & -1 & -\alpha \\ \alpha & 2 & 2\alpha \end{vmatrix}\right|}{\left|\begin{vmatrix} i & j & k \\ \alpha & -1 & -\alpha \\ \alpha & 2 & 2\alpha \end{vmatrix}\right|} = \sqrt{2} \text{ (given)}$$

$$\frac{|-1(-2\alpha+2\alpha)-1(2\alpha^2+\alpha^2)+3(2\alpha+\alpha)|}{|(-2\alpha+2\alpha)\hat{i}-(2\alpha^2+\alpha^2)\hat{j}+(2\alpha+\alpha)\hat{k}|} = \sqrt{2}$$

$$\frac{|-\alpha+3|}{\sqrt{\alpha^2+1}} = \sqrt{2}$$

$\Rightarrow \alpha^2 + 9 - 6\alpha = 2\alpha^2 + 2 \Rightarrow \alpha^2 + 6\alpha - 7 = 0$

Therefore, the required sum is $= -7 + 1 = -6$

Numerical Value Type

28. (6) Line perpendicular to the given two lines is parallel to the vector

$$\begin{vmatrix} \hat{i} & \hat{j} & \hat{k} \\ 4 & 1 & 1 \\ 1 & 1 & 0 \end{vmatrix} = -\hat{i}+\hat{j}+3\hat{k}$$

So, the equation of line L is $\frac{x-1}{-1} = \frac{y-1}{1} = \frac{z-1}{3}$

Line L intersects yz-plane at $Q(0, 2, 4)$

Another line passing through the point $S(1, 0, -1)$ and parallel to L is

$$\frac{x-1}{-1} = \frac{y}{1} = \frac{z+1}{4}$$

It intersects the yz-plane at R(0, 1, 2)

Area of parallelogram with adjacent vectors $\overrightarrow{PQ}$ and $\overrightarrow{PS}$

$\overrightarrow{PQ} = -\hat{i}+\hat{j}+3\hat{k}$

$\overrightarrow{PS} = -\hat{j}-2\hat{k}$

Area of parallelogram

$$\overrightarrow{PQ} \times \overrightarrow{PS} = \begin{vmatrix} \hat{i} & \hat{j} & \hat{k} \\ -1 & 1 & 3 \\ 0 & -1 & -2 \end{vmatrix} = \hat{i}-2\hat{j}+\hat{k}$$

So, Area of parallelogram $= \sqrt{1^2+(-2)^2+1^2} = \sqrt{6}$

29. (170)

The equation of the line through the point P and having direction ratios 3, –1, 0 is

$$\frac{x-43}{3} = \frac{y-\alpha}{-1} = \frac{z-\beta}{0}$$

Let the point $Q(43 + 3\lambda, \alpha - \lambda, \beta)$ on this line lies on the second line.

Putting this point in the given line, we get

$$\frac{43+3\lambda-4}{2} = \frac{\alpha-\lambda}{0} = \frac{\beta+1}{3}$$

$\lambda = \alpha$ and $2\beta + 2 = 117 + 9\lambda$

$\Rightarrow \lambda = \alpha$ and $2\beta = 115 + 9\lambda$

Now $PQ^2 = 1690$

$9\lambda^2 + \lambda^2 = 1690$ or $\lambda = -13$, as $\beta < 0$

$\alpha = -13$ and $\beta = -1$

So, $\alpha^2 + \beta^2 = 170$

30. (3)

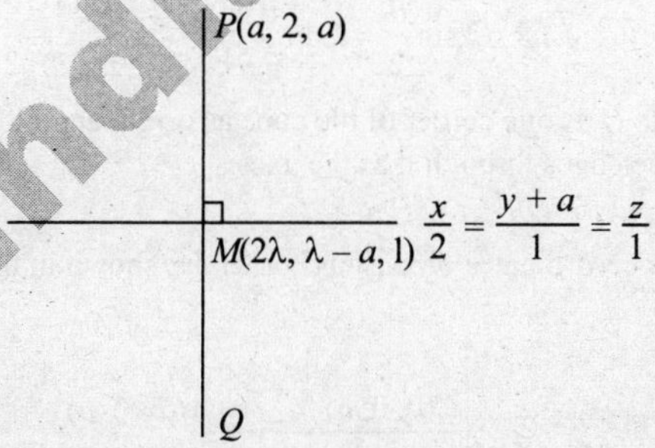

$PM \perp (2\hat{i}+\hat{j}+\hat{k})$

$\therefore 2(2\lambda - a) + 1(\lambda - a - 2) + 1(\lambda - a) = 0$

$\Rightarrow (2\hat{i}+\hat{j}+\hat{k}) = M(2\lambda, \lambda - a, \lambda)$

$$\Rightarrow M \equiv \left(\frac{4a+2}{3}, \frac{1-a}{3}, \frac{2a+1}{3}\right)$$

$$\Rightarrow Q \equiv \left(\frac{5a+4}{3}, \frac{-4-2a}{3}, \frac{a+2}{3}\right)$$

Similarly, the image of Q in the second line is

$$\left(\frac{32b-25a-24}{15}, \frac{40a-14b+18}{15}, \frac{2b-a}{3}\right) \equiv (a, 2, a)$$

Solving, we get $a = 1, b = 2$

$\therefore a + b = 1 + 2 = 3$

JEE (MAIN) MOCK TEST

1. (2) Since F is the middle point of BD, therefore

$$\vec{AB}+\vec{AD}=2\vec{AE} \quad \ldots\text{(i)}$$

Similarly $\vec{CB}+\vec{CD}=2\vec{CE}$ ...(ii)

Adding (i) and (ii), we get

$$\vec{AB}+\vec{AD}+\vec{CB}+\vec{CD} = 2(\vec{AF}+\vec{CF}) = -2(\vec{FA}+\vec{FC})$$

$$= -2(\,2(\vec{FE}) \qquad [\because E \text{ is the midpoint of } AC]$$

$$= 4\vec{EF}$$

2. (2) Any point on the given line is $(k+1, 2k+1, k+1)$.

but $x = 0 \Rightarrow A(0, -1, 0)$

and $y = 0 \Rightarrow B(1/2, 0, 1/2)$

3. (1) $|\vec{a}-\vec{b}| = \sqrt{1^2+1^2-2.1^2\cos\theta} = \sqrt{2(1-\cos\theta)}$

$$= \sqrt{2}\times\sqrt{2}\sin\frac{\theta}{2} = 2\sin\frac{\theta}{2} \Rightarrow \sin\frac{\theta}{2} = \frac{|\vec{a}-\vec{b}|}{2}$$

4. (2) Take O as one corner of the cube as origin and OA, OB, OC the three edges through it as the axes.

Let, $OA = OB = OC = a$

The co-ordinates of various corner are shown in the figure.

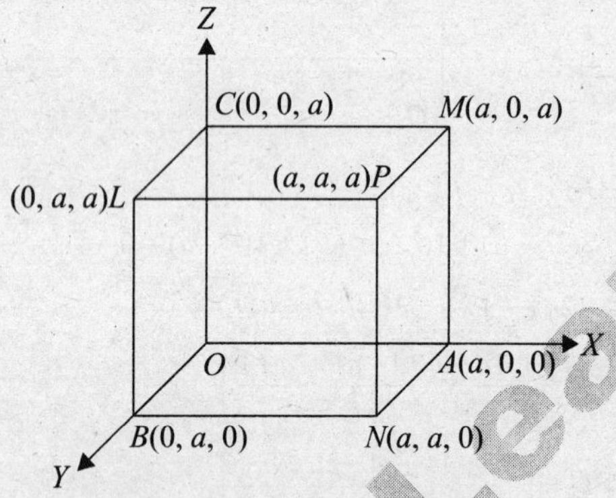

The four diagonals are AL, BM, CN and OP

The DRs of diagonal AL are $0-a, a-0, a-0$, i.e., $-a, a, a$

The DRs of diagonal OP are $a-0, a-0, a-0$ i.e., a, a, a

If θ is the angle between them, then

$$\cos\theta = \frac{|(-a)(a)+(a)(a)+(a)(a)|}{\sqrt{a^2+a^2+a^2}\sqrt{a^2+a^2+a^2}}$$

$$\cos\theta = \frac{a^2}{3a^2} = \frac{1}{3} \Rightarrow \theta = \cos^{-1}\left(\frac{1}{3}\right)$$

Hence, the correct option is (2).

5. (4) $\vec{BA}\times\vec{F} = (2\vec{i}+4\vec{j}+8\vec{k})\times(2\vec{i}+2\vec{j}+5\vec{k})$

$$= \begin{vmatrix} \vec{i} & \vec{j} & \vec{k} \\ 2 & 4 & 8 \\ 2 & 2 & 5 \end{vmatrix} = 4\vec{i}+6\vec{j}-4\vec{k}$$

$$= 4\vec{i}+6\vec{j}+2\lambda\vec{k} \text{ (given)}$$

$\therefore \quad \lambda = -2$

6. (3) $3l + m + 5n = 0$ ...(i)

$6mn - 2nl + 5ml = 0$ ...(ii)

Substituting the value of n from Equation (i) in Equation (ii), then

$$6l^2 + 9lm - 6m^2 = 0$$

$$\Rightarrow 6\left(\frac{l}{m}\right)^2 + 9\left(\frac{l}{m}\right) - 6 = 0$$

$$\therefore \frac{l_1}{m_1} = \frac{1}{2} \text{ and } \frac{l_2}{m_2} = -2$$

From Equation (i), we get,

$$\frac{l_1}{n_1} = -1 \text{ and } \frac{l_2}{n_2} = -2$$

$$\therefore \frac{l_1}{1} = \frac{m_1}{2} = \frac{n_1}{-1} = \frac{\sqrt{l_1^2+m_1^2+n_1^2}}{\sqrt{1+4+1}} = \frac{1}{\sqrt{6}}$$

$$\text{and } \frac{l_2}{2} = \frac{m_2}{-1} = \frac{n_2}{-1} = \frac{\sqrt{l_2^2+m_2^2+n_2^2}}{\sqrt{4+1+1}} = \frac{1}{\sqrt{6}}$$

If θ be the angle between the lines

$$\therefore \cos\theta = \left(\frac{1}{\sqrt{6}}\right)\left(\frac{2}{\sqrt{6}}\right)+\left(\frac{2}{\sqrt{6}}\right)\left(-\frac{1}{\sqrt{6}}\right)+\left(-\frac{1}{\sqrt{6}}\right)\left(-\frac{1}{\sqrt{6}}\right)$$

$$= \frac{1}{6}$$

$$\therefore \theta = \cos^{-1}\left(\frac{1}{6}\right)$$

7. (4) $[\vec{a}-2\vec{b} \;\; \vec{b}-3\vec{c} \;\; \vec{c}-4\vec{a}]$

$$= (\vec{a}-2\vec{b})\times(\vec{b}-3\vec{c}).(\vec{c}-4\vec{a})$$

$$= (\vec{a}\times\vec{b} - 3\vec{a}\times\vec{c} + 6\vec{b}\times\vec{c}).(\vec{c}-4\vec{a})$$

$$= \vec{a}\times\vec{b}.\vec{c} - 24\vec{b}\times\vec{c}.\vec{a}$$

$$= -23(\vec{a}\times\vec{b}).\vec{c}$$

$$= -23.|\vec{a}||\vec{b}|\sin 90^\circ \;\hat{n}.\vec{c}$$

$$= -23(1)(3)|\vec{c}|$$

$$= -69\times 5 = -345$$

8. (4) Any line through the point (1, 1, 1) is

$$\frac{x-1}{a} = \frac{y-1}{b} = \frac{z-1}{c} \quad \ldots(1)$$

This line intersects the line $\frac{x-1}{2} = \frac{y-2}{3} = \frac{z-3}{4}$

If $a:b:c \neq 2:3:4$ and $\begin{vmatrix} 1-1 & 2-1 & 3-1 \\ a & b & c \\ 2 & 3 & 4 \end{vmatrix} = 0$

i.e. if $0 - 1(4a-2c) + 2(3a-2b) = 0$

i.e. if $a - 2b + c = 0$ ...(2)

Again the line (1) intersects the line

$$\frac{x-(-2)}{1}=\frac{y-3}{2}=\frac{z-(-1)}{4}$$

If $a : b : c \neq 1 : 2 : 4$ and

$$\begin{vmatrix} -2-1 & 3-1 & -1-1 \\ a & b & c \\ 1 & 2 & 4 \end{vmatrix}=0$$

i.e. if $-3(4b-2c)-2(4a-c)-2(2a-b)=0$

i.e. if $6a+5b-4c=0$ ...(3)

From (2) and (3) by cross multiplication, we have

$$\frac{a}{8-5}=\frac{b}{6+4}=\frac{c}{5+12}$$

$$\Rightarrow \frac{a}{3}=\frac{b}{10}=\frac{c}{17}$$

So, the equation of required line is $\frac{x-1}{3}=\frac{y-1}{10}=\frac{z-1}{17}$

Hence, the correct option is (4).

9. (1) $[(\vec{a}\times\hat{i})\times\hat{j}]^2=(\vec{a}\cdot\hat{j})\,\hat{i}$

$[(\vec{a}\times\hat{j})\times\hat{k}]^2=(\vec{a}\cdot\hat{k})\,\hat{j}$

and $[(\vec{a}\times\hat{k})\times\hat{i}]^2=(\vec{a}\cdot\hat{i})\,\hat{k}$

$\Rightarrow [(\vec{a}\times\hat{i})\times\hat{j}]^2+[(\vec{a}\times\hat{j})\times\hat{k}]^2+[(\vec{a}\times\hat{k})\times\hat{i}]^2$

$=a_2^2+a_3^2+a_1^2=1+4+1=6$

10. (2) The lines $\vec{r}=\vec{a}+\lambda(\vec{b}\times\vec{c})$ and $\vec{r}=\vec{b}+\mu(\vec{c}\times\vec{a})$ pass through points $\vec{a}$ and $\vec{b}$ respectively and are parallel to the vectors $\vec{b}\times\vec{c}$ and $\vec{c}\times\vec{a}$ respectively. Therefore, they intersect if $\vec{a}-\vec{b}, \vec{b}\times\vec{c}$ and $\vec{c}\times\vec{a}$ are coplanar and so

$(\vec{a}-\vec{b}).\{(\vec{b}\times\vec{c})\times(\vec{c}\times\vec{a})\}=0$

$\Rightarrow (\vec{a}-\vec{b}).([\vec{b}\vec{c}\vec{a}]\vec{c}-[\vec{b}\vec{c}\vec{c}]\vec{a})=0$

$\Rightarrow ((\vec{a}-\vec{b}).\vec{c})[\vec{b}\vec{c}\vec{a}]=0$

$\Rightarrow \vec{a}.\vec{c}-\vec{b}.\vec{c}=0$

$\Rightarrow \vec{a}.\vec{c}=\vec{b}.\vec{c}$

11. (3) Vector are coplanar if $\begin{vmatrix} 2 & -1 & 1 \\ 1 & 2 & -3 \\ 3 & a & 5 \end{vmatrix}=0$

$\Rightarrow 2(10+3a)+(5+9)+a-6=0$

$\Rightarrow 7a+28=0$

$\Rightarrow a=-4$

12. (3) The straight line $\vec{r}=\vec{a}+\lambda\vec{b}$ meets the plane $\vec{r}.\vec{n}=0$ in P for which λ is given by

$(\vec{a}+\lambda\vec{b}).\vec{n}=0$

$$\Rightarrow \lambda=-\frac{\vec{a}.\vec{n}}{\vec{b}.\vec{n}}$$

Thus, the position vector of P is

$$\vec{r}=\vec{a}-\frac{\vec{a}.\vec{n}}{\vec{b}.\vec{n}}\vec{b}$$

[Putting the value of λ in $\vec{r}=\vec{a}+\lambda\vec{b}$]

Hence, the correct option is (3).

13. (1) It is given that the vectors $\hat{i}-2x\hat{j}-3y\hat{k}$ and $\hat{i}+3x\hat{j}+2y\hat{k}$ are orthogonal. Therefore,

$(\hat{i}-2x\hat{j}-3y\hat{k})\cdot(\hat{i}+3x\hat{j}+2y\hat{k})=0$

$\Rightarrow 1-6x^2-6y^2=0$

$\Rightarrow 6x^2+6y^2=1$, which is circle.

14. (4) Let AD be perpendicular and D be foot of perpendicular which divide BC in ratio $\lambda : 1$, then

$$D\left(\frac{10\lambda-9}{\lambda+1},\frac{4}{\lambda+1},\frac{-\lambda+5}{\lambda+1}\right) \quad ...(i)$$

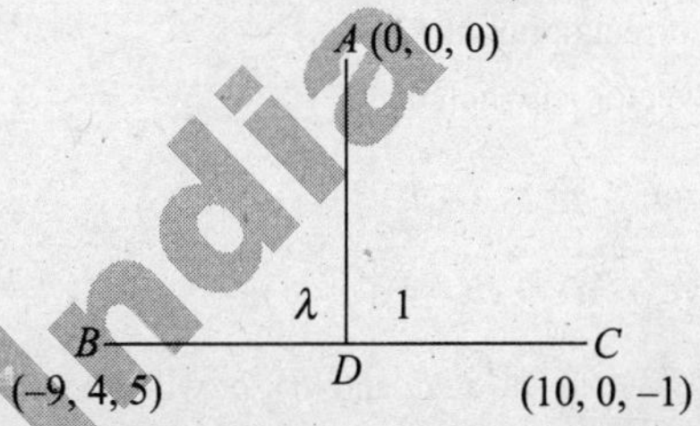

The direction ratio of AD are $\frac{10\lambda-9}{\lambda+1},\frac{4}{\lambda+1},\frac{-\lambda+5}{\lambda+1}$ and direction ratio of BC are 19, – 4 and – 6.

Since $AD \perp BC$

$$\Rightarrow 19\left(\frac{10\lambda-9}{\lambda+1}\right)-4\left(\frac{4}{\lambda+1}\right)-6\left(\frac{-\lambda+5}{\lambda+1}\right)=0$$

$$\Rightarrow \lambda=\frac{31}{28}.$$

Hence on putting the value of λ in (i), we get required foot of the perpendicular i.e., $\left(\frac{58}{59},\frac{112}{59},\frac{109}{59}\right)$.

Trick: The line passing through these points is

$$\frac{x+9}{19}=\frac{y-4}{-4}=\frac{z-5}{-6}.$$

Now co-ordinates of the foot lie on this line, so they must satisfy the given line. But here no point satisfies the line, hence answer is (4).

15. (2) $\Delta=\frac{1}{2}|\vec{a}\times\vec{b}|$

But $\vec{a}\times\vec{b}=\begin{vmatrix} \hat{i} & \hat{j} & \hat{k} \\ 3 & 1 & -2 \\ 1 & -3 & 4 \end{vmatrix}=-2\hat{i}-14\hat{j}-10\hat{k}$

Hence $\Delta=\frac{1}{2}|\vec{a}\times\vec{b}|=\frac{1}{2}\sqrt{4+196+100}=5\sqrt{3}$

16. (2) $\Delta\, OPM$ and $\Delta\, OAC$ are similar

$$\frac{OP}{OA} = \frac{OM}{OC}$$

$$\frac{\sqrt{x^2+y^2+z^2}}{\sqrt{2}} = \frac{x}{1}$$

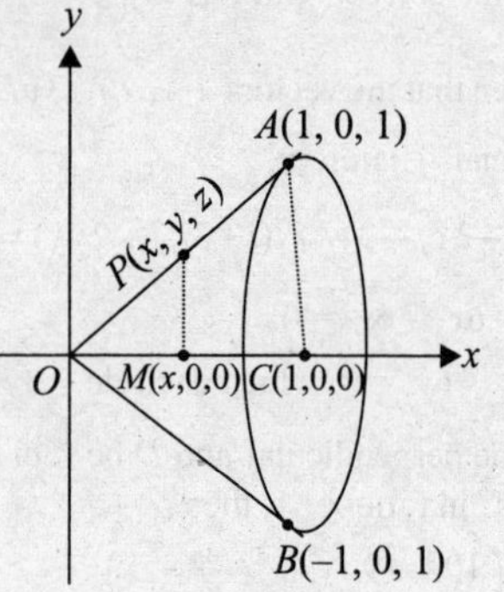

$\Rightarrow\ x^2+y^2+z^2 = 2x^2 \ \Rightarrow\ x^2-y^2-z^2=0$

Hence, the correct option is (2).

17. (3) Let required vector $\vec{r}$ is such that

$$\vec{r} = x_1\vec{a} + x_2\vec{b} + x_3\,\vec{a}\times\vec{b}$$

We must have $\vec{r}.\vec{a} = \vec{r}.\vec{b} = \vec{r}.(\vec{a}\times\vec{b})$

(as $\vec{r}, \vec{a}, \vec{b}$ and $\vec{a}\times\vec{b}$ will are unit vectors)

$\vec{r}.\vec{a} = x_1,\ \vec{r}.\vec{b} = x_2,\ \vec{r}.\vec{c} = x_3$

$\Rightarrow\ \vec{r} = \lambda(\vec{a}+\vec{b}+(\vec{a}\times\vec{b}))$

Also, $\vec{r}.\vec{r} = 1$

$\Rightarrow\ \lambda^2 = (\vec{a}+\vec{b}+\vec{a}\times\vec{b}).(\vec{a}+\vec{b}+(\vec{a}\times\vec{b})) = 1$

$\Rightarrow\ \lambda^2\,(\vec{a}.\vec{a}+\vec{a}.\vec{b}+\vec{a}.(\vec{a}\times\vec{b})+\vec{b}.\vec{a}$

$+\ \vec{b}.\vec{b}+\vec{b}.(\vec{a}\times\vec{b})+(\vec{a}\times\vec{b}).\vec{a}$

$+\ (\vec{a}\times\vec{b}).\vec{b}+(\vec{a}\times\vec{b}).(\vec{a}\times\vec{b}) = 1$

$\Rightarrow\ \lambda^3 = \frac{1}{3}\ \Rightarrow\ \lambda = \pm\frac{1}{\sqrt{3}}$

$\Rightarrow\ \vec{r} = \pm\frac{1}{\sqrt{3}}\,(\vec{a}+\vec{b}+\vec{a}\times\vec{b})$

18. (2) Direction ratios of normal to the plane are (2, 3, – 6)

$\therefore$ Direction cosines are:

$$\left(\frac{2}{\sqrt{49}}, \frac{3}{\sqrt{49}}, -\frac{6}{\sqrt{49}}\right) \text{ or } \left(\frac{2}{7}, \frac{3}{7}, \frac{-6}{7}\right).$$

19. (2) $\vec{d}.(\vec{a}\times(\vec{b}\times(\vec{c}\times\vec{d})))$

$= \vec{d}.(\vec{a}\times((\vec{b}.\vec{d})\vec{c}-(\vec{b}.\vec{c})\vec{d}))$

$= \vec{d}.((\vec{b}.\vec{d})(\vec{a}\times\vec{c})-(\vec{b}.\vec{c})(\vec{a}\times\vec{d}))$

$= (\vec{b}.\vec{d})(\vec{d}.(\vec{a}\times\vec{c}))-(\vec{b}.\vec{c})(\vec{d}.(\vec{a}\times\vec{d}))$

$= (\vec{b}.\vec{d})[\vec{d}\ \vec{a}\ \vec{c}] = (\vec{b}.\vec{d})[\vec{a}\ \vec{c}\ \vec{d}]$

20. (1) Trick : Both lines are satisfied by (–1,–1, –1).

21. (8) Obviously one in each octant

22. (28) $\vec{a} = a_1\hat{i}+a_2\hat{j}+a_3\hat{k}$;

$\vec{b} = b_1\hat{i}+b_2\hat{j}+b_3\hat{k}$;

$\vec{c} = c_1\hat{i}+c_2\hat{j}+c_3\hat{k}$;

LHS = $[3\vec{a}+\vec{b}\ \ 3\vec{b}+\vec{c}\ \ 3\vec{c}+\vec{a}]$

$= [3\vec{a}\ \ 3\vec{b}\ \ 3\vec{c}] + [\vec{b}\ \vec{c}\ \vec{a}]$

$= 3^3[\vec{a}\ \vec{b}\ \vec{c}]+[\vec{a}\ \vec{b}\ \vec{c}] = 28\ [\vec{a}\ \vec{b}\ \vec{c}]$

23. (7) Clearly minimum value of $a^2+b^2+c^2$

$$= \left(\frac{|(3(0)+2(0)+(0)-7|}{\sqrt{(3)^2+(2)^2+(1)^2}}\right)^2 = \frac{49}{14} = \frac{7}{2} \text{ units.}$$

24. (3) $\vec{c}\times\vec{a}=\vec{b}$

$\Rightarrow\ |\vec{c}\times\vec{a}| = |\vec{b}|$

$\Rightarrow\ |\vec{c}||\vec{a}|\sin\theta = 3\ \Rightarrow\ |\vec{c}| = \frac{3}{2\sin\theta}$

$|\vec{c}-\vec{a}|^2 = |\vec{c}|^2+|\vec{a}|^2-2\vec{c}.\vec{a}$

$= |\vec{c}|^2+4-2|\vec{c}|.|\vec{a}|\cos\theta$

$= \frac{9}{4\sin^2\theta}+4-2.\frac{3}{2\sin\theta}.2.\cos\theta$

$= 4+\frac{9}{4}\operatorname{cosec}^2\theta-6\cot\theta$

$= \frac{9}{4}+\left(\frac{3}{2}\cot\theta-2\right)^2$

$\Rightarrow\ |\vec{c}-\vec{a}|^2 \ge \frac{9}{4} \Rightarrow |\vec{c}-\vec{a}| \ge \frac{3}{2}$

25. (5) $\frac{x-2}{3} = \frac{y+1}{2} = \frac{z-1}{-1} = \lambda$

General point on the line is $(3\lambda+2,\ 2\lambda-1,\ -\lambda+1)$.

This point is on the curve $xy = c^2, z = 0$

$\Rightarrow\ \lambda = 1$ and $(3\lambda+2)\,(2\lambda-1) = c^2$

$\Rightarrow\ c^2 = 5$

JEE (ADVANCED) MOCK TEST (PAPER 1)

1. (2) $A(-1, 1, 2)$, $B(1, 2, 3)$ and $C(t, 1, 1)$

$\Rightarrow \quad \overrightarrow{AB} = 2\hat{i} + \hat{j} + \hat{k}$

$\overrightarrow{AC} = (t+1)\hat{i} + 0\hat{j} - \hat{k}$

$$\overrightarrow{AB} \times \overrightarrow{AC} = \begin{vmatrix} \hat{i} & \hat{j} & \hat{k} \\ 2 & 1 & 1 \\ t+1 & 0 & -1 \end{vmatrix} = -\hat{i} + (t+3)\hat{j} - (t+1)\hat{k}$$

$$|\overrightarrow{AB} \times \overrightarrow{AC}| = \sqrt{1+(t+3)^2+(t+1)^2} = \sqrt{2t^2+8t+11}$$

Area of $\Delta ABC = \frac{1}{2}|\overrightarrow{AB} \times \overrightarrow{AC}|$

$$= \frac{1}{2}\sqrt{2t^2+8t+11}$$

$$= \frac{1}{2}\sqrt{2(t+2)^2+3} \geq \frac{\sqrt{3}}{2}$$

2. (3) The perpendicular distance of the origin (0, 0, 0) from the plane $x + y + z = p$ is $\left|\frac{-p}{\sqrt{1+1+1}}\right| = \frac{|p|}{\sqrt{3}}$.

If the coordinates of P are (x, y, z), then we must have

$$\left|\frac{x+y+z-p}{\sqrt{3}}\right| = \frac{|p|}{\sqrt{3}}$$

$\Rightarrow \quad |x+y+z-p| = |p|$

3. (3) Here, $\overrightarrow{OA} = 2\hat{i} - 3\hat{j} = \vec{a}$ (say), $\overrightarrow{OB} = \hat{i} + \hat{j} - \hat{k} = \vec{b}$ (say), and $\overrightarrow{OC} = 3\hat{i} - \hat{k} = c$ (say)

Hence volume is $[\vec{a}\,\vec{b}\,\vec{c}] = \vec{a}\cdot(\vec{b} \times \vec{c}) = \begin{vmatrix} 2 & -3 & 0 \\ 1 & 1 & -1 \\ 3 & 0 & -1 \end{vmatrix} = 4$

4. (1) Line L is $\frac{x-1}{1} = \frac{y-2}{2\sqrt{2}} = \frac{z}{0}$

This line L makes 45° with plane $\sqrt{2}x + y - z = 1$

Required distance (d) of PQ is perpendicular distance of plane from P.

$$\therefore \quad d = \frac{|\sqrt{2}+7-\sqrt{2}-1|}{\sqrt{2+1+1}} = 3.$$

5. (2), (3), (4)

Line is parallel to $\hat{i} - 3\hat{j}$.

Also put $\vec{r}_1 = 3\hat{i} + 3\hat{j}$ for which $\lambda = 1$

and $\vec{r}_1 = \hat{i} + 9\hat{j}$ for which $\lambda = -1$

6. (1), (3)

$\vec{a} \times \vec{b} = 4(\vec{a} \times \vec{c})$

$\Rightarrow \quad \vec{a} \times (\vec{b} - 4\vec{c}) = \vec{0}$

$\Rightarrow \quad \vec{a}$ is parallel to $\vec{b} - 4\vec{c}$

$\therefore \quad \vec{a} = t(\vec{b} - 4\vec{c})$, where t is scalar.

$\therefore \quad |\vec{a}|^2 = t^2\left\{|b|^2 + 16|c|^2 - 8(\vec{b}\cdot\vec{c})\right\}$

$\therefore \quad 1 = t^2\left\{1 + 16\cdot\frac{1}{16} - 8.1.\frac{1}{4}.1\right\}2$

$\therefore \quad 1 = t^2[1] \quad \therefore \quad t = \pm 1$

$\therefore \quad \vec{a} = \vec{b} - 4\vec{c}$ or $\vec{a} = -\vec{b} + 4\vec{c}$

$\Rightarrow \quad \vec{b} = \vec{a} + 4\vec{c}$ or $\vec{b} = 4\vec{c} - \vec{a}$

7. (1), (2), (3)

Treating the three equations as a 3×3 linear system, we have

$$D = \begin{vmatrix} a & 4 & 1 \\ 0 & 2 & 3 \\ 3 & 0 & -b \end{vmatrix} = 30 - 2ab$$

$$D_x = \begin{vmatrix} 0 & 4 & 1 \\ 1 & 2 & 3 \\ -2 & 0 & -b \end{vmatrix} = 4b - 20$$

$D_y = -ab + 6a - 3$,

$D_z = -4a + 12$

If $ab \neq 15$ the system will have unique solution.

Hence, planes will meet in a point.

If $ab = 15$,

$$D_x = \frac{60}{a} - 20, D = 6a - 18, D_z = -4a + 12$$

We know that there is a value of a for which D_x, D_y, D_z vanish simultaneously.

Which is obviously $a = 3$.

For $a \neq 3$ or $b \neq 5$ at least one of the determinants D_x, D_y, D_z become non-zero.

So, planes will have no common point.

8. (7)

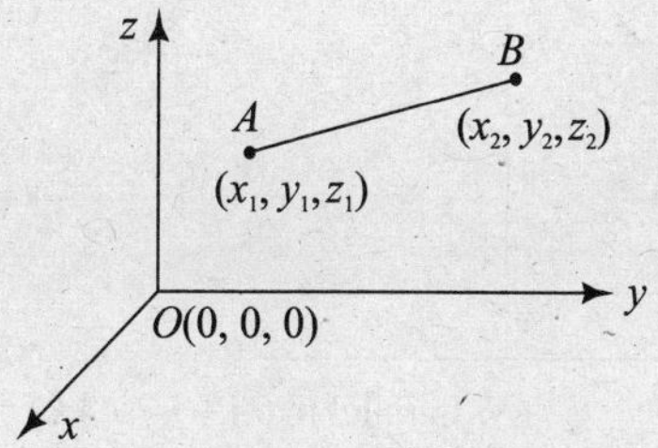

$$\therefore \quad AB = \sqrt{(x_1-x_2)^2+(y_1-y_2)^2+(z_1-z_2)^2}$$

$$= \sqrt{4+9+36} = \sqrt{49} = 7$$

9. (1) If image of point $(2, -3, 3)$ in the plane $x - 2y - z + 1 = 0$ is (a, b, c) then

We have

$$\frac{a-2}{1} = \frac{b+3}{-2} = \frac{c-3}{-1} = \frac{-2(2-2(-3)-3+1)}{(1)^2+(-2)^2+(-1)^2} = -2$$

Hence the image is (0, 1, 5)

Obviously distance of image of the point from z axis is 1.

10. (36) Let the angle between $\vec{a}$ and $\vec{b}$ be α and that between $\vec{a} \times \vec{b}$ and $\vec{c}$ be β

$\therefore \quad \left|[\vec{a}\vec{b}\vec{c}]\right| = 6$

$\Rightarrow \quad \sin\alpha\cos\beta = 1$

$\Rightarrow \quad \sin\alpha = 1, \cos\beta = 1 \quad \Rightarrow \quad \alpha = 90°, \beta = 0°$

$\Rightarrow \quad \vec{a}, \vec{b}, \vec{c}$ are mutually perpendicular

$$|(\vec{a}\times\vec{c}).\vec{b}|^2 + |(\vec{a}\times\vec{c})\times\vec{b}|^2 = |\vec{a}\times\vec{c}|^2|\vec{b}|^2$$

$$= |\vec{a}|^2|\vec{c}|^2|\vec{b}|^2 = 36$$

11. (2) Vector normal to the plane is $\vec{n} = \hat{i} - 3\hat{j} + 2\hat{k}$ and vector along the line is $\vec{v} = 2\hat{i} + \hat{j} - 3\hat{k}$

Now $\sin\theta = \dfrac{\vec{x}\cdot\vec{v}}{|\vec{x}||\vec{v}|} = \left|\dfrac{2-3-6}{\sqrt{14}\ \sqrt{14}}\right| = \left|\dfrac{7}{14}\right|$

Hence cosec $\theta = 2$

12. (5) Let $\vec{c}$ be inclined at angles θ with $\vec{a}$ and 2θ with $\vec{b}$

Now $\vec{a} + \vec{b} = \vec{c}$

$\Rightarrow \quad \vec{a}.\vec{a} + \vec{a}.\vec{b} = \vec{a}.\vec{c}$

$\Rightarrow \quad 36 + 24\cos 3\theta = 6c\cos\theta$

$\Rightarrow \quad 6 + 4\cos 3\theta = c\cos\theta$ ...(i)

and $\vec{b}.\vec{a} + \vec{b}.\vec{b} = \vec{b}.\vec{c}$

$\Rightarrow \quad 24\cos 3\theta + 16 = 4c\cos 2\theta$

$\Rightarrow \quad 6\cos 3\theta + 4 = c\cos 2\theta$ ...(ii)

Eliminating c from (i) and (ii),

$(6 + 4\cos 3\theta)\cos 2\theta = (6\cos 3\theta + 4)\cos\theta$

$\Rightarrow \quad \cos\theta = \dfrac{3}{4}$

Putting in (i), we get $c = 5$

13. (7) From the figure

$\overrightarrow{AP} = -3\vec{i} - \vec{j} + 10\vec{k}$

$|\overrightarrow{AP}| = \sqrt{110}$

$\overrightarrow{PN}$ is perpendicular to the line

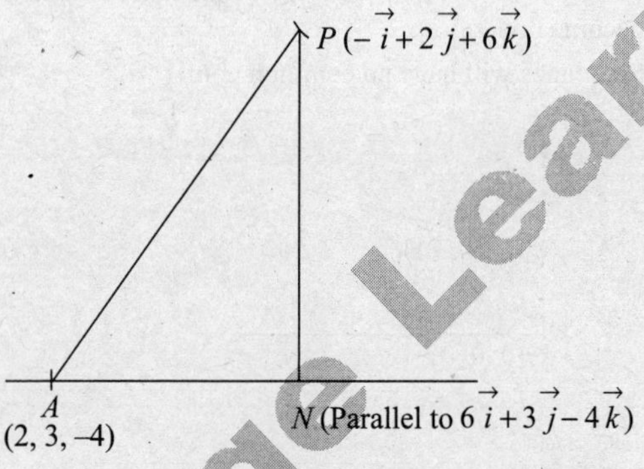

$\overrightarrow{AN}$ = projection of $\overrightarrow{AP}$ on $6\vec{i} + 3\vec{j} - 4\vec{k}$

$$= \left|\frac{\overrightarrow{AP}.(6\vec{i}+3\vec{j}-4\vec{k})}{(6\vec{i}+3\vec{j}-4\vec{k})}\right| = \left|\frac{-18-3-40}{\sqrt{61}}\right| = \sqrt{61}$$

$\therefore \quad \overrightarrow{PN}^2 = \overrightarrow{AP}^2 - \overrightarrow{AN}^2 = 110 - 61 = 49$

$\therefore \quad PN = 7$

14. (3)

(P) We know, distance between parallel planes $ax + by + cz + d_1 = 0$ and $ax + by + cz + d_2 = 0$ is,

$$\frac{|d_1 - d_2|}{\sqrt{a^2+b^2+c^2}}$$

$\therefore$ distance between $x + 2y - 2z + 1 = 0$

and $x + 2y - 2z + \dfrac{5}{2} = 0$ is: $\dfrac{|5/2-1|}{\sqrt{1+4+4}} = \dfrac{1}{2}$

(Q) Radius of the sphere = 5

Given plane is $x + y - z = 4\sqrt{3}$

Length of the $\perp$ from the centre (0, 0, 0) of the sphere on the plane $= \dfrac{4\sqrt{3}}{\sqrt{1+1+1}} = 4$

Hence radius of the circular section $= \sqrt{25-16} = \sqrt{9} = 3$

(R) Line is along the vector $\vec{a} = -3\hat{i} + 2\hat{j} + \hat{k}$ and plane is normal to the vector $\vec{b} = \hat{i} + \hat{j} + \hat{k}$

Since $\vec{a}\cdot\vec{b} = 0$, line is parallel to the plane.

Hence distance between line and plane is distance of point (–1, 3, 2) from the plane,

$$= \frac{|-1+3+2|}{\sqrt{1+1+1}} = \frac{4}{\sqrt{3}}$$

(S) Any point on the line $\dfrac{x-2}{3} = \dfrac{y+1}{4} = \dfrac{z-2}{12} = \lambda$ is

$(3\lambda + 2, 4\lambda - 1, 12\lambda + 2)$

This lies on $x - y + z = 5$

if $\quad 3\lambda + 2 - 4\lambda + 1 + 12\lambda + 2 = 5$

$\Rightarrow \quad \lambda = 0$

$\therefore \quad$ point is (2, –1, 2)

Its distance from (–1, –5, – 10) is

$$= \sqrt{(2+1)^2 + (-1+5)^2 + (2+10)^2}$$

$$= \sqrt{9+16+144} = 13$$

15. (2)

(P) If $\vec{a}, \vec{b}, \vec{c}$ are mutually perpendicular

$$[\vec{a}\times\vec{b}\ \ \vec{b}\times\vec{c}\ \ \vec{c}\times\vec{a}] = \left[\vec{a}\ \vec{b}\ \vec{c}\right]^2$$

$$= (|\vec{a}||\vec{b}||\vec{c}|)^2 = 16$$

(Q) Given $\vec{a}, \vec{b}$ are two unit vectors, i.e., $|\vec{a}| = |\vec{b}| = 1$ and angle between them is $\pi/3$

$$\sin\theta = \frac{|\vec{a}\times\vec{b}|}{|\vec{a}||\vec{b}|} \Rightarrow \sin\frac{\pi}{3} = |\vec{a}\times\vec{b}|$$

$$\frac{\sqrt{3}}{2} = |\vec{a}\times\vec{b}|$$

Now $\left[\vec{a}\ \ \vec{b} + \vec{a}\times\vec{b}\ \ \vec{b}\right] = \left[\vec{a}\ \vec{b}\ \vec{b}\right] + \left[\vec{a}\ \ \vec{a}\times\vec{b}\ \ \vec{b}\right]$

$$= 0 + \left[\vec{a}\ \ \vec{a}\times\vec{b}\ \ \vec{b}\right] = \left(\vec{a}\times\vec{b}\right).\left(\vec{b}\times\vec{a}\right)$$

$$= -\left(\vec{a}\times\vec{b}\right).\left(\vec{a}\times\vec{b}\right)$$

$$= -|\vec{a}\times\vec{b}|^2 = -\frac{3}{4}$$

(R) If $\vec{b}, \vec{c}$ are orthogonal $\vec{b}.\vec{c} = 0$ and also given $\vec{b} \times \vec{c} = \vec{a}$

Now $\left[\vec{a}+\vec{b}+\vec{c} \quad \vec{a}+\vec{b} \quad \vec{b}+\vec{c}\right]$

$= \left[\vec{a} \quad \vec{a}+\vec{b} \quad \vec{b}+\vec{c}\right] + \left[\vec{b}+\vec{c} \quad \vec{a}+\vec{b} \quad \vec{b}+\vec{c}\right]$

$= \left[\vec{a}\,\vec{b}\,\vec{c}\right] = \vec{a}.\left(\vec{b}\times\vec{c}\right)$

$= \vec{a}.\vec{a} = |\vec{a}|^2 = 1$ [$\because \vec{a}$ is unit vector]

(S) $\left[\vec{x}\,\vec{y}\,\vec{a}\right] = 0$

$\Rightarrow \vec{x}, \vec{y}, \vec{a}$ coplanar ...(i)

$\left[\vec{x}\,\vec{y}\,\vec{b}\right] = 0$

$\Rightarrow \vec{x}, \vec{y}, \vec{b}$ are coplanar ...(ii)

also $\left[\vec{a}\,\vec{b}\,\vec{c}\right] = 0$

$\therefore \vec{a}, \vec{b}, \vec{c}$ are coplanar ...(iii)

from (i) (ii) and (iii)

$\vec{x}\,\vec{y}\,\vec{c}$ are coplanar

$\therefore \left[\vec{x}\,\vec{y}\,\vec{c}\right] = 0$

16. (4)

(P) $\begin{vmatrix} 2 & 1 & 1 \\ 1 & -1 & 1 \\ \alpha & -1 & 3 \end{vmatrix} = 0 \Rightarrow \alpha = 4$

(Q) Put $z = 0$ in P_1 and P_2

$2x + y = 1$ and $x - y = 2$

$\therefore P \equiv (1, -1, 0)$

Put $x = 0$ in P_1 and P_2

$y + z = 1$ and $-y + z = 2$

$\therefore Q \equiv (0, -1/2, 3/2)$

$\vec{PQ} = -\hat{i} + \frac{1}{2}\hat{j} + \frac{3}{2}\hat{k}$

Projection of PQ on x-axis $= \left|\vec{PQ}\cdot\hat{i}\right| = 1$

(R)

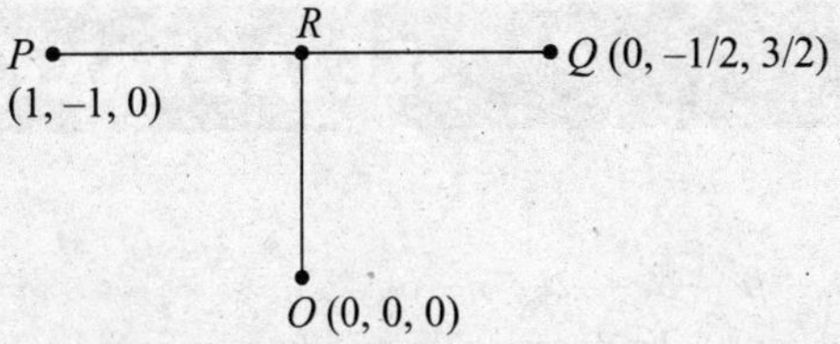

Equation of line PQ is

$$\frac{x-1}{-1} = \frac{y+1}{1/2} = \frac{z}{3/2} = 1$$

let $R \equiv (-\lambda + 1, \lambda/2 - 1, \frac{3\lambda}{2})$

$\vec{OR} \cdot \vec{PQ} = 0$

$\Rightarrow -1(-\lambda + 1) + \frac{1}{2}\left(\frac{\lambda}{2} - 1\right) + \frac{9\lambda}{4} = 0$

$\Rightarrow \lambda = 3/7$

$\therefore R \equiv \left(\frac{4}{7}, \frac{-11}{14}, \frac{9}{14}\right)$

$\Rightarrow 7a + 14b + 14c = 2$

(S) Area $= \frac{1}{2}|\vec{PQ} \times \vec{OQ}| = \sqrt{\frac{19}{16}}$

$\therefore a - b = 3$

17. (3)

(P) $\vec{a} \times \vec{b} = \begin{vmatrix} \hat{i} & \hat{j} & \hat{k} \\ -1 & 1 & 2 \\ -1 & -2 & -1 \end{vmatrix} = -3\hat{i} - 3\hat{j} + 3\hat{k}$

Hence area of triangle is $\frac{1}{2} \times 3\sqrt{3}$.

(Q) Area of parallelogram is $3\sqrt{3}$

(R) Area of parallelogram whose diagonals area $\vec{a}$ *and* $\vec{b}$ is

$\frac{1}{2}|2\vec{a} \times 4\vec{b}| = 12\sqrt{3}$

(S) Volume of parallelepiped $= |(\vec{a} \times \vec{b}) \cdot \vec{c}| = \sqrt{9 + 36 + 9} = 3\sqrt{6}$

JEE (ADVANCED) MOCK TEST (PAPER 2)

1. (1) $\overrightarrow{AB} = \vec{\beta} - \vec{\alpha} = -2i - 3j - 6k$

Equation of plane passing through B and perpendicular to AB is

$(\vec{r} - \overrightarrow{OB}).\overrightarrow{AB} = 0$

$\Rightarrow \quad \vec{r}\,.(2i + 3j + 6k) + 28 = 0$

hence required distance from ($\vec{c} = -i + j + k$)

$= \left|\dfrac{\vec{c}.(2i+3j+6k)+28}{|2i+3j+6k|}\right|$

$= \left|\dfrac{-2+3+6+28}{7}\right| = 5$ units.

2. (3) $\vec{a} \times \vec{b} \cdot \vec{c} = \vec{c} \cdot \vec{c} = |\vec{c}|^2 \Rightarrow \left[\vec{a}\ \vec{b}\ \vec{c}\right] = |\vec{c}|^2$

Similarly, $[\vec{a}\ \vec{b}\ \vec{c}] = |\vec{a}|^2$ and $[\vec{a}\ \vec{b}\ \vec{c}] = |\vec{b}|^2$

$\Rightarrow \quad |\vec{a}| = |\vec{b}| = |\vec{c}|$

Also $\left[\vec{b} \times \vec{c},\ \vec{c} \times \vec{a},\ \vec{a} \times \vec{b}\right] = \left[\vec{a}\ \vec{b}\ \vec{c}\right]^2$

$\Rightarrow \quad \left[\vec{a}\ \vec{b}\ \vec{c}\right] = \left[\vec{a}\ \vec{b}\ \vec{c}\right]^2$

$\Rightarrow \quad \left[\vec{a}\ \vec{b}\ \vec{c}\right] = 1$ as $\left[\vec{a}\ \vec{b}\ \vec{c}\right] \neq 0$

Since $\vec{a}, \vec{b}, \vec{c}$ are non –coplanar.

$\therefore \quad |\vec{a}| = |\vec{b}| = |\vec{c}| = 1$

$\therefore \quad |\vec{a}| + |\vec{b}| + |\vec{c}| = 3.$

3. (4) We have $Q(-1, 0, 1)$ and $R(1, 0, -1)$

$\therefore \quad QR = 2\sqrt{2}$

Now mid point of QR is $M(0, 0, 0)$

Triangle PQR is equilateral

$\therefore \quad PM = MQ \times \tan 60° = \sqrt{2} \times \sqrt{3} = \sqrt{6}$

$\therefore$ P can lie on the plane whose distance from point M is less than or equal to $\sqrt{6}$

Hence P can lie on the plane $x + y + z + 3\sqrt{2} = 0$

4. (2) Since $\vec{a}, \vec{b}$ and $\vec{c}$ are inclined to each other at an angle of $\dfrac{2\pi}{3}$,

vectors $\vec{a}, \vec{b}, \vec{c}$ are coplanar.

$\Rightarrow \quad \vec{p}, \vec{q}$ and $\vec{r}$ are also coplanar

$\Rightarrow \quad [\vec{p}\ \vec{q}\ \vec{r}] = 0$

5. (2), (3)

6. (1), (4)

Equation of plane through the intersection of the plane

$2x - y = 0$ and $3z - y = 0$ is

$2x - y + \lambda(3z - y) = 0 \qquad \ldots(i)$

or $\quad 2x - y(\lambda + 1)\,3\lambda^2$

Plane (i) is perpendicular to $4x + 5y - 3z = 8$

$\therefore \quad 4 \times 2 - 5(\lambda + 1) - 9\lambda = 0$

$8 - 5\lambda - 5 - 9\lambda = 0$

$\Rightarrow \quad 3 - 14\lambda = 0$

$\lambda = 3/14$

Putting in (i), we get

$\therefore \quad 2x + y + \dfrac{3}{14}(3z - y) = 0$

$28x - 17y + 9z = 0$

7. (2), (4)

$\bar{a} \times (\bar{r} \times \bar{a}) = \bar{a} \times \bar{b}$

$3\bar{r} - (\bar{a}\,.\bar{r})\ \bar{a} = \bar{a} \times \bar{b}$

Also $|\bar{r} \times \bar{a}| = |\bar{b}|$

$\Rightarrow \quad 3|\bar{r}|^2 \sin^2\theta = 2$

$\Rightarrow \quad (1 - \cos^2\theta) = \dfrac{2}{3}$

$\Rightarrow \quad \dfrac{1}{3} = \cos^2\theta$

$\Rightarrow \quad \bar{a}\,.\bar{r} = \pm 1$

$\Rightarrow \quad \bar{r} \pm \bar{a} = \bar{a} \times \bar{b}$

$\Rightarrow \quad \bar{r} = \dfrac{1}{3}(\bar{a} \times \bar{b} \pm \bar{a})$

8. (1) $\because$ Angle between $\vec{u}$ and $\hat{i}$ is 60°

$\vec{u}.i = |\vec{u}||\hat{i}|\cos 60° = \dfrac{|\vec{u}|}{2}$

given that $|\vec{u}|, |\vec{u} - \hat{i}|$ and $|\vec{u} - 2\hat{i}|$ are in G.P.

So $\quad |\vec{u} - \hat{i}|^2 = |\vec{u}| \cdot |\vec{u} - 2\hat{i}|$

Squaring both sides, we get

$[|\vec{u}|^2 + |\hat{i}|^2 - 2\vec{u} \cdot \hat{i}]^2 = |\vec{u}|^2 [|\vec{u}|^2 + 4|\hat{i}|^2 - 4\vec{u} \cdot \hat{i}]$

$\left[|\vec{u}|^2 + 1 - \dfrac{2|\vec{u}|}{2}\right]^2 = |\vec{u}|^2 \left[|\vec{u}|^2 + 4 - 4\dfrac{|\vec{u}|}{2}\right]$

$\Rightarrow \quad |\vec{u}|^2 + 2|\vec{u}| - 1 = 0 \quad \Rightarrow \quad |\vec{u}| = -\dfrac{2 \pm 2\sqrt{2}}{2}$

$\Rightarrow \quad |\vec{u}| = \sqrt{2} - 1$

9. (8) Any plane through the line

$3x - y - 2z + 4 = 0 = 2x + y + z + 1$

is $\quad (3x - y - 2z + 4) + \lambda(2x + y + z + 1) = 0$

or $\quad (3 + 2\lambda) + (\lambda - 1)y + (\lambda - 2)z + 4 + \lambda = 0$

If this plane is parallel to the first line then its normal must be at right angle to first line.

$\Rightarrow \quad (3 + 2\lambda)\,2 + (\lambda - 1)\,4 + (\lambda - 2) = 0$

$\Rightarrow \quad \lambda = 0$

$\Rightarrow$ equation of plane through the second line and parallel to the first line is

$3x - y - 2z + 4 = 0$

Now the required shortest distance,

d = perpendicular distance of a point (1, 3, – 2) on the first line to the plane $3x - y - 2z + 4 = 0$

$= \dfrac{8}{\sqrt{14}}$

$\Rightarrow \quad \sqrt{14}\,d = 8.$